Lecture Notes in Mathematics

Edited by A. Dold and B. Eckmann

960

Multigrid Methods

Proceedings of the Conference
Held at Köln-Porz, November 23–27, 1981

Edited by W. Hackbusch and U. Trottenberg

Springer-Verlag
Berlin Heidelberg New York 1982

Editors
W. Hackbusch
Institut für Informatik und Praktische Mathematik
Christian-Albrechts-Universität Kiel
Olshausenstr. 40–60, D-2300 Kiel 1

U. Trottenberg
Institut für Angewandte Mathematik
Rheinische Friedrich-Wilhelms-Universität Bonn
Wegelerstr. 6, D-5300 Bonn 1

1st Edition 1982
2nd Printing 1986

AMS Subject Classifications (1980): 65 N XX; 65-02, 65-06;
65 N 20, 65 N 30; 65 F 10, 65 H 10; 65 B 05; 35 J XX, 76 XX; 68 B XX, 68 C 25

ISBN 3-540-11955-8 Springer-Verlag Berlin Heidelberg New York
ISBN 0-387-11955-8 Springer-Verlag New York Heidelberg Berlin

Printing and binding: Beltz Offsetdruck, Hemsbach/Bergstr.
2146/3140-543210

PREFACE

These proceedings contain the introductory and specific scientific papers presented at the international

Conference on Multigrid Methods

which was held at Cologne-Porz from 23rd to 27th November, 1981.

The introductory part describes basic methods, theoretical approaches and practical aspects in a systematic way. Furthermore, some simple applications are discussed, and an exemplary multigrid program for a simple model problem is presented. The four authors of the introductory papers have tried to use a uniform notation. This has been rather difficult since in the meantime several systems of notations have come into general use, and good arguments can be found for all of them. The uniformity reached despite of these difficulties does not concern all occurring quantities but the essential ones.

The specific papers deal with the fields of theory, applications and software development. Most studies concern elliptic problems and their solution by means of difference methods. The conference and the papers reflect an increasing interest in the combination of multigrid techniques with defect correction methods as well as in the solution of singularly perturbed and (indefinite) nonlinear problems. Apart from introductory and specific papers this volume also contains a multigrid bibliography.

120 scientists from 12 countries participated in the Conference. Thanks to the generous financial support by the organizing institutions it was not necessary to charge a conference fee.

The organizers are as follows:

- Gesellschaft für Mathematik und Datenverarbeitung (GMD, St. Augustin)
- Sonderforschungsbereich (SFB) 72 "Approximation und mathematische Optimierung" at the University of Bonn, funded by the Deutsche Forschungsgemeinschaft
- Fachausschuß "Effiziente numerische Verfahren für partielle Differentialgleichungen" of the Gesellschaft für Angewandte Mathematik und Mechanik (GAMM).

Within the cooperation with the GMD the Deutsche Forschungs- und Versuchsanstalt für Luft- und Raumfahrt (DFVLR) provided the conference rooms. On this occasion the editors like to thank the mentioned institutions and their representatives, Prof. Dr. Norbert Szyperski (chairman of the Executive Board of the GMD),

Prof. Dr. Stephan Hildebrandt (speaker of SFB 72) and Prof. Dr. Hermann L. Jordan (chairman of the Executive Board of the DFVLR) for the immaterial and material support of the conference.

The practical organization was carried out by Kurt Brand and Heinz Reutersberg (Institut für Mathematik of the GMD). They were supported by Margarete Donovang, Elisabeth Harf and Reinhild Schwarz. Furthermore, the Abteilung für Informationswesen of the GMD provided substantial assistance to the completion of this volume. We like to express our gratitude to all persons involved.

Finally we like to thank all conference participants and especially the lecturers for their contributions to the success of the conference.

Wolfgang Hackbusch

Ulrich Trottenberg

CONTENTS

MULTIGRID METHODS: FUNDAMENTAL ALGORITHMS, MODEL PROBLEM ANALYSIS AND APPLICATIONS

Klaus Stüben*

Ulrich Trottenberg**

*Gesellschaft für Mathematik und Datenverarbeitung
Postfach 1240, D-5205 St. Augustin 1, Germany

**Institut für Angewandte Mathematik, Universität Bonn,
Wegeler Straße 6, D-5300 Bonn, Germany

Contents

1. Introduction

This paper gives a systematic introduction to multigrid methods for the solution of elliptic differential equations. The paper is based on the two introductory lectures held by the authors on the occasion of the "Conference on Multigrid Methods". It includes basic ideas (Part I) and fundamental methodical approaches (Part II), theoretical approaches (Part III) and simple applications (Part IV). The paper is to be seen in the context of the two other introductory papers in which Wolfgang Hackbusch outlines his general theory of multigrid methods and Achi Brandt gives a guide to the practical realization of multigrid methods. Brandt's paper deals, in particular, with problems of a more general type (systems of differential equations in higher dimensions) than that of the problems we consider in our paper. Brandt also discusses more sophisticated multigrid techniques.

Although our description of the multigrid principle and of the fundamental methodic approaches is quite general, the concrete considerations in this paper refer - in accordance with its introductory character - to a limited class of problems: We explicitly treat only scalar equations in two dimensions; the underlying discretizations are based on finite difference methods. Mostly we are concerned with second order Dirichlet boundary value problems. Most of these restrictions, in particular the restriction to two dimensions, are mainly for the sake of technical simplification.

In this introduction we give a short survey of the development of multigrid methods and on the state of the art (Section 1.1). We will then describe contents and objectives of this paper in some more detail (Section 1.2). In Section 1.3, we will introduce some fundamental notation which is needed.

1.1 Historical remarks and current perspectives

Multigrid history. The multigrid principle (for discrete elliptic boundary value problems) is extremely simple: Approximations with *smooth errors* are obtained very efficiently by applying suitable *relaxation* methods. Because of the error smoothness, corrections of these approximations can be calculated on *coarser grids*. This basic idea can be used *recursively* employing coarser and coarser grids. This leads then to "*(asymptotically) optimal*" iterative methods, i.e. methods for which the computational work required for achieving a fixed accuracy is proportional to the number of discrete unknowns. If the multigrid methods are then combined with the idea of *nested iteration* (use of coarser grids to obtain good initial approximations on finer grids), a suitable algorithmization even yields methods for which the computational work required for achieving the discretisation accuracy is still proportional to the number of discrete unknowns.

Consequently, we may distinguish three elements (stages):
(1) error smoothing by relaxation,
(2) calculation of corrections on coarser grids and recursive application,
(3) combination with nested iteration.

Looking back on the development of multigrid methods we see that the above elements, if considered separately, have already been known or used for a long time before they were combined to efficient multigrid methods. Especially the error smoothing effect of relaxation methods belongs to the classical inventory of numerical knowledge. The idea to use this effect for convergence acceleration can already be found in the early literature (e.g. Southwell [92], [93]; Stiefel [94]). However, the recursive use of coarser grids is not yet elaborate there. But it is only this recursion which gives the above mentioned "optimality".

On the other hand, the recursive application of coarser grids for an efficient solution of specific discrete elliptic boundary value problems was used in the context of "*reduction methods*" introduced by Schröder [86] (see also [85], [87], [88]). Here, however, no explicit error smoothing is performed. Elimination techniques are used instead which transform the original problem "equivalently" to coarser grids. (These elimination techniques restrict the range of direct application of reduction methods to a small class of problems.)

Finally, the self-suggesting idea of nested iterations has in principle been known for a long time.

The first studies introducing and investigating multigrid methods in a narrow

sense (elements (1) and (2)) are those by Fedorenko [34], [35] and then that of Bakhvalov [6]. While in [35] Fedorenko restricts the convergence investigation to the Poisson equation in the unit square, Bakhvalov [6] discusses general elliptic boundary value problems of second order with variable coefficients (in the unit square). Bakhvalov also indicates the possibility of combining multigrid methods with nested iteration (element (3)).

Though the studies published by Fedorenko and Bakhvalov have, in principle, shown the asymptotic optimality of the multigrid approach (and to a certain extent its generality as well), their actual efficiency is first recognised only by Achi Brandt (by 1970). Studying adaptive grid refinements and their relation to fast solvers, Brandt has been led to the papers of Fedorenko and Bakhvalov through information given by Olof Widlund. In the first two papers [15], [16] and later on summarised in the systematic work [17], Brandt shows the actual efficiency of multigrid methods. His essential contributions (in the early studies) concern the introduction of non-linear multigrid methods ("FAS-scheme") and adaptive techniques ("MLAT"), the discussion of general domains and local grid refinements, the systematic application of the nested iteration idea ("full multigrid" FMG) and - last but not least - the provision of the tool of the "local Fourier analysis" for theoretical investigation and method optimisation.

Representative for the further multigrid development are the following papers which we would like to mention as being "historically" relevant contributions.

In [4] Astrakhantsev generalises Bakhvalov's convergence result to general boundary conditions; like Bakhvalov he uses a variational formulation in his theoretical approach. - In [39], Frederickson introduces an approximate multigrid-like solver which can be regarded as a forerunner of the "MGR methods", which were developed later on. - After a first study of multigrid methods for Poisson's equation in a square [75], Nicolaides discusses multigrid ideas in connection with finite element discretisations systematically in [76]. -

In the years 1975/76, Hackbusch develops the fundamental elements of multigrid methods anew without having knowledge of the existing literature. It is again Olof Widlund who informs Hackbusch about the studies which are already available. Hackbusch's first systematic report [42] contains many theoretical and practical investigations which have been taken up and developed further by several authors. So one finds considerations of the "model problem analysis" type, the use of "red black" and "four colour" relaxation methods for smoothing, the treatment of non-rectangular domains and of nonlinear problems etc. In the papers [43], [45], [49], Hackbusch then presents a general convergence theory of multigrid methods.

The recent development. Since about 1977 multigrid methods have increasingly gained broad acceptance. This more recent development shall not be described here in detail. (A survey of the literature presently available is given by the multigrid bibliography in this Proceedings.) However, we want to mention some important fields of applications and mathematical areas to which multigrid methods have been applied and extended. The field of finite elements which has first been of a more theoretical interest to multigrid methods (see, for example, [76], [43], [8]) is now undergoing an intensive practical investigation (see, for example, [9], [32]). Apart from linear and non-linear boundary value problems (scalar equations and systems) eigenvalue problems and bifurcation problems (see, for example, [44], [27], [73]) are treated as well. Parabolic (see, for example, [33], [90], [63]) and other time-dependent and non-elliptic problems (see e.g. [23], [22], [84]) are attracting more and more interest. All these situations occur in numerical fluid dynamics, probably the most challenging field for multigrid methods. Here the studies are now concentrating on singular perturbation phenomena, transonic flow, shocks, the treatment of Euler equations and of the full Navier Stokes equations.

Apart from differential equations, integral equations can also be efficiently solved by multigrid methods (see e.g. [25] and the whole complex of multigrid methods "of the second kind" [48], [57]). Furthermore, multigrid-like methods are also being suggested for the solution of special systems of equations without continuous background [25]. A certain amount of multi-level structure (at least the nested iteration idea) can also be found in algorithms used in pattern recognition.

Perhaps as important as the extension of the field of applications of multigrid methods is the combination of the multigrid idea with other numerical and more general mathematical principles. In this context we would like to mention the combination with *extrapolation* and *defect correction methods* (see e.g. [25], [5], [51], [56]). Finally, there are considerations which refer to the optimal use of multigrid methods on *vector* and *parallel computers* (and the construction of corresponding multigrid components) (see, for example, [24]) as well as to approaches within computer architecture concerning a direct mapping of the multigrid principle onto a suitable - perhaps *pyramidal* - multiprocessor structure (see corresponding remarks in [103]).

Delayed acceptance, resentments. The historical survey has shown that the acceptance of multigrid methods was first a rather troublesome process. Only the rapid development of recent years has convinced most people working in the field of numerical methods for partial differential equations of the sensational possibilities provided by the multigrid principle.

Nevertheless, even today's situation is still unsatisfactory in several respects. If this is true for the development of standard methods, it applies all the more to the area of really difficult, complex applications. With respect to standard applications, we would like to discuss this in some detail (since this area is in the center of this introductory paper) and with respect to the complex applications, for example in fluid dynamics, we would like to confine ourselves to some remarks.

As far as standard problems (simple elliptic 2D problems of second order) are concerned, the opinion prevailed for a long time - even and just among experts - that, despite of their "asymptotic optimality", multigrid methods were in reality far from being as efficient as the "direct fast solvers" (such as the Buneman algorithm [29] or the method of total reduction [88]) and their combination with capacitance matrix techniques [81]. Only by providing generally available programs (such as MG00, MG01, see chapter 10), has it been proved in practice that suitable multigrid methods are at least competitive in these areas as well. The decisive advantage of multigrid methods is however that they can be applied easily to problems which do not meet - or do not fully meet - the requirements demanded by direct fast solvers and capacitance matrix techniques.

Doubts in the high efficiency of multigrid methods were also fed by the multigrid convergence theories. The abstract theories are often far too pessimistic and do usually not provide constructive criteria for the construction of optimal methods for concrete situations (see also Section 9.3).Only the *model problem analysis* (see Chapters 3, 7 and 8) and *local Fourier analysis* (see Sections 9.1, 9.2) yield quantitative results to be used for the construction of algorithms. On the other hand, these theoretical approaches, being relatively simple from the mathematical viewpoint, also have disadvantages: The model problem analysis can be applied directly to a small class of problems only, and local Fourier analysis is based on idealising assumptions.

As a consequence, even in the field of standard applications the disagreement about which approach would really supply the "best" or the "most robust" algorithms, is not completely settled as yet. For example, as far as the smoothing methods are concerned, each expert recommends "his" method and emphasises its benefits (A.Brandt recommends standard relaxation techniques - pointwise, linewise and "distributed"; Wesseling the ILU smoothing, Jameson smoothing methods of the ADI type, we recommend MGR methods....). Since so far systematic and fair comparisons were hardly available, it was also impossible, until recently, to obtain reliable statements on which method should be preferred in which situation. Among users this confusion has led to misunderstandings and false conclusions.

While in the field of standard problems the differences in efficiency shown by the various algorithms are, after all, not very large and the disagreement previously mentioned is therefore of a more or less academic nature, the disagreement in the field of non-elementary applications is of direct practical importance and it has especially unpleasant consequences there.

Such a controversy exists, for example, in the field of fluid dynamics between many numerical practitioners who like to take up multigrid methods and multigrid experts (even among the practically oriented experts) who like to develop "optimal" methods from a more fundamental viewpoint. With respect to more complex problems the experts usually supply efficient algorithms for simplified situations only and do not go to the work of solving full fledged industrial problems. The practitioners are therefore sceptical about the full applicability of multigrid methods. They mostly prefer to include single multigrid components in certain parts of available software. Thus, they obtain improvements which are possibly rather impressing, but, on the other hand, they regard their scepsis as being justified since the improvements obtained are far from being as large as predicted for "optimal" methods. However the multigrid experts also feel justified since they regard the stepwise inclusion of multigrid elements in the available "non-multigrid software" as being unsatisfactory in any case. This discrepancy can be found in many publications and comments and it was also reflected on the conference which is the subject of these proceedings. There is some hope, that these proceedings contribute towards bridging the gap between multigrid experts and practitioners.

1.2. Contents of this paper, acknowledgements

In part I, we describe the multigrid idea (Chapter 2) and give a first analysis of a sample method for Poisson's equation. For both chapters we have intentionally chosen a very detailed and elementary representation. The sample method considered in Chapter 3 is a rather inefficient method (since Jacobi relaxation is used for smoothing), but it has the advantage of being particularly theoretically transparent. The theoretical considerations and the tools introduced in Chapter 3 are characteristic for the model problem analysis which is discussed more systematically in part III.

Part II (Chapters 4,5,6) describes the well-known fundamental multigrid techniques: the *recursively defined complete multigrid cycle* (Chapter 4), the *nonlinear full approximation scheme* (Chapter 5), and the *full multigrid method* (Chapter 6).

Parts III and IV, in particular Chapters 7,8 (together with Chapter 3) and 10, 11, inform about results which are largely new and have not been published as yet.

Part III discusses the concepts of the so-called *model problem analysis* and *local Fourier analysis*. For a certain class of model problems and a certain class of multigrid algorithms, it is possible to give exact statements (not estimates) on the convergence behaviour of the method in question using basic tools of discrete Fourier analysis. In Chapter 7, we introduce the required formalism. In this context, various cases of the coarse grid definition are discussed.

Readers who are interested in concrete results rather than in the technically quite complicated formalism should proceed to Chapter 8. All results in this chapter refer to *standard coarsening* (doubling the meshwidths); the emphasis lies on the discussion of efficient smoothing methods, namely on *RB (= red black)*, *ZEBRA*, and *alternating ZEBRA relaxation*. Within the class of methods discussed, the model problem analysis allows the construction of optimal multigrid components.

Problems and methods which can no longer be treated rigorously by model problem analysis may possibly be studied by means of Fourier analysis (Chapter 9). In this context, however, no exact statements on the problem given are obtained but only statements on an idealised problem (and thus on an idealised method) where, in particular, the influence of the boundary and the boundary conditions are neglected. The exact statements on the idealised problem (and method) are then regarded as approximate statements on the original problem (and method). Subjects of this idealizing local Fourier analysis are, for example, the usual *Gauß-Seidel-relaxation* method (with *lexicographic ordering* of the grid points) and *ILU-smoothing*. Among other things, we make a short comparison of ILU-smoothing with ZEBRA relaxation in Section 9.2. - In Section 9.3., we make some remarks on more abstract convergence theories.

On the basis of the model problem and local Fourier analysis, the programs MG00 and MG01 for elliptic "standard problems" have been developed. MG01 is described in Chapter 10. - Chapter 11 describes the possibility of applying multigrid methods in combination with simultaneous use of various coordinate systems to a given problem (composite mesh system).

This is not the first introductory paper to multigrid methods (see [17], [55], [52]). In our presentation, the emphasis lies on the theoretical and practical discussion of the following central problem: How are the different multigrid components to be chosen in concrete situations? Clearly, there are different possible objectives which can be persued in answering this question, e.g. efficiency, simplicity or/and robustness of the respective algorithms. In our paper we tend toward demonstrating the efficiency of multigrid methods (for standard applications) rather than their generality. This shall, however, by no means modify or question the generality of the principle.

Acknowledgements

For various pleasant discussions concerning this paper (or certain parts of its contents) we would like to thank Achi Brandt, Wolfgang Hackbusch, Theodor Meis, Olof Widlund, and Kristian Witsch.

Christoph Börgers and Clemens August Thole read the manuscript and checked proofs and examples. Kristian Witsch supported us in providing the sample program. Kurt Brand and Horst Schwichtenberg accomplished several technical tasks.

Rudolph Lorentz corrected our use of the English language and also Ursula Bernhard supported us in this respect.

Gertrud Jacobs typed the manuscript, never tiring in making subsequent changes. She was supported partly by Elisabeth Harf. Maria Heckenbach drew the illustrating figures.

We owe sincere thanks to all of them.

1.3 Some notation

In this section, we want to list the basic notation needed for our description of discrete elliptic problems and their multigrid treatment. (Most of the notation below will - for clarity - be shortly explained once more when it occurs in the paper for the first time.)

1.3.1 Continuous boundary value problems

Linear boundary value problems are denoted by

$$\begin{aligned} L^{\Omega}u &= f^{\Omega}(x) \qquad (x \in \Omega) \\ L^{\Gamma}u &= f^{\Gamma}(x) \qquad (x \in \Gamma := \partial\Omega). \end{aligned} \tag{1.1}$$

Here $x=(x_1,\ldots,x_d)$ and Ω is a given domain with boundary Γ. L^{Ω} is a linear (elliptic) differential operator on Ω and L^{Γ} stands for one or several linear boundary operators. f^{Ω} denotes a given function on Ω and f^{Γ} one or several functions on Γ. Solutions of (1.1) are always denoted by $u=u(x)$. For brevity, we also write simply $Lu=f$ instead of (1.1). All concrete considerations refer to the case $d=2$.

Nonlinear differential operators are denoted by $\mathbf{L}$ rather than L.

1.3.2 Discrete boundary value problems

For discrete problems, we use the terminology of *grid functions, grid operators* and *grid equations* (rather than matrix terminology). The discrete analog of (1.1) is denoted by

$$\begin{aligned} L_h^{\Omega}u_h &= f_h^{\Omega}(x) \qquad (x \in \Omega_h) \\ L_h^{\Gamma}u_h &= f_h^{\Gamma}(x) \qquad (x \in \Gamma_h). \end{aligned} \tag{1.2}$$

h is a (formal) discretization parameter here. The discrete solution u_h is a grid-function defined on $\Omega_h \cup \Gamma_h$. f_h^{Ω} and f_h^{Γ} are discrete analogs of f^{Ω} and f^{Γ}. L_h^{Ω} and L_h^{Γ} are grid operators, i.e. mappings between spaces of grid functions. (L_h^{Ω} is also called a *discrete* or *difference operator,* L_h^{Γ} a *discrete boundary operartor.*)

For simplicity, we will usually assume that the discrete boundary equations are eliminated from (1.2). This is, for example, quite natural in case of second order equations with Dirichlet boundary conditions (for an example, see Section 1.3.3). We then simply write

$$L_h u_h = f_h \quad (\Omega_h). \tag{1.3}$$

Here u_h and f_h are grid functions on Ω_h and L_h is a linear operator

$$L_h : \mathbb{G}(\Omega_h) \to \mathbb{G}(\Omega_h) \tag{1.4}$$

where $\mathbb{G}(\Omega_h)$ denotes the linear space of grid functions on Ω_h. Clearly, (1.3) represents a system of linear algebraic equations. We consider it, however, as one *grid equation*.

At many places in this paper, Ω is a rectangular domain $\Omega=(0,A_1)\times(0,A_2)$ and Ω_h a rectangular grid "matching well" with Ω. In this case, h stands for a vector of meshsizes: $h=(h_{x_1},h_{x_2})$ and Ω_h is described by

$$\Omega_h := \Omega \cap G_h \tag{1.5}$$

where G_h denotes the infinite grid

$$G_h := \{x = \kappa\cdot h : \kappa \in \mathbb{Z}^2\}, \qquad h_{x_j} = A_j/N_j, \quad N_j \in \mathbb{N} \quad (j = 1,2). \tag{1.6}$$

Here $\kappa\cdot h := (\kappa_1 h_{x_1}, \kappa_2 h_{x_2})$. In the special case of square grids, we simply write $h=h_{x_1}=h_{x_2}$.

For (1.5) and (1.6), the space of grid functions $\mathbb{G}(\Omega_h)$ is canonically endowed with the *Euclidian* (more precisely, the *discrete* L_2*-) inner product*

$$(u_h,w_h)_2 := \frac{1}{N_1 N_2} \sum_{x\in\Omega_h} u_h(x)\overline{w}_h(x), \qquad \| u_h \|_2 := \sqrt{(u_h,u_h)_2}. \tag{1.7}$$

The corresponding operator norm is the *spectral norm*, denoted by $\|\cdot\|_S$. For L_h symmetric and positive definite, we also consider the *energy inner product*

$$(u_h,w_h)_E := (L_h u_h,w_h)_2 \tag{1.8}$$

and the corresponding operator norm $\|\cdot\|_E$.

1.3.3 Model problem (P)

For demonstration purposes, we sometimes refer to the simple case of Poisson's equation with Dirichlet boundary conditions on the unit square, namely

$$\begin{aligned} L^\Omega u &:= -\Delta u = f^\Omega(x) \quad (x \in \Omega := (0,1)^2) \\ L^\Gamma u &:= \quad u = f^\Gamma(x) \quad (x \in \Gamma). \end{aligned} \tag{1.9}$$

We speak of *model problem (P)*, if this problem is discretized on a square h-grid using the ordinary 5-point approximation (with order of consistency 2). In particular, we then have

$$L_h^{\Omega} = -\Delta_h \mathrel{\hat{=}} \frac{1}{h^2}\begin{bmatrix} & -1 & \\ -1 & 4 & -1 \\ & -1 & \end{bmatrix}_h ,$$

$$\Omega_h = \Omega \cap G_h, \quad G_h = \{x = \kappa\cdot h : \kappa \in \mathbb{Z}^2\}, \quad h = 1/N \quad (N \in \mathbb{N}). \tag{1.10}$$

(The notation of difference stars used here is described in more detail below.)

Eliminating the discrete boundary conditions leads to a grid equation (1.3). More precisely, L_h in (1.3) is then given by the difference star in (1.10) degenerating to a 4-point star near the edges and to a 3-point star near the corners. Clearly, the right hand side f_h then also includes the boundary values: certain terms of the form f_h^{Γ}/h^2 have to be added to f_h^{Ω} at grid points near the boundary Whenever we refer to model problem (P), we mean the corresponding grid equation (1.3)

Another important discrete problem considered in this paper, is the *anisotropic model problem*. The difference to model problem (P) is that L^{Ω} in (1.9) is replaced by

$$L^{\Omega}u := -\varepsilon u_{x_1x_1} - u_{x_2x_2}. \tag{1.11}$$

The standard 5-point approximation of L^{Ω} is given by

$$L_h^{\Omega} \mathrel{\hat{=}} \frac{1}{h^2}\begin{bmatrix} & -1 & \\ -\varepsilon & 2(1+\varepsilon) & -\varepsilon \\ & -1 & \end{bmatrix}_h . \tag{1.12}$$

1.3.4 General difference stars on rectangular grids

For the concrete definition of discrete operators L_h^{Ω} (on rectangular grids) the terminology of *difference stars* is convenient. We will make use of this terminology throughout this paper. For a simplified introduction of this terminology, we make use of the infinite grid in (1.6). (Note that the following definitions can also be understood locally for fixed $x \in \Omega_h$.)

A general difference approximation at some $x \in G_h$ is of the form

$$L_h^{\Omega}w_h(x) = \sum_{\kappa\in V} s_\kappa w_\kappa(x+\kappa\cdot h) \tag{1.13}$$

where $\mathbb{V}$ denotes a certain finite subset of $\mathbb{Z}^2$ (containing (0,0)). In the terminology of difference stars this is written as

$$L_h^{\Omega} w_h(x) = [s_\kappa]_h w_h(x) := \begin{bmatrix} & & \cdot & \cdot & \cdot & & \\ & & \cdot & \cdot & \cdot & & \\ \cdot & \cdot & s_{-1,1} & s_{0,1} & s_{1,1} & \cdot & \cdot \\ \cdot & \cdot & s_{-1,0} & s_{0,0} & s_{1,0} & \cdot & \cdot \\ \cdot & \cdot & s_{-1,-1} & s_{0,-1} & s_{1,-1} & \cdot & \cdot \\ & & \cdot & \cdot & \cdot & & \\ & & \cdot & \cdot & \cdot & & \end{bmatrix}_h w_h(x) \qquad (1.14)$$

The coefficients s_κ depend, of course, on h and possibly also on x. In our theoretical considerations, however, the s_κ do not depend on x. We will mainly consider discrete operators L_h^{Ω} on rectangular domains which can be described by *5-point* and *compact 9-point stars*

$$\begin{bmatrix} & s_{0,1} & \\ s_{-1,0} & s_{0,0} & s_{1,0} \\ & s_{0,-1} & \end{bmatrix}_h , \quad \begin{bmatrix} s_{-1,1} & s_{0,1} & s_{1,1} \\ s_{-1,0} & s_{0,0} & s_{1,0} \\ s_{-1,-1} & s_{0,-1} & s_{1,-1} \end{bmatrix}_h . \qquad (1.15)$$

L_h^{Ω} is then identified with its difference star:

$$L_h^{\Omega} \mathrel{\hat{=}} [s_\kappa]_h. \qquad (1.16)$$

We sometimes also identify L_h in (1.3) with the difference star corresponding to L_h^{Ω}. This always means that near boundaries the star is assumed to be properly modified (as described above for model problem (P)).

1.3.5 Restriction and interpolation operators

Apart from discrete operators L_h, we need restriction and interpolation operators for the intergrid transfer of grid functions in the multigrid context. For their description, we use a star terminology also. We illustrate this for the grid transfer between the grid G_h and the grid corresponding to the meshsize 2h, namely

$$G_{2h} = \{x = 2\kappa\cdot h : \kappa \in \mathbb{Z}^2\}. \qquad (1.17)$$

A *restriction operator* I_h^{2h} maps h-grid functions into 2h-grid functions:

$$(I_h^{2h} w_h)(x) = \sum_{\kappa\in\mathbb{V}} \hat{t}_\kappa w_h(x+\kappa\cdot h) \quad (x \in G_{2h}). \qquad (1.18)$$

Here $\mathbb{V}$ again denotes some finite subset of $\mathbb{Z}^2$. The coefficients $\hat{t}_\kappa$ may depend on h and x. Throughout this paper, however, the $\hat{t}_\kappa$ are constants (depending neither on h nor on x). For (1.18) we write

$$I_h^{2h} \mathrel{\hat{=}} [\hat{t}_\kappa]_h^{2h} := \begin{bmatrix} & & \cdot & \cdot & \cdot & & \\ & & \cdot & \cdot & \cdot & & \\ \cdot & \cdot & \hat{t}_{-1,1} & \hat{t}_{0,1} & \hat{t}_{1,1} & \cdot & \cdot \\ \cdot & \cdot & \hat{t}_{-1,0} & \hat{t}_{0,0} & \hat{t}_{1,0} & \cdot & \cdot \\ \cdot & \cdot & \hat{t}_{-1,-1} & \hat{t}_{0,-1} & \hat{t}_{1,-1} & \cdot & \cdot \\ & & \cdot & \cdot & \cdot & & \\ & & \cdot & \cdot & \cdot & & \end{bmatrix}_h^{2h} \tag{1.19}$$

The most frequently used restriction operator is the operator of *full weighting* (FW):

$$\frac{1}{16}\begin{bmatrix} 1 & 2 & 1 \\ 2 & 4 & 2 \\ 1 & 2 & 1 \end{bmatrix}_h^{2h} . \tag{1.20}$$

Similarly, an *interpolation (prolongation) operator* maps 2h-grid functions into h-grid functions. For the description of such operators, we introduce the following notation:

$$I_{2h}^{h} \mathrel{\hat{=}}]\check{t}_\kappa[_{2h}^{h} := \left]\begin{matrix} & & \cdot & \cdot & \cdot & & \\ & & \cdot & \cdot & \cdot & & \\ \cdot & \cdot & \check{t}_{-1,1} & \check{t}_{0,1} & \check{t}_{1,1} & \cdot & \cdot \\ \cdot & \cdot & \check{t}_{-1,0} & \check{t}_{0,0} & \check{t}_{1,0} & \cdot & \cdot \\ \cdot & \cdot & \check{t}_{-1,-1} & \check{t}_{0,-1} & \check{t}_{1,-1} & \cdot & \cdot \\ & & \cdot & \cdot & \cdot & & \\ & & \cdot & \cdot & \cdot & & \end{matrix}\right[_{2h}^{h} \tag{1.21}$$

The meaning of this star terminology is that coarse-grid values are "distributed" to the fine grid weighted by $\check{t}_\kappa$. More precisely, (1.21) means that a 2h-grid function w_{2h} is mapped into the h-grid function w_h defined by

$$w_h := \sum_{y \in G_{2h}} w_{h,y} \tag{1.22}$$

where $w_{h,y}$ is the h-grid function (with finite support)

$$w_{h,y}(x) := \begin{cases} w_{2h}(y)\check{t}_\kappa & \text{for } x = y+\kappa\cdot h \text{ with } \kappa \in \mathbb{V} \\ 0 & \text{for } x = y+\kappa\cdot h \text{ with } \kappa \notin \mathbb{V} . \end{cases} \tag{1.23}$$

Note that, for fixed $x \in G_h$, only a finite number of summands $w_{h,y}(x)$ in (1.22) is nonzero.

The most frequently used interpolation method is the one of *bilinear interpolation* from G_{2h} to G_h. The corresponding star is given by

$$\frac{1}{4}\left]\begin{matrix} 1 & 2 & 1 \\ 2 & 4 & 2 \\ 1 & 2 & 1 \end{matrix}\right[_{2h}^{h} \tag{1.24}$$

Obviously, this interpolation operator corresponds to the FW restriction operator (1.20): these two operators are *adjoint* to each other (see [50],[109]). Without going into details, we mention that - in a certain sense - such a correspondence holds for general restriction and interpolation operators if $\hat{t}_\kappa = \check{t}_{-\kappa}/4$.

1.3.6 Some remarks on the parameter h and admissible meshsizes

So far, we have sometimes considered h as a formal parameter (also used for non-rectangular domains and meshes) and at other places as a vector of meshsizes $h=(h_{x_1},h_{x_2})$ or as a scalar parameter of square meshes. Similarly, an index H is used to denote quantities on (or involving) any coarser grid Ω_H, e.g., L_H, I_h^H, I_H^h. In concrete cases, H denotes the meshsize of this coarser grid, e.g., $H=2h$ or $H=(2h_{x_1},h_{x_2})$ etc. Even then, H is often considered as a formal parameter: For example, if L_h is given on Ω_h, then L_{2h} does <u>not</u> necessarily mean the 2h-discretization of L which corresponds to L_h; in general, L_{2h} may be <u>any</u> discrete operator on Ω_{2h}. This use of the index $2h$ is somewhat inconsistent. It is, however, common and very convenient for reasons of technical simplicity. In concrete cases, it will always be clear which coarse grid quantities are actually meant.

In many places within this paper we consider certain important quantities as a function of h, e.g. asymptotic convergence factors $\rho(h)$. Of particular interest are their suprema with respect to h. These suprema have to be taken over those h which are meaningful and of interest within the multigrid context and the respective considerations. We denote the corresponding range of *admissible* h by $\mathcal{H}^*$. Thus we write, e.g.,

$$\rho^* := \sup \{\rho(h) : h \in \mathcal{H}^*\}. \tag{1.25}$$

Assuming rectangular meshes, the general form of $\mathcal{H}^*$ is

$$\mathcal{H}^* := \{h : h \le h^*, \; h_{x_2}/h_{x_1} = q^*\}. \tag{1.26}$$

Here q^* denotes a given fixed meshsize ratio and h^* denotes the (vector of) maximal meshsizes of interest. The inequality in (1.26) has to be understood componentwise. Furthermore, "≤" implicitly restricts the meshsizes h to those for which the process of constructing coarser grid at hand is meaningful. For example, if Ω is a rectangle and $H=2h$, we admit only those h for which both Ω_h and Ω_{2h} match well with the given domain. For square meshsizes (i.e. $q^*=1$), both h and h^* are scalar quantities. In the case of model problem (P), $\mathcal{H}^*$ is defined by

$$\mathcal{H}^* := \{h : h \le h^*, \; h = 1/N \quad (N \text{ even})\}, \; h^* := 1/4. \tag{1.27}$$

In all concrete cases considered in this paper, the precise meaning of $\mathcal{H}^*$ will be clear.

2. The multigrid idea, multigrid components

In this section we describe the fundamental multigrid idea. For this purpose, we consider a discrete linear elliptic boundary value problem, formally given by

$$L_h u_h = f_h \quad (\Omega_h). \tag{2.1}$$

Here L_h is a linear operator

$$L_h : \mathbb{G}(\Omega_h) \to \mathbb{G}(\Omega_h), \tag{2.2}$$

and $\mathbb{G}(\Omega_h)$ denotes the linear space of grid functions on Ω_h. We assume Ω_h to consist of $\mathcal{N} = \mathcal{N}_h$ grid points corresponding to the unknown grid values of u_h. Thus $\mathbb{G}(\Omega_h)$ is of dimension $\mathcal{N}$. Furthermore, we assume L_h^{-1} to exist.

As we are not going to give concrete quantitative results in this section, we do not make precise assumptions on the discrete operator L_h, the right hand side f_h and the grid Ω_h. A simple but characteristic example will be treated explicitly in Chapter 3.

In (2.1), we have assumed that the boundary conditions of the corresponding continuous problem (1.1) have been "eliminated". Thus f_h consists of some discrete analogue of the right hand side f^Ω as well as of the boundary values f^Γ. The assumption about the elimination of boundary conditions allows a simple description of the multigrid idea. We point out, however, that this treatment of the boundary conditions may not give the best multigrid methods in general cases. In certain simple cases (e.g. Dirichlet boundary conditions for second order equations) this approach is indeed suitable. In general, a separate treatment of the differential equation and the boundary conditions by multigrid techniques should be taken into consideration (see Section 2.4.5).

2.1 Iteration by approximate solution of the defect equation

Let u_h^j be any approximation of the solution u_h of (2.1). Then by

$$v_h^j := u_h - u_h^j \tag{2.3}$$

we denote the *error* of u_h^j (also regarded as *correction* of u_h^j), and by

$$d_h^j := f_h - L_h u_h^j \tag{2.4}$$

the *defect* (or *residual*) of u_h^j. Trivially, the *defect equation*

$$L_h v_h^j = d_h^j \tag{2.5}$$

is equivalent to the original equation, yielding

$$u_h = u_h^j + v_h^j.$$

The defect equation and approximations to it play an essential role in our description of the multigrid idea.

We begin the description by pointing out that most of the classical iterative methods for the solution of (2.1) can also be interpreted as approximations to (2.5): If in (2.5) L_h is replaced by any "simpler" operator $\hat{L}_h$ such that $\hat{L}_h^{-1}$ exists, the solution $\hat{v}_h^j$ of

$$\hat{L}_h \hat{v}_h^j = d_h^j \tag{2.6}$$

gives a new approximation

$$u_h^{j+1} = u_h^j + \hat{v}_h^j.$$

Starting with some u_h^o, the successive application of this process defines an iterative procedure. Obviously, the *iteration operator* of this method is given by

$$I_h - B_h L_h : \mathbb{G}(\Omega_h) \to \mathbb{G}(\Omega_h), \tag{2.7}$$

where $B_h := \hat{L}_h^{-1}$ and I_h denotes the identity on $\mathbb{G}(\Omega_h)$. We have

$$v_h^{j+1} = (I_h - B_h L_h)\, v_h^j \quad (j = 0,1,2,\ldots) \tag{2.8}$$

for the errors and

$$d_h^{j+1} = L_h (I_h - B_h L_h) L_h^{-1}\, d_h^j = (I_h - L_h B_h)\, d_h^j \quad (j = 0,1,2,\ldots) \tag{2.9}$$

for the defects.

It is a well-known fact that the asymptotic convergence properties of the above iterative process are characterized by the *spectral radius (asymptotic convergence factor)* of the iteration operator, i.e

$$\rho(I_h - B_h L_h) = \max\{|\lambda| : \lambda \text{ eigenvalue of } I_h - B_h L_h\}. \tag{2.10}$$

If some norm $\|.\|$ is introduced in $\mathbb{G}(\Omega_h)$, the corresponding operator norms

$$\| I_h - B_h L_h \|, \quad \| I_h - L_h B_h \| \tag{2.11}$$

give the *error reducing factor* and the *defect reducing factor*, respectively, per iteration step.

The following examples of iterative methods are of direct relevance to the multigrid method explained below:

Example 2.1a: The classical *Jacobi method* (or *simultaneous displacement procedure* or *total step procedure*) is characterized by replacing L_h by its "diagonal" part (in matrix terminology).

Example 2.1b: Similarly, the classical *Gauss-Seidel method* (or *successive displacement procedure* or *single step procedure*) is obtained by replacing L_h by its "upper triangular" part (in matrix terminology). As the structure of the matrix associated with L_h depends on the enumeration (ordering) of the grid points of Ω_h, $\hat{L}_h$ also depends essentially on this enumeration,

Example 2.2: A quite different choice of $\hat{L}_h$ (or, more precisely, of B_h), which will lead us to the multigrid idea, consists in using an appropriate approximation L_H of L_h on a coarser grid Ω_H. This means that the defect equation (2.5) is replaced by an equation

$$L_H \hat{v}_H^j = d_H^j. \tag{2.12}$$

Here we assume

$$L_H : \mathbb{G}(\Omega_H) \to \mathbb{G}(\Omega_H), \quad \dim \mathbb{G}(\Omega_H) << \dim \mathbb{G}(\Omega_h) \tag{2.13}$$

and L_H^{-1} to exist. As d_H^j and $\hat{v}_H^j$ are grid functions on the coarser grid Ω_H, we assume two (linear) transfer operators

$$I_h^H : \mathbb{G}(\Omega_h) \to \mathbb{G}(\Omega_H), \quad I_H^h : \mathbb{G}(\Omega_H) \to \mathbb{G}(\Omega_h). \tag{2.14}$$

to be given. I_h^H is used to *restrict* d_h^j to Ω_H:

$$d_H^j := I_h^H d_h^j, \tag{2.15}$$

and I_H^h is used to *interpolate* (or *prolongate*) the correction $\hat{v}_H^j$ to Ω_h:

$$\hat{v}_h^j := I_H^h \hat{v}_H^j. \tag{2.16}$$

Altogether, one iteration step (calculating u_h^{j+1} from u_h^j) proceeds as follows:

- Compute the defect: $d_h^j := f_h - L_h u_h^j$.
- Restrict the defect (fine-to-coarse transfer): $d_H^j := I_h^H d_h^j$.
- Solve exactly on Ω_H: $L_H \hat{v}_H^j = d_H^j$.
- Interpolate the correction (coarse-to-fine transfer): $\hat{v}_h^j := I_H^h \hat{v}_H^j$.
- Compute new approximation: $u_h^{j+1} := u_h^j + \hat{v}_h^j$.

This process is illustrated in Figure 2.1. The associated iteration operator is given by

$$I_h - B_h L_h \quad \text{with} \quad B_h = I_H^h L_H^{-1} I_h^H. \tag{2.17}$$

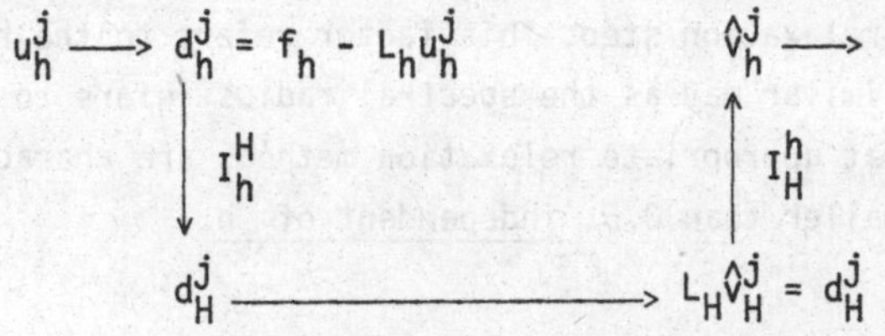

Figure 2.1: Structure of a coarse-grid correction process.

2.2 Relaxation and coarse-grid correction

If the methods in Example 2.1 and 2.2 are regarded as iterative methods for the solution of (2.1), they turn out to have very unsatisfactory convergence properties (or they are even not convergent at all). We are going to discuss this in more detail. In particular, we shall recognize that both types of methods behave very differently. This difference in behavior can be exploited: suitable combinations of both types of methods have very good convergence properties. Such combinations yield so-called *two-grid methods* which are the basis of *multigrid methods*.

Relaxation methods. Methods as considered in Example 2.1 are called *relaxation methods* in this paper. We also use this term for Jacobi- and Gauss-Seidel-type methods with under- or overrelaxation parameters, for Jacobi and Gauss-Seidel block-relaxation and related techniques.

The convergence properties of relaxation methods when applied to h-discrete elliptic equations (2.1) are known to become very bad if $h \to 0$. For Poisson's equation (model problem (P)), for instance, the spectral radii of the usual Jacobi and Gauss-Seidel methods behave like $1-O(h^2)$; SOR with optimal overrelaxation parameter has a spectral radius $1-O(h)$ [105].

The error reducing properties of these methods may be analyzed expanding the errors v_h^j into discrete Fourier series. In terms of this Fourier expansion one may roughly distinguish between *smooth (low-frequency)* and *non-smooth (high-frequency)* error components. It is already well-known from classical investigations on relaxation methods that especially the smooth error components are responsible for the slow asymptotic convergence (also see Section 3.2).

On the other hand, suitable relaxation methods are very efficient in smoothing the errors, i.e. in reducing the high-frequency error components (see Figure 2.2). The smoothing properties of relaxation methods are measured by a *smoothing factor* (see Sections 3.2 and 7.5), i.e. the worst (largest) factor, by which high-frequency error components are reduced per relaxation step. This factor refers to the high-frequency error components in a similar way as the spectral radius refers to all error components. We shall see that appropriate relaxation methods are characterized by smoothing factors which are smaller than 0.5, independent of h.

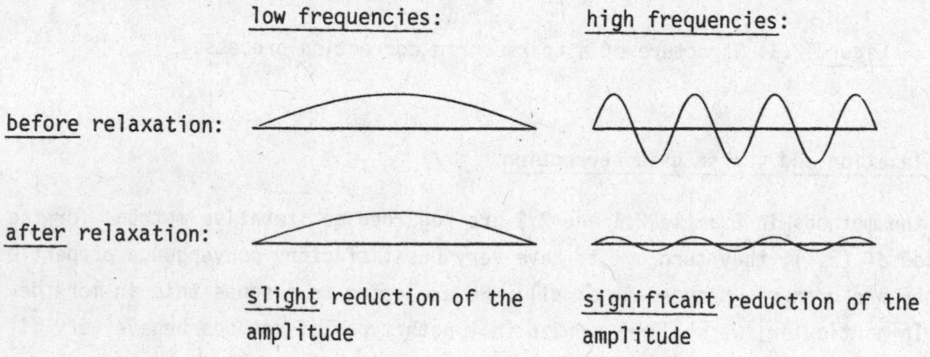

Figure 2.2: Typical error smoothing behavior of appropriate relaxation methods.

For convenience, we introduce the notation

$$\bar{w}_h = \mathrm{RELAX}^{\nu}(w_h, L_h, f_h) \tag{2.18}$$

to denote the result $\bar{w}_h$ of ν relaxation steps applied to (2.1) starting with w_h as first approximation. If $\nu=1$ we omit the upper index.

Coarse-grid correction. Processes as considered in Example 2.2 are called *coarse grid correction* (CGC) processes. Our first observation is that such a process (with linear operators I_h^H, I_H^h), used as an iterative method by itself, is not convergent

Lemma 2.1: $$\rho(I_h - I_H^h L_H^{-1} I_h^H L_h) \geq 1.$$

This follows directly from the fact that any $w_h \in \mathbb{G}(\Omega_h)$ $(w_h \neq 0)$ such that $L_h w_h$ lies in the kernel of I_h^H, is reproduced by the corresponding iteration operator.

This shows that the coarse-grid defect equation (2.12) in general is not a reasonable approximation for the original defect equation (2.5). In particular, those components of v_h^j which cannot be represented on the coarse grid Ω_H (which are - so to say - "not visible" on the H-grid) can, of course, not be reduced by use of this grid. We illustrate this in Figure 2.3 for $H=2h$.

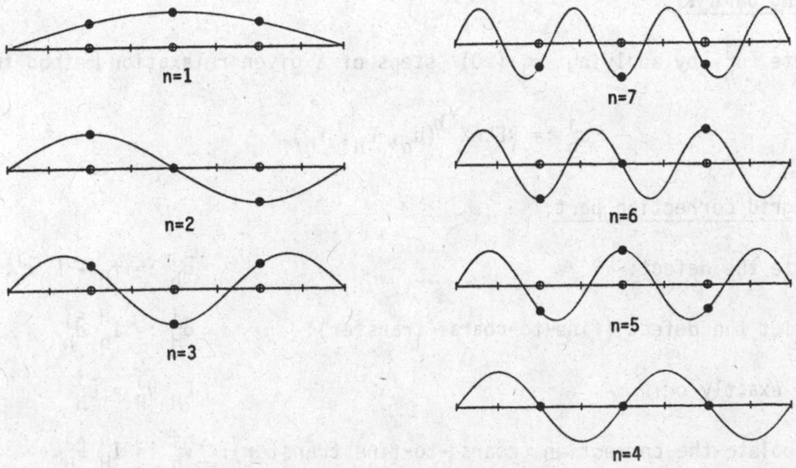

Components which are visible on Ω_H (wavelength $> 4h$)

Components which are not visible on Ω_H (wavelength $\leq 4h$)

Figure 2.3: $\sin(n\pi x)$, low $(n=1,2,3)$ and high $(n=4,5,6,7)$ frequency components for $h=1/8$ and $H=1/4$

We conclude from the above considerations:

The error v_h^j which is determined by the original defect equation (2.5), can be approximated well by use of the H-grid only if this error is "smooth", i.e. if its high-frequency components are small compared to its low-frequency components. This is, however, exactly what can be achieved by suitable relaxation methods very efficiently. (Of course, the terms "high" and "low" have to be related to the given h- and H-grid, see Chapter 7.)

2.3 Structure of an (h,H) two-grid iteration operator

Summarizing the above considerations, it is reasonable to combine the two processes of relaxation and of coarse-grid correction. In this way we obtain an iterative (h,H) *two-grid method*. Each iteration step of such a method consists of a *smoothing* and a *coarse-grid correction* part.

One step of such an (h,H)-method (calculating u_h^{j+1} from u_h^j) proceeds as follows:

(1) Smoothing part I:

- Compute $\bar{u}_h^j$ by applying ν_1 (≥ 0) steps of a given relaxation method to u_h^j:

$$\bar{u}_h^j := \mathrm{RELAX}^{\nu_1}(u_h^j, L_h, f_h).$$

(2) Coarse-grid correction part:

- Compute the defect: $\bar{d}_h^j := f_h - L_h \bar{u}_h^j$.
- Restrict the defect (fine-to-coarse transfer): $\bar{d}_H^j := I_h^H \bar{d}_h^j$.
- Solve exactly on Ω_H: $L_H \hat{v}_H^j = \bar{d}_H^j$. (2.19)
- Interpolate the correction (coarse-to-fine transfer): $\hat{v}_h^j := I_H^h \hat{v}_H^j$.
- Compute the corrected approximation: $\bar{u}_h^j + \hat{v}_h^j$.

(3) Smoothing part II:

- Compute u_h^{j+1} by applying ν_2 (≥ 0) steps of the given relaxation method to $\bar{u}_h^j + \hat{v}_h^j$:

$$u_h^{j+1} := \mathrm{RELAX}^{\nu_2}(\bar{u}_h^j + \hat{v}_h^j, L_h, f_h).$$

This process is illustrated in Figure 2.4. The difference to Figure 2.1 lies in the additional relaxation steps before and after the coarse-grid correction.

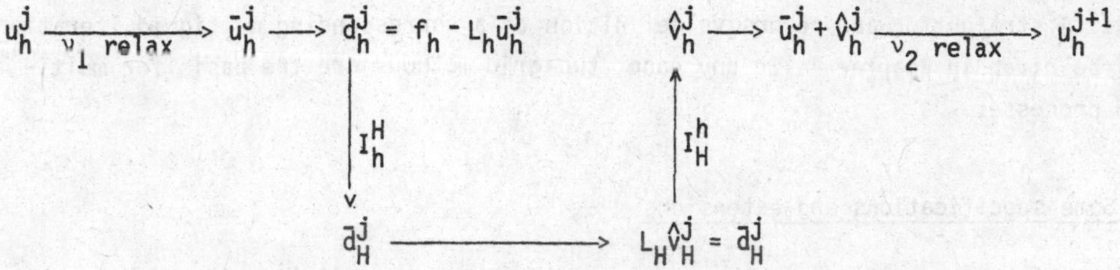

Figure 2.4: Structure of an (h,H) two-grid method

From the above description, one immediately obtains the iteration operator M_h^H of the (h,H) two-grid method:

Lemma 2.2: $$M_h^H = S_h^{\nu_2} K_h^H S_h^{\nu_1} \quad \text{with} \quad K_h^H := I_h - I_H^h L_H^{-1} I_h^H L_h .$$

Here S_h denotes the iteration operator corresponding to the relaxation process used. Obviously, the following individual *components of the (h,H)-method* have to be specified:

- the relaxation procedure, characterized by S_h;
- the numbers ν_1, ν_2 of relaxation steps;
- the coarse grid Ω_H;
- the fine-to-coarse restriction operator I_h^H;
- the coarse-grid operator L_H;
- the coarse-to-fine interpolation operator I_H^h.

Experience with multigrid methods shows that the choice of these components has - on the one hand - a strong influence on the efficiency of the resulting algorithms. On the other hand, there seem to be no general rules on how to choose the individual components in order to construct optimal algorithms. One can, however, recommend certain choices for certain situations. Whenever possible, such recommendations should, of course, be theoretically founded. The main objective of the *model problem* and the *local mode analysis* is to determine the asymptotic convergence factor $\rho(M_h^H)$ or suitable norms $\|M_h^H\|$ and to investigate the influence of the above mentioned choices on $\rho(M_h^H)$, $\|M_h^H\|$. In Chapter 3, we will have a preliminary discussion of this question for the special case of Poisson's equation.

The (h,H)-method is not yet a real *multi*grid method as only one coarser grid is used so far. In practice, the exact solution of the defect equation (2.19) on Ω_H is replaced by an approximate solution, which is obtained by using still coarser grids. A straightforward recursive definition of a corresponding multigrid iteration will be given in Chapter 4. In any case, two-grid methods are the basis for multigrid processes.

2.4 Some specifications and extensions

Some of the assumptions in the previous sections were made in order to keep the description of the basic multigrid idea simple. Not all multigrid methods used in practice satisfy these assumptions. We want to mention some important modifications in this section. First, however, we want to specify some of the quantities used above and to introduce the corresponding notation.

2.4.1 Choice of the coarser grid

The most important and most frequently used choice of Ω_H is characterized by doubling the given meshsize h, i.e. $H=2h$. Most of the results and considerations in this paper refer to this choice which will be called *standard coarsening*.

If the meshsize h is doubled in one direction only, i.e. $H=(2h_{x_1},h_{x_2})$ or $H=(h_{x_1},2h_{x_2})$, we speak of *semi-coarsening*. This is of interest for anisotropic and certain singularly perturbed differential operators [25],[23]. Furthermore, semi-coarsening is natural for the so-called MG-AR methods [82],[111].

We speak of *red-black coarsening*, if the coarse grid points are distributed in the fine grid in a checkerboard manner. We will consider this coarsening only for square grids ($h_{x_1}=h_{x_2}$). In this case, Ω_H can obviously be identified with a rotated grid of meshsize $\sqrt{2}h$. In particular, red-black coarsening is characteristic for the so-called MG-TR methods [82],[111].

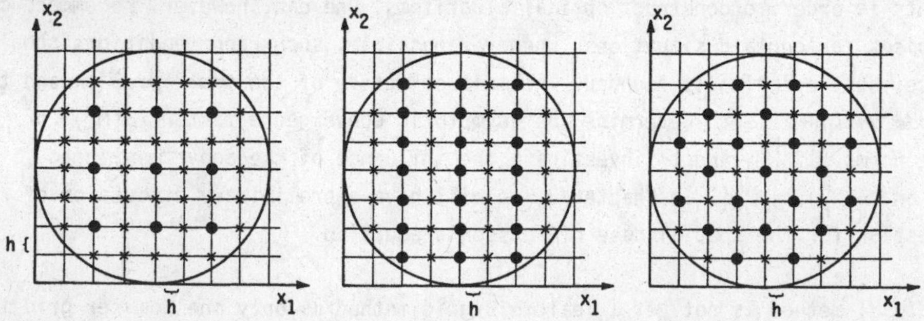

Figure 2.5: Example of standard, semi and red-black coarsening. The grid points of Ω_H are marked by dots. The grid points of Ω_h are just the union of points marked by crosses or dots.

2.4.2 Choice of the coarse-grid difference operator

In this paper, we mainly consider multigrid methods in which L_H is constructed with respect to Ω_H in a way analogous to the construction of L_h with respect to Ω_h. This is, however, not necessary: In principle, L_H may be any reasonable difference operator approximating L_h in some sense. In particular, we want to mention the important case of the *Galerkin approach* [109],[70],[42], which has a natural background in the finite-element area. Here the transfer operators I_h^H and I_H^h are used to define the coarse-grid operator L_H:

$$L_H := I_h^H L_h I_H^h. \tag{2.21}$$

2.4.3 More general smoothing procedures

Instead of applying the same relaxation operators S_h ν_1 times before and ν_2 times after the coarse-grid correction step, one may, of course, use different operators in the different smoothing steps. This straightforward extension takes, for example, the possibility into account that different relaxation parameters are chosen in different relaxation steps (see Section 3.6). M_h^H is then given by

$$M_h^H = S_{h,\nu} \cdot \ldots \cdot S_{h,\nu_1+1} \; K_h^H \; S_{h,\nu_1} \cdot \ldots \cdot S_{h,1}$$

where K_h^H is defined in Lemma 2.2.

Furthermore, not only relaxation methods may be used for error smoothing: any (iterative) procedure which has good smoothing properties and requires little computational work per iteration step, can, in principle, be used as a smoother in the multigrid context. In particular, certain pre-conditioning methods (various *incomplete LU-decomposition* methods) have been shown to yield good and robust smoothers [54],[60],[109],[110], also see Section 9.2.

2.4.4 Coarse-to-fine transfer using the grid equation

In certain efficient multigrid algorithms, the right hand side of the (original) grid equation (2.1) is used within the coarse-to-fine transfer (instead of performing a pure interpolation) [82],[111]. Note that such a coarse-to-fine tranfer can, in general, not be described by a linear operator but rather by an affine operator. Clearly, in such cases, also Lemma 2.1 is no longer true: The coarse-grid correction may now have a spectral radius < 1 (see [82]).

Coarse-to-fine transfers as mentioned above can often be interpreted as an interpolation (in the usual sense) followed by one (or one half) step of a suitable relaxation process [82]. They are of particular interest for the full multigrid method (FMG, see Chapter 6; [36]) also.

2.4.5 More general treatment of boundary conditions

In Section 2.1, we have assumed for simplicity that the (discrete) boundary conditions have been "eliminated". Because of this assumption we had to consider only one grid equation

$$L_h u_h = f_h \quad (\Omega_h) \tag{2.22}$$

with $u_h, f_h \in \mathbb{G}(\Omega_h)$, $L_h : \mathbb{G}(\Omega_h) \to \mathbb{G}(\Omega_h)$. In this introductory paper, we will always make this assumption. For example, for the multigrid treatment of second order equations with Dirichlet boundary conditions in general bounded domains (see Chapter 10) or with Neumann boundary conditions in rectangular domains [37],[36], the elimination of boundary conditions is a well established technique. In more general cases, however, a different treatment of boundary conditions may be necessary.

For its formal description, (2.22) has to be replaced by a system of grid equations (1.2) where Ω_h denotes an *interior grid* and Γ_h a *boundary grid.* In principle, all considerations and explanations of the previous sections can be interpreted with respect to this more general situation. All multigrid components $(S_h, I_h^H, I_H^h,..)$ have of course, to be defined separately for Ω_h and for Γ_h. For a more detailed description of these techniques, we refer to Brandt [25], Section 5. In particular, the question of suitable *boundary relaxation methods* is discussed there.

3. Analysis of a sample (h,2h) two-grid method for Poisson's equation

In this chapter, we consider a very simple two-grid method for the model problem (P). This is a concrete example for the general description of two-grid methods as given in Section 2.3. By determining the *spectral radius* $\rho(M_h^{2h})$ of the associated iteration operator M_h^{2h}, we prove the h-independency of its convergence factor. This result is valid even if only one relaxation step $(\nu=\nu_1=1)$ is performed per two-grid iteration. The considerations of this section are representative for what we call *model problem analysis* (see Chapters 7 and 8).

In Section 3.1 we define the algorithm. Section 3.2 refers to the relaxation operator S_h which is used for smoothing; on the basis of Fourier analysis, we give a first definition of a smoothing factor there. The Fourier representation of the coarse-grid correction operator K_h^{2h} is given in Section 3.3. Using the representations of S_h and K_h^{2h}, we determine the spectral radius $\rho(M_h^{2h})$ (Section 3.4). In Section 3.5 we are concerned with norms for M_h^{2h}, namely with its *spectral* and its *energy norm*. In Section 3.6 we outline algorithmic variants and their influence on the theoretical results.

Within the two-grid method which is considered here, we use a *Jacobi (under-) relaxation method* for smoothing. This smoothing method is by far not the most efficient one; it is, however, theoretically transparent and allows a simple rigorous and quantitative analysis. A disadvantage of Jacobi's method is the need of a relaxation parameter for good smoothing; this is not typical for smoothing techniques in general. (Other relaxation methods without this disadvantage will be discussed, e.g., in the Chapters 7 and 8.)

3.1 An (h,2h)-algorithm

For the solution of the discrete Poisson equation on the unit square, model problem (P) (see (1.10)), we consider an algorithm as described in Section 2.3, the j-th iteration of which (computing u_h^{j+1} from u_h^j) consists of the following components:

Smoothing part I: Apply ν_1 iteration steps of Jacobi ω-relaxation to u_h^j with fixed relaxation parameter ω (the choice of ω will be discussed in the following section):

$$\bar{u}_h^j := \mathrm{RELAX}^{\nu_1}(u_h^j, L_h, f_h; \omega). \tag{3.1}$$

Coarse-grid correction on

$$\Omega_{2h} = (0,1)^2 \cap G_{2h} = \{x = (2ih, 2jh) : i,j = 1,2,\ldots,N/2-1\} \tag{3.2}$$

proceeding as follows:

- Compute the defect $\bar{d}_h^j := f_h - L_h\bar{u}_h^j$.
- Restrict the defect $\bar{d}_h^j$ (fine-to-coarse transfer) using the *full weighting* (FW) operator, i.e.

$$I_h^{2h} \hat{=} \frac{1}{16}\begin{bmatrix} 1 & 2 & 1 \\ 2 & 4 & 2 \\ 1 & 2 & 1 \end{bmatrix}_h^{2h}. \tag{3.3}$$

- Compute the exact solution $\hat{v}_{2h}^j$ of the Ω_{2h}-defect equation $L_{2h}\hat{v}_{2h}^j = \bar{d}_{2h}^j$. Here L_{2h} is defined analogously to L_h.
- Interpolate the correction $\hat{v}_{2h}^j$ (coarse-to-fine transfer) using *bilinear interpolation*, i.e.

$$I_{2h}^h \hat{=} \frac{1}{4}\left]\begin{matrix} 1 & 2 & 1 \\ 2 & 4 & 2 \\ 1 & 2 & 1 \end{matrix}\right[_{2h}^h. \tag{3.4}$$

- Compute the corrected approximation on Ω_h: $\bar{u}_h^j + \hat{v}_h^j$.

Smoothing part II: Compute u_h^{j+1} by applying ν_2 smoothing steps of Jacobi ω-relaxation to $\bar{u}_h^j + \hat{v}_h^j$, i.e.

$$u_h^{j+1} = \text{RELAX}^{\nu_2}(\bar{u}_h^j + \hat{v}_h^j, L_h, f_h; \omega). \tag{3.5}$$

By Lemma 2.2, the iteration operator of this $(h,2h)$ two-grid method is given by

$$M_h^{2h} = M_h^{2h}(\nu_1,\nu_2;\omega) = S_h^{\nu_2}(\omega)\; K_h^{2h}\; S_h^{\nu_1}(\omega) \quad \text{with} \quad K_h^{2h} = I_h - I_{2h}^h\, L_{2h}^{-1}\, I_h^{2h}\, L_h \tag{3.6}$$

where $S_h = S_h(\omega)$ denotes the iteration operator which corresponds to Jacobi ω-relaxation.

3.2 The relaxation operator

One step of Jacobi ω-relaxation applied to problem (P) with first approximation w_h, i.e.

$$\bar{w}_h = \text{RELAX}(w_h, L_h, f_h; \omega) \tag{3.7}$$

is defined by

$$\bar{w}_h = w_h + \omega(z_h - w_h), \quad \frac{4}{h^2} z_h(x) + L_h^- w_h(x) = f_h(x) \quad (x \in \Omega_h). \tag{3.8}$$

Here L_h^- denotes the "off-diagonal" part of L_h, namely

$$L_h^- w_h(x) := L_h w_h(x) - \frac{4}{h^2} w_h(x) \quad (x \in \Omega_h). \tag{3.9}$$

Clearly, the corresponding iteration operator is given by

$$S_h = S_h(\omega) = I_h - \frac{\omega h^2}{4} L_h. \tag{3.10}$$

Let us first recall some well-known facts about the convergence properties of Jacobi ω-relaxation. These facts can be derived easily by considering the eigenfunctions of S_h, which are the same as those of L_h, namely

$$\varphi_n(x) = 2\sin(n_1\pi x_1)\sin(n_2\pi x_2) \quad (x \in \Omega_h;\ |n| \le N-1) \tag{3.11}$$

where $n=(n_1,n_2) \in \mathbb{N}^2$ and $|n|=\max(n_1,n_2)$. The corresponding eigenvalues of S_h are

$$\chi_n = \chi_n(\omega) = 1 - \frac{\omega}{2}(2 - \cos(n_1\pi h) - \cos(n_2\pi h)). \tag{3.12}$$

For the spectral radius $\rho(S_h) = \max\{|\chi_n| : |n| \le N-1\}$ we obtain

$$\begin{aligned} &\text{for } 0 < \omega \le 1: && \rho(S_h) = |\chi_{1,1}| = 1 - \omega(1-\cos\pi h) = 1 - O(h^2); \\ &\text{for } \omega \le 0 \text{ or } \omega > 1: && \rho(S_h) \ge 1 \quad \text{(if } h \text{ small enough).} \end{aligned} \tag{3.13}$$

In particular, with respect to the (very unsatisfactory) asymptotic convergence, there is no use in introducing the relaxation parameter: $\omega=1$ is the best choice.

The situation is quite different with respect to the smoothing properties of Jacobi ω-relaxation. For $0 < \omega \le 1$, we first observe by (3.13) that it is the smoothest eigenfunction $\varphi_{1,1}$ which is responsible for the slow convergence of Jacobi's method. Highly oscillating eigenfunctions are reduced much faster if ω is chosen properly. To see this, we expand the errors before and after one relaxation step, namely

$$v_h := u_h - w_h \quad \text{and} \quad \bar{v}_h := u_h - \bar{w}_h,$$

into discrete eigenfunction series:

$$v_h = \sum_{|n| \le N-1} \alpha_n \varphi_n, \quad \bar{v}_h = \sum_{|n| \le N-1} \chi_n \alpha_n \varphi_n. \tag{3.14}$$

The smoothing properties of $S_h(\omega)$ are measured by distinguishing low and high frequencies (with respect to the coarser grid Ω_{2h} used). As motivated in Section 2.2, it is reasonable to define as

$$\begin{aligned} &\textit{low frequencies:} \quad \varphi_n \ \text{with} \ |n| < N/2, \\ &\textit{high frequencies:} \quad \varphi_n \ \text{with} \ N/2 \le |n| \le N-1. \end{aligned} \tag{3.15}$$

In other words: The low frequencies are those eigenfunctions of L_h, which are representable also on the coarser grid Ω_{2h}. The high frequencies are "not visible" on Ω_{2h} at all. (Cf. Figure 2.3, where this distinction was illustrated for the corresponding 1D-case.)

We now define the *smoothing factor* $\mu(h;\omega)$ of S_h (and its supremum $\mu^*(\omega)$ over h) as the worst factor by which high frequency error components are reduced per relaxation step, i.e.

$$\begin{aligned} \mu(h;\omega) &:= \max\{|\chi_n| : N/2 \le |n| \le N-1\}, \\ \mu^*(\omega) &:= \sup\{\mu(h;\omega) : h \le 1/4\}. \end{aligned} \tag{3.16}$$

Remark: This or similar definitons of the smoothing factor can also be used for some other simple smoothing methods. It has, however, to be substantially refined for smoothing methods like RB and ZEBRA relaxation which are much more efficient in smoothing than Jacobi's method. We give a refined definition in Section 7.5.

Inserting (3.12), we get from (3.16)

$$\begin{aligned} \mu(h;\omega) &= \max\{|1-\omega(2-\cos\pi h)/2|, |1-\omega(1+\cos\pi h)|\}, \\ \mu^*(\omega) &= \max\{|1-\omega/2|, |1-2\omega|\}. \end{aligned} \tag{3.17}$$

This shows that Jacobi's relaxation has no smoothing properties for $\omega \le 0$ or $\omega > 1$:

$$\mu(h;\omega) \ge 1 \ \text{if} \ \omega \le 0 \ \text{or} \ \omega > 1 \quad (\text{and } h \text{ is sufficiently small}).$$

For $0 < \omega < 1$, however, the smoothing factor is smaller than 1 and bounded away from 1, independently of h. For $\omega=1$, we have a smoothing factor of $1-O(h^2)$ only. In particular, we find by (3.17):

$$\mu(h;\omega) = \begin{cases} \cos\pi h & \text{if } \omega = 1 \\ (2+\cos\pi h)/4 & \text{if } \omega = 1/2 \\ (1+2\cos\pi h)/5 & \text{if } \omega = 4/5 \end{cases} \qquad \mu^*(\omega) = \begin{cases} 1 & \text{if } \omega = 1 \\ 3/4 & \text{if } \omega = 1/2 \\ 3/5 & \text{if } \omega = 4/5\,. \end{cases}$$

The choice $\omega=4/5$ is optimal in the following sense:

$$\inf\{\mu^*(\omega) : 0 \le \omega \le 1\} = \mu^*(4/5) = 3/5. \tag{3.18}$$

With respect to $\mu(h,\omega)$, one obtains

$$\inf\{\mu(h;\omega) : 0 \le \omega \le 1\} = \mu(h;\frac{4}{4+\cos\pi h}) = \frac{3\cos\pi h}{4+\cos\pi h} = \frac{3}{5} - O(h^2).$$

3.3 The coarse-grid correction operator

For the coarse-grid correction operator

$$K_h^{2h} = I_h - I_{2h}^h L_{2h}^{-1} I_h^{2h} L_h,$$

it turns out that the (at most) 4-dimensional subspaces of $\mathbb{G}(\Omega_h)$

$$E_{h,n} := \text{span}\,\{\varphi_{n_1,n_2};\ \varphi_{N-n_1,N-n_2};\ -\varphi_{N-n_1,n_2};\ -\varphi_{n_1,N-n_2}\} \quad (|n|\le N/2) \qquad (3.19)$$

are invariant under K_h^{2h}, i.e.

$$K_h^{2h} : E_{h,n} \to E_{h,n} \quad (|n| \le N/2). \qquad (3.20)$$

Consequently, as the φ_n $(|n| \le N-1)$ form an orthonormal basis of $\mathbb{G}(\Omega_h)$ (with respect to (1.7)), K_h^{2h} is orthogonally equivalent to a block-diagonal matrix consisting of (at most) (4,4)-blocks $\hat{K}_{h,n}^{2h}$. This is a characteristic feature of what we call *model problem analysis*.

For a detailed description of the matrices $\hat{K}_{h,n}^{2h}$, we also need a basis of eigenfunctions of L_{2h}. A suitable basis is given by

$$\Phi_n(x) := 2\sin(n_1\pi x_1)\sin(n_2\pi x_2) \qquad (x \in \Omega_{2h};\ |n| \le N/2-1). \qquad (3.21)$$

On Ω_{2h}, the Φ_n and the basis functions of $E_{h,n}$ coincide:

$$\varphi_{n_1,n_2}(x) = \varphi_{N-n_1,N-n_2}(x) = -\varphi_{N-n_1,n_2}(x) = -\varphi_{n_1,N-n_2}(x) = \Phi_{n_1,n_2}(x) \qquad (3.22)$$

$$(x \in \Omega_{2h};\ |n| \le N/2-1).$$

For $n_1 = N/2$ and/or $n_2 = N/2$, the spaces $E_{h,n}$ are 1-/2-dimensional, respectively, and their basis functions coincide on Ω_{2h} with the zero grid function.

The transfer operators I_{2h}^h and I_h^{2h} have the characteristic properties

$$\left.\begin{array}{ll} I_{2h}^h : \text{span}\,\{\Phi_n\} \to E_{h,n} & (|n| \le N/2-1), \\ I_h^{2h} : E_{h,n} \to \text{span}\,\{\Phi_n\} & (|n| \le N/2-1), \\ I_h^{2h}\varphi_n = 0 & (n_1 = N/2 \text{ and/or } n_2 = N/2). \end{array}\right\} \qquad (3.23)$$

In more detail, we have for fixed n ($|n| \le N/2-1$)

$$I_{2h}^{h}\,\Phi_{n_1,n_2} = (1-\xi)(1-\eta)\varphi_{n_1,n_2} + \xi\eta\varphi_{N-n_1,N-n_2} - \xi(1-\eta)\varphi_{N-n_1,n_2} - (1-\xi)\eta\varphi_{n_1,N-n_2} \tag{3.24}$$

and

$$I_h^{2h}\begin{Bmatrix}\varphi_{n_1,n_2}\\ \varphi_{N-n_2,N-n_2}\\ -\varphi_{N-n_1,n_2}\\ -\varphi_{n_1,N-n_2}\end{Bmatrix} = \begin{Bmatrix}(1-\xi)(1-\eta)\\ \xi\eta\\ \xi(1-\eta)\\ (1-\xi)\eta\end{Bmatrix}\Phi_{n_1,n_2}. \tag{3.25}$$

Here we use the abbreviations

$$\xi = \sin^2(n_1\pi/2N), \qquad \eta = \sin^2(n_2\pi/2N). \tag{3.26}$$

Together with the fact that the φ_n and the Φ_n are eigenfunctions of L_h and L_{2h}, respectively, (3.20) follows immediately. In particular, we obtain

$$\hat{K}_{h,n}^{2h} = \begin{cases} I - \frac{1}{\Lambda}\begin{bmatrix} b_i c_j \end{bmatrix}_{4,4} & (\text{if } |n| < N/2)\\ (2,2)\text{-identity matrix} & (\text{if } n_1 \text{ or } n_2=N/2)\\ (1,1)\text{-identity matrix} & (\text{if } n_1=n_2=N/2)\end{cases} \tag{3.27}$$

with $\Lambda = \xi(1-\xi) + \eta(1-\eta)$ and

$$b_1 = (1-\xi)(1-\eta),\quad b_2 = \xi\eta,\quad b_3 = \xi(1-\eta),\quad b_4 = (1-\xi)\eta,$$
$$c_1 = b_1(\xi+\eta),\quad c_2 = b_2(2-\xi-\eta),\quad c_3 = b_3(1-\xi+\eta),\quad c_4 = b_4(1+\xi-\eta).$$

3.4 Spectral radius of the two-grid operator

The invariance of the spaces $E_{h,n}$ ($|n| \le N/2$) under K_h^{2h} and under S_h also imply their invariance under M_h^{2h}. Using (3.12), (3.27) and the abbreviations (3.26) one immediately obtains the $E_{h,n}$-representation $\hat{M}_{h,n}^{2h}$ of M_h^{2h}:

$$\hat{M}_{h,n}^{2h} = \hat{M}_{h,n}^{2h}(\nu_1,\nu_2;\omega) = \hat{S}_{h,n}^{\nu_2}(\omega)\;\hat{K}_{h,n}^{2h}\;\hat{S}_{h,n}^{\nu_1}(\omega) \tag{3.28}$$

where

$$\hat{S}_{h,n}(\omega) = \begin{cases} \begin{bmatrix} 1-\omega(\xi+\eta) & & & \\ & 1-\omega(2-\xi-\eta) & & \\ & & 1-\omega(1-\xi+\eta) & \\ & & & 1-\omega(1+\xi-\eta) \end{bmatrix}_{4,4} & (\text{if } |n| < N/2) \\ \begin{bmatrix} 1-\omega(\xi+\eta) & \\ & 1-\omega(2-\xi-\eta) \end{bmatrix}_{2,2} & (\text{if } n_1 \text{ or } n_2 = N/2) \\ \begin{bmatrix} 1-\omega(\xi+\eta) \end{bmatrix}_{1,1} & (\text{if } n_1 = n_2 = N/2) \end{cases} \tag{3.29}$$

Thus the determination of the spectral radius $\rho(M_h^{2h})$ has been reduced to the calculation of the spectral radii of (at most) (4,4)-matrices:

$$\rho(M_h^{2h}) = \max\{\rho(\hat{M}_{h,n}^{2h}) : |n| \le N/2\}. \tag{3.30}$$

This quantity depends in particular on the parameter ω and on $\nu := \nu_1 + \nu_2$. (Since $\rho(AB) = \rho(BA)$ for any linear operators A and B, $\rho(M_h^{2h})$ does not depend on ν_1 and ν_2 individually.) In the following, we shall use the notation

$$\rho(h,\nu;\omega) := \rho(M_h^{2h}(\nu_1,\nu_2;\omega)). \tag{3.31}$$

Usually, one is more interested in

$$\rho^*(\nu;\omega) := \sup\{\rho(h,\nu;\omega) : h \le 1/4\} \tag{3.32}$$

than in $\rho(h,\nu;\omega)$ for fixed h. From the representation (3.28) (with (3.27) and (3.29)), one recognizes that $\rho(\hat{M}_{h,n}^{2h})$ can be written as a certain function $f(\xi,\eta)$. Using this function, ρ^* is conveniently computed as

$$\rho^*(\nu,\omega) = \sup\{f(\xi,\eta) : 0 < \xi,\eta \le 1/2\}. \tag{3.33}$$

(For simplicity, we use the term "convergence factor" for both ρ and ρ^*. It will be clear from the context, which quantity is actually considered.)

In Table 3.1, we have listed ρ^* as a function of ν for the two relaxation parameters $\omega=0.5$ and $\omega=0.8$. First, we recognize that - as already suggested by the respective smoothing factors, see Section 3.2 - the parameter $\omega=0.8$ indeed yields better convergence factors ρ^* than $\omega=0.5$. One sees that the convergence factors ρ^* decrease for increasing ν. This does not mean, however, that large values of ν are suitable with respect to efficiency, as also the computational work increases with ν. We postpone the question of efficiency: In Chapter 4 we will discuss the computational work in connection with complete multigrid iterations; results concerning the efficiency of several methods will be given, e.g., in Chapter 8.

That too large values of ν are useless, can already be seen from a comparison of ρ^* and $(\mu^*)^\nu$. For small values of ν we observe a remarkable accordance between these quantities. If ν increases, however, $(\mu^*)^\nu$ is no longer a good prediction for ρ^*: the high smoothing effect is not fully exploited as the reduction of low error frequencies by one coarse-grid correction step is not good enough, or the smoothing effect is even partly destroyed by the coarse-grid correction (which introduces new high frequencies by itself). Typically, one has $\rho^*(\nu;\omega) \sim \text{const}/\nu$ $(\nu \to \infty)$ (see, for instance, Theorem 8.1). The difference between $(\mu^*)^\nu$ and ρ^* occurs all the sooner, the better the smoothing properties of S_h are ($\nu \geq 4$ for $\omega = 0.8$, $\nu \geq 8$ for $\omega = 0.5$).

ν	$\omega = 0.5$		$\omega = 0.8$	
	$(\mu^*(\omega))^\nu$	$\rho^*(\nu;\omega)$	$(\mu^*(\omega))^\nu$	$\rho^*(\nu;\omega)$
1	0.750	0.750	0.600	0.600
2	0.563	0.563	0.360	0.360
3	0.422	0.422	0.216	0.216
4	0.316	0.316	0.130	0.137
5	0.237	0.237	0.078	0.113
6	0.178	0.178	0.047	0.097
7	0.133	0.133	0.028	0.085
8	0.100	0.118	0.017	0.076
9	0.075	0.106	0.010	0.068
10	0.056	0.097	0.006	0.062

Table 3.1: Comparison of smoothing factors μ^* and two-grid convergence factors ρ^* for different ν and ω.

Finally, we give some values for $\rho(h,\nu;\omega)$ as a function of h. The corresponding results in Table 3.2 show that ρ tends to ρ^* rather quickly. Thus, in the cases considered, the main information about the two-grid convergence is contained in ρ^*.

h	$\omega = 0.5$				$\omega = 0.8$			
	ν=1	ν=2	ν=3	ν=4	ν=1	ν=2	ν=3	ν=4
1/4	0.677	0.458	0.310	0.217	0.483	0.233	0.171	0.130
1/8	0.731	0.534	0.391	0.285	0.570	0 324	0.185	0.130
1/16	0.745	0.555	0.414	0.308	0.592	0.351	0.208	0.135
1/32	0.749	0.561	0.420	0.314	0.598	0.358	0.214	0.137
1/64	0.750	0.562	0.421	0.316	0.600	0.359	0.215	0.137
1/128	0.750	0.562	0.422	0.316	0.600	0.360	0.216	0.137
$\rho^*(\nu;\omega)$	0.750	0.563	0.422	0.316	0.600	0.360	0.216	0.137

Table 3.2: The two-grid convergence factor $\rho(h,\nu;\omega)$ as a function of h

3.5 Norms of the two-grid operator

Whereas the spectral radius $\rho(M_h^{2h})$ gives insight into the asymptotic convergence behavior of a two-grid method, norms are needed to measure the actual error (or defect) reducing per iteration step. In particular, essential use of norms of M_h^{2h} is made in the theoretical investigations of complete multigrid iterations (Chapter 4) and of the full multigrid method (Chapter 6).

There are many reasonable possibilities to choose norms. Of course, different choices of norms will in general lead to very different results. A general observation is that the spectral radius ρ is less sensitive with respect to algorithmical details than norms usually are. For example, norms considerably depend on ν_1 and on ν_2, whereas ρ depends only on the sum $\nu=\nu_1+\nu_2$.

In this paper, we mainly consider the operatornorm $\|\cdot\|_S$ corresponding to the Euclidian inner product (1.7) on $\mathbb{G}(\Omega_h)$, i.e. the *spectralnorm*

$$\|M\|_S = \sqrt{\rho(MM^*)}, \tag{3.34}$$

where M denotes any linear operator $M : \mathbb{G}(\Omega_h) \to \mathbb{G}(\Omega_h)$. Apart from the error reduction $(M=M_h^{2h})$, we sometimes also consider the defect reduction $(M=L_hM_h^{2h}L_h^{-1})$.

For positive-definite symmetric operators L_h, the *energy norm* (which is induced by the inner product (1.8)) is also of - mainly theoretical - interest. The corresponding operatornorm is given by

$$\|M\|_E = \|L_h^{1/2}\, M\, L_h^{-1/2}\|_S = \sqrt{\rho(L_hML_h^{-1}M^*)}. \tag{3.35}$$

(Here M^* denotes the operator adjoint to M with respect to the Euclidian inner product.) We introduce the following notations:

$$\sigma_S := \|M_h^{2h}\|_S,\quad \sigma_E := \|M_h^{2h}\|_E,\quad \sigma_d := \|L_hM_h^{2h}L_h^{-1}\|_S. \tag{3.36}$$

In particular, these quantities depend on h,ν_1,ν_2 and ω. By σ_S^*, σ_E^* and σ_d^* we denote the suprema of σ_S, σ_E and σ_d with respect to h, e.g.

$$\sigma_S^*(\nu_1,\nu_2;\omega) := \sup\,\{\sigma_S(h,\nu_1,\nu_2;\omega) : h \le 1/4\}. \tag{3.37}$$

All the above norms can be determined from the representation (3.28) in much the same way as ρ and ρ^*. In particular, one obtains

$$\sigma_S = \max \{ \| \hat{M}^{2h}_{h,n} \|_S : |n| \le N/2 \},$$
$$\sigma_E = \max \{ \| \hat{L}^{1/2}_{h,n} \hat{M}^{2h}_{h,n} \hat{L}^{-1/2}_{h,n} \|_S : |n| \le N/2 \}, \tag{3.38}$$
$$\sigma_d = \max \{ \| \hat{L}_{h,n} \hat{M}^{2h}_{h,n} \hat{L}^{-1}_{h,n} \|_S : |n| \le N/2 \},$$

where $\hat{L}_{h,n}$ denotes the (diagonal) matrix representation of L_h with respect to $E_{h,n}$. The computation of σ^*_S, σ^*_E and σ^*_d can be performed analogously as for ρ^* (cf. (3.33)).

For our sample (h,2h)-method, we have listed several values σ^*_S, σ^*_E and σ^*_d in Tables 3.3a and 3.3b. For comparison, we also recall the corresponding ρ^*-values already given in Table 3.1. It is, of course, a general aim to have not only a small spectral radius of M^{2h}_h but also small norms. In both tables, 3.3a and 3.3b, we have underlined those norm-values which are optimal (=spectral radius). According to these results, it seems to be reasonable to choose ν_1 and ν_2 not very different from each other (and rather $\nu_1 \ge \nu_2$ than $\nu_1 \le \nu_2$).

(ν_1,ν_2)	$\rho^*(\nu;\omega)$	$\sigma^*_S(\nu_1,\nu_2;\omega)$	$\sigma^*_E(\nu_1,\nu_2;\omega)$	$\sigma^*_d(\nu_1,\nu_2;\omega)$
(1,0)	0.750	$\underline{0.750}$	$\underline{0.750}$	1.118
(0,1)		1.118	$\underline{0.750}$	$\underline{0.750}$
(2,0)	0.563	$\underline{0.563}$	$\underline{0.563}$	1.031
(1,1)		$\underline{0.563}$	$\underline{0.563}$	$\underline{0.563}$
(0,2)		1.031	$\underline{0.563}$	$\underline{0.563}$
(3,0)	0.422	$\underline{0.422}$	$\underline{0.422}$	1.008
(2,1)		$\underline{0.422}$	$\underline{0.422}$	0.515
(1,2)		0.515	$\underline{0.422}$	$\underline{0.422}$
(0,3)		1.008	$\underline{0.422}$	$\underline{0.422}$
(4,0)	0.316	$\underline{0.316}$	0.323	1.002
(3,1)		$\underline{0.316}$	$\underline{0.316}$	0.504
(2,2)		$\underline{0.316}$	$\underline{0.316}$	$\underline{0.316}$
(1,3)		0.504	$\underline{0.316}$	$\underline{0.316}$
(0,4)		1.002	0.323	$\underline{0.316}$

Table 3.3a: Spectral radii and norms for $\omega = 0.5$

(ν_1,ν_2)	$\rho^*(\nu;\omega)$	$\sigma_S^*(\nu_1,\nu_2;\omega)$	$\sigma_E^*(\nu_1,\nu_2;\omega)$	$\sigma_d^*(\nu_1,\nu_2;\omega)$
(1,0)	0.600	$\underline{0.600}$	$\underline{0.600}$	1.020
(0,1)		1.020	$\underline{0.600}$	$\underline{0.600}$
(2,0)		$\underline{0.360}$	$\underline{0.360}$	1.000
(1,1)	0.360	$\underline{0.360}$	$\underline{0.360}$	$\underline{0.360}$
(0,2)		1.000	$\underline{0.360}$	$\underline{0.360}$
(3,0)		$\underline{0.216}$	0.269	1.000
(2,1)	0.216	$\underline{0.216}$	$\underline{0.216}$	0.239
(1,2)		0.239	$\underline{0.216}$	$\underline{0.216}$
(0,3)		1.000	0.269	$\underline{0.216}$
(4,0)		0.148	0.233	1.000
(3,1)		$\underline{0.137}$	0.140	0.209
(2,2)	0.137	$\underline{0.137}$	$\underline{0.137}$	$\underline{0.137}$
(1,3)		0.209	0.140	$\underline{0.137}$
(0,4)		1.000	0.233	0.148

Table 3.3b: Same as Table 3.3a for $\omega = 0.8$

The following equalities hold between the quantities considered:

$$\sigma_S^*(\nu_1,\nu_2;\omega) = \sigma_d^*(\nu_2,\nu_1;\omega), \quad \sigma_E^*(\nu_1,\nu_2;\omega) = \sigma_E^*(\nu_2,\nu_1;\omega),$$
$$\sigma_E^*(\nu_1,\nu_2;\omega) = \rho^*(\nu;\omega) \quad (\text{if } \nu_1 = \nu_2). \tag{3.39}$$

They are an immediate consequence of the relations

$$(M_h^{2h}(\nu_1,\nu_2;\omega))^* = L_h \, M_h^{2h}(\nu_2,\nu_1;\omega) \, L_h^{-1},$$
$$(L_h^{1/2} \, M_h^{2h}(\nu_1,\nu_2;\omega) \, L_h^{-1/2})^* = L_h^{1/2} \, M_h^{2h}(\nu_2,\nu_1;\omega) \, L_h^{-1/2} \tag{3.40}$$

which hold in our particular example and can easily be verified.

3.6 Algorithmic variants

We want to mention two modifications of the (h,2h)-method considered and to show their influence on the quantities introduced above.

3.6.1 Use of straight injection for the fine-to-coarse transfer

In practice the FW operator (3.3) may often be replaced by simpler restriction operators. The simplest (and cheapest) fine-to-coarse transfer is given by the operator of *straight injection* (INJ)

$$I_h^{2h} = [1]_h^{2h}, \text{ i.e. } (I_h^{2h} w_h)(x) = w_h(x) \quad (x \in \Omega_{2h}). \tag{3.41}$$

Heuristically, it is clear that this operator should give similar results as the FW operator (3.3) as long as the defects (to which I_h^{2h} is applied) are really smooth.

If the INJ operator is used in our sample method, the theoretical considerations have to be modified only slightly: Instead of (3.25), we now have

$$I_h^{2h} \begin{Bmatrix} \varphi_{n_1,n_2} \\ \varphi_{N-n_1,N-n_2} \\ -\varphi_{N-n_1,n_2} \\ -\varphi_{n_1,N-n_2} \end{Bmatrix} = \Phi_{n_1,n_2} . \tag{3.42}$$

With this modification, one can calculate ρ^*, σ_S^*, σ_E^*, σ_d^* as the previous sections. Calculating ρ^*, it turns out that the asymptotic convergence properties are not influenced significantly by this exchange of the fine-to-coarse transfer operator: One obtains the same ρ^*-values as shown in Table 3.1 if ν is not too large ($\nu \le 7$ for $\omega = 0.5$ and $\nu \le 3$ for $\omega = 0.8$). For larger values of ν, the asymptotic convergence factor is even slightly better if INJ is used instead of FW.

The behavior of the norms σ_S^* and σ_E^*, however, is quite different now: We find for all ν and ω:

$$\sigma_S^*(\nu;\omega) = \sigma_E^*(\nu;\omega) = \infty. \tag{3.43}$$

The reason for this can easily be seen by applying M_h^{2h} to one of the <u>highest</u> frequencies, e.g. to $\varphi_{N-1,1}$: This frequency is mapped into a grid function which contains the <u>low</u> frequency component $O(1/h^2)\ \varphi_{1,1}$.

The above behavior of σ_S^* and σ_E^* is characteristic for the use of straight injection within multigrid processes. For many <u>theoretical</u> approaches, where the above norms are needed, the INJ operator is therefore useless. On the other hand, in <u>practice</u>, straight injection gives often similar (or even better) results as full weighting. One should be aware, however, that errors which contain significant highest frequency components (see above), may be enlarged considerably if only <u>one</u> multigrid iteration step in performed. (This is the usual application in the FMG method, see Chapter 6!)

3.6.2 Jacobi ω-relaxation with several parameters

As we have seen above, the smoothing properties of Jacobi ω-relaxation significantly depend on the choice of ω. If $\nu \geq 2$ relaxation steps are carried out (per two-grid iteration), one can try to use different parameters $\omega_1, \ldots, \omega_\nu$ in each step in order to improve the total smoothing effect. A straightforward extension of the definition of the smoothing factor (3.16) to this more general case is given by

$$\mu(h,\nu;\omega_1,\ldots,\omega_\nu) := \sqrt[\nu]{\max\{|\chi_n(\omega_1)\cdot\ldots\cdot\chi_n(\omega_\nu)| : N/2 \leq |n| \leq N-1\}}. \tag{3.44}$$

Instead of (3.17), one now gets:

$$\mu^*(\nu;\omega_1,\ldots,\omega_\nu) = \sqrt[\nu]{\max\{|(1-\omega_1 t)\cdot\ldots\cdot(1-\omega_\nu t)| : 1/2 \leq t \leq 2\}}.$$

Minimizing μ^* with respect to $\omega_1,\ldots,\omega_\nu$ (for fixed ν) gives the optimal parameters

$$\omega_j = \left(\frac{5}{4} + \frac{3}{4}\cos\left(\frac{2j-1}{2\nu}\pi\right)\right)^{-1} \quad (j = 1,\ldots,\nu) \tag{3.45}$$

(zeros of Chebyshev polynomials).

These parameters are used in Table 3.4 where some values for μ^* and ρ^* are given. As one can see, the use of different relaxation parameters gives some improvement in the case considered (cf. Table 3.1). One should, however, take the following into account: Firstly, the explicit determination of optimal parameters is restricted to rather special situations. Secondly, as we have already pointed out previously, there are more efficient smoothing methods (for Poisson-like equations) than Jacobi ω-relaxation, which do not even need a parameter (for example, RB relaxation, see Section 8.2).

ν	$(\mu^*(\nu;\omega_1,\ldots,\omega_\nu))^\nu$	$\rho^*(\nu;\omega_1,\ldots,\omega_\nu)$
1	0.600	0.600
2	0.220	0.220
3	0.074	0.126
4	0.025	0.110

Table 3.4: Jacobi ω-relaxation with optimal parameters (3.45)

4. Complete multigrid cycle

Up to now, we have described the multigrid principle only in its two-grid version. We have, however, already pointed out that two-grid methods - usually - are not used in practice: they serve only as the (theoretical) basis for the real multigrid method.

The multigrid idea starts from the observation that in a convergent two-grid method it is not necessary to solve the coarse-grid defect equation (2.19)

$$L_H \hat{v}_H^j = \bar{d}_H^j \tag{4.1}$$

exactly. Instead, without essential loss of convergence speed, one may replace $\hat{v}_H^j$ by a suitable approximation. A natural way to obtain such an approximation is to apply an analogous two-grid method to (4.1) also, where an even coarser grid than Ω_H is used. Clearly, if the convergence factor of this two-grid method is small enough, it is sufficient to perform only a few, say γ (see Figure 4.1), iteration steps to obtain a good enough approximation to the solution of (4.1). This idea can, in a straightforward manner, be applied recursively, using coarser and coarser grids, down to some coarsest grid. On this coarsest grid any solution method may be used (e.g. a direct method or the smoothing process itself if it has sufficiently good convergence properties on the coarsest grid).

Most parts of the considerations in this chapter are independent of the way in which coarser grids are constructed. Usually, however, we have standard coarsening in mind. In particular, in this case, the asymptotic optimality of multigrid methods follows easily from a very simple result on their h-independent convergence (see Section 4.3) and on the computational work needed (see Section 4.4).

4.1 Notation, sequence of grids and operators

Before we provide the notation for a formal description of the multigrid recursion, let us illustrate the structure of one iteration step *(cycle)* of a multigrid method with a few pictures which are given in Figure 4.1. Here $\circ$, $\square$, \ and / mean smoothing, solving exactly, fine-to-coarse and coarse-to-fine transfer, respectively. With respect to the computational work (see Section 4.4), mainly the case $\gamma \le 2$ is of practical interest. For obvious reasons, we refer to the cases $\gamma=1$ and $\gamma=2$ as to *V-cycles* and *W-cycles*, respectively.

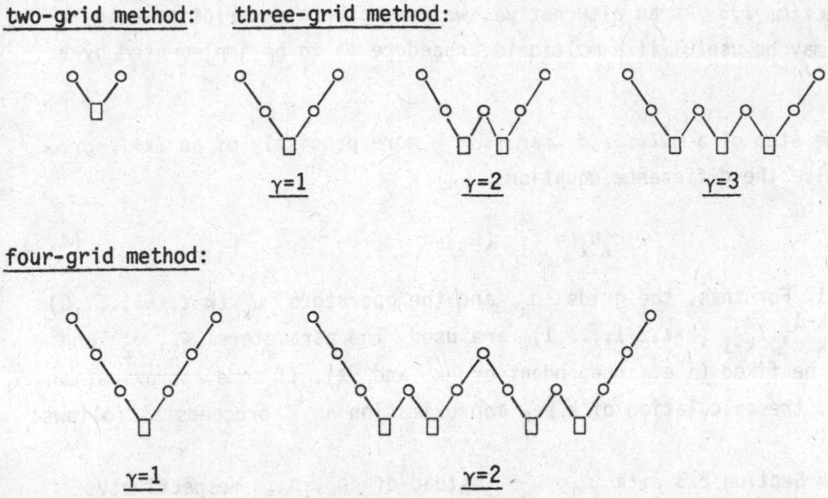

Figure 4.1: Structure of one multigrid cycle for different numbers of grids and different values of γ.

For a formal description of multigrid methods we now use a sequence of increasingly finer grids Ω_{h_ℓ}, characterized by a sequence of meshsizes h_ℓ ($\ell=0,1,2,\ldots$). For simplicity, we replace the index h_ℓ by ℓ (for grids, grid functions and grid operators) in the following. For each Ω_ℓ, we assume linear operators

$$
\begin{aligned}
&L_\ell : \mathbb{G}(\Omega_\ell) \to \mathbb{G}(\Omega_\ell), &\qquad &S_\ell : \mathbb{G}(\Omega_\ell) \to \mathbb{G}(\Omega_\ell),\\
&I_\ell^{\ell-1} : \mathbb{G}(\Omega_\ell) \to \mathbb{G}(\Omega_{\ell-1}), &\qquad &I_{\ell-1}^{\ell} : \mathbb{G}(\Omega_{\ell-1}) \to \mathbb{G}(\Omega_\ell)
\end{aligned}
\tag{4.2}
$$

and discrete equations

$$L_\ell u_\ell = f_\ell \qquad (\Omega_\ell) \tag{4.3}$$

(with L_ℓ invertible) to be given. Here $\mathbb{G}(\Omega_\ell)$ denotes the space of gridfunctions on Ω_ℓ. The operators S_ℓ denote the linear iteration operators corresponding to given relaxation methods. The result $\bar{w}_\ell$ of ν relaxation steps (applied to $L_\ell u_\ell = f_\ell$ with first approximation w_ℓ) will be denoted by

$$\bar{w}_\ell = \mathrm{RELAX}^{\nu}(w_\ell, L_\ell, f_\ell). \tag{4.4}$$

4.2 Recursive definition of a complete multigrid cycle

A convenient way to define a complete multigrid iteration step (*cycle*) is to use an Algol-like description. The recursive definition of a multigrid cycle then can easily be established using a self-calling procedure. A description of this type is given in [50], Section 1.3. As an alternative, we here give a description using a flow chart. This may be useful if a multigrid procedure is to be implemented by a FORTRAN program.

We describe one step of a *multigrid iteration* - more precisely of an *$(\ell+1)$-grid iteration* - to solve the difference equations

$$L_\ell u_\ell = f_\ell \quad (\Omega_\ell) \tag{4.5}$$

for a fixed $\ell \geq 1$. For this, the grids Ω_k and the operators L_k $(k=\ell,\ell-1,\ldots,0)$ as well as S_k, I_k^{k-1}, I_{k-1}^k $(k=\ell,\ell-1,\ldots,1)$ are used. The parameters ν_1, ν_2 and γ are assumed to be fixed (i.e. independent of k and ℓ). If some approximation u_ℓ^j of u_ℓ is given, the calculation of a new approximation u_ℓ^{j+1} proceeds as follows:

If $\ell = 1$: Like in Section 2.3 with Ω_1, Ω_o instead of Ω_h, Ω_H, respectively.

If $\ell > 1$:

(1) Smoothing part I:

- Compute $\bar{u}_\ell^j$ by applying ν_1 (≥ 0) smoothing steps to u_ℓ^j:

$$\bar{u}_\ell^j := \text{RELAX}^{\nu_1}(u_\ell^j, L_\ell, f_\ell).$$

(2) Coarse-grid correction:

- Compute the defect: $\qquad \bar{d}_\ell^j := f_\ell - L_\ell \bar{u}_\ell^j.$

- Restrict the defect: $\qquad \bar{d}_{\ell-1}^j := I_\ell^{\ell-1} \bar{d}_\ell^j.$

- Compute an approximate solution $\tilde{v}_{\ell-1}^j$ of the defect equation on $\Omega_{\ell-1}$

$$L_{\ell-1} \hat{v}_{\ell-1}^j = \bar{d}_{\ell-1}^j \tag{4.6}$$

by performing $\gamma \geq 1$ iterations of the *ℓ-grid method* (using the grids $\Omega_{\ell-1}, \Omega_{\ell-2}, \ldots, \Omega_o$ and the corresponding grid operators) applied to (4.6) with the zero grid function as first approximation.

- Interpolate the correction: $\qquad \tilde{v}_\ell^j := I_{\ell-1}^\ell \tilde{v}_{\ell-1}^j.$

- Compute the corrected approximation on Ω_ℓ: $\qquad \bar{u}_\ell^j + \tilde{v}_\ell^j.$

(3) Smoothing part II:

- Compute u_ℓ^{j+1} by applying ν_2 (≥ 0) smoothing steps to $\bar{u}_\ell^j + \tilde{v}_\ell^j$:

$$u_\ell^{j+1} := \text{RELAX}^{\nu_2}(\bar{u}_\ell^j + \tilde{v}_\ell^j, L_\ell, f_\ell).$$

The same process is described in the flow-chart below. There a switching parameter $0 \le C(k) \le \gamma$ is introduced to control when to go to a coarser grid and when to go back to a finer grid.

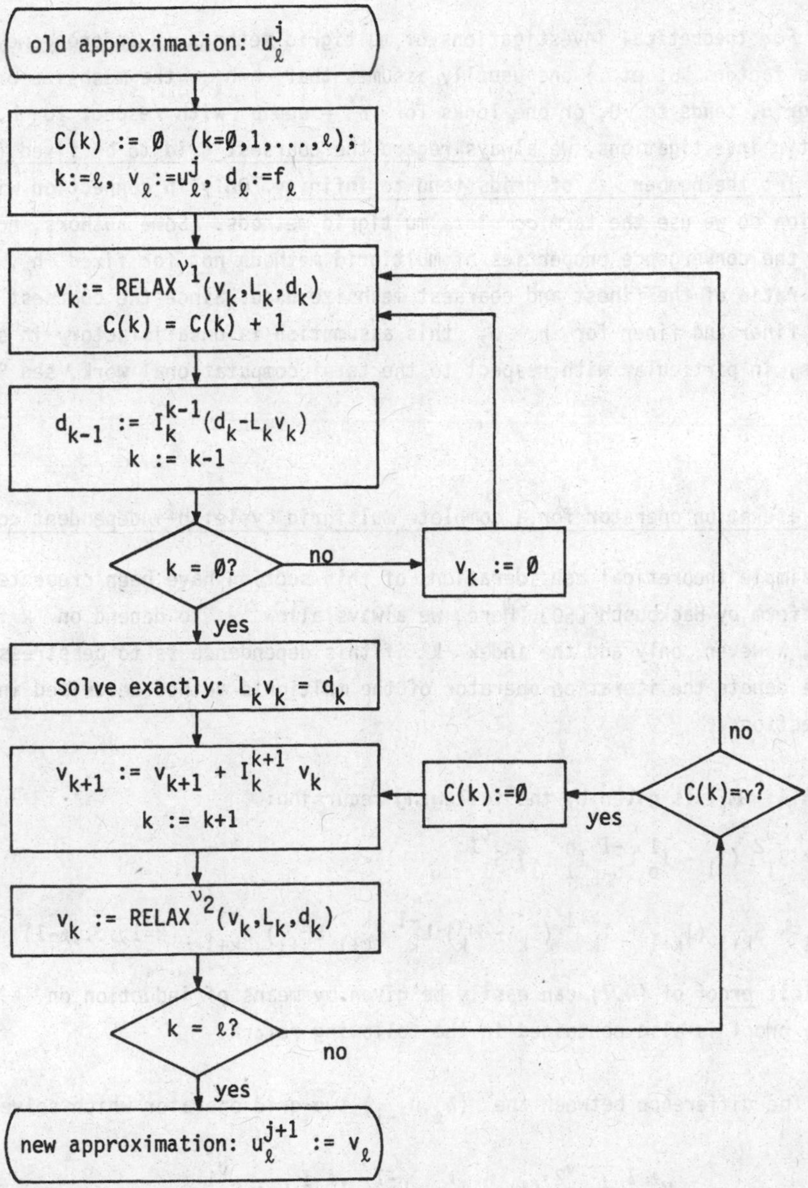

Figure 4.2: Flow-chart for one multigrid iteration step to solve $L_\ell u_\ell = f_\ell$ $(\ell \ge 1)$.

So far, we have assumed that the parameters ν_1, ν_2 and the recursion parameter γ are fixed numbers. This is, of course, not necessary. In particular, γ may depend on k (and possibly on ℓ). Certain combinations of $\gamma=1$ and $\gamma=2$ are indeed used in practice. We will discuss one such choice in the following section.

Remark: For theoretical investigations of multigrid methods (h-independency of convergence factors ρ, etc.) one usually assumes that $h=h_\ell$, the meshsize of the finest grid, tends to 0, or one looks for $\rho^* = \sup \rho$ with respect to h. For such asymptotic investigations, we always regard the coarsest grid to be fixed (meshsize h_0) and let the number ℓ of grids tend to infinity. Only in connection with this conception do we use the term *complete* multigrid methods. (Some authors, however, discuss the convergence properties of multigrid methods not for fixed h_0, but for a fixed ratio of the finest and coarsest meshsize used. Since the coarsest grid then becomes finer and finer for $h \to 0$, this assumption is unsatisfactory in several respects, in particular with respect to the total computational work, see Section 4.4.)

4.3 The iteration operator for a complete multigrid cycle; h-independent convergence

The simple theoretical considerations of this section have been presented in a similar form by Hackbusch [50]. Here, we always allow γ to depend on $k : \gamma = \gamma_k$. We will, however, only add the index k if this dependence is to be stressed. By $\mathbf{M}_\ell$, we denote the iteration operator of the multigrid method described in the previous section.

Lemma 4.1: $\mathbf{M}_\ell$ is given by the following recursion:

$$\begin{aligned} \mathbf{M}_1 &= S_1^{\nu_2} (I_1 - I_0^1 L_0^{-1} I_1^0 L_1) S_1^{\nu_1} \\ \mathbf{M}_{k+1} &= S_{k+1}^{\nu_2} (I_{k+1} - I_k^{k+1} (I_k - \mathbf{M}_k^\gamma) L_k^{-1} I_{k+1}^k L_{k+1}) S_{k+1}^{\nu_1} \quad (k=1,\dots,\ell-1) \end{aligned} \tag{4.7}$$

An explicit proof of (4.7) can easily be given by means of induction on ℓ. Implicitly, a proof is also contained in the following remark.

Remark: The difference between the $(h_\ell, h_{\ell-1})$ two-grid operator which solves (4.5), namely

$$M_\ell^{\ell-1} = S_\ell^{\nu_2} (I_\ell - I_{\ell-1}^\ell L_{\ell-1}^{-1} I_\ell^{\ell-1} L_\ell) S_\ell^{\nu_1}, \tag{4.8}$$

and the above multigrid operator $\mathbf{M}_\ell$ is obviously that

$$L_{\ell-1}^{-1} \text{ is replaced by } (I_{\ell-1} - \mathbf{M}_{\ell-1}^\gamma) L_{\ell-1}^{-1}. \tag{4.9}$$

This reflects the fact that the coarse-grid equation (4.6) is solved approximately by γ ($=\gamma_{\ell-1}$) multigrid steps on the grid $\Omega_{\ell-1}$ starting with an initial approximation = 0. (Here we use the simple consideration: If any non-singular system of linear equations $Aw = r$ is solved approximately by γ steps of an iterative method $w^{j+1} = Mw^j + s$ with $w^0 = 0$, then the γ-th iterate can be represented as $w^\gamma = (I - M^\gamma) A^{-1} r$.)

For the following norm estimations, it is convenient to write $\mathbb{M}_\ell$ as a perturbation of $M_\ell^{\ell-1}$. Lemma 4.1 yields

<u>Corollary</u> 4.2: For $k=1,\ldots,\ell-1$ the equations

$$\mathbb{M}_{k+1} = M_{k+1}^k + A_k^{k+1} \mathbb{M}_k^\gamma A_{k+1}^k \tag{4.10}$$

hold, where

$$\begin{aligned} A_k^{k+1} &:= S_{k+1}^{\nu_2} I_k^{k+1} &&: \mathbb{G}(\Omega_k) \to \mathbb{G}(\Omega_{k+1}), \\ A_{k+1}^k &:= L_k^{-1} I_{k+1}^k L_{k+1} S_{k+1}^{\nu_1} &&: \mathbb{G}(\Omega_{k+1}) \to \mathbb{G}(\Omega_k) \end{aligned} \tag{4.11}$$

and M_{k+1}^k is as in (4.8) with $k+1$ instead of ℓ.

From this representation, one can immediately derive an estimate for $\|\mathbb{M}_\ell\|$, provided that estimates for $\|M_{k+1}^k\|$, $\|A_k^{k+1}\|$ and $\|A_{k+1}^k\|$ ($k \le \ell-1$) are known. Here $\|.\|$ denotes any reasonable operator norm.

<u>Lemma</u> 4.3: Let the following estimates hold uniformly with respect to k ($\le \ell-1$):

$$\|M_{k+1}^k\| \le \sigma^*, \qquad \|A_k^{k+1}\| \cdot \|A_{k+1}^k\| \le C. \tag{4.12}$$

Then we have $\|\mathbb{M}_\ell\| \le \eta_\ell$ where η_ℓ is recursively defined by

$$\eta_1 := \sigma^*, \quad \eta_{k+1} := \sigma^* + C\eta_k^\gamma \quad (k=1,\ldots,\ell-1). \tag{4.13}$$

<u>Remark</u>: Clearly, we could also have admitted bounds σ^* and C in (4.12) which depend on k. In particular, this may be advantageous if these bounds achieve their maximal values for small k (which is typical for certain indefinite problems).

From (4.13), one can already conclude the h-(ℓ-)independent convergence of multigrid methods. Mainly, one has to assume that the corresponding (h_{k+1}, h_k) two-grid methods converge for all k with σ^* sufficiently small. Furthermore, one has to make a decision on the choice of γ. We consider two cases:

(1) $\quad \gamma_k \equiv 2 \quad (k=1,2,\ldots;$ W-cycle) (4.14)

(2) $$\gamma_k = \begin{cases} 1 & (k \text{ odd}) \\ 2 & (k \text{ even}) \end{cases} \tag{4.15}$$

The second choice (see Figure 4.3) is of particular interest in connection with semi or red-black coarsening (see Section 4.4).

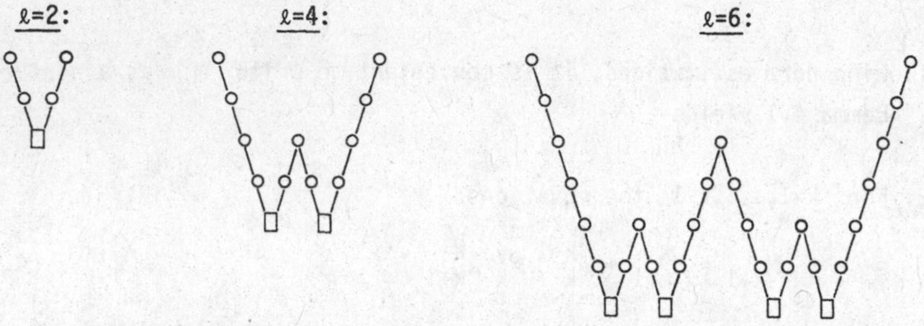

Figure 4.3: Structure of one multigrid cycle for ℓ even and γ as defined in (4.15)

Corollary 4.4: In the case (1) we have the following uniform estimate provided that $4C\sigma^* \le 1$:

$$\|\mathbb{M}_\ell\| \le \eta := (1 - \sqrt{1 - 4C\sigma^*})/2C \le 2\sigma^* \quad (\ell \ge 1). \tag{4.16}$$

Similarly, in the case (2) we obtain

$$\|\mathbb{M}_\ell\| \le \begin{cases} (1 - \sqrt{1 - 4C^2(1+C)\sigma^*})/2C^2 \le 2\sigma^*(1+C) & (\ell \text{ even}) \quad (4.17a) \\ (1 - 2C^2\sigma^* - \sqrt{1 - 4C^2(1+C)\sigma^*})/2C^3 \le \sigma^*(1+2C)/C & (\ell \text{ odd}) \quad (4.17b) \end{cases}$$

if $4C^2(1+C)\sigma^* \le 1$.

Remarks:

(1) If $\gamma=2$, i.e. if W-cycles are used, we obviously obtain, under the sole assumption that σ^* is small enough, $\eta \approx \sigma^*$ for the bound η given in (4.16). For example, if $C=1$, then we get from (4.16)

$$\eta \le 0.113 \quad \text{if} \quad \sigma^* \le 0.1.$$

(Typically, the constant C is ≥ 1, but not very large. For instance, if we use $\|.\| = \|.\|_E$ in the sample method treated in Sections 3.1 - 3.5, we obtain $C = 1$, independent of ν_1 and ν_2. For $\|.\| = \|.\|_S$ we have $C \le \sqrt{2}$ for all choices of ν_1 and ν_2 and we have $C \searrow 1$ if $\nu_1 \to \infty$.) In this sense we may say: <u>If a given two-grid method converges sufficiently well (small enough σ^*), then the corresponding multigrid method with $\gamma=2$ will have similar convergence pro-</u>

perties. In this respect, for the construction of multigrid methods, it is usually sufficient to analyze only the corresponding two-grid method. Furthermore, there usually is no need to work with $\gamma>2$.

(2) If γ is defined as in (4.15), the bounds for $\|M_\ell\|$ become somewhat worse than those for $\gamma=2$. For example, from (4.17a) we obtain that the upper bound of $\|M_\ell\|$ (ℓ even) approaches $\sigma^*(1+C)$ if σ^* is small enough (instead of $\eta \approx \sigma^*$ in the case $\gamma=2$). On the other hand, a smaller amount of numerical work is needed for one cycle (cf. Section 4.4) than for the W-cycle.

(3) If $\gamma=1$, i.e. if V-cycles are used, Lemma 4.3 gives no ℓ-independent upper bound for $\|M_\ell\|$ if $C \geq 1$. However, instead of $\gamma=1$ one could use, e.g., $\gamma=\gamma_k$ with $\gamma_k=1$ if $\ell \geq k \leq \ell-\ell_o$ and $\gamma_k=2$ otherwise. For larger values of ℓ_o, this would result in only a very slight increase of the computational work compared to the V-cycle. For a cycle of this type, Lemma 4.3 could, in principle, be used to derive ℓ-independent bounds for $\|M_\ell\|$. If ℓ_o is large, we would then, however, have to assume σ^* to be very small and the estimate would become completely unrealistic from a practical point of view.

There is another approach in proving ℓ-independent convergence of multigrid methods which is also applicable to V-cycles. This approach was first presented by Braess [13]. Hackbusch has also incorporated the corresponding idea into his theory (see [50], Section 4.3). The approach is based on the energy norm and makes essential use of the following assumptions:

- L_ℓ is symmetric and positive definite.
- The restriction operators I_k^{k-1} and the interpolation operators I_{k-1}^k are adjoint to each other: $I_k^{k-1} = (I_{k-1}^k)^*$ $(k = 1,\dots,\ell)$.
- The coarse-grid difference operators L_k $(k = 0,1,\dots,\ell-1)$ are defined to be the "Galerkin operators" (see Section 4.5): $L_{k-1} := I_k^{k-1} L_k I_{k-1}^k$ $(k=1,\dots,\ell)$.
- The difference operators and the smoothing operators are supposed to commute. (This assumption can easily be weakened up to a certain extend [100]).

We remark that under these assumptions the h-independent convergence of V-cycle methods can also be shown in the framework of local Fourier analysis [100].

In practice, if σ^* is small enough such that W-cycles have good convergence properties, usually also V-cycles may be used (even if the above listed assumptions are not satisfied). Often, the convergence properties of V-cycles are somewhat worse than those of W-cycles, but with respect to effiency nevertheless competitive (also see Section 10.2).

4.4 Computational work and efficiency

The fact that a certain method has an h-independent convergence factor says nothing about its efficiency as long as the computational work is not taken into account. In the following, we will estimate the computational work of a multigrid method. It will turn out that the number of arithmetic operations needed for one multigrid cycle is proportional to the number of grid points of the finest grid (under quite natural assumptions which are satisfied for reasonable multigrid methods). Together with the h-independent convergence, this means that multigrid methods are asymptotically optimal. The constant of proportionality depends on the type of the cycle, i.e. on γ, the type of coarsening and the other multigrid components. For reasonable choices of these components, the constants of proportionality are small.

From the recursive definition of a multigrid cycle as given in Section 4.2, it immediately follows that the *computational work* W_ℓ per multigrid cycle Ω_ℓ is recursively given by

$$W_1 = W_1^0 + W_0, \quad W_{k+1} = W_{k+1}^k + \gamma_k W_k \qquad (k = 1,\dots,\ell-1). \tag{4.18}$$

Here W_{k+1}^k denotes the computational work of one (h_{k+1},h_k) two-grid cycle excluding the work needed to solve the defect equations on Ω_k, and W_0 denotes the work needed to compute the exact solution on the coarsest grid Ω_0. By "computational work", we always denote some reasonable measure, for example, the number of arithmetic operations needed. If γ is independent of k, we obtain from (4.18)

$$W_\ell = \sum_{k=1}^{\ell} \gamma^{\ell-k} W_k^{k-1} + \gamma^{\ell-1} W_0 \qquad (\ell \geq 1). \tag{4.19}$$

Let us first discuss the case of standard coarsening with γ independent of k. Obviously, we have in this case

$$\mathcal{N}_k \doteq 4\,\mathcal{N}_{k-1} \qquad (k = 1,2,\dots) \tag{4.20}$$

where $\mathcal{N}_k = \#\,\Omega_k$ (number of gridpoints on Ω_k) and " $\doteq$ " means equality up to lower order terms (boundary effects). Furthermore, we assume that the multigrid components (relaxation, computation of defects, fine-to-coarse and coarse-to-fine transfers) require a number of arithmetic operations per point of the respective grids which is bounded by a constant C, independent of k:

$$W_k^{k-1} \stackrel{\cdot}{\leq} C\,\mathcal{N}_k \qquad (k = 1,2,\dots). \tag{4.21}$$

(As above, " $\stackrel{\cdot}{\leq}$ " means " $\leq$ " up to lower order terms.) This is a typical feature of multigrid methods. In particular, (4.21) is satisfied with $\doteq$ instead of $\stackrel{\cdot}{\leq}$ if all multigrid components are constructed in the same way on all grids.

Under these assumptions, one immediately obtains from (4.19) the following estimate for the total computational work W_ℓ of one complete multigrid cycle:

$$W_\ell \doteq \begin{cases} \frac{4}{3} C \mathcal{N}_\ell & \text{(for } \gamma=1) \\ 2 C \mathcal{N}_\ell & \text{(for } \gamma=2) \\ 4 C \mathcal{N}_\ell & \text{(for } \gamma=3) \\ O(\mathcal{N}_\ell \log \mathcal{N}_\ell) & \text{(for } \gamma=4) \end{cases} \tag{4.22}$$

This estimate of W_ℓ together with the h-independent convergence as discussed in the previous section shows the asymptotic optimality of iterative multigrid methods if $\gamma \le 3$ and standard coarsening is used. (As mentioned in the previous section, for V-cycles h-independent convergence has been proved - so far - only under certain additional assumptions. In practice, however, this convergence behavior can be observed in much more general situations. In this respect, we have asymptotic optimality also for $\gamma=1$.)

Remark: W_k^{k-1} in (4.18) is determined by the computational work needed for the individual multigrid components of the (h_k, h_{k-1}) two-grid method, namely

$$W_k^{k-1} \doteq (\nu w_o + w_1 + w_2) \mathcal{N}_k . \tag{4.23}$$

Here $\nu=\nu_1+\nu_2$ is the number of relaxation steps used; w_o, w_1 and w_2 are measures for the computational work per grid point of Ω_k needed for the single components, namely

- w_o: one relaxation step on Ω_k;
- w_1: computation of the defect and its transfer to Ω_{k-1};
- w_2: interpolation of the correction to Ω_k and its addition to the previous approximation.

Usually (in particular, when the multigrid components are constructed in the same way on all grids), w_o, w_1 and w_2 are independent of k. In general, however, they may depend on k.

Example 4.1: If the multigrid algorithm which corresponds to our sample method (Section 3.1) is arranged suitably, we obtain the following operation count

	w_o	w_1	w_2	W_k^{k-1}
+/-	5	25/4	7/4	$(5\nu + 8)\mathcal{N}_k$
*	1	5/4	3/4	$(\nu + 2)\mathcal{N}_k$

If we count additions and multiplications in the same way, we obtain (4.22) with $C=6\nu+10$. Further examples, namely for more efficient methods and more general problems, will be given in Chapters 8, 9 and 10.

For other grid coarsenings than the standard coarsening,

$$\mathcal{N}_k \doteq \tau \mathcal{N}_{k-1} \quad (k = 1,2,\ldots) \quad \text{with} \quad \tau > 1.$$

and for γ independent of k we obtain

$$\boxed{W_\ell \overset{\cdot}{\leq} \begin{cases} \frac{\tau}{\tau-\gamma} C\mathcal{N}_\ell & (\text{for } \gamma < \tau) \\ O(\mathcal{N}_\ell \log \mathcal{N}_\ell) & (\text{for } \gamma = \tau) \end{cases}} \tag{4.24}$$

instead of (4.22).

If we consider, for example, <u>red-black coarsening</u> or <u>semi-coarsening</u>, we have $\tau=2$. In this case, we already see that W-cycles do not yield an asymptotically optimal multigrid method: For fixed γ, only $\gamma=1$ yields a cycle for which W_ℓ is proportional to $\mathcal{N}_\ell$. Because of the theoretical restrictions in proving the h-independent convergence of pure V-cycle iterations, the choice $\gamma=\gamma_k$ as given in (4.15) is of particular interest for $\tau=2$. We obtain for $\tau=2$:

$$\boxed{W_\ell \overset{\cdot}{\leq} \begin{cases} 2C\mathcal{N}_\ell & (\text{for } \gamma = 1) \\ 3C\mathcal{N}_\ell & (\text{for } \ell \text{ even and } \gamma \text{ as defined in (4.15)}) \\ 4C\mathcal{N}_\ell & (\text{for } \ell \text{ odd and } \gamma \text{ as defined in (4.15)}) \\ O(\mathcal{N}_\ell \log \mathcal{N}_\ell) & (\text{for } \gamma = 2). \end{cases}}$$

Clearly, for $\gamma=\gamma_k$ we have to use the more general formula

$$W_\ell = \sum_{k=1}^{\ell} \Big(\prod_{j=k}^{\ell-1} \gamma_j \Big) W_k^{k-1} + \Big(\prod_{j=1}^{\ell-1} \gamma_j \Big) W_0 \qquad (\ell \geq 1)$$

instead of (4.19).

<u>Remark</u>: There are, of course, many other possible choices of $\gamma=\gamma_k$ and still more general ways to construct a cycle. We mention here just the so-called *F-cycle* [25] which is illustrated in Figure 4.4. The corresponding iteration operator $\mathbb{M}_\ell^F$ is recursively defined by

$$\mathbb{M}_1^F = \mathbb{M}_1 \quad \text{(as in (4.7))}$$

$$\mathbb{M}_{k+1}^F = S_{k+1}^{\nu_2} (I_{k+1} - I_k^{k+1} (I_k - \mathbb{M}_k^V \mathbb{M}_k^F) L_k^{-1} I_{k+1}^k L_{k+1}) S_{k+1}^{\nu_1} \quad (k=1,\ldots,\ell-1).$$

Here $\mathbb{M}_k^V$ is the corresponding V-cycle iteration operator (i.e. (4.7) with $\gamma=1$ and k instead of $k+1$).

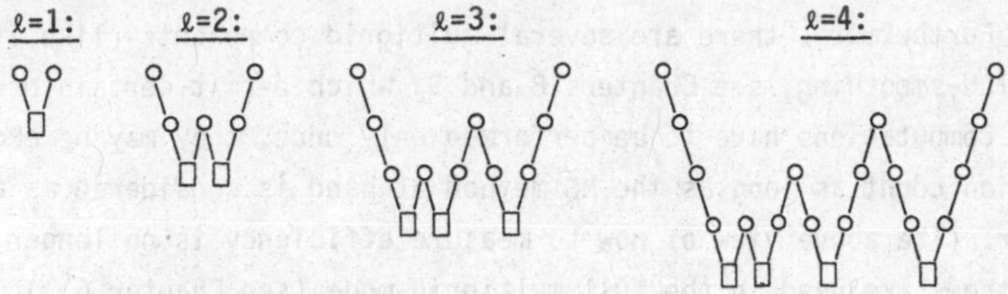

Figure 4.4: Structure of an F-cycle

In *self-adaptive* algorithms as proposed by Brandt [25],[18], no fixed cycles are used: Switching from one grid to another (to a finer or a coarser one) is controlled by suitable accommodative criteria.

Efficiency of multigrid iterations. Let us make some comments on the questions of efficiency and of efficiency measures for multigrid methods. Trivially, the efficiency of an iterative method is determined by both its convergence factor and the computational work needed per iteration step. Reasonable measures of efficiency are

$$op(\varepsilon) := w \frac{\log\varepsilon}{\log\rho}, \tag{4.25}$$

the number of numerical operations (per point of the finest grid) required to reduce the error by a factor of ε, or

$$\rho_{eff} := \rho^{1/w}. \tag{4.26}$$

Here ρ characterizes the convergence factor of the method at hand and w the work needed per iteration step and per point of the finest grid.

Although (4.25) and (4.26), at first sight, look simple and well-defined, there is little uniformity in the multigrid literature with respect to the definition of ρ and w: As for ρ, one might use (empirical) asymptotic convergence factors or (empirical) error reducing factors (norms) for a suitable multigrid cycle (e.g., V- or W-cycle). Similarly, there are different possible definitions of w.

A choice which is usually reasonable and which avoids the arbitrariness in the decision about the type of cycle and the norms, is for ρ to use the asymptotic convergence factor of the corresponding two-grid operator $M_\ell^{\ell-1}$ (or its supremum

ρ^* with respect to ℓ). This choice corresponds to the work quantity

$$w := W_\ell^{\ell-1} / \mathcal{N}_\ell \tag{4.27}$$

in a natural way.

One still has to decide about the question of how to count the different arithmetic operations. Furthermore, there are several multigrid components (like linewise relaxation and ILU-smoothing, see Chapters 8 and 9) which permit certain precomputations. As these computations have to be performed only once, they may be excluded from the operation count as long as the MG method at hand is considered as a purely iterative solver. (The above view of how to measure efficiency is no longer correct if multigrid methods are used in the full multigrid mode (see Chapter 6).)

In Chapters 8 and 9 we give results on ρ^*, σ_S^*, σ_E^* etc. for several methods along with an operation count in terms of (4.27).

4.5 Other coarse-grid operators, extensions

In Section 4.1 we have assumed a fixed sequence of difference operators L_ℓ (ℓ=0,1,2,...) to be given. We had in mind there that the L_ℓ were chosen in a uniform manner on all grid Ω_ℓ, e.g. always using the same discretization. For fixed ℓ, the grid equation (4.5) was solved by using the L_k ($k=\ell-1,\ell-2,\ldots,0$) as coarse-grid difference operators. Another way of defining these coarse-grid operators (maintaining the sequence of transfer operators) has already been mentioned in Section 2.4.2:

- *The Galerkin approach:* Here, for a given fine-grid operator L_ℓ, the grid equation (4.5) is solved using coarse-grid operators L_k ($k=\ell-1,\ldots,0$) which are recursively defined by

$$L_k := I_{k+1}^k L_{k+1} I_k^{k+1} \quad (k = \ell-1,\ \ell-1,\ldots,0).$$

From this recursion one sees that no fixed sequence of coarse-grid operators is defined. Instead, the L_k depend on the operator L_ℓ which is given on the finest grid. This means that we have to work with a "triangular" scheme of operators

$$L_k^{(\ell)} \quad (k = \ell,\ell-1,\ldots;\ \ell = 0,1,2,\ldots)$$

where

$$L_\ell^{(\ell)} = L_\ell \quad \text{and} \quad L_k^{(\ell)} := I_{k+1}^k\, L_{k+1}^{(\ell)}\, I_k^{k+1} \quad (k = \ell-1,\ell-2,\ldots,0). \tag{4.28}$$

The description of the multigrid method given in Section 4.2 carries over to this more general case: One simply has to replace L_ℓ by $L_\ell^{(\ell)}$ and L_k by $L_k^{(\ell)}$ $(k = \ell-1,\ldots,0)$ for all ℓ. (Of course, relaxation processes now have to be given with smoothing operators $S_k^{(\ell)}$ corresponding to the $L_k^{(\ell)}$.) All further results of this chapter still hold after a few obvious changes. In particular, all operators occuring in (4.10) now depend not only on k but also on ℓ. Lemma 4.3 carries over to this situation. One now, however, needs norm estimates (4.12) which hold for <u>all</u> operators within the above "triangular" scheme of operators.

The main practical difference to the approach in the previous sections is that the $L_k^{(\ell)}$ are not known in advance but have to be calculated from the recursion formula (4.28):

<u>Example</u> 4.2: (a) If $L_\ell^{(\ell)}$ $(\ell = 0,1,2,\ldots)$ and $I_\ell^{\ell-1}$, $I_{\ell-1}^{\ell}$ $(\ell = 1,2,\ldots)$ are defined as in the sample method (see Section 3.1, in particular (3.3),(3.4)), we obtain

$$L_{\ell-1}^{(\ell)} = I_\ell^{\ell-1} L_\ell^{(\ell)} I_{\ell-1}^{\ell} \mathrel{\hat{=}} \frac{1}{h_{\ell-1}^2} \begin{bmatrix} -1/4 & -1/2 & -1/4 \\ -1/2 & 3 & -1/2 \\ -1/4 & -1/2 & -1/4 \end{bmatrix}_{h_{\ell-1}}. \tag{4.29}$$

For $\ell\to\infty$ and k fixed, $L_k^{(\ell)}$ tends to a difference operator which is characterized by the difference star

$$\frac{1}{h_k^2} \begin{bmatrix} -1/3 & -1/3 & -1/3 \\ -1/3 & 8/3 & -1/3 \\ -1/3 & -1/3 & -1/3 \end{bmatrix}_{h_k}. \tag{4.30}$$

This is a well-known approximation for the Laplace operator which occurs in connection with bilinear finite elements.

(b) If, for $k=\ell$, (4.30) is used as difference operator $L_\ell^{(\ell)}$ on the finest grid Ω_ℓ (and if $I_\ell^{\ell-1}$, $I_{\ell-1}^{\ell}$ are chosen as above), $L_\ell^{(\ell)}$ is "reproduced" by the Galerkin-recursion (4.28): $L_{\ell-1}^{(\ell)}$ ist just (4.30) with $k=\ell-1$.

(c) The 5-point Laplace difference operator $L_\ell^{(\ell)}$ from (a) is reproduced by the Galerkin recursion, if we use different transfer operators, namely the 7-point operators [109]

$$I_{\ell}^{\ell-1} \hat{=} \frac{1}{8}\begin{bmatrix} 1 & 1 & \\ 1 & 2 & 1 \\ & 1 & 1 \end{bmatrix}_{h_\ell}^{h_{\ell-1}}, \qquad I_{\ell-1}^{\ell} \hat{=} \frac{1}{2}\Bigg] \begin{matrix} 1 & 1 & \\ 1 & 2 & 1 \\ & 1 & 1 \end{matrix} \Bigg[_{h_{\ell-1}}^{h_\ell}. \tag{4.31}$$

- *Reduction-type approach:* For 1D-problems (discrete ordinary boundary value problems), it usually is possible to define operators I_h^{2h} and L_{2h} such that the coarse-grid equation is equivalent to the original grid equation for $x \in \Omega_{2h}$. If, in the coarse-to-fine transfer, the original grid equation is used (for $x \in \Omega_h \setminus \Omega_{2h}$, cf. Section 2.4.4), then one obtains the exact discrete solution on Ω_h after one cycle only: These methods degenerate to direct solvers (without performing smoothing steps), and they coincide with so-called 1D-reduction methods [87],[104] (also called cyclic reduction [99].)

Such a transfomation of a fine-grid equation to an equivalent equation on some coarser grid can also be carried out for certain 2D- (and 3D-) problems. Corresponding methods are known as *total, cyclic* and *alternating reduction methods* [87],[29], [88]. These methods differ mainly with respect to the coarsening (successive red-black coarsening, successive semi-coarsening and alternating semi-coarsening, respectively). All of these methods have the disadvantage of being directly applicable only to a rather small class of problems. Moreover, the coarse-grid operators (and the fine-to-coarse operators) become more and more complicated. For the total and alternating reduction method, however, the corresponding difference stars can be "truncated". This possibility (in combination with certain smoothing processes) is the basis of the *MGR principle* [82],[37] yielding particularly efficient multigrid solvers (MG-TR and MG-AR methods).

The reduction-type approach shows a connection between multigrid methods and certain direct solvers for a given discrete problem. The following approach is, on the contrary, closer related to the original continuous problem and not to a fixed discrete problem.

- *Double discretization:* In principle, multigrid methods may be applied with different operators in the relaxation process (L_ℓ) and in the process of the calculation of defects (L_ℓ^*). Brandt [25],[23] recommends such "double discretization" multigrid methods for certain applications. For example, the operators L_ℓ^* may be of a higher order of consistency than the operators L_ℓ. As a consequence, one can expect to obtain higher order accuracy although only low order operators are employed for smoothing. In particular, it is possible to use unstable higher order operators L_0^* for the defect computation.

The latter choice is of particular interest in connection with singular perturbation problems. For instance, the L_ℓ^* may be (unstable) operators of second order consistency based on central differencing, whereas the L_ℓ are defined by introducing a certain amount of artificial ellipticity (leading to first order consistency). This concept is discussed in [25] and, in detail, in [21],[12].

The double discretization idea is clearly related to the defect correction principle [96],[5]. Possibilities to combine multigrid and defect correction methods are discussed in several papers (see, e.g., [25],[5],[51],[56]).

5. Nonlinear multigrid methods, the full approximation scheme (FAS)

So far, we have discussed multigrid methods only in connection with linear problems. Clearly, if a linear multigrid method is combined with some iterative (global) linearization process like Newton's method, it can also be used for the solution of nonlinear problems. This "indirect" application of (linear) multigrid methods to nonlinear problems is more or less straightforward (see Section 5.1).

The multigrid idea can, however, also be applied directly to nonlinear problems. Again, we only need a procedure for smoothing errors and a procedure for approximating corrections on coarser grids. For error smoothing, suitable relaxation methods for nonlinear equations now have to be used. This "direct" approach leads to nonlinear multigrid methods in form of the so-called *full approximation technique* (*full approximation scheme*, "FAS", introduced by Brandt [16],[17]). In this approach, no global linearization has to be carried out explicitly (except perhaps on the coarsest grid). We will describe the nonlinear multigrid methods in Section 5.2.

In Section 5.3 we point out the close relationship between the nonlinear multigrid method and the indirect approach, giving some numerical results. This relationship can be exploited for a convergence theory of the nonlinear methods. We do not give such proofs here (see Hackbusch [50]). However, some simple theoretical considerations concerning the appropriate choice of relaxation methods for nonlinear problems are given in Section 5.4.

In Section 5.5.1, we make some remarks on the multigrid treatment of an exemplary bifurcation problem, in which a global constraint has to be taken into account. Furthermore, there are several specific features of the full approximation scheme which are the starting point for more sophisticated multigrid techniques. In this respect, FAS is of interest for linear problems also (although it is then theoretically equivalent to the usual linear scheme). Some of the more sophisticated techniques will be sketched in Section 5.5.2.

5.1 Indirect application of multigrid methods to nonlinear problems

In the following, we consider a discrete elliptic equation

$$L_h u_h = f_h \quad (\Omega_h). \tag{5.1}$$

Here $L_h : \mathbb{G}(\Omega_h) \to \mathbb{G}(\Omega_h)$ is assumed to be a nonlinear operator; $f_h \in \mathbb{G}(\Omega_h)$ is a given grid function (which is introduced for technical reasons only). In order not to have too many formal requirements and restrictions, we assume explicitly only that this equation has at least one isolated solution u_h. All other assumptions are implicitly contained in the following considerations.

For the solution of (5.1), an iterative (global) linearization method

$$\mathbf{L}_h u_h^j + L_h^j v_h^j = f_h, \quad u_h^{j+1} = u_h^j + v_h^j \quad (j = 0,1,2,\ldots) \tag{5.2}$$

may be used. Here L_h^j is some linear approximation of $\mathbf{L}_h'(u_h^j)$ which characterizes the iteration process. In particular, we consider Newton's method $(L_h^j = \mathbf{L}_h'(u_h^j))$. In each step of the iteration (5.2), a (linear) multigrid method can be applied to solve the linear equations

$$L_h^j v_h^j = d_h^j := f_h - \mathbf{L}_h u_h^j. \tag{5.3}$$

One way to combine Newton's method with an iterative linear multigrid method for (5.3) (of the type shown in Figure 4.2), is to adapt the number of multigrid iterations in each Newton step. Here the aim is to exploit the convergence speed of Newton's method as far as possible. For example, if Newton's method converges quadratically, the number of MG iterations should roughly be doubled from one Newton step to the next. The main problem in this approach is that one has to use an appropriate control technique in order to obtain the information needed about the convergence of Newton's method. We shall refer to this approach as to *method I*.

Another possibility is to fix the number of multigrid iterations per Newton step. For example, one may perform only one multigrid iteration per Newton step. As a consequence, Newton's method is, of course, truncated to a linearly convergent method. A disadvantage of this approach, which we will refer to as *method II*, is the larger amount of linearization work. On the other hand, no control technique is needed as in I.

A few numerical results and a short comparison with a nonlinear multigrid method of FAS type will be given in Section 5.3. This comparison refers to the special case that (5.1) is the 5-point discretization of

$$\mathbf{L}^\Omega u = -\Delta u + g(x,u) = f^\Omega(x) \quad (x \in \Omega) \tag{5.4}$$

with Dirichlet boundary conditions on a bounded region Ω. In terms of (5.2), (5.3), Newton's method then reads as

$$L_h v_h^j + c_h^j(x) v_h^j = d_h^j(x) \quad (x \in \Omega_h) \quad \text{with} \quad c_h^j(x) = \frac{\partial g}{\partial u}(x, u_h^j(x)). \tag{5.5}$$

Here L_h is given by the 5-point descretization of $-\Delta$. Thus, in each Newton step, a discrete Helmholtz-like equation has to be solved. With respect to algorithmic simplifications, let us add the following

Remark: There are several reasons, why, in practice, Newton's method is often replaced by - only linearly convergent - "approximate" Newton's methods. Very simple approximate methods are obtained when, for instance, $c_h^j(x)$ in (5.5) is replaced by

$$c_h^o(x) := \frac{\partial g}{\partial u}(x, u_h^o(x)) \quad \text{(modified Newton's method)}$$

or by a constant, e.g.

$$\hat{c}_h^j := 1/2\, (\min_x c_h^j(x) + \max_x c_h^j(x)). \tag{5.6}$$

The latter simplification is of particular interest in connection with the application of so-called direct Fast Elliptic Solvers (like Buneman's algorithm) for which a constant Helmholtz-c is required. We point out that a simplification of the type (5.6) is not needed if multigrid methods are used. (The application of Fast Elliptic Solvers has been studied systematically in [90] in connection with a nonlinear parabolic problem.)

5.2 The full approximation scheme

Similar to the linear case, the nonlinear FAS multigrid method can be recursively defined on the basis of an FAS two-grid method. Thus we again start with the description of one iteration step of the (h,H) two-grid method for (5.1), computing u_h^{j+1} from u_h^j. An illustration of this step, which is analogous to the one given in Figure 2.4 for the linear two-grid method, is given in Figure 5.1.

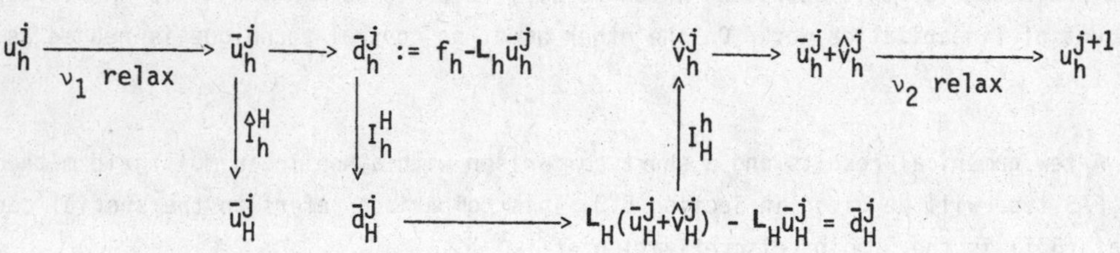

Figure 5.1: FAS (h,H) two-grid method

In this description, "relax" stands for a nonlinear relaxation procedure which has suitable error smoothing properties. As in the linear case, ν_1 smoothing steps are performed before and ν_2 smoothing steps are performed after the coarse-grid correction. In contrast to the linear case, not only is the defect $\bar{d}_h^j$ transfered to the coarse grid (by some linear operator I_h^H), but also the relaxed approximation $\bar{u}_h^j$ itself (by some linear operator $\hat{I}_h^H$, which may be different from I_h^H). This is necessary, as in the nonlinear case the Ω_h-defect equation is given by

$$L_h(\bar{u}^j_h + v^j_h) - L_h\bar{u}^j_h = \bar{d}^j_h. \tag{5.7}$$

This equation is approximated on Ω_H by

$$L_H(\bar{u}^j_H + \hat{v}^j_H) - L_H\bar{u}^j_H = \bar{d}^j_H, \tag{5.8}$$

or equivalently by

$$L_H w^j_H = \bar{d}^j_H + L_H\bar{u}^j_H, \quad \hat{v}^j_H := w^j_H - \bar{u}^j_H. \tag{5.9}$$

This means that in the FAS mode on the coarse grid, one does not solve for the correction $\hat{v}^j_H$, but rather for the "full approximation" w^j_H. Of course, transfered back to the fine grid Ω_h, is not w^j_H but the correction $\hat{v}^j_H$. This is important since only correction (and defect) quantities are smoothed by relaxation processes and can therefore be approximated well on coarser grids (see the explanations in Section 2, which in principle apply also to the nonlinear case). Clearly, if L_h is a linear operator, the FAS two-grid method is equivalent to the linear method, see Figure 2.4, which is called the *correction scheme* (CS) by Brandt.

In the corresponding nonlinear multigrid process, the nonlinear coarse-grid equation in (5.9) is not solved exactly, but approximately by several multigrid steps using still coarser grids. This leads to the following algorithmic description of one step of the FAS multigrid method. Here we use notations analogous to those in Sections 4.1, 4.2. In particular, we assume a sequence of grids Ω_ℓ and grid operators L_ℓ, $I^{\ell-1}_\ell$, $\hat{I}^{\ell-1}_\ell$, $I^\ell_{\ell-1}$ etc. to be given. One *FAS multigrid* (more precisely: *$(\ell+1)$-grid*) *step* for the solution of

$$L_\ell u_\ell = f_\ell \quad (\ell \geq 1, \text{ fixed}) \tag{5.10}$$

proceeds as follows:

If $\ell = 1$, we just have the two-grid method described above with Ω_0 and Ω_1 instead of Ω_H and Ω_h, respectively.

If $\ell > 1$:

(1) Smoothing part I:

- Compute $\bar{u}^j_\ell$ by applying ν_1 (≥ 0) smoothing steps to u^j_ℓ:

$$\bar{u}^j_\ell := \text{RELAX}^{\nu_1}(u^j_\ell, L_\ell, f_\ell).$$

(2) Coarse-grid correction:

- Compute the defect: $\bar{d}^j_\ell := f_\ell - L_\ell\bar{u}^j_\ell$.

- Restrict the defect: $\bar{d}^j_{\ell-1} := I^{\ell-1}_{\ell}\, \bar{d}^j_{\ell}$.

- Restrict $\bar{u}^j_{\ell}$: $\bar{u}^j_{\ell-1} := \hat{I}^{\ell-1}_{\ell}\, \bar{u}^j_{\ell}$.

- Compute an approximate solution $\tilde{w}^j_{\ell-1}$ of

$$\mathbf{L}_{\ell-1} w^j_{\ell-1} = \bar{d}^j_{\ell-1} + \mathbf{L}_{\ell-1}\bar{u}^j_{\ell-1} \tag{5.11}$$

by applying $\gamma \geq 1$ steps of the FAS ℓ-grid method (using the grids $\Omega_0,\ldots,\Omega_{\ell-1}$) to (5.11) with $\bar{u}^j_{\ell-1}$ as first approximation. Then compute the correction

$$\tilde{v}^j_{\ell-1} := \tilde{w}^j_{\ell-1} - \bar{u}^j_{\ell-1}.$$

- Interpolate the correction: $\tilde{v}^j_{\ell} := I^{\ell}_{\ell-1}\, \tilde{v}^j_{\ell-1}$.

- Compute the corrected approximation on Ω_ℓ: $\bar{u}^j_{\ell} + \tilde{v}^j_{\ell}$.

(3) Smoothing part II:

- Compute u^{j+1}_{ℓ} by applying ν_2 (≥ 0) smoothing steps to $\bar{u}^j_{\ell} + \tilde{v}^j_{\ell}$:

$$u^{j+1}_{\ell} := \mathrm{RELAX}^{\nu_2}(\bar{u}^j_{\ell} + \tilde{v}^j_{\ell}, \mathbf{L}_\ell, f_\ell).$$

One sees from this description that no global linearization is needed in the FAS multigrid process, except perhaps on the coarsest grid. Apart from that, only (nonlinear) relaxation methods are required as well as (linear) fine-to-coarse and coarse-to-fine transfer operators.

Concerning the concrete choice of the occuring multigrid components, one can orient oneself to the corresponding linearized problem. For the latter, techniques such as given in Chapters 3,7,8 and 9, can be applied. As to the relaxation methods, there usually exist (several) nonlinear analogs to a given linear relaxation method (see, e.g., [80]). We will make some remarks about the smoothing properties of a simple nonlinear relaxation method in Section 5.4.

5.3 A simple example

The nonlinear multigrid method as described in the previous section and the indirect multigrid approaches as outlined in Section 5.1 are quite different algorithmically, but closely related from a theoretical point of view. In particular, consider one iteration step of method II (cf. Section 5.1; one linear multigrid cycle per linearization step) and one FAS cycle. Without going into details, we only mention that - apart from the solution process on the coarsest grid - the main difference between these two cycles lies in the relaxation process (which in the one case refers to L_h and in the other case refers to its current linearization L_h^j). To make this clear, it is useful to write the linear multigrid cycle in the FAS form also.

The similarity of these approaches is reflected by the numerical results for the following

Example:

$$L^{\Omega}u = -\Delta u + e^{u} = f^{\Omega}(x) \quad (x \in \Omega), \qquad u = f^{\Gamma}(x) \quad (x \in \Gamma) \tag{5.12}$$

with solution $u(x) = \sin 3(x_1+x_2)$. The domain Ω is composed of semicircles and straight lines as shown in Figure 5.2. This problem is discretized with the usual 5-point formula (and $h=h_{x_1}=h_{x_2}$) except for grid points near the boundary, where the Shortley-Weller approximation is used (cf. Section 10.1).

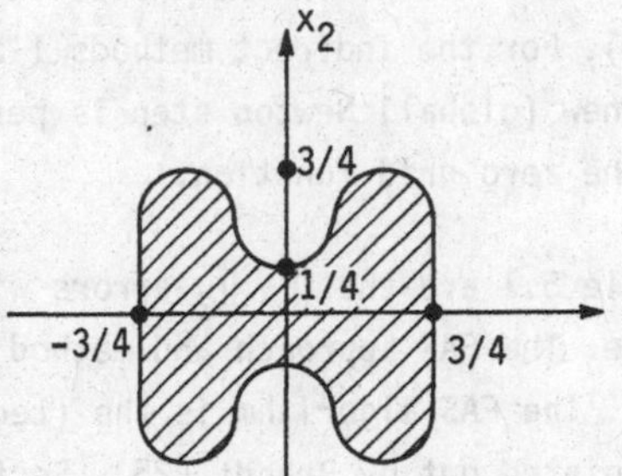

Figure 5.2: Domain Ω treated in (5.12)

Table 5.1 shows some numerical results if multigrid methods are applied to this problem. Here the columns I and II refer to the indirect methods I and II as described in Section 5.1. In II, one MG cycle is performed per (global) Newton-step, in I the number of MG cycles is doubled from one Newton-step to the next. The concrete (linear) MG program used is a version of the MGØ1 program collection described in Sections 10.1 and 10.2 (with $\nu_1=2$, $\nu_2=1$ and $\gamma=2$). The last column in Table 5.1 shows the corresponding FAS results. The concrete algorithm used here is a nonlinear analog of the linear one described in Section 10.2: The RB relaxation is replaced by a corresponding nonlinear relaxation method (performing one Newton step for each single equation in relaxing at the corresponding grid point). The operator $\hat{I}_h^H$ which occurs in the description of the FAS method, is chosen to be straight injection (cf. Section 3.6). For all methods, the zero grid function is used as first approximation.

multigrid steps	method		
	I	II	FAS
1	0.18(+2)	0.18(+2)	0.14(+2)
2	0.20(0)	0.20(0)	0.20(0)
3	0.86(-2)	0.55(-2)	0.54(-2)
4	0.14(-3)	0.14(-3)	0.14(-3)
5	0.43(-5)	0.42(-5)	0.42(-5)
6	0.13(-6)	0.13(-6)	0.13(-6)
7	0.47(-8)	0.39(-8)	0.39(-8)
8	0.13(-9)	0.12(-9)	0.12(-9)
9	0.42(-11)	0.40(-11)	0.39(-11)

<u>Table</u> 5.1: Behavior of the $\| \cdot \|_2$-error (with respect to the discrete solution, h=1/32) in case of direct and indirect applications of multigrid methods to problem (5.12). For the indirect methods I and II, horizontal lines indicate that a new (global) Newton step is performed. The first approximation used is the zero grid function.

The numbers shown in Table 5.1 are the $\| \cdot \|_2$-errors with respect to the discrete solution after each MG cycle. The FAS approach and method II give indeed very similar results in this example. The FAS algorithm is the (technically) simplest of the three algorithms used. As pointed out by Brandt [25], Section 8.3, there are several other advantages of FAS over the indirect methods.

The numerical similarity between method II and the FAS algorithm can, of course, only be expected if the first approximation used is sufficiently close to the solution (so that the convergence of Newton's method is sufficiently good). If we replace L^Ω in (5.12) by

$$L^\Omega u = -\Delta u + \lambda e^u \tag{5.13}$$

it turns out that - for $\lambda \geq 0$ - the dependence of the FAS method on the first approximation is much less sensitive than that of method II. For an example see Table 5.2 where results analogous to those in Table 5.1 are given for $\lambda = 100$. (The corresponding solution and the first approximation are chosen as above.)

MG steps:	1	2	3	4	5	6
method II	0.26(+3)	0.33(+2)	0.38(0)	0.67(-2)	0.11(-3)	0.19(-5)
FAS	0.14(+2)	0.21(0)	0.39(-2)	0.75(-4)	0.17(-5)	0.41(-7)

Table 5.2: Results corresponding to those in Table 5.1 for $\lambda = 100$ (see (5.13))

5.4 A remark on nonlinear relaxation methods

Relaxation methods for linear problems usually have several analogs for nonlinear problems (see [80],[101]). In the MG context, we are mainly interested in the smoothing properties of such nonlinear relaxation methods. We want to discuss this question briefly for a simple nonlinear problem of the type (5.4). For simplicity, we consider only Jacobi's method. A corresponding analysis can, however, also be made for other relaxation methods.

Let, in particular, a nonlinear counterpart of model problem (P) be given, namely

$$\mathbf{L}_h u_h := L_h u_h + g(x,u_h) = f_h(x) \quad (x \in \Omega_h) \tag{5.14}$$

where L_h and Ω_h are given as in model problem (P). Using the same notation as in Section 3.2, in particular.

$$L_h^- w_h := L_h w_h - \frac{4}{h^2} w_h,$$

one complete step of the *nonlinear Jacobi ω-relaxation* is defined by $\bar{w}_h = w_h + \omega(z_h - w_h)$ and

$$\frac{4}{h^2} z_h(x) + L_h^- w_h(x) + g(x,z_h(x)) = f_h(x) \quad (x \in \Omega_h). \tag{5.15}$$

In practice, one may replace (5.15) by one Newton-step for each single equation (*Jacobi-Newton ω-relaxation*):

$$\frac{4}{h^2} z_h(x) + L_h^- w_h(x) + g(x,w_h(x)) + g_u(x,w_h(x))(z_h(x) - w_h(x)) = f_h(x) \quad (x \in \Omega_h). \tag{5.16}$$

An even simpler linearized version of (5.15) which does not use any derivatives at all, is mentioned by Hackbusch [50], Section 7.1. Here (5.15) is simply replaced by

$$\frac{4}{h^2} z_h(x) + L_h^- w_h(x) + g(x,w_h(x)) = f_h(x) \quad (x \in \Omega_h). \tag{5.17}$$

We call this method *Jacobi-Picard ω-relaxation*.

The latter relaxation method should, however, be used with care. One difficulty which arises in connection with this method can already be demonstrated by looking at the special case of g being a linear function of u, namely

$$g(x,u) = cu, \quad \text{with constant } c > 0. \tag{5.18}$$

In this linear case, the relaxation operators of the Jacobi-Newton and the Jacobi-Picard methods are given by, respectively,

$$S_h^N = (1 - \frac{\omega ch^2}{4+ch^2}) I_h - \frac{\omega h^2}{4+ch^2} L_h \qquad (\omega = \omega^N), \tag{5.19}$$

$$S_h^P = (1 - \frac{\omega ch^2}{4}) I_h - \frac{\omega h^2}{4} L_h \qquad (\omega = \omega^P).$$

Obviously, both operators coincide if

$$\omega^N = \frac{4+ch^2}{4} \omega^P. \tag{5.20}$$

It is therefore sufficient to analyze $S_h^N = S_h^N(\omega)$.

By considerations similar to those in Section 3.2, we obtain the eigenvalues χ_n^N and the smoothing factor μ^N of $S_h^N(\omega)$:

$$\chi_n^N(\omega) = \frac{4}{4+ch^2} \chi_n(\omega) + \frac{(1-\omega)ch^2}{4+ch^2} \qquad (\chi_n \text{ in } (3.12)), \tag{5.21}$$

$$\mu^N(h;\omega,c) = \max \{ |1-\omega(1 - \frac{2\cos\pi h}{4+ch^2})|, |1-\omega(1 + \frac{4\cos\pi h}{4+ch^2})| \}. \tag{5.22}$$

From this, we see that, for any fixed $0<\omega<1$ (and any fixed h), the smoothing properties of S_h^N become better for increasing c:

$$\mu^N(h;\omega,c) \searrow 1 - \omega \qquad (0 \le c \to \infty). \tag{5.23}$$

The situation is quite different for the Jacobi-Picard method: From the relation (5.20), we see that for any fixed $\omega=\omega^P$, $0<\omega^P<1$, the Jacobi-Picard method has no smoothing properties (in fact, strong divergence occurs), if ch^2 is sufficiently large.

In our special case (5.18), of course, one could overcome this difficulty by choosing ω^P small enough, as a function of ch^2. This is, however, no longer possible for the more general case (5.14). Although the above analysis connot be applied directly to this general case, it is clear that the unfavorable behavior of the Jacobi-Picard method carries over:

The Jacobi-Picard method, with fixed $\omega=\omega^P$, cannot be used for smoothing purposes whenever $h^2 g_u(x,u_h(x))$ is large compared to 1 for certain $x \in \Omega_h$. If, however, $0 \le h^2 g_u(x,u_h(x)) < 1$ $(x \in \Omega_h)$ and if w_h is sufficiently close to u_h, the Jacobi-Picard method should give results similar to those of the Jacobi-Newton method.

We finally remark that the application of one Newton step in relaxing each single equation, seems to be a reasonable possibility for rather general nonlinear smoothing methods. This has been confirmed by several numerical experiments.

Remark: As a by-product, in formula (5.22) we have determined the smoothing factor μ of Jacobi ω-relaxation for the Helmholtz equation (with Helmholtz constant $c \ge 0$). For any fixed h and c one can easily determine the optimal relaxation parameter. One obtains:

$$\omega_{opt} = \frac{4+ch^2}{4+ch^2+\cos\pi h}, \qquad \mu_{opt} = \frac{3\cos\pi h}{4+ch^2+\cos\pi h}.$$

In particular, the optimal relaxation parameter and the corresponding smoothing factor approach 1 and 0, respectively, if c tends to infinity.

5.5 Some additional remarks

In this section, we want to mention only two further complexes which are connected with FAS multigrid methods:

- the application of multigrid methods to bifurcation problems,

- the dual view of multigrid methods.

5.5.1 An exemplary bifurcation problem

In the general description of the FAS multigrid method in Section 5.2, we have considered a discrete nonlinear elliptic operator L_h without making particular assumptions on it. Even in linear cases, but clearly even more in nonlinear cases, there are many situations where the fundamental algorithms have to be modified to take special features of a given problem into account.

For demonstration purposes, we want to mention only one such case here. Consider the following discrete problem for $\lambda>0$:

$$L_h^{\Omega} u_h = -\Delta_h u_h - \lambda e^{u_h} = 0 \quad (\Omega_h), \quad u_h = 0 \quad (\partial\Omega_h) \qquad (5.24)$$

on the unit square $\Omega=(0,1)^2$ with $h=h_{x_1}=h_{x_2}$. It is well known that the corresponding continuous problem has solutions which behave - as a function of λ - as illustrated in Figure 5.3. Methods of multigrid type have been used by several authors [30], [73],[74] to treat this problem.

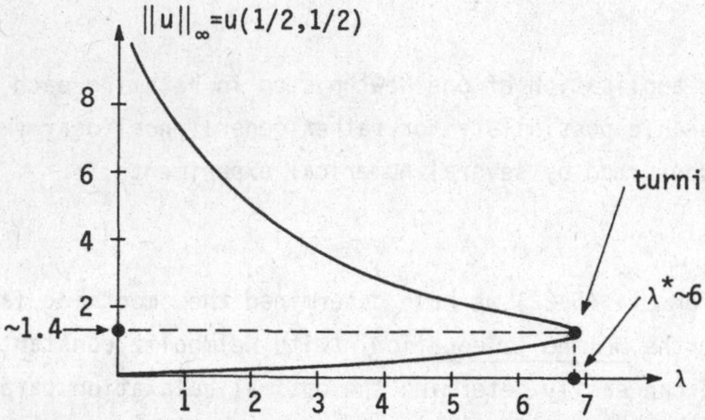

Figure 5.3: "Bifurcation diagram" for $-\Delta u - \lambda e^u = 0$ (Ω), $u=0$ $(\partial\Omega)$ showing $u(P)$ (for $P=(1/2,1/2)$) as a function of λ

For fixed $\lambda<\lambda^*$ (with λ not too close to λ^*), any reasonable algorithm of the type described in Section 5.2, e.g. the one mentioned briefly in Section 5.3 (with zero first approximation), will succeed in computing the "lower" solution (cf. Figure 5.3). Clearly, if λ approaches the critical value λ^*, the multigrid method will fail to converge (cf. [30]). Without going into details, we point out that this effect is caused by certain low frequency error components which are not reduced by coarse-grid corrections. This effect is very similar to the one which occurs if Helmholtz' equation is treated by multigrid methods for a fixed Helmholtz constant which is close to an eigenvalue of the discrete Laplacian.

This difficulty can easily be overcome by combining the multigrid approach with a suitable continuation method. In this particular example, one can, for instance, regard λ as an additional variable and prescribe another parameter ($\|u_h\|_2$ or $u_h(P)$ at the center point $P=(1/2,1/2)$ of Ω_h). Thus we consider (5.24) no longer as a boundary value problem with λ fixed, but as a nonlinear eigenvalue problem of the form

$$B_h(u_h,\lambda) = 0 \quad (\Omega_h), \quad u_h = 0 \quad (\partial\Omega_h),$$

together with an additional constraint, e.g., $u_h(P) = \delta$.

The FAS multigrid method (i.e. mainly its smoothing part) has to be modified with respect to this problem. We briefly describe one possibility how this modification can be done. (For other possibilities see, for example, [73].) We assume that the constraint is given by $u_h(P)=\delta$. One smoothing step is defined such that it takes both the variable λ and the constraint into account, e.g., by the following partial steps:

- (1) Apply one (nonlinear) relaxation step to the actual approximation of u_h (including the value at P).
- (2) Multiply the relaxed approximation by a factor such that the constraint is satisfied afterwards.
- (3) Compute a new value for λ by using the difference equations (e.g., using a suitable average over all equations).

If a reasonable FAS method is constructed which uses this kind of a smoothing process on all grids, one can march along the curve in Figure 5.3, starting somewhere on the lower "branch". Taking the previous solution (and the corresponding value of λ) as a first approximation and enlarging δ, one will come around the turning point without any difficulty. (Corresponding experiments have been carried out in cooperation with A. Brandt.)

However, somewhere beyond the turning point one has to be careful with the use of standard pointwise relaxation methods. Difficulties will arise if the coarsest grid used is too coarse. This is due to the fact that the "diagonal elements" of the corresponding relaxation operator (by which one has to divide) may be close to 0 or change sign on the coarsest grids. This difficulty can be overcome by simple modifications of the relaxation method used. In practice a combination of the (nonlinear) Gauß-Seidel method with Kaczmarz' method (see [21]), for instance, turned out to be suitable.

Already from the 1D-case it is known [11] that a <u>discrete</u> problem of the type considered here may have a very large variety of solutions. In our numerical experiments, we indeed found several solutions. In particular, for large values of δ (roughly $\delta\sim 9$ and $h=1/32$), the constraint used seems to be no longer suitable in order to distinguish between different discrete solutions. What happens in practice (convergence, divergence, jump to another branch of discrete solutions) depends somewhat on the algorithmical details (e.g., on the coarsest grid used). One should try to use more sophisticated continuation methods in this critical field (e.g. the arclength continuation method [30]).

5.5.2 The (h,H)-relative truncation error and the dual view of multigrid methods

The coarse-grid defect equation (5.11) of the FAS scheme can be written in the form

$$L_H w_H^j = I_h^H f_h + \tau_h^H[\bar{u}_h^j] \tag{5.25}$$

where

$$\tau_h^H[z_h] := L_H \hat{I}_h^H z_h - I_h^H L_h z_h .$$

Trivially, the following identity holds for the discrete solution u_h:

$$L_H(\hat{I}_h^H u_h) = I_h^H f_h + \tau_h^H[u_h] . \tag{5.26}$$

$\tau_h^H[u_h]$ is called the *(h,H)-relative truncation error* (with respect to I_h^H, $\hat{I}_h^H$). With respect to the grids Ω_h and Ω_H, τ_h^H plays a similar role as the *truncation error* (*local discretization error*)

$$\tau_h[u] := L_h I^h u - I^h Lu \tag{5.27}$$

of the continuous solution u with respect to Ω and Ω_h. (Here I^h denotes the straight injection operator from Ω to Ω_h.) If, in particular, $\Omega_H \subset \Omega_h$ and $\hat{I}_h^H$ is the straight injection operator, we see from (5.26) that $\tau_h^H[u_h]$ can be interpreted in the following way: τ_h^H is that quantity which has to be added to the right hand side $I_h^H f_h$ to obtain the values of the fine-grid solution u_h (on Ω_H) by solving the coarse-grid equation (5.26).

The quantity τ_h^H (and approximations to it) are the starting point for several more sophisticated multigrid techniques which we only want to list here (for details and further references, see [25]):

- τ-estimation;
- τ-extrapolation;
- frozen-τ techniques, especially for the treatment of parabolic problems [63];
- adaptive multigrid techniques, where the switching from fine to coarse and from coarse to fine is controlled by defects and τ-quantities;
- small storage algorithms.

All these techniques can be regarded from the *dual point of view* - as Brandt calls it. Here not the fine grid Ω_h is regarded as the primary grid, but the coarse grid Ω_H. The objective is to obtain - by treating coarse-grid equations - a solution which has the fine-grid accuracy. For this purpose, of course, a suitable approximation for $\tau_h^H[u_h]$ has to be provided.

6. The full multigrid method (nested iteration)

So far, we have considered only iterative multigrid methods for the solution of discrete elliptic boundary value problems. Although suitable versions of these methods are very efficient, their efficiency can still be improved essentially if they are used in form of the *full multigrid* (FMG) technique [25],[19], also called *nested iteration* [50],[49]. Although this approach can be applied to linear as well as to nonlinear problems (in connection with FAS), we restrict ourselves to considering the linear case.

6.1 Idea and purpose

The full multigrid approach can be regarded from different points of view. We here prefer a quite narrow interpretation of this method, namely as an "approximate" direct solver (for a given h-discrete elliptic problem (2.1)) which is characterized by the following features:

(1) An approximation $\tilde{u}_h$ of the discrete solution u_h is computed up to an error $\| \tilde{u}_h - u_h \|$ which is smaller than the discretization error $\| u - u_h \|$.

(2) The number of arithmetic operations needed is proportional to the number of grid-points of Ω_h (with only a small constant of proportionality).

We would like to point out that the idea of nested iteration is a well-known numerical principle which has also been used successfully independently of the multigrid concept. Within an arbitrary iterative process (e.g. SOR) for the solution of a given discrete problem, this principle simply means that lower (coarser) discretization levels are used in order to provide good initial approximations for the iteration on the next higher (finer) discretization level [64],[65]. With classical methods like SOR, however, it is not possible to achieve both, (1) and (2).

6.2 Structure of the full multigrid method

The FMG method, as we understand it here, is most easily explained by a flow chart. We refer to a general elliptic boundary value problem

$$L^{\Omega}u = f^{\Omega} \quad (\Omega), \qquad L^{\Gamma}u = f^{\Gamma} \quad (\Gamma := \partial\Omega) \tag{6.1}$$

on a bounded domain Ω. As in Section 4.2., let

$$L_{\ell}u_{\ell} = f_{\ell} \quad (\Omega_{\ell}) \qquad (\ell = 0,1,2,\ldots) \tag{6.2}$$

be a sequence of discrete approximations to (6.1). For all $\ell \geq 1$, we denote by

$$MGI^r(\cdot,\ell,L_\ell,f_\ell) : \mathbb{G}(\Omega_\ell) \to \mathbb{G}(\Omega_\ell) \tag{6.3}$$

a procedure consisting of r iteration steps of a suitable iterative multigrid method for (6.2) (using grids $\Omega_o,\ldots,\Omega_\ell$; cf. Figure 4.2). In the following, standard values for r are $r=1$ and $r=2$ (fixed). In general, however, r may depend on ℓ.

Now let an arbitrary but <u>fixed</u> ℓ be chosen with $h=h_\ell$. Then the full multigrid method proceeds as shown in Figure 6.1. The final FMG approximation is denoted by $\tilde{u}_\ell$.

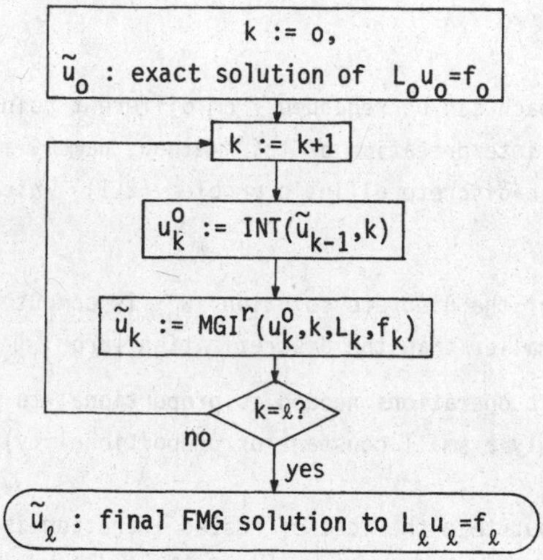

<u>Figure</u> 6.1: FMG method for the approximate solution of $L_\ell u_\ell = f_\ell$ (Ω_ℓ)

In Figure 6.1, the notation

$$INT(\cdot,k) : \mathbb{G}(\Omega_{k-1}) \to \mathbb{G}(\Omega_k) \tag{6.4}$$

is used for a suitable interpolation procedure. In general, we assume INT to be an affine operator of the form

$$INT(u_{k-1},k) = \Pi_{k-1}^{k} u_{k-1} + w_k \tag{6.5}$$

with a linear operator

$$\Pi_{k-1}^{k} : \mathbb{G}(\Omega_{k-1}) \to \mathbb{G}(\Omega_k)$$

and a certain fixed gridfunction $w_k \in \mathbb{G}(\Omega_k)$. This representation of INT allows one to take the given boundary conditions into account within the interpolation process or to make use of L^Ω and of f^Ω as well ("interpolation using the grid equation" as mentioned already in Section 2.4; also see [36]). In order to distinguish the interpolation (6.5) from the interpolation used within the MG iterations, we call it *FMG interpolation*.

6.3 A simple theoretical result

Under reasonable assumptions, the FMG method asymptotically (i.e. for sufficiently large ℓ) has indeed the properties (1) and (2) which were pointed out in Section 6.1. This can be shown by very simple estimations. Several such estimates have been given, which differ only slightly (see, e.g., [50] and reference "[11]" in [36]). For completeness, we give an estimate, too.

By $\mathbb{M}_\ell : \mathbb{G}(\Omega_\ell) \to \mathbb{G}(\Omega_\ell)$ we denote the iteration operator corresponding to (6.3). Let any reasonable vector norm $\|\,.\,\|$ with corresponding operator norm be given on $\mathbb{G}(\Omega_\ell)$ $(\ell=0,1,2,\ldots)$. For simplicity, we restrict ourselves to square grids in this section. (Otherwise, in the following inequalities, h_ℓ would have to be replaced by its maximal component.) Now we make the following assumptions.

(1) Let the norms of $\mathbb{M}_\ell$ and $\Pi^\ell_{\ell-1}$ be uniformly bounded:

$$\|\mathbb{M}_\ell\| \le \eta < 1, \quad \|\Pi^\ell_{\ell-1}\| \le C \qquad (\ell = 1,2,\ldots). \tag{6.6}$$

(2) The discretization error and the FMG interpolation error are assumed to be of order κ_1 and κ_2, respectively:

$$\|u-u_\ell\| \le K_1 h_\ell^{\kappa_1}, \quad \|u-\Pi^\ell_{\ell-1}u-w_\ell\| \le K_2 h_\ell^{\kappa_2} \tag{6.7}$$

with $\kappa_1, \kappa_2, K_1, K_2$ independent of ℓ.

(3) We assume a fixed meshsize ratio $\zeta > 1 : h_{\ell-1} = \zeta h_\ell$ $\quad (\ell = 1,2,\ldots)$.

Under these assumptions we obtain the following

<u>Lemma</u> 6.1: Let $\kappa_2 \ge \kappa_1$ and $\eta^r A < 1$ with $A := C\zeta^{\kappa_1}$. Then the following estimate holds for all $\ell \ge 1$:

$$\|\tilde{u}_\ell - u_\ell\| \le \delta^* h_\ell^{\kappa_1} \quad \text{with} \quad \delta^* := \frac{\eta^r K}{1-\eta^r A} + o(1) \quad (\ell \to \infty) \tag{6.8}$$

and

$$K := \begin{cases} K_1(1+A) + K_2 & (\text{if } \kappa_2 = \kappa_1) \\ K_1(1+A) & (\text{if } \kappa_2 > \kappa_1). \end{cases} \tag{6.9}$$

If we additionally have a lower bound for the discretization error

$$\| u - u_\ell \| \geq \hat{K}_1 h_\ell^{\kappa_1} \qquad (\hat{K}_1 > o, \text{ independent of } \ell),$$

then we obtain by (6.8)

$$\| \tilde{u}_\ell - u_\ell \| \leq \beta^* \| u - u_\ell \| \quad \text{with} \quad \beta^* := \frac{K}{\hat{K}_1} \frac{\eta^r}{1-\eta^r A} + o(1) \quad (\ell \to \infty). \tag{6.10}$$

Proof: By definition of the FMG method, we have for all $\ell \geq 1$

$$\tilde{u}_\ell - u_\ell = M_\ell^r (u_\ell^o - u_\ell), \quad u_\ell^o = \Pi_{\ell-1}^\ell \tilde{u}_{\ell-1} + w_\ell .$$

Using the identity

$$u_\ell^o - u_\ell = \Pi_{\ell-1}^\ell (\tilde{u}_{\ell-1} - u_{\ell-1}) + \Pi_{\ell-1}^\ell (u_{\ell-1} - u) + (\Pi_{\ell-1}^\ell u + w_\ell - u) + (u - u_\ell),$$

we get the recursive estimation

$$\delta_\ell \leq \eta^r (A \, \delta_{\ell-1} + K_1(1+A) + K_2 \, h_\ell^{\kappa_2 - \kappa_1}), \quad \delta_\ell := \| \tilde{u}_\ell - u_\ell \| / h_\ell^{\kappa_1}.$$

From this, (6.8) follows by a simple calculation.

Remarks:

(1) In the case $\kappa_2 = \kappa_1$, one sees from the definition of K in (6.9) that the contribution of the FMG interpolation constant K_2 to the bounds δ^* and β^* may become arbitrarily large (depending on u). In particular, in (6.10), $K_1/\hat{K}_1$ often can be assumed to be bounded independently of u (see below), not, however, $K_2/\hat{K}_1$. If, on the other hand, $\kappa_2 > \kappa_1$, then the contribution of K_2 asymptotically vanishes. Thus, it might be advantageous to choose an FMG interpolation of an order which is higher than that of the discretization.

(2) If $\kappa_2 > \kappa_1$ and we additionally assume an expansion

$$u_\ell = u + h_\ell^{\kappa_1} e + o(h_\ell^{\kappa_1}), \tag{6.11}$$

then we can replace $K/\hat{K}_1$ in (6.10) by $A+1$. Moreover, under certain stronger assumptions (see Hackbusch [50], Chapter 5, where the FMG interpolation itself is assumed to be described by a linear operator), one can replace $A+1$ by $\zeta^{\kappa_1}-1$. (Notice that the assumption (5.2) in [50] does not distinguish between κ_1 and κ_2.)

6.4 Computational work, some practical remarks

The computational work W_ℓ^{FMG} needed for the FMG method (as described in Figure 6.1) can easily be estimated. By arguments similar to those in Section 4.4, one immediately obtains in terms of W_ℓ (cf. (4.22)), for example in case of standard coarsening ($\zeta=2$):

$$W_\ell^{FMG} \lessdot \frac{4}{3}\, r\, W_\ell + \frac{4}{3} W_{\ell-1}^{INT} \tag{6.12}$$

(neglecting lower order terms). Here W_{k-1}^{INT} denotes the work needed for the FMG interpolation process from grid Ω_{k-1} to grid Ω_k ($k=1,\ldots,\ell$).

Whether (1) in Section 6.1 is satisfied (for sufficiently large ℓ and fixed r, usually $r=1$ or $r=2$), clearly depends essentially on the size of η. Let us assume in the following that $\zeta=2$, $\kappa_2>\kappa_1=2$ and that (6.11) is satisfied. According to Remark (2) above, β^* in (6.10) can then be replaced by

$$\beta^* = \eta^r \frac{1+4C}{1-4C\eta^r} \tag{6.13}$$

which now depends on C, η and r only. Here $C\geq 1$, but for reasonable FMG interpolation processes, C usually can be assumed not to be very large.

Example: Let us consider model problem (P). An appropriate and especially cheap FMG interpolation is described in the following. This interpolation is an example of an interpolation "using the grid equation". (It is used in the sample program listed in the appendix.) The process $u_k^o = INT(\tilde{u}_{k-1},k)$ (cf. Figure 6.1) proceeds in three partial steps (1), (2), (3):

(1) At points $x \in \Omega_k \cap \Omega_{k-1}$ define $u_k^o(x) := \tilde{u}_{k-1}(x)$.

(2) At points $x \in \Omega_k \backslash \Omega_{k-1}$, $x=\kappa\cdot h$ with $\kappa_1+\kappa_2$ even, we define $u_k^o(x)$ by

$$\frac{1}{2h_k^2}\begin{bmatrix} -1 & & -1 \\ & 4 & \\ -1 & & -1 \end{bmatrix}_{h_k} u_k^o(x) = f^\Omega(x). \tag{6.14}$$

Note that this is an explicit equation (corresponding to a diagonal matrix). At boundary points, $u_k^o(x)$ is assumed to be replaced by the corresponding boundary values given.

(3) At points $x \in \Omega_k$, $x=\kappa\cdot h$ with $\kappa_1+\kappa_2$ odd, we define $u_k^o(x)$ by

$$\frac{1}{h_k^2}\begin{bmatrix} & -1 & \\ -1 & 4 & -1 \\ & -1 & \end{bmatrix}_{h_k} u_k^o(x) = f^\Omega(x). \tag{6.15}$$

Again, this is an explicit equation. (We remark that this partial step can be interpreted as one half-step of RB relaxation (see Section 7.3)).

For this FMG interpolation one has $C=1$ (with respect to $\|\cdot\|_2$) and $\kappa_2=4$. Table 6.1 shows the dependence of β^* on η in this case. We see that, for $r=1$, $\eta\le0.1$ is sufficient to guartantee a value of $\beta^*<1$ (and by this also (1) in Section 6.1). η-values of this size are typical for efficient multigrid methods for standard applications. This is confirmed, e.g., by the numerical results of Chapter 10. For the underlying example of model problem (P), one can construct multigrid methods with $\eta\le0.1$ by using the theoretical results of Chapter 8 together with the considerations of Section 4.3 (see, for example, Corollary 8.4).

r \ η	0.5	0.25	0.20	0.10	0.05	0.01
1	-	∞	>1	0.833	0.313	0.052
2	∞	0.417	0.238	0.052	0.013	0.001

Table 6.1: Values of β^* (6.13) as a function of η if $C=1$

Remark: Instead of the FMG interpolation described above, one could also use, e.g., cubic interpolation. This is, however, somewhat more expensive. (Cubic interpolation is used within the program MG01 described in Chapter 10.)

Clearly, the efficiency of an FMG method depends on the numerical work needed to ensure (1) in Section 6.1. Some concrete examples on computational work and accuracy will be given in Chapter 10. We here only want to make the following point clear: The efficiency of an FMG process is not solely connected to the efficiency of the (iterative) multigrid process which it is based upon. It is, for example, not worthwile to construct an FMG method using a multigrid cycle which is characterized by a convergence factor η, much smaller than needed to achieve the discretization accuracy. Such a multigrid method will usually be too expensive per cycle.

Finally, we want to point out that the FMG concept is only the starting point for several more sophisticated techniques (see Part II in Brandt [25] for some details and references). The common idea of these more advanced approaches is the orientation to the continuous solution u rather than to the discrete solution u_h.

7. The concept of model problem analysis, smoothing and two-grid convergence factors

So far, we have described the fundamental multigrid methods in a rather general form. The practically very important question of how the individual multigrid components should be chosen for concrete situations has, however, not yet been discussed systematically. As pointed out in the introduction, there are two related approaches in treating this question: the *model problem* and the *local Fourier analysis*.

A first example of model problem analysis has been already considered in Chapter 3. The main characteristics of this example which carry over to the general concept of both, the model problem and local Fourier analysis are the following:

- For the given linear difference operator L_h a basis of discrete eigenfunctions is known.
- The $(h,2h)$ two-grid method is constructed in such a way, that the corresponding iteration operator M_h^{2h} can be represented with respect to the L_h-eigenbasis by a block-matrix consisting of small blocks only. (In the particular example at most (4,4)-blocks occured.)

Because of these properties, the determination of the corresponding two-grid convergence factors turned out to be relatively easy. Furthermore, we were able to give a straightforward definition of a smoothing factor by making use of the fact that the eigenfunctions of L_h and the smoothing operator S_h were the same.

In this chapter, the analysis of Chapter 3 is generalized in several respects. In Section 7.1, we describe the class of problems under consideration. Section 7.2 refers to the Fourier representation of coarse-grid correction operators. Here other coarse grids than Ω_{2h} are admitted. In Section 7.3, we give Fourier representations of efficient smoothing methods like RB and ZEBRA relaxation. A general definition of smoothing factors is given in Section 7.5. Some other classes of problems which can be treated similarly are sketched in Section 7.6.

Concrete results of the model problem analysis are presented in Chapter 8. In Chapter 9, the local Fourier analysis and its relation to the model problem analysis are described. Furthermore some other theoretical approaches are outlined.

7.1 Assumptions on the difference operator

Let

$$L_h u_h = f_h \quad (\Omega_h) \tag{7.1}$$

be a given grid equation on a rectangular grid $\Omega_h = \Omega \cap G_h$

$$\Omega = (0,A_1)\times(0,A_2), \quad G_h = \{x=\kappa\cdot h : \kappa \in \mathbb{Z}^2\}; \quad h_{x_j}=A_j/N_j,\ N_j \text{ even } (j=1,2), \tag{7.2}$$

$h = (h_{x_1},h_{x_2})$, $\kappa\cdot h := (\kappa_1 h_{x_1},\kappa_2 h_{x_2})$. We consider invertible operators L_h for which the following grid functions form a basis of eigenfunctions:

$$\varphi_n(x) := 2\sin(n_1\pi x_1/A_1)\sin(n_2\pi x_2/A_2) \quad (x \in \Omega_h;\ 0 < n < N) \tag{7.3}$$

with $n=(n_1,n_2)$, $N=(N_1,N_2)$. (The inequalities are to be understood componentwise.) These φ_n are orthonormal with respect to (1.7). Examples of operators L_h with this property are given below.

For the formalism used in the following, it is convenient to write the grid functions (7.3) in the form

$$\varphi(\theta,x) := 2\sin(\theta_1 x_1/h_{x_1})\sin(\theta_2 x_2/h_{x_2}) \quad (x \in \Omega_h;\ \theta \in T_h) \tag{7.4}$$

with

$$T_h = \{\theta = (\frac{n_1}{N_1}\pi, \frac{n_2}{N_2}\pi) : 0 < n < N\} \subset (0,\pi)^2 . \tag{7.5}$$

(For technical reasons, we sometimes consider $\varphi(\theta,x)$ in (7.4) also for certain $\theta \in (-\pi,\pi]^2$. Whenever such a θ occurs in the following - see (7.22), for example - then there exists a $\tilde{\theta} \in T_h$ such that $\varphi(\theta,x) = \pm\,\varphi(\tilde{\theta},x)$ $(x \in \Omega_h)$. So this only means that some of the above basis functions are used with a negative sign.)

The following continuous problems typically lead to operators L_h of the above type:

<u>Example 7.1</u>: Let a linear boundary value problem (1.1) be given on Ω with

$$L^{\Omega}u = -a_1 u_{x_1x_1} - a_2 u_{x_2x_2} + cu \quad (a_1,a_2 > 0,\ c \in \mathbb{R}) \tag{7.6}$$

and Dirichlet boundary conditions

$$u = f^{\Gamma}(x) \quad (x \in \Gamma). \tag{7.7}$$

We consider discretizations of L^{Ω} on Ω_h which can be described by a symmetric compact 9-point star, i.e.

$$L_h^{\Omega} \triangleq \begin{bmatrix} s_{-1,1} & s_{0,1} & s_{1,1} \\ s_{-1,0} & s_{0,0} & s_{1,0} \\ s_{-1,-1} & s_{0,-1} & s_{1,-1} \end{bmatrix}_h \quad \text{with} \quad \begin{cases} s_{1,1} = s_{-1,1} = s_{1,-1} = s_{-1,-1}, \\ s_{1,0} = s_{-1,0},\ s_{0,1} = s_{0,-1}. \end{cases} \tag{7.8}$$

(Clearly, these coefficients depend on h.) Every discretization of this type leads after elimination of the boundary conditions - to a discrete system (7.1). The grid functions (7.4) are eigenfunctions with corresponding eigenvalues

$$\lambda(\theta,h) = \sum_{|\kappa|\leq 1} s_\kappa \cos(\theta\kappa) \quad (\theta \in T_h), \qquad (\kappa=(\kappa_1,\kappa_2), \quad |\kappa|=\max(|\kappa_1|,|\kappa_2|)). \tag{7.9}$$

In particular, higher-order discretizations of the "Mehrstellen"-type ("Hermitian methods") [31] are admitted here.

<u>Example 7.2</u>: Let a fourth-order boundary value problem (1.1) be given on Ω with

$$L^\Omega u = \Delta\Delta u - b\Delta u + cu \quad (b,c \in \mathbb{R}) \tag{7.10}$$

and boundary conditions

$$u = f_1^\Gamma(x), \; -\Delta u = f_2^\Gamma(x) \quad (x \in \Gamma). \tag{7.11}$$

Approximating all occuring Δ-operators by symmetric 5- or compact 9-point difference stars and eliminating the boundary conditions, again leads to a discrete operator L_h which satisfies the above assumptions. (L_h^Ω is then described by a 13- or 25-point star, respectively.)

Although for the following considerations we mainly need the assumption (7.3), for the derivation of concrete results we shall additionally assume that L_h is given by a compact 9-point star (7.8). This is done, in order to avoid some technical complications which are introduced by larger stars (e.g. within the relaxation process), see also Section 7.6.

7.2 <u>The (h,H) coarse-grid correction operator</u>

For the definition of the coarse-grid correction (CGC) operator

$$K_h^H = I_h - I_H^h L_H^{-1} I_h^H L_h \tag{7.12}$$

we first need a coarser grid Ω_H with $\mathcal{N}_H \ll \mathcal{N}_h$. In the following, we consider four choices of Ω_H (cf. Section 2.4.1). In all these cases we have $\Omega_H = \Omega \cap G_H$. For the cases (1),(2) and (4), G_H is defined analogously to G_h in (7.2) with different meshsizes only. For the case (3), G_H is explicitly defined below.

(1) <u>Standard coarsening</u>: H=2h.

(2) <u>Semi-coarsening</u>: x_1-coarsening, $H=(2h_{x_1},h_{x_2})$ or x_2-coarsening, $H=(h_{x_1},2h_{x_2})$.

(3) <u>Red-black coarsening</u>: For simplicity, we consider this type of coarsening only

for square grids Ω_h ($h=h_{x_1}=h_{x_2}$). Then $\Omega_H = \Omega \cap G_H$, where G_H can be identified with a rotated grid of meshsize $H=\sqrt{2}h$:

$$G_H = \{x = \kappa \cdot h : \kappa \in \mathbb{Z}^2, \kappa_1 + \kappa_2 \text{ even}\}.$$

(4) Quadrupling h: H=4h. In this case we assume that N_j is a multiple of 4.

In all these cases there is a natural distinction between *low* and *high (h-)frequencies*. As before, the low frequencies are those which can be represented (are visible) on the coarser grid Ω_H also. We will define the low and high frequencies for each of the above cases separately. Corresponding to this distinction, the space $\mathbb{G}(\Omega_h)$ is written as an orthogonal sum

$$\mathbb{G}(\Omega_h) = \mathbb{G}^{low}(\Omega_h) \overset{\perp}{\oplus} \mathbb{G}^{high}(\Omega_h). \tag{7.13}$$

This splitting of $\mathbb{G}(\Omega_h)$ can simply be described by a corresponding splitting of the index set T_h in (7.5)

$$T_h = T_h^{low} \,\dot{\cup}\, T_h^{high}, \quad \#T_h^{low} = \mathcal{N}_H \tag{7.14}$$

(Here $\#T_h^{low}$ denotes the number of elements in T_h^{low}.) In all cases considered, there is a natural identification of $\mathbb{G}^{low}(\Omega_h)$ and $\mathbb{G}(\Omega_H)$: If we denote by Φ the restrictions of low frequencies φ to Ω_H

$$\Phi(\theta,x) := \varphi(\theta,x) \quad (x \in \Omega_H;\ \theta \in T_h^{low}) \tag{7.15}$$

then these Φ form a basis of $\mathbb{G}(\Omega_H)$:

$$\mathbb{G}(\Omega_H) = \text{span}\,\{\Phi(\theta,\cdot) : \theta \in T_h^{low}\}. \tag{7.16}$$

The high frequencies have the property that for all $\theta' \in T_h^{high}$

$$\text{either} \quad \varphi(\theta',x) = \pm\,\Phi(\theta,x) \quad (x \in \Omega_H) \quad \text{for a suitable } \theta \in T_h^{low}, \tag{7.17}$$

$$\text{or} \quad \varphi(\theta'x) \equiv 0 \quad (x \in \Omega_H). \tag{7.18}$$

As in Section 3.3, we will introduce low-dimensional subspaces $E_{h,\theta}^H$ of $\mathbb{G}(\Omega_h)$. Each of these spaces is defined as the span of functions (7.4) which coincide on Ω_H (up to sign). The $E_{h,\theta}^H$ are called spaces of *H-harmonics*. Because of (7.17),(7.18), their basis functions coincide on Ω_H either with $\pm\Phi(\theta,x)$ (for a certain $\theta \in T_h^{low}$) or with 0. For each of the coarser grids Ω_H considered, the corresponding spaces $E_{h,\theta}^H$ of H-harmonics yield a splitting of $\mathbb{G}(\Omega_h)$ into an orthogonal sum

$$\mathbb{G}(\Omega_h) = \overset{\perp}{\bigoplus_{\theta \in T_h^H}} E_{h,\theta}^H \tag{7.19}$$

Here the index set $T_h^H \subset T_h$ is a superset of T_h^{low} taking the degenerate cases (7.18) into account (see the examples below).

In the following, we will specify the above quantities for the different coarsenings (1)-(4). We use the notation $|\theta| := \max(|\theta_1|,|\theta_2|)$. Figure 7.1 illustrates the respective θ-ranges of the low frequencies (hatched part). Furthermore, exemplary θ-values of frequencies (7.4) are shown there which coincide on Ω_H with a low frequency (up to sign). This low frequency is marked by "•" and the corresponding high frequencies are marked by "o".

- <u>case (1), standard coarsening, H=2h:</u>

$$T_h^{low} = \{\theta \in T_h : |\theta| < \pi/2\}; \quad T_h^H = \{\theta \in T_h : |\theta| \le \pi/2\}; \tag{7.20}$$

$$E_{h,\theta}^H = \text{span}\{\varphi(\theta^\alpha,x) : \alpha=(0,0);(1,1);(1,0);(0,1)\}, \ \theta \in T_h^H \tag{7.21}$$

with

$$\theta^\alpha = \theta - \pi\alpha, \quad \alpha = (\alpha_1,\alpha_2). \tag{7.22}$$

The spaces $E_{h,\theta}^H$ are of dimension 4 if $\theta \in T_h^{low}$. Otherwise they are of dimension 2 or 1.

- <u>case (2), x_1-coarsening, $H=(2h_{x_1},h_{x_2})$ (analogous for x_2-coarsening):</u>

$$T_h^{low} = \{\theta \in T_h : \theta_1 < \pi/2\}; \quad T_h^H = \{\theta \in T_h : \theta_1 \le \pi/2\}; \tag{7.23}$$

$$E_{h,\theta}^H = \text{span}\ \{\varphi(\theta^\alpha,x) : \alpha = (0,0);(1,0)\}, \quad \theta \in T_h^H \tag{7.24}$$

with θ^α as defined in (7.22). The spaces $E_{h,\theta}^H$ are of dimension 2 and 1 for $\theta \in T_h^{low}$ and $\theta \notin T_h^{low}$, respectively.

- <u>case (3), red-black coarsening, "$H=\sqrt{2}h$"</u> $(h=h_{x_1}=h_{x_2})$:

$$T_h^{low} = \{\theta \in T_h : \theta_2 \le \pi-\theta_1 \ (\text{if } \theta_1 \le \pi/2),\ \theta_2 < \pi-\theta_1 \ (\text{if } \theta_1 > \pi/2)\} = T_h^H; \tag{7.25}$$

$$E_{h,\theta}^H = \text{span}\ \{\varphi(\theta^\alpha,x) : \alpha = (0,0);(1,1)\}, \quad \theta \in T_h^H \tag{7.26}$$

with θ^α as defined in (7.22). The spaces $E_{h,\theta}^H$ are of dimension 2 if $\theta \ne (\pi/2,\pi/2)$. For $\theta = (\pi/2,\pi/2)$, $E_{h,\theta}^H$ is of dimension 1.

<u>Remark</u>: The above definition of low frequencies is in accordance with the natural interpretation of them being representable on the coarse grid Ω_H. This can easily be seen by use of a transformation to the rotated coordinates

$$\xi_1 = (x_1+x_2)/\sqrt{2}, \quad \xi_2 = (x_1-x_2)/\sqrt{2}.$$

- case (4), quadrupling h, H=4h:

$$T_h^{low} = \{\theta \in T_h : |\theta| < \pi/4\}; \quad T_h^H = \{\theta \in T_h : |\theta| \le \pi/4\}; \tag{7.27}$$

$$E_{h,\theta}^H = \text{span}\,\{\varphi(\theta^\alpha,x) : \alpha_1,\alpha_2 \in \{0,-1/2,1/2,1\}\}, \quad \theta \in T_h^H \tag{7.28}$$

with θ^α as defined in (7.22). The spaces $E_{h,\theta}^H$ are of dimension 16 if $\theta \in T_h^{low}$. Otherwise they are of dimension 8 or 4.

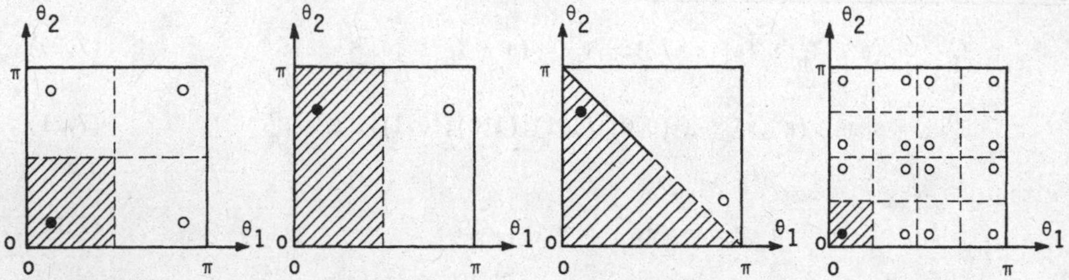

Figure 7.1: θ-range of low frequencies for standard coarsening, x_1-coarsening, red-black coarsening and quadrupling, respectively

In the model problem analysis we consider (h,H) two-grid methods such that the corresponding coarse-grid correction operator (7.12) leaves the spaces of H-harmonics invariant:

$$K_h^H : E_{h,\theta}^H \to E_{h,\theta}^H \quad (\theta \in T_h^H). \tag{7.29}$$

For this, we first assume that L_H^{-1} exists. Furthermore, the individual components I_h^H, L_H and I_H^h of K_h^H are supposed to have the following properties

$$\left.\begin{array}{llll} I_h^H : & E_{h,\theta}^H \to \text{span}\,\{\Phi(\theta,x)\} & (\theta \in T_h^H), \\ L_H : & \text{span}\,\{\Phi(\theta,x)\} \to \text{span}\,\{\Phi(\theta,x)\} & (\theta \in T_h^{low}), \\ I_H^h : & \text{span}\,\{\Phi(\theta,x)\} \to E_{h,\theta}^H & (\theta \in T_h^{low}). \end{array}\right\} \tag{7.30}$$

Here we use the formal notation

$$\Phi(\theta,x) :\equiv 0 \quad (x \in \Omega_H) \quad \text{for} \quad \theta \in T_h^H \setminus T_h^{low}.$$

Obviously, the assumptions (7.30) imply (7.29). By this, K_h^H is orthogonally equivalent to a block matrix. The corresponding blocks - represented with respect to the basis of $E^H_{h,\theta}$ given above - will be denoted by $\hat{K}^H_{h,\theta}$:

$$\hat{K}^H_{h,\theta} \mathrel{\hat{=}} K^H_h\Big|_{E^H_{h,\theta}} \qquad (\theta \in T^H_h). \tag{7.31}$$

The dimensions of these blocks clearly are equal to the dimensions of the respective spaces $E^H_{h,\theta}$. We remark that $\hat{K}^H_{h,\theta}$ is the identity matrix if $\theta \in T^H_h \setminus T^{low}_h$.

As an example, we give the representation of K^H_h for the case of <u>standard coarsening</u> and <u>9-point operators</u>:

<u>Lemma 7.1</u>: Let L_h and L_{2h} be given by symmetric compact 9-point stars with coefficients s_κ (cf. (7.8)) and $\hat{s}_\kappa$, respectively. We assume L^{-1}_{2h} to exist on $G(\Omega_{2h})$. Furthermore, I^{2h}_h and I^h_{2h} are supposed to be defined by symmetric 9-point stars

$$I^{2h}_h \mathrel{\hat{=}} \begin{bmatrix} \hat{t}_{-1,1} & \hat{t}_{0,1} & \hat{t}_{1,1} \\ \hat{t}_{-1,0} & \hat{t}_{0,0} & \hat{t}_{1,0} \\ \hat{t}_{-1,-1} & \hat{t}_{0,-1} & \hat{t}_{1,-1} \end{bmatrix}^{2h}_{h}, \qquad I^h_{2h} \mathrel{\hat{=}} \left]\begin{matrix} \check{t}_{-1,1} & \check{t}_{0,1} & \check{t}_{1,1} \\ \check{t}_{-1,0} & \check{t}_{0,0} & \check{t}_{1,0} \\ \check{t}_{-1,-1} & \check{t}_{0,-1} & \check{t}_{1,-1} \end{matrix}\right[^{h}_{2h} \tag{7.32}$$

with $\hat{t}_\kappa = \hat{t}_{\kappa'}$, $\check{t}_\kappa = \check{t}_{\kappa'}$ if $|\kappa_1| = |\kappa_1'|$, $|\kappa_2| = |\kappa_2'|$.

Then the properties (7.29), (7.30) are valid. In particular, we have for all $\theta \in T^{low}_h$ and all α as given in (7.21) (and θ^α as in (7.22)):

$$\left.\begin{aligned} &L_h\, \varphi(\theta^\alpha,x) = \lambda(\theta^\alpha,h)\varphi(\theta^\alpha,x), \qquad L_{2h}\Phi(\theta,x) = \Lambda(\theta,h)\Phi(\theta,x), \\ &I^{2h}_h\, \varphi(\theta^\alpha,x) = q(\theta^\alpha)\Phi(\theta,x), \qquad I^h_{2h}\Phi(\theta,x) = \sum_\alpha p_\alpha(\theta)\varphi(\theta^\alpha,x) \end{aligned}\right\} \tag{7.33}$$

with

$$\left.\begin{aligned} &\lambda(\theta^\alpha,h) = \sum_{|\kappa|\le 1} s_\kappa \cos(\theta^\alpha\kappa), \qquad \Lambda(\theta,h) = \sum_{|\kappa|\le 1} \hat{s}_\kappa \cos(2\theta\kappa) \\ &q(\theta^\alpha) \;=\; \sum_{|\kappa|\le 1} \hat{t}_\kappa \cos(\theta^\alpha\kappa), \qquad p_\alpha(\theta) = \frac{1}{4} \sum_{|\kappa|\le 1} \check{t}_\kappa \cos(\theta^\alpha\kappa). \end{aligned}\right\} \tag{7.34}$$

From this, we obtain the following matrix representation for K^{2h}_h with respect to $E^{2h}_{h,\theta}$ $(\theta \in T^{low}_h)$:

$$\hat{K}^{2h}_{h,\theta} = I - \frac{1}{\Lambda}\Big(B_i C_j\Big)_{4,4} \tag{7.35}$$

with $\Lambda = \Lambda(\theta,h)$ and

$$B_1 = p_{00}(\theta), \qquad C_1 = q(\theta^{00})\lambda(\theta^{00},h),$$

$$B_2 = p_{11}(\theta), \qquad C_2 = q(\theta^{11})\lambda(\theta^{11},h),$$

$$B_3 = p_{10}(\theta), \qquad C_3 = q(\theta^{10})\lambda(\theta^{10},h),$$

$$B_4 = p_{01}(\theta), \qquad C_4 = q(\theta^{01})\lambda(\theta^{01},h).$$

The statements of the Lemma can be verified easily. Corresponding results can be derived for the other types of coarsening considered: In all these cases, the essential assumption needed is that the respective operators can be described by symmetric difference stars.

Example 7.3: If L_h, L_{2h}, I_h^{2h} and I_{2h}^h are defined as in the sample method in Section 3.1, we obtain for $\theta \in T_h^{low}$ and all α:

$$\lambda(\theta^\alpha,h) = 2(2 - \cos\theta_1^\alpha - \cos\theta_2^\alpha)/h^2, \qquad \Lambda(\theta,h) = \lambda(2\theta,2h),$$

$$q(\theta^\alpha) = p_\alpha(\theta) = (1 + \cos\theta_1^\alpha)(1 + \cos\theta_2^\alpha)/4.$$

Remark: For the above representation of coarse-grid operators K_h^H we have made essential use of the space of H-harmonics. These are the minimal invariant spaces of K_h^H occuring because of the natural assumptions (7.30). Often, however, it is convenient to consider matrix representations of K_h^H with respect to different (larger) invariant spaces. For instance, operators K_h^H corresponding to red-black or semi-coarsening may - under the assumption (7.29) - be represented with respect to the spaces of 2h-harmonics as well. (This has technical advantages if K_h^H is combined with one of the smoothing operators considered in the next section.) Adding a '$\wedge$' to any of the occuring grid operators always denotes their representation with respect to suitable spaces of harmonics (more precisely: with respect to the respective basis, see (7.21), (7.24), (7.26) and (7.28)). From the context, it will be clear which representations are actually used.

7.3 Smoothing operators

If a suitable representation of K_h^H by block matrices is given (cf. the remark above), one can obtain corresponding representations also for two-grid operators M_h^H if the smoothing operator used can be decomposed in a similar way. In this section, we give Fourier representations for some important smoothing methods, namely for *RB* and *ZEBRA relaxation*. From the latter a corresponding representation for *alternating ZEBRA relaxation* can be derived easily. Other important smoothing methods like Gauß-Seidel relaxation with lexicographic ordering of the grid points or ILU-smoothing do not have comparable invariance properties. We will treat them in connection with local Fourier analysis in Chapter 9.

Before we give the Fourier representations mentioned above, we make a few comments on RB and ZEBRA relaxation. In order to avoid some technical complications, we restrict our considerations to 9-point difference operators (7.8).

One complete step of *RB relaxation* (on Ω_h) consists of two half-steps. For this, the grid points of Ω_h are divided - in a checkerboard manner - into two sets of points, the "red" and the "black" points. In the first half-step all red grid points are relaxed in a Jacobi-like manner (i.e. the actual change of grid values is not done until all red points are passed through). Using the new values at the red points, the second half-step relaxes all black grid points in an analogous way. For both half-steps, a relaxation parameter ω may be used. The above partial relaxation steps are described by "partial step operators" which will be denoted by S_h^{red} and S_h^{black}, respectively.

Notice that only for 5-point difference operators does RB relaxation coincide with Gauß-Seidel ω-relaxation with red-black ordering of the grid points. For "Poisson-like" equations (discretized by 5-point stars) this method of relaxation (with relaxation parameter ω=1) probably yields the most efficient smoother of all.

ZEBRA relaxation (for compact 9-point difference operators) simply means Gauß-Seidel line-relaxation for which the lines (rows or columns) are not passed through one after the other but in a zebra-wise manner (first the even lines and then the odd lines or vice versa). We use the notation *x_1-ZEBRA* for ZEBRA row-relaxation (*x_2-ZEBRA* for ZEBRA column-relaxation). The respective partial step operators are denoted by $S_h^{x_1\text{-even}}$, $S_h^{x_1\text{-odd}}$ ($S_h^{x_2\text{-even}}$, $S_h^{x_2\text{-odd}}$). ZEBRA relaxation is of particular interest for anisotropic operators $-\varepsilon u_{x_1x_1} - u_{x_2x_2}$: x_1-(x_2-) ZEBRA has very efficient smoothing properties if $1 \ll \varepsilon$ ($0 < \varepsilon \ll 1$).

Alternating ZEBRA relaxation is defined by a combination of x_1- and x_2-ZEBRA. This yields a robust smoother, which is, for example, suitable for anisotropic operators with $\varepsilon=\varepsilon(x)$ such that $0 < \varepsilon(x) \ll 1$ and $\varepsilon(x) \gg 1$ for different x.

The partial relaxation steps mentioned above, both of RB and ZEBRA relaxation, can be written in the following way: For any approximation w_h to the solution u_h of (7.1), the new approximation $\bar{w}_h$ is given by

$$\bar{w}_h = w_h + \omega(z_h - w_h) \qquad (7.36)$$

(with some relaxation parameter ω) where z_h is determined by

$$L_h^o z_h(x) + L_h^- w_h(x) = f_h(x) \quad (x \in \tilde{\Omega}_h)$$
$$z_h(x) = w_h(x) \quad (x \in \Omega_h \setminus \tilde{\Omega}_h). \tag{7.37}$$

Here $\tilde{\Omega}_h$ is the subset of Ω_h consisting of those grid points, which are to be relaxed, and L_h^o and L_h^- are connected by

$$L_h = L_h^o + L_h^-.$$

L_h^o is assumed to be invertible. Both $\tilde{\Omega}_h$ and L_h^o are characteristic for the respective partial step. They are specified in Table 7.1.

partial relaxation step		$\tilde{\Omega}_h$	L_h^o
RB	red points	$\{x\in\Omega_h : x=\kappa\cdot h,\ \kappa_1+\kappa_2 \text{ even}\}$	$\begin{bmatrix} 0 & 0 & 0 \\ 0 & s_{0,0} & 0 \\ 0 & 0 & 0 \end{bmatrix}_h$
	black points	$\{x\in\Omega_h : x=\kappa\cdot h,\ \kappa_1+\kappa_2 \text{ odd}\}$	
x_1-ZEBRA	even rows	$\{x\in\Omega_h : x=\kappa\cdot h,\ \kappa_2 \text{ even}\}$	$\begin{bmatrix} 0 & 0 & 0 \\ s_{-1,0} & s_{0,0} & s_{1,0} \\ 0 & 0 & 0 \end{bmatrix}_h$
	odd rows	$\{x\in\Omega_h : x=\kappa\cdot h,\ \kappa_2 \text{ odd}\}$	
x_2-ZEBRA	even columns	$\{x\in\Omega_h : x=\kappa\cdot h,\ \kappa_1 \text{ even}\}$	$\begin{bmatrix} 0 & s_{0,1} & 0 \\ 0 & s_{0,0} & 0 \\ 0 & s_{0,-1} & 0 \end{bmatrix}_h$
	odd columns	$\{x\in\Omega_h : x\in\kappa\cdot h,\ \kappa_1 \text{ odd}\}$	

Table 7.1: Definition of $\tilde{\Omega}_h$ and L_h^o in (7.37)

For the derivation of Fourier representations for the partial step operators, we consider the errors $v_h=u_h-w_h$, $\bar{v}_h=u_h-\bar{w}_h$. From (7.36),(7.37) one obtains

$$\bar{v}_h(x) = \begin{cases} (I_h - \omega(L_h^o)^{-1}L_h)v_h(x) & (x \in \tilde{\Omega}_h) \\ v_h(x) & (x \in \Omega_h \setminus \tilde{\Omega}_h).. \end{cases} \tag{7.38}$$

In particular, any frequency $\varphi(\theta,x)$ $(\theta \in T_h)$ is mapped into

$$\bar{\varphi}(\theta,x) := \begin{cases} A(\theta,h;\omega)\ \varphi(\theta,x) & (x \in \tilde{\Omega}_h) \\ \varphi(\theta,x) & (x \in \Omega_h \setminus \tilde{\Omega}_h). \end{cases} \tag{7.39}$$

Here

$$A(\theta,h;\omega) = 1 - \omega\,\frac{\lambda(\theta,h)}{\lambda^o(\theta,h)} \tag{7.40}$$

and $\lambda(\theta,h)$, $\lambda^{0}(\theta,h)$ denote the eigenvalues of L_h and L_h^{0}, respectively, with respect to $\varphi(\theta,x)$ $(\theta \in T_h)$. In particular, we have:

$$\lambda^{0}(\theta,h) = \begin{cases} s_{00} & \text{(RB)} \\ \sum\limits_{\kappa_2=0} s_{\kappa}\cos(\theta\kappa) & (x_1\text{-ZEBRA}) \\ \sum\limits_{\kappa_2=0} s_{\kappa}\cos(\theta\kappa) & (x_2\text{-ZEBRA}) \end{cases} \tag{7.41}$$

(as for $\lambda(\theta,h)$, see (7.9)).

Clearly, (7.39) is not yet a Fourier representation of the partial step operator defined by (7.38). However, $\bar{\varphi}$ in (7.39) can always be written as a linear combination of only a few of the eigenfrequencies (7.4), depending on the concrete "relaxation pattern" used. It turns out that all of the above partial step operators - and by this also the complete relaxation operators - have invariant subspaces which already occured in connection with the coarse-grid correction operators in the previous section. The *minimal invariant subspaces* actually depend on the relaxation pattern used. In particular, the iteration operators of

RB, x_1-ZEBRA, x_2-ZEBRA and alternating ZEBRA

leave the spaces of harmonics invariant which correspond to

red-black, x_2-, x_1- and standard coarsening,

respectively. Instead of giving representations with respect to these minimal spaces, we prefer to use the spaces of 2h-harmonics throughout. This is mainly for two reasons:

(1) Most concrete results given refer to the case of standard coarsening.

(2) If, e.g., semi-coarsening is combined with RB relaxation, the minimal invariant spaces of the corresponding operators K_h^H and S_h are different. Therefore, the respective representations are not compatible. Both operators - and therefore also M_h^H - can, however, be represented with respect to the 2h-harmonics.

The Fourier representations of all partial step operators mentioned above are given in Table 7.2. The distribution of zero entries in the respective matrices show that they can be decomposed still further. As one easily recognizes, this reflects just the invariance properties with respect to the minimal spaces mentioned above.

In the case of alternating ZEBRA, four partial steps are connected (e.g. x_2-odd, x_2-even, x_1-even, x_1-odd). The corresponding product matrices do not allow a further decomposition

relaxation	Fourier representation of partial step operators with respect to $E^{2h}_{h,\theta}$: first partial step	second partial step	$\theta \in T^{2h}_h$
RB 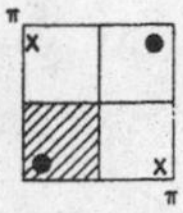	$\hat{S}^{red}_{h,\theta} = \frac{1}{2}\left(\begin{array}{cc\|cc} A_{00}+1 & A_{11}-1 & & \\ A_{00}-1 & A_{11}+1 & \multicolumn{2}{c}{0} \\ \hline & & A_{10}+1 & A_{01}-1 \\ \multicolumn{2}{c\|}{0} & A_{10}-1 & A_{01}+1 \end{array}\right)$	$\hat{S}^{black}_{h,\theta} = \frac{1}{2}\left(\begin{array}{cc\|cc} A_{00}+1 & -A_{11}+1 & & \\ -A_{00}+1 & A_{11}+1 & \multicolumn{2}{c}{0} \\ \hline & & A_{10}+1 & -A_{01}+1 \\ \multicolumn{2}{c\|}{0} & -A_{10}+1 & A_{01}+1 \end{array}\right)$	$\|\theta\| < \pi/2$
	$\hat{S}^{red}_{h,\theta} = \frac{1}{2}\begin{pmatrix} A_{00}+1 & A_{11}-1 \\ A_{00}-1 & A_{11}+1 \end{pmatrix}$	$\hat{S}^{black}_{h,\theta} = \frac{1}{2}\begin{pmatrix} A_{00}+1 & -A_{11}+1 \\ -A_{00}+1 & A_{11}+1 \end{pmatrix}$	$\|\theta\| = \pi/2$, $\theta \neq (\pi/2,\pi/2)$
	$\hat{S}^{red}_{h,\theta} = (A_{00}) = (1-\omega)$	$\hat{S}^{black}_{h,\theta} = (1)$	$\theta = (\pi/2,\pi/2)$
x_1-ZEBRA 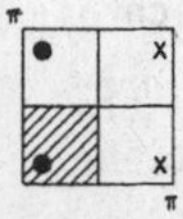	$\hat{S}^{x_1\text{-even}}_{h,\theta} = \frac{1}{2}\begin{pmatrix} A_{00}+1 & 0 & 0 & A_{01}-1 \\ 0 & A_{11}+1 & A_{10}-1 & 0 \\ 0 & A_{11}-1 & A_{10}+1 & 0 \\ A_{00}-1 & 0 & 0 & A_{01}+1 \end{pmatrix}$	$\hat{S}^{x_1\text{-odd}}_{h,\theta} = \frac{1}{2}\begin{pmatrix} A_{00}+1 & 0 & 0 & -A_{01}+1 \\ 0 & A_{11}+1 & -A_{10}+1 & 0 \\ 0 & -A_{11}+1 & A_{10}+1 & 0 \\ -A_{00}+1 & 0 & 0 & A_{01}+1 \end{pmatrix}$	$\|\theta\| < \pi/2$
	$\hat{S}^{x_1\text{-even}}_{h,\theta} = \frac{1}{2}\begin{pmatrix} A_{00}+1 & A_{01}-1 \\ A_{00}-1 & A_{01}+1 \end{pmatrix}$	$\hat{S}^{x_1\text{-odd}}_{h,\theta} = \frac{1}{2}\begin{pmatrix} A_{00}+1 & -A_{01}+1 \\ -A_{00}+1 & A_{01}+1 \end{pmatrix}$	$\|\theta\| = \pi/2$, $\theta \neq (\pi/2,\pi/2)$
	$\hat{S}^{x_1\text{-even}}_{h,\theta} = (1)$	$\hat{S}^{x_1\text{-odd}}_{h,\theta} = (A_{00}) = (1-\omega)$	$\theta = (\pi/2,\pi/2)$
x_2-ZEBRA 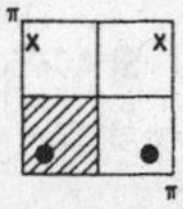	$\hat{S}^{x_2\text{-even}}_{h,\theta} = \frac{1}{2}\begin{pmatrix} A_{00}+1 & 0 & A_{10}-1 & 0 \\ 0 & A_{11}+1 & 0 & A_{01}-1 \\ A_{00}-1 & 0 & A_{10}+1 & 0 \\ 0 & A_{11}-1 & 0 & A_{01}+1 \end{pmatrix}$	$\hat{S}^{x_2\text{-odd}}_{h,\theta} = \frac{1}{2}\begin{pmatrix} A_{00}+1 & 0 & -A_{10}+1 & 0 \\ 0 & A_{11}+1 & 0 & -A_{01}+1 \\ -A_{00}+1 & 0 & A_{10}+1 & 0 \\ 0 & -A_{11}+1 & 0 & A_{01}+1 \end{pmatrix}$	$\|\theta\| < \pi/2$
	$\hat{S}^{x_2\text{-even}}_{h,\theta} = \frac{1}{2}\begin{pmatrix} A_{00}+1 & A_{10}-1 \\ A_{00}-1 & A_{10}+1 \end{pmatrix}$	$\hat{S}^{x_2\text{-odd}}_{h,\theta} = \frac{1}{2}\begin{pmatrix} A_{00}+1 & -A_{10}+1 \\ -A_{00}+1 & A_{10}+1 \end{pmatrix}$	$\|\theta\| = \pi/2$, $\theta \neq (\pi/2,\pi/2)$
	$\hat{S}^{x_2\text{-even}}_{h,} = (1)$	$\hat{S}^{x_2\text{-odd}}_{h,\theta} = (A_{00}) = (1-\omega)$	$\theta = (\pi/2,\pi/2)$

Table 7.2: Fourier representations with respect to $E^{2h}_{h,\theta}$, $\theta \in T^{2h}_h$ (see 7.21)). Here we use the abbreviation (for any fixed θ, h, ω and all α as in (7.21)):

$$A_\alpha := A(\theta^\alpha, h; \omega) \tag{7.42}$$

with A as defined in (7.40), (7.41). The leftmost column illustrates which basis functions of $E^{2h}_{h,\theta}$ are actually coupled (cf. Figure 7.1, "standard coarsening").

We conclude this section with some remarks concerning simplifications and extensions:

Remarks: (1) In the case of RB relaxation with $\omega=1$ for 5-point operators L_h, the relations

$$A_{11} = -A_{00}, \quad A_{01} = -A_{10} \tag{7.43}$$

hold. Therefore, the representation $\hat{S}_{h,\theta}$ of the complete RB relaxation operator $S_h = S_h^{\text{black}} S_h^{\text{red}}$ becomes

$$\hat{S}_{h,\theta} = \begin{cases} \frac{1}{2}\left[\begin{array}{cc|cc} A_{00}(1+A_{00}) & -A_{00}(1+A_{00}) & & \\ A_{00}(1-A_{00}) & -A_{00}(1-A_{00}) & \multicolumn{2}{c}{0} \\ \hline & & A_{10}(1+A_{10}) & -A_{10}(1+A_{10}) \\ \multicolumn{2}{c|}{0} & A_{10}(1-A_{10}) & -A_{10}(1-A_{10}) \end{array}\right] & (|\theta| < \pi/2) \\ \frac{1}{2}\begin{bmatrix} A_{00}(1+A_{00}) & -A_{00}(1+A_{00}) \\ A_{00}(1-A_{00}) & -A_{00}(1-A_{00}) \end{bmatrix} & (|\theta| = \pi/2,\ \theta \neq (\pi/2,\pi/2)) \\ (0) & (\theta = (\pi/2,\pi/2)) \end{cases} \tag{7.44}$$

Obviously, all (2,2)-blocks of $\hat{S}_{h,\theta}$ are of rank 1. The range of S_h therefore has only about half the dimension of $\mathcal{G}(\Omega_h)$. This fact can be exploited for the computation of two-grid convergence factors (see, for example, Section 8.1).

(2) Similar relations hold in the case of ZEBRA relaxation with $\omega=1$ (for 9-point operators L_h):

$$\begin{aligned} A_{01} &= -A_{00}, \quad A_{10} = -A_{11} \qquad \text{(for } x_1\text{-ZEBRA)}, \\ A_{10} &= -A_{00}, \quad A_{01} = -A_{11} \qquad \text{(for } x_2\text{-ZEBRA)}. \end{aligned} \tag{7.45}$$

They lead to corresponding simplifications as in (7.44).

(3) One complete step of Jacobi ω-relaxation (both point and line relaxation) is a special case of (7.38), characterized by $\tilde{\Omega}_h = \Omega_h$. For the three cases point, x_1-line and x_2-line relaxation, L_h^o is here defined in the same way as for RB, x_1-ZEBRA and x_2-ZEBRA, respectively (see Table 7.1). The corresponding relaxation operators S_h are represented by diagonal matrices: We have

$$S_h\, \varphi(\theta,x) = A(\theta,h;\omega)\, \varphi(\theta,x) \qquad (\theta \in T_h) \tag{7.46}$$

with A defined in (7.40), (7.41).

(4) Analogous to RB relaxation ("two-colour relaxation"), one can define a general <u>k-colour relaxation</u> [24]. Here Ω_h is divided into k natural subsets each of which is passed through in a Jacobi-like manner. Thus one complete step of k-colour relaxation consists of k partial steps of the form (7.37). For compact 9-point operators as considered here, <u>4-colour relaxation</u> coincides with Gauß-Seidel relaxation with a proper enumeration of grid points. The above Fourier analysis can immediately be applied to the corresponding partial step operators (giving full (4,4) matrices now). See [98] for details.

7.4 Two-grid operator

In Chapter 8, we will give results for several fundamental model problems and two-grid methods in terms of the following quantities

$$\rho(h,\nu) := \rho(M_h^H(\nu_1,\nu_2)), \qquad \rho^*(\nu) := \sup\{\rho(h,\nu) : h \in \mathcal{H}^*\};$$
$$\sigma_S(h,\nu_1,\nu_2) := \| M_h^H(\nu_1,\nu_2)\|_S, \qquad \sigma_S^*(\nu_1,\nu_2) := \sup\{\sigma_S(h,\nu_1,\nu_2) : h \in \mathcal{H}^*\};$$
$$\sigma_d(h,\nu_1,\nu_2) := \| L_h M_h^H(\nu_1,\nu_2) L_h^{-1}\|_S, \qquad \sigma_d^*(\nu_1,\nu_2) := \sup\{\sigma_d(h,\nu_1,\nu_2) : h \in \mathcal{H}^*\}.$$

In addition, we will consider

$$\sigma_E(h,\nu_1,\nu_2) := \| M_h^H(\nu_1,\nu_2)\|_E, \qquad \sigma_E^*(\nu_1,\nu_2) := \sup\{\sigma_E(h,\nu_1,\nu_2) : h \in \mathcal{H}^*\}$$

which is a meaningful norm if L_h is positive definite. Here $\mathcal{H}^*$ is defined as in (1.26) with some vector of coarsest meshsizes h^* and fixed meshsize ratio q^*. Because of space limitations, we will confine ourselves to the detailed discussion of methods using standard coarsening in Chapter 8. (For methods which use red-black or semi-coarsening, see, for example, [82],[111]).

The actual computation of the above quantities is performed by use of Fourier representations for M_h^H (as considered in the previous sections). In the case of standard coarsening, for instance, we have

$$\left.\begin{aligned}
\rho(h,\nu) &= \max\{\rho(\hat{M}_{h,\theta}^{2h}) : \theta \in T_h^{2h}\},\\
\sigma_S(h,\nu_1,\nu_2) &= \max\{\| \hat{M}_{h,\theta}^{2h}\|_S : \theta \in T_h^{2h}\},\\
\sigma_d(h,\nu_1,\nu_2) &= \max\{\| \hat{L}_{h,\theta}\, \hat{M}_{h,\theta}^{2h}\, \hat{L}_{h,\theta}^{-1}\|_S : \theta \in T_h^{2h}\},\\
\sigma_E(h,\nu_1,\nu_2) &= \max\{\| \hat{L}_{h,\theta}^{1/2}\, \hat{M}_{h,\theta}^{2h}\, \hat{L}_{h,\theta}^{-1/2}\|_S : \theta \in T_h^{2h}\}.
\end{aligned}\right\} \qquad (7.47)$$

Here '$\wedge$' denotes the representation of all operators occuring with respect to the spaces of 2h-harmonics.

7.5 General definition of smoothing factors

Clearly, the asymptotic convergence behavior of a two-grid method is characterized by the spectral radius $\rho(M_h^H)$ of the corresponding iteration operator. A priori, however, it is not clear how to measure the smoothing properties of a general relaxation process.

The purpose of introducing such a smoothing measure is to separate the influences of the two main parts of a two-grid method, the smoothing and the coarse-grid correction part, on the error reduction. A reasonable "smoothing factor" should - at least - give some information on the reduction of the high-frequency error part which is achieved by one smoothing step. The following definition does not take into account any transfer or coarse-grid difference operators. Nevertheless, the smoothing factor yields some information on the behavior of a two-grid method when some "idealized" assumptions on the coarse-grid correction part are made (see below).

In Section 3.2, we defined such a smoothing factor μ (or μ^*) to be the worst factor by which high-frequency error components are reduced per relaxation step. This was possible as all high frequencies were eigenfunctions of S_h. The situations is the same here, as long as we use Jacobi ω-relaxation (point- or line-wise, see Remark (3) in Section 7.3). For relaxation processes with more complicated smoothing operators like RB or ZEBRA relaxation, we have to extend the definition of a smoothing factor.

For the general definition of a smoothing factor, we now assume that only a coarser grid Ω_H and a corresponding splitting

$$\mathbb{G}(\Omega_h) = \mathbb{G}^{low}(\Omega_h) \oplus \mathbb{G}^{high}(\Omega_h) \tag{7.48}$$

are given (several examples have been discussed in Section 7.2). As motivated above, the "real" coarse-grid operator K_h^H is replaced by an "ideal" operator Q_h^H which has the following properties: Q_h^H annihilates the low-frequency error components and leaves the high-frequency components unchanged. Thus Q_h^H is a projection operator onto the space of high frequencies:

$$Q_h^H w_h := \begin{cases} w_h & (w_h \in \mathbb{G}^{high}(\Omega_h)) \\ 0 & (w_h \in \mathbb{G}^{low}(\Omega_h)) \, . \end{cases} \tag{7.49}$$

Then the quantity

$$\rho(S_h^{\nu_2}\, Q_h^H\, S_h^{\nu_1}) = \rho(Q_h^H\, S_h^{\nu})$$

should be a good measure for the <u>total</u> smoothing effect of applying ν smoothing steps of S_h. In addition, one can hope that this quantity gives a realistic prediction of the spectral radius of M_h^H $(= \rho(K_h^H\, S_h^{\nu}))$ as long as the transfer and coarse-grid difference operators are "sufficiently good".

This leads to the following

<u>Definition</u>: We define the *smoothing factor* μ of S_h by

$$\mu(h,\nu) := \sqrt[\nu]{\rho(Q_h^H\, S_h^{\nu})} \tag{7.50}$$

which measures the average smoothing effect per smoothing step. As usual, we denote by μ^* the corresponding supremum with respect to the admissable meshsizes h:

$$\mu^*(\nu) := \sup\,\{\mu(h,\nu) :\ h \in \mathcal{H}\}. \tag{7.51}$$

<u>Remark</u>: As one can easily see, the above definition is indeed an extension of the one given in Section 3.2: In the case considered there, S_h had, in particular, the property

$$S_h : \mathfrak{G}^{high}(\Omega_h) \rightarrow \mathfrak{G}^{high}(\Omega_h). \tag{7.52}$$

In this case, (7.50) can be simplified to

$$\mu(h,\nu) := \rho(S_h^{high}) \quad \text{with} \quad S_h^{high} := S_h\Big|_{\mathfrak{G}^{high}(\Omega_h)}\ . \tag{7.53}$$

This coincides with the definition (3.16) given in Section 3.2. In particular, μ is independent of ν. In general, however, low and high frequencies are intermixed by S_h, and μ does depend on ν.

As an elucidation, we now consider operators S_h corresponding to any of the complete relaxation methods considered in Section 7.3. For the actual computation of smoothing factors of these methods, we can take advantage of their respective invariance properties. In Table 7.2 we have given the matrix representations $\hat{S}_{h,\theta}$ of S_h with respect to the spaces of 2h-harmonics. Trivially, Q_h^H is represented by diagonal matrices $\hat{Q}_{h,\theta}^H$ $(\theta \in T_h^{2h})$ with respect to the corresponding basis (see (7.21)). These matrices are given in Table 7.3 (for all different coarsenings). Using these representations, we obtain (for any fixed type of coarsening):

$$\mu(h,\nu) = \max\,\{\sqrt[\nu]{\rho(\hat{Q}_{h,\theta}^H\, \hat{S}_{h,\theta}^{\nu})} : \theta \in T_h^{2h}\}. \tag{7.54}$$

coarsening	matrix representation of Q_h^H with respect to $E_{h,\theta}^{2h}$ $(\theta \in T_h^{2h})$				
	$\lvert\theta\rvert < \pi/2$		$\lvert\theta\rvert = \pi/2,\ \theta \neq (\pi/2,\pi/2)$		$\theta = (\pi/2,\pi/2)$
standard coarsening	$\begin{bmatrix}0&&&\\&1&&\\&&1&\\&&&1\end{bmatrix}$		$\begin{bmatrix}1&\\&1\end{bmatrix}$		(1)
	$\lvert\theta\rvert < \pi/2$		$\lvert\theta\rvert = \pi/2,\ \theta_1 < \pi/2$	$\lvert\theta\rvert = \pi/2,\ \theta_2 < \pi/2$	$\theta=(\pi/2,\pi/2)$
x_1-coarsening	$\begin{bmatrix}0&&&\\&1&&\\&&1&\\&&&0\end{bmatrix}$		$\begin{bmatrix}0&\\&1\end{bmatrix}$	$\begin{bmatrix}1&\\&1\end{bmatrix}$	(1)
	$\lvert\theta\rvert < \pi/2$		$\lvert\theta\rvert = \pi/2,\ \theta_1 < \pi/2$	$\lvert\theta\rvert = \pi/2,\ \theta_2 < \pi/2$	$\theta=(\pi/2,\pi/2)$
x_2-coarsening	$\begin{bmatrix}0&&&\\&1&&\\&&0&\\&&&1\end{bmatrix}$		$\begin{bmatrix}1&\\&1\end{bmatrix}$	$\begin{bmatrix}0&\\&1\end{bmatrix}$	(1)
	$\lvert\theta\rvert < \pi/2,\ \theta_2 \geq \theta_1$	$\lvert\theta\rvert < \pi/2,\ \theta_2 < \theta_1$	$\lvert\theta\rvert = \pi/2,\ \theta \neq (\pi/2,\pi/2)$		$\theta=(\pi/2,\pi/2)$
red-black coarsening	$\begin{bmatrix}0&&&\\&1&&\\&&1&\\&&&0\end{bmatrix}$	$\begin{bmatrix}0&&&\\&1&&\\&&0&\\&&&1\end{bmatrix}$	$\begin{bmatrix}0&\\&1\end{bmatrix}$		(0)
	$\lvert\theta\rvert < \pi/4$	$\pi/4 \leq \lvert\theta\rvert < \pi/2$	$\lvert\theta\rvert = \pi/2,\ \theta \neq (\pi/2,\pi/2)$		$\theta=(\pi/2,\pi/2)$
quadrupling	$\begin{bmatrix}0&&&\\&1&&\\&&1&\\&&&1\end{bmatrix}$	$\begin{bmatrix}1&&&\\&1&&\\&&1&\\&&&1\end{bmatrix}$	$\begin{bmatrix}1&\\&1\end{bmatrix}$		(1)

Table 7.3: Matrix representations $\hat{Q}_{h,\theta}^H$ of the projection operator Q_h^H with respect to $E_{h,\theta}^{2h}$ $(\theta \in T_h^{2h})$ for different coarsenings.

Table 7.4 gives a comparison of smoothing factors $\mu^*(\nu)$ for all combinations of smoothers and coarsenings treated in this chapter. Table 7.4a refers to the model problem (P) and Table 7.4b to the anisotropic model operator with $\varepsilon=0.1$ (cf. Section 1.3.3). The range $\mathcal{H}^*$ of admissable h is

$$\mathcal{H}^* := \{h : h \leq h^*,\ h=1/N,\ N \text{ even}\},\quad h^* := 1/4.$$

We are not going to discuss all of the results in Table 7.4 and their significance for the construction of multigrid methods in detail here. We mainly pick out some exemplary cases which are of interest for the discussions of two-grid methods in Chapters 8 and 9. In all of the examples below, the following quantity $\chi(\nu)$ occurs:

$$\chi(\nu) := \left(\frac{2\nu-1}{2\nu}\right)^2 \Big/ \sqrt[\nu]{2(2\nu-1)}. \tag{7.55}$$

Some particular values are

$$\chi(1) = 0.125,\quad \chi(2) = 0.229..,\quad \chi(3) = 0.322..,\quad \chi(\nu) \sim \sqrt[\nu]{1/4e\nu}\ (\nu \text{ "large"}). \tag{7.56}$$

coarsening	ν	RB	x_1-ZEBRA	x_2-ZEBRA	alt.ZEBRA
		type of relaxation			
standard coarsening	1 2 3	0.250 0.250 0.322	0.250 0.250 0.322	0.250 0.250 0.322	0.048 0.118 -
quadrupling	1 2 3	0.729 0.729 0.729	0.598 0.598 0.598	0.598 0.598 0.598	0.380 0.380 -
x_1-coarsening	1 2 3	0.375 0.306 0.322	0.250 0.250 0.250	0.125 0.230 0.322	0.028 0.092 -
x_2-coarsening	1 2 3	0.375 0.306 0.322	0.125 0.230 0.322	0.250 0.250 0.250	0.028 0.092 -
red-black coarsening	1 2 3	0.125 0.230 0.322	0.125 0.230 0.322	0.125 0.230 0.322	0.012 0.078 -

<u>Table</u> 7.4a: Smoothing factors $\mu^*(\nu)$ for model problem (P)

coarsening	ν	RB	x_1-ZEBRA	x_2-ZEBRA	alt.ZEBRA
		type of relaxation			
standard coarsening	1 2 3	0.826 0.826 0.826	0.826 0.826 0.826	0.125 0.230 0.322	0.102 0.203 -
quadrupling	1 2 3	0.947 0.947 0.947	0.944 0.944 0.944	0.500 0.500 0.500	0.496 0.496 -
x_1-coarsening	1 2 3	0.868 0.847 0.840	0.826 0.826 0.826	0.125 0.230 0.322	0.101 0.203 -
x_2-coarsening	1 2 3	0.125 0.230 0.322	0.125 0.230 0.322	0.008 0.032 0.100	0.002 0.032 -
red-black coarsening	1 2 3	0.744 0.706 0.693	0.694 0.694 0.694	0.125 0.230 0.322	0.084 0.195 -

<u>Table</u> 7.4b: Analogous to Table 7.4a for the anisotropic model operator (ε=0.1)

Examples: (1) We consider RB relaxation (with relaxation parameter $\omega=1$) in connection with model problem (P). The matrix representation of one complete RB step with respect to $E^{2h}_{h,\theta}$ ($\theta \in T^{2h}_h$) is given in (7.44). We have for fixed h and $\theta \in T^{2h}_h$

$$A_{00} = A(\theta^{00},h) = (\cos\theta_2 + \cos\theta_1)/2, \quad A_{10} = A(\theta^{10},h) = (\cos\theta_2 - \cos\theta_1)/2. \qquad (7.57)$$

With respect to standard coarsening, the smoothing factor $\mu(h,\nu)$ can be computed by (7.54) using the corresponding matrix representation of Q^{2h}_h in Table 7.3. For $\mu^*(\nu)$ one obtains

$$\mu^*(\nu) = \sqrt[\nu]{\sup \{\max \{|A_{00}^{2\nu-1}(1-A_{00})|/2,\ A_{10}^{2\nu}\} : |\theta| \le \pi/2\}} \qquad (7.58)$$

$$= \max \{0.25,\ \chi(\nu)\} = \begin{cases} 0.25 & (\nu \le 2) \\ \chi(\nu) & (\nu \ge 3)\ . \end{cases} \qquad (7.59)$$

A comparison with Jacobi ω-relaxation (cf. (3.18)) shows that the smoothing properties of RB are - for relevant (small) values of ν - considerably better. For large values of ν, however, the smoothing factor of RB approaches 1 (cf. (7.56)). This is due to the coupling of low and high frequencies within one RB step. Thus, here already the smoothing factor indicates that the performance of too many smoothing steps ($\nu>3$, say) is not recommendable.

(2) For RB relaxation and the 5-point discretization (1.12) of the anisotropic model equation, we have

$$A_{00} = (\cos\theta_2 + \varepsilon \cos\theta_1)/(1+\varepsilon), \quad A_{10} = (\cos\theta_2 - \varepsilon \cos\theta_1)/(1+\varepsilon) \qquad (7.60)$$

instead of (7.57). Evaluating (7.58), we obtain the smoothing factor with respect to standard coarsening

$$\mu^*(\nu) = \max \{(\varepsilon^+/(1+\varepsilon))^2,\ \chi(\nu)\}, \qquad \varepsilon^+ := \max \{1,\varepsilon\}. \qquad (7.61)$$

In particular, we obtain (for fixed ν) $\mu^*(\nu) \to 1$ if either $\varepsilon \to 0$ or $\varepsilon \to \infty$, showing that the smoothing properties of RB relaxation become very bad if ε is considerably different from 1. For some concrete values, see the Tables 7.4 and 8.3. Heuristically, the reason for this is clear: If ε is, for example, very small, there is nearly no coupling of grid values in x_1-direction (cf. (1.12)). In particular, there is hardly a smoothing effect on the error in this direction. The error is, however, smoothed in the x_2-direction. This can be seen from the next example:

(3) Let the problem and the smoother be the same as in Example (2). Now, however, we consider the smoothing effect of RB relaxation with respect to x_2-coarsening. With A_{00} and A_{10} as in (7.60) we now obtain

$$\mu^*(\nu) = \sqrt[\nu]{\sup\{\max\{|A_{00}^{2\nu-1}(1-A_{00})|/2,\ |A_{10}^{2\nu-1}(1-A_{10})|/2\} : |\theta|\le\pi/2\}} \qquad (7.62)$$

and

$$\mu^*(\nu) \to \begin{cases} \chi(\nu) & (\varepsilon\to 0) \\ 1 & (\varepsilon\to\infty). \end{cases} \qquad (7.63)$$

Thus RB relaxation has, with respect to x_2-coarsening and for small ε, the same smoothing properties as in Example (1) if $\nu\ge 3$. (For $\nu=1$ or 2 they are even better.) For large ε, however, RB cannot be used in connection with x_2-coarsening: Then x_1-coarsening has to be used instead. This result is in full accordance with the heuristic explanation given in (2).

(4) We have seen above that - for the anisotropic model equation with ε considerably different from 1 - RB has good smoothing properties only if it is combined with semi-coarsening. (In fact, this is true for <u>any</u> pointwise relaxation method by the same heuristic argument which was given in Example (2).) In order to use <u>standard coarsening</u>, the smoothing process has to be changed: One can use, for instance, <u>ZEBRA relaxation</u>. The matrix representation of ZEBRA relaxation can be computed from Table 7.2 (also see (7.45)). Let us consider x_2-ZEBRA. With

$$A_{00} = \varepsilon\cos\theta_1/(1+\varepsilon-\cos\theta_2),\quad A_{11} = -\varepsilon\cos\theta_1/(1+\varepsilon+\cos\theta_2), \qquad (7.64)$$

one obtains the same formula as (7.58) with A_{10} replaced by A_{11} and by that

$$\mu^*(\nu) = \max\{(\varepsilon/(1+\varepsilon))^2,\ \chi(\nu)\} \to \begin{cases} \chi(\nu) & (\varepsilon\to 0) \\ 1 & (\varepsilon\to\infty). \end{cases} \qquad (7.65)$$

In particular, x_2-ZEBRA has, in connection with standard coarsening and for small ε, the same good smoothing properties as RB had in connection with x_2-coarsening (7.63). For large ε, however, x_2-ZEBRA is not suitable: x_1-ZEBRA has to be used instead. If one smoothing step is defined by one step of <u>alternating ZEBRA</u>, one can show that this kind of a smoother has very good smoothing properties, independent of the size of ε. Some explicit values are given in Table 7.4 (also see Table 8.4b). In the judgement of alternating ZEBRA one has, of course, to take into account that this smoother needs twice the work per step as one single ZEBRA step.

We want to make one final remark on coarsening by <u>quadrupling h</u>. This kind of coarsening leads to multigrid algorithms which are (slightly) cheaper per cycle than corresponding ones obtained by, for example, standard coarsening (cf. (4.24)). On the other hand, the smoothing factors shown in Table 7.4 indicate that this saving of computational work does not pay: the smoothing factors which correspond to coarsening by quadrupling h are much worse than those which correspond to standard coarsening.

7.6 Modifications and extensions

The concept of model problem analysis was so far described on the basis of the discrete sine functions (7.4). For second order differential operators (7.6) (on rectangular domains), these sine functions naturally occur as eigenfunctions of L_h in connection with *Dirichlet boundary conditions*.

The whole concept carries over, if, instead of (7.4), cosine or (complex) exponential functions (or certain combinations of them) are used. More precisely, we may consider functions

$$\varphi(\theta,x) = \varphi_1(\theta_1,x_1)\,\varphi_2(\theta_2,x_2) \tag{7.66}$$

where φ_1 and φ_2 are functions of the form

$$\varphi_j(\theta_j,x_j) = \begin{cases} \sin(\theta_j x_j/h_{x_j}) \\ \cos(\theta_j x_j/h_{x_j}) \\ e^{i\theta_j x_j/h_{x_j}} \end{cases}, \quad (j = 1,2). \tag{7.67}$$

Examples: (1) Let the same differential equation (7.6) be given as in Example 7.1, but with *Neumann boundary conditions*

$$\partial u/\partial n = f^{\Gamma}(x) \qquad (x \in \Gamma) \tag{7.68}$$

instead of (7.7). For L^{Ω}, we consider any discretization of the form (7.8) applied at all points of $\bar{\Omega} \cap G_h$ using auxiliary grid points outside Ω; these points are also used in discretizing the Neumann boundary conditions by two-point central differences. The grid values corresponding to auxiliary points are assumed to be eliminated then. Thus, we obtain a grid equation (7.1) on $\Omega_h := \bar{\Omega} \cap G_h$ the eigenfunctions of which are (up to normalization)

$$\varphi(\theta,x) = \cos(\theta_1 x_1/h_{x_1})\,\cos(\theta_1 x_2/h_{x_2}) \quad (\theta \in T_h) \tag{7.69}$$

with T_h as in (7.5) but with $0 \le n \le N$ instead of $0<n<N$.

(2) Similarly, (7.6) with *periodic boundary conditions* can be discretized in a straightforward manner such that one obtains a grid equation (7.1) on

$$\Omega_h := \{x \in G_h : 0 \le x_j < A_j \quad (j=1,2)\} \tag{7.70}$$

with eigenfunctions

$$\varphi(\theta,x) = e^{i\theta_1 x_1/h_{x_1}}\, e^{i\theta_2 x_2/h_{x_2}} \quad (\theta \in T_h),\; T_h := \{\theta = (\frac{2n_1}{N_2}\pi, \frac{2n_2}{N_2}\pi) : -N/2<n\le N/2\}. \tag{7.71}$$

Clearly, mixtures of the above functions correspond to certain mixed boundary conditions. All these problems can be analyzed in much the same way as shown in Sections 7.1 through 7.5. There are only a few technical changes necessary. In particular, changes are caused by the grid Ω_h (which is slightly different now) and by the different θ-ranges. For demonstration purposes, we make some comments on the Neumann case as described above in Example (1).

In this case, the definition of high and low frequencies given in Section 7.2 is maintained except that now additional frequencies occur. If we include these additional frequencies, the definition of the spaces of harmonics remains also unchanged. As the functions (7.69) are not orthogonal with respect to (1.7), the definition of the inner product has to be modified: We replace (1.7) by

$$(u_h,w_h)_2 := \frac{1}{N_1N_2} \sum_{x\in\Omega_h} u_h(x)\, \overline{w}_h(x)\, c_h(x) \tag{7.72}$$

where

$$c_h(x) := \begin{cases} 1 & (x \in \Omega) \\ 1/2 & (x \in \partial\Omega, \text{ edges}) \\ 1/4 & (x \in \partial\Omega, \text{ corners}). \end{cases}$$

With respect to this inner product, the functions in (7.69) are orthogonal (and can be normalized).

Taking these formal changes into account, the results of the previous sections carry over in a straightforward manner. In particular, (7.33) and (7.34) of Lemma 7.1 remain valid if we extend the definition of I_h^{2h} and I_{2h}^h to grid points at the boundary of Ω. This extension has to be done in a way which takes the symmetric properties of the above cosine functions into account. For instance, at the left boundary (away from the corners), I_h^{2h} and I_{2h}^h are defined by

$$I_h^{2h}: \begin{bmatrix} 0 & \hat{t}_{0,1} & 2\hat{t}_{1,1} \\ 0 & \hat{t}_{0,0} & 2\hat{t}_{1,0} \\ 0 & \hat{t}_{0,-1} & 2\hat{t}_{1,-1} \end{bmatrix}_h^{2h}, \qquad I_{2h}^h: \left]\begin{matrix} 0 & \check{t}_{0,1} & \check{t}_{1,1} \\ 0 & \check{t}_{0,0} & \check{t}_{1,0} \\ 0 & \check{t}_{0,-1} & \check{t}_{1,-1} \end{matrix}\right[_{2h}^h \tag{7.73}$$

and at the upper left corner by

$$I_h^{2h}: \begin{bmatrix} 0 & 0 & 0 \\ 0 & \hat{t}_{0,0} & 2\hat{t}_{1,0} \\ 0 & 2\hat{t}_{0,-1} & 4\hat{t}_{1,-1} \end{bmatrix}_h^{2h}, \qquad I_{2h}^h: \left]\begin{matrix} 0 & 0 & 0 \\ 0 & \check{t}_{0,0} & \check{t}_{1,0} \\ 0 & \check{t}_{0,-1} & \check{t}_{1,-1} \end{matrix}\right[_{2h}^h . \tag{7.74}$$

Also, the analysis of relaxation operators given in Section 7.3 and their matrix representations do not change essentially. The definition of the smoothing factor applies directly. Altogether, one can compute spectral radii and norm quantities of two-grid methods just as before and one obtains nearly the same results. (The minor differences are caused only by the fact that the number of spaces of harmonics is slightly larger now. These differences vanish if one maximizes with respect to h.)

Remark: The model problems treated in Chapter 8 (5-point discretization of the Poisson equation and the anisotropic model equation) become singular in connection with pure Neumann conditions. Such cases were excluded so far as we always assumed L_h^{-1} and L_{2h}^{-1} to exist. It is quite easy to see, however, how the analysis has to be changed in order to include the above singular cases: One can use slightly shrunken spaces $\tilde{\mathbb{G}}(\Omega_h)$ and $\tilde{\mathbb{G}}(\Omega_{2h})$ defined by

$$\tilde{\mathbb{G}}(\Omega_h) := \mathbb{G}(\Omega_h) \setminus E_{h,0}^{2h}, \quad \tilde{\mathbb{G}}(\Omega_{2h}) := \mathbb{G}(\Omega_{2h}) \setminus \text{span}\,\{1\}.$$

where $E_{h,0}^{2h}$ denotes that space of harmonics which contains the lowest (constant) frequency. With respect to these spaces, L_h^{-1} and L_{2h}^{-1} exist and the analysis can be performed as usual. (One can easily arrange things so that the remaining three non-constant harmonics in $E_{h,0}^{2h}$ give no significant contribution.) These changes in the analysis also have direct analogs in the algorithmic construction of a corresponding two- (and multi-) grid method. Without going into details, we only mention that the results for Neumann boundary conditions are exactly the same as for Dirichlet conditions as far as ρ^*, σ_S^* (now defined with respect to (7.72)) are concerned. For more details and, in particular, for numerical investigations, we refer to [10].

Similar analogs as in the Dirichlet and Neumann case do hold also for the periodic case. We are not going to discuss the model problem analysis with respect to (7.71) here. We want to point out, however, that the technical details of an analysis with respect to (7.71) are implicitly contained in the local Fourier analysis as described in Chapter 9.

In most of our concrete investigations in the previous sections we have restricted our considerations to grid operators (L_h, I_h^H, I_H^h) which are described by 5- or compact 9-point stars. Technically, however, the model problem analysis also can be applied in connection with "larger" stars. For this, all larger stars have to be changed near the boundary in such a way that the functions (7.4) - or more generally (7.66) - remain eigenfunctions of L_h and M_h^H has still the required invariance properties (see also the changes of I_h^H and I_H^h in (7.73) and (7.74)). Instead of performing such changes explicitly, there is an equivalent way of treating such situations which is theoretically more transparent: One considers the given problem not on Ω_h, but extends it (equivalently) to the infinite grid G_h. For this, for all grid functions occuring one has to perform certain continuation processes. The kind of continuation is, of course, given by the type of boundary conditions at hand. Without going into details, we only mention that there is a direct correspondence between an antisymmetric, symmetric and periodic continuation of grid functions and Dirichlet, Neumann and periodic boundary conditions, respectively. (Such continuation processes have been studied extensively in the framework of reduction methods [87], [71].)

Remark: Although larger stars can be included in the concept of model problem analysis, one has to be cautious with the interpretation of corresponding results. If changes near the boundary are performed in the way as sketched above, these operators may loose some of their properties. For example, a higher order discretization or interpolation becomes of lower order near the boundary. On the other hand, if changes near the boundary, e.g. for L_h, are done such that the order of approximation is maintained, the model problem analysis can, in general, not be applied directly as the functions (7.4) or (7.66) may possibly no longer be eigenfunctions.

8. Applications of model problem analysis

In this Chapter, we give some exemplary results derived by the analysis introduced in the previous chapter. We consider some of the most fundamental problems and algorithms and discuss them briefly.

In Section 8.1, we treat Poisson's equation, namely model problem (P), in detail. We consider MG components which lead to very efficient MG algorithms. The main intention here is to show how model problem analysis can be used to obtain explicit analytic expressions for convergence factors. Sections 8.2 and 8.3 give several additional results in form of tables. Section 8.2 again refers to Poisson's equation. In particular, we make some efficiency considerations. Besides the usual 5-point discretization, we consider the 9-point "Mehrstellen" discretization, also. Section 8.3 refers to the anisotropic model operator.

Other operators which can be treated by means of the model problem analysis as well, are not discussed here in detail. Such problems are: Helmholtz' equation (with positive or negative Helmholtz' constant), 4th order equations (cf. Example 7.2), problems with Neumann or periodic boundary conditions (cf. Section 7.6), certain systems of elliptic equations, higher order discretizations, etc.

All results of this chapter refer, for simplicity, to standard coarsening. As for results concerning red-black and semi-coarsening (MGR methods), we refer to [82],[111] Furthermore, we restrict ourselves to square meshsizes $h=h_{x_1}=h_{x_2}$. This can be done without loss of generality, as the discretization of

$$- \varepsilon u_{x_1x_1} - u_{x_2x_2} \quad \text{with} \quad h_{x_1} = h/q^*, h_{x_2} = h \tag{8.1}$$

is clearly equivalent to the discretization of

$$- \tilde{\varepsilon} u_{x_1x_1} - u_{x_2x_2} \quad \text{with} \quad h_{x_1} = h_{x_2} = h \text{ and } \tilde{\varepsilon} = \varepsilon(q^*)^2. \tag{8.2}$$

As already pointed out in the introduction, we regard the model problem analysis (and the local Fourier analysis) as the theoretical basis for the construction of efficient multigrid solvers. In particular, the results given in this chapter have been used in developing the programs MGØØ and MGØ1 for the solution of certain standard elliptic equations of second order (see Chapter 10 and [36]). The connection between model problem analysis and standard applications is - roughly - given by the following:

Each of the model problems represents a considerably larger class of more general "standard" problems (with variable coefficients, on more general domains etc.) to which the quantitative theoretical results obtained for the corresponding model problem carry over in practice. This "stable behavior" of multigrid methods is a general experience familiar to multigrid experts and has been demonstrated by a great number of systematical experiments. It is because of this behavior that the model problem analysis becomes really worthwhile for practical purposes.

The above behavior is heuristically explained by the fact that the spectral properties of M_h^{2h}, by which the two-grid convergence behavior is determined, are "robust" with respect to small changes of a given model problem (for example, with respect to changes of L_h or the domain). In particular, the influence of a given relaxation technique is nearly the same for neighboring problems; this is mainly due to the local nature of relaxation processes.

8.1 Analytic results for an efficient two-grid method

We consider an $(h,2h)$ two-grid method for model problem (P) which is very similar to the sample method treated in Chapter 3. The only difference is that now RB relaxation (with relaxation parameter $\omega=1$) is used instead of Jacobi ω-relaxation. This small change of the algorithm will prove to be essential for the resulting efficiency. We obtain for the asymptotic convergence factor $\rho^*(\nu)=\sup\{\rho(h,\nu):h\in\mathcal{H}^*\}$, $h^*=1/4$:

Theorem 8.1: Let

$$M_h^{2h} = S_h^{\nu_2} (I_h - I_{2h}^h L_{2h}^{-1} I_h^{2h} L_h) S_h^{\nu_1} \tag{8.3}$$

where L_h, I_h^{2h}, I_{2h}^h, L_{2h} are defined as in Section 3.1 and S_h characterizes one complete step of RB relaxation (cf. Section 7.3). Then we have for $\nu=\nu_1+\nu_2$

$$\rho^*(\nu) = \begin{cases} 1/4 & (\nu = 1) \\ \frac{1}{2\nu}\left(\frac{\nu}{\nu+1}\right)^{\nu+1} & (\nu \geq 2). \end{cases} \tag{8.4}$$

In particular,

$$\rho^*(2) = 2/27 = 0.074\ldots, \quad \rho^*(3) = 0.052\ldots; \quad \nu\rho^*(\nu) \to 1/2e \quad (\nu \to \infty). \tag{8.5}$$

Remark: We obtain the same result on ρ^* if Poisson's equation is considered on a rectangular domain (with $h_{x_1} = h_{x_2}$) rather than on a square. The necessary changes in the proof below are obvious.

The asymptotic behavior of ρ^* shown in (8.5) already indicates that - for reasons of efficiency - it is not reasonable to choose too large a value of ν. Taking computational work into account, we will see in Section 8.2 that $\nu=2$ is the optimal value. The corresponding convergence factor shows that the mere replacement of Jacobi ω-relaxation by RB relaxation leads to a considerably faster method (cf. Table 3.1). In addition, it requires less numerical operations per iteration step, see Section 8.2.

Before we give a proof for the above theorem, we would like to formulate related statements with respect to norms (Supplements 8.2 and 8.3). The norms σ_S^* and σ_E^* can be determined in a similar way as ρ^*. (We give some explicit values in Section 8.2.) For the method at hand, however, certain modified norm values turn out to be more interesting from a theoretical and practical point of view. In the following supplement concerning the <u>spectral norm</u>, we take advantage of the fact that S_h has a nullspace of a high dimension (cf. Remark (1) in Section 7.3). For $\nu_2 \geq 1$, we have

$$M_h^{2h} : \mathbb{G}(\Omega_h) \to \tilde{\mathbb{G}}(\Omega_h) \quad \text{where} \quad \tilde{\mathbb{G}}(\Omega_h) := S_h(\mathbb{G}(\Omega_h)). \tag{8.6}$$

We therefore consider

$$\tilde{\sigma}_S(h,\nu_1,\nu_2) := \| \tilde{M}_h^{2h} \|_S \quad \text{where} \quad \tilde{M}_h^{2h} := M_h^{2h}\Big|_{\tilde{\mathbb{G}}(\Omega_h)} \tag{8.7}$$

For the corresponding supremum $\tilde{\sigma}_S^*(\nu_1,\nu_2) := \sup \{\tilde{\sigma}_S(h,\nu_1,\nu_2) : h \in \mathcal{H}\}$, $h^* = 1/4$, we obtain the

<u>Supplement 8.2</u>: Let the same assumptions as in Theorem 8.1 be satisfied. Then

$$\tilde{\sigma}_S^*(\nu_1,\nu_2) = \begin{cases} 1/\sqrt{2} & \text{for} \quad \nu_1 = 0,\ \nu_2 \geq 1 \\ \rho^*(\nu) & \text{for} \quad \nu_1 \geq 1,\ \nu_2 \geq 1. \end{cases} \tag{8.8}$$

The practical meaning of this result is that (for $\nu_1 \geq 1$, $\nu_2 \geq 1$) the actual $\|\cdot\|_2$-error reduction per two-grid step is described by ρ^* if the error of the very first approximation is an element of $\tilde{\mathbb{G}}(\Omega_h)$. Such a first approximation, however, can easily be obtained by applying one RB step to <u>any</u> first approximation. (In fact, applying only one "black" partial step is sufficient.)

We have already mentioned before that $\nu=2$ is the optimal choice for ν. By (8.8) we see that one should use $\nu_1=\nu_2=1$.

A somewhat more essential modification of the original algorithm yields the following statement on the <u>energy norm</u>.

Supplement 8.3: If, in contrast to (8.3), the two-grid method is arranged in a "symmetric way", namely if

$$M_h^{2h} = S_h^{(2)} \left(I_h - I_{2h}^h L_{2h}^{-1} I_h^{2h} L_h\right) S_h^{(1)} \tag{8.9}$$

where $S_h^{(1)}$ denotes any product of partial step operators S_h^{red}, S_h^{black} and $S_h^{(2)}$ denotes the corresponding product taken in the reversed order, then we obtain

$$\| M_h^{2h} \|_E = \rho(M_h^{2h}). \tag{8.10}$$

In the following, we give the main steps of the proof of Theorem 8.1. Several technical details of the rather involved considerations have been omitted.

Proof of Theorem 8.1: Let $h=1/N$ (N even, $N\geq 4$). Without loss of generality we assume $\nu_1=0$, $\nu_2=\nu$. With $\tilde{M}_h^{2h}$ as defined in (8.7), we then have

$$\rho(M_h^{2h}) = \rho(\tilde{M}_h^{2h}).$$

Because of

$$\tilde{M}_h^{2h} : \tilde{E}_{h,\theta}^{2h} \to \tilde{E}_{h,\theta}^{2h} \quad (\theta \in T_h^{2h}) \quad \text{with} \quad \tilde{E}_{h,\theta}^{2h} := S_h(E_{h,\theta}^{2h}), \quad S_h = S_h^{black} S_h^{red},$$

we need to consider only the restrictions $\tilde{M}_{h,\theta}^{2h}$ of $\tilde{M}_h^{2h}$ to $\tilde{E}_{h,\theta}^{2h}$:

$$\rho(\tilde{M}_h^{2h}) = \max \{\rho(\tilde{M}_{h,\theta}^{2h}) : \theta \in T_h^{2h}\}.$$

Let us now assume a fixed $\theta \in T_h^{2h}$, $|\theta|<\pi/2$ to be given. (Omitting the case $|\theta|=\pi/2$, does not influence the result.) From (7.44) we see that $\tilde{E}_{h,\theta}^{2h}$ is of dimension 2 and is spanned by $\psi_1, \psi_2 \in \mathbf{G}(\Omega_h)$:

$$\psi_1(x):=(1+a)\varphi(\theta^{00},x)+(1-a)\varphi(\theta^{11},x), \quad \psi_2(x):=(1+b)\varphi(\theta^{10},x)+(1-b)\varphi(\theta^{01},x) \quad (x\in\Omega_h). \tag{8.11}$$

Here we use the abbreviations

$$a := 1-\xi-\eta, \quad b := \xi-\eta; \quad \xi := \sin^2(\theta_1/2), \quad \eta := \sin^2(\theta_2/2). \tag{8.12}$$

The matrix representation of $\tilde{M}_{h,\theta}^{2h}$ with respect to (8.11) can be shown to be

$$\tilde{M}_{h,\theta}^{2h} \mathrel{\hat{=}} M(a,b;\nu) := \begin{bmatrix} m_{ij}(a,b;\nu) \end{bmatrix}_{2,2}$$

where

$$m_{11}(a,b;\nu) = \frac{a^{2\nu}}{2}\left(1-a^2 - \frac{2a^2b^2}{1-a^2-b^2}\right), \quad m_{22}(a,b;\nu) := m_{11}(b,a;\nu),$$

$$m_{12}(a,b;\nu) = -\frac{a^{2\nu}}{2}\left(1+b^2 + \frac{2a^2b^2}{1-a^2-b^2}\right), \quad m_{21}(a,b;\nu) := m_{12}(b,a;\nu).$$

Allowing θ to range continuously in $(0,\pi/2)$, we obtain (cf. (8.12))

$$\rho^*(\nu) = \sup\{F(a,b;\nu) : 0 < a < 1,\ 0 \le b < \min\{a,1-a\}\} \tag{8.13}$$

with $F(a,b;\nu):=\rho(M(a,b;\nu))\ (=\rho(\tilde{M}^{2h}_{h,\theta}))$. Thus we have to compute the supremum of the function F with respect ot (a,b). (This function is plotted in Figure 8.1 with respect to the coordinates $0\le\xi,\eta\le 1/2$.)

We first note that

$$m_{1,2}m_{2,1} \ge 0,\ m_{11} \ge 0,\ m_{22} \ge 0.$$

In particular, the eigenvalues of $M(a,b;\nu)$ are real for all a,b,ν and we obtain

$$F(a,b;\nu) = (T + \sqrt{T^2 - 4D})\ /\ 2$$

with

$$T = T(a,b;\nu) = m_{11} + m_{22},\quad D = D(a,b;\nu) = m_{11}m_{22}-m_{12}m_{21}$$

As the behavior of $F(a,b;\nu)$ is different for $\nu=1$ and $\nu>1$ (cf. Figure 8.1), we treat these cases separately.

If $\nu=1$: Using polar coordinates $a=r\cos(\vartheta)$, $b=r\sin(\vartheta)$, one computes

$$F(a,b;1) = (r^2(1-r^2) + r^4 \sin^2(2\vartheta))/2.$$

The right hand side is monotonically increasing with respect to ϑ $(0\le\vartheta\le\pi/4)$. Thus F achieves its maximum at the upper boundary of the (a,b)-range, i.e. for either $b=a$, $a\le 1/2$ or $b=1-a$, $a\ge 1/2$. Because $F(a,a)=a^2$ and $F(a,1-a)=a(1-a)$, F achieves its maximum exactly for $a=b=1/2$ (corresponding to $\xi=1/2$, $\eta=0$ or $\theta_1=\pi/2$, $\theta_2=0$). We obtain $\rho^*(1)=1/4$.

If $\nu=2$: For this case the use of the coordinates r and $t:=\sin^2(2\vartheta)$ turns out to be suitable. Note that t has the ranges

$$0 \le t \le 1 \qquad \text{if} \qquad 0 \le r \le 1/\sqrt{2},$$

$$0 \le t \le (1/r^2-1)^2 \quad \text{if} \quad 1/\sqrt{2} \le r \le 1.$$

Some technical calculations yield

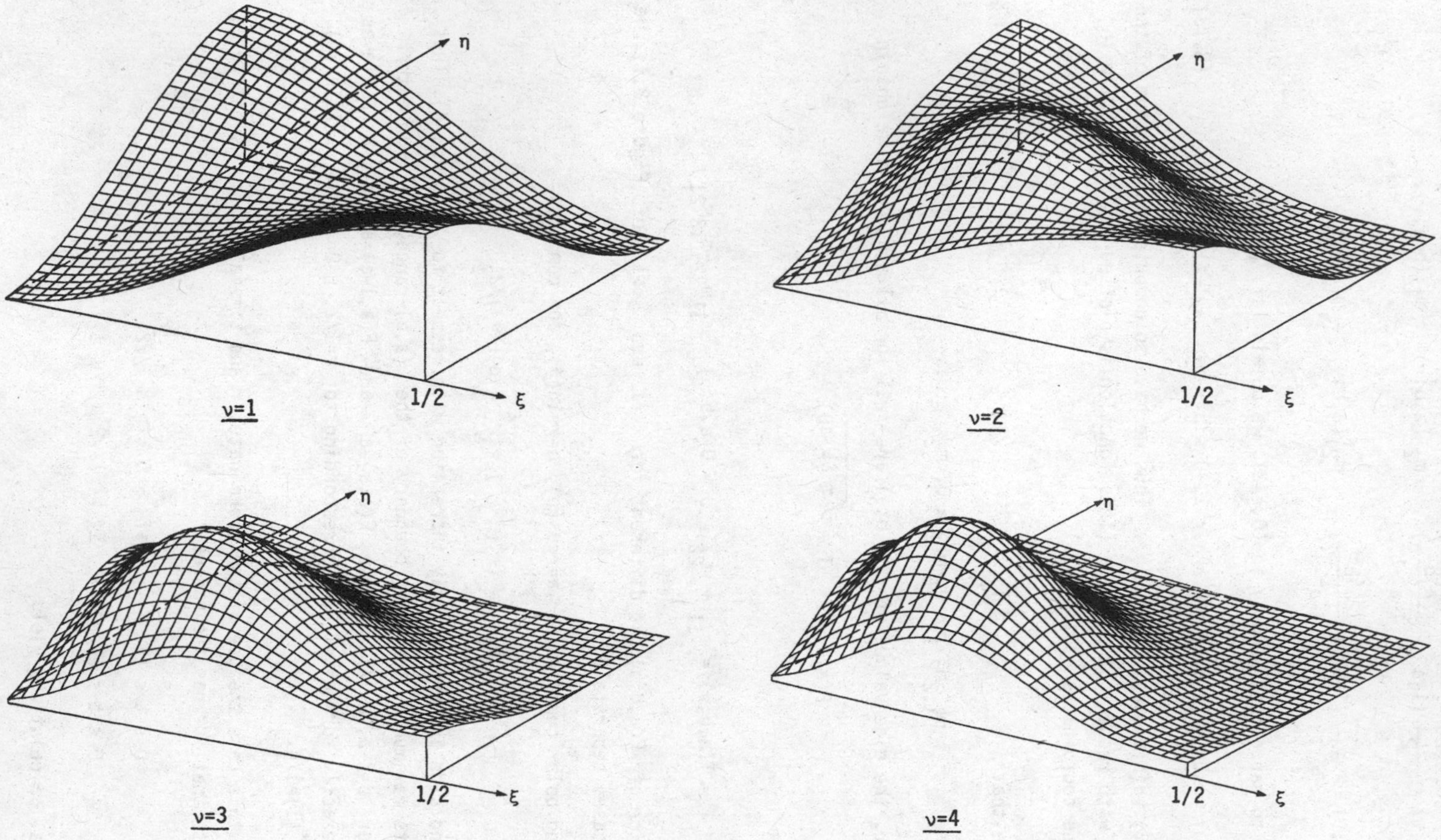

Figure 8.1: F in (8.13) as a function of ξ,η $(0 \le \xi,\eta \le 1/2)$. The maximum of F gives ρ^* for the two-grid method considered in Theorem 8.1 (in particular, I_h^{2h} : FW).

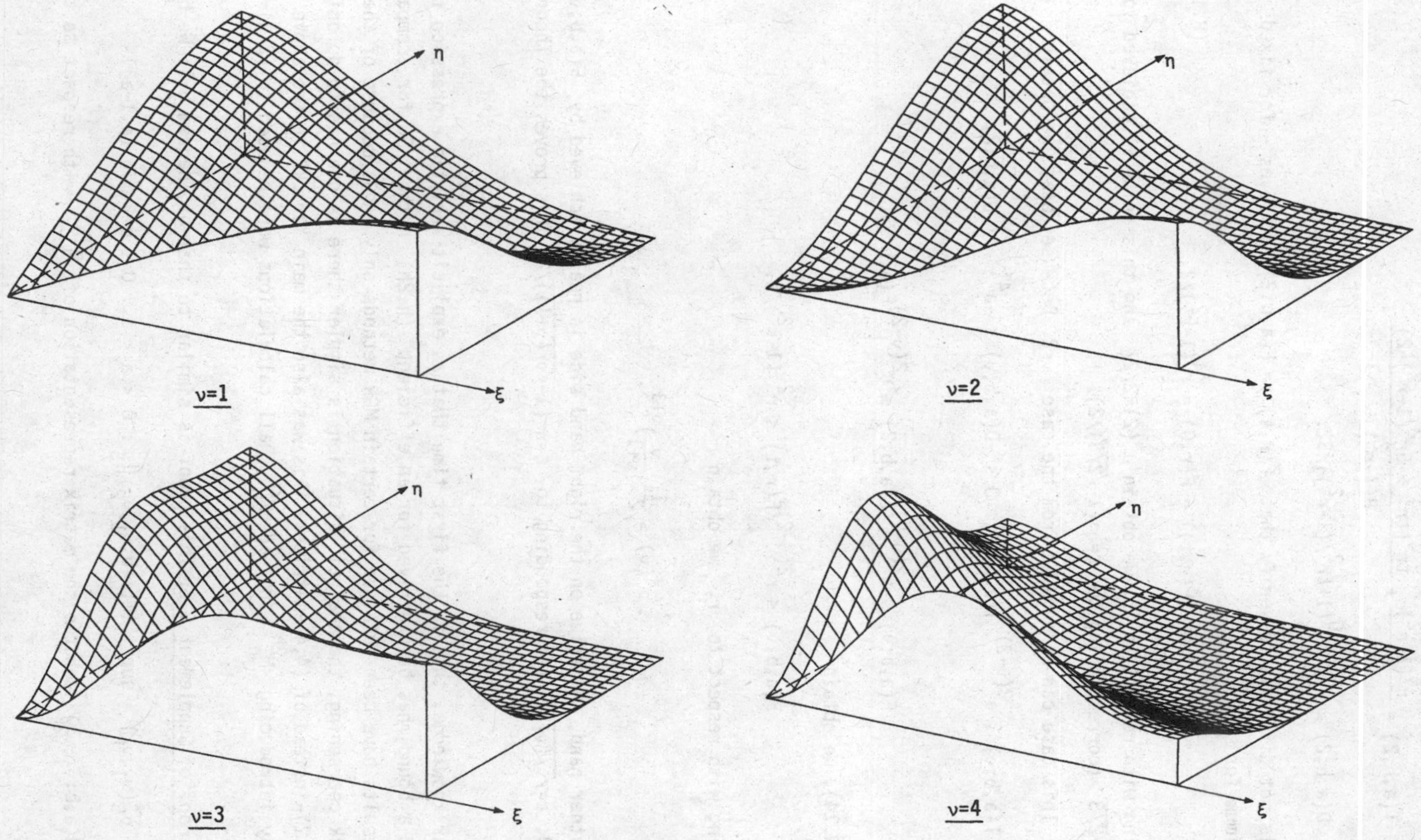

Figure 8.2: Analogous to Figure 8.1. The difference is now that the FW operator is replaced by the HW operator.

$$T(a,b;2) = \frac{r^4(1-r^2)}{2} + \frac{tr^4(tr^4 + 5r^2(1-r^2)-2)}{8(1-r^2)},$$

$$D(a,b;2) = -\, t^2r^{10}(1+tr^2/(1-r^2))/32.$$

Now it is not difficult to verify that $\tilde{F}(r,t) := F(a,b;2)$ achieves - for fixed r - its maximum for $t=0$:

$$\max_t \{\tilde{F}(r,t)\} = \tilde{F}(r,0) = r^4(1-r^2)/2. \tag{8.14}$$

Maximizing with respect to r, we obtain $\rho^*(2)=2/27$ and this value is achieved for $b=0$, $a=\sqrt{2/3}$ (corresponding to $\xi=\eta=(1-\sqrt{2/3})/2$).

<u>If $\nu>2$</u>: This case can be deduced from the case $\nu=2$. Because $b\le a$, we obtain

$$0 \le T(a,b;\nu) \le a^{2(\nu-2)}\, T(a,b;2), \quad 0 \le -D(a,b;\nu) \le -a^{4(\nu-2)}\, D(a,b;2)$$

and thus

$$F(a,b;\nu) \le a^{2(\nu-2)}F(a,b;2) \le r^{2(\nu-2)}F(a,b;2).$$

Using (8.14), we obtain

$$F(a,b;\nu) \le r^{2(\nu-2)}\tilde{F}(r,t) \le r^{2\nu}(1-r^2)/2.$$

Maximizing with respect to r, we obtain

$$\rho^*(\nu) \le \frac{1}{2\nu}\left(\frac{\nu}{\nu+1}\right)^{\nu+1}.$$

On the other hand, the value on the right hand side is really achieved by $F(a,b;\nu)$ for $b=0$, $a=\sqrt{\nu/(\nu+1)}$ (corresponding to $\xi=\eta=(1-\sqrt{\nu/(\nu+1)})/2$. This proves the Theorem.

□

To our knowledge, this is the first time that an explicit analytic expression for ρ^* (not a bound) has been derived for an efficient (h,2h) method. So far, comparable results have been obtained for certain MGR methods only [82]. Because of the red-black coarsening, the analytic situation is simpler there as one has to do only with (2,2)-instead of (4,4)-matrices. This was also the main "trick" in the above proof: By introducing $\tilde{M}_h^{2h}$ instead M_h^{2h}, all calculations were reduced to (2,2)-matrices.

<u>On the proof of Supplement 8.2</u>: This proof is similar to that of Theorem 8.1: We have

$$\tilde{\sigma}_S^*(\nu_1,\nu_2) = \sup\,\{\|M(a,b;\nu_1,\nu_2)\|_S : 0 < a < 1,\ 0 \le b < \min\{a,1-a\}\}$$

where $M(a,b;\nu_1,\nu_2)$ denotes the matrix representation of $\tilde{M}_{h,\theta}^{2h}$ with respect to the normalized basis $\bar{\psi}_1(x) := \psi_1(x)/\|\psi_1\|_2$, $\bar{\psi}_2(x) := \psi_2(x)/\|\psi_2\|_2$ (cf. 8.11)). We do not give the details of the complicated considerations here. We only want to mention that for $\nu_1 \geq 1$, $\nu_2 \geq 1$ the above supremum is achieved for $b=0$. For $b=0$, however, M turns out to be a symmetric matrix. This is the reason why $\tilde{\sigma}_S^* = \rho^*$ then.

□

Proof of Supplement 8.3: We first note that both S_h^{red} and S_h^{black} are orthogonal projection operators with respect to the energy inner product (1.8) (cf. [13]). In particular, they are self-adjoint with respect to this inner product meaning that

$$L_h S_h^{red} = (S_h^{red})^* L_h, \quad L_h S_h^{black} = (S_h^{black})^* L_h.$$

Here "*" denotes the adjoint with respect to the Euclidian inner product (1.7). From this, we obtain (8.10) because

$$(L_h^{1/2} M_h^{2h} L_h^{-1/2})^* = L_h^{1/2} M_h^{2h} L_h^{-1/2}.$$

□

Both results (8.8) and (8.10) can be used to obtain norm estimates for complete cycles based on the respective two-grid methods. Compared to the general estimation in Lemma 4.3, we have the advantage here that the spectral radius ρ^* can be used instead of a norm σ^*. For both cases, however, the estimate used in Lemma 4.3 has to be modified slightly. For (8.10) it is immediately clear from the definition of M_h^{2h} in (8.9) how this modification has to be done. If one wants to apply (8.8), one has to guarantee that the first approximation used on the coarser grids (after every fine-to-coarse transfer) is such that the $\|\cdot\|_2$-error reduction can be estimated by ρ^*. (This is not true if the zero grid function is used as first approximation.) It is sufficient, however, to modify the recursion of MG cycle (cf. (4.10)) in the following way: Apply one extra "black" partial relaxation step after each fine-to-coarse transfer. This means that the MG iteration operator is now - different from (4.10) - recursively defined by

$$\mathbf{M}_{k+1} := M_{k+1}^k + A_k^{k+1} \mathbf{M}_k^\gamma S_k^{black} A_{k+1}^k \quad (k = 1,2,\ldots,\ell-1) \tag{8.15}$$

with two-grid operators corresponding to (8.3). We obtain the

Corollary 8.4: Let $\gamma=2$ and $\nu_1 \geq 1$, $\nu_2 \geq 1$. Then one obtains the following bound on $\rho(\mathbf{M}_\ell)$ for $\mathbf{M}_\ell$ as defined in (8.15)

$$\rho(\mathbf{M}_\ell(\nu_1,\nu_2)) \leq \|\tilde{\mathbf{M}}_\ell(\nu_1,\nu_2)\|_S \leq (1-\sqrt{1-4\rho^*(\nu)})/2 \quad (\ell = 1,2,\ldots). \tag{8.16}$$

Here $\tilde{\mathbf{M}}_\ell$ denotes the restriction of $\mathbf{M}_\ell$ to $\tilde{\mathbf{G}}(\Omega_\ell)$. In particular, for $\nu_1=\nu_2=1$

$$\rho(\mathbb{M}_\ell) < 0.081 \qquad (\ell = 1,2,\ldots).$$

Proof: The estimate in (8.16) follows by a recursive estimation analogous to the one used in Corollary 4.4 but now applied to (8.15). One only has to observe that $\| B_k^{k+1}\|_S \cdot \|B_{k+1}^k\|_S \le 1$ (k=1,2,...) where

$$B_k^{k+1} := A_k^{k+1}\Big|_{\tilde{\mathbb{G}}(\Omega_k)}, \qquad B_{k+1}^k := S_k^{black} A_{k+1}^k\Big|_{\tilde{\mathbb{G}}(\Omega_{k+1})}.$$

This estimate can be verified by use of the Fourier representations of B_k^{k+1} and B_{k+1}^k.

□

Remarks: (1) If we replace the FW operator used above by the operator of *half weighting* (HW), i.e.

$$I_h^{2h} \mathrel{\hat{=}} \frac{1}{8}\begin{bmatrix} & 1 & \\ 1 & 4 & 1 \\ & 1 & \end{bmatrix}_h^{2h}, \tag{8.17}$$

we obtain a method which is still more efficient than the one using FW. For the corresponding two-grid operator M_h^{2h}, we can make a similar analysis with respect to ρ^* and σ_S^* as above. We do not give the details of the corresponding analysis here, but show the most important values in Table 8.1 of the next section. (See also Figure 8.2: This Figure corresponds to Figure 8.1 and it shows how the function F in (8.13) changes, if the FW operator used there is replaced by the HW operator.) The FORTRAN program listed in the appendix is the MG version corresponding to this two-grid method.

(2) In contrast to the HW operator, the operator of *straight injection* (INJ, see (3.41)) does not lead to a reasonable two-grid method here. This follows immediately from the observation that the defect after one RB step is zero at all black points and nonzero otherwise.

8.2 Further results for Poisson's equation

In this section, we give some more (h,2h)-results on Poisson's equation with Dirichlet boundary conditions on rectangular domains. As before, we restrict ourselves to the case of square grids $h=h_{x_1}=h_{x_2}$. For the case of non-square grids, see Section 8.3.

The first Table 8.1a recalls results on μ^* and ρ^* already given in Sections 7.5 and 8.1. In addition, corresponding convergence factors are given for the case that the FW operator (3.3) is replaced by the HW operator (8.17). One can see that the use of the HW operator (for $\nu \geq 3$) leads to a method which is both more rapidly convergent and cheaper. In the table, the operation count is given in terms of $W_h^{2h}/\mathcal{N}_h$ (cf. Section 4.4). (Details on the operation count of the single MG components are given in Table 8.1c.) Using the efficiency measure (4.26), (4.27) (and counting additions and multiplications weighted equally), we compute from Table 8.1a that the optimal values ν for FW and HW are given by $\nu=2$ and $\nu=3$, respectively.

Remark: Note that after one RB relaxation step the defect is zero at all black points. This means that the fine-to-coarse transfers become fairly simple. In particular, the application of the HW operator simply means that residuals are transfered from the finer to the coarser grid by multiplying them by a factor of 0.5. (This kind of transfer operator is also called *half injection* (HI) in the following.) Similarly, the performance of the coarse-to-fine transfer is simplified if it is followed by one step of RB relaxation: interpolation then has to be carried out explicitly only at black points.

Table 8.1b shows the quantities σ_S^*, σ_E^* and σ_d^* for the respective methods. In order to obtain values less than one for all these quantities, we see that one should choose neither $\nu_1=0$ nor $\nu_2=0$. As for the optimal ν-values mentioned above, one should choose $\nu_1=1$, $\nu_2=1$ and $\nu_1=2$, $\nu_2=1$ in case of FW and HW, respectively, in order to obtain the smallest norm values.

Remark: For $\nu_1=0$ and HW, we see from the table that $\sigma_S^*=\sigma_E^*=\infty$. The reason for this can easily be seen by a similar consideration as for Jacobi's method in connection with the INJ operator (cf. (3.43)). In contrast to Jacobi's method, this effect vanishes here if $\nu_1 \geq 1$. A similar behavior does not occur in connection with the FW restriction operator; this is due to the fact that the FW operator acts like a "filter": High frequencies are essentially damped by the FW operator, more precisely, we have for $\alpha \neq (0,0)$ (cf. (7.33), (7.34))

$$q(\theta^\alpha) = O(|\theta|^2) \qquad (|\theta| \to 0). \tag{8.18}$$

		I_h^{2h} : FW			I_h^{2h} : HW		
ν	$(\mu^*)^\nu$	ρ^*	# Add	# Mult	ρ^*	# Add	# Mult
1	0.250	0.250	6.75	2.25	0.500	5.5	1.75
2	0.063	0.074	9.75	3.25	0.125	8.5	2.75
3	0.034	0.053	12.75	4.25	0.034	11.5	3.75
4	0.025	0.041	15.75	5.25	0.025	14.5	4.75

Table 8.1a: μ^*, ρ^* and computational work $W_h^{2h}/\mathcal{N}_h$ in case of smoothing by RB relaxation (for 5-point Laplace discretization)

	I_h^{2h} : FW			I_h^{2h} : HW		
(ν_1,ν_2)	σ_S^*	σ_E^*	σ_d^*	σ_S^*	σ_E^*	σ_d^*
(1,0)	0.559	0.500	1.414	0.707	0.514	1.414
(0,1)	1.414	0.500	0.559	∞	∞	1.000
(2,0)	0.200	0.285	1.414	0.191	0.300	1.414
(1,1)	0.141	0.125	0.141	0.707	0.257	0.280
(0,2)	1.414	0.285	0.212	∞	∞	1.000
(3,0)	0.137	0.226	1.414	0.115	0.233	1.414
(2,1)	0.081	0.062	0.081	0.070	0.064	0.096
(1,2)	0.081	0.062	0.081	0.707	0.165	0.108
(0,3)	1.414	0.226	0.144	∞	∞	1.000
(4,0)	0.105	0.193	1.414	0.082	0.197	1.414
(3,1)	0.062	0.046	0.062	0.039	0.036	0.063
(2,2)	0.062	0.046	0.062	0.038	0.029	0.037
(1,3)	0.062	0.046	0.062	0.707	0.131	0.073
(0,4)	1.414	0.193	0.109	∞	∞	1.000

Table 8.1b: Norm values corresponding to the method in Table 8.1a

MG component	# Add	# Mult
one RB step	3	1
I_h^{2h}/FW (if preceeded by RB step)	2.75	0.75
I_h^{2h}/HW (if preceeded by RB step)	1.5	0.25
I_{2h}^h (if followed by RB step)	1	0.5

Table 8.1c: Operation count for the individual MG components used in Tables 8.1a and 8.1b (number of operations per point of Ω_h). The numbers given for I_h^{2h} and I_{2h}^h include the work needed for the computation of the defect and adding the correction, respectively.

As in the case of Jacobi ω-relaxation, one could try to introduce one (or more) parameters into the process of RB relaxation. In contrast to Jacobi's method, however, this results only in a non-significant improvement of the quantities in Table 8.1. Taking also work into account, the corresponding methods are even less efficient: $\omega=1$ is indeed the best value here. Also see Figure 9.1, where the dependence on ω is shown for the case of Gauß-Seidel relaxation with lexicographic ordering of the grid points. The dependence shown there is typical for RB relaxation, also.

We point out once more that for Jacobi's method the use of a parameter is unavoidable to ensure good smoothing. This is a typical feature of Jacobi relaxation which makes it inefficient in variable coefficient cases. But even if the optimal parameter is known, Jacobi's method usually is considerably less efficient in smoothing than, e.g., RB relaxation (cf. [25], Section 3.5).

Remark: For the Helmholtz operator

$$L^{\Omega}u = -\Delta u + cu \tag{8.19}$$

one obtains - for non-negative c - similar results as shown in Table 8.1 (for the methods considered there). For any fixed h, all two-grid quantities shown become even better for increasing c. For $c<0$, the above results still hold if the coarsest mesh size used (h^*) is assumed to be sufficiently fine.

Table 8.2 refers to the 9-point "Mehrstellen"-discretization

$$L_h^{\Omega} \hat{=} \frac{1}{h^2}\begin{bmatrix} -1 & -4 & -1 \\ -4 & 20 & -4 \\ -1 & -4 & -1 \end{bmatrix}_h . \tag{8.20}$$

Again, we use I_{2h}^{h} defined by linear interpolation and we define L_{2h} in the same way as L_h. We compare RB with Jacobi (JAC) and "four-colour" relaxation (FC, see Remark (4) in Section 7.3). As before we have to use a parameter ω for Jacobi relaxation. In Table 8.2 we used $\omega=0.5$ and $\omega=10/11$ (which is optimal with respect to the smoothing factor). As for the fine-to-coarse transfer, we compare several operators which are meaningful for the respective relaxation methods.

Table 8.2 shows only two choices of ν_1 and ν_2, namely $\nu_1=\nu_2=1$ and $\nu_1=2$, $\nu_2=1$ (other choices lead to less efficient methods). The best method considered is the one using FC relaxation followed by the one using RB relaxation (and FW restriction). As in the 5-point discretization, Jacobi's method leads to less efficient algorithms, even with the parameter ω chosen for optimal smoothing. We point out that all norm values σ_S^* and σ_E^* are infinite except for FW. The "Mehrstellen" discretization and corresponding two-grid methods have been studied systematically in [12].

Remark: Instead of using the 9-point operator for L_h and L_{2h}, one could have equally well used the usual 5-point operator on the coarser grid. This would not change the results given in Table 8.2 considerably.

Relax	$(\mu^*)^\nu$	I_h^{2h}	ρ^*	σ_S^*	σ_E^*	σ_d^*
RB	0.040	FW	0.066	0.205	0.078	0.205
		HW	0.152	∞	∞	0.205
		HI	0.241	∞	∞	0.317
FC	0.048	FW	0.039	0.185	0.069	0.185
JAC (ω=0.5)	0.490	FW	0.490	0.490	0.490	0.490
		HW	0.490	∞	∞	0.490
		INJ	0.490	∞	∞	0.648
JAC (ω=10/11)	0.207	FW	0.207	0.207	0.207	0.207
		HW	0.207	∞	∞	0.207
		INJ	0.207	∞	∞	0.515

Table 8.2a: Results for the 9-point "Mehrstellen" operator and $(\nu_1,\nu_2) = (1,1)$

Relax	$(\mu^*)^\nu$	I_h^{2h}	ρ^*	σ_S^*	σ_E^*	σ_d^*
RB	0.024	FW	0.036	0.052	0.048	0.202
		HW	0.063	∞	∞	0.202
		HI	0.144	∞	∞	0.273
FC	0.030	FW	0.023	0.036	0.037	0.181
JAC (ω=0.5)	0.343	FW	0.343	0.343	0.343	0.405
		HW	0.343	∞	∞	0.403
		INJ	0.343	∞	∞	0.414
JAC (ω=10/11)	0.094	FW	0.101	0.101	0.106	0.121
		HW	0.094	∞	∞	0.104
		INJ	0.094	∞	∞	0.228

Table 8.2b: Analogous to Table 8.2a for $(\nu_1,\nu_2) = (2,1)$

8.3 Results for the anisotropic model equation

We now consider the anisotropic model equation

$$L^\Omega u = -\varepsilon u_{x_1x_1} - u_{x_2x_2} = f^\Omega(x) \quad (x \in \Omega) \tag{8.21}$$

with Dirichlet boundary conditions on a rectangular domain Ω. We assume this problem to be discretized by the 5-point operator

$$L_h^\Omega \hat{=} \frac{1}{h^2}\begin{bmatrix} & -1 & \\ -\varepsilon & 2(1+\varepsilon) & -\varepsilon \\ & -1 & \end{bmatrix}_h \tag{8.22}$$

on a square grid $h=h_{x_1}=h_{x_2}$.

As pointed out in Section 7.5, pointwise relaxation methods - if applied in connection with standard coarsening - have very unsatisfactory smoothing properties for $\varepsilon>>1$ or $0<\varepsilon<<1$. This deficiency is fully reflected by the two-grid convergence properties. We have listed some of the corresponding quantities in Table 8.3 for the case of RB relaxation with $\omega=1$, $\nu_1=2$, $\nu_2=1$. Note that the half weighting operator yields unbounded norms σ_S^* and σ_E^* as soon as $\varepsilon\neq1$. This shows the special role of the half weighting operator in connection with model problem (P).

		I_h^{2h} : FW			I_h^{2h} : HW		
ε	$(\mu^*)^\nu$	ρ^*	σ_S^*	σ_E^*	ρ^*	σ_S^*	σ_E^*
1.0	0.034	0.053	0.081	0.062	0.034	0.070	0.064
0.5	0.088	0.088	0.158	0.134	0.132	∞	∞
0.1	0.564	0.564	0.849	0.621	0.621	∞	∞
0.01	0.942	0.942	1.829	0.951	0.951	∞	∞

Table 8.3: Results for the anisotropic model operator using RB relaxation (I_{2h}^h : linear interpolation; $(\nu_1,\nu_2) = (2,1)$)

According to our considerations on smoothing factors in Section 7.5, ZEBRA relaxation (with $\omega=1$) has very good smoothing properties for problem (8.22) if applied in the "correct" direction. Table 8.4a gives results for the case of x_2-ZEBRA relaxation (using the FW restriction operator). This direction of lines is correct for $\varepsilon\leq1$. In full accordance with the prediction obtained from its smoothing factor, x_2-ZEBRA does not yield a good method if $\varepsilon>>1$. In this case x_1-ZEBRA would be appropriate.

	$(\nu_1,\nu_2) = (1,0)$				$(\nu_1,\nu_2) = (1,1)$			
ε	$(\mu^*)^\nu$	ρ^*	σ_S^*	σ_E^*	$(\mu^*)^\nu$	ρ^*	σ_S^*	σ_E^*
1000	0.998	0.998	1.99	0.999	0.996	0.996	1.99	0.997
100	0.980	0.980	1.94	0.990	0.961	0.961	1.903	0.971
10	0.826	0.826	1.53	0.909	0.683	0.683	1.316	0.751
2	0.444	0.444	0.801	0.666	0.198	0.198	0.548	0.296
1	0.250	0.250	0.559	0.500	0.063	0.063	0.317	0.125
0.5	0.125	0.112	0.351	0.368	0.053	0.028	0.170	0.049
0.1	0.125	0.111	0.181	0.289	0.053	0.047	0.082	0.065
0.01	0.125	0.123	0.177	0.274	0.053	0.052	0.089	0.073
0.001	0.125	0.124	0.175	0.272	0.053	0.053	0.090	0.074
$W_h^{2h}/\mathcal{N}_h$	#Add: 8.25, #Mult: 4.5				#Add: 12.25, #Mult: 7.5			

Table 8.4a: Results for the anisotropic model operator using x_2-ZEBRA relaxation (I_h^{2h}: FW restriction, I_{2h}^h: linear interpolation)

In order to get rid of the above dependence on the size of ε, we can use alternating ZEBRA relaxation. Corresponding results are given in Table 8.4b. Here, one smoothing step is arranged in the following "symmetric" way

$$S_h = S_h^{x_1\text{-odd}} \; S_h^{x_1\text{-even}} \; S_h^{x_2\text{-even}} \; S_h^{x_2\text{-odd}} \; . \tag{8.23}$$

This arrangement of one smoothing step is more favourable than others: in particular, it leads to values of ρ^* which are symmetric with respect to ε and $1/\varepsilon$.

Table 8.4b clearly shows that the convergence properties of the (h,2h)-method using alternating ZEBRA relaxation are very good independent of the size of ε. If $\varepsilon\to 0$ or $\varepsilon\to\infty$, the asymptotic convergence behavior becomes the same as that of ZEBRA relaxation with lines in the correct direction. This is clear heuristically, as only the "correct" partial steps contained in one alternating ZEBRA step are really "active" then.

	$(\nu_1,\nu_2) = (1,0)$				$(\nu_1,\nu_2) = (1,1)$			
ε	$(\mu^*)^\nu$	ρ^*	σ_S^*	σ_E^*	$(\mu^*)^\nu$	ρ^*	σ_S^*	σ_E^*
1000	0.125	0.124	0.249	0.273	0.053	0.053	0.126	0.074
100	0.122	0.119	0.242	0.271	0.051	0.051	0.119	0.071
10	0.102	0.082	0.198	0.266	0.041	0.038	0.075	0.049
2	0.061	0.019	0.171	0.279	0.020	0.013	0.020	0.016
1	0.048	0.023	0.198	0.305	0.014	0.009	0.031	0.009
0.5	0.061	0.019	0.242	0.347	0.020	0.013	0.041	0.016
0.1	0.102	0.082	0.337	0.448	0.041	0.038	0.062	0.049
0.01	0.122	0.119	0.379	0.493	0.051	0.051	0.085	0.074
0.001	0.125	0.124	0.384	0.5	0.053	0.053	0.089	0.074
$W_h^{2h}/\mathcal{N}_h$	#Add: 12, #Mult: 7.25				#Add: 20, #Mult: 13.25			

Table 8.4b: Analogous to Table 8.4a with alternating ZEBRA instead of x_2-ZEBRA

MG component	# Add	# Mult
one ZEBRA step	4	3
I_h^{2h} (if preceeded by odd partial step):	3.25	1
I_{2h}^h (if followed by odd partial step):	0.75	0.5
I_{2h}^h (if followed by even partial step):	1	0.5

Table 8.4c: Operation count for the individual MG components used in Tables 8.4a and 8.4b (number of operations per point of Ω_h). The numbers given for I_h^{2h} and I_{2h}^h include the work needed for the computation of the defect and adding the correction, respectively.

Remarks: (1) With respect to a relaxation parameter ω, we have a similar behavior as already mentioned for RB relaxation: In general, the use of a parameter $\omega \neq 1$ does not pay.

(2) As for the operation count given in Table 8.4, the work needed for the matrix decompositions (in solving the occuring tridiagonal systems), is not counted: It has to be performed only once and the correponding amount of operations is only proportional to $\sqrt{\mathcal{N}_h}$.

(3) Clearly, with respect to efficieny, one should take into account that one step of alternating ZEBRA relaxation requires about twice the work of one single ZEBRA step. Thus, if we have to do with a fixed constant ε, alternating relaxation does not pay. In practice, the same holds if $\varepsilon=\varepsilon(x)$ with either $\varepsilon(x)\geq 1$ or $0<\varepsilon(x)\leq 1$. On the other hand, if $\varepsilon(x)\gg 1$ <u>and</u> $0<\varepsilon(x)\ll 1$ in different parts of a given domain, alternating relaxation is needed (for examples, see Section 10.3).

9. Local Fourier analysis and some general theoretical approaches

The model problem analysis gives realistic quantitative results on the convergence behavior of certain multigrid methods. Its main disadvantages is, however, that it can directly be applied only to a rather small class of problems. In this chapter we review some more general theoretical approaches, namely

- local Fourier analysis (Brandt) which can be regarded as an extension of the model problem analysis;
- more abstract convergence theories.

In Section 9.1 we describe the purpose of local Fourier analysis, provide the formal tools needed and point out the connections to the model problem analysis. Section 9.2 gives some concrete results and comparisons. In particular, multigrid methods based on Gauß-Seidel relaxation (with lexicographic ordering of the grid points) and ILU smoothing are considered. As far as general theories are concerned, we will shortly discuss only two of them (Section 9.3).

All considerations in this section directly refer only to two-grid methods of (h,2h)-type.

9.1 Purpose and formal tools of local Fourier analysis

Local Fourier analysis (or *local mode analysis*) was introduced by Brandt in [16] and developed further in several papers (see [25] and the references given there). This theoretical approach can be regarded from different points of view and has several objectives. We here consider it as a general tool for providing quantitative and realistic insight into the smoothing properties of relaxation methods and the convergence properties of two-grid methods. As for more sophisticated aspects of the local Fourier analysis (FMG mode analysis, connections to h-ellipticity and interior stability etc.) and for extensions (to systems of differential equations etc.), see Brandt [25] and the references cited there.

As mentioned already in Section 7.6, there are close connections between the local Fourier analysis and the model problem analysis as presented in Chapter 7 and 8. First of all, local mode analysis can, in a sense, be regarded as an analysis for special model problems, namely those with periodic boundary conditions (on rectangular domains). On the other hand, all results obtained by the model problem analysis in Chapters 7 and 8 can be derived by means of local Fourier analysis also. This application of local results to "global" situations is possible since the model problems considered were required to satisfy certain essential assumptions. Roughly speaking, these assumptions were: the "symmetry" of all grid operators occuring in terms of their corresponding difference stars (cf. Lemma 7.1) as well as certain "continuation properties" for the Ω_h-grid functions occuring with respect to the infinite grid G_h (9.1). The latter assumption has been used in Chapter 7 only implicitly; the three types of boundary conditions - Dirichlet, Neumann and periodic - are associated with an antisymmetric, symmetric and periodic continuation of grid functions, respectively. An analysis of model problems derived from the concept of local Fourier analysis has been sketched in [98].

The more important objective of local Fourier analysis is, however, to give approximative predictions or estimations of μ^*, ρ^* etc. even in those cases for which a rigorous determination of these quantities is very difficult (or practically impossible). The "local" character of this theoretical approach consists in the following: A given discrete operator L_h with variable coefficients (on a regular grid) is considered in the neighborhood of a fixed $x_o \in \Omega_h$. Here L_h is replaced by a formal discrete operator $\mathcal{L}_h$ with constant coefficients (obtained by "freezing" those of L_h at x_o). The resulting operator $\mathcal{L}_h$ is then defined on the infinite grid

$$G_h := \{x = \kappa\cdot h : \kappa \in \mathbb{Z}^2\} \quad \text{with} \quad \kappa\cdot h = (\kappa_1 h_{x_1}, \kappa_2 h_{x_2}). \tag{9.1}$$

As domain of $\mathcal{L}_h$ we consider the (complex) linear space $\mathcal{E}_h$ spanned by all frequencies

$$\varphi(\theta,x) := e^{i\theta x/h} \ (x \in G_h) \quad \text{for} \quad \theta \in \mathbb{R}^2, \ -\pi < \theta \le \pi. \tag{9.2}$$

Here the inequality has to be understood with respect to θ_1 and θ_2, and x/h is defined by $x/h := (x_1/h_{x_1}, x_2/h_{x_2})$. Obviously, for any $\theta \in \mathbb{R}^2$, $e^{i\theta x/h}$ can be identified with one of the basis elements in (9.2) because

$$e^{i\theta x/h} = e^{i\theta' x/h} \ (x \in G_h) \quad \text{if} \quad \theta = \theta' (\mathrm{mod}\ 2\pi). \tag{9.3}$$

Note that $\mathcal{E}_h$ is a space with a non-denumerable basis. This is the main formal difference compared to the model problem analysis where the discrete operators were defined on finite dimensional spaces $(\theta \in T_h)$. The use of infinite-dimensional spaces and operators (with θ varying continuously in $-\pi < \theta \le \pi$) gives some technical simplifications in the analysis. Clearly, we are, after all, only interested in two-grid methods for finite-dimensional discrete operators, in practice.

As with $\mathcal{L}_h$, we replace all other operators by the corresponding infinite-grid operators. We denote them by

$$\mathcal{S}_h, \mathcal{I}_h, \mathcal{I}_h^{2h}, \mathcal{L}_{2h}, \mathcal{I}_{2h}^{h}, \mathcal{K}_h^{2h}, \mathcal{M}_h^{2h}. \tag{9.4}$$

(For simplicity, we restrict ourselves to the case of standard coarsening. All other types of coarsenings considered in Chapter 7 can be treated analogously however.)

On the coarser grid G_{2h}, the space $\mathcal{E}_{2h}$ is defined analogously to $\mathcal{E}_h$ with h replaced by $2h$. By $\mathcal{L}_h$ and $\mathcal{M}_h^{2h}$, both an "ideal" difference operator and an "ideal" two-grid operator are defined. In particular, boundary effects are not taken into account by considering these operators. This ideal situation can be analyzed in much the same way as was described in Chapter 7, if complex Fourier frequencies (9.2) are used instead of sine functions (7.4). In the following, we are going to list the necessary tools only briefly without going into details.

Let $\mathcal{L}_h$ be any <u>difference operator</u> on G_h described by a general difference star (1.14):

$$\mathcal{L}_h w_h(x) := \sum_{\kappa\in V} s_\kappa w_h(x+\kappa\cdot h) \quad (x \in G_h). \tag{9.5}$$

Obviously, all basis functions (9.2) are eigenfunctions of $\mathcal{L}_h$:

$$\mathcal{L}_h\, \varphi(\theta,x) = \lambda(\theta,h)\varphi(\theta,x) \quad (-\pi < \theta \le \pi), \quad \lambda(\theta,h) = \sum_{\kappa\in V} s_\kappa e^{i\theta\kappa}. \tag{9.6}$$

In direct analogy to Section 7.2, we define low and high frequencies (with respect to the coarser grid G_{2h}):

$$\varphi(\theta,\cdot) \text{ is called } \begin{cases} \textit{low frequency}, & \text{if } -\pi/2 < \theta \le \pi/2 \\ \textit{high frequency}, & \text{otherwise.} \end{cases} \tag{9.7}$$

Corresponding to this definition, the space $\mathcal{E}_h$ is split into an orthogonal sum

$$\mathcal{E}_h = \mathcal{E}_h^{\text{low}} \overset{\perp}{\oplus} \mathcal{E}_h^{\text{high}}. \tag{9.8}$$

Here "orthogonal" is meant with respect to the inner product

$$(v_h,w_h) := \lim_{n\to\infty} \frac{1}{4n^2} \sum_{|\kappa|\le n} v_h(\kappa\cdot h)\, \overline{w_h(\kappa\cdot h)}. \tag{9.9}$$

The basis functions (9.2) of $\mathcal{E}_h$ are orthonormal with respect to (9.9). Again, there is a natural identification of $\mathcal{E}_h^{\text{low}}$ and the space $\mathcal{E}_{2h}$. This becomes obvious if $\mathcal{E}_{2h}$ is written in the form

$$\mathcal{E}_{2h} = \text{span}\,\{\Phi(\theta,\cdot) : -\pi/2 < \theta \le \pi/2\}, \quad \Phi(\theta,x) := \varphi(\theta,x) \quad (x \in G_{2h}). \tag{9.10}$$

As in Section 7.2, we define spaces $\mathcal{E}_{h,\theta}$ of (2h-)*harmonics* to be spanned by just those basis functions of $\mathcal{E}_h$ which coincide on the coarser grid used. Because of

$$\varphi(\theta,x) = \varphi(\theta',x) \quad (x \in G_{2h}) \quad \text{iff} \quad \theta = \theta' (\text{mod } \pi), \tag{9.11}$$

we have for all $-\pi/2 < \theta \le \pi/2$

$$\mathcal{E}_{h,\theta} = \text{span}\,\{\varphi(\theta',\cdot) : -\pi < \theta' \le \pi, \quad \theta' = \theta \ (\text{mod } \pi)\}$$
$$= \text{span}\,\{\varphi(\theta^\alpha,\cdot) : \alpha = (0,0); (1,1); (1,0); (0,1)\} \tag{9.12}$$

with

$$\theta^\alpha := \theta - \begin{pmatrix} \alpha_1 \,\text{sign}(\theta_1) \\ \alpha_2 \,\text{sign}(\theta_2) \end{pmatrix} \pi, \qquad \text{sign}(t) := \begin{cases} 1 \ (t > 0) \\ -1 \ (t \le 0). \end{cases} \tag{9.13}$$

The <u>transfer operators</u> $\mathcal{I}_h^{2h}$, $\mathcal{I}_{2h}^h$ and the <u>coarse-grid difference operator</u> $\mathcal{L}_{2h}$ are assumed to be represented by difference stars with coefficients $\hat{t}_\kappa$, $\check{t}_\kappa$ and $\hat{s}_\kappa$, respectively. We obtain for all $-\pi/2 < \theta \le \pi/2$ and all α, θ^α as defined in (9.12), (9.13) (cf. Lemma 7.1):

$$\begin{aligned} \mathcal{L}_{2h}\,\Phi(\theta,x) &= \Lambda(\theta,h)\Phi(\theta,x), & \Lambda(\theta,h) &= \sum_{\kappa\in\mathbf{V}} \hat{s}_\kappa e^{2i\theta\kappa}\,; \\ \mathcal{I}_h^{2h}\,\varphi(\theta^\alpha,x) &= q(\theta^\alpha)\Phi(\theta,x), & q(\theta^\alpha) &= \sum_{\kappa\in\mathbf{V}} \hat{t}_\kappa e^{i\theta^\alpha\kappa}\,; \\ \mathcal{I}_{2h}^h\,\Phi(\theta,h) &= \sum_\alpha p_\alpha(\theta)\varphi(\theta^\alpha,x), & p_\alpha(\theta) &= \frac{1}{4}\sum_{\kappa\in\mathbf{V}} \check{t}_{-\kappa} e^{i\theta^\alpha\kappa}. \end{aligned} \tag{9.14}$$

For the definition of <u>relaxation operators</u> $\mathcal{S}_h$, we use the following notation (which is a generalization of the one used in Section 7.3). The difference operator $\mathcal{L}_h$ is split into a sum

$$\mathcal{L}_h = \mathcal{L}_h^+ + \mathcal{L}_h^0 + \mathcal{L}_h^- \tag{9.15}$$

defined by a corresponding splitting of the index set $\mathbf{V}$ in (9.5) into three disjoint subsets $\mathbf{V}^+$, $\mathbf{V}^0$, $\mathbf{V}^-$:

- $\mathbf{V}^0$: characterizes those grid points where grid values are solved for simultaneously within the relaxation process;
- $\mathbf{V}^-$: characterizes those grid points where previous grid values are used within the relaxation process;
- $\mathbf{V}^+$: characterizes those grid points where already new grid values are available and are used instead of the previous values.

Using these notations, all common relaxation operators can be described in a formally simple manner. Considering errors v_h and $\bar{v}_h$ before and after one complete or one partial relaxation step, we have (with some relaxation parameter ω)

$$(\mathcal{L}_h^o + \omega\mathcal{L}_h^+)\,\bar{v}_h(x) - ((1-\omega)\,\mathcal{L}_h^o - \omega\mathcal{L}_h^-)\,v_h(x) = 0 \quad (x \in \tilde{G}_h)$$
$$\bar{v}_h(x) = v_h(x) \quad (x \in G_h \backslash \tilde{G}_h). \tag{9.16}$$

Here $\tilde{G}_h$ is either G_h (for complete relaxation steps) or a certain subset of G_h (for partial relaxation steps). Corresponding to the splitting of $\mathcal{L}_h$, we have a splitting of its eigenvalues $\lambda(\theta,h) = \lambda^+(\theta,h) + \lambda^o(\theta,h) + \lambda^-(\theta,h)$:

$$\lambda^+(\theta,h) = \sum_{\kappa\in\mathbf{V}^+} s_\kappa e^{i\theta\kappa}, \quad \lambda^o(\theta,h) = \sum_{\kappa\in\mathbf{V}^o} s_\kappa e^{i\theta\kappa}, \quad \lambda^-(\theta,h) = \sum_{\kappa\in\mathbf{V}^-} s_\kappa e^{i\theta\kappa}. \tag{9.17}$$

We assume throughout that

$$\lambda^o(\theta,h) + \omega\,\lambda^+(\theta,h) \neq 0 \quad (-\pi < \theta \leq \pi).$$

The following examples show how common relaxation methods are included in the above concept. As for more examples, see [98],[16].

Examples: (1) Gauß-Seidel point relaxation with lexicographic ordering of the grid points (*GS*; here: from left to right and from bottom to top) is characterized by $\tilde{G}_h = G_h$ and

$$\mathbf{V}^o = \{(0,0)\}, \quad \mathbf{V}^+ = \{\kappa \in \mathbf{V} : \kappa_1 < 0 \text{ or } (\kappa_1 = 0 \text{ and } \kappa_2 < 0)\}.$$

(2) Gauß-Seidel line relaxation with line-by-line ordering (*line GS*; here: lines in x_2-direction marching from left to right) is characterized by $\tilde{G}_h = G_h$ and

$$\mathbf{V}^o = \{\kappa \in \mathbf{V} : \kappa_1 = 0\}, \quad \mathbf{V}^+ = \{\kappa \in \mathbf{V} : \kappa_1 < 0\}.$$

(3) Jacobi relaxation is characterized by $\tilde{G}_h = G_h$ and

$$\mathbf{V}^o = \{(0,0)\}, \quad \mathbf{V}^+ = \emptyset.$$

(4) RB relaxation proceeds in two partial steps. For both steps the splitting of $\mathbf{V}$ is the same as in (3). In the first partial step, $\tilde{G}_h$ consists of the "red" and in the second step $\tilde{G}_h$ consists of the "black" points (cf. Table 7.1).

(5) Gauß-Seidel point relaxation with red-black ordering of the grid points *(red-black GS)* is obtained by choosing $\tilde{G}_h$ as in (4) and

$$\mathbf{V}^o = \{(0,0)\}, \quad \mathbf{V}^+ = \{\kappa \in \mathbf{V} : \kappa_1 < 0 \text{ or } (\kappa_1 = 0 \text{ and } \kappa_2 < 0)\} \cap \{\kappa \in \mathbf{V} : \kappa_1 + \kappa_2 \text{ even}\}.$$

(As pointed out before, this method is equivalent to RB relaxation iff $\mathcal{L}_h$ is described by a 5-point star.)

It is quite easy to construct Fourier representations for all common relaxation operators (with respect to the basis (9.2) of $\mathcal{E}_h$). In particular, if $\tilde{G}_h = G_h$, then the relaxation operator $\mathcal{S}_h$ defined by (9.16) is represented by

$$\mathcal{S}_h \varphi(\theta,x) = A(\theta,h;\omega)\,\varphi(\theta,x) \qquad (-\pi < \theta \le \pi) \tag{9.18}$$

with

$$A(\theta,h;\omega) := \frac{(1-\omega)\lambda^0(\theta,h) - \omega\,\lambda^-(\theta,h)}{\lambda^0(\theta,h) + \omega\,\lambda^+(\theta,h)}\,. \tag{9.19}$$

For methods like RB or ZEBRA, the $\varphi(\theta,\cdot)$ are no longer eigenfunctions of the respective operators $\mathcal{S}_h$. However, the spaces $\mathcal{E}_{h,\theta}$ of harmonics are still invariant:

$$\mathcal{S}_h : \mathcal{E}_{h,\theta} \to \mathcal{E}_{h,\theta} \qquad (-\pi/2 < \theta \le \pi/2) \tag{9.20}$$

This is analogous to the finite-grid situation studied in Section 7.3. In fact, <u>for RB and ZEBRA (on the infinite grid) one obtains formally the same (4,4)-matrix representations (with respect to $\mathcal{E}_{h,\theta}$) as we have shown in Table 7.2.</u> Clearly, θ ranges in $-\pi/2 < \theta \le \pi/2$ instead of T_h^{2h}. (The "degenerate" cases shown in Table 7.2 do not occur here.) Also the definiton of A_α in Table 7.2 has to be understood with respect to all $-\pi/2 < \theta \le \pi/2$ and with A, θ^α as defined in (9.19), (9.13), respectively.

<u>Remark</u>: In contrast to RB and ZEBRA, there is no comparable equivalence of finite-grid analysis (concerning S_h) and infinite-grid analysis (concerning $\mathcal{S}_h$) in the cases of, for instance, GS relaxation (1) and line GS relaxation (2). In fact, with respect to the basis (7.4), all frequencies are intermixed by the relaxation operator S_h (leading to a representation of S_h by a full matrix). Therefore these methods cannot be directly analyzed in the framework of the model problem analysis. The reason for the different structure of S_h and $\mathcal{S}_h$ lies in the fact that $\mathbf{V}^+ \ne \emptyset$ for (1) and (2) (meaning that there is - so to say - no natural starting point for the relaxation on the infinite grid). For RB and ZEBRA we have $\mathbf{V}^+ = \emptyset$; here any point (or line) can be used to start the relaxation. In general, the analyses of S_h and $\mathcal{S}_h$ differ due to algorithmic changes near the boundary. Concerning the <u>high</u> frequencies, however, the boundary influence is not essential. Therefore, the <u>smoothing properties</u> of $\mathcal{S}_h$ should be a good approximation to those of S_h in all cases.

The general definition of a *smoothing factor* given in Section 7.5 carries over to the infinite-grid operators in a straightforward manner. In particular, whenever a matrix representation $\hat{\mathcal{S}}_{h,\theta}$ of $\mathcal{S}_h$ (with respect to $\mathcal{E}_{h,\theta}$) is available, we can compute μ by

$$\mu(h,\nu) = \sup\,\{\sqrt[\nu]{\rho(\hat{Q}^{2h}_{h,\theta}\,\hat{\mathcal{S}}^{\nu}_{h,\theta})} : -\pi/2 < \theta \le \pi/2\} \tag{9.21}$$

where

$$\hat{Q}^{2h}_{h,\theta} = \begin{pmatrix} 0 & & & \\ & 1 & & \\ & & 1 & \\ & & & 1 \end{pmatrix} .$$

(This quantitiy is denoted by $\bar{\mu}_\nu$ in Brandt [25].) As usual, we define

$$\mu^*(\nu) := \sup \{\mu(h,\nu) : h \in \mathcal{H}^*\} \qquad (9.22)$$

We give exemplary values of $\mu^*(\nu)$ in the following section. For more examples, see [98],[16].

Let us now consider a <u>two-grid operator</u> for $\mathcal{L}_h$. For the relaxation operator $\mathcal{S}_h$ used, we assume (9.20) to hold. In order for the two-grid operator to be well-defined, we replace the domain $\mathcal{E}_h$ of $\mathcal{L}_h$ by a slightly shrunk subspace $\tilde{\mathcal{E}}_h$ such that $\mathcal{L}_h^{-1}$ exists and also $\mathcal{L}_{2h}^{-1}$ can be reasonably defined on the coarser grid. We choose

$$\tilde{\mathcal{E}}_h := \mathcal{E}_h \setminus \bigoplus_{\theta\in\Psi} \mathcal{E}_{h,\theta} \qquad (9.23)$$

Here the set Ψ contains just those $-\pi/2 < \theta \le \pi/2$ with

- either $\Lambda(\theta,h) = 0$,
- or $\lambda(\theta',h) = 0$ for some $-\pi < \theta' \le \pi$ with $\theta' = \theta \pmod{\pi}$.

Clearly, $\mathcal{L}_h^{-1}$ exists on $\tilde{\mathcal{E}}_h$. The transfer operators have the properties (cf. (9.14))

$$\mathcal{I}_h^{2h} : \tilde{\mathcal{E}}_h \to \tilde{\mathcal{E}}_{2h}, \quad \mathcal{I}_{2h}^h : \tilde{\mathcal{E}}_{2h} \to \tilde{\mathcal{E}}_h$$

with

$$\tilde{\mathcal{E}}_{2h} := \mathcal{E}_{2h} \setminus \text{span}\,\{\Phi(\theta,\cdot) : \theta \in \Psi\}.$$

$\mathcal{L}_{2h}$, regarded as an operator

$$\mathcal{L}_{2h} : \tilde{\mathcal{E}}_{2h} \to \tilde{\mathcal{E}}_{2h},$$

is invertable. Thus $\mathcal{M}_h^{2h}$ is well-defined as an operator on $\tilde{\mathcal{E}}_h$. In particular, we have

$$\mathcal{M}_h^{2h} : \mathcal{E}_{h,\theta} \to \mathcal{E}_{h,\theta} \qquad (-\pi/2 < \theta \le \pi/2,\ \theta \notin \Psi). \qquad (9.24)$$

Using these invariance properties, we can compute

$$\rho(h,\nu) := \sup_\theta \{\rho(\hat{\mathcal{M}}^{2h}_{h,\theta})\}, \qquad \sigma_d(h,\nu_1,\nu_2) := \sup_\theta \{\|\hat{\mathcal{L}}_{h,\theta} \hat{\mathcal{M}}^{2h}_{h,\theta} \hat{\mathcal{L}}^{-1}_{h,\theta}\|_S\},$$

$$\sigma_S(h,\nu_1,\nu_2) := \sup_\theta \{\|\hat{\mathcal{M}}^{2h}_{h,\theta}\|_S\}, \qquad \sigma_E(h,\nu_1,\nu_2) := \sup_\theta \{\|\hat{\mathcal{L}}^{1/2}_{h,\theta} \hat{\mathcal{M}}^{2h}_{h,\theta} \hat{\mathcal{L}}^{-1/2}_{h,\theta}\|_S\}, \qquad (9.25)$$

where θ ranges as in (9.24) and $\hat{\mathcal{M}}_{h,\theta}^{2h}$ and $\hat{\mathcal{L}}_{h,\theta}$ denote the matrix representations of $\mathcal{M}_h^{2h}$ and $\mathcal{L}_h$ with respect to $\mathcal{E}_{h,\theta}$. Clearly, σ_E only makes sense if $\mathcal{L}_h$ is positive definite. In the next section we give exemplary results on ρ^*, σ_S^*, σ_E^* and σ_d^* which are, as usual, defined by taking the supremum of the respective values ρ, σ_S, σ_E, σ_d over the admissable values $h \in \mathcal{H}^*$.

Remarks: (1) Note that, in general, ρ in (9.25) is not equal to the spectral radius of $\mathcal{M}_h^{2h}$ (with respect to the norm $\|\cdot\|_2$ induced by (9.9)). If $\mathcal{M}_h^{2h}$ is not bounded with respect to $\|\cdot\|_2$, the spectrum of $\mathcal{M}_h^{2h}$ may, for example, be the whole complex plane, whereas ρ may have a finite value. Clearly, this phenomenon connot occur for finite-dimensional operators, which we are actually interested in (cf. the corresponding remark above). We have allowed θ to range continuously over $-\pi/2 < \theta \leq \pi/2$ only for reasons of technical simplifications. Therefore, ρ as defined in (9.25) is the practically relevant quantity and not the spectral radius of $\mathcal{M}_h^{2h}$.

(2) If $\mathcal{L}_h$ and $\mathcal{L}_{2h}$ consist of "principal terms" only, more precisely, if for fixed meshsize ratio q^* and some $m \in \mathbb{N}$

$$s_\kappa h_{x_1}^m, \quad \hat{s}_\kappa h_{x_1}^m \qquad (\kappa \in \mathbb{V}) \tag{9.26}$$

are independent of h (and if the star coefficients of $\mathcal{I}_h^{2h}$ and $\mathcal{I}_{2h}^h$ are independent of h) then $\mathcal{M}_h^{2h}$ also does not depend explicitly on h. In particular, we then have

$$\rho = \rho^*, \quad \sigma_S = \sigma_S^*, \quad \sigma_E = \sigma_E^*, \quad \sigma_d = \sigma_d^*. \tag{9.27}$$

Clearly, if $\mathcal{L}_h$ and $\mathcal{L}_{2h}$ do contain lower order terms, for $h \to 0$ only their principal parts are relevant for the two-grid convergence properties.

9.2 Applications of local Fourier analysis

In this section, we give some results obtained by the application of local Fourier analysis. The main objective here is to treat some methods which are similar to those of Chapter 8 but which cannot be analyzed by the model problem analysis and to compare the corresponding results with those of Chapter 8. As in Chapter 8, we again restrict ourselves to square grids $h=h_{x_1}=h_{x_2}$.

All examples considered satisfy (9.27). The list of applications could easily be extended as the concept of local Fourier analysis is very general. Many more applications can be found in the papers of Brandt (see [25] and the references cited there). For a detailed discussion of operators also containing lower order terms, in particular of singularly perturbed problems, we refer to [12].

Tables 9.1, 9.2 and 9.4 directly correspond to Tables 8.1, 8.2 and 8.4, respectively. We consider the same difference operators and similar MG components as in Chapter 8 (now, of course, on the infinite grid G_h). As in Chapter 8, we define the coarse-grid difference operator $\mathcal{L}_{2h}$ in the same way as the fine-grid operator $\mathcal{L}_h$; the coarse-to-fine grid transfer $\mathcal{I}_{2h}^{h}$ again is assumed to be given by bilinear interpolation. Differences in the MG components are in the smoothing and the fine-to-coarse transfer components. In particular, we treat GS and line GS relaxation (cf. Examples (1) and (2) in the previous Section). Table 9.5 refers to the equation $-\Delta u-\varepsilon u_{x_1x_2}$. As the (standard) 9-point discretization of this problem leads to a non-symmetric difference star, the analysis of Chapter 7 is not applicable to this problem. We conclude this section with some considerations on ILU smoothing. As the ILU smoothers are not included in the concept of relaxation methods given so far, we give a short description of the (local) ILU Fourier representation.

In Table 9.1, we give results for the 5-point discretization of the Laplace operator. The relaxation method used is GS with relaxation parameter $\omega=1$. As for $\mathcal{I}_h^{2h}$, we compare the FW restriction operator (3.3) with the simplest transfer operator possible, namely the operator of straight injection (INJ). Of course, we could also have used the HW restriction operator. This operator leads to results somewhere between FW and INJ.

Comparing Table 9.1a with Table 8.1a, one sees that the use of RB relaxation leads to considerably more efficient methods than GS. If one takes the numerical work into account, Table 9.1a shows that the restriction operator INJ is more favourable than FW. The application of the INJ operator has, however, the disadvantage that the spectral and the energy norms of the corresponding local two-grid operators are not bounded (see Table 9.1b). Although we obtain this result only in the framework of infinite-grid analysis, practically the same effect occurs for finite-grid algorithms (if $h\to 0$). Thus one has to be careful with this method (both if it is used in a complete cycle or in the FMG version). As already observed before, FW can be regarded as more robust, at least from a theoretical point of view.

		I_h^{2h} : FW			I_h^{2h} : INJ		
ν	$(\mu^*)^\nu$	ρ^*	# Add	# Mult	ρ^*	# Add	# Mult
1	0.500	0.400	12	3	0.447	7	2.25
2	0.250	0.193	16	4	0.200	11	3.25
3	0.125	0.119	20	5	0.089	15	4.25
4	0.063	0.084	24	6	0.042	19	5.25

Table 9.1a: μ^*, ρ^* and computational work in case of smoothing by GS relaxation (for 5-point Laplace discretization)

	I_h^{2h} : FW			I_h^{2h} : INJ		
(ν_1,ν_2)	σ_S^*	σ_E^*	σ_d^*	σ_S^*	σ_E^*	σ_d^*
(1,0)	0.447	0.447	1.000	∞	∞	1.067
(0,1)	1.000	0.447	0.447	∞	∞	1.743
(2,0)	0.208	0.232	1.000	∞	∞	1.006
(1,1)	0.203	0.202	0.203	∞	∞	0.356
(0,2)	1.000	0.232	0.208	∞	∞	1.733
(3,0)	0.128	0.183	1.000	∞	∞	1.001
(2,1)	0.119	0.119	0.131	∞	∞	0.132
(1,2)	0.131	0.119	0.119	∞	∞	0.335
(0,3)	1.000	0.183	0.128	∞	∞	1.732
(4,0)	0.091	0.156	1.000	∞	∞	1.000
(3,1)	0.084	0.084	0.106	∞	∞	0.103
(2,2)	0.084	0.084	0.084	∞	∞	0.112
(1,3)	0.106	0.084	0.084	∞	∞	0.334
(0,4)	1.000	0.156	0.091	∞	∞	1.732

Table 9.1b: Norm values corresponding to the methods in Table 9.1a.

Using a relaxation parameter $\omega \neq 1$ in the methods considered above does not improve the two-grid results. The reason for introducing a parameter in the classical approach to SOR is to speed up the relaxation method as an iterative solver by itself: mainly the reduction of low frequencies is improved by a parameter. The reduction of high frequencies (and by that the smoothing property) usually becomes even worse. Figure 9.1 shows graphically the dependence of μ^* and ρ^* (for FW) on the choice of ω. The optimal value of ω is not exactly 1, but very close to 1. (The gain in convergence speed by using the optimal ω would be only very small and does not pay if the additional work (2 operations per point and smoothing step) is taken into account.)

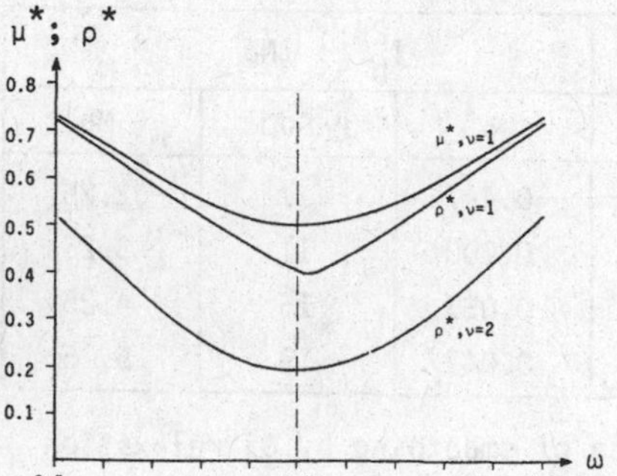

Figure 9.1: μ^*, ρ^* as a function of ω for the method as considered in Table 9.1a (using FW)

Table 9.2 shows two-grid quantities for the 9-point "Mehrstellen" discretization (8.20) using GS and red-black GS (see Example (6) in the previous section) as smoothing methods. Both methods are combined with reasonable fine-to-coarse transfer operators. From the numbers given in Table 9.2, we see that it is advisable to order the grid points in a red-black manner rather than simply lexicographically and use either the HW operator (with $\nu_1=2$, $\nu_2=1$) or the FW operator. Except for FW, we again obtain $\sigma_S^*=\sigma_E^*=\infty$. Compared to the corresponding methods in Table 8.2, we see that red-black GS is slightly preferable to RB. RB has, however, the advantage of being analyzable in the framework of model problem analysis. FC relaxation still yields the most efficient method of all and is also analyzable by the model problem analysis. Many more considerations on the Mehrstellenverfahren, both theoretical and practical, can be found in [12]. In particular, the influence of relaxation parameters is investigated there. Again, the main result is that the use of relaxation parameters in combination with RB, FC, GS and red-black GS does not pay if numerical work is taken into account.

		$(\nu_1,\nu_2)=(1,1)$			$(\nu_1,\nu_2)=(2,1)$		
Relax	I_h^{2h}	$(\mu^*)^\nu$	ρ^*	σ_S^*	$(\mu^*)^\nu$	ρ^*	σ_S^*
GS	FW	0.223	0.134	0.179	0.105	0.075	0.076
	HW		0.141	∞		0.061	∞
	INJ		0.178	∞		0.075	∞
red-black GS	FW	0.060	0.044	0.12	0.027	0.029	0.042
	HW		0.10	∞		0.029	∞
	HI		0.13	∞		0.097	∞

Table 9.2: Results for the 9-point "Mehrstellen"-operator

Tables 9.4a and 9.4b refer to the anisotropic model operator in (8.22) and contain results obtained by using line GS and alternating line GS as smoothers. They are to be compared with Tables 8.4a and 8.4b, respectively. They demonstrate that the use of ZEBRA and alternating ZEBRA is considerably preferable to line GS and alternating line GS, respectively, both with respect to convergence speed and to numerical work. In fact, the relation of ZEBRA to line GS is here quite similar to that of RB to GS (for the Laplace operator). In particular, the use of relaxation parameters does not improve the methods essentially any further.

	$(\nu_1,\nu_2)=(1,0)$				$(\nu_1,\nu_2)=(1,1)$			
ε	$(\mu^*)^\nu$	ρ^*	σ_S^*	σ_E^*	$(\mu^*)^\nu$	ρ^*	σ_S^*	σ_E^*
1000	0.998	0.998	1.41	0.998	0.996	0.996	1.41	0.996
100	0.980	0.980	1.36	0.980	0.961	0.961	1.33	0.961
10	0.833	0.833	1.00	0.833	0.694	0.694	0.878	0.694
2	0.500	0.500	0.500	0.500	0.250	0.250	0.351	0.250
1	0.447	0.333	0.471	0.447	0.200	0.134	0.351	0.200
0.5	0.447	0.333	0.471	0.447	0.200	0.134	0.351	0.200
0.1	0.447	0.333	0.471	0.447	0.200	0.134	0.351	0.200
0.01	0.447	0.333	0.471	0.447	0.200	0.134	0.351	0.200
0.001	0.447	0.333	0.471	0.447	0.200	0.134	0.351	0.200
$W_h^{2h}/\mathcal{N}_h$	# Add: 11.75, # Mult: 5.5				# Add: 15.75, # Mult: 8.5			

Table 9.4a: Results for the anisotropic model operator using x_2-line GS relaxation ($\mathcal{I}_h^{2h}$: FW restriction, $\mathcal{I}_{2h}^h$: linear interpolation)

	$(\nu_1,\nu_2)=(1,0)$				$(\nu_1,\nu_2)=(1,1)$			
ε	$(\mu^*)^\nu$	ρ^*	σ_S^*	σ_E^*	$(\mu^*)^\nu$	ρ^*	σ_S^*	σ_E^*
1000	0.446	0.332	0.469	0.446	0.199	0.133	0.35	0.199
100	0.438	0.321	0.453	0.438	0.192	0.126	0.337	0.192
10	0.373	0.238	0.373	0.373	0.139	0.088	0.245	0.139
2	0.223	0.111	0.223	0.236	0.050	0.047	0.112	0.055
1	0.149	0.086	0.149	0.183	0.022	0.042	0.067	0.042
0.5	0.223	0.111	0.223	0.236	0.050	0.047	0.112	0.055
0.1	0.373	0.238	0.373	0.373	0.139	0.088	0.245	0.139
0.01	0.438	0.321	0.453	0.438	0.192	0.126	0.337	0.192
0.001	0.446	0.332	0.469	0.446	0.199	0.133	0.35	0.199
$W_h^{2h}/\mathcal{N}_h$	# Add: 15.75, # Mult: 8.5				# Add: 23.75, # Mult: 14.5			

Table 9.4b: As in Table 9.4a with alternating line GS instead of x_2-line GS

The discrete problems considered so far are model problems in the sense of Section 7.1. This means that appropriate multigrid methods can be constructed for them which allow the application of the analysis of Chapter 7. This is different for the differential operator

$$L^{\Omega} = -\Delta u - \varepsilon u_{x_1 x_2}, \tag{9.28}$$

since this operator cannot be consistently discretized by a symmetric difference star. In Table 8.5, we show results for the discretization

$$\mathcal{L}_h \hat{=} \frac{1}{h^2} \begin{bmatrix} \frac{\varepsilon}{4} & -1 & -\frac{\varepsilon}{4} \\ -1 & 4 & -1 \\ -\frac{\varepsilon}{4} & -1 & \frac{\varepsilon}{4} \end{bmatrix}_h \tag{9.29}$$

using local Fourier analysis. For $\varepsilon=0$, we have just the Laplace operator; if $|\varepsilon|\geq 2$ the operator (9.28) is no longer elliptic. We see from the table that - for the values $\varepsilon\neq 0$ shown - RB and GS behave much the same way. In particular, $\nu=1$ already yields the best values. The efficiency of both methods deteriorates considerably for $|\varepsilon|\to 2$. To improve the efficiency for this problem, one should take the characteristic unsymmetry of the difference star into account. Also, the use of ZEBRA or alternating ZEBRA relaxation does not lead to considerably more efficient methods as these smoothers are primarily suitable for unsymmetries with respect to the axes. We are not going to discuss the problem of optimal smoothing in more detail here. See, however, the use of ILU-smoothing below.

	$(\nu_1,\nu_2)=(1,0)$				$(\nu_1,\nu_2)=(1,1)$			
ε	$(\mu^*)^\nu$	ρ^*	σ_S^*	σ_E^*	$(\mu^*)^\nu$	ρ^*	σ_S^*	σ_E^*
-1.7	0.678	0.623	0.656	0.676	0.460	0.543	0.563	0.546
-1.5	0.651	0.569	0.600	0.620	0.424	0.461	0.474	0.464
-1.0	0.591	0.480	0.514	0.523	0.350	0.325	0.332	0.329
0.0	0.500	0.400	0.447	0.447	0.250	0.193	0.203	0.202
1.0	0.538	0.467	0.557	0.538	0.289	0.260	0.361	0.302
1.5	0.568	0.509	0.675	0.619	0.342	0.420	0.550	0.450
1.7	0.610	0.588	0.760	0.678	0.372	0.516	0.676	0.540

Table 9.5a: Results for the operator (9.29) using GS relaxation ($\mathcal{I}_h^{2h}$: FW restriction; $\mathcal{I}_{2h}^{h}$: linear interpolation)

	$(\nu_1,\nu_2)=(1,0)$				$(\nu_1,\nu_2)=(1,1)$			
ε	$(\mu^*)^\nu$	ρ^*	σ_S^*	σ_E^*	$(\mu^*)^\nu$	ρ^*	σ_S^*	σ_E^*
-1.7	0.568	0.596	0.772	0.698	0.324	0.539	0.696	0.561
-1.5	0.520	0.521	0.686	0.637	0.271	0.443	0.576	0.471
-1.0	0.405	0.363	0.559	0.513	0.163	0.226	0.364	0.301
0.0	0.250	0.250	0.559	0.500	0.063	0.074	0.141	0.125
1.0	0.405	0.363	0.559	0.513	0.163	0.266	0.364	0.301
1.5	0.520	0.521	0.686	0.637	0.271	0.443	0.576	0.471
1.7	0.568	0.596	0.772	0.698	0.324	0.539	0.696	0.561

Table 9.5b: Analogous to Table 9.5a with RB instead of GS.

There are smoothing methods which are not contained in the general description of relaxation methods given in Section 9.1. For example, *ILU-smoothing* has attracted attention recently and is supposed to be distinguished by its "robustness" [110],[60]. We shortly describe how ILU-smoothing can be treated in the framework of local Fourier analysis. As for concrete algorithms, see [110].

For a given discrete difference operator $\mathcal{L}_h$ (9.5), an imcomplete ILU decomposition is generally described by a splitting

$$\mathcal{L}_h = \mathcal{L}_h^L \mathcal{L}_h^U + \mathcal{L}_h^R \tag{9.30}$$

with certain "lower" and "upper" difference operators $\mathcal{L}_h^L$, $\mathcal{L}_h^U$ and a "remainder" operator $\mathcal{L}_h^R$.

Examples: For a given 5-point operator

$$\mathcal{L}_h \mathrel{\hat{=}} \begin{bmatrix} 0 & * & 0 \\ * & * & * \\ 0 & * & 0 \end{bmatrix}_h$$

it is reasonable to use a "7-point" ILU-decomposition [110]:

$$\mathcal{L}_h^L \mathrel{\hat{=}} \begin{bmatrix} 0 & 0 & 0 \\ * & * & 0 \\ 0 & * & * \end{bmatrix}_h, \quad \mathcal{L}_h^U \mathrel{\hat{=}} \begin{bmatrix} * & * & 0 \\ 0 & * & * \\ 0 & 0 & 0 \end{bmatrix}_h, \quad \mathcal{L}_h^R \mathrel{\hat{=}} \begin{bmatrix} * & 0 & 0 & 0 & 0 \\ 0 & 0 & 0 & 0 & 0 \\ 0 & 0 & 0 & 0 & * \end{bmatrix}_h. \tag{9.31}$$

For a 9-point operator

$$\mathcal{L}_h \hat{=} \begin{bmatrix} * & * & * \\ * & * & * \\ * & * & * \end{bmatrix}_h$$

it is convenient to use a "9-point" ILU-decomposition:

$$\mathcal{L}_h^L \hat{=} \begin{bmatrix} 0 & 0 & 0 \\ * & * & 0 \\ * & * & * \end{bmatrix}_h, \quad \mathcal{L}_h^U \hat{=} \begin{bmatrix} * & * & * \\ 0 & * & * \\ 0 & 0 & 0 \end{bmatrix}_h, \quad \mathcal{L}_h^R \hat{=} \begin{bmatrix} * & 0 & 0 & 0 & 0 \\ * & 0 & 0 & 0 & * \\ 0 & 0 & 0 & 0 & * \end{bmatrix}_h. \tag{9.32}$$

Here " * " stands for certain non-zero elements. These decompositions are unique if, for example, the center element of $\mathcal{L}_h^L$ is scaled to one.

We denote the eigenvalues of $\mathcal{L}_h, \mathcal{L}_h^L$ and $\mathcal{L}_h^U$ (cf. (9.6)) by

$$\lambda(\theta,h), \quad \lambda^L(\theta,h), \quad \lambda^U(\theta,h) \quad (-\pi < \theta \le \pi),$$

respectively. In terms of errors v_h, $\bar{v}_h$ (before and after one ILU smoothing step), we obtain

$$\mathcal{L}_h^L \mathcal{L}_h^U \bar{v}_h = -\mathcal{L}_h^R v_h. \tag{9.33}$$

Thus, the corresponding smoothing operator is given by

$$\mathcal{S}_h = -(\mathcal{L}_h^L \mathcal{L}_h^U)^{-1} \mathcal{L}_h^R. \tag{9.34}$$

With respect to the basis (9.2), $\mathcal{S}_h$ is represented by

$$\mathcal{S}_h \varphi(\theta,x) = A(\theta,h)\varphi(\theta,x) \ (-\pi<\theta\le\pi) \quad \text{with} \quad A(\theta,h) := \frac{\lambda^L(\theta,h)\lambda^U(\theta,h) - \lambda(\theta,h)}{\lambda^L(\theta,h)\lambda^U(\theta,h)}. \tag{9.35}$$

Using this representation, local mode analysis can be applied in connection with ILU-smoothing in a straightforward manner.

Table 9.6a shows results for the <u>anisotropic model operator</u> in (8.22) using 7-point ILU for smoothing. As transfer operators, the 7-point operators (4.31) are used. The coarse-grid difference operator $\mathcal{L}_{2h}$ is assumed to be the same as $\mathcal{L}_h$ with $2h$ instead of h. (We note that this operator coincides with the Galerkin operator $\mathcal{L}_{2h} := \mathcal{I}_h^{2h} \mathcal{L}_h \mathcal{I}_{2h}^h$ here.)

The results of Tables 9.6a show, in particular, that the convergence properties of the above method are not "symmetric" with respect to the size of ε: We have good convergence for $\varepsilon \le 1$ but ρ^* tends to 1 for $\varepsilon \to \infty$. This behavior is similar to the one observed in connection with x_2-ZEBRA relaxation in Table 8.4a. There, however, the dependence of ρ^* on the size of ε was more critical: even for values of ε which are only slightly larger than 1, one has to use x_1-ZEBRA relaxation instead of x_2-ZEBRA relaxation. If ZEBRA relaxation is used with lines in the appropriate direction, we see that ZEBRA clearly is superior to ILU.

On the other hand, in practice the less sensitive ε-dependence of ILU-smoothing has advantages if ε depends on x and if $\varepsilon(x)>1$ and $\varepsilon(x)<1$ in different parts of Ω: If ε does not become too large, ILU-smoothing will still be satisfactory. If we wanted to use ZEBRA relaxation, we would have to apply it in the alternating version. Table 9.6b gives a comparison of the computational work for ZEBRA and ILU in case of arbitrary 5-point difference operators (with variable coefficients). From this operation count and the convergence behavior shown in Tables 8.4b and 9.6a, one sees that for values of ε which are not too large both methods, ILU and alternating ZEBRA, are comparable in efficiency. The asymptotic convergence behavior of alternating ZEBRA, however, is completely symmetric with respect to ε. (Of course, ILU could be applied in an alternating manner also.)

For the FMG method (cf. Chapter 6) applied to anisotropic problems, we note that the use of alternating ZEBRA is slightly preferable over ILU: For FMG the pre-computations contained in Table 9.6b can no longer be neglected. This results in more efficient algorithms if alternating ZEBRA is used (in particular, if it is compared with "alternating ILU" which needs twice the amount of pre-computations as shown in Table 9.6b), [100].

	$(\nu_1,\nu_2)=(1,0)$			$(\nu_1,\nu_2)=(1,1)$		
ε	$(\mu^*)^\nu$	ρ^*	σ_S^*	$(\mu^*)^\nu$	ρ^*	σ_S^*
1000	0.843	0.843	1.20	0.710	0.710	1.10
100	0.607	0.607	0.858	0.368	0.368	0.710
10	0.273	0.273	0.386	0.075	0.095	0.287
2	0.146	0.128	0.146	0.021	0.034	0.119
1	0.126	0.126	0.131	0.016	0.027	0.085
0.5	0.144	0.144	0.144	0.021	0.022	0.071
0.1	0.165	0.165	0.165	0.027	0.027	0.067
0.01	0.171	0.171	0.171	0.029	0.029	0.068
0.001	0.171	0.172	0.172	0.029	0.029	0.068
$W_h^{2h}/\mathcal{N}_h$	# Add: 12.5, # Mult: 11			# Add: 20.5, # Mult: 19		

Table 9.6a: Results for the anisotropic model operator using 7-point ILU (9.31) for smoothing (and 7-point restriction and interpolation (4.31))

	pre-computations per point of fine grid			$W_h^{2h}/\mathcal{N}_h$ for $(\nu_1,\nu_2)=(1,0)$		$W_h^{2h}/\mathcal{N}_h$ for $(\nu_1,\nu_2)=(1,1)$	
method	# Add	# Mult	# Div	# Add	# Mult	# Add	# Mult
ZEBRA	1	2	1	8.25	8.5	12.25	13.5
alt.ZEBRA	2	4	2	12.25	13.5	20.5	23.5
ILU	5	12	1	12.5	12	20.5	21

Table 9.6b: Comparison of computational work for arbitrary 5-point difference operators (variable coefficients) and the two-grid methods used in Table 8.4a, 8.4b and 9.6a, respectively.

In Table 9.7, we show results for (9.29) corresponding to Table 9.5: the smoothing methods used there have simply been replaced by 9-point ILU smoothing (9.32). The results of Table 9.7 show that this method behaves considerably better for $\varepsilon \geq 0$. For $\varepsilon < 0$, it behaves similarly to those of Table 9.5. This "unsymmetric behavior" is due to a corresponding unsymmetry in the ILU-decomposition: The convergence behavior shown in Table 9.7 is just reversed if we use a 9-point decomposition of the form

$$\mathcal{L}_h^L \triangleq \begin{bmatrix} * & * & 0 \\ * & * & 0 \\ * & 0 & 0 \end{bmatrix}_h, \quad \mathcal{L}_h^U \triangleq \begin{bmatrix} 0 & 0 & * \\ 0 & * & * \\ 0 & * & * \end{bmatrix}_h, \quad \mathcal{L}_h^R \triangleq \begin{bmatrix} 0 & * & * \\ 0 & 0 & 0 \\ 0 & 0 & 0 \\ 0 & 0 & 0 \\ * & * & 0 \end{bmatrix}_h$$

instead of (9.32). Clearly, both decompositions can be used for smoothing in an alternating manner, giving a robust method for boundary value problems involving (9.29).

	$(\nu_1,\nu_2)=(1,0)$		$(\nu_1,\nu_2)=(1,1)$	
ε	ρ^*	σ_S^*	ρ^*	σ_S^*
-1.9	0.573	2.41	0.513	2.40
-1.7	0.402	0.775	0.307	0.740
-1.5	0.300	0.451	0.200	0.417
-1.0	0.165	0.214	0.083	0.172
0.0	0.126	0.126	0.021	0.021
1.0	0.075	0.076	0.030	0.038
1.5	0.083	0.088	0.047	0.049
1.7	0.107	0.108	0.061	0.062
1.9	0.142	0.142	0.084	0.084

Table 9.7: Results for the operator (9.29) using 9-point ILU-smoothing (9.32) (J_h^{2h}: FW restriction, $\mathcal{I}_{2h}^h$: linear interpolation)

9.3 A short discussion of other theoretical approaches

As mentioned already in the introduction, all theoretical approaches to multigrid methods known so far have some specific disadvantages:

- Whereas the model problem analysis can be directly applied only to a rather small class of problems, the local Fourier analysis refers to certain "ideal" situations.
- For general problems, rigorous results are obtained by several more abstract approaches. These show the h-independent convergence behavior of multigrid methods in principle, but only under algorithmic assumptions or with constants, which are (very) unrealistic in concrete cases. Usually, these general approaches do not give the full impression about the high efficiency of multigrid methods, nor can they be used for the optimal (or even a reasonable) arrangement of algorithms.

As in the model problem and local Fourier analysis, in any proof for h-independent convergence of two-grid methods one has to distinguish the effects of the methods on low- and on high-frequency error components. The proofs differ, among other things, with respect to the definition of "low" and "high" and in the way how the above effects are measured.

We want to sketch two of the general approaches here and make some additional remarks. For simplicity, we restrict ourselves to (h,2h) two-grid methods.

9.3.1 Splitting of the two-grid operator norm into a product

In Hackbusch's approach (see [50] and the references cited there) the above effects are measured implicitly by use of a certain splitting of $\| M_h^{2h} \|$ (for a suitably chosen norm) into a product. For $\nu_1=\nu$, $\nu_2=0$ this splitting is of the form:

$$\| M_h^{2h} \| \le \| K_h^{2h} L_h^{-1} \| \cdot \| L_h S_h^{\nu} \| . \qquad (9.36)$$

The following properties are assumed to hold:

(1) approximation property:

$$\| K_h^{2h} L_h^{-1} \| = \| L_h^{-1} - I_{2h}^{h} L_{2h}^{-1} I_h^{2h} \| \le Ch^{\delta} \quad (\delta > 0), \qquad (9.37)$$

(2) smoothing property:

$$\| L_h S_h^{\nu} \| \le \eta(\nu) h^{-\delta} \quad \text{with} \quad \eta(\nu) \to 0 \quad (\nu \to \infty). \qquad (9.38)$$

Under these assumptions, the h-independent boundedness of $\| M_h^{2h} \|$ (by a constant less than one) immediately follows for sufficiently large ν. Hackbusch was able to verify the assumptions (1) and (2) (or suitable generalizations) for quite general classes of problems. In particular, for second order problems with sufficient regularity, (1) and (2) can be fulfilled - by a suitable choice of MG components - with $\|\cdot\| = \|\cdot\|_S$, $\delta=2$ and $\eta(\nu) \sim 1/\nu$. For details see [50], Chapter 3.

Thus Hackbusch's approach is very satisfactory from a theoretical point of view. With respect to its quantitative part, however, there are several limitations and disadvantages in this theory.

- Firstly, in its simplest form the theory is an asymptotic theory, in particular, with respect to ν. If the constant C in (9.37) is large, this has to be compensated by choosing ν suffiently large. Thus, in general, this theory does not allow one to prove the h-independent convergence of a given two-grid method (with fixed, small ν).
- In its general form, the theory does not distinguish between very different situations (e.g. Poisson-like operators, anisotropic operators, singular perturbed operators, highly indefinite cases, etc.). We know, however, that multigrid algorithms have to be arranged quite differently for such different situations in order to become efficient. In Hackbusch's theory, these differences may be reflected by the size of the constants occuring. If one wants to obtain results which take specific situations into account, one has to choose - often very complicated - special norms, discrete spaces etc. (For a rather simple example, see [50], Chapter 6.)
- The theory directly applies only to norms, not to the spectral radius of a two-grid iteration operator. Thus, in any case, the suitable choice of norms is a substantial requirement of this approach. There are, however, practically important multigrid methods for which $\| M_h^{2h} \|_S$ is unbounded as $h \to 0$ (see Chapter 8 and the previous section). These methods cannot be treated with Hackbusch's approach (at least not in the spectral or energy norm). In particular, the use of "straight injection" in the fine-to-coarse transfer causes this difficulty.
- In (1) and (2) above, only $\nu_2=0$ (or $\nu_1=0$) are suitable choices of ν_1, ν_2. In practice, however, multigrid methods with $\nu_1>0$ <u>and</u> $\nu_2>0$ usually have advantages: see, e.g., Tables 3.3, 8.1 and [25], Section 6.1 (also cf. [50], Section 3.5).

<u>Example</u>: We consider model problem (P) and the two-grid method which was treated in Theorem 8.1. If we were to specialize the general estimates of Hackbusch's theory to this situation, the constants occuring would become by far too large. But even if we compute the optimal norm bounds in (1) and (2) (with $\delta=2$ and $\|\cdot\| = \|\cdot\|_S$) by use of Fourier representations, the resulting bounds for $\| M_h^{2h} \|_S$ differ considerably from ρ^* (cf. Section 8.1). Table 9.8 compares corresponding norm bounds for $\|M_h^{2h}\|_S$

with ρ^* for different values of ν. The optimal value of ν for the method at hand was found to be $\nu=2$ in Section 8.2 giving $\rho^*=0.074$. Table 9.8 shows that 10 relaxation steps are needed to achieve a bound for $\| M_h^{2h} \|_S$ of this size.

ν	1	2	3	4	...	10
$\rho^*(\nu)$	0.250	<u>0.074</u>	0.053	0.041	...	0.018
$\| M_h^{2h} \|_S \leq$	0.770	0.372	0.247	0.185	...	<u>0.074</u>

Table 9.8: Comparison of $\rho^*(\nu)$ with bounds of $\| M_h^{2h} \|_S$ obtained by Hackbusch's estimation

If in the above two-grid method the FW restriction operator is replaced by the HW restriction operator, we obtain norm values σ_S^* as shown in Table 8.1b. Property (1), however, is not at all satisfied (for $\|\cdot\|=\|\cdot\|_S$ and $\delta=2$).

We finally remark that the above smoothing property (2) (and the factor $\eta(\nu)$ occuring there) is not directly related to the definition of the smoothing factor given in Section 7.5.

9.3.2 Splitting of the two-grid operator norm into a sum

Another theoretical approach starts from a splitting

$$\mathbb{G}(\Omega_h) = \mathbb{G}^{low}(\Omega_h) \oplus \mathbb{G}^{high}(\Omega_h) \tag{9.39}$$

and a corresponding splitting of I_h into a sum of projection operators

$$I_h = Q_h^{low} + Q_h^{high}. \tag{9.40}$$

(Here "low" and "high" do not need to have the same meaning as in the model problem or in the local mode analysis.)

In order to derive norm estimates for $M_h^{2h} = K_h^{2h} S_h^\nu$, the above splitting is used in the following way:

$$M_h^{2h} = K_h^{2h} (Q_h^{low} + Q_h^{high}) S_h^\nu \tag{9.41}$$

which is estimated by

$$\| M_h^{2h} \| \leq \| K_h^{2h} Q_h^{low} \| \cdot \| S_h^\nu \| + \| K_h^{2h} \| \cdot \| Q_h^{high} S_h^\nu \|. \tag{9.42}$$

This estimate can be regarded as the basis of Wesseling's approach [108]. Under reasonable circumstances, $\| S_h^\nu \|$ and $\| K_h^{2h} \|$ can be assumed to be bounded by constants

C and C' (independent of h). The idea then is to make - again independently of h - $\| K_h^{2h} Q_h^{low} \|$ small by a suitable definition of "low" and to make $\| Q_h^{high} S_h^{\nu} \|$ small by choosing a sufficiently large number ν of relaxation steps.

One obvious disadvantage of this approach is the following. Once the term "low" (i.e. Q_h^{low}) is defined, the first summand in (9.42) (more precisely: its supremum over h for fixed meshsize ratio q^*) is bounded from below, independently of ν. In particular, an h-independent bound for $\| M_h^{2h} \|$ derived from (9.42) then is bounded from below, independently of ν, also. This is, however, not realistic: typically $\| M_h^{2h} \|$ is proportional to $1/\nu$ for $\nu \to \infty$ (see the corresponding remark in the previous section).

<u>Example</u>: Consider model problem (P) and the sample method in Section 3.1. If we use the natural definition of "low" given in (3.15) one can compute explicitly (using the Fourier representations given in Chapter 3):

$$\sup_h \| K_h^{2h} Q_h^{low} \|_S = 0.901\ldots, \qquad \sup_h \| S_h^{\nu} \|_S = 1 \quad (\nu=1,2,\ldots). \tag{9.43}$$

Wesseling gives bounds for the single terms in (9.42) which are valid for rather general situations. For special situations, however, these bounds turn out to be very unrealistic. Starting from (9.42), Wesseling estimates

$$\| K_h^{2h} Q_h^{low} \| \le \| K_h^{2h} L_h^{-1} \| \cdot \| L_h Q_h^{low} \| , \; \| K_h^{2h} \| \le \| K_h^{2h} L_h^{-1} \| \cdot \| L_h \| . \tag{9.44}$$

By this, the approximation property (9.37) enters again. The further estimates of Wesseling's proof - which correspond to a very restrictive definition of "low" - are aimed at obtaining a small bound for the right hand side of the first inequality in (9.44). Applied to model problem (P) and the sample method in Section 3.1 (with the exception that the Galerkin operator (2.21) is used on the coarser grid), this definition means that only frequencies (7.4) with - roughly - $|\theta| < 10^{-3}$ are regarded as low frequencies. As a consequence, millions of relaxation steps have to be carried out per cycle in order to make $\| Q_h^{high} S_h^{\nu} \|$ sufficiently small (cf. the corresponding remark in [108]).

This extremely unsatisfactory result in due to the fact that the definition of "low" frequencies is not natural here: The definition used is not oriented to the ratio of the meshsizes h and 2h, but rather to the technique of the proof.

9.3.3 <u>Further remarks on the definition of "low" and "high"</u>

(1) Another interesting way to measure low and high frequencies is contained in Braess' approach [13]. This approach refers to red-black coarsening $(H=\sqrt{2}h)$. Apart

from the energy norm $\|\cdot\|_E$ on $\mathbb{G}(\Omega_h)$, Braess introduces a certain semi-norm $|\cdot|$ on $\mathbb{G}(\Omega_h)$ which takes only grid points of Ω_H into account (see [13], Section 2). Then the quotient $\lambda := |\cdot|/\|\cdot\|_E$ turns out to be a measure for the smoothness of Ω_h-grid functions with respect to Ω_H ($\lambda \sim 0$ corresponds to "high" and $\lambda \sim 1$ corresponds to "low" frequencies).

Using this measure in connection with a strenghtened Cauchy inequality, Braess obtains (not optimal, but) very good h-independent norm bounds for an MGR-method for Poisson equation on certain convex domains. These results were extended to methods using standard coarsening by Verfürth [106].

(2) Finally, we mention that in McCormick's algebraic approach to multigrid methods [70], the elements of the nullspace of I_h^{2h} can essentially be regarded as "high" frequencies.

10. Multigrid programs for standard applications

On the basis of the model problem analysis, two collections of programs for certain second order elliptic equations have been developed. We refer to these collections as MGØØ and MGØ1.

The MGØØ programs are restricted to rectangular domains (with different types of boundary conditions). The sample program listed in the Appendix (for Poisson's equation with Dirichlet boundary conditions in the unit square) is a very special program of the MGØØ collection. More information about MGØØ and, in particular, a systematic comparison with several direct Fast Elliptic Solvers and related methods is given by Foerster, Witsch [36].

The MGØ1 programs refer to second order equations with Dirichlet boundary conditions on non-rectangular bounded domains. The main objective in the development of these programs was to study the influence of the shape of the domain on the behavior of several multigrid algorithms. Because of the complexity of programs for problems on general domains and with respect to possible generalizations, special emphasis was laid upon the structural transparency of these programs; efficiency was only our second objective. (The technical structure of the MGØ1 programs is partially adopted from a similar program contained on MUGTAPE, see [25]).

Current versions of the MGØ1 collection refer to the equations

(1) $-\Delta u + c(x)u = f(x),$

(2) $-a_1(x)u_{x_1x_1} - a_2(x)u_{x_2x_2} + c(x)u = f(x),$

(3) $-a_1(x)u_{x_1x_1} - a_2(x)u_{x_2x_2} + b_1(x)u_{x_1} + b_2(x)u_{x_2} + c(x)u = f(x),$

(4) generalizations to $f(x,u)$.

In Section 10.1, we give a short description of admissable domains and the discretization. Sections 10.2 and 10.3 refer to the cases (1) and (2), respectively. As for (2), we are especially interested in the case of strongly differing coefficients a_1 and a_2. In Section 10.4, only a few remarks are made with respect to the technically more complicated case (3). Nonlinear applications (4) have been reported upon already in Chapter 5.

10.1 Description of domains and discretization

For technical reasons only, MGØ1 has, up to now, been restricted to domains Ω which can be described by two values A, B and two functions $FL(x_1)$ (F-low) and

$FH(x_1)$ (F-high) in the following way:

$$\Omega = \{(x_1,x_2) : A < x_1 < B,\ FL(x_1) < x_2 < FH(x_1)\}.$$

Here the functions FL and FH may have a finite number of jumps (cf. Figure 10.1). For the discretization, a grid $\Omega_h \cup \Gamma_h$ is used consisting of the set Ω_h of *interior grid points* and the set Γ_h of *boundary grid points*. Here Ω_h is defined by

$$\Omega_h := \Omega \cap G_h(x^*),\quad G_h(x^*) := \{x = x^* + h\kappa : \kappa \in \mathbb{Z}^2\},\quad h = h_{x_1} = h_{x_2}$$

with grid origin $x^* \in \mathbb{R}^2$. The boundary grid points are just the intersection of Γ with grid lines of $G_h(x^*)$.

We denote by

- *regular interior grid points:* all points of Ω_h, whose neighboring grid points in northern, southern, western and eastern direction are also points of Ω_h;
- *irregular interior grid points:* all interior grid points which are not regular.

At regular interior grid points, the differential operator L is approximated by the usual 5-point discretization (order of consistency = 2), replacing e.g.

$$u_{x_1x_1} \text{ by } \frac{1}{h^2}[1\ \ -2\ \ 1]_h, \qquad u_{x_1} \text{ by } \frac{1}{2h}[-1\ \ 0\ \ 1]_h.$$

Near the boundary, i.e. in the irregular grid points, a 5-point Shortley-Weller approximation [89] is used. This means, for example, that

$$a_1(x)u_{x_1x_1} + a_2(x)u_{x_2x_2}$$

is replaced by

$$2\begin{bmatrix} & \dfrac{a_2(x)}{h_N(h_N+h_S)} & \\ \dfrac{a_1(x)}{h_W(h_W+h_E)} & -\dfrac{a_1(x)}{h_Wh_E}+\dfrac{a_2(x)}{h_Nh_S} & \dfrac{a_1(x)}{h_E(h_W+h_E)} \\ & \dfrac{a_2(x)}{h_S(h_N+h_S)} & \end{bmatrix} u(x).$$

Here h_N, h_S, h_W, and h_E denote the distances to the neighboring grid points in northern, southern, western and eastern direction, respectively.

So far, we have not yet made any assumptions about the coefficients a_1, a_2, b_1, b_2, c. Since only elementary multigrid techniques will be considered in the following sections, certain degeneracies have to be excluded. We want to mention these restrictions here only qualitatively:

- the problem is assumed to be elliptic and "nicely definite";
- all coefficients should have some "smoothness", e.g. considerable jumps in size are not admitted;
- the ratios of a_1, b_1 and of a_2, b_2, respectively, should be such that a singularly perturbed behavior (of the discrete problem) is not likely and that the central differencing of the first derivatives leads to no instability.

If one of these assumptions is not fulfilled, in general a more sophisticated multigrid approach will be necessary (which is adapted to the particular situation at hand, see Section 10.4).

10.2 Helmholtz equation (with variable c)

The discrete problem

$$\begin{aligned} -\Delta_h u_h + c(x)u_h &= f^{\Omega}(x) \quad (x \in \Omega_h) \\ u_h &= f^{\Gamma}(x) \quad (x \in \Gamma_h) \end{aligned} \tag{10.1}$$

is solved by MGØ1 [97]

(1) either by using the underlying multigrid method as an iterative solver (which is especially suitable to investigate the influence of the shape of the domain Ω on the convergence behavior),

(2) or in the full multigrid mode as explained in Chapter 6 (which is the usual way to apply the program in practice).

The main components of MGØ1 used in solving (10.1) iteratively are the following:

- Coarsening: The grid-coarsening is done by doubling the (interior) meshsize in both directions, i.e. $\Omega_{2h} = \Omega \cap G_{2h}(x^*)$. On all grids the discrete operators are constructed in the same way as on the finest grid. The coarsest grid may be "very coarse" (but should be geometrically reasonable, see Figure 10.1).
- Smoothing: For smoothing RB relaxation is used (usually with $\nu_1=2$, $\nu_2=1$).
- Grid transfer: For the fine-to-coarse and coarse-to-fine transfers we use half

injection (HI, see Section 8.1) and linear interpolation, respectively.

- Structure of cycles: Alternatively, V- or W-cycles (see Section 4.1) may be used.

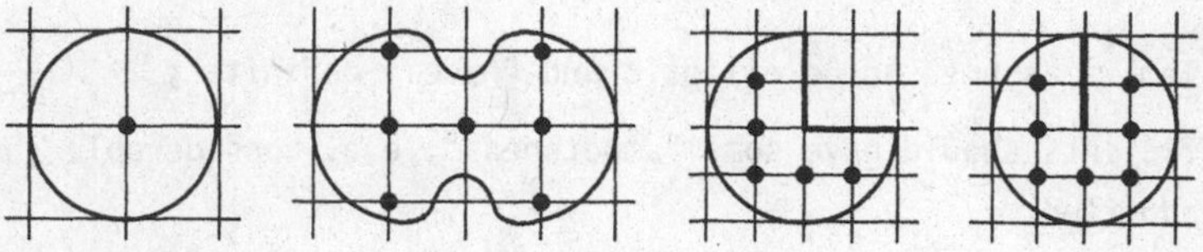

Figure 10.1: "Reasonable" coarsest grids

All following quantitative results refer to Poisson's equation and the MGØ1 version described above with

$$\nu_1 = 2, \nu_2 = 1. \tag{10.2}$$

If $\mathcal{N}$ denotes the number of grid points of Ω_h, the total computational work for one iteration step of the corresponding method is less than

$$\boxed{\begin{array}{lll} 15\mathcal{N} \text{ additions}, & 5\mathcal{N} \text{ multiplications} & \text{(for V-cycles)}, \\ 23\mathcal{N} \text{ additions}, & 7.5\mathcal{N} \text{ multiplications} & \text{(for W-cycles)}, \end{array}} \tag{10.3}$$

neglecting lower order terms. These numbers are independent of the shape of the domain.

Table 10.1 shows some numerically calculated asymptotic convergence factors of the multigrid iteration (for both V- and W-cycles) for several domains. All domains are comparable in size with the unit square. The convergence factors have been computed by a v. Mises vector iteration.

We recall that on rectangular domains for $h_{x_1}=h_{x_2}$ the corresponding asymptotic two-grid convergence factor is given by $\rho^*(3) \approx 0.034$ (see Section 8.2). As far as W-cycles (γ=2) are concerned, Table 10.1 shows that the multigrid convergence factors (for h=1/128) are nearly the same as ρ^*. This is also true for general domains, as long as there are no reentrant corners. The worst convergence factor, namely 0.097 instead of ~0.03, can be observed for the domain with a cut. Here a singular behavior like $\sqrt{R}$ is typical for the solution $u(x)$ near the singular point (where R denotes the distance of x to the singular point). Clearly, for such problems Poisson's equation on a rectangle is no longer a model case. Nevertheless, the W-cycle convergence turns out to be very satisfactory even in such cases. We point out that the given convergence factors remain essentially unchanged if γ is increased ($\gamma \geq 3$). This means that these factors can practically be regarded as the asymptotic two-grid convergence factors.

As expected, the multigrid convergence factors for V-cycles (γ=1) are worse than those for W-cycles. For domains without reentrant corners, the convergence factors are nevertheless so good that the V-cycle efficiency (convergence factor versus computational work) is slightly better than the W-cycle efficiency. In the cases with reentrant corners, however, the V-cycle convergence deteriorates considerably. Altogether, V-cycles may be used just as well as W-cycles at least for "harmless" problems. To be on the safe side, however, we usually prefer to use the more robust W-cycles.

domain	convergence factor	domain	convergence factor
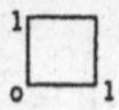	V: 0.059 W: 0.033		V: 0.056 W: 0.033
	V: 0.059 W: 0.033		V: 0.122 W: 0.049
	V: 0.063 W: 0.032		V: 0.182 W: 0.075
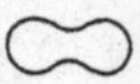	V: 0.058 W: 0.033		V: 0.230 W: 0.097
	V: 0.088 W: 0.033		V: 0.221 W: 0.095

Table 10.1: Numerically computed asymptotic convergence factors per multigrid iteration step for different domains (ν_1=2, ν_2=1, h=1/128) and both V- and W-cycles.

For the full multigrid version we used r=1 W-cycles per level and cubic FMG-interpolation (cf. Section 6.3). In Table 10.2, we give FMG results for various domains. For some values of h, both the error $\| \tilde{u}_h - u_h \|_2$ of the computed FMG solution $\tilde{u}_h$ compared to the solution u_h of the discrete problem (upper values) and the exact discretization error $\| u_h - u \|_2$ (lower values) are shown.

For the examples treated in Table 10.2, the continuous solution is given by

$$u(x) = \sin\pi(x_1+x_2) \qquad \text{for the upper 5 domains}$$

$$u(R,\phi) = \sqrt[\alpha]{R}\,\sin(\phi/\alpha) \qquad \text{for the lower 4 domains}$$

In the latter case, u reflects the singular behavior of solutions which is typical for respective reentrant corners. Here (R,ϕ) denote polar coordinates with respect to the singular point; $\alpha\pi$ is the inner angle of the domain at this point (see Table 10.2).

As we have seen in Table 10.1, the convergence factors of the multigrid iteration deteriorate with increasing α for the domains with reentrant corners $(1 < \alpha \le 2)$. The question arises, whether the corresponding convergence speed is still sufficient for the satisfactory performance of FMG (cf. the influence of η^r in the estimation (6.10)). For this, one should notice that the discretization error becomes larger for increasing α also. This means that the value of κ_1 in (6.7) is smaller than 2 in these cases. More precisely, the following estimate is valid [66]: For any $1 \le \alpha \le 2$ and any $\varepsilon > 0$ there exists a constant C such that

$$|u(R,\phi) - u_h(R,\phi)| \le C\, h^{\frac{2}{\alpha} - 2\varepsilon} R^{-\frac{1}{\alpha} + \varepsilon} = \begin{cases} O(h^{\frac{2}{\alpha} - 2\varepsilon}) & \text{if } R \text{ fixed} \\ O(h^{\frac{1}{\alpha} - \varepsilon}) & \text{if } R = O(h). \end{cases} \tag{10.4}$$

Therefore, the loss of MG convergence speed is - so to say - compensated by a loss of discretization accuracy. The errors given in Table 10.2 show indeed that the main objective of the FMG method, namely to obtain approximate solutions $\tilde{u}_h$ with $\| \tilde{u}_h - u_h \|_2 \le \| u_h - u \|_2$, is achieved for all examples considered. The same is true for highly oscillatory solutions, see Table 10.3.

All results in Tables 10.2 and 10.3 refer to W-cycles. We have computed corresponding errors $\| \tilde{u}_h - u_h \|_2$ for V-cycles also (maintaining $r = 1$). The ratio $\| \tilde{u}_h - u_h \|_2 / \| u_h - u \|_2$ is larger then, but in all cases still ≤ 1. This means, that V-cycles may also be employed in the cases considered.

The <u>total computational work</u> of MGØ1 in the FMG version $(r=1)$ is less than

$22\mathcal{N}$ additions, $8\mathcal{N}$ multiplications (if V-cycles are used),	(10.5)
$32.5\mathcal{N}$ additions, $11.5\mathcal{N}$ multiplications (if W-cycles are used)	

(neglecting lower order terms), where $\mathcal{N}$ is the number of grid points on the finest grid. These numbers are independent of the shape of the domain. In particular, they are the same as for a corresponding special program for rectangular domains. Concerning the real computing times, this special program is, of course, faster than MGØ1 (for a given reasonable $\mathcal{N}$), as in MGØ1 additional work has to be performed due to the more complicated grid structure. (As for computing times concerning programs on rectangular domains, see MGØØ [36].)

domain	h=1/32	h=1/64	h=1/128	solution:
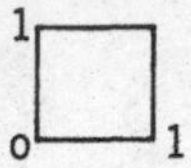	0.157(-5)	0.114(-6)	0.789(-8)	$u(x)=\sin\pi(x_1+x_2)$
	0.194(-3)	0.486(-4)	0.121(-4)	
	0.119(-5)	0.264(-6)	0.300(-7)	
	0.145(-3)	0.368(-4)	0.922(-5)	
	0.562(-6)	0.606(-7)	0.945(-8)	
	0.274(-4)	0.672(-5)	0.166(-5)	
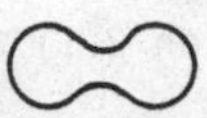	0.445(-5)	0.520(-6)	0.853(-7)	
	0.339(-3)	0.849(-4)	0.213(-4)	
	0.246(-5)	0.453(-6)	0.701(-7)	
	0.129(-3)	0.331(-4)	0.841(-5)	
	0.123(-4)	0.168(-4)	0.174(-5)	$u(x)=\sqrt[\alpha]{R}\,\sin(\phi/\alpha)$
$\alpha=5/4$	0.198(-3)	0.694(-4)	0.239(-4)	
	0.792(-4)	0.363(-4)	0.156(-4)	
$\alpha=3/2$	0.816(-3)	0.337(-3)	0.137(-3)	
	0.234(-3)	0.122(-3)	0.590(-4)	
$\alpha=7/4$	0.201(-2)	0.938(-3)	0.432(-3)	
	0.471(-3)	0.269(-3)	0.144(-3)	
$\alpha=2$	0.384(-2)	0.196(-2)	0.989(-3)	

Table 10.2: FMG results by MGØ1 ($\nu_1=2$, $\nu_2=1$, W-cycles, r=1) for smooth solutions and singular solutions (upper value: $\| \tilde{u}_h - u_h \|_2$, lower value: $\| u_h - u \|_2$

domain	$n_1=9, n_2=6$	$n_1=18, n_2=12$	$n_1=30, n_2=20$	
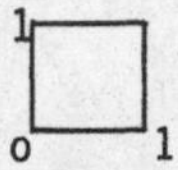	0.634(-5)	0.723(-4)	0.290(-3)	$u(x)=$ $\sin(n_1\pi x_1)\sin(n_2\pi x_2)$
	0.169(-2)	0.679(-2)	0.191(-1)	

Table 10.3: Method as in Table 10.2 for highly oscillatory solutions (h = 1/128)

10.3 Anisotropic operators

The structure of the MGØ1-version for anisotropic problems

$$\begin{aligned} -a_1(x)u_{x_1x_1} - a_2(x)u_{x_2x_2} + c(x)u &= f^{\Omega}(x) \quad (x \in \Omega) \\ u &= f^{\Gamma}(x) \quad (x \in \Gamma) \end{aligned} \tag{10.6}$$

$(a_1(x) > o,\ a_2(x) > o)$ is the same as for the Helmholtz equation with the following exceptions:

- RB relaxation is replaced by ZEBRA or - if necessary - by alternating ZEBRA relaxation;
- The HI restriction operator is replaced by the FW operator (see Section 3.1) which becomes particulary simple in connection with ZEBRA relaxation.

If $a_1(x)-a_2(x)$ does not change its sign, ZEBRA relaxation is used with lines in a fixed direction (with $\nu_1=1,\ \nu_2=0$ or $\nu_1=\nu_2=1$; cf. Section 8.3), namely

$$\begin{aligned} &x_1\text{-ZEBRA} \quad \text{if } a_1(x) \geq a_2(x), \\ &x_2\text{-ZEBRA} \quad \text{if } a_1(x) \leq a_2(x). \end{aligned} \tag{10.7}$$

If $a_1(x)-a_2(x)$ changes its sign on Ω, alternating ZEBRA relaxation of type (8.23) is used instead (with $\nu_1=1,\ \nu_2=0$ or $\nu_1=\nu_2=1$). (Notice that one smoothing step of alternating ZEBRA costs twice the work of one single x_2- or x_1-ZEBRA step.)

In the following we give some convergence factors for MGØ1 using alternating ZEBRA relaxation: The convergence behavior of this method is very good for any size of a_1 and a_2 and independent of the domain.

Table 10.4 shows some numerically calculated W-cycle convergence factors for (10.6) with constant coefficients

$$a_1(x) \equiv \varepsilon, \quad a_2(x) \equiv 1, \quad c(x) \equiv 0 \tag{10.8}$$

on the unit square $\Omega=(0,1)^2$. For comparison, also the corresponding smoothing factors $(\mu^*)^\nu$ and two-grid convergence factors ρ^* are given in Table 10.4 (cf. Section 8.3). We see that in particular ρ^* predicts the behavior of the multigrid method very well. For $\varepsilon \to \infty$ or $\varepsilon \to o$ all three values approach 0.125 (in the case $\nu_1=1,\ \nu_2=0$) and 0.052 (in the case $\nu_1=1,\ \nu_2=1$).

ε	$\nu_1 = 1, \nu_2 = 0$			$\nu_1 = 1, \nu_2 = 1$		
	$\mu^*(1)$	$\rho^*(1)$	W-cycle	$(\mu^*(2))^2$	$\rho^*(2)$	W-cycle
100	0.122	0.119	0.117	0.051	0.051	0.047
10	0.102	0.082	0.079	0.041	0.038	0.036
2	0.061	0.019	0.018	0.020	0.013	0.012
1	0.048	0.023	0.019	0.014	0.009	0.008
0.5	0.061	0.019	0.018	0.020	0.013	0.012
0.1	0.102	0.082	0.079	0.041	0.038	0.036
0.01	0.122	0.119	0.117	0.051	0.051	0.047

Table 10.4: Comparison of smoothing factors, two-grid convergence factors ρ^* and numerically computed W-cycle convergence factor for alternating ZEBRA relaxation and constant coefficients (10.8) $(\Omega = (0,1)^2,\ h = 1/64)$.

Table 10.5 shows convergence factors for

$$a_1(x) = K^{x_1+x_2-1}, \quad a_2(x) \equiv 1, \quad c(x) \equiv 0 \tag{10.9}$$

instead of (10.8). Obviously, $1/K \le a_1(x) \le K \ (x \in \Omega)$ holds. The given results for $K = 1,2,10$ and 100 show that even for variable $a_1(x)$ the convergence behavior is well predicted by the two-grid convergence factors for constant a_1, if the respective "worst cases" (namely: $\max a_1(x)$ or $\min a_1(x)\ (x \in \Omega)$) are considered: For example, in the case $K = 100$ the corresponding convergence factor 0.097 (for $\nu_1=1, \nu_2=0$) is smaller than the two-grid convergence factor corresponding to the "worst" constant a_1-values (namely $a_1 \equiv 100$ and $a_1 \equiv 1/100$) which is 0.117 (see Table 10.4).

K	$\nu_1 = 1, \nu_2 = 0$	$\nu_1 = 1, \nu_2 = 1$
1	0.019	0.008
2	0.014	0.008
10	0.058	0.025
100	0.097	0.038

Table 10.5: Numerically computed W-cycle convergence factors for alternating ZEBRA relaxation and variable coefficients (10.9) $(\Omega = (0,1)^2,\ h = 1/64)$

Finally, Table 10.6 shows W-cycle convergence factors for some non-rectangular domains (again for $a_2(x) \equiv 1,\ c(x) \equiv 0$). The convergence factors given for the upper three domains behave as in the case of the unit square treated in Table 10.5. The differences in the convergence factors are due to the fact that $a_1(x)$ varies over different scales for the different domains. Again, in all cases the convergence factors listed are smaller than the two-grid convergence factors ρ^* corresponding to the worst constant a_1. The situation for the lower three domains (reentrant corners) is quite similar in spite of the singularity. Only for the last domain (with a cut) does the convergence behavior slow down slightly in comparison to the first domain.

domain	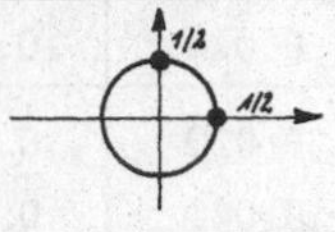	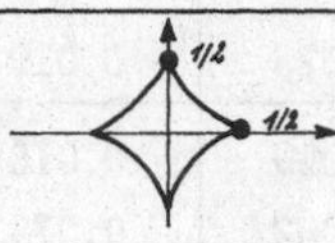	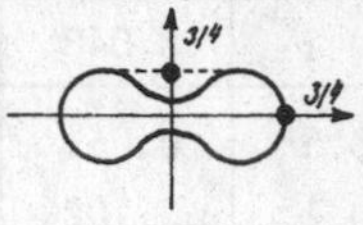
$a_1(x)$	$10^{2(x_1+x_2)}$	$10^{2(x_1+x_2)}$	$10^{4(x_1+x_2)/3}$
$\nu_1=1,\ \nu_2=0$	0.073	0.063	0.093
$\nu_1=1,\ \nu_2=1$	0.031	0.017	0.039
domain	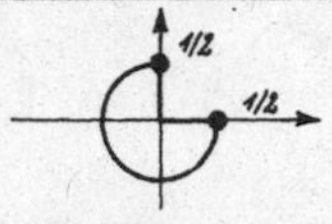		
$a_1(x)$	$10^{2(x_1+x_2)}$	$10^{2(x_1+x_2)}$	$10^{2(x_1+x_2)}$
$\nu_1=1,\ \nu_2=0$	0.065	0.069	0.091
$\nu_1=1,\ \nu_2=1$	0.035	0.041	0.056

Table 10.6: Numerically computed W-cycle convergence factors for alternating ZEBRA relaxation, variable coefficients and different domains

Table 10.7 shows the amount of computational work needed per MG iteration step in the case of arbitrary 5-point difference approximations. The numbers given do not include the computational work required for the decomposition of the occuring tridiagonal systems. These decompositions are regarded as precomputations which have to be performed only once before starting the MG iteration. They amount to

$$2\tfrac{2}{3}\mathcal{N} \text{ additions}, \quad 5\tfrac{1}{3}\mathcal{N} \text{ multiplications}, \quad 2\tfrac{2}{3}\mathcal{N} \text{ divisions}. \tag{10.10}$$

	$\nu_1 = 1,\ \nu_2 = 0$		$\nu_1 = \nu_2 = 1$	
	+/-	*	+/-	*
V-cycle	$16\frac{1}{3}\mathcal{N}$	$18\mathcal{N}$	$27\frac{1}{3}\mathcal{N}$	$31\frac{1}{3}\mathcal{N}$
W-cycle	$24\frac{1}{2}\mathcal{N}$	$27\mathcal{N}$	$41\mathcal{N}$	$47\mathcal{N}$

Table 10.7: Operation count for one step of MG iteration with alternating ZEBRA relaxation for general 5-point discretizations with variable coefficients (neglecting lower order terms).

The FMG version of the iterative MG method described above behaves similarly as shown in Tables 10.2 and 10.3 for Poisson's equation. Thus we do not give detailed results here. In the FMG mode, it is in practice sufficient for most cases to use

$$\text{V-cycles;}\quad r = 1;\quad \nu_1 = 1, \nu_2 = 0.$$

To be on the safe side, however, one could use W-cycles instead (with $\nu_1=1$, $\nu_2=0$ or even $\nu_1=\nu_2=1$).

The total operation count for the FMG versions (corresponding to Table 10.7) is given in Table 10.8, assuming cubic FMG-interpolation. (Here, of course, we take the set-up operations (10.10) into account.)

	$\nu_1 = 1, \nu_2 = 0$			$\nu_1 = \nu_2 = 1$		
	+/-	*	÷	+/-	*	÷
V-cycle, r = 1	$25\frac{4}{9}\mathcal{N}$	$30\mathcal{N}$	$2\frac{2}{3}\mathcal{N}$	$40\frac{1}{9}\mathcal{N}$	$47\frac{5}{9}\mathcal{N}$	$2\frac{2}{3}\mathcal{N}$
W-cycle, r = 1	$36\frac{1}{3}\mathcal{N}$	$42\mathcal{N}$	$2\frac{2}{3}\mathcal{N}$	$58\frac{1}{3}\mathcal{N}$	$68\frac{2}{3}\mathcal{N}$	$2\frac{2}{3}\mathcal{N}$

Table 10.8: Operation count for the FMG method (corresponding to Table 10.7).

10.4 More general situations

For the numerical treatment of problems involving the more general differential operator

$$-a_1(x)u_{x_1x_1} - a_2(x)u_{x_2x_2} + b_1(x)u_{x_1} + b_2(x)u_{x_2} + c(x)u = f(x),$$

one has to take the following complications into account.

For fixed h, the discretization of first derivatives by central differencing leads to instabilities if the coefficients a_1, a_2 are very small compared to b_1, b_2 roughly if

$$|a_1| \leq |b_1|h/2, \quad |a_2| \leq |b_2|h/2. \tag{10.11}$$

But even if such *singularly perturbed* cases are excluded, one must be aware of similar difficulties which may still occur on the coarser grids used in the MG process.

In many cases these difficulties may be overcome, e.g., by introducing some *artificial viscosity* on the respective grids, i.e. the actual values of a_1 and/or a_2 are increased in a suitable way. This possibility (and its relevance for multigrid methods) has been described by Brandt in several papers [23],[26]. A systematic study for a model equation was made by Börgers [12].

In the singularly perturbed case (10.11), artificial viscosity is already needed on the finest grid. Through this, the order of discretization accuracy is also reduced. The MG treatment of such problems has been studied, e.g., in [23],[12],[56], [3].

If, on the other hand,

$$|a_1| > |b_1|h_0/2, \quad |a_2| > |b_2|h_0/2 \tag{10.12}$$

holds (where h_0 denotes the meshsize of the coarsest grid), the multigrid methods described above may also be used without any change for these more general problems. Exceptions are given by *indefinite* and *highly indefinite problems* (for which we refer to [28]) and problems with *(strongly) discontinuous coefficients*. Problems of the latter kind are treated in [1],[60].

11. Multigrid methods on composite meshes

11.1 Composite mesh discretization and a "naive" multigrid approach

The numerical treatment of elliptic equations with general boundary conditions on general domains is known to be technically rather complicated. One approach is to use different coordinate systems in the "main part" of the interior of the given domain and near the boundary. Advantages of this approach are due to the fact that suitably chosen local coordinates (with the boundary line being a grid line) allow the use of regular discretizations of the boundary conditions as well as higher order discretizations near the boundary. Furthermore, mesh refinement (orthogonally or tangentially to the boundary) can be performed in a technically simple way, for example for the treatment of boundary layers.

In the following description, we assume for simplicity that the given domain Ω is bounded and simply connected and that it has a smooth boundary Γ. For such a situation, in [95], a *composite mesh discretization method* has been considered. Here the given domain Ω is divided into two overlapping parts, Ω_I and Ω_0: Ω_I is an *interior domain* ($\overline{\Omega}_I \subset \Omega$) and $\Omega_0 \subset \Omega$ is an "annulus-shaped" region along the boundary Γ *(boundary domain)*. Ω_0 is assumed to be the image of a rectangular domain Ω_R under an orthogonal transformation Φ (see Figure 11.1).

The given problem on Ω is now discretized in both Ω_I and Ω_R (using the transformed equations in Ω_R), by use of, for example, rectangular grids. The grids are connected to each other by a suitable interpolation scheme. We use the notation *composite mesh system* for the resulting system which consists of discrete problems on Ω_I, Ω_R and of the interpolation relations.

This composite mesh system may be solved iteratively using a discrete analogue of Schwarz' alternating method. In each step of this method, the two discrete elliptic problems on Ω_I and Ω_R are solved in an alternating manner. Clearly, for each of these problems, multigrid methods may be used separately. This possibility is straightforward and has been studied in [68] for a model problem. Though the efficiency of this method is much better than, e.g., that of the corresponding SOR application, the total efficiency is limited by the convergence properties of Schwarz' method. The convergence of Schwarz' method, however, depends on the geometrical situation, e.g. on the overlapping of Ω_I and Ω_0 (roughly: the smaller the overlapping, the slower the convergence). On the other hand, a large region of overlap involves many extra grid points resulting in more computational work.

Instead of the "naive" combination of Schwarz' method with multigrid techniques, we propose a more direct multigrid approach to the composite mesh system. In this method, which has been investigated systematically in [68] for a model problem, a multigrid hierarchy of composite meshes in used: The principle of Schwarz' alternating method is applied here only within the relaxation process for smoothing. It turns out that the efficiency of this smoothing does not depend sensitively on the geometrical situation as, e.g., the overlapping.

Clearly, the composite mesh idea can be used not only in such simple geometrical situations as assumed above. In general, one will have to compose not only two, but several meshes (for example, if Ω has a boundary which is only piecewise smooth).

11.2 A "direct" multigrid method for composite meshes

Let a linear elliptic boundary value problem (1.1) be given on a simply connected bounded domain Ω with smooth boundary Γ. As described in the previous section, we assume

$$\Omega = \Omega_I \cup \Omega_0, \quad \overline{\Omega}_I \subset \Omega, \quad \Omega_I \cap \Omega_0 \neq \emptyset. \tag{11.1}$$

Here Ω_0 denotes an annulus-shaped "boundary domain" with "outer" boundary Γ and "inner" boundary Γ_0. Following the lines of Schwarz' alternating method, the original problem (1.1) is replaced by the two boundary value problems (0) and (I):

(0) *boundary problem* (on Ω_0):

$$\begin{aligned} L^\Omega u_0 &= f^\Omega(x) \quad (x \in \Omega_0), \\ L^\Gamma u_0 &= f^\Gamma(x) \quad (x \in \Gamma), \quad u_0 = f^{\Gamma_0}(x) \quad (x \in \Gamma_0); \end{aligned} \tag{11.2}$$

(I) *interior problem* (on Ω_I):

$$\begin{aligned} L^\Omega u_I &= f^\Omega(x) \quad (x \in \Omega_I), \\ u_I &= f^{\Gamma_I}(x) \quad (x \in \Gamma_I := \partial\Omega_I), \end{aligned} \tag{11.3}$$

where the connection to (1.1) is given by

$$f^{\Gamma_0}(x) = u_I(x) \quad (x \in \Gamma_0), \qquad f^{\Gamma_I}(x) = u_0(x) \quad (x \in \Gamma_I). \tag{11.4}$$

We assume that Ω_0 is the image of a rectangular domain

$$\Omega_R := \{(s,t) : o < s \le S, \ o < t < T\} \tag{11.5}$$

under a suitable orthogonal transformation Φ (see Figure 11.1). For example, if Γ is parametrized with respect to arclength, i.e.

$$\Gamma = \{(x_1^o(s), x_2^o(s)) : o \le s \le S\} \quad (\dot{x}_1^o(s))^2 + (\dot{x}_2^o(s))^2 \equiv 1,$$

a suitable domain Ω_0 can be defined using the mapping [18]

$$\Phi(s,t) = \begin{bmatrix} x_1(s,t) \\ x_2(s,t) \end{bmatrix} := \begin{bmatrix} x_1^o(s) - t\,\dot{x}_2^o(s) \\ x_2^o(s) + t\,\dot{x}_1^o(s) \end{bmatrix} . \tag{11.6}$$

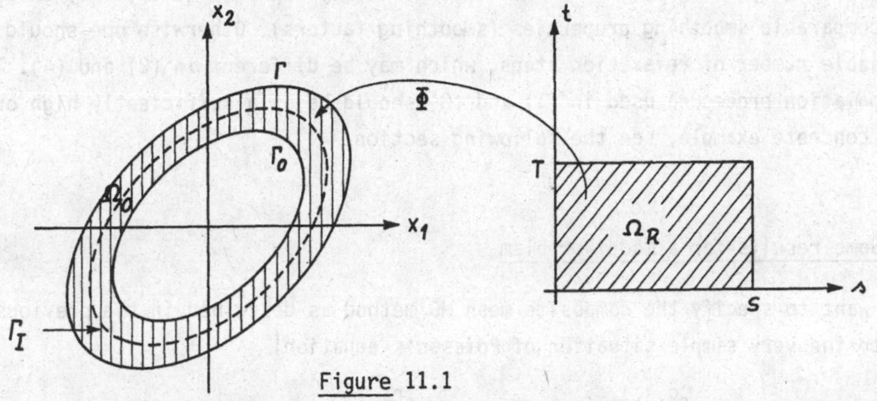

Figure 11.1

By (R) we denote problem (O) in terms of the local coordinates (s,t). Clearly, periodic boundary conditions are prescribed for s=0 and s=S (o ≤ t ≤ T).

We now assume that appropriate discretizations for both problems (R) and (I) are given. We use the formal parameter h to characterize this composite mesh discretization. In particular, (R_h), (I_h) denote the discrete problems and $\Omega_{R,h}$, $\Omega_{I,h}$ the corresponding grids. In the following, we do not distinguish between $\Omega_{R,h}$-gridfunctions $u_{R,h}$ and the corresponding transformed gridfunctions $u_{0,h}$ on $\Omega_{0,h}$. The *composite mesh multigrid method* now is applied to composite mesh grid functions U_h (which consist of both, $u_{I,h}$ and $u_{R,h}$). In principle, this method proceeds as follows.

Apart from the h-discretization, a sequence of coarser composite meshes and corresponding discretizations has to be given. For the fine-to-coarse and the coarse-to-fine transfer, the usual operators are applied, but now individually to $\Omega_{I,h}$ and $\Omega_{R,h}$. The only essential difference compared to the usual MG methods consists in the smoothing part, in which the idea of Schwarz' alternating method is used. One smoothing step, e.g. for the h-grid, consists of the following four parts (assuming a first approximation $\tilde{U}_h$ to be given):

(1) interpolate the grid values $\tilde{u}_{I,h}$ to the discrete boundary points of Γ_0;
(2) apply one relaxation step to $\tilde{u}_{R,h}$ with respect to the discrete problem (R_h);
(3) interpolate the grid values $\tilde{u}_{R,h}$ to the discrete boundary points of Γ_I;
(4) apply one relaxation step to $\tilde{u}_{I,h}$ with respect to the discrete problem (I_h).

Here the relaxation methods used in (2) and (4) for the discrete problems (R_h) and (I_h), respectively, have to be chosen suitably. In particular, they should have comparable smoothing properties (smoothing factors). Otherwise one should allow a variable number of relaxation steps, which may be different in (2) and (4). The interpolation procedure used in (1) and (3) should be of a sufficiently high order. For a concrete example, see the following section.

11.3 Some results for a model problem

We want to specify the composite mesh MG method as described in the previous section to the very simple situation of Poisson's equation

$$-\Delta u = f^{\Omega}(x) \quad (x \in \Omega), \qquad u = f^{\Gamma}(x) \quad (x \in \Gamma)$$

on the unit disk Ω. Clearly, for this utterly simple problem, one would not use the composite mesh approach in practice. For the purpose of demonstrating typical properties of the composite mesh MG method, however, this problem is quite suitable. In particular, one can discuss the question of overlapping and its influence on the convergence speed, the question of how to interpolate between the grids and the smoothing techniques. The results of these considerations are of a more general relevance. They are not restricted to the above problem.

We define Ω_0 using the orthogonal transformation Φ given by (11.6), i.e.

$$\Phi(s,t) = \begin{bmatrix} x_1(s,t) \\ x_2(s,t) \end{bmatrix} := \begin{bmatrix} (1-t)\cos(s) \\ (1-t)\sin(s) \end{bmatrix}$$

and the domain Ω_I to be an octagon. (An octagon has been chosen because it can easily be matched by a rectangular grid.) A composite mesh (for given $h=(h_I,h_s,h_t)$) is defined by the two grids $\Omega_{I,h}$ and $\Omega_{R,h}$ as shown in Figure 11.2. On $\Omega_{I,h}$ the Laplace-operator is discretized using the standard 5-point formula. We have Dirichlet boundary conditions along Γ_I. On $\Omega_{R,h}$ the transformed Laplace-operator

$$-\Delta^* := -\frac{1}{1-t}\left(\frac{1}{1-t}\frac{\partial^2}{\partial^2 s} + \frac{\partial}{\partial t}\left((1-t)\frac{\partial}{\partial t}\right)\right)$$

is discretized at a point $P = (s_i,t_j) = (ih_s,jh_t) \in \Omega_{R,h}$ by

$$\frac{1}{h_s^2(1-jh_t)}\begin{bmatrix} & -q^2(1-jh_t-h_t/2) & \\ \frac{-1}{1-jh_t} & -\Sigma & \frac{-1}{1-jh_t} \\ & -q^2(1-jh_t+h_t/2) & \end{bmatrix}$$

where $q := h_s/h_t$ and Σ denotes the sum of the four neighboring coefficients. We have Dirichlet boundary conditions along ∂_N, ∂_S and periodic boundary conditions at ∂_W and ∂_E (see Figure 11.2).

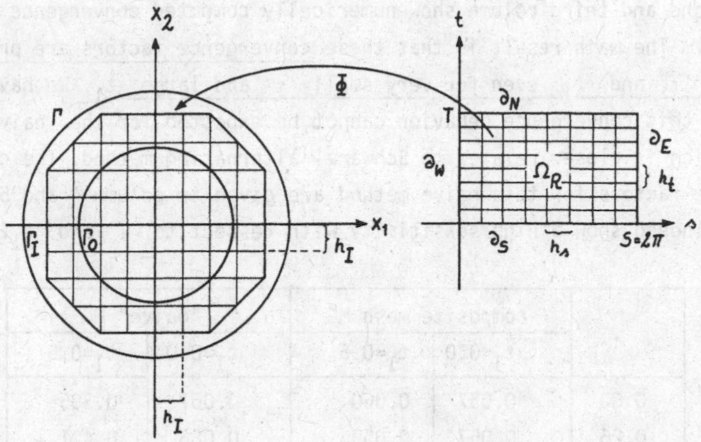

Figure 11.2: Composite mesh

The components of the MG algorithm used are the following:

- Coarsening, grid transfer, type of cycle: We apply standard coarsening (for both grids) and coarse-grid operators using the same discretization as on the fine grids. The transfer are done by full weighting and linear interpolation, respectively. The results given below are based on W-cycles.

- Smoothing: On smoothing step is performed as was described in the previous section. For (R_h) ZEBRA relaxation is used, with lines in the appropriate direction (depending on T and h_s/h_t). In the results given below, for (I_h) ZEBRA relaxation was used, also. (Here one could apply RB relaxation as well.) To connect the grids $\Omega_{R,h}$ and $\Omega_{I,h}$ (steps (1) and (3) in the previous section), cubic interpolation is used. This turns out to be necessary: using, for example, linear interpolation results in a much slower multigrid convergence.

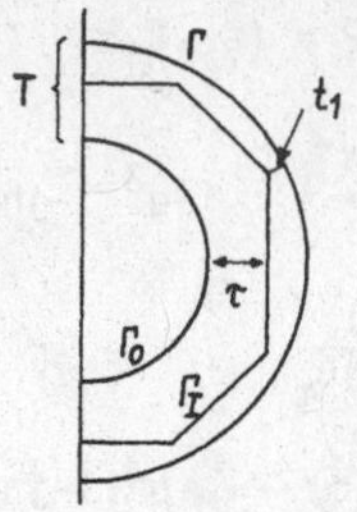

Figure 11.3:

In Table 11.1 we show the dependence of the multigrid algorithm on the size τ of the overlapping and on t_1, the distance of Γ_I to the boundary Γ (see Figure 11.3). The second and third column show numerically computed convergence factors (for $\nu_1=\nu_2=1$). The main result is that these convergence factors are practically independent of τ and t_1 even for very small τ and large t_1. We have already mentioned that this convergence behavior cannot be expected for the "naive" multigrid method which is closer related to Schwarz' alternating method. The corresponding convergence factors for this naive method are given in column 4 and 5 of Table 11.1: they do indeed show a high sensitivity with respect to τ and t_1.

τ	composite mesh MG		"naive" MG	
	t_1=0.0	t_1=0.5	t_1=0.0	t_1=0.5
0.30	0.057	0.060	0.067	0.385
0.26	0.057	0.059	0.086	0.411
0.22	0.057	0.059	0.119	0.462
0.18	0.057	0.059	0.159	0.527
0.14	0.055	0.059	0.215	0.602
0.10	0.057	0.059	0.303	0.687
0.08	0.058	0.059	0.368	0.734
0.06	0.067	0.062	0.448	0.783

Table 11.1: Numerically computed convergence factors ($\nu_1=\nu_2=1$)

In Table 11.2 we finally compare the convergence of the above composite mesh multigrid method with the convergence of the SOR method (applied in a straightforward alternating manner following the lines of Schwarz' alternating method). The results given refer to the case $t_1 = 0$ and $T_2 = 0.4$ (cf. Figure 11.3) and to different meshsizes $h_I = 1/N$, $h_s = 2\pi/N_s$, $h_t = T/N_t$. To compare the numbers given, one has to take into account that the total computational work for one multigrid iteration step is larger than that of one SOR step by a factor of about 4.

N, N_t, N_s	composite mesh MG	SOR
8,16,16	0.057	0.837
16,32,32	0.057	0.906
32,64,64	0.057	0.968
64,128,128	0.057	0.988

Table 11.2: Numerically computed multigrid convergence factors compared with SOR convergence factors (per iteration step) for $t_1 = 0$, $T_2 = 0.4$.

Remark: The composite mesh approach may also be used for the multigrid treatment of boundary value problems on unbounded domains. For example, let a differential equation be given in $\Omega = \mathbb{R}^2 \setminus \bar{D}$, where D is some bounded domain; on ∂D and at infinity boundary conditions are assumed to be prescribed.

Clearly, there are several well-known ways to handle the unbounded domain and the boundary condition at infinity numerically (transformation techniques, e.g. of conformal mapping type, replacement of the unbounded domain by a bounded one etc.). In the composite mesh approach, the unbounded domain Ω is divided into two parts Ω_I and Ω_0 with (11.1). Here Ω_I is some bounded domain "around $\bar{D}$" and Ω_0 is some geometrically simple unbounded domain. For example, Ω_0 may be the exterior of a circle (which can immediately be represented as the image of a rectangle), see Figure 11.4. A composite mesh MG method can then be applied to this combined system on Ω_I and Ω_0 as described above.

Figure 11.4:

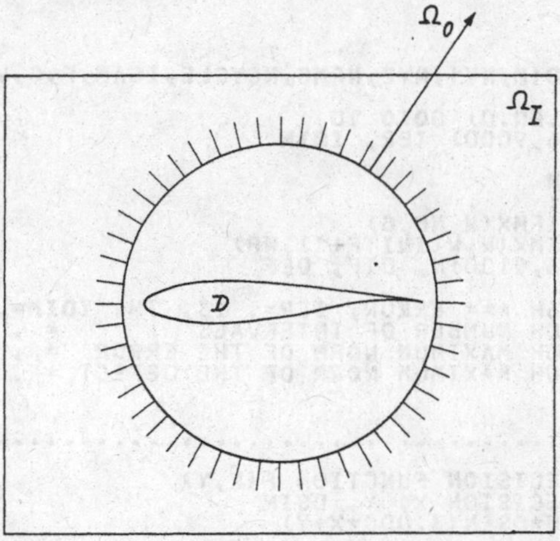

APPENDIX

In this appendix we list a very specialized program MGØØD from the MGØØ program collection (see Foerster, Witsch [36]). It solves model problem (P) by using an efficient multigrid method (see Section 8.1), which can be applied either as an iterative solver or in the FMG mode.

The following program is an exemplary driving routine for calling the multigrid subroutine MGØØD:

```
C++++++++++++++++++++++++++++++++++++++++++++++++++++++++++++++++++++++++++
C
C     MAIN PROGRAM FOR DEMONSTRATION OF MGOOD
C
C++++++++++++++++++++++++++++++++++++++++++++++++++++++++++++++++++++++++++
      DOUBLE PRECISION F, G, DIFMX, DEFMX, DIF, DEF
      INTEGER IDIM, IER, M, N, NP, NY1, NY2, NCYCLE, NFMG, IGAM, INITF
      DOUBLE PRECISION W(12000)
      EXTERNAL F, G
C
      IDIM = 12000
      M = 6
      NY1 = 2
      NY2 = 1
      NCYCLE = 1
      NFMG = 1
      IGAM = 1
C
      CALL MGOOD(M,NY1,NY2,NFMG,NCYCLE,IGAM,F,G,W,IDIM,INITF,IER)
C
        IF (IER.EQ.0) GOTO 10
        WRITE (6,9000) IER, IDIM
        STOP
 10     N = 2**M
        NP = N+1
        DIF = DIFMX(W,NP,G)
        DEF =DEFMX(W,W(INITF+1),NP)
        WRITE (6,9100)N, DIF, DEF
      STOP
 9000 FORMAT (16H *** ERROR, IER=, I3, 7H, IDIM=, I6, 4H ***)
 9100 FORMAT (30H NUMBER OF INTERVALS          = , I4 /
     *         30H MAXIMUM NORM OF THE ERROR   = , D12.4 /
     *         30H MAXIMUM NORM OF THE DEFECT = , D12.4)
      END
C
C..........................................................................
C
      DOUBLE PRECISION FUNCTION F(X,Y)
      DOUBLE PRECISION X, Y, DSIN
      F = 10.0D0*DSIN(3.0D0*X+Y)
      RETURN
      END
C
C
C..........................................................................
C
      DOUBLE PRECISION FUNCTION G(X,Y)
      DOUBLE PRECISION X, Y, DSIN
      G = DSIN(3.0D0*X+Y)
      RETURN
      END
```

```
C***********************************************************************
C
C     M G O O D     VERSION  20/04/82
C
C***********************************************************************
C
      SUBROUTINE MGOOD(M, NY1, NY2, NFMG, NCYCLE, IGAM, F, G, W, IDIM,
     *                 INITF, IER)
C
C     MULTIGRID MODULE FOR THE FAST SOLUTION OF POISSON'S EQUATION
C     WITH DIRICHLET BOUNDARY CONDITIONS ON THE UNIT SQUARE
C
C          DIFFERENTIAL EQUATION :    -DELTA U(X,Y) = F(X,Y)
C          BOUNDARY CONDITION :              U(X,Y) = G(X,Y)
C
C     CYCLE STRUCTURE:
C
C                COARSENING: H, 2H, 4H, ...
C                DIFFERENCE OPERATORS: USUAL 5-POINT STARS ON ALL GRIDS
C                RELAXATION: RED-BLACK
C                FINE-TO-COARSE: HALF INJECTION
C                COARSE-TO-FINE: BILINEAR INTERPOLATION
C
C     FULL MULTIGRID MODE:
C
C                FULL MULTIGRID INTERPOLATION USES GRID EQUATION
C                (4-TH ORDER)
C
C     INPUT:
C
C        M       NUMBER OF GRIDS (0 < M < 11). FOR GIVEN M:
C                NUMBER OF POINTS ON THE FINEST GRID (INCLUDING BOUNDARY
C                POINTS) = 2**M + 1 IN BOTH DIRECTIONS
C        NY1     NUMBER OF RELAXATIONS BEFORE COARSE-GRID CORRECTION
C                (NY1 > 0)
C        NY2     NUMBER OF RELAXATIONS AFTER COARSE-GRID CORRECTION
C                (NY2 > 0)
C        NCYCLE  NUMBER OF MULTIGRID ITERATIONS (NCYCLE > 0)
C        NFMG    .EQ. 0 : NCYCLE MULTIGRID ITERATIONS ARE PERFORMED
C                .NE. 0 : FMG-VERSION IS PERFORMED PLUS NCYCLE-1
C                         ADDITIONAL MG ITERATIONS AFTERWARDS
C        IGAM    TYPE OF CYCLING (IGAM > 0).
C                E.G.: IGAM=1 FOR V-CYCLES, IGAM=2 FOR W-CYCLES
C        W       DOUBLE PRECISION WORK ARRAY OF DIMENSION IDIM
C        IDIM    DIMENSION OF W. APPROXIMATELY: IDIM > 2.8*4**M
C
C     EXTERNALS:
C
C        F       DOUBLE PRECISION FUNCTION F(X,Y), RIGHT HAND SIDE OF
C                THE DIFFERENTIAL EQUATION
C        G       DOUBLE PRECISION FUNCTION G(X,Y), BOUNDARY VALUES
C
C     REMARK:    GRID #1 AND #M ARE THE FINEST AND COARSEST GRID USED,
C                RESPECTIVLY
C
C     OUTPUT:
C
C        IER     ERROR INDICATOR
C                = 0  NO ERRORS
C                = 1  INSUFFICIENT MEMORY , I.E. IDIM TOO SMALL.
C                     IN THIS CASE IDIM IS USED AS OUTPUT PARAMETER
C                     TO SHOW THE MINIMAL DIMENSION
C                = 2  M, NY1, NY2, NCYCLE OR IGAM WRONG
C        IDIM    ONLY IN CASE IER=1 : MINIMAL LENGTH OF W
C        INITF   CF. DESCRIPTION OF W
C        W       W CONTAINS THE DISCRETE APPROXIMATION TO THE GIVEN
C                BOUNDARY VALUE PROBLEM ON THE FINEST GRID. THE GRID
C                VALUES ARE STORED ROWWISE FROM LEFT TO RIGHT AND
C                FROM BOTTOM TO TOP.
C                I.E. THE GRID VALUE CORRESPONDING TO THE GRID POINT
C                (XI,YJ) :
C
C                     XI = (I-1)*H,  YJ = (J-1)*H    (H=1/N, N=2**M)
C
C                IS STORED AT
C
C                             W((J-1)*(N+1)+I)      (0 < I,J < N+2)
C
C                THE CORRESPONDING VALUES OF THE RIGHT HAND SIDE ARE
C                STORED IN THE SAME MANNER AT
C
C                             W((J-1)*(N+1)+I+INITF).
C
C                THE REMAINING STORAGE CONTAINS COARSE GRID VALUES.
```

```
      INTEGER IDC, IDF, IDR, IT, ITM, INITF, L, L1, LEV, LIN, NP,
     *NY1, NY2, IGAM, IDIM, IER, NCYCLE, NFMG, M, ID(11), NPK(10)
      DOUBLE PRECISION W(    1), HK(10), DFLOAT
      EXTERNAL F, G
      DFLOAT(L1)=DBLE(FLOAT(L1))
      IER = 0
C
      IF (M.LE.0 .OR. M.GE.11 .OR. NY1.LE.0 .OR. NY2.LE.0 .OR.
     *    NCYCLE.LE.0 .OR. IGAM.LE.0 ) IER=2
      IF (IER.GT.0) GOTO 50
C
C        DETERMINATION OF COARSER GRIDS
C
      NP = 2**M + 1
      HK(1) = 1.0D0/DFLOAT(NP-1)
      ID(1) = 1
      ID(2) = NP*NP + 1
      NPK(1) = NP
      DO 10 L=2,M
        NPK(L) = (NPK(L-1)+1)/2
        HK(L) = HK(L-1) + HK(L-1)
        ID(L+1) = ID(L) + NPK(L)**2
   10 CONTINUE
C
C        CHECK OF DIMENSIONS
C
      INITF = ID(M+1)
      IF (2*INITF.LE.IDIM) GOTO 20
      IER = 1
      IDIM = 2*INITF
      GO TO 50
   20 LIN = 1
      IF (NFMG.NE.0) LIN = M
C
C        SET UP ALL GRID VALUES NEEDED
C
      CALL INIT1(NP, HK(1), F, G, W(1), W(INITF+1))
      CALL INIT2(M, LIN, NPK, ID, W(1), W(INITF+1), INITF)
C
C        (FULL) MULTIGRID PROCEDURE
C
      DO 40 L=1,LIN
        LEV = LIN - L + 1
        ITM = 1
        IF (LEV.EQ.1) ITM = NCYCLE
        DO 30 IT=1,ITM
        CALL MGI(LEV, M, NY1, NY2, IGAM, NPK, ID, W, W(INITF+1), INITF)
   30   CONTINUE
        IF (L.GE.LIN) GO TO 50
        IDF = ID(LEV-1)
        IDC = ID(LEV)
        IDR = IDF+INITF
        CALL INT4(NPK(LEV), NPK(LEV-1), W(IDC), W(IDF), W(IDR))
        CALL PUTZB(NPK(LEV), W(IDC))
   40 CONTINUE
   50 RETURN
      END
C
C..........................................................................
C
      SUBROUTINE INIT1(NP, H, F, G, U, FR)
C
C        COMPUTES INITIAL VALUES ON THE FINEST GRID
C
      INTEGER I, J, N, NP
      DOUBLE PRECISION FR(NP,NP), U(NP,NP), F, G, DFLOAT, H, H2, X, Y
      DFLOAT(K) = DBLE(FLOAT(K))
C
      H2 = H*H
      N = NP - 1
      DO 20 J=2,N
        Y = DFLOAT(J-1)*H
        U(1,J) = G(0.0D0,Y)
        DO 10 I=2,N
          X = DFLOAT(I-1)*H
          U(I,J) = 0.0D0
          FR(I,J) = H2*F(X,Y)
   10   CONTINUE
        U(NP,J) = G(1.0D0,Y)
   20 CONTINUE
      DO 30 I=1,NP
        X = DFLOAT(I-1)*H
        U(I,1) = G(X,0.0D0)
        U(I,NP) = G(X,1.0D0)
```

```
   30 CONTINUE
      RETURN
      END
C
C..........................................................................
C
      SUBROUTINE INIT2(M, LIN, NPK, ID, U, FR, IDIM)
C
C        COMPUTES INITIAL VALUES ON THE COARSER GRIDS
C
      INTEGER IDIM, LIN, M, IDC, IDF, L, LH, ID(11), NPK(10)
      DOUBLE PRECISION FR(IDIM), U(IDIM)
C
      IF (LIN.EQ.1) GO TO 20
C
C        TRANSFER OF F AND U TO COARSER GRIDS
C
      DO 10 L=2,LIN
         IDC = ID(L)
         IDF = ID(L-1)
         CALL TRANS(NPK(L), NPK(L-1), U(IDC), FR(IDC), U(IDF), FR(IDF))
   10 CONTINUE
C
C        PUT ZERO TO BOUNDARY VALUES FOR MULTIGRID CORRECTIONS
C
   20 IF (LIN.EQ.M) GO TO 40
      LH = LIN + 1
      DO 30 L=LH,M
         IDC = ID(L)
         CALL PUTZB(NPK(L), U(IDC))
   30 CONTINUE
   40 RETURN
      END
C
C..........................................................................
C
      SUBROUTINE TRANS(NPC, NPF, UC, FC, UF, FF)
C
C       TRANSFER OF F AND U FROM GRID NPF TO NPC
C
      INTEGER NC, NPC, NPF, I, IF, J, JF
      DOUBLE PRECISION FC(NPC,NPC), FF(NPF,NPF), UC(NPC,NPC),
     *                 UF(NPF,NPF)
C
      NC = NPC - 1
C
      DO 20 J=2,NC
         JF = J + J - 1
         UC(1,J) = UF(1,JF)
         DO 10 I=2,NC
            IF = I + I - 1
            UC(I,J) = UF(IF,JF)
            FC(I,J) = 4.0D0*FF(IF,JF)
   10    CONTINUE
         UC(NPC,J) = UF(NPF,JF)
   20 CONTINUE
      DO 30 I=1,NPC
         IF = I + I - 1
         UC(I,1) = UF(IF,1)
         UC(I,NPC) = UF(IF,NPF)
   30 CONTINUE
      RETURN
      END
C
C..........................................................................
C
      SUBROUTINE PUTZB(NPC, UC)
C
C       PUTS ZERO TO BOUNDARY OF COARSER GRIDS
C
      INTEGER NC, NPC, I, J
      DOUBLE PRECISION UC(NPC,NPC)
C
      NC = NPC - 1
      DO 10 J=2,NC
         UC(1,J) = 0.0D0
         UC(NPC,J) = 0.0D0
   10 CONTINUE
      DO 20 I=1,NPC
         UC(I,1) = 0.0D0
         UC(I,NPC) = 0.0D0
   20 CONTINUE
      RETURN
      END
```

```
C.......................................................................
C
      SUBROUTINE MGI(LEV, M, NY1, NY2, IGAM, NPK, ID, U, FR, IDIM)
C
C         ONE MULTIGRID ITERATION STEP (ON ACTUAL FINEST GRID LEV)
C
      INTEGER IDIM, LEV, M, NY1, NY2, ID(11), NPK(10), IDC, IDF, IZ,
     *        K, IGAM, ICGAM(10)
      DOUBLE PRECISION FR(IDIM), U(IDIM)
C
      IZ = 1
      DO 10 K=LEV,M
        ICGAM(K) = 0
   10 CONTINUE
      K = LEV
      IF (K.EQ.M) GO TO 30
   20   IDF = ID(K)
C
C         RELAXATIONS BEFORE CGC
C
        IZ = 1
        IF (K.GT.LEV .AND. ICGAM(K).EQ.0) IZ = 0
        CALL RELAX(NY1+NY1, IZ, NPK(K), U(IDF), FR(IDF), FR(1))
        ICGAM(K) = ICGAM(K) + 1
C
C         RESIDUAL TRANSFER TO NEXT COARSER GRID
C
        IDC = ID(K+1)
        CALL RESTR(NPK(K+1), NPK(K), FR(IDC), U(IDF), FR(IDF))
        K = K + 1
        IF (K.LT.M) GOTO 20
      IZ = 0
   30 IDC = ID(M)
C
C         EXACT SOLUTION ON COARSEST GRID
C
      CALL RELAX(1,IZ,NPK(M),U(IDC),FR(IDC),FR(1))
      IF (K.EQ.LEV) GOTO 50
   40   K = K - 1
        IDF = ID(K)
        IDC = ID(K+1)
C
C         LINEAR INTERPOLATION TO NEXT FINER GRID
C
        CALL INT2A(NPK(K+1), NPK(K), U(IDC), U(IDF))
C
C         RELAXATION AFTER CGC
C
        CALL RELAX(NY2+NY2, 1, NPK(K), U(IDF), FR(IDF), FR(1))
      IF (K.EQ.LEV) GOTO 50
      IF (ICGAM(K).LT.IGAM) GOTO 20
      ICGAM(K) = 0
      GOTO 40
   50 RETURN
      END
C
C.......................................................................
C
      SUBROUTINE RELAX(ITM, IZ, NPF, UF, FF, W)
C
C       RED-BLACK RELAXATION ON GRID NPF
C
      INTEGER ITM, IZ, NF, NPF, I, IS, IT, ITMAX, J
      DOUBLE PRECISION FF(NPF,NPF), UF(NPF,NPF), W(NPF)
C
      NF = NPF - 1
      ITMAX = IABS(ITM)
      IS = 2
      IF (ITM.LT.0) IS = 3
C
      DO 70 IT=1,ITMAX
C
C
C
C         RELAXATION OF EVEN POINTS FOR IS=2, OF ODD POINTS FOR IS=3
C
        IF (IZ.NE.0 .OR. IT.GT.1) GO TO 30
C
C         ZERO STARTING VALUES
C
        DO 20 J=2,NF
          DO 10 I=IS,NF,2
            UF(I,J) = 0.25D0*FF(I,J)
   10     CONTINUE
```

```
            IS = 5 - IS
   20     CONTINUE
          GO TO 70
C
C          NON-ZERO STARTING VALUES
C
   30     DO 40 I=IS,NF,2
            W(I) = UF(I,1) + UF(I-1,2)
   40     CONTINUE
          DO 60 J=2,NF
            IF (IS.EQ.3) W(2) = UF(2,J) + UF(1,J+1)
            DO 50 I=IS,NF,2
              W(I+1) = UF(I+1,J) + UF(I,J+1)
              UF(I,J) = 0.25D0*(FF(I,J)+W(I)+W(I+1))
   50       CONTINUE
            IS = 5 - IS
   60     CONTINUE
   70 CONTINUE
C
      RETURN
      END
C
C.....................................................................
C
      SUBROUTINE RESTR(NPC, NPF, FC, UF, FF)
C
C         COMPUTATION OF THE DEFECT AND FINE-TO-COARSE TRANSFER
C         (HALF-INJECTION)
C
      INTEGER NC, NPC, NPF, I, IF, J, JF
      DOUBLE PRECISION FC(NPC,NPC), FF(NPF,NPF), UF(NPF,NPF), H
C
      NC = NPC - 1
      DO 20 J=2,NC
        JF = J + J - 1
        DO 10 I=2,NC
          IF = I + I - 1
          H = FF(IF,JF) - 4.0D0*UF(IF,JF) + UF(IF,JF-1) + UF(IF-1,JF)
     *          + UF(IF+1,JF) + UF(IF,JF+1)
          FC(I,J) = H + H
   10   CONTINUE
   20 CONTINUE
      RETURN
      END
C
C.....................................................................
C
      SUBROUTINE INT2A(NPC, NPF, UC, UF)
C
C         COARSE-TO-FINE TRANSFER (BILINEAR INTERPOLATION) AND CORRECTION
C
      INTEGER NC, NPC, NPF, I, IF, J, JF
      DOUBLE PRECISION UC(NPC,NPC), UF(NPF,NPF)
C
      NC = NPC - 1
      DO 20 J=2,NC
        JF = J + J - 1
        DO 10 I=1,NC
          IF = I + I
          UF(IF,JF) = UF(IF,JF) + 0.5D0*(UC(I,J)+UC(I+1,J))
   10   CONTINUE
   20 CONTINUE
      DO 40 J=1,NC
        JF = J + J
        DO 30 I=1,NC
          IF = I + I - 1
          UF(IF,JF) = UF(IF,JF) + 0.5D0*(UC(I,J)+UC(I,J+1))
   30   CONTINUE
   40 CONTINUE
      RETURN
      END
C
C.....................................................................
C
      SUBROUTINE INT4(NPC, NPF, UC, UF, FRF)
C
C         4-TH ORDER FMG-INTERPOLATION
C
      INTEGER NC, NPC, NPF, I, IF, J, JF, JM
      DOUBLE PRECISION FRF(NPF,NPF), UC(NPC,NPC), UF(NPF,NPF), HA, HN
```

```
C     TRANSFER OF COARSE GRID VALUES TO THE FINE GRID
C
      NC = NPC - 1
      JF = 3
      DO 20 J=2,NC
        IF = 3
        DO 10 I=2,NC
          UF(IF,JF) = UC(I,J)
          IF = IF + 2
   10   CONTINUE
        JF = JF + 2
   20 CONTINUE
C
C     COMPUTATION OF THE REMAINING EVEN POINTS BY USING THE
C     ROTATED 5-POINT STAR
C
      JF = 2
      JM = 1
      DO 40 J=2,NPC
        IF = 2
        HA = UC(1,JM) + UC(1,J)
        DO 30 I=2,NPC
          HN = UC(I,JM) + UC(I,J)
          UF(IF,JF) = 0.25D0*(FRF(IF,JF)+FRF(IF,JF)+HA+HN)
          HA = HN
          IF = IF + 2
   30   CONTINUE
        JM = JM + 1
        JF = JF + 2
   40 CONTINUE
C
C     COMPUTATION OF THE ODD POINTS BY ONE HALF (ODD) RELAXATION STEP
C
      CALL RELAX(-1, 1, NPF, UF, FRF, FRF)
      RETURN
      END
C
C..............................................................................
C
      DOUBLE PRECISION FUNCTION DIFMX(UC, NP, SOL)
C
C       COMPUTES THE MAXIMUM NORM OF THE DIFFERENCE BETWEEN
C       SOL (=SOLUTION OF THE BVP) AND THE VALUES IN UC
C
      INTEGER NP, I, J
      DOUBLE PRECISION UC(NP,NP), SOL, DFLOAT, H, X, Y
      DFLOAT(K) = DBLE(FLOAT(K))
C
      H = 1.0D0/DFLOAT(NP-1)
      DIFMX = 0.0D0
      DO 20 J=1,NP
        Y = DFLOAT(J-1)*H
        DO 10 I=1,NP
          X = DFLOAT(I-1)*H
          DIFMX = DMAX1(DIFMX,DABS(SOL(X,Y)-UC(I,J)))
   10   CONTINUE
   20 CONTINUE
      RETURN
      END
C
C..............................................................................
C
      DOUBLE PRECISION FUNCTION DEFMX(UC, FC, NP)
C
C       COMPUTES THE MAXIMUM NORM OF THE DEFECT
C
      INTEGER N, NP, I, J
      DOUBLE PRECISION UC(NP,NP), FC(NP,NP), D, HSQR
C
      N = NP - 1
      HSQR = DBLE(FLOAT(N*N))
      DEFMX = 0.0D0
      DO 20 J=2,N
        DO 10 I=2,N
          D = (FC(I,J)-4.0D0*UC(I,J)+UC(I-1,J)+UC(I,J-1)+UC(I,J+1)
     *          +UC(I+1,J))*HSQR
          DEFMX = DMAX1(DEFMX,DABS(D))
   10   CONTINUE
   20 CONTINUE
      RETURN
      END
```

References:

1. Alcouffe, R.E.; Brandt, A.; Dendy, J.E.(Jr.); Painter, J.W.: *The multi-grid methods for the diffusion equation with strongly discontinuous coefficients.* Siam J. Sci. Stat. Comput., 2, pp. 430-454, 1981.

2. Allgower, E.L.; Böhmer, K.; McCormick, S.F.: *Discrete correction methods for operator equations.* Numerical Solution of Nonlinear Equations. Proceedings, Bremen 1980 (E.L. Allgower, K. Glashoff, H.O. Peitgen, eds.). Lecture Notes in Mathematics, 878, pp. 30-97. Springer-Verlag, Berlin, 1981.

3. Asselt, E.J. van: *The multi grid method and artificial viscosity.* This Proceedings.

4. Astrakhantsev, G.P.: *An iterative method of solving elliptic net problems.* U.S.S.R. Computational Math. and Math. Phys., 11 no. 2, pp. 171-182, 1971.

5. Auzinger, W.; Stetter, H.J.: *Defect correction and multigrid iterations.* This Proceedings.

6. Bakhvalov, N.S.: *On the convergence of a relaxation method with natural constraints on the elliptic operator.* U.S.S.R. Computational Math. and Math. Phys., 6 no. 5, pp. 101-135, 1966.

7. Bank, R.E.: *A multi-level iterative method for nonlinear elliptic equations.* Elliptic Problem Solvers (M.H. Schultz, ed.), pp. 1-16. Academic Press, New York, NY, 1981.

8. Bank, R.E.; Dupont, T.F.: *Analysis of a two-level scheme for solving finite element equations.* Report CNA-159, Center for Numerical Analysis, University of Texas at Austin, 1980.

9. Bank, R.E.; Sherman, A.H.: *An adaptive multi-level method for elliptic boundary value problems.* Computing, 26, pp. 91-105, 1981.

10. Becker, K.: *Mehrgitterverfahren zur Lösung der Helmholtz-Gleichung im Rechteck mit Neumannschen Randbedingungen.* Diplomarbeit, Institut für Angewandte Mathematik, Universität Bonn, 1981.

11. Beyn, W.J.; Lorenz, J.: *Spurious solution for discrete superlinear boundary value problems*. Computing, 28, pp. 43-51, 1982.

12. Börgers, C.: *Mehrgitterverfahren für eine Mehrstellendiskretisierung der Poissongleichung und für eine zweidimensionale singulär gestörte Aufgabe*. Diplomarbeit, Institut für Angewandte Mathematik, Universität Bonn, 1981.

13. Braess, D.: *The convergence rate of a multigrid method with Gauss-Seidel relaxation for the Poisson equation*. This Proceedings.

14. Brand, K.: *Multigrid bibliography*. Gesellschaft für Mathematik und Datenverarbeitung, St. Augustin, 1982.

15. Brandt, A.: *Multi-level adaptive technique (MLAT) for fast numerical solution to boundary value problems*. Proceedings Third International Conference on Numerical Methods in Fluid Mechanics, Paris 1972 (H. Cabannes, R. Teman, eds.). Lecture Notes in Physics, 18, pp. 82-89. Springer-Verlag, Berlin, 1973.

16. Brandt, A.: *Multi-level adaptive techniques (MLAT). I. The multi-grid method.* Research Report RC 6026, IBM T.J. Watson Research Center, Yorktown Heights, NY, 1976.

17. Brandt, A.: *Multi-level adaptive solutions to boundary-value problems*. Math. Comp. , 31, pp. 333-390, 1977.

18. Brandt, A.: *Multi-level adaptive techniques (MLAT) for partial differential equations: ideas and software*. Mathematical Software III. (J.R. Rice, ed.), pp. 277-318. Academic Press, New York, NY, 1977.

19. Brandt, A.: *Multi-level adaptive finite-element methods. I. Variational problems.* Special Topics of Applied Mathematics (J. Frehse, D. Pallaschke, U. Trottenberg, eds.), pp. 91-128. North-Holland Publishing Company, Amsterdam, 1979.

20. Brandt, A.: *Multi-level adaptive techniques (MLAT) for singular-perturbation problems*. Numerical Analysis of Singular Perturbation Problems (P.W. Hemker, J.J.H. Miller, eds.), pp. 53-142. Academic Press, London 1979.

21. Brandt, A.: *Numerical stability and fast solutions to boundary value problems.* Boundary and Interior Layers - Computational and Asymptotic Methods (J.J.H. Miller, ed.), pp. 29-49. Boole Press, Dublin, 1980.

22. Brandt, A.: *Multigrid solutions to steady-state compressionable Navier-Stokes equations. I.* Preprint no. 492, Sonderforschungsbereich 72, Universität Bonn, 1981.

23. Brandt, A.: *Multigrid solvers for non-elliptic and singular-perturbation steady-state problems.* Research Report, Dept. of Applied Mathematics, Weizmann Institute of Science, Rehovot, 1981.

24. Brandt, A.: *Multigrid solvers on parallel computers.* Elliptic Problem Solvers (M.H. Schultz, ed.), pp. 39-84. Academic Press, New York, NY, 1981.

25. Brandt, A.: *Guide to multigrid development.* This Proceedings.

26. Brandt, A.; Dinar, N.: *Multi-grid solutions to elliptic flow problems.* Numerical Methods for Partial Differential Equations (S.V. Parter, ed.), pp. 53-147. Academic Press, New York, NY, 1979.

27. Brandt, A.; McCormick, S.F.; Ruge, J.: *Multigrid methods for differential eigenproblems.* Report, Dept. of Mathematics, Colorado State University, Ft. Collins, CO, 1981.

28. Brandt, A.; Ta'asan, S.: *Multi-grid methods for highly oscillatory problems.* Research Report, Dept. of Applied Mathematics, Weizmann Institute of Science, Rehovot, 1981.

29. Buzbee, B.L.; Golub, G.H.; Nielson, C.W.: *On direct methods for solving Poisson's equation.* SIAM J. Numer. Anal., 7, pp. 627-656, 1970.

30. Chan, T.F.; Keller, H.B.: *Arc-length continuation and multi-grid techniques for nonlinear elliptic eigenvalue problems.* Technical Report no. 197, Computer Science Dept., Yale University, New Haven, CT, 1981.

31. Collatz, L.: *The Numerical Treatment of Differential Equations.* Springer Verlag, Berlin, 1966.

32. Deconinck, H.; Hirsch, C.: *A multigrid finite element method for the transonic potential equation.* This Proceedings.

33. Dinar, N.: *Fast methods for the numerical solution of boundary-value problems.* PH.D. Thesis, Dept. of Applied Mathematics, Weizmann Institute of Science, Rehovot, 1978.

34. Fedorenko, R.P.: *A relaxation method for solving elliptic difference equations.* U.S.S.R. Computational Math. and Math. Phys., 1 no. 5, pp. 1092-1096, 1962.

35. Fedorenko, R.P.: *The speed of convergence of an iterative process.* U.S.S.R. Computational Math. and Math. Phys., 4 no. 3, pp. 227-235, 1964.

36. Foerster, H.; Witsch, K.: *Multigrid software for the solution of elliptic problems on rectangular domains: MGOO (Release 1).* This Proceedings.

37. Foerster, H.; Stüben, K.; Trottenberg, U.: *Non standard multigrid techniques using checkered relaxation and intermediate grids.* Elliptic Problems Solvers (M.H. Schultz, ed.), pp. 285-300. Academic Press, New York, NY, 1981.

38. Forsythe, G.E.; Wasow, W.R.: *Finite Difference methods for Partial Differential Equations.* John Wiley, New York-London, 1960.

39. Frederickson, P.O.: *Fast approximate inversion of large sparse linear systems.* Mathematics Report no. 7-75, Dept. of Mathematical Sciences, Lakehead University, Ontario, 1975.

40. Fuchs, L.: *Transonic flow computation by a multi-grid method.* Numerical Methods for the Computation of Inviscid Transonic Flows with Shock Waves (A. Rizzi, H. Viviand, eds.), pp. 58-65. Vieweg, Braunschweig, 1981.

41. Gary, J.: *The multigrid method applied to the collocation method.* SIAM J. Numer. Anal., 18, pp. 211-224, 1981.

42. Hackbusch, W.: *Ein iteratives Verfahren zur schnellen Auflösung elliptischer Randwertprobleme.* Report 76-12, Institut für Angewandte Mathematik, Universität Köln, 1976.

43. Hackbusch, W.: *On the convergence of a multi-grid iteration applied to finite element equations.* Report 77-8, Institut für Angewandte Mathematik, Universität Köln, 1977.

44. Hackbusch, W.: *On the computation of approximate eigenvalues and eigenfunctions of elliptic operators by means of a multi-grid method.* SIAM J. Numer. Anal., 16, pp. 201-215, 1979.

45. Hackbusch, W.: *Convergence of multi-grid iterations applied to difference equations*. Math. Comp., 34, pp. 425-440, 1980.

46. Hackbusch, W.: *Survey of convergence proofs for multigrid iterations*. Special Topics of Applied Mathematics (J. Frehse, D. Pallaschke, U. Trottenberg, eds.), pp. 151-164. North-Holland Publishing Company, Amsterdam, 1980.

47. Hackbusch, W.: *Bemerkungen zur iterierten Defektkorrektur und zu ihrer Kombination mit Mehrgitterverfahren*. Rev. Roumaine Math. Pures Appl., 26, pp. 1319-1329, 1981.

48. Hackbusch, W.: *Die schnelle Auflösung der Fredholmschen Integralgleichung zweiter Art*. Beiträge Numer. Math., 9, pp. 47-62, 1981.

49. Hackbusch, W.: *On the convergence of multi-grid iterations*. Beiträge Numer. Math., 9, pp. 213-239, 1981.

50. Hackbusch, W.: *Multi-grid convergence theory*. This Proceedings.

51. Hackbusch, W.: *On multi-grid iterations with defect correction*. This Proceedings.

52. Hackbusch, W.: *Introduction to multi-grid methods for the numerical solution of boundary value problems*. Computational Methods for Turbulent, Transonic and Viscous Flows (J.A. Essers, ed.). Hemisphere, to appear.

53. Hemker, P.W.: *Fourier analysis of gridfunctions, prolongations and restrictions*. Preprint NW 93/80, Dept. of Numerical Mathematics, Mathematical Centre, Amsterdam, 1980.

54. Hemker, P.W.: *The incomplete LU-decomposition as a relaxation method in multigrid algorithms*. Boundary and Interior Layers - Computational and Asymptotic Methods (J.J.H. Miller, ed.), pp. 306-311. Boole Press, Dublin, 1980.

55. Hemker, P.W.: *Introduction to multigrid methods*. Nieuw Archief voor Wiskunde (3), 29, pp. 71-101, 1981.

56. Hemker, P.W.: *Mixed defect correction iteration for the accurate solution of the convection diffusion equation*. This Proceedings.

57. Hemker, P.W.; Schippers, H.: *Multiple grid methods for the solution of Fredholm integral equations of the second kind.* Math. Comp., 36, pp. 215-232, 1981.

58. Hockney, R.W.; Eastwood, J.W.: *Computer Simulation Using Particles.* McGraw-Hill, New York, 1981.

59. Jameson, A.: *Acceleration of transonic potential flow calculations on arbitrary meshes by the multiple grid method.* Paper AIAA-79-1458, AIAA Fourth Computational Fluid Dynamics Conference, New York, NY, 1979.

60. Kettler, R.: *Analysis and comparison of relaxation schemes in robust multigrid and preconditioned conjugate gradient methods.* This proceedings.

61. Kettler, R.; Meijerink, J.A.: *A multigrid method and a combined multigrid-conjugate gradient method for elliptic problems with strongly discontinuous coeffiencients in general domains.* Shell Publication 604, KSEPL, Rijswijk, 1981.

62. Klunker, E.B.: *Contribution to methods for calculating the flow about thin lifting wings at transonic speeds - analytic expressions for the far field.* NASA Technical Note D-6530, Langley Research Center, 1971.

63. Kroll, N.: *Direkte Anwendungen von Mehrgittertechniken auf parabolische Anfangsrandwertaufgaben.* Diplomarbeit, Institut für Angewandte Mathematik, Universität Bonn, 1981.

64. Kronsjö, L.: *A note on the "nested iterations" method.* BIT, 15, pp. 107-110, 1975.

65. Kronsjö, L.; Dahlquist, G.: *On the design of nested iterations for elliptic difference equations.* BIT, 11, pp. 63-71, 1972.

66. Laasonen, P.: *On the discretization error of the Dirichlet problem in a plane region with corners.* Ann. Acad. Sci. Fenn. Ser. AI Math. Dissertationes, 408, pp. 1-16, 1967.

67. Linden, J.; Trottenberg, U.; Witsch, K.: *Multigrid computation of the pressure of an incompressible fluid in a rotating spherical gap.* Proceedings of the Fourth GAMM-Conference on Numerical Methods in Fluid Mechanics (H. Viviand, ed.), pp. 183-193. Vieweg, Braunschweig, 1982.

68. Linden, J.: *Mehrgitterverfahren für die Poisson-Gleichung in Kreis und Ringgebiet unter Verwendung lokaler Koordinaten.* Diplomarbeit, Institut für Angewandte Mathematik, Universität Bonn, 1981.

69. Magnus, R.J.; Gallaher, W.H.: *Flow over airfoils in the transonic regime - computer programs.* AFFDL-TR-70-16, Vol. II, U.S. Air Force, 1970.

70. McCormick, S.F.; Ruge, J.: *Multigrid methods for variational problems.* Preprint, Dept. of Mathematics, Colorado State University, Ft. Collins, CO, 1981 (submitted to: SIAM J. Num. Anal.).

71. Meis, T.; Marcowitz, U.: *Numerical Solution of Partial Differential Equations.* New York-Heidelberg-Berlin, Springer, 1981.

72. Meis, T.; Branca, H.W.: *Schnelle Lösung von Randwertaufgaben.* Z. Angew. Math. Mech., 62, 1982.

73. Meis, T.; Lehmann, H.; Michael, H.: *Application of the multigrid method to a nonlinear indefinite problem.* This Proceedings.

74. Mittelmann, H.D.: *Multi-grid methods for simple bifurcation problems.* This Proceedings.

75. Nicolaides, R.A.: *On multiple grid and related techniques for solving discrete elliptic systems.* J. Comput. Phys., 19, pp. 418-431, 1975.

76. Nicolaides, R.A.: *On the* l^2 *convergence of an algorithm for solving finite element equation.* Math. Comp., 31, pp. 892-906, 1977.

77. Nicolaides, R.A.: *On multi-grid convergence in the indefinite case.* Math. Comp., 32, pp. 1082-1086, 1978.

78. Nicolaides, R.A.: *On some theoretical and practical aspects of multigrid methods.* Math. Comp., 33, pp. 933-952, 1979.

79. Ophir, D.: *Language for processes of numerical solutions to differential equations.* PH.D. Thesis, Dept. of Applied Mathematics, Weizmann Institute of Science, Rehovot, 1978.

80. Ortega, J.M.; Rheinboldt, W.C.: *Iterative Solution of Nonlinear Equations in Several Variables.* New York-London, Academic Press, 1970.

81. Proskurowski, W.; Widlund, O.: *On the numerical solution of Helmholtz' equation by the capacitance matrix method.* Math. Comp., 30, pp. 433-468, 1976.

82. Ries, M.; Trottenberg, U.; Winter, G.: *A note on MGR methods.* Linear Algebra Appl., to appear, 1982.

83. Ruge, J.W.: *Multigrid methods for differential eigenvalue and variational problems and multigrid simulation.* PH.D. Thesis, Dept. of Mathematics, Colorado State University, Ft. Collins, CO, 1981.

84. Schmidt, W.; Jameson, A.: *Applications of multi-grid methods for transonic flow calculations.* This Proceedings.

85. Schröder, J.: *Zur Lösung von Potentialaufgaben mit Hilfe des Differenzenverfahrens.* ZAMM, 34, pp. 241-253, 1954.

86. Schröder, J.: *Beiträge zum Differenzenverfahren bei Randwertaufgaben.* Habilitationsschrift , Hannover 1955.

87. Schröder, J.; Trottenberg, U.: *Reduktionsverfahren für Differenzengleichungen bei Randwertaufgaben I.* Numer. Math., 22, pp. 37-68, 1973.

88. Schröder, J.; Trottenberg, U.; Witsch, K.: *On fast Poisson solvers and applications.* Numerical Treatment of Partial Differential Equations (R. Bulirsch, R.D. Grigorieff, J. Schröder, eds.). Proceedings of a Conference held at Oberwolfach, July 4-10, 1976. Lecture Notes in Mathematics, 631, pp. 153-187, Springer-Verlag, Berlin, 1978.

89. Shortley, G.H.; Weller, R.: *Numerical solution of Laplace's equation.* J. Appl. Phys., 9, pp. 334-348, 1938.

90. Solchenbach, K.; Stüben, K.; Trottenberg, U.; Witsch, K.: *Efficient solution of a nonlinear heat conduction problem by use of fast reduction and multigrid methods.* Preprint no. 421, Sonderforschungsbereich 72, Universität Bonn, 1980.

91. South, J.C.(Jr.); Brandt, A.: *Application of a multi-level grid method to transonic flow calculations.* Transonic Flow Problems in Turbomachinery (T.C. Adamson, M.F. Platzer, eds.). Hemisphere, Washington, DC, 1977.

92. Southwell, R.V.: *Stress calculation in frameworks by the method of systematic relaxation of constraints. I, II*, Proc. Roy. Soc. London Ser. A, 151, pp. 56-95, 1935.

93. Southwell, R.V.: *Relaxation Methods in Theoretical Physics*. Clarendon Press, Oxford, 1946.

94. Stiefel, E.: *Über einige Methoden der Relaxationsrechnung*. Z. Angew. Math. Phys., 3, pp. 1-33, 1952.

95. Starius, G.C.: *Composite mesh difference methods for elliptic boundary value problems*. Numer. Math., 28, pp. 242-258, 1977.

96. Stetter, H.J.: *The defect correction principle and discretization methods*. Numer. Math., 29, pp. 425-443, 1978.

97. Stüben, K.: *MGØ1: A multi-grid program to solve* $\Delta U - c(x,y)U = f(x,y)$ *(on* Ω*),* $U = g(x,y)$ *(on* $\partial\Omega$*), on nonrectangular bounded domains* Ω. IMA-Report no. 82.02.02, Gesellschaft für Mathematik und Datenverarbeitung, St. Augustin, 1982.

98. Stüben, K.; Trottenberg, U.: *On the construction of fast solvers for elliptic equations*. Computational Fluid Dynamics. Lecture Series 1982-04, von Karman Institute for Fluid Dynamics, Rhode-Saint-Genese, 1982.

99. Temperton, C.: *Algorithms fot the solution of cyclic tridiagonal systems*. J. Comput. Phys., 19, pp. 317-323, 1975.

100. Thole, C.A.: *Beiträge zur Fourieranalyse von Mehrgittermethoden: V-cycle, ILU-Glättung, anisotrope Operatoren*. Diplomarbeit, Institut für Angewandte Mathemathik, Universität Bonn, to appear.

101. Törnig, W.: *Numerische Mathematik für Ingenieure und Physiker*, Band 1, Springer-Verlag, Berlin, 1979.

102. Trottenberg, U.: *Reduction methods for solving discrete elliptic boundary value problems - an approach in matrix terminology*. Fast Elliptic Solvers (U. Schumann, ed.). Advance Publications, London, 1977.

103. Trottenberg, U.: *Schnelle Lösung partieller Differentialgleichungen - Idee und Bedeutung des Mehrgitterprinzips*. Jahresbericht 1980/81, Gesellschaft für Mathematik und Datenverarbeitung, pp. 85-95, Bonn, 1981.

104. Trottenberg, U.; Witsch, K.: *Zur Kondition diskreter elliptischer Randwertaufgaben*. GMD-Studien no. 60. Gesellschaft für Mathematik und Datenverarbeitung, St. Augustin, 1981.

105. Varga, R.S.: *Matrix Iterative Analysis*. Englewood Cliffs, Prentice Hall, 1962.

106. Verfürth, R.: *The contraction number of a multigrid method with mesh ratio 2 for solving Poisson's equation*. Report, Institut für Angewandte Mathematik, Ruhr-Universität Bochum, 1982.

107. Wesseling, P.: *Numerical solution of stationary Navier-Stokes equation by means of a multiple grid method and Newton iteration*. Report NA-18, Dept. of Mathematics, Delft University of Technology, Delft, 1977.

108. Wesseling, P.: *The rate of convergence of a multiple grid method*. Numerical Analysis. Proceedings, Dundee 1979 (G.A. Watson, ed.). Lecture Notes in Mathematics, 773, pp. 164-184. Springer-Verlag, Berlin, 1980.

109. Wesseling, P.: *Theoretical and practical aspects of a multigrid method*. Report NA-37, Dept. of Mathematics, Delft University of Technology, Delft, 1980.

110. Wesseling, P.: *A robust and efficient multigrid method*. This Proceedings.

111. Winter, G.: *Fourieranalyse zur Konstruktion schneller MGR-Verfahren*. Dissertation, Institut für Angewandte Mathematik, Universität Bonn, to appear.

MULTI-GRID CONVERGENCE THEORY

W. Hackbusch

Mathematisches Institut, Ruhr-Universität Bochum,

Postfach 102148, D-4630 Bochum 1, Germany

Contents.

1. Algorithms

1.1 Problem and Notations

Let Ω be a domain in $\mathbb{R}^d$. The boundary value problem consists of a differential equation

(1.1a) $$L^{\Omega} u = f^{\Omega} \quad \text{in } \Omega,$$

where L^{Ω} is an elliptic differential operator, and of some boundary condition

(1.1b) $$L^{\Gamma} u = f^{\Gamma} \quad \text{on } \Gamma = \partial\Omega.$$

Here L^{Γ} is a boundary operator, e.g., $L^{\Gamma} u = u$ (Dirichlet boundary condition) or $L^{\Gamma} u = \partial u/\partial n$ (Neumann condition). In the sequel both equations (1.1a,b) are denoted by one abstract equations:

(1.2) $$L\,u = f.$$

Let

(1.3) $$h_o > h_1 > h_2 > \ldots > h_{l-1} > h_l > \ldots$$

be a sequence of decreasing *discretization parameters*, e.g., a sequence of grid sizes. The discretization corresponding to the parameter h_l (*level* l) is denoted by

(1.4) $$L_l u_l = f_l.$$

L_l is the discretization matrix, u_l and f_l are 'grid functions'. We denote the linear space of grid functions u_l by $\mathcal{U}_l$ and the space of the right-hand sides f_l by $\mathcal{F}_l$. Of course, $\mathcal{U}_l$ and $\mathcal{F}_l$ have same dimension:

(1.5) $$n_l = \dim(\mathcal{U}_l) = \dim(\mathcal{F}_l).$$

But $\mathcal{U}_l$ and $\mathcal{F}_l$ will be equipped with possibly different norms. In case of difference schemes, n_l is the number of grid points, whereas in case of finite element methods, n_l is the dimension of the finite element subspace.

For the prolongation (interpolation) from the coarse to the fine grid we use the symbol

$$I_{l-1}^{l} : \mathcal{U}_{l-1} \rightarrow \mathcal{U}_l,$$

while

$$I_l^{l-1} : \mathcal{F}_l \rightarrow \mathcal{F}_{l-1}$$

denotes some restriction from the fine to the coarse grid (cf. [16]).

Finally we have to introduce a symbol for the *smoothing procedure* used in the multi-grid iteration. This (linear) smoothing iteration is denoted by $\mathcal{S}_l$:

(1.6) $$u_l^{j+1} = \mathcal{S}_l(u_l^j, f_l) := S_l u_l^j + T_l f_l, \quad I = S_l + T_l L_l$$

Examples for $\mathcal{S}_l$ are given in § 3.3. The equation $I = S_l + T_l L_l$ ensures that the solution of Eq. (1.4) is a stationary point of iteration (1.6).

1.2. Two-Grid Iteration

As described in the tutorial contribution of Stüben and Trottenberg [16] one iteration of the two-grid algorithm consists of a smoothing step and a coarse-grid correction. The j+1st iterate u_l^{j+1} is computed from the jth iterate by

(1.7a) compute $\overline{u}_l$ as result of ν steps of the smoothing procedure $\mathcal{S}_l$:

$$u_l^{j,o} := u_l^j; \; u_l^{j,\mu} := \mathcal{S}_l(u_l^{j,\mu-1}, f_l) \quad (\mu = 1,2,..,\nu), \quad \overline{u}_l := u_l^{j,\nu};$$

(1.7b) $$u_l^{j+1} := \overline{u}_l + I_{l-1}^l L_{l-1}^{-1} I_l^{l-1} (f_l - L_l \overline{u}_l).$$

The algorithm depends on the choice of $\nu, I_{l-1}^l, I_l^{l-1}$, and L_{l-1}. In the sequel the algorithms will be formulated in quasi-ALGOL. The iteration (1.7a,b) becomes

procedure TGM(l,u,f);

(1.8a) <u>if</u> l = o <u>then</u> u := L_o^{-1} * f <u>else</u>

<u>begin</u> <u>integer</u> j; <u>array</u> d,v;

(1.8b) <u>for</u> j := 1 <u>step</u> 1 <u>until</u> ν <u>do</u> u := $\mathcal{S}_l$(u,f);

(1.8c) d := I_l^{l-1}* (f - L_l * u);

(1.8d) v := L_{l-1}^{-1} * d;

(1.8e) u := u + I_{l-1}^l * v

<u>end</u>;

The parameters l, u, f of TGM (<u>two</u>-<u>grid</u> <u>method</u>) have the following meaning. l is the level number. f is the right-hand side f_l of Eq. (1.4). The input value u is a given iterate u_l^j, while the output value u is the next iterate u_l^{j+1}. The two-grid method can be applied only for $l \geq 1$. The statement (1.8a) is added to include the case l = o. The smoothing step (1.8b) corresponds to (1.7a), while (1.8 c-e) is the coarse-grid correction (1.7b).

Instead of applying ν smoothing iterations before the coarse-grid correction, one can perform ν_1 iterations before and ν_2 iterations after the coarse-grid correction, where $\nu_1 + \nu_2 = \nu$. Algorithm (1.8) corresponds to $\nu_1 = \nu$, $\nu_2 = o$. The other extrem case is $\nu_1 = o$, $\nu_2 = \nu$:

```
        procedure TGM' (l, u, f);
        if l = o then u := L_o^-1 * f else
        begin integer j;
(1.9)        u := u + I_{l-1}^l * L_{l-1}^-1 * I_l^{l-1} * (f - L_l * u);
             for j := 1 step 1 until ν do u := S_l(u,f)
        end;
```

1.3 Multi-Grid Iteration

In order to avoid the exact solving of the problem $L_{l-1}\, v_{l-1} = d_{l-1}$ in (1.8d) we use the same iteration to approximate the solution v_{l-1}. This approach yields the following recursive program MGM:

```
          procedure MGM (l, u, f);
(1.10a)   if l = o then u := L_o^-1 * f else
          begin integer j; array v,d;
(1.10b)        for j := 1 step 1 until ν do u := S_l(u,f);
(1.10c)        d := I_l^{l-1} * (f - L_l * u);
(1.10d)        v := o; for j := 1 step 1 until γ do MGM (l-1, v, d);
(1.10e)        u := u + I_{l-1}^l * v
          end;
```

Here, $\gamma \geq 1$ is a further parameter (number of coarse-grid iterations). Modifications according to (1.9) are possible, but they are not discussed in the following.

1.4 Nested Iteration

One can solve Eq. (1.4) by starting with $u_l^o = o$ and applying a certain number of iterations. Usually, one can save computational work by replacing u_l^o with a better initial guess. A very good initial value can be obtained from the approximate result of the coarser grid equation $L_{l-1}\, u_{l-1} = f_{l-1}$. This leads us to the nested iteration. The combination of the multi-grid iteration with the nested iteration is also called 'full multi-grid method' (cf. [16]). The ALGOL program reads as follows:

$\tilde{u}_o := L_o^{-1} * f_o;$

<u>for</u> k := 1 <u>step</u> 1 <u>until</u> l <u>do</u>

(1.11) <u>begin</u> $\tilde{u}_k := \tilde{I}_{k-1}^{k} \tilde{u}_{k-1};$

<u>for</u> j := 1 <u>step</u> 1 <u>until</u> i <u>do</u> MGM $(k, \tilde{u}_k, f_k)$

<u>end</u>;

Here, $\tilde{I}_{l-1}^{l} : \mathcal{U}_{l-1} \to \mathcal{U}_l$ is some interpolation. It may be the same as used in the multi-grid iteration ($\tilde{I}_{l-1}^{l} = I_{l-1}^{l}$) or a more accurate one. i is the number of MGM-iterations per level, independent of l.

The *computational work* of the multi-grid iteration is discussed by Stüben and Trottenberg [16].

2. Outline of the Proofs

2.1 Iterative Analysis

We recall some elementary propositions about iterations. Any linear iteration has the representation

$$u_l^{j+1} = M_l\, u_l^j + N_l\, f_l, \tag{2.1}$$

where M_l and N_l are $n_l \times n_l$ - matrices (cf. (1.5)). M_l is called *iteration matrix*. Obviously, the solution u_l of (1.4), $L_l u_l = f_l$, is a stationary point (i.e. $u_l = M_l u_l + N_l f_l$) if and only if

$$I = M_l + N_l L_l \qquad \text{(I: identity matrix)}. \tag{2.2}$$

For regular L_l, (2.2) becomes

$$N_l = (I - M_l)\, L_l^{-1}. \tag{2.3}$$

The behaviour of the iteration (2.1) depends on M_l only. It is well-known that the process (2.1) converges for any starting value u_l^o if and only if

$$\rho(M_l) < 1 \tag{2.4}$$

holds for the *spectral radius* $\rho(M_l) = \max\{|\lambda| : \lambda \text{ eigenvalue of } M_l\}$. The spectral radius describes the asymptotic behaviour as $j \to \infty$. Since fast iterative processes should be terminated for small j, $\rho(M_l)$ is not a sufficient description of the iteration. Instead, we have to determine suitable *norm estimates* of the iteration matrix M_l.

$\mathcal{U}_l$ was introduced in § 1.1 as linear space of grid functions. This space is equipped with some norm $\|\cdot\|_u$. An example is the Euclidean norm

$$\| u_l \|_u = \sqrt{h^d \sum_P |u_l(P)|^2}\ , \tag{2.5}$$

where Σ_P is the summation over all grid points P (all components of u_l) and d is the dimension of the domain $\Omega \subset \mathbb{R}^d$. The norm of the space $\mathcal{F}_l$ of right-hand sides f_l is denoted by $\|\cdot\|_F$.

Let $A : \mathcal{U}_l \to \mathcal{U}_k$ $(l,k \geq o)$ be some matrix, then the *matrix norm* of A induced by the (vector) norms of $\mathcal{U}_l$ and $\mathcal{U}_k$ is

$$\| A \|_{u \leftarrow u} = \sup\{\| Au_l \|_u \,/\, \| u_l \|_u \ : o \neq u_l \in \mathcal{U}_l\}.$$

Similarly we define

$$\| A\|_{F\leftarrow u} = \sup \{\| Au_l\|_F / \| u_l\|_u : o \neq u_l \in \mathcal{U}_l\},$$

$$\| A\|_{u\leftarrow F} = \sup \{\| Af_l\|_u / \| f_l\|_F : o \neq f_l \in \mathcal{F}_l\},$$

$$\| A\|_{F\leftarrow F} = \sup \{\| Af_l\|_F / \| f_l\|_F : o \neq f_l \in \mathcal{F}_l\}$$

for matrices $A : \mathcal{U}_l \to \mathcal{F}_k$, $A : \mathcal{F}_l \to \mathcal{U}_k$, $A : \mathcal{F}_l \to \mathcal{F}_k$, resp.

The error $u_l^j - u_l$ (u_l solution of (1.4)) satisfies

$$u_l^j - u_l = M_l \,(u_l^{j-1} - u_l). \tag{2.6}$$

Therefore, the iteration converges monotoneously ($\| u_l^j - u_l\|_u \searrow o$) if $\| M_l\|_{u\leftarrow u} < 1$. Note that this condition is sufficient, but not necessary for convergence.

2.2 Two-Grid Iteration

It will turn out that the convergence proof of the two-grid iteration is the key for the analysis of the multi-grid iteration. However, it is not easy to estimate the norm $\| M_l\|_{u\leftarrow u}$ of the iteration matrix. Even for the discrete Poisson equation in a square (cf. [16]) it is complicated to compute the precise value or good estimates of $\| M_l\|_{u\leftarrow u}$ (or $\rho(M_l)$). In § 3 we propose a technique to provide for a bound of $|M_l\|_{u\leftarrow u}$, that explains the characteristic features to the multi-grid convergence. The convergence follows from two conditions called 'smoothing property' and 'approximation property'. The former expresses that the smoothing procedure is really smoothing, while the latter describes that the fine-grid correction is sufficiently well approximated by the coarse-grid correction.

2.3 Multi-Grid Iteration

The multi-grid iteration (1.10) is obtained from the two-grid iteration (1.8) by replacing the exact solution of the coarse-grid equation with the recursive application of the same procedure. The recursive structure gives rise to a recursive definition of the iteration matrix of the multi-grid algorithm. The iteration matrix can be regarded as the two-grid iteration matrix plus a perturbation. Therefore, the two-grid convergence together with some technical conditions implies the multi-grid convergence .

2.4 Nested Iteration

It can be shown that the iteration error $\tilde{u}_k - u_k$ of the results $\tilde{u}_k$ of the nested iteration (1.11) is of the order of the discretization error $u - u_k$. The ratio of both errors can be described explicitly in terms of $\| M_l \|_{U \leftarrow U}$ (norm of multi-grid iteration matrix).

2.5 Bibliographical Comments

Fedorenko [8] considers the case of a discrete Poisson equation in a square. The convergence proof is formulated by means of sin-functions, which are the eigenfunctions of this model problem (cf. [16]).

Bachvalov [3] extends the foregoing proof to difference operators of second order with varying coefficients. He uses a decompositon of $\mathcal{U}_l$ into $\mathcal{U}_l^1 \oplus \mathcal{U}_l^2$, where $\mathcal{U}_l^1$ ($\mathcal{U}_l^2$) are spanned by smooth (non-smooth) eigenfunctions.

Astrachancev [1] discusses the convergence of the multi-grid iteration in the case of a finite element discretization in a general domain (with an assumption similar to (3.15)).

Also Nicolaides [15] considers the case of finite element discretizations for second order equations. He discusses the convergence for symmetric and positive definite L_l and extends the proof to the indefinite case, too.

Bank and Dupont [4] give a more general proof for the finite element case, that is very similar to our approach in § 3.4.2.

Wesseling's proof [17] applies to second order difference equations in a rectangle. It is related to the approach of Bachvalov. He formulates conditions that imply multi-grid convergence in the general case. One of the conditions is the approximation property of § 3, while another is similar to the smoothing property we use in § 3.

Braess [5,6] and Maitre and Musy [12] give another proof that applies to a special finite element discretization of the Poisson equation. This approach is based on a sharpened Cauchy inequality. In some cases the convergence proof is independent of the regularity of the problem (cf. [5]).

3. Convergence of the Two-Grid Iteration

3.1 Smoothing Property and Approximation Property

The two-grid iteration (1.8) is a linear iteration with representation (2.1). The iteration matrix M_l depends on the number $\nu : M_l = M_l(\nu)$. Sometimes, we add the superscript TGM to indicate that M_l^{TGM} is the *two*-grid iteration matrix. From (1.8), (1.9), and the definition of S_l in (1.6) we obtain

<u>Note 3.1</u> The iteration matrices M_l and M_l' of the iterations (1.8) and (1.9), resp., are

(3.1) $\quad M_l = (I - I_{l-1}^{l} L_{l-1}^{-1} I_l^{l-1} L_l) S_l^{\nu}, \; M_l' = S_l^{\nu}(I - I_{l-1}^{l} L_{l-1}^{-1} I_l^{l-1} L_l) \quad (l \geq 1).$

Assuming L_l regular, we can rewrite (3.1) as

(3.1') $M_l = (L_l^{-1} - I_{l-1}^{l} L_{l-1}^{-1} I_l^{l-1}) L_l S_l^{\nu}, \; \hat{M}_l := L_l M_l' L_l^{-1} = L_l S_l^{\nu}(L_l^{-1} - I_{l-1}^{l} L_{l-1}^{-1} I_l^{l-1}).$

Note that the iteration matrix M_l is the amplification matrix for the error:

$$\| u_l^j - u_l \|_u \leq \| M_l \|_{u \leftarrow u}^j \; \| u_l^o - u_l \|_u \qquad \text{(cf. (2.6))},$$

whereas in the case of the second variant (1.9), the matrix $\hat{M}_l$ (similar to M_l') is the amplification matrix of the defect:

$$\| L_l u_l^j - f_l \|_F \leq \| \hat{M}_l \|_{F \leftarrow F}^j \, \| L_l u_l^o - f_l \|_F .$$

By the submultiplicativity of the matrix norms of § 2.1, a product AB ($A : \mathcal{U}_l \to \mathcal{F}_l$, $B : \mathcal{F}_l \to \mathcal{U}_l$) can be estimated by $\| AB \|_{u \leftarrow u} \leq \| A \|_{u \leftarrow F} \; \| B \|_{F \leftarrow u}$. Hence the representation (3.1') implies

(3.2a) $\quad \| M_l \|_{u \leftarrow u} \leq \| L_l^{-1} - I_{l-1}^{l} L_{l-1}^{-1} I_l^{l-1} \|_{u \leftarrow F} \, \| L_l S_l^{\nu} \|_{F \leftarrow u},$

(3.2b) $\quad \| \hat{M}_l \|_{F \leftarrow F} \leq \| L_l S_l^{\nu} \|_{F \leftarrow u} \, \| L_l^{-1} - I_{l-1}^{l} L_{l-1}^{-1} I_l^{l-1} \|_{u \leftarrow F}.$

Note that the right-hand sides of (3.2a) and (3.2b) coincide. Both factors $\| L_l S_l^{\nu} \|_{F \leftarrow u}$ and $\| L_l^{-1} - I_{l-1}^{l} L_{l-1}^{-1} I_l^{l-1} \|_{u \leftarrow F}$ will be considered separately.

A vector $w_l \in \mathcal{U}_l$ can be called 'smooth' if the product of the difference operator L_l and w_l satisfies $\| L_l w_l \|_F \ll \| L_l \|_{F \leftarrow u} \; \| w_l \|_u$. Therefore, the smoothing procedure $\mathcal{S}_l$ is really smoothing if $\| L_l S_l^{\nu} \|_{F \leftarrow u} \ll \| L_l \|_{F \leftarrow u} \; \| S_l^{\nu} \|_{u \leftarrow u}$. Usually,

$\| L_l \|_{F\leftarrow u}$ is of order $O(h_l^{-\alpha})$, e.g. $O(h_l^{-2m})$ if L is a differential operator of order 2m and if $\|\cdot\|_u = \|\cdot\|_F$ = Enclidean norm. The norm $\| S_l^\nu \|_{u\leftarrow u}$ is of order O(1), often $\| S_l^\nu \|_{u\leftarrow u} \le 1$. These considerations lead us to the following condition:

Smoothing property. There are some functions $\eta(\nu)$ and $\nu_{max}(h)$ and some number α such that

(3.3a) $\quad \| L_l S_l^\nu \|_{F\leftarrow u} \le \eta(\nu)\, h_l^{-\alpha}$ for all $l \ge 1$, $\; 1 \le \nu < \nu_{max}(h_l)$,

(3.3b) $\quad \eta(\nu) \searrow 0 \quad (\nu \to \infty)$

(3.3c) $\quad \nu_{max}(h) = \infty \quad$ or $\quad \nu_{max}(h) \nearrow \infty \quad (h \to 0)$.

This condition will be discussed in § 3.3. Note that $\eta(\nu)$ does not depend on l or h_l.

The coarse grid is involved only in the second factor $L_l^{-1} - I_{l-1}^{l} L_{l-1}^{-1} I_l^{l-1}$. The following condition guarantees that L_l^{-1} is approximated by L_{l-1}^{-1}:

Approximation property. (3.4) holds with the same α as in (3.3a):

(3.4) $\quad \| L_l^{-1} - I_{l-1}^{l} L_{l-1}^{-1} I_l^{l-1} \|_{u\leftarrow F} \le C h_l^{\alpha} \qquad$ for all $l \ge 1$.

C denotes a generic constant not depending on l or h_l. α from (3.3a) and (3.4) depends on the choice of the norms $\|\cdot\|_u$ and $\|\cdot\|_F$. Its value is not important, since $h_l^{\alpha} h_l^{-\alpha} = 1$ in any case.

The estimates (3.2a,b), (3.3a), and (3.4) yield

(3.5) $\quad \| M_l \|_{u\leftarrow u} \le C\,\eta(\nu), \qquad \| \hat{M}_l \|_{F\leftarrow F} \le C\,\eta(\nu) \qquad (1 \le \nu < \nu_{max}(h_l))$.

Note that the right-hand sides $C\eta(\nu)$ are independent of the grid size h_l. A sufficient condition for convergence is $C\,\eta(\nu) < 1$. Let $\rho \in (0,1)$ be arbitrary. By (3.3b), $C\,\eta(\nu) \le \rho$ holds for $\nu \ge \nu_{min}$ for some $\nu_{min} = \nu_{min}(\rho)$ (independent of h_l). Therefore, we have convergence provided that $\nu_{min} < \nu_{max}(h_l)$; otherwise (3.3a) is not valid. By (3.3b) there is some h_{max} (coarsest grid size) such that $\nu_{min} < \nu_{max}(h_{max}) \le \nu_{max}(h_l)$ if $h_l < h_{max}$. Hence, we obtain the following result:

Theorem 3.2 Let $\rho \in (0,1)$ be fixed. Suppose that the smoothing and approximation properties are valid. Then there are bounds h_{max} and ν_{min} so that the two-grid iterations (1.8) and (1.9) converge with rate ρ if $\nu_{min} \le \nu < \nu_{max}(h_l)$, $h_l \le h_{max}$:

(3.6) $\quad \| M_l \|_{u\leftarrow u} \le C\,\eta(\nu) \le \rho < 1, \qquad \| \hat{M}_l \|_{F\leftarrow F} \le C\,\eta(\nu) \le \rho < 1.$

We recall that $h_l \le h_{max}$ is needed to avoid that the intervall $[\nu_{min}, \nu_{max}(h_l))$ is empty. This restriction is not necessary if $\nu_{max}(h) = \infty$ (cf. (3.3c)):

Corollary 3.3 If $\nu_{max} = \infty$ (i.e. (3.3a) holds for all $\nu \ge 1$), then the restrictions $h_l \le h_{max}$ and $\nu \le \nu_{max}(h_l)$ can be omitted.

Theorem 3.2 shows that the two-grid convergence can be reduced to the proof of the smoothing property and of the approximation property. In the following sections 3.2 and 3.3 we discuss in detail the approximation property (3.4) and the smoothing property (3.3). In particular we want to show that these conditions are very natural and can be verified in general situations.

3.2 Discussion of the Approximation Property

3.2.1 Finite Element Equation (Simple Case)

In the case of a difference operator L_l in a general region, the proof of the approximation property is not quite easy (cf. [10]). But for a finite element equation one can prove this condition directly. In this subsection we consider the case, where both norms $\|\cdot\|_U = \|\cdot\|_F$ may be chosen as Euclidean norm.

First, we recall some details of the finite element discretization and introduce some notations. We restrict our considerations to a scalar differential equation (1.1a) of second order with Dirichlet boundary condition (1.1b). For higher orders or other boundary conditions we refer to [9].

The weak formulation of (1.2), Lu = f, is

$$\text{(3.7a)} \qquad a(u,v) = (f,v)_{L^2(\Omega)} \quad \text{for all } v \in H_0^1(\Omega)$$

where $a(\cdot,\cdot)$ is the bilinear form

$$\text{(3.7b)} \qquad a(u,v) = \int_\Omega \Big[\sum_{i,j=1}^{d} a_{ij} u_{x_i} \overline{v}_{x_j} + \sum_{i=1}^{d} (b_i u_{x_i} \overline{v} + c_i u \overline{v}_{x_i}) + du\overline{v} \Big] dx.$$

and $(\cdot,\cdot)_{L^2(\Omega)}$ is the scalar product of $L^2(\Omega)$ extended to the dual form of $H^{-1}(\Omega) \times H_0^1(\Omega)$.

Assume that there is a hierarchy of finite element subspaces:

$$\text{(3.8)} \qquad \mathcal{H}_0 \subset \mathcal{H}_1 \subset \ldots \subset \mathcal{H}_{l-1} \subset \mathcal{H}_l \subset \ldots \subset H_0^1(\Omega).$$

Let h_l be the corresponding discretization parameter satisfying

(3.9) $\inf\{\|u-v\|_1 : v \in \mathcal{H}_l\} \le C\, h_l \,\|u\|_2$ for all $u \in H^2(\Omega) \cap H^1_o(\Omega)$.

Here, $\|\cdot\|_k$ is the norm of the Sobolev space $H^k(\Omega)$ of order k. In particular, $\|\cdot\|_o$ is the $L^2(\Omega)$ norm.

The discrete problem of level l is to seek $u_{\mathcal{H}_l} \in \mathcal{H}_l$ fulfilling

(3.10) $$a(u_{\mathcal{H}_l}, v) = (f,v)_{L^2(\Omega)} \text{ for all } v \in \mathcal{H}_l.$$

However, we do not compute the function $u_{\mathcal{H}_l}$ but a certain coefficient vector $u_l \in \mathcal{U}_l = \mathcal{F}_l$, where $\dim(\mathcal{U}_l) = \dim(\mathcal{H}_l) = n_l$. Usually, the coefficients (components of u_l) are the values of $u_{\mathcal{H}_l}$ at nodal points. The bijective mapping from $\mathcal{U}_l$ onto $\mathcal{H}_l \subset H^1_o(\Omega)$ is denoted by I_l:

$$u_{\mathcal{H}_l} = I_l\, u_l.$$

In order to formulate the finite element equation $L_l\, u_l = f_l$ being equivalent to (3.10), we have to introduce the scalar product

$$\langle u_l, v_l\rangle = h_l^d \sum_P u_l(P)\,\overline{v_l(P)} \qquad (P\text{: nodal points in } \Omega \subset \mathbb{R}^d)$$

and the adjoint

$$I^l = I_l^* \qquad (\text{i.e., } \langle I^l v, w_l\rangle = (v, I_l w_l)_{L^2(\Omega)}).$$

Then the equation $L_l\, u_l = f_l$ with $I_l u_l = u_{\mathcal{H}_l}$ satisfying (3.10) is given by

(3.11) $$L_l = I^l\, L\, I_l, \qquad f_l = I^l f.$$

Here, $L : H^1_o(\Omega) \to H^{-1}(\Omega)$ is defined by $a(u,v) = (Lu,v)_{L^2(\Omega)}$.

In the case of a finite element hierarchy (3.8) there is a *canonical choice* of the prolongation I^l_{l-1} and the restriction I^{l-1}_l for the multi-grid algorithm. I^l_{l-1} is uniquely defined by

(3.12a) $$I_{l-1} = I_l \cdot I^l_{l-1}.$$

That means that for any coefficient vector $v_{l-1} \in \mathcal{U}_{l-1}$ the corresponding function $I_{l-1}\, v_{l-1} \in \mathcal{H}_{l-1} \subset \mathcal{H}_l$ coincides with the function $I_l(I^l_{l-1}\, v_{l-1})$ corresponding to $I^l_{l-1}\, v_{l-1} \in \mathcal{U}_l$. $I^l_{l-1}\, v_{l-1}$ is the representation of $I_{l-1}\, v_{l-1}$ in $\mathcal{U}_l$; v_{l-1} and $I^l_{l-1}\, v_{l-1}$ describe the same finite element function. The canonical choice of the restriction I^{l-1}_l is

(3.12b) $$I^{l-1}_l = (I^l_{l-1})^* \quad (\text{equivalent to } I^{l-1} = I^{l-1}_l\, I^l).$$

We shall prove the approximation property for the choice $\|\cdot\|_u = \|\cdot\|_F$ = Euclidean norm. Note that the Euclidean norm (2.5) can be redefined by

(3.13a) $$\|v_l\| = \sqrt{\langle v_l, v_l\rangle}.$$

The symbol $\|\cdot\|$ will always denote the Euclidean norm if it is used for vectors. The scaling by h_l^d in the definition of $\langle\cdot,\cdot\rangle$ is applied to ensure that the Euclidean norm of a vector v_l and the $L^2(\Omega)$-norm of the corresponding function $I_l\, v_l \in \mathcal{U}_l$ are uniformly equivalent:

(3.14) $$\frac{1}{C}\,\|v_l\| \le \|I_l\, v_l\|_o \le C\,\|v_l\| \quad \text{for all } v_l \in \mathcal{U}_l,\ l \ge o.$$

In § 3.4 we shall discuss the condition implying the usual error estimate for the finite element solution $u_{\mathcal{U}_l}$ of (3.10):

(3.15) $$\|u_{\mathcal{U}_l} - u\|_o \le C' h_l^2\,\|u\|_2 \le C\,h_l^2\,\|f\|_o.$$

Using the operator norm

(3.13b) $$\|A\|_{o\leftarrow o} = \sup\{\|Au\|_o / \|u\|_o : o \ne u \in L^2(\Omega)\} \text{ for } A : L^2(\Omega) \to L^2(\Omega),$$

we can rewrite the inequality (3.15) as

(3.15') $$\|I_l\, L_l^{-1}\, I^l - L^{-1}\|_{o\leftarrow o} \le C\,h_l^2,$$

since $u_{\mathcal{U}_l} = I_l\, L_l^{-1}\, I^l\, f$, $u = L^{-1} f$. For level l-1 (3.15') becomes

(3.15'') $$\|I_{l-1}\, L_{l-1}^{-1}\, I^{l-1} - L^{-1}\|_{o\leftarrow o} \le C\,h_{l-1}^2.$$

By (3.12a,b), $I_{l-1} = I_l\, I_{l-1}^l$ and $I^{l-1} = I_l^{l-1}\, I^l$, one obtains

(3.15''') $$\|I_l\,[L_l^{-1} - I_{l-1}^l\, L_{l-1}^{-1}\, I_l^{l-1}]\, I^l\|_{o\leftarrow o} \le C\,[h_l^2 + h_{l-1}^2].$$

Inequality (3.14) implies $\|I_l A I^l\|_{o\leftarrow o} \le C^2\,\|A\|$, where now $\|\cdot\|$ denotes the spectral norm. Since $\|\cdot\|_u = \|\cdot\|_F$ = Euclidean norm, we have $\|\cdot\| = \|\cdot\|_{u\leftarrow F}$. Hence, (3.15''') together with

(3.16) $$h_{l-1} / h_l \le C \qquad \text{for all } l \ge 1$$

yield the approximation property (3.4) with $\alpha = 2$. We summarize:

<u>Theorem 3.4</u> Choose I_{l-1}^l and I_l^{l-1} canonically and set $\|\cdot\|_u = \|\cdot\|_F$ = Euclidean norm (3.13). Assume (3.14), (3.15), and (3.16). Then the approximation property (3.4)

holds with $\alpha = 2$:

$$\| L_l^{-1} - I_{l-1}^{l} L_{l-1}^{-1} I_l^{l-1} \| \le C\, h_l^2 \qquad (\|\cdot\| = \|\cdot\|_{u\leftarrow F} \text{ spectral norm}).$$

3.2.2 Finite Element Equation (More General Case)

In the foregoing section the simplest choice of the norms, $\|\cdot\|_u = \|\cdot\|_F = \|\cdot\|$ (Euclidean norm), was possible. Unfortunately, this choice does not apply in general. In this section we consider a case, where $\|\cdot\|_u$ and $\|\cdot\|_F$ are different though $\mathcal{U}_l = \mathcal{F}_l$ have the same elements.

In (3.15) we used $\|u\|_2 \le C\, \|f\|_o$ (H^2-regularity of L). E.g., for domains with re-entrant corners (L-shaped region), estimate (3.15) does not hold, but

$$\text{(3.17)} \qquad \| u_{\mathcal{T}_l} - u \|_s \le C\, h_l^{2-2s}\, \|f\|_{-s} \qquad (Lu = f \in H^s(\Omega))$$

for some $o < s < 1$ (cf. § 3.4). Here, $\|\cdot\|_{\pm s}$ are the norms of the Sobolev spaces $H^s(\Omega)$ and $H^{-s}(\Omega)$ = dual space of $H_o^s(\Omega)$.

The Euclidean norm $\|\cdot\|$ is the discrete analogue of the L^2-norm. Now we need a counterpart of $\|\cdot\|_{\pm s}$. This can be done as follows. By ellipticity of L, the operator

$$\text{(3.18)} \qquad \Lambda = \tfrac{1}{2}\,(L + L^*) + c\, I \qquad \text{(c sufficiently large)}$$

is $H_o^1(\Omega)$ - coercive. Set

$$\Lambda_l = I^l\ \Lambda\ I_l$$

according to $L_l = I^l\, L\, I_l$ (cf. (3.11)). Obviously. Λ_l is symmetric and positive definite, so that powers Λ_l^t ($t \in \mathbb{R}$) are well-defined. Suitable norms for $\mathcal{U}_l$ and $\mathcal{F}_l$ are

$$\text{(3.19)} \qquad \| u_l \|_u = \| \Lambda_l^{s/2}\, u_l \|, \quad \| f_l \|_F = \| \Lambda_l^{-s/2}\, f_l \| \qquad (\|\cdot\| \text{ from (3.13a)})$$

with s from (3.17). For $s = o$ we regain $\|\cdot\|_u = \|\cdot\|_F = \|\cdot\|$.

In general, the product $I^l \cdot I_l$ is not the identity, but thanks to (3.14) the inverse $(I^l\ I_l)^{-1} : \mathcal{U}_l \to \mathcal{U}_l$ exists and is bounded. The product

$$\text{(3.20)} \qquad Q_l = I_l\ (I^l\ I_l)^{-1}\ I^l$$

is the orthogonal L^2-projection of $H_o^1(\Omega)$ onto the finite element subspace $\mathcal{T}_l$. We have $\| Q_l \|_{o\leftarrow o} = 1$. Assume in addition

(3.21) $$\| Q_l \|_{s\leftarrow s} \le C \quad \text{for all } l,$$

where $\|\cdot\|_{s\leftarrow s}$ is defined analogously to (3.13b). A sufficient condition for (3.21) is $\| Q_l \|_{1\leftarrow 1} \le C$.

In [9] we proved

Theorem 3.5 Define I_{l-1}^{l} and I_l^{l-1} canonically and choose $\|\cdot\|_u$, $\|\cdot\|_F$ as in (3.19). Under the assumptions of this section 3.2.2, the approximation property (3.4) holds with $\alpha = 2 - 2s$.

In the next section we have to prove the smoothing property for the *same* choice of the norms and for the *same* value of α.

3.3 Discussion of the Smoothing Property

3.3.1 Jacobi-like Iteration for Positive Definite L_l

The smoothing property was the estimate $\| L_l S_l^{\nu} \|_{F\leftarrow u} \le \eta(\nu) h_l^{-\alpha}$ for $1 \le \nu < \nu_{max}(h_l)$. In § 3.1 we motivated the right-hand side $\eta(\nu)\, h_l^{-\alpha}$. But the bound $\nu_{max}(h_l)$ of ν still deserves a comment. It is know that the multi-grid method works for indefinite problems, too. Usual smoothing iterations as those discussed below are divergent: $\| S_l^{\nu} \|_{u\leftarrow u} \ge \exp(c\nu h_l^2) > 1$. Therefore, the inequality (3.3a), $\| L_l S_l^{\nu} \|_{F\leftarrow u} \le \eta(\nu) h_l^{-\alpha}$, cannot hold for all ν as a consequence of

Note 3.6 If (3.3a,b) holds for all $\nu \ge 1$, the smoothing iteration is convergent: $\rho(S_l) < 1$.

The simplest example for a smoothing procedure is a Jacobi-like iteration. The original Jacobi iteration

$$u_l^{j+1} = u_l^j - D_l^{-1}\,(L_l u_l^j - f_l) \qquad (D_l\text{: diagonal of } L_l).$$

does not satisfy the smoothing property. However, this condition holds for ωD_l^{-1} instead of D_l^{-1}, where e.g. $\omega = 1/2$. In many cases the diagonal is $D_l = d\cdot h_l^{-2}\cdot I\,(d\in\mathbb{R})$. Replacing D_l^{-1} by $\omega h_l^2 I$ (ω suitable), we obtain

(3.22) $$\mathcal{S}_l(u_l,f_l) = u_l - \omega h_l^2\,(L_l\, u_l - f_l),\; S_l = I - \omega h_l^2\, L_l,\; T_l = \omega h_l^2\, I.$$

A possible choice of ω is

(3.23) $$\omega = 1/C_L$$

where C_L is a (good) bound for $h_l^2 L_l$:

(3.24) $$\| h_l^2 L_l \| \le C_L \qquad (l \ge o).$$

Here, $\|\cdot\|$ is the *spectral norm* for matrices. This norm coincides with $\|\cdot\|_{F\leftarrow u}$, since in this section we shall use $\|\cdot\|_u = \|\cdot\|_F$ = Euclidean norm as in § 3.2.1. We recall two properties of the spectral norm:

(3.25a) $$\| A\| = \sup \{ \sqrt{\lambda} : \lambda \text{ eigenvalue of } A^*A\},$$

(3.25b) $$\| A\| = \rho(A) = \sup \{|\lambda| : \lambda \text{ eigenvalue of } A\} \quad \text{if } A = A^*.$$

The result for positive semi-definite L_l is given by

Theorem 3.7 Assume that L_l is symmetric and positive semi-definite. Then the modified Jacobi iteration (3.22) with ω from (3.23) satisfies

(3.26) $$\| L_l S_l^\nu \| \le \frac{3}{8} C_L \frac{1}{\nu+1/2} h_l^{-2} \quad \text{for all } \nu \ge 1,\ l \ge o.$$

That means, the smoothing property (3.3) holds with $\|\cdot\|_u = \|\cdot\|_F$ = Euclidean norm and

$$\eta(\nu) = \frac{3}{8} C_L \,/\, (\nu + \tfrac{1}{2}), \quad \alpha = 2, \quad \nu_{max}(h) = \infty .$$

Note that $\alpha = 2$ is the same value as in § 3.2.1 for the approximation property.

Proof. The matrix $L_l S_l^\nu = L_l(I - \omega h_l^2 L_l)^\nu$ is symmetric. Its eigenvalues μ are $\lambda(1 - \omega h_l^2 \lambda)^\nu$ with λ eigenvalue of L_l. By (3.25b) we have

$$\| L_l S_l^\nu \| = \sup \{ |\lambda(1 - \omega h_l^2 \lambda)^\nu| : \lambda \text{ eigenvalue of } L_l\}.$$

λ is non-negative since L_l is positive semi-definite; $1 - \omega h_l^2 \lambda$ is non-negative by definition of ω. As all eigenvalues of L_l are in $[o, C_L h_l^{-2}]$, the estimate

$$\| L_l S_l^\nu \| \le \sup \{\lambda(1 - \omega h_l^2 \lambda)^\nu : o \le \lambda \le C_L h_l^{-2}\}$$

follows. Set $x = \omega h_l^2 \lambda$. x varies in $[0,1]$. Hence, we have

$$\| L_l S_l^\nu \| \le \sup \{C_L h_l^{-2} x(1-x)^\nu : o \le x \le 1\}.$$

The maximum of $x(1-x)^\nu$ in $[0,1]$ is taken at $x = 1/(\nu+1)$. A very close upper bound for this maximum is $3/[8(\nu + 1/2)]$. Hence, (3.26) is proved.

□

The result of Theorem 3.7 corresponds to the approximation property in Theorem 3.4, since both estimates hold for $\|\cdot\|_u = \|\cdot\|_F$ = Euclidean norm and $\alpha = 2$. If we use the approximation property of Theorem 3.5 with norms (3.19), we have to show the corresponding smoothing property. The norms are defined by means of Λ_l. If L_l is positive definite, we may choose $\Lambda_l = L_l$ (cf. (3.18)). $\|\cdot\|_{F\leftarrow u}$ can expressed by means of the spectral norm:

$$\|A\|_{F\leftarrow u} = \|L_l^{-s/2}\, A\, L_l^{-s/2}\| \qquad (s \text{ from } (3.19)).$$

Hence, the smoothing property (3.3a) requires an estimate of

$$\|L_l^{-s/2}\, L_l S_l^{\nu}\, L_l^{-s/2}\| = \|L_l^{1-s}\, S_l^{\nu}\|.$$

Here, we used the fact that S_l and L_l commute. $L_l^{1-s} S_l^{\nu}$ can be estimated in the same way as in the proof of Theorem 3.7. The result is

$$(3.27) \quad \|L_l^{1-s}\, S_l^{\nu}\| \le \eta(\nu)\, h_l^{2s-2}, \quad \eta(\nu) = C_L^{1-s}\left(\frac{1-s}{\nu+1-s}\right)^{1-s}\left(1 - \frac{1-s}{\nu+1-s}\right)^{\nu} \quad (\nu \ge 1).$$

A simpler upper bound of $\eta(\nu)$ is $[C_L/(\frac{2}{1-s}\nu - 1)]^{1-s}$. Therefore, we have

Theorem 3.8 Let $\|\cdot\|_u$ and $\|\cdot\|_F$ defined by (3.19) with $\Lambda_l = L_l$ and $s < 1$. Assume (3.24) for the symmetric and positive definite matrix L_l. Then the modified Jacobi iteration (3.22) with ω from (3.23) satisfies the smoothing property (3.3) with

$$\eta(\nu) = [C_L/(\tfrac{2\nu}{1-s} - 1)]^{1-s}, \quad \alpha = 2 - 2s, \quad \nu_{max} = \infty.$$

Note that $s < 1$ is necessary for (3.3b) : $\eta(\nu) \searrow 0$ $(\nu \to \infty)$. Again, the value $\alpha = 2 - 2s$ coincides with α from the approximation property in Theorem 3.5.

Later we shall need the following result:

Note 3.9 The modified Jacobi iteration satisfies $\|S_l^{\nu}\|_{u\leftarrow u} \le 1$ ($\|\cdot\|_u$ Euclidean norm or defined by (3.19)).

3.3.2 Modified Jacobi Iteration for General L_l

Often, the matrix L_l is not positive definite, but it can be split into

$$L_l = L_l' + L_l'',$$

where L_l' is the positive definite principal part and L_l'' is of lower order. Let S_l' be the smoothing iteration matrix corresponding to L_l' and set $S_l'' = S_l - S_l'$. As seen in the foregoing section 3.3.1, L_l' and S_l' fulfil the smoothing property. The folloing lemma shows, that under suitable assumptions, L_l and S_l have the smoothing property, too. The conditions (3.29) will be discussed below.

Lemma 3.10 Let L_l and S_l be split into $L_l' + L_l''$ and $S_l' + S_l''$, resp. Assume that L_l' and S_l' satisfy the smoothing property:

$$(3.28)\qquad \begin{aligned} &\| L_l' S_l'^{\nu} \|_{F\leftarrow u} \le \eta'(\nu) h_l^{-\alpha} \quad \text{for all } \nu \in [o, \nu'_{max}(h_l)) \text{ and } l \ge 1, \\ &\eta'(\nu) \searrow o \quad (\nu\to\infty), \ \nu'_{max}(h) \nearrow \infty \ (h\to o) \quad \text{or } \nu'_{max}(h) = \infty. \end{aligned}$$

Furthermore, suppose that there is some $\beta > o$ such that

$$(3.29)\qquad \| L_l'' \|_{F\leftarrow u} \le C_L'' h_l^{\beta-\alpha}, \quad \| S_l' \|_{u\leftarrow u} \le C_S', \quad \| S_l'' \|_{u\leftarrow u} \le C_S'' h_l^{\beta}.$$

Then the smoothing property (3.3a-c) holds for L_l with the same α. $\eta(\nu)$ can be chosen as $c\eta'(\nu)$ for any $c > 1$. $\nu_{max}(h) \le \nu'_{max}(h)$ satisfies (3.3c): $\nu_{max}(h) \nearrow \infty$ $(h\to o)$.

Proof. The straight-forward estimation yields

$$\| L_l S_l^{\nu} \|_{F\leftarrow u} \le \eta(\nu,h_l) h_l^{-\alpha} := \{\eta'(\nu) + h_l^{\beta}[C_L''(C_S' + C_S'' h_l^{\beta})^{\nu} + C_S''\eta'(o)(C_S' + C_S'' h_l^{\beta})^{\nu-1}]\}\, h_l^{-\alpha}$$

for $\nu < \nu'_{max}(h_l)$. Because of $\eta(\nu,o) = \eta'(\nu)$, $\eta(\nu,h_l) \le c\eta'(\nu) =: \eta(\nu)$ $(c > 1)$ holds for $\nu \in [1,\nu_{max}]$ if h_l is sufficiently small. Denote the largest possible h_l by $h_l(\nu_{max})$. $h_l(\nu)$ is positive and monotoneously increasing. Hence, there exists the inverse function $\nu_{max}(h)$ with $\nu_{max}(h(\nu)) = \nu$ and $\nu_{max}(h) \to \infty$ for $h \to o$.

□

Note that (3.28) with $\nu'_{max}(h) = \infty$ does not imply $\nu_{max}(h) = \infty$. For example, a positive definite principal part L_l' satisfies (3.28) with $\nu'_{max} = \infty$ as shown in § 3.3.1. But $L_l = L_l' + L_l''$ may be indefinite (i.e. there is also a negative eigenvalue of L_l). Then $\mathcal{S}_l$ is divergent, and Note 3.6 proves $\nu_{max}(h) < \infty$.

In the sequel we describe the application of the perturbation Lemma 3.10 to finite element discretizations. The bilinear form (3.7b) corresponds to the diffential operator $L = L' + L''$ with

$$L' = \sum_{i,j=1}^{d} \frac{\partial}{\partial x_i} a_{ij} \frac{\partial}{\partial x_j}, \quad L'' = \sum_{i=1}^{d} (b_i \frac{\partial}{\partial x_i} - \frac{\partial}{\partial x_i} c_i) + d.$$

$L_l = I^l L I_l$ (cf. (3.11)) can be split into $L_l' + L_l''$, where $L_l' = I^l L' I_l$, $L_l'' = I^l L'' I_l$. S_l' is $I - \omega h_l^2 L_l'$ (cf. (3.22)) and, therefore, $S_l'' = S_l - S_l' = \omega h_l^2 L_l''$. Note that L'' is of lower order (at most first order). From this fact we obtain

Note 3.11 Assume that the coefficients are bounded:

$$(3.30)\qquad a_{ij},\ b_i,\ c_i,\ d \in L^\infty(\Omega).$$

Suppose ellipticity

$$(3.31)\qquad \sum_{i,j=1}^{d} a_{ij}(x)\,\xi_i\,\xi_j \geq \varepsilon \sum_{i=1}^{d} \xi_i^2,\ \varepsilon > 0, \text{ for all } \xi \in \mathbb{R}^d$$

and the inverse assumption

$$(3.32)\qquad \|v\|_1 \leq C\, h_l^{-1}\, \|v\|_0 \quad \text{for all } v \in \mathcal{U}_l.$$

Then the conditions (3.28) and (3.29) of Lemma 3.9 are satisfied for $\|\cdot\|_u = \|\cdot\|_F$ = Euclidean norm with $\beta = 1$, $\alpha = 2$.

Proof. L_l' fulfils (3.28) as demonstrated in § 3.3.1. (3.31) and (3.32) imply $\|L_l''\| \leq C\, h_l^{-1} = C\, h_l^{\beta-\alpha}$ and consequently $\|S_l''\| \leq C'\, h_l = C'\, h_l^{\beta}$.
□

Note 3.9 shows $\|S_l'^{\nu}\|_{u\leftarrow u} \leq C'$ $(1 \leq \nu < \nu'_{max}(h_l))$. The following lemma states the corresponding estimate for S_l.

Lemma 3.12 Assume $\|S_l'^{\nu}\|_{u\leftarrow u} \leq C'$ for $\nu \in [1, \nu'_{max}(h_l))$ with $\nu'_{max}(h) \nearrow \infty$ $(h \to o)$ and let the supposition (3.29) of Lemma 3.10 hold. Then

$$\|S_l^{\nu}\|_{u\leftarrow u} \leq C \quad \text{for all } \nu \in [1, \nu_{max}(h_l)), \quad l \geq 1$$

is valid for some $\nu_{max}(h) \nearrow \infty$ $(h \to o)$.

3.3.3 Smoothing Property for Gauß-Seidel Iteration

In the previous subsections we studied the modified Jacobi iteration, since the proof of its smoothing property is very simple but typical for more general cases. Usually, Gauß-Seidel iteration (= relaxation) is the preferred smoothing procedure. In particular, Gauß-Seidel iteration with 'red-black' ordering of the grid points turns out to be very efficient (cf. Stüben and Trottenberg [16]). Here, we prove the smoothing property of the Gauß-Seidel iteration for the case of a positive definite matrix L_l. The discussion of a general L_l follows the considerations of § 3.3.2 and is given in [10].

Define some ordering $(u_l)_\mu$ $(\mu = 1,\ldots,n_l)$ of the components of u_l and f_l. Then, the matrix L_l is assumed to have the block structure

$$L_l = \begin{bmatrix} d_1 & -b \\ -a & d_2 \end{bmatrix} = D_l - A_l - B_l,$$

(3.33)

$$D_l = \begin{bmatrix} d_1 & o \\ o & d_2 \end{bmatrix}, \quad A_l = \begin{bmatrix} o & o \\ a & o \end{bmatrix}, \quad B_l = \begin{bmatrix} o & b \\ o & o \end{bmatrix},$$

where D_l is diagonal or block-diagonal. Examples for matrices with structure (3.33) are

- 'red-black' ordering for five-point formulae,
- 'zebra-line' ordering for nine-point formulae

(cf. [16]). The estimate (3.24) of L_l implies

(3.34) $\qquad \| D_l \| \le C_D \, h_l^{-2} \qquad (\|\cdot\|$: spectral norm).

The Gauß-Seidel iteration is $u_l^{j+1} = (D_l - A_l)^{-1} (B_l u_l^j + f_l)$; hence,

(3.35) $\qquad \mathcal{S}_l(u_l, f_l) = (D_l - A_l)^{-1}(B_l u_l^j + f_l), \; S_l = (D_l - A_l)^{-1} B_l, \; T_l = (D_l - A_l)^{-1}.$

We prove the following result:

<u>Theorem 3.13</u> Let L_l be a symmetric and positive definite matrix with decomposition (3.33). Then, the Gauß-Seidel iteration (3.35) satisfies

(3.36) $\qquad \| L_l S_l^\nu \| \le \| D_l \| \dfrac{2}{3\sqrt{3}\nu} \qquad$ for all $\nu \ge 1$.

Together with estimate (3.34) the smoothing property (3.3a-c) holds with $\alpha = 2$, $\eta(\nu) = 2C_D/(3\sqrt{3}\nu)$, $\nu_{max}(h) = \infty$, $\|\cdot\|_U = \|\cdot\|_F$ = Euclidean norm.

<u>Proof</u>. Since L_l is positive definite, D_l is positive definite, too. The matrices S_l and $L_l S_l^\nu$ are

$$S_l = \begin{bmatrix} o & d_1^{-1} b \\ o & d_2^{-1} a d_1^{-1} b \end{bmatrix}, \quad L_l S_l^\nu = \begin{bmatrix} o & b\{[d_2^{-1} a d_1^{-1} b]^{\nu-1} - [d_2^{-1} a d_1^{-1} b]^{\nu}\} \\ o & o \end{bmatrix}.$$

Introducing $B = d_2^{-1/2} a d_1^{-1} b d_2^{-1/2}$ and $C = d_1^{-1/2} b d_2^{-1/2} B^{\nu-1}(I-B)$, we have $D_l^{-1/2} L_l S_l^\nu D^{-1/2} = \begin{bmatrix} o & C \\ o & o \end{bmatrix}$. By symmetry of L_l, $b^* = a$ holds implying $B^* = B$. Hence, $C^*C = B^{*\nu-1} (I-B^*) \, B \, B^{\nu-1}(I-B) = B^{2\nu-1}(I-B)^2$ shows

$$\| D_1^{-1/2} L_1 S_1^{\nu} D_1^{-1/2} \| = \rho(B^{2\nu-1}(I-B)^2)^{1/2}$$

(cf. (3.25a)). Note that $\rho(S_1) = \rho(d_2^{-1} a d_1^{-1} b) = \rho(B)$. Under our assumptions the Gauß-Seidel iteration converges, i.e. $\rho(S_1) < 1$ (cf. Meis and Marcowitz [13]). Hence, the eigenvalues of B are contained in the interval [0,1]. As in the proof of Theorem 3.7 we have

$$\| D_1^{-1/2} L_1 S_1^{\nu} D_1^{-1/2} \|^2 = \rho(B^{2\nu-1}(I-B)^2) \le \sup\{\lambda^{2\nu-1}(1-\lambda)^2 : 0 \le \lambda \le 1\} \le \left(\frac{2}{3\sqrt{3}\nu}\right)^2$$

for $\nu \ge 1$. The assertion (3.35) follows from

$$\| L_1 S_1^{\nu} \| \le \| D_1^{1/2} \| \, \| D_1^{-1/2} L_1 S_1^{\nu} D_1^{-1/2} \| \, \| D_1^{1/2} \| = \| D_1 \| \, \| D_1^{-1/2} L_1 S_1^{\nu} D_1^{-1/2} \| .$$

□

It is also possible to prove the smoothing property of the Gauß-Seidel iteration in case of the norms (3.19) (cf. [10]). For general L_1 one can apply Lemma 3.10.

3.4 Two-Grid Convergence for Finite Element Equations

3.4.1 Case of H^2-Regular Problems

The Dirichlet problem $Lu = f$ has the weak formulation (3.7a), $a(u,v) = (f,v)$. The *adjoint problem* $L^* w = f$ is defined by

$$a(v,w) = (v,f)_{L^2(\Omega)} \quad \text{for all } v \in H_0^1(\Omega).$$

L and L^* are called H^2-*regular* if the solutions of $Lu = f$ and $L^* w = f$ satisfy

(3.37) $\quad \| u \|_2 \le C \, \| f \|_0 , \; \| w \|_2 \le C \, \| f \|_0 \quad$ for all $f \in L^2(\Omega)$.

The existence of L^{-1} and L_1^{-1} is guaranteed by

Note 3.14 The original problem (3.7a) and the finite element problem (3.10) are uniquely solvable if for some $\varepsilon > 0$ (3.38a,b) hold:

(3.38a) $\quad \sup\{|a(u,v)| : v \in H_0^1(\Omega), \|v\|_1 = 1\} \ge \varepsilon \, \|u\|_1$ for all $u \in H_0^1(\Omega)$,

(3.38b) $\quad \sup\{|a(u,v)| : v \in \mathcal{H}_1, \|v\|_1 = 1\} \ge \varepsilon \, \|u\|_1$ for all $u \in \mathcal{H}_1$ and $1 \ge 0$.

Furthermore, the conditions (3.9), (3.30), (3.31), (3.37) imply the error estimate (3.15): $\| u_{\mathcal{H}_1} - u \|_0 \le C \, h_1^2 \, \|f\|_0$.

The estimate (3.15) was the main condition for the approximation property (cf.

Theorem 3.4). The Theorems 3.7 or 3.13 together with Lemma 3.10 and Note 3.11 imply the smoothing property for the same choice of norms: $\|\cdot\|_u = \|\cdot\|_F$ = Euclidean norm. Therefore, Theorem 3.2 proves the two-grid convergence for a general finite element equation. Collecting all suppositions we obtain

Theorem 3.15 (two-grid convergence for f.e.m) If the assumptions mentioned above [(3.8), (3.9), (3.12a,b), (3.14), (3.16), (3.30), (3.31), (3.32), (3.37), (3.38a,b)] hold and if $\mathcal{S}_l$ is either the modified Jacobi iteration (3.22) with ω from (3.23) or the Gauß-Seidel iteration (3.35) with decomposition (3.33), then the two-grid iteration (1.8) converges:

(3.39) $\quad \| M_l \| \le C/\nu \le \rho < 1 \qquad (\nu_{min} \le \nu < \nu_{max}(h_l),\ h_l \le h_{max}$, cf. Theorem 3.2)

Here, we used that for both smoothing procedures the function $\eta(\nu)$ behaves like C'/ν. The estimation of $\| M_l(\nu)\|$ by C/ν is asymptotically optimal. The exact evaluation of $\| M_l(\nu)\|$ in the one-dimensional case yields $O(1/\nu)$, too (cf. Stüben and Trottenberg [16]).

3.4.2 Less Regular Problems

As mentioned in § 3.2.2 the H^2-regularity (3.37) does not hold if for instance re-entrant corners are present. Such domains having boundaries locally described by Lipschitz continuous functions are called Lipschitz domains.

Theorem 3.16 (Nečas [14]) Let Ω be a Lipschitz domain. Assume that L satisfies (3.30), (3.31) and that a_{ij} are real and Hölder continuous: $a_{ij} \in \overline{C}^{\sigma}(\Omega)$. If $\lambda = o$ is no eigenvalue of L, then

(3.40) $\quad \|u\|_{1+t} \le C\, \|f\|_{t-1},\ \|w\|_{1+t} \le \|f\|_{t-1}$ for all $f \in H^{t-1}(\Omega)$

holds for all $|t| < \min(\sigma, 1/2)$, where $u,w \in H_o^{1+t}(\Omega)$ are the solutions of the Dirichlet boundary value problems $Lu = L^*w = f$, $w = u = o$ on Γ.

With H^{1+t}-regularity (3.40) instead of H^2-regularity we obtain nearly the same convergence result:

Theorem 3.17 (two-grid convergence) Suppose the assumptions of Theorem 3.15 but replace the H^2-regularity (3.37) by H^{2-s}-regularity (3.40) [$t = 1-s$] for some $s < 1$. In addition assume (3.21) and choose the norms $\|\cdot\|_u$, $\|\cdot\|_F$ by (3.19). Then we have for suitable ν_{min} and h_{max} that

(3.41) $\quad \| M_l(\nu)\|_{u \leftarrow u} \le C/\nu^{1-s} \le \rho < 1$

for $\nu_{min} \le \nu < \nu_{max}(h_l)$, $h_l \le h_{max}$.

Note that the order 2-s of regularity influences the asymptotic behaviour of $\| M_l(\nu) \|_{U \leftarrow U}$ as $\nu \to \infty$. A proof of Theorem 3.17 can be found in [9].

3.5 Quantitative Estimates for Symmetric and Positive Definite Problems

The foregoing results show convergence for number ν sufficiently large. In the case of symmetric and positive problems,

(3.42a) $$L_l = L_l^* > 0,$$

it is possible to prove convergence of the two-grid iteration for *all* $\nu > 0$, if the restriction and prolongation satisfy

(3.42b) $$I_l^{l-1} = (I_{l-1}^l)^*,$$

if L_{l-1} is constructed by

(3.42c) $$L_{l-1} = I_l^{l-1} \, L_l \, I_{l-1}^l$$

and if the two-grid iteration (1.7a,b) is followed by further smoothing steps of the modified Jacobi iteration (3.22). Thus, the algorithm is

u_l^j : j^{th} iterate

$\bar{u}_l$: result of $\nu/2$ smoothing steps of (3.22),

(3.43) $$\hat{u}_l := \bar{u}_l + I_{l-1}^l \, L_{l-1}^{-1} \, I_l^{l-1} \, (f_l - L_l \bar{u}_l),$$

u_l^{j+1} : result of $\nu/2$ smoothing steps of (3.22).

Note that the conditions (3.42b,c) hold for the canonical choice of I_{l-1}^l and I_l^{l-1} in the case of finite element discretizations (cf. (3.12b)).

The iteration matrix of the process (3.43) is

$$M_l = S_l^{\nu/2} \, (I - I_{l-1}^l \, L_{l-1}^{-1} \, I_l^{l-1} \, L_l) \, S_l^{\nu/2} \quad \text{with } S_l = I - \omega h_l^2 L_l, \ \omega = 1/C_L \text{ from (3.23).}$$

Since S_l and L_l are commutative, a matrix similar to M_l is

$$\hat{M}_l = L_l^{1/2} \, M_l \, L_l^{-1/2} = S_l^{\nu/2} \, Q_l \, S_l^{\nu/2} \quad \text{with } Q_l = L_l^{1/2} \, (L_l^{-1} - I_{l-1}^l \, L_{l-1}^{-1} \, I_l^{l-1}) \, L_l^{1/2}.$$

The symmetry of $\hat{M}_l$ is evident. The spectral norm of $\hat{M}_l$ coincides with the $\|\cdot\|_{U \leftarrow U}$ norm of M_l,

(3.44) $$\| M_l \|_{U \leftarrow U} = \| \hat{M}_l \| \qquad (\|\cdot\| \text{ spectral norm}),$$

where $\|\cdot\|_U$ is the 'energy norm':

(3.45) $$\| u_l \|_U = \| L_l^{1/2} u_l \| \qquad (\|\cdot\| \text{ Euclidian norm}).$$

<u>Theorem 3.18</u> Assume (3.42a-c) and the approximation property

(3.46) $$\| L_l^{-1} - I_{l-1}^{l} L_{l-1}^{-1} I_l^{l-1} \| \le C_A h_l^2$$

with respect to the spectral norm (cf. Theorem 3.4). Then the two-grid iteration (3.43) converges for $\nu > o$. The $\|\cdot\|_{U \leftarrow U}$ norm of the iteration matrix $M_l = M_l(\nu)$ is

(3.47) $$\| M_l(\nu) \|_{U \leftarrow U} \le \begin{cases} C_M \nu^\nu/(1+\nu)^{1+\nu} < 1 & \text{if } o \le C_M \le 1+\nu \\ (1 - 1/C_M)^\nu < 1 & \text{if } C_M \ge 1+\nu \end{cases}$$

where $C_M = C_A C_L$ ($C_A C_L$ from (3.46), (3.23)).

The asymptotic behaviour of $\| M_l(\nu) \|_{U \leftarrow U}$ is again as $C/(1+\nu)$ for $\nu \to \infty$.

<u>Proof:</u> In the sequel the notation $A \le B$ means that A and B are symmetric and that B - A is positive semi-definite. Because of (3.42c) the matrix $Q_l = I - L_l^{1/2} I_{l-1}^{l} L_{l-1}^{-1} I_l^{l-1} L_l^{1/2}$ satisfies $0 \le Q_l \le I$, implying

(3.48) $$0 \le Q_l \le \alpha Q_l + (1-\alpha) I \qquad \text{for all } o \le \alpha \le 1.$$

The approximation property (3.46) yields $L_l^{-1} - I_{l-1}^{l} L_{l-1}^{-1} I_l^{l-1} \le C_A h_l^2 I$ and

$$Q_l = L_l^{1/2} (L_l^{-1} - I_{l-1}^{l} L_{l-1}^{-1} I_l^{l-1}) L_l^{1/2} \le C_A h_l^2 L_l.$$

This inequality together with (3.48) shows

$$0 \le Q_l \le \alpha C_A h_l^2 L_l + (1-\alpha) I \qquad \text{for all } o \le \alpha \le 1.$$

By definition of $\hat{M}_l$ and $\omega = 1/C_L$ we have

(3.49) $$0 \le \hat{M}_l = S_l^{\nu/2} Q_l S_l^{\nu/2} \le S_l^{\nu/2} [\alpha C_M \omega h_l^2 L_l + (1-\alpha) I] S_l^{\nu/2}.$$

The spectral norm of the right-hand side is

$$\max \{ (1 - \omega h_l^2 x)^\nu (\alpha C_M \omega h_l^2 x + 1 - \alpha) : x \in \text{spectrum of } L_l \}$$

as can be shown by the same argument as in the proof of Theorem 3.7. This value is bounded by

$$m(\alpha) = \max_{0\le\xi\le 1} (1-\xi)^{\nu}(\alpha C_M \xi + 1 - \alpha).$$

Thus, (3.49) becomes

(3.50) $$0 \le \hat{M}_l(\nu) \le m(\alpha)\, I \qquad \text{for all } 0 \le \alpha \le 1.$$

If $0 \le C_M \le 1 + \nu$, choose $\alpha = 1$. Then the maximum $m(1) = C_M\, \nu^{\nu}/(1+\nu)^{1+\nu}$ is taken at $\xi = 1/(1+\nu)$. Otherwise, choose $\alpha = \nu/(C_M - 1) \in [0,1]$. The maximum $m(\alpha) = (1 - 1/C_M)^{\nu}$ is taken at $\xi = 1/C_M$. Hence, (3.47) is proved.

□

Note that the approximation property is explicitly used, whereas the smoothing property $\| L_l^{1/2} S_l^{\nu/2} \| = \| L_l S_l^{\nu} \|^{1/2} \le \ldots$ is changed into an estimate of $\| [a L_l + b I]^{1/2} S_l^{\nu/2} \|$. A similar technique will be used in Section 4.3.

Note 3.19 Assume (3.42a-c) and (3.46). If we use the two-grid iteration (1.7a,b) with ν smoothing steps (3.22) before and no smoothing after the coarse-grid correction, the spectral radius of the iteration matrix $M_l = (I - I_{l-1}^{l} L_{l-1}^{-1} I_l^{l-1} L_l) S_l^{\nu}$ satisfies the same estimate as $\|M_l(\nu)\|_{u\leftarrow u}$ in Theorem 3.18. Therefore convergence is guaranteed.

Proof. $M_l(\nu)$ as defined in Note 3.19 and $\hat{M}_l(\nu)$ from Theorem 3.18 have the same spectral radius. Note that, formally, Theorem 3.18 holds for odd ν, too.

□

Note 3.20 For the proof of the estimate (3.47) the approximation property (3.46) with respect to the spectral norm is needed. If we require the weaker approximation property with respect to the norms (3.19) for some fixed $s \in [0,1)$:

(3.51) $$\| L_l^{s/2} (L_l^{-1} - I_{l-1}^{l} L_{l-1}^{-1} I_l^{l-1}) L_l^{s/2} \| \le (C_A h_l^2)^{1-s},$$

then the conditions (3.42a-c) still imply convergence. Instead of (3.47) we have

(3.52) $$\| M_l(\nu) \|_{u\leftarrow u} \le \begin{cases} C_M^{1-s} (1-s)^{1-s} \nu^{\nu}/(\nu+1-s)^{\nu+1-s} < 1 & \text{if } 0 \le C_M \le (\nu+1-s)/(1-s) \\ (1 - 1/C_M)^{\nu} < 1 & \text{if } C_M \ge (\nu+1-s)/(1-s), \end{cases}$$

where $\|\cdot\|_u$ is again the energy norm (3.45).

By the results of § 3.2.2, the estimate (3.17) (for some $s < 1$) ensures (3.51) and, therefore, the estimate (3.52) that is independent of h_l. Even for $s = 1$, $\| M_l(\nu) \|_{u\leftarrow u} \le [1 - h_l^2 / (C_L \| L_l^{-1} \|)]^{\nu} < 1$ can be shown.

4. Convergence of the Multi-Grid Iteration

4.1 Iteration Matrix

Let

$$(4.1)\qquad u_l^{j+1} = M_l^{MGM} u_l^j + N_l^{MGM} f_l \quad (M_l^{MGM} = M_l^{MGM}(\nu),\ N_l^{MGM} = N_l^{MGM}(\nu))$$

be the representation (2.1) of the multi-grid iteration (1.10). Since the multi-grid algorithm is defined recursively, also the iteration matrix M_l^{MGM} is defined by a recursive formula.

For $l = 1$, the multi-grid iteration coincides with the two-grid iteration (1.8); thus

$$(4.2a)\qquad M_1^{MGM} = M_1^{TGM}, \quad N_1^{MGM} = N_1^{TGM}.$$

Let $l > 1$. The smoothing step (1.10b) yields the intermediate function

$$\bar{u}_l = S_l^\nu u_l^j + [\ldots]f_l.$$

The γ-fold application of the multi-grid iteration at level $l-1$ to $L_{l-1} v_{l-1} = d_{l-1} := I_l^{l-1}(f_l - L_l\bar{u}_l)$ with starting guess $v_{l-1}^0 = o$ results in

$$v_{l-1}^\gamma = \sum_{\mu=o}^{\gamma-1} (M_{l-1}^{MGM})^\mu N_{l-1}^{MGM} I_l^{l-1} (-L_l S_l^\nu u_l^j) + [\ldots] f_l.$$

The coarse-grid correction (1.10e) produces

$$u_l^{j+1} = [I - I_{l-1}^l \sum_{\mu=o}^{\gamma-1} (M_{l-1}^{MGM})^\mu N_{l-1}^{MGM} I_l^{l-1} L_l] S_l^\nu u_l^j + [\ldots] f_l.$$

Obviously, any solution of $L_l u_l = f_l$ is a stationary point of (4.1). Assuming L_{l-1} to be regular and applying Eq. (2.3) to N_{l-1}^{MGM} we obtain

$$\sum_{\mu=o}^{\gamma-1} (M_{l-1}^{MGM})^\mu N_{l-1}^{MGM} = \sum_{\mu=o}^{\gamma-1} (M_{l-1}^{MGM})^\mu (I - M_{l-1}^{MGM}) L_{l-1}^{-1} = [I - (M_{l-1}^{MGM})^\gamma] L_{l-1}^{-1}.$$

Inserting this expression into the foregoing equation, we conclude that

$$M_l^{MGM} = \{I - I_{l-1}^l [I - (M_{l-1}^{MGM})^\gamma] L_{l-1}^{-1} I_l^{l-1} L_l\} S_l^\nu$$

is the multi-grid iteration matrix. The comparison with the two-grid iteration matrix M_l^{TGM} (cf. (3.1)) demonstrates

Note 4.1 The iteration matrix of the multi-grid algorithm (1.10) is defined by (4.2a) for $l = 1$ and

$$(4.2b)\qquad M_l^{MGM} = M_l^{TGM} + I_{l-1}^l (M_{l-1}^{MGM})^\gamma L_{l-1}^{-1} I_l^{l-1} L_l S_l^\nu \quad \text{for } l \geq 2.$$

4.2 Multi-Grid Convergence

The representation (4.2b) leads to

$$(4.3)\qquad \| M_l^{MGM}\|_{U\leftarrow U} \le \| M_l^{TGM}\|_{U\leftarrow U} + \| I_{l-1}^{l}\|_{U\leftarrow U}\, \| M_{l-1}^{MGM}\|_{U\leftarrow U}^{\gamma}\, \| L_{l-1}^{-1}\, I_l^{l-1}\, L_l\, S_l^{\nu}\|_{U\leftarrow U}.$$

The norms of I_{l-1}^{l} and $L_{l-1}^{-1}\, I_l^{l-1}\, L_l\, S_l^{\nu}$ are expected to be uniformly bounded. Unfortunately, the naive estimate $\|L_{l-1}^{-1} I_l^{l-1} L_l S_l^{\nu}\|_{U\leftarrow U} \le \| L_{l-1}^{-1}\|_{U\leftarrow F}\, \|I_l^{l-1}\|_{F\leftarrow F}\, \|L_l S_l^{\nu}\|_{F\leftarrow U}$ would result in a bound $O(h_l^{-\alpha})$ with $\alpha > o$ instead of $O(1)$. However, the next lemma shows that the desired estimate follows from estimates of the two-grid iteration matrix and simple additional assumptions on I_{l-1}^{l} and S_l^{ν}.

Lemma 4.2 Assume

$$(4.4)\qquad \| S_l^{\nu}\|_{U\leftarrow U} \le C \quad \text{for all } l \ge 1 \text{ and all } o < \nu < \nu_{max}(h_l),$$

$$(4.5)\qquad \| I_{l-1}^{l}\, v_{l-1}\|_{U} \ge \frac{1}{C}\, \| v_{l-1}\|_{U} \quad \text{for all } v_{l-1} \in \mathcal{U}_{l-1}.$$

Then the estimate (4.6) holds:

$$(4.6)\qquad \| L_{l-1}^{-1}\, I_l^{l-1}\, L_l\, S_l^{\nu}\|_{U\leftarrow U} \le C'(1 + \| M_l^{TGM}(\nu)\|_{U\leftarrow U}) \text{ for } l\ge 1,\ o < \nu < \nu_{max}(h_l).$$

Proof: Inequality (4.5) implies $\| L_{l-1}^{-1}\, I_l^{l-1}\, L_l\, S_l^{\nu}\|_{U\leftarrow U} \le C\, \| I_{l-1}^{l}\, L_{l-1}^{-1}\, I_l^{l-1} L_l S_l^{\nu}\|_{U\leftarrow U}$. Therefore, the assertion follows from

$$I_{l-1}^{l}\, L_{l-1}^{-1}\, I_l^{l-1}\, L_l\, S_l^{\nu} = S_l^{\nu} - [L_l^{-1} - I_{l-1}^{l}\, L_{l-1}^{-1}\, I_l^{l-1}]\, L_l\, S_l^{\nu} = S_l^{\nu} - M_l^{TGM}.$$

□

The estimate (4.4) is very natural. For a convergent smoothing procedure and for a suitable norm one would expect $\| S_l^{\nu}\|_{U\leftarrow U} \le 1$ ($l \ge 1$, $\nu \ge o$, cf. Note 3.9). Moreover, a criterion for (4.4) is given by Lemma 3.12: If S_l' satisfies (4.4), then $S_l = S_l' + S_l''$ with lower order term S_l'' does, too. Also inequality (4.5) is not restrictive for an injective prolongation I_{l-1}^{l}. Assuming $\| I_{l-1}^{l}\|_{U\leftarrow U}\, \| L_{l-1}^{-1}\, I_l^{l-1}\, L_l\, S_l^{\nu}\|_{U\leftarrow U} \le C^{*}$ ($l \ge 2$) we obtain from (4.3) the recursive estimate

$$(4.7)\qquad \rho_l \le \rho + C^{*}\rho_{l-1}^{\gamma}, \qquad \rho_1 \le \rho$$

for $\rho_k = \| M_k^{MGM}\|_{U\leftarrow U}$ and $\rho = \sup_k \| M_k^{TGM}\|_{U\leftarrow U}$. We need $\overline{\lim}\, \rho_l < 1$ for convergence. An elementary analysis shows

Lemma 4.3 Let $\gamma \geq 2$. If ρ satisfies

(4.8) $$\rho < 1 - 1/\gamma, \quad C^*\rho^{\gamma-1} \leq (\gamma - 1)^{\gamma-1}\,\gamma^{-\gamma},$$

then any solution of (4.7) satisfies $\rho_1 \leq \rho^* < 1$. Furthermore, $\rho_1 \leq C(\rho)\rho$ holds with $C(\rho) \to 1$ $(\rho \to o)$. In the case of $\gamma = 2$, condition (4.8) becomes

(4.8') $$\rho < 1/2, \quad 4\rho\, C^* \leq 1, \quad \gamma = 2$$

and ρ_1 is bounded by

(4.9) $$\rho_1 \leq C(\rho)\,\rho < 1 \quad \text{with } C(\rho) = 2/(1 + \sqrt{1 - 4\rho\, C^*}).$$

Now we are able to prove the multi-grid convergence.

Theorem 4.4 Assume the smoothing property (3.3), the approximation property (3.4), and in addition (4.4) and

(4.10) $$\frac{1}{C}\,\| v_{l-1}\|_u \leq \| I_{l-1}^{l}\, v_{l-1}\|_u \leq C\,\| v_{l-1}\|_u \quad \text{for all } v_{l-1} \in \mathcal{U}_{l-1}.$$

For any $\rho^* \in (0,1)$ there are ν_{min} and h_{max} such that

(4.11) $$\| M_l^{MGM}(\nu)\|_{u \leftarrow u} \leq C'\eta(\nu) \leq \rho^* < 1 \quad (l \geq 1)$$

if $\gamma \geq 2$, $\nu_{min} \leq \nu < \nu_{max}(h_l)$, $h_l \leq h_{max}$. Hence, the multi-grid iteration (1.10) converges.

Proof: Choose ρ such that $C(\rho)\,\rho \leq \rho^*$ with $C(\rho)$ from Lemma 4.3. Theorem 3.2 shows $\| M_l^{TGM}\|_{u \leftarrow u} \leq C\,\eta(\nu) \leq \rho$ $(\nu_{min} \leq \nu < \nu_{max}(h_l) \leq \nu_{max}(h_1),\ h_1 \leq h_{max})$ for suitable ν_{min} and h_{max}. From Note 4.1, estimate (4.3), and Lemma 4.2 we obtain (4.7). Lemma 4.3 proves $\rho_l \leq \rho^*$. The constant C' of (4.11) is $C \cdot C(\rho)$ (C from (3.4), $C(\rho)$ from Lemma 4.3).

□

Theorem 4.4 shows that two-grid convergence (implied by the smoothing and approximation property) is almost sufficient for multi-grid convergence. Conditions (4.4) and (4.10) are usually satisfied. The requirement $\gamma \geq 2$ is necessary for this convergence proof. For $\gamma = 1$, inequality (4.7) allows a similar convergence proof if and only if $C^* < 1$. But usually one expects $C^* \geq 1$ and $C^* \approx 1$. Another kind of proof for the case of $\gamma = 1$ ('V-cycle') is described in the following section.

4.3 Estimates for the V-Cycle in the Symmetric and Positive Definite Case

Consider again the situation of § 3.5. L_l is assumed to be symmetric and positive definite. The restriction is the adjoint of the prolongation. The smoothing by the modified Jacobi iteration (3.22) is performed $\nu/2$-times before and $\nu/2$-times after the coarse-grid correction. The multi-grid iteration with the choice $\gamma = 1$ is called 'V-cycle'.

In this case the technique of Braess [6] can be extended.

Theorem 4.5 Assume (3.42a-c), (3.46) as in Theorem 3.18. Then the multi-grid version of the two-grid iteration (3.43) with $\gamma = 1$ and $\nu > o$ converges. The iteration matrix $M_l'(\nu)$ satisfies

(4.12) $$\| M_l'(\nu)\|_{u\leftarrow u} \le C_M / (\nu + C_M) < 1,$$

where $\|\cdot\|_u$ is the energy norm (3.45) and $C_M = C_A C_L$ as in Theorem 3.18.

Proof. Define $\hat{M}_l'$ by $L_l^{1/2} M_l' L_l^{-1/2}$. We want to prove

(4.13) $$0 \le \hat{M}_l'(\nu) \le \varepsilon I, \quad \varepsilon = C_M / (\nu + C_M)$$

by induction. As in (4.2b) we have

$$M_l'(\nu) = M_l(\nu) + S_l^{\nu/2} I_{l-1}^{l} M_{l-1}' L_{l-1}^{-1} I_l^{l-1} L_l S_l^{\nu/2}$$

Introducing $\hat{I}_l^{l-1} = L_{l-1}^{-1/2} I_l^{l-1} L_l^{1/2}$ and $\hat{I}_{l-1}^{l} = L_l^{1/2} I_l^{l-1} L_{l-1}^{-1/2}$ we get

$$\hat{M}_l'(\nu) = \hat{M}_l(\nu) + S_l^{\nu/2} I_{l-1}^{l} \hat{M}_{l-1}' \hat{I}_l^{l-1} S_l^{\nu/2} = S_l^{\nu/2} [Q_l + \hat{I}_{l-1}^{l} \hat{M}'_{l-1} \hat{I}_l^{l-1}] S_l^{\nu/2}$$

with $Q_l = I - \hat{I}_{l-1}^{l} \hat{I}_l^{l-1}$ and $\hat{M}_l(\nu)$ as in § 3.5. Obviously, $\hat{M}_l' \ge 0$ is fulfilled. The representation of Q_l yields

$$0 \le \hat{M}_l' \le S_l^{\nu/2} [(1-\varepsilon) Q_l + \hat{I}_{l-1}^{l} (\hat{M}_{l-1}' - \varepsilon I)\hat{I}_l^{l-1} + \varepsilon I] S_l^{\nu/2}.$$

Since $\hat{M}_{l-1}' - \varepsilon I \le 0$ by induction, the inequality

$$0 \le \hat{M}_l' \le S_l^{\nu/2} [(1-\varepsilon)Q_l + \varepsilon I] S_l^{\nu/2}$$

holds. As in the proof of Theorem 3.18, the right-hand matrix is bounded by $\max \{[(1-\varepsilon) C_M \xi + \varepsilon] (1-\xi)^{\nu} : 0 \le \xi \le 1\}$. Thanks to the choice of $\varepsilon = C_M/(\nu + C_M)$, the maximum is taken at $\xi = 0$ and its value is ε. Thus, (4.13) is proved for l, too.

□

Even without the approximation property (3.46) it is possible to show convergence:

$\| M_l'(\nu)\|_{u\leftarrow u} \le 1/[1 + \nu h_l^2 / (C_L \| L_l^{-1}\|)] < 1.$

5. Analysis of the Nested Iteration

In § 1.4 we described the nested iteration (1.11) that produces approximations $\tilde{u}_k$ for the levels $k = 0,1,\ldots, l$. Of course we have to assume that the multi-grid iteration converges. Theorem 4.4 yields the following estimate with $\rho < 1$:

$$(5.1) \qquad \| M_k^{MGM} \|_{U\leftarrow U} \le \rho \qquad \text{for } 1 \le k \le l.$$

The *relative discretization error* is the difference of the exact solutions u_k, u_{k-1} of two consecutive levels. Using the prolongation $\tilde{I}_{k-1}^k$ from (1.11), we may form the difference $\tilde{I}_{k-1}^k u_{k-1} - u_k$. Assuming some smoothness of the solution, one expects the difference of the order $O(h_k^\kappa)$ (κ: order of consistency):

$$(5.2) \qquad \| \tilde{I}_{k-1}^k u_{k-1} - u_k \|_U \le C_1 h_k^\kappa \quad \text{for } 1 \le k \le l.$$

Assumption (4.10) implies $\| I_{k-1}^k \|_{U\leftarrow U} \le C$. The same condition on $\tilde{I}_{k-1}^k$ and the boundedness of the ratio h_{k-1} / h_k (cf. (3.16)) give

$$(5.3) \qquad \| \tilde{I}_{k-1}^k \|_{U\leftarrow U} \, (h_{k-1} / h_k)^\kappa \le C_2 \quad \text{for } 1 \le k \le l.$$

Now we are able to formulate the error estimates of $\tilde{u}_k$:

<u>Theorem 5.1</u> Suppose (5.1), (5.2), (5.3) with

$$(5.4) \qquad C_2 \rho^i < 1 \qquad \text{(i from algorithm (1.11))}.$$

Then the nested iteration (1.11) with i multi-grid iterations per level yields functions $\tilde{u}_k$ ($o \le k \le l$) with

$$(5.5) \qquad \| \tilde{u}_k - u_k \|_U \le C_3(\rho)\, C_1 h_k^\kappa \quad \text{with } C_3(\rho) = \rho^i / (1 - C_2 \rho^i)$$

for $k = 0,1,\ldots, l$, where u_k is the exact solution.

<u>Proof:</u> Since $\tilde{u}_o = u_o$, (5.5) holds for $k = o$. Let (5.5) be valid for $k - 1$. The estimates

$$\| \tilde{u}_k - u_k \|_U = \| u_k^i - u_k \|_U \le \rho^i \| u_k^o - u_k \|_U,$$

$$\| u_k^o - u_k \|_U = \| \tilde{I}_{k-1}^k \tilde{u}_{k-1} - u_k \|_U = \| (\tilde{I}_{k-1}^k u_{k-1} - u_k) + \tilde{I}_{k-1}^k (\tilde{u}_{k-1} - u_{k-1}) \|_U$$

$$\le \| \tilde{I}_{k-1}^k u_{k-1} - u_k \|_U + \| \tilde{I}_{k-1}^k \|_{U\leftarrow U} \| \tilde{u}_{k-1} - u_{k-1} \|_U$$

imply

$$\| \tilde{u}_k - u_k \|_U \le \rho^i C_1 h_k^\kappa (1 + C_2 C_3(\rho) \rho^i) = C_3(\rho) C_1 h_k^\kappa.$$

□

Inequality (5.5) ensures that the iteration error $\tilde{u}_k - u_k$ is a certain percentage of the (upper bound for the) relative discretization error. Its ratio is bounded by $C_3(\rho)$. For piecewise linear, quadratic or cubic interpolation $\| \tilde{I}^k_{k-1} \| \le 1$ holds for the spectral norm. Assuming the standard case of $\kappa = 2$ and $h_{k-1} = 2h_k$, we find

(5.6) $$C_2 = 4, \quad C_3(\rho) = \rho^i / (1-4\rho^i)$$

<u>Note 5.2</u> Assume (5.6) and $\rho < 1/4$. Then the estimate (5.5) from Theorem 5.1 holds for $i = 1$ (only one multi-grid iteration per level).

In order to compare $\tilde{u}_k - u_k$ with the (absolute) discretization error assume an error expansion

(5.7) $$u_l = u + h_l^2 e + o(h_l^2).$$

This shows $C_1 = 3 \|e\|_u + o(1)$, provided that the interpolation error of $\tilde{I}^k_{k-1}$ is $o(h_l^2)$. Estimate (5.5) becomes

(5.8) $$\| \tilde{u}_k - u_k \|_u \le C_4(\rho) \cdot h_k^2 \|e\|_u + o(h_k^2) \quad \text{for } o \le k \le l$$

where $C_4(\rho) := 3\rho^i / (1 - 4\rho^i)$. Note that $h_k^2 \|e\|_u$ is the discretization error.

The following table of $C_4(\rho)$ in the case of $i = 1$

ρ	0.2	0.175	1/7=.14...	0.1	0.07	0.05
$C_4(\rho)$	3.0	1.75	1.0	0.5	0.292	0.188

shows that for the typical size of ρ the iteration errors of $\tilde{u}_k$ are of the same order as the discretization errors.

6. Anisotropic Problems

The model problem for an anisotropic problem is the equation

$$-\varepsilon\, u_{xx} - u_{yy} = f \text{ in } \Omega, \quad u = o \text{ on } \Omega,$$

where $\varepsilon > o$ is small. The multi-grid iteration with the pointwise Gauß-Seidel iteration or Jacobi-like iteration turns out to be slow (cf. [16]). Nevertheless, for fixed ε the convergence theorems apply, but its constants are depending on ε. In particular, the minimal number ν_{min} of smoothing iterations may increase with $1/\varepsilon$.

It is easy to see that the constant $C(\varepsilon)$ of the approximation property behaves like C/ε:

$$\text{(6.1)} \qquad \| L_l^{-1} - I_{l-1}^{l} L_{l-1}^{-1} I_l^{l-1} \| \le C\, \varepsilon^{-1} h_l^2 .$$

As described in [16] the multi-grid iteration has fast convergence if a linewise Gauß-Seidel iteration is used for smoothing. The purpose of this section is to demonstrate that the presented convergence theory can be applied to parameter dependent problems, too. It is possible to analyse the line-Gauß-Seidel iteration, but to simplify the proof we discuss a modified line-Jacobi iteration.

The discretization matrix L_l can be split into $L_l = \varepsilon L_l^x + L_l^y$. We consider the smoothing iteration

$$u_l^{j+1} = u_l^j - [L_l^y + \varepsilon\, \omega\, h_l^{-2} I]^{-1} (L_l u_l^j - f_l).$$

The iteration matrix S_l of this smoothing procedure is

$$S_l = B^{-1} A, \text{ where } B = L_l^y + \varepsilon\, \omega\, h_l^{-2} I,\ A = \varepsilon(\omega\, h_l^{-2} I - L_l^x).$$

Because of (6.1), the smoothing property $\| L_l S_l^\nu \| \le C\, \eta(\nu)$ does not suffice. The suitable estimate is provided by

<u>Lemma 6.1</u> Assume L_l^x and L_l^y to be symmetric and positive semi-definite. Choose $\omega > o$ such that $\omega\, h_l^{-2} I - L_l^x$ is positive semi-definite. Then the smoothing property

$$\text{(6.2)} \qquad \| L_l S_l^\nu \| \le \varepsilon\, \eta(\nu)\, h_l^{-2} \text{ with } \eta(\nu) = C/\nu \qquad (\nu \ge 1)$$

holds with C independent of ε.

<u>Proof:</u> Since $L_l = B - A$ we have

$$L_l S_l^\nu = (B - A)(B^{-1}A)^\nu = A(B^{-1}A)^{\nu-1} - A(B^{-1}A)^\nu .$$

A is positive semi-definite. Thus, $A^{1/2}$ and $E := A^{1/2} B^{-1} A^{1/2}$ can be introduced yielding

$$L_1 S_1^{\nu} = A^{1/2} E^{\nu-1} (I-E) A^{1/2}$$

and

$$\| L_1 S_1^{\nu} \| \le \| A^{1/2} \|^2 \ \| E^{\nu-1} (I-E) \| = \|A\| \ \| E^{\nu-1} (I-E) \| .$$

Since $B \ge A \ge o$, the spectrum of E is contained in [0,1]. As in the proof of Theorem 3.7 one obtains $\| E^{\nu-1}(I-E) \| \le C/\nu$ for $\nu \ge 1$ with C independent of ε. Since the matrix A has a factor ε, the estimate $\|A\| = \varepsilon \, \| \omega \, h_1^{-2} I - L_1^{x} \| \le \varepsilon \, C \, h_1^{-2}$ concludes the proof.

□

Now, we can proceed with the convergence proofs of the foregoing sections. For example, the two-grid iteration matrix can be estimated using (6.1) and (6.2):

$$\| M_1^{TGM} \| \le C \, \eta(\nu) .$$

The right-hand side (i.e. the convergence rate) is independent of ε.

7. Nonlinear Multi-Grid Iteration

7.1 Notations

The nonlinear boundary value problem is denoted by

$$\mathcal{L}(u) = o. \tag{7.1}$$

Its discretization of grid size h_l is written as

$$\mathcal{L}_l(u_l) = o \qquad (u_l \in \mathcal{U}_l). \tag{7.2}$$

In the following we assume that (7.2) has at least one isolated solution u_l^*. The derivative

$$L_l := \partial \mathcal{L}_l (u_l) / \partial u_l \quad \text{at } u_l = u_l^* \text{ (solution of (7.2))} \tag{7.3}$$

is supposed to be regular. Under some additional conditions, there is a neighbourhood $\mathcal{O}_l \subset \mathcal{F}_l$ of zero such that the more general problem.

$$\mathcal{L}_l(u_l) = f_l \qquad (f_l \in \mathcal{O}_l \subset \mathcal{F}_l) \tag{7.4}$$

has a solution which is the only solution in a neighbourhood of the solution u_l^* of (7.2). For simplicity assume that $\mathcal{O}_l$ is a ball of radius δ:

$$\mathcal{O}_l = \{f_l \in \mathcal{F}_l : \| f_l \|_F \le \delta\} \qquad (\delta > o). \tag{7.5}$$

Denote the inverse mapping of $\mathcal{L}_l$ by ϕ_l:

$$u_l = \phi_l(f_l) \quad \text{is solution of (7.4)} \qquad (f_l \in \mathcal{O}_l) \tag{7.6}$$

Let $\mathcal{V}_l$ be the range of ϕ_l:

$$\mathcal{V}_l := \phi_l(\mathcal{O}_l) = \{u_l = \phi_l(f_l) : f_l \in \mathcal{O}_l\}. \tag{7.7}$$

$\mathcal{V}_l$ is a neighbourhood of $u_l^* = \phi_l(o)$.

In the following sections we shall describe an algorithm for solving Eq. (7.4). Evidently, such an algorithm will apply to (7.2), too. The treatment of the generalized problem (7.4) will be necessary because of the auxiliary problems at lower levels.

Also the nonlinear algorithm consists of a smoothing step and a coarse-grid correction. The nonlinear smoothing procedure is denoted by $\mathcal{S}_l(u_l,f_l)$ as before. Since $u_l = \phi_l(f_l)$ is required to be a fixed point of $\mathcal{S}_l$, we have

(7.8) $$u_l = \mathcal{S}_l(u_l, \mathcal{L}_l(u_l)) \qquad \text{for all } u_l \in \mathcal{V}_l.$$

Example. The nonlinear analogue of the (linear) modified Jacobi iteration (3.22) is

(7.9) $$\mathcal{S}_l(u_l, f_l) := u_l - \omega\, h_l^2\, (\mathcal{L}_l(u_l) - f_l).$$

Also the Gauß-Seidel iteration has an obvious nonlinear counterpart (cf. [16]).

The linearization L_l is already defined (cf. (7.3)). The matrix S_l is the derivative

(7.10) $$S_l := \partial\, \mathcal{S}_l(u_l^*, o) \,/\, \partial u_l \qquad (u_l^* = \phi_l(o)).$$

Instead of Taylor expansions of a matrix $A_l(u_l)$ we introduce the difference DA_l (not uniquely) defined by $A_l(u_l) - A_l(v_l) = DA_l(u_l, v_l) \cdot (u_l - v_l)$. If A_l is differentiable, DA_l can be represented by $DA_l(u_l,v_l) = \int_o^1 A_l'(v_l + \lambda(u_l - v_l))\, d\lambda$. Continuity of the derivative A_l' implies continuity of $DA_l(u_l,v_l)$. $A_l'(u_l)$ can be regarded as limit of $DA_l(u_l,v_l)$ as $v_l \to u_l$.

The differences of $\mathcal{L}_l$ and $\mathcal{S}_l$ are denoted by

(7.11a) $$\mathcal{L}_l(v_l) - \mathcal{L}_l(w_l) = DL_l(v_l,w_l) \cdot (v_l - w_l),$$

(7.11b) $$\mathcal{S}_l(v_l,f_l) - \mathcal{S}_l(w_l,f_l) = DS_l(v_l,w_l,f_l) \cdot (v_l - w_l).$$

For the example (7.9) the matrices S_l and DS_l are

$$S_l = I - \omega\, h_l^2\, L_l, \qquad DS_l = I - \omega\, h_l^2\, DL_l.$$

The matrices DL_l and DS_l are assumed to be continuous (uniformly with respect to l) at $u_l^* = \phi_l(o)$:

(7.12a) $$\| L_l^{-1} DL_l(v_l,w_l) - I \|_{u\leftarrow u} \le \varepsilon(\| v_l - u_l^*\|_u + \| w_l - u_l^*\|_u) \quad (v_l,w_l \in \mathcal{V}_l)$$

(7.12b) $$\| DS_l(v_l,w_l,\mathcal{L}_l(w_l)) - S_l \|_{u\leftarrow u} \le \varepsilon\,(\| v_l - u_l^*\|_u + \| w_l - u_l^*\|_u)\ (v_l,w_l \in \mathcal{V}_l),$$

where $\varepsilon(\cdot)$ vanishes at zero:

(7.12c) $$\varepsilon(\xi) \to o \qquad \text{as } \xi \to o.$$

Note that we do not require continuous differentiability of $\mathcal{L}_l$ for all $u_l \in \mathcal{V}_l$.

7.2 Nonlinear Two-Grid Iteration

The smoothing step is the same as in the linear case, but the nonlinear coarse-grid correction differs from the linear one. Let $\bar{u}_l \in \mathcal{U}_l$ be the result of the smoothing step. We want to correct $\bar{u}_l$ by solving a certain nonlinear coarse-grid equation. But in order to approximate a directional derivative we need some coarse-grid function

(7.13a) $$\tilde{u}_{l-1} \in \mathcal{U}_{l-1}$$

and its defect

(7.13b) $$\tilde{f}_{l-1} = \mathcal{L}_{l-1}(\tilde{u}_{l-1}) \in \mathcal{F}_{l-1}.$$

$\tilde{u}_{l-1}$ may be regarded as an approximation to $u^*_{l-1} = \phi_{l-1}(o)$.

The exact correction v_l of the smoothing result $\bar{u}_l$ is given by

(7.14) $$\phi_l(f_l) = \bar{u}_l + v_l.$$

The defect of $\bar{u}_l$ is d_l:

(7.15) $$d_l = f_l - \mathcal{L}_l(\bar{u}_l) \qquad \text{(equivalent: } \bar{u}_l = \phi_l(f_l - d_l)).$$

Comparing (7.14) and (7.15), we are led to the representation

(7.16) $$v_l = \phi_l(f_l) - \phi_l(f_l - d_l) = D\phi_l(f_l, f_l - d_l) \cdot d_l$$

of the exact correction. We try to approximate $v_l = D\phi_l \cdot d_l$ by means of a coarse-grid function. For this purpose set

(7.17) $$d_{l-1} = \tilde{f}_{l-1} + \sigma I_l^{l-1} d_l \qquad (\sigma \in \mathbb{R})$$

with $\tilde{f}_{l-1}$ from (7.13b) and solve the nonlinear coarse-grid equation

(7.18) $$\mathcal{L}_{l-1}(w_{l-1}) = d_{l-1} \qquad \text{(equivalent:} w_{l-1} = \phi_{l-1}(d_{l-1})).$$

From (7.18) and (7.13b) one concludes

$$w_{l-1} - \tilde{u}_{l-1} = \phi_{l-1}(d_{l-1}) - \phi_{l-1}(\tilde{f}_{l-1}) = \sigma\, D\phi_{l-1} I_l^{l-1} d_l.$$

Set

$$\tilde{v}_l = \frac{1}{\sigma} I_{l-1}^{l}(w_{l-1} - \tilde{u}_{l-1}) = I_{l-1}^{l}\, D\phi_{l-1} I_l^{l-1}\, d_l.$$

Since d_l is the defect of the smoothed function $\bar{u}_l$ we expect

$$\tilde{v}_l = I_{l-1}^{l}\, D\phi_{l-1}\, I_l^{l-1}\, d_l \approx D\phi_l\, d_l = v_l.$$

Note that $D\phi_k = (DL_k)^{-1} \approx L_k^{-1}$. The foregoing statement corresponds to $I_{l-1}^{l} L_{l-1}^{-1} I_l^{l-1} d_l \approx L_l^{-1} d_l$ (d_l smooth) in the linear case. The coarse-grid correction of $\bar{u}_l$ is $\bar{u}_l + \tilde{v}_l$ (instead of $\bar{u}_l + v_l = \phi_l(f_l)$). Thus, the nonlinear two-grid iteration reads as

(7.19a) $\quad u_l^{j,o} := u_l^j,\ u_l^{j,\mu} := \mathcal{S}_l(u_l^{j,\mu-1}, f_l)\ (\mu = 1,\dots,\nu),\ \bar{u}_l := u_l^{j,\nu}$

(7.19b) $\quad d_l := f_l - \mathcal{L}_l(\bar{u}_l)$

(7.19c) $\quad$ choose $\tilde{u}_{l-1}$ and σ

(7.19d) $\quad \tilde{f}_{l-1} := \mathcal{L}_{l-1}(\tilde{u}_{l-1})$

(7.19e) $\quad d_{l-1} := \tilde{f}_{l-1} + \sigma\, I_l^{l-1}\, d_l$

(7.19f) $\quad w_{l-1} := \phi_{l-1}(d_{l-1})$, i.e. solve $\mathcal{L}_{l-1}(w_{l-1}) = d_{l-1}$

(7.19g) $\quad u_l^{j+1} := \bar{u}_l + I_{l-1}^{l}(w_{l-1} - \tilde{u}_{l-1}) / \sigma.$

There are two different choices of $\tilde{u}_{l-1}$ and σ in step (7.19c) among other possible ones.

1st variant: Set

(7.20a) $\quad \tilde{u}_{l-1} = I_l^{l-1}\, \bar{u}_l \qquad$ ($\bar{u}_l$: result of smoothing step (7.19a)),

(7.20b) $\quad \sigma = 1.$

In this case $\tilde{f}_{l-1}$ has to be computed again in every iteration of (7.19) since $\tilde{u}_{l-1}$ changes. d_{l-1} has the representation

$$d_{l-1} = \tilde{f}_{l-1} + d_l = I_l^{l-1} f_l + \mathcal{L}_{l-1}(I_l^{l-1}\bar{u}_l) - I_l^{l-1}\, \mathcal{L}_l(\bar{u}_l).$$

If $I_l^{l-1} f_l$ is small enough, e.g. $\| I_l^{l-1} f_l \|_F \le \delta/2$ (cf. (7.5)), and if by consistency $\mathcal{L}_{l-1}(I_l^{l-1}\, \bar{u}_l) - I_l^{l-1}\, \mathcal{L}_l(\bar{u}_l)$ is sufficiently small, too, the right-hand side d_{l-1} belongs to $\mathcal{O}_{l-1}$. Under these assumptions the equation $\mathcal{L}_{l-1}(w_{l-1}) = d_{l-1}$ in (7.19f) makes sense.

It is easier to ensure $d_{l-1} \in \mathcal{O}_{l-1}$ under the assumption

(7.21a) $$\| \tilde{f}_{l-1} \|_F \le \delta/2$$

if we apply the

2nd variant. Let $\tilde{u}_{l-1} \in \mathcal{U}_{l-1}$ be any *fixed* value with defect satisfying (7.21a) and define σ by

(7.21b) $$\sigma = \delta / [2\| I_l^{l-1} d_l \|_F] \qquad (\sigma = 1 \text{ if } I_l^{l-1} d_l = o).$$

Note 7.1 The second variant (7.21a,b) requires the evaluation of $\tilde{f}_{l-1}$ in (7.19d) only once and ensures $d_{l-1} \in \mathcal{F}_{l-1}$, i.e. $\| d_{l-1} \|_F \le \delta$.

The two-grid iteration (7.19) is not practical since it requires the exact solution of the coarse-grid equation in (7.19f). Let

$$u_k^{j+1} = \tilde{\phi}_k (u_k^j, f_k)$$

$(k = l-1)$ be any iteration converging to the solution $u_k = \phi_k(f_k)$. Examples are $\tilde{\phi}_k(u_l, f_l) = \phi_k(f_l)$ and $\tilde{\phi}_k = \mathcal{S}_k$. The convergence of $\tilde{\phi}_k$ can be described by

(7.22) $$\| \tilde{\phi}_k(v_k, f_k) - \phi_k(f_k) \|_u \le \kappa_k \| v_k - \phi_k(f_k) \|_u, \ \kappa_k < 1$$

for all $f_k \in \mathcal{F}_k$, $v_k \in \mathcal{V}_k$. Replacing the exact solution with an approximation by means of $\tilde{\phi}_{l-1}$ we obtain the following modification of the second two-grid variant:

(7.23a) choose some $\tilde{u}_{l-1}$ with $\tilde{f}_{l-1} := \mathcal{L}_{l-1}(u_{l-1})$, $\| \tilde{f}_{l-1} \|_F \le \delta/2$,

(7.23b) $u_l^{j,o} := u_l^j$, $u_l^{j,\mu} := \mathcal{S}_l(u_l^{j,\mu-1}, f_l)$ $(\mu = 1, \ldots, \nu)$, $\bar{u}_l := u_l^{j,\nu}$,

(7.23c) $d_l := f_l - \mathcal{L}_l(\bar{u}_l)$,

(7.23d) $\sigma := \delta / [2\| I_l^{l-1} d_l \|_F]$ if $I_l^{l-1} d_l \ne o$, $\sigma = 1$ otherwise,

(7.23e) $w_{l-1} := \tilde{\phi}_{l-1}(\tilde{u}_{l-1}, \tilde{f}_{l-1} + \sigma I_l^{l-1} d_l)$,

(7.23f) $u_l^{j+1} := \bar{u}_l + I_{l-1}^l (w_{l-1} - \tilde{u}_{l-1}) / \sigma$.

Identifying $\tilde{\phi}_{l-1}$ with γ steps of the two-grid iteration at level $l-1$ we are led to the multi-grid iteration.

7.3 Nonlinear Multi-Grid Iteration and Nested Iteration

Assume that there is some iteration $\tilde{\phi}_o$ at the lowest level. At the levels $k = o,1,\ldots, l-1$ we need approximations $\tilde{u}_k$ (cf. (7.13a), (7.23a)). Then we can apply the two-grid iteration recursively for approximating the solution in (7.23e). The first variant (7.20a,b) results in the nonlinear multi-grid method (7.24) that Brandt [7] calls FAS method.

procedure NMGM (l,u,f);

if $l = o$ *then* $u := \tilde{\phi}_o\,(u,f)$ *else*

begin integer j; *array* w,d;

for j := 1 *step* 1 *until* ν *do* $u := \mathcal{S}_l(u,f)$;

(7.24) $d := \mathcal{L}_{l-1}\,(I_l^{l-1}\,u) + I_l^{l-1}(f - \mathcal{L}_l(u))$;

$w := I_l^{l-1}\,u$; *for* j := 1 *step* 1 *until* γ *do* NMGM $(l-1,w,d)$;

$u := u + I_{l-1}^{l}\,(w - I_l^{l-1}\,u)$

end;

The meaning of the parameters is the same as in the linear procedures (1.8), TGM, or (1.10), MGM. Note that we use the starting guess $w := I_l^{l-1}\,u$ (not $w = o$). This w is the exact solution of $\mathcal{L}_{l-1}\,(w) = d$ if the defect $f - \mathcal{L}_l\,(u)$ vanishes.

For the second variant (cf. (7.21a,b)) we need grid functions $\tilde{u}_k (o \le k \le l-1)$ such that $\tilde{f}_k = \mathcal{L}_k(\tilde{u}_k)$ satisfies (7.21a):

$$\text{(7.25)} \qquad \|\tilde{f}_k\|_F \le \delta/2 \qquad\qquad (o \le k \le l-1).$$

If $\tilde{u}_k$ and $\tilde{f}_k$ $(o \le k \le l-1)$ are given, the following nonlinear multi-grid iteration (2nd variant) is well-defined:

procedure NMGM (l,u,f);

if $l = o$ *then* $u := \tilde{\phi}_o\,(u,f)$ *else*

begin integer j; *real* σ; *array* w,d;

for j := 1 *step* 1 *until* ν *do* $u := \mathcal{S}_l\,(u,f)$;

$$d := I_l^{l-1} \, (f - \mathcal{L}_l(u));$$

(7.26) $\sigma :=$ *if* $d = o$ *then* 1 *else* $\delta / (2 * \|d\|_F)$;

$$d := \tilde{f}_{l-1} + \sigma * d;$$

$w := \tilde{u}_{l-1}$; *for* $j := 1$ *step* 1 *until* γ *do* NMGM $(l-1, w, d)$;

$$u := u + I_{l-1}^{l} * (w - \tilde{u}_{l-1}) \, / \, \sigma$$

end;

A consequence of Note 7.1 is

Note 7.2 Condition (7.25) implies $\|d\|_F \le \delta$ for all d appearing in (7.26); thus, the auxiliary problems are always uniquely solvable in $\mathcal{V}_{l-1}$.

In order to provide for values $\tilde{u}_k$, $\tilde{f}_k$ we combine (7.26) with the nested iteration from § 1.4:

$\tilde{u}_o :=$ approximation of $u_o^* = \phi_o(o)$ (solution of $\mathcal{L}_o(u_o) = o$);

for $k := 1$ *step* 1 *until* l *do*

begin $\tilde{f}_{k-1} := \mathcal{L}_{k-1}(\tilde{u}_{k-1})$;

(7.27) $\tilde{u}_k := \tilde{I}_{k-1}^{k} \, \tilde{u}_{k-1}$;

for $j := 1$ *step* 1 *until* i *do* NMGM $(k, \tilde{u}_k, o)$

end;

Here, we inserted $f_k = o$, since the original problem (7.2) with $f_k = o$ is to be solved. Note that $\tilde{f}_{k-1}$ and $\tilde{u}_{k-1}$ are computed *before* NMGM is called at level k.

7.4 Convergence of the Nonlinear Two-Grid Iteration

The relationship of the nonlinear two grid iterations and the linear algorithm is given by

Note 7.3 If $\mathcal{L}_l(u_l)$ is the affine mapping $\mathcal{L}_l(u_l) = L_l \, u_l - g_l$, then the two-grid iterations (7.19) (nonlinear) and (1.7) (linear) yield identical results. Moreover, the nonlinear and linear multi-grid algorithm (7.24), (7.26), and (1.10) produce the same results.

Therefore, one may expect that the nonlinear iteration behaves asymptotically as the linear iteration applied to the linearized problem. Indeed, the following theorem can be proved (cf. [9]):

Theorem 7.4 *(Convergence of the nonlinear two-grid iteration (7.23)).* Suppose $f_l \in \mathcal{O}_l$ [cf. (7.5)]. Let the matrices L_l and S_l [defined by (7.3) and (7.10)] satisfy the smoothing property (3.3) and the approximation property (3.4). In addition assume (4.4), (4.10), (7.12a-c), (7.21a), and (7.22) ($k = l-1$). Then the iterates of (7.23) satisfy

$$(7.28) \qquad \| u_l^{j+1} - u_l \|_u \le C[\eta(\nu) + \kappa_{l-1} + \delta] \, \| u_l^j - u_l \|_u$$

$[u_l = \phi_l(f_l)$, κ_{l-1} from (7.22), δ from (7.5)], whenever $u_l^j \in \mathcal{U}_l$, $\nu < \nu_{max}(h_l)$. Hence, the nonlinear two-grid iteration converges if $\nu_{min} \le \nu < \nu_{max}(h_l)$, $h_l \le h_{max}$, $\delta \le \delta_{max}$, and if κ_{l-1} is small enough.

The convergence rate depends on the function $\eta(\nu)$ of the smoothing property, on the convergence rate κ_{l-1} of $\tilde{\phi}_{l-1}$, and on δ. Usually, δ can be chosen very small and does not deteriorate the result. Note that $\kappa_{l-1} = o$ if and only if $\tilde{\phi}_l(v_l, f_l) = \phi_l(f_l)$

Neglecting the influence of κ_{l-1}, we see from Theorem 7.4 that the convergence rate of the nonlinear two-grid iteration is asymptotically (as $\delta \to o$) the same as for the linear iteration (1.8) applied to the linearized problem $L_l w_l = g_l$. A similar result will be true for the multi-grid case.

7.5 Convergence of the Nonlinear Multi-Grid Iteration

In Theorem 7.4 we state the convergence of a perturbed two-grid iteration, since $\tilde{\phi}_{l-1}$ is used instead of the exact inverse ϕ_{l-1}. The following proof shows a second variant of demonstrating the multi-grid convergence by means of the two-grid convergence.

Theorem 7.5 *(Convergence of the Nonlinear Multi-Grid Iteration (7.26))*

Suppose $f_l \in \mathcal{O}_l$, $\tilde{u}_k$ ($o \le k \le l-1$) with $\tilde{f}_k = \mathcal{L}_k(\tilde{u}_k)$ fulfilling (7.25). Let the matrices L_k, S_k ($o \le k \le l$, cf. (7.3), (7.10)) satisfy all conditions required in Theorem 4.4 for the linear multi-grid convergence. In addition assume (7.12a-c) and (7.22) for $\tilde{\phi}_o$ with $\kappa_o \le \rho^* \in (o,1)$. Then there are numbers ν_{min}, h_{max}, δ_{max} such that the iterates of the nonlinear multi-grid iteration (7.26) satisfy

$$(7.29) \qquad \| u_l^{j+1} - u_l \|_u \le \rho^* \, \| u_l^j - u_l \|_u \qquad (u_l = \phi_l(f_l)),$$

whenever $\gamma \geq 2$, $\nu_{min} \leq \nu < \nu_{max}(h_l)$, $h_l \leq h_{max}$, $\delta \leq \delta_{max}$, $u_l^j \in \mathcal{V}_l$.

Proof. Estimate (7.29) holds for $l = o$ because of (7.22) and $\kappa_o \leq \rho^*$. Let (7.29) be valid for l-1. Denote the γ-fold application of iteration (7.26) at level l-1 by the symbol $\tilde{\phi}_{l-1}$. Then the multi-grid iteration (7.26) and the two-grid iteration (7.23) coincide. The rate is bounded by $C \cdot [\eta(\nu) + \kappa_{l-1} + \delta]$ (cf. Theorem 7.4). Use $\kappa_{l-1} \leq (\rho^*)^\gamma$ and apply Lemma 4.3. For suitable ν_{min} and δ_{max} we conclude $C[\eta(\nu) + \rho^{*\gamma} + \delta] \leq \rho^*$ proving (7.29).

□

References:

[1] ASTRACHANCEV, G.P.: An iterative method of solving elliptic net problems. Ž. vyčisl. Mat. mat. Fiz 11,2,439-448 (1971)

[2] ASTRACHANCEV, G.: and L.A. RUCHOVEC: A relaxation method in a sequence of grids for elliptic equations with natural boundary condition. Ž. vyčisl. Mat. mat Fiz. 21,4,926-944 (1981)

[3] BACHVALOV, N.S.: On the convergence of a relaxation method with natural constraints on the elliptic operator. Ž. vyčisl. Mat. mat. Fiz. 6,5,861-883 (1966)

[4] BANK, R.E. and T. DUPONT: An optimal order process for solving elliptic finite element equations. Math. Comp. 36, 35-51 (1981)

[5] BRAESS, D.: The contraction number of a multi-grid method for solving the Poisson equation. Numer. Math. 37, 387-404 (1981)

[6] ----------: The convergence rate of a multi-grid method with Gauß-Seidel relaxation for the Poisson equation. This volume

[7] BRANDT, A.: Multi-level adaptive solutions to boundary value problems. Math. Comp. 31, 333-390 (1977)

[8] FEDORENKO, R.P.: A relaxation method for solving elliptic difference equations. Ž. vyčisl. Mat. mat. Fiz. 1,5,922-927 (1961) - The speed of convergence of one iterative process. Ž. vyčisl. Mat. mat. Fiz. 4,3, 559-564 (1964)

[9] HACKBUSCH, W.: On the convergence of multi-grid iterations. Beiträge Numer. Math. 9, 213-329 (1981)

[10] -------------: Convergence of multi-grid iterations applied to difference equations. Math. Comp. 34, 425-440 (1980)

[11] -------------: Introduction to multi-grid methods for the numerical solution of boundary value problems. In. Computational Fluid Dynamics, Lecture Series 1981-5, von Karman Institute for Fluid Dynamics, Rhode Saint Genese, Belgium 1981

[12] MAITRE. J.F. and F. MUSY: The contraction number of a class of two-level methods - An exact evaluation for some finite element subspaces and model problems. This volume

[13] MEIS, Th. and U. MARCOWITZ: Numerische Behandlung partieller Differentialgleichungen. Berlin-Heidelberg-New York: Springer 1978. English translation: Numerical solution of partial differential equations. New York-Heidelberg-Berlin: Springer 1981

[14] NEČAS, J.: Sur la coercivité des formes sesqui-linéaires elliptiques. Rev. Roumaine Math. Pure Appl. 9, 47-69 (1964)

[15] NICOLAIDES, R.A.: On the l^2 convergence of an algorithm for solving finite element equations. Math. Comp. 31, 892-906 (1977) - On multigrid convergence in the indefinite case. Math. Comp. 32, 1082-1086 (1978)

[16] STÜBEN, K. and U. TROTTENBERG: Multi-grid method: fundamental algorithm, model problem analysis, and applications. This volume

[17] WESSELING, P.: The rate of convergence of a multiple grid method. In: Numerical Analysis, Proceedings, Dundee 1979 (ed.: G.A. Watson), Lecture Notes in Math. 773, Springer-Verlag, Berlin 1980, pp. 164-180

GUIDE TO MULTIGRID DEVELOPMENT

Achi Brandt*

Department of Applied Mathematics
The Weizmann Institute of Science
Rehovot, Israel 76100

CONTENTS

---

*This research is sponsered by the Air Force Wright Aeronautical Laboratories, Air Force Systems Command, United States Air Force, under Grant AFOSR 82-0063.

1. INTRODUCTION

1.1 Purpose of the paper

On starting to write this paper, I have in mind the following situation, so often encountered in the last few years: A good numerical analyst tries a multigrid solver on a new problem. He knows the basics, he has seen it implemented on another problem, so he has no trouble writing the program. He gets results, showing a certain rate of convergence, perhaps improving a former rate obtained with a one-grid program. Now he is confronted with the question: Is this the real multigrid efficiency? Or is it many times slower, due to some conceptual error or programming bug? The algorithm has many parts and aspects: relaxation sweeps and coarse-to-fine and fine-to-coarse transfers at interior points and at points near boundaries; relaxation and transfers of the boundary conditions themselves; treatment of boundary and interior singularities and/or discontinuities; choosing the coarse-grid variables and defining its equations; the method of solving on the coarsest grid; the general flow of the algorithm; etc. A single error (a wrong scheme or a bug) in any of these parts may degrade the whole performance very much, but is still likely to give an improvement over a one-grid method, misleading the analyst to believe he has done a good job. How can an error be suspected and detected? How can one distinguish between various possible troubles? And what improved techniques are available?

The key to a fully successful code is to know in advance what efficiency is ideally obtainable, and then to construct the code gradually in a way which ensures approaching that ideal, telling us at each stage which process may be responsible for a slowdown. It is important to work in that spirit: Do not just observe what efficiency is obtained by a given multigrid algorithm, but ask yourself what is the ideal efficiency and find out how to obtain it. To guide inexperienced multigridders in that spirit is the main purpose of this paper.

We believe that any discrete system derived from a continuous problem is solvable "to the level of truncation errors" in just few "work units" (see Sec. 7.3). To obtain this performance, the first crucial step is to construct a relaxation scheme with a high "smoothing rate" (see Sec. 3). Then the interior inter-grid transfers and coarse-grid operator should be designed (Sec. 4), and full numerical experiments can be started with cycling algorithms, aiming at obtaining the interior rate (Secs. 5.6). Finally, "Full Multi-Grid" (FMG) algorithm can then be implemented, and "solvability in just few work units" can be tested (Sec. 7). These stages of development are outlined in Part I below, pointing out many possibilities and technical points, together with theoretical tools needed for quantitative insights into the main processes.

The quantitative aspect in these theoretical tools is important, since we want to distinguish between the efficiency of several candidate multigrid algorithms, all of which may be "asymptotically optimal" (i.e., solving the problem in a uniformly bounded number of work units), but some of which may still be several orders of magnitude faster than others. Except for some model problems, the present-day rigorous mathematical theories of multigrid algorithms do not give us accurate enough insights (see Sec. 14), hence the present guide will emphasize the role of "local mode analyses". These analyses (see Secs. 2.1, 3.1, 4.1, 7.4, 7.5) neglect some of the less work-consuming processes so as to obtain a clear and precise picture of the efficiency of the more important processes. The predictions so obtained can be made accurate enough to serve in program optimization and debugging. Experience has taught us that careful incorporation of such theoretical studies is essential for producing reliable programs which fully utilize the potential of the method.

Part II of this article summarizes more advanced multigrid techniques and insights. Mainly, it is intended to show how to use the multilevel techniques far beyond their more familiar capacity as fast linear algebraic solvers. See the survey in Sec. 1.2 and the list of contents.

I have also used this opportunity to mention several topics which have not appeared in the literature before. These include some new relaxation schemes such as "Box Gauss-Seidel" (Sec. 3.4) and relaxation with only sub-principal terms (Sec. 10.3); the general rule of block relaxation (Sec. 3.3); an analysis of the orders of interpolation and residual transfers which should be used in solving <u>systems</u> of differential equations (Secs. 4.3, 7.1); the multigrid treatment of global constraints (Sec. 5.6); an application of FAS to obtain much more efficient discretization to <u>integral</u> equations, leading sometimes to solutions in $O(n)$ operations, where n is the number of discrete unknowns (Sec. 8.6); some innovations in higher-order techniques (Sec. 10.2); unified switching and adaptation criteria (Sec. 9.6); multi-level approach to optimization (Sec. 13); and the Algebraic Multi-Grid (AMG) method (Sec. 13.1). Results from a recent work, still in press, on non-elliptic and singular perturbations [B17] are also mentioned, including a summary of stability requirements (Sec. 2.1); the double-discretization scheme (Sec. 10.2); the two-level FMG mode analysis, which tends to replace the usual two-level mode analysis (Secs. 7.4, 7.5); the F cycle (a hybrid of V and W cycles; Sec. 6.2). (The main topic from [B17] hardly mentioned here is the multigrid treatment of discontinuities, in which research has barely started. But see Secs. 2.2 and 8.5.) The discussion of the real role of relaxation (Sec. 12) and the general approach to coarsening questions (Sec. 11) are useful new viewpoints.

1.2 <u>Where and why multigrid can help</u>

The starting point of the multigrid method (or, more generally, the Multi-Level Adaptive Technique — MLAT), and indeed also its ultimate upshot, is the following "<u>golden rule</u>":

<u>The amount of computational work should be proportional to the</u>

amount of real physical changes in the computed system. Stalling numerical processes must be wrong.

That is, whenever the computer grinds very hard for very small or slow real physical effect, there must be a better computational way to achieve the same goal. Common examples of such stalling are the usual iterative processes for solving the algebraic equations arising from discretizing partial-differential, or integro-differential, boundary-value (steady-state) problems, in which the error has relatively small changes from one iteration to the next. Another example is the solution of time-dependent problems with time-steps (dictated by stability requirements) much smaller than the real scale of change in the solution. Or, more generally, the use of too-fine discretization grids, where in large parts of the computational domain the meshsize and/or the timestep are much smaller than, again, the real scale of solution changes. Etc.

If you have such a problem, multi-level techniques may help. The trouble is usually related to some "stiffness" in your problem; i.e., to the existence of several solution components with different scales, which conflict with each other. For example, smooth components, which are efficiently approximated on coarse grids but are slow to converge in fine-grid processes, conflict with high-frequency components which must be approximated on fine grids. By employing interactively several scales of discretization, multilevel techniques resolve such conflicts, avoid stalling and do away with the computational waste.

In fully developed MLAT processes the amount of computations should be determined only by the amount of real physical information.

The main development of multilevel techniques has so far been limited to their role as fast solvers of the algebraic equations arising in discretizing boundary-value problems (steady-state problems or implicit steps in evolution problems). The multigrid solution of such problems usually requires just few (four to ten) work units, where a work unit is the amount of computational work involved in expressing the algebraic equations (see Sec. 7.3). This efficiency is obtained for all problems on which sufficient research has been made, from simple model problems to complicated nonlinear systems on general domains, including diffusion problems with strongly discontinuous coefficients, integral equations, minimization problems with constraints; from regular elliptic to singular-perturbation and non-elliptic boundary-value problems. Due to the iterative nature of the method, nonlinear problems require no more work than the corresponding linearized problems. (No outer Newton-like steps are required, and no global linearization is needed. See Sec. 8.3).

Problems with global constraints are solved as fast as the corresponding unconstrained difference equations, using a technique of enforcing the constraints only at the coarse-grid stages of the algorithm (Sec. 5.6). Few work units are also all the work required in calculating each eigenfunction of discretized eigenproblems (Sec. 8.3.1).

Beyond the fast solvers, multilevel techniques can be very useful in other ways related to stiffness. They can provide very efficient grid-adaptation procedures for problems (either boundary-value or evolution problems) in which different scales of discretization are needed in different parts of the domain (see Sec. 9). They can give new dimension of efficiency to stiff evolution problems (Sec. 16). And they can resolve the conflict between higher accuracy and stability in cases of non-elliptic and singular perturbation boundary-value problems. (Sec. 10.2). In addition, multi-level techniques can enormously reduce the amount of discrete relations employed in solving chains of similar boundary-value problems (as in processes of continuation, and in optimization problems; see Secs. 15, 13), or in solving integral equations (see Sec. 8.6). They can also be used to vastly cut the required computer storage (Sec. 8.7). All these topics are now under active research.

Multilevel processes can also introduce new dimensions of efficiency to solving or treating some large systems which do not originate from partial-differential or integral equations. The common feature in those systems is that they involve many unknowns related in a low-dimensional space; i.e., each unknown u_P is defined at a point $P = (x_1,\dots,x_d)$ of a low-dimensional space (d is usually 2 or 3), and the equations are given in terms of these coordinates x_j. Moreover, the coupling between two values u_P and u_Q generally becomes weaker or smoother as the distance between P and Q increases, except perhaps for a small number of particular pairs (P,Q). Examples are the equations of multivariate interpolation of scattered data [M1], geodetic problems of finding the locations of many stations that best fit a large set of local observations [M2] and pattern-recognition systems [N1], [T1].

1.3 Elementary acquaintance with multigrid

If you have no multigrid experience, the easiest way to acquire the elementary acquaintance with the method is to go through a simple example. Thus, reading Secs. 2, 3.1, 4 and Appendix B in [B7], just 12 pages, would acquaint you with the most basic concept, with elementary mode analysis, and with a simple algorithm together with its short Fortran code and output.

The sections below are intended as a guide for developing less elementary algorithms for less elementary problems. For a comprehensive treatment of some model problems by a variety of multigrid algorithms, mode analyses and numerical experiments, for a model program, and for general introduction to the basic algorithms, see [S4]. The model program of [B7, App. B] and several other model programs and multigrid software are available on [M3].

PART I. STAGES IN DEVELOPING FAST SOLVERS

The intention of this part is to organize existing multigrid approaches in an order which corresponds to actual stages in developing fast multigrid solvers. Each section (Secs. 2 through 7) represents a separate stage. To get an overview of these stages, the reader may first go through the opening remark of all the sections, skipping the subsections. The actual sequence of development may correspond to the actual order of the sections; but Secs. 5, 6, and 7.4 represent three independent stages, which can be taken in any order following Sec. 4. In fact, an increasing current tendency is to replace the usual two-level mode analysis (Sec. 4) by the two-level FMG mode analysis (Sec. 7.4). Generally, one can skip a stage, risking a lesser control of potential mistakes. Even when one does, the information and advice contained in the corresponding subsections are still important.

This part emphasizes the linear solver: relaxation of nonlinear equations is described, but the Full Approximation Scheme (FAS) used in inter-grid transfers of nonlinear solvers is described in the next part (Sec. 8).

2. STABLE DISCRETIZATION

The formulation of good discretization schemes is of course the first step in any numerical solution of continuous equations. For multigrid solutions some additional considerations enter. First, discrete equations should be written for general meshsizes h, including large ones (to be used on coarse grids). Also, the multigrid processes offer several simplifications of the discretization procedures.

For example, only uniform-grid discretization is needed. Non-uniform discretization can be effected through multigrid interactions between uniform grids (see Sec. 9). In fact, if the basic grids are non-uniform, various structural and algorithmic complications are introduced, in multigrid as well as, and even more than, in unigrid processes. The uniform discretization can be made either with finite element or finite difference formulations. Finite elements with general non-uniform partitions, on the other hand, require an assembly process so costly as to make the use of fast multigrid solvers almost pointless, except perhaps in an adaptive process. It is preferable to use either piecewise uniform partitions as in [B2] and [B7, §7.3], or uniform partitions modified at special parts (e.g., at boundaries) as in [B11, Fig. 2], and produce local refinements by the multigrid process (Sec. 9).

Another simplification is that one has to write low-order (first or second order) discretization only. Higher-order schemes can later be superposed for little extra programming effort and computer time (Sec. 10). The low order makes it easier to write stable equations, easier to devise and analyze relaxation schemes, and cheaper to operate them.

Low order finite elements on uniform grids yield kinds of difference equations. The description below will therefore be in terms of finite difference formulations only. It should be emphasized, however, that variational formulations, where appropriate, automatically yield good prescriptions for the main multigrid processes. See [B7, App. A.5], [N3] and [B11]. This is especially useful in some complicated situations, as in [A1]. We return to this issue in Sec. 11 below.

To be solvable by a fast multigrid algorithm, the discretization of a boundary-value problem should be suitably stable. More precisely, the type of stability determines the kind of multigrid algorithm that can be efficient. The simplest differencing of regular elliptic equations yield equations which are stable in every respect, so the reader interested in only such problems can skip the rest of this section. But remember:

An important <u>advantage</u> of multigird solvers is that bad (unstable and/or inaccurate) discretizations cannot usually be passed unnoticed; they must show up as slow convergence of the algebraic solver. This may however turn into a disadvantage if one is not aware of this possible source of slowdown.

2.1 Interior stability measures: h-ellipticity

Numerical stability is a <u>local</u> property, i.e., a property significant only for non-smooth components of the solution (components that change significantly over a meshsize), whereas the smooth-components' stability depends, by consistency, on the <u>differential</u> system, not on its discretization. Indeed, in multigrid solvers, stability of the discrete operator is needed only in the <u>local</u> process of relaxation (cf. Sec. 10.2). Moreover, what counts is not really the stability of the static difference operator itself, but the overall efficiency with which the dynamic process of relaxation smoothes the differential error (cf. Sec. 12); numerical stability of the operator is just a necessary condition for achieving that smoothing.

Because of the local character of the required stability (corresponding to the local task of relaxation), it is easy to get a very good quantitative idea about it, for any given difference operator L^h, by <u>local mode analysis</u>, analogous to the Von-Neumann stability analysis for time-dependent problems. It turns out, however, that for steady-state problems, especially non-elliptic or singular perturbation ones, the distinction between stable and unstable discrete operators is not enough. More important is the measure of stability. When that measure, for the given meshsize, is low the scheme is still formally stable, but its actual behavior can be intolerably bad (see example in [B17, §3]).

Briefly, the basic relevant measure of stability of an interior (not at boundaries) linear difference operator L^h with constant coefficients is its h-ellipticity measure $E^h(L^h)$ [B19, §3.9], defined for example by

$$E^h(L^h) = \min_{\hat{\rho}\pi \leq |\underline{\theta}| \leq \pi} |\tilde{L}^h(\underline{\theta})| \, / \, |L^h| , \tag{2.1}$$

where the complex function $\tilde{L}^h(\underline{\theta})$ is the "symbol" of L^h, i.e., $L^h e^{i\underline{\theta}\cdot\underline{x}/\underline{h}} = \tilde{L}^h(\underline{\theta})e^{i\underline{\theta}\cdot\underline{x}/\underline{h}}$; $\underline{\theta} = (\theta_1,\ldots,\theta_d)$, $\underline{\theta}\cdot\underline{x}/\underline{h} = \theta_1 x_1/h_1 + \ldots + \theta_d x_d/h_d$; $|\underline{\theta}| = \max(|\theta_1|,\ldots,|\theta_d|)$; d is the dimension; h_j is the meshsize in direction x_j; and $|L^h|$ is any measure of the size of L^h , e.g., $|L^h| = \max |\tilde{L}^h(\underline{\theta})|$. The constant $0 < \hat{\rho} < 1$ is in fact arbitrary, but for convenient multigrid applications a natural choice is the meshsize ratio, hence usually $\hat{\rho} = \frac{1}{2}$. The range $\hat{\rho}\pi \leq |\underline{\theta}| \leq \pi$ is then the range of "high frequency" components on grid h , i.e., components $\exp(i\underline{\theta}\cdot\underline{x}/\underline{h})$ which on the next coarser grid, with meshsize $h/\hat{\rho}$, coincide (alias) with lower components. In case of systems of equations, L^h and $\tilde{L}^h(\underline{\theta})$ are matrices and $|\tilde{L}^h(\underline{\theta})|$ should then be understood as a measure of the non-singularity of $\tilde{L}^h(\underline{\theta})$ (e.g., its smallest eigenvalue, or its determinant). See more details and explanations in [B12, §3], [B17, §3.1].

In case the differential operator L , and hence also L^h , have variable coefficients, L^h is called h-elliptic if $E^h(L^h) = O(1)$ for each combination of coefficients appearing in the domain. If L is nonlinear , L^h is called h-elliptic if its linearizations, around any approximate solution encountered in the calculations, are all h-elliptic.

A major simplification in selecting the discretization scheme for complicated systems is the fact that, being interested in local properties only, we can confine our considerations (not the actual process, to be sure) to the "scaled principal terms" of the operator [B19, §3.8], i.e., to the discretization of the subprincipal terms of L. These are defined as the principal terms (the terms contributing to the highest order terms in the determinant of the linearized operator) plus the principal terms of the reduced operator (the operator without singular perturbation terms). See an example in [B16].

Regular discretizations of elliptic systems should, and usually do have good (i.e., O(1)) h-ellipticity measures. (But see a counter-example in [B19, §5.2]). Singular perturbation or non-elliptic systems can also have such good measures, e.g., by using artificial viscosity or by upwind (upstream) differencing. (Note that a regular elliptic system with lower-order terms may be a singular-perturbation problem on a sufficiently coarse grid). If, however, characteristic or subcharacteristic directions (i.e., characteristic directions of the reduced equations, in case of singular perturbation problems) coincide with grid directions, upwind differencing schemes are only semi h-elliptic. They have, that is, bad h-ellipticity measure E^h, but they still have a good <u>semi</u> h-ellipticity measure in the characteristic direction. This measure is defined as follows.

Let $S \subset \{1,\dots,d\}$ be a subset of grid directions. The measure of <u>semi h-ellipticity in directions S</u>, or briefly <u>S-h-ellipticity</u>, of a difference operator L^h is defined by

$$E_S^h(L^h) = \min_{\hat{\rho}\pi \leq |\underline{\theta}|_S \leq \pi} |\tilde{L}^h(\underline{\theta})| \,/\, |L^h| \,, \qquad (2.2)$$

where $|\underline{\theta}|_S = \max_{j\in S} |\theta_j|$. Full h-ellipticity is the special case $S = \{1,\dots,d\}$. If $S_2 \subset S_1$ then clearly $E^h_{S_1} \leq E^h_{S_2}$, hence S_1-h-ellipticity entails S_2-h-ellipticity.

In case (sub)characteristics are aligned with grid directions, full h-ellipticity is not needed for stability. The corresponding S-h-ellipticity is enough; it allows large local oscillations perpendicular to the characteristics, but those oscillations are also allowed by the differential equations.

Fully h-elliptic approximations can be constructed even for non-elliptic equations by using isotropic artificial viscosity. In various cases, however, semi h-elliptic approximations are preferable, since they entail much less cross-stream smearing. These are mainly cases of <u>strong alignment</u>, that is, cases where (sub)characteristic lines are non-locally (i.e., for a length of many meshsizes) aligned with a grid-line, and where this non-local alignment occurs either for many grid-lines, or even for one gridline if that line is adjacent to a boundary layer or a similar layer of sharp change in the solution. (For a method to obtain strong alignments and thus avoid smearing — see Sec. 9.3).

A convenient way of constructing h-elliptic and semi h-elliptic operators is by term-by-term R-elliptic or semi R-elliptic approximations [B10, §5.2], [B19, §3.6]. Another, more physical way is to regard the given boundary-value problem as a limit of an elliptic problem (usually this is physically so anyway), and enlarge the elliptic singular perturbation to serve as artificial-viscosity terms [B17], [B16]. With proper amount of anisotropic artificial viscosity this gives the correct upstream differencing whenever desired.

The optimal amount of artificial viscosity (either isotropic or anisotropic) is mainly determined not by stability considerations but by the smoothing properties of relaxation. Below a certain level of viscosity, more costly, distributive relaxation will have to be used. Increasing the artificial viscosity slightly beyond the miminum required for convergence of the simplest scheme makes the relaxation ordering-free (see Sec. 3.6 and [B17, §5.7], [B12, §4.2]), which is desirable, except perhaps near discontinuities. Considerably larger artificial viscosity makes the algebraic smoothing faster, but impedes the differential smoothing (cf. Sec. 12).

Interior difference equations which are not even semi h-elliptic should rarely be used. Their solutions may show large numerical oscillations (giving nice solutions only in the average), and their fast multigrid solvers must have more complicated fine-to-coarse interactions (see [B19, §3.4], [B15, §3.2], [B21] and Sec. 4.2.2 below).

2.2 Boundaries, discontinuities

We have so far discussed the stability conditions related to the interior difference equations, away from boundaries. To gain overall stability some additional conditions should be placed at the boundaries. These can be analyzed by mode analysis in case the boundaries are parallel to grid directions (cf. Sec. 7.5). But more general boundaries are difficult to analyze. Usually, however, h-elliptic approximations consistent with a well-posed problem and employing low-order approximations to boundary conditions are stable. The order can then be raised in a stable way by one of the methods of Sec. 10. At any rate, the boundary stability is not related to the stages of developing the main (interior) multigrid processes.

More critical than discretization near boundaries is the treatment of discontinuities, whether at boundaries (e.g., boundary layers) or in the interior (e.g., shocks). The basic rule, in multigrid as in unigrid processes, is to try not to straddle the discontinuity by any difference operator (in relaxation as well as in residual transfers.

The rule should also be applied to the interpolation operators.) More precisely, the rule is not to difference any quantity which is discontinuous in the interval of differencing. This can fully be achieved only in cases where the location of the discontinuity is known or traced (at such discontinuities the above rule overrides upstream differencing if they happen to conflict), or when the discontinuity is more or less parallel to grid directions (so that upstream differencing will automatically satisfy the rule). Captured discontinuities which are not in grid directions must be smeared, and the only way to get high accuracy is by local refinements (see Sec. 9). Generally, multigrid procedures for discontinuities are now under active investigation (see [B17, §4] and a remark in Sec. 8.5 below).

3. INTERIOR RELAXATION AND SMOOTHING FACTORS

The crucial step in developing multigrid solvers is the design of interior relaxation schemes with high error-smoothing rates. Namely, the crucial question is how to reduce non-smooth error components for as little computational work as possible, neglecting interactions with boundaries. This is the crucial question, from the point of view of solution efficiency, since reducing smooth error components will require less computational work (being done on coarser grids), and since reducing non-smooth error components near boundaries will require much less work because it involves local work only near the boundary (which is a lower-dimensional manifold). Also, relaxation is the most problem-dependent part of the algorithm -- other parts are usually quite standard. (The relaxation of boundary conditions is discussed in Sec. 5.3).

3.1 Local analysis of smoothing

To reduce non-smooth error components is basically a local task; it can be done in a certain neighborhood independently of other parts of the domain. This is why it can efficiently be performed through relaxation, which is basically a local process (the information propagates just few meshsizes per sweep). Hence also, the efficiency of this process can accurately be measured by local mode analysis.

That is, one can assume the problem to be in an unbounded domain, with constant (frozen) coefficients, in which case the algebraic error $u^h - \tilde{u}^h$ (where u^h is the exact solution of the discrete equations and $\tilde{u}^h$ is the computed approximation) is a combination of Fourier components $\exp(i\underline{\theta}\cdot\underline{x}/\underline{h})$. For each such Fourier component and any proposed relaxation scheme one can easily calculate the amplification factor $\mu(\underline{\theta})$, defined as the factor by which the amplitude of that component is multiplied as a result of a relaxation sweep (see simple examples in [B7, §3.1] and in [S4, §3.2]). The smoothing factor $\bar{\mu}$ of the relaxation scheme, defined by

$$\bar{\mu} = \max_{\pi/2 \leq |\underline{\theta}| \leq \pi} |\mu(\underline{\theta})| \tag{3.1}$$

can then easily be computed (usually by a standard computer routine, such as the SMORATE routine in [M3]). This is indeed the measure we need: $\bar{\mu}$ is the worst (largest) factor by which all high-frequency error components are reduced per sweep, where we define the frequency to be high if the component is not visible (aliases with a lower component) on the next coarser grid (grid 2h).

In case of a system of q grid equations in q unknown grid functions (i.e., q unknowns and q algebraic equations are defined per mesh cell), each Fourier amplitude is a q-vector, hence $\mu(\underline{\theta})$ is a $q \times q$ amplification matrix. $\bar{\mu}$ is still defined as in (3.1), except that $|\mu(\underline{\theta})|$ is replaced by $\rho(\mu(\underline{\theta}))$, where $\rho(\mu)$ is the spectral radius of μ.

For L^h with non-constant coefficients, $\bar{\mu}$ defined by (3.1) depends on the location. In case of nonlinear L^h, the analysis is made for the linearized operator, hence $\bar{\mu}$ depends also on the solution around which linearization is made. The quality of relaxation is then determined by the worst $\bar{\mu}$, i.e., the maximum of $\bar{\mu}$ over all possible coefficients of L^h for any solution which may evolve in the calculations. (One may disregard $\bar{\mu}$ at isolated points: See Sec. 3.3.)

A major simplification in calculating $\bar{\mu}$ is to look at subprincipal terms only (see Sec. 2.1 and an example in [B17, §4]).

Some relaxation schemes do not transform each Fourier component of the error to a multiple of itself. Instead, they couple several (ℓ, say) Fourier components at a time (even in the infinite domain). For example, if relaxation is performed in red-black (checker-board) ordering (cf. Sec. 3.6), the $\underline{\theta}$ component is coupled to the $\underline{\theta} + (\pi, \ldots, \pi)$ component. Instead of the $q \times q$ amplification matrix $\mu(\underline{\theta})$ we then have the $(q\ell) \times (q\ell)$ matrix $\mu(\underline{\theta}^1, \ldots, \underline{\theta}^\ell)$, describing the transformation of the ℓ q-vector amplitudes corresponding to the coupled components $\underline{\theta}^1, \ldots, \underline{\theta}^\ell$. Definition (3.1) is extended in such cases to

$$\bar{\mu}_\nu = \max \left[\rho\left(C(\underline{\theta}^1, \ldots, \underline{\theta}^\ell)\, \mu(\underline{\theta}^1, \ldots, \underline{\theta}^\ell)^\nu\right) \right]^{1/\nu} \tag{3.2}$$

where the max is taken over all coupled ℓ-tuples $(\underline{\theta}^1, \ldots, \underline{\theta}^\ell)$, C is an $\ell \times \ell$ matrix of $q \times q$ blocks C_{ij} such that $C_{ij} = 0$ for $i \neq j$, $C_{jj} = I_q$ (the $q \times q$ identity matrix) if $|\underline{\theta}^j| \geq \pi/2$, and $C_{jj} = 0$ otherwise. ν is the number of sweeps performed on the finest grid per multigrid cycle; only in the simple case $(\ell = 1)$ $\overline{\mu}$ does not depend on ν. Examples with $\ell > 1$ see in [B14, §3.3], [L2, §2.3.1].

The smoothing factor is the first and simplest quantitative predictor of the obtainable multigrid efficiency: $\overline{\mu}^\nu$ (or $\overline{\mu}_\nu^{\,\nu}$) is an approximation to the asymptotic convergence factor obtainable per multigrid cycle. Usually this prediction is more accurate than needed. There are still more accurate predictors (see Sec. 4.1). But the main importance of $\overline{\mu}$ is that it separates the design of the interior relaxation from all other algorithmic questions. Moreover, it sets an ideal figure against which the performance of the full algorithm can later be judged (see Secs. 4, 5).

The analysis of relaxation within multigrid is thus much easier than the analysis of relaxation as an independent iterative solver. The latter is not a local process, and its speed depends on smooth components, which are badly approximated by mode analysis, due to boundaries and variable coefficients. For multigrid purposes, however, wherever the equations (or their linearized version) do not change too much within few meshsizes, the smoothing factor can be used as a standard measure of performance. A general routine for calculating $\overline{\mu}$, called SMORATE, is available [M3].

3.2 Work, robustness and other considerations

In comparing several candidate relaxation schemes we should of course take into account not only their smoothing factors, but also the amount of work per sweep. The aim is generally to have the best high-frequency convergence rate per operation, i.e., the largest $w_o^{-1} \log(1/\overline{\mu})$, where w_o is the number of operations per gridpoint per sweep. But other considerations should enter as well: The rate should be robust, that is, $\overline{\mu}$ should not depend too sensitively on problem parameters or on a precise choice of various relaxation parameters. Also, between two schemes with similar values of $w_o^{-1} \log(1/\overline{\mu})$ but with very different w_o, the simpler scheme (where w_o is smaller) should be preferred, because very small factors $\overline{\mu}$ cannot fully be obtained in practice (owing to the inability of the coarse-grid correction to obtain such small factors for the smooth components, and owing to interactions with boundaries).

Moreover, large values of w_o leave us with less flexibility as to the amount of relaxation work to be performed per cycle. Very small $\bar{\mu}$ may in fact be below what we need in the Full Multigrid (FMG) algorithm (see Sec. 7).

An important consideration, sometimes overlooked, is that each relaxation sweep should of course be stable. The most familiar schemes are stable, but distributive schemes (Sec. 3.4) for example, can be unstable exactly in cases showing the best $\bar{\mu}$. A trivial example: satisfy each difference equation in its turn by changing its latest unknown (in the sweeping ordering) instead of its usual corresponding unknown. $\bar{\mu}$ will then vanish, but the process will be unstable. Stability analysis can in each case be performed as Von-Neumann analysis for time dependent problems, taking the main relaxation marching direction as the time-like direction.

Also, do not forget that relaxation has a certain effect on smooth (low frequency) components, too. Usually this effect is slow: $\mu(\underline{\theta})$ is close to 1 for small $|\underline{\theta}|$. But sometimes schemes which show spectacularly small values of $\bar{\mu}$ also show either bad divergence $(|\mu(\underline{\theta})| >> 1)$ or fast convergence $(|\mu(\underline{\theta})| << 1)$ for low frequencies. This for example may happen in relaxing hyperbolic equations (relative to some time-like direction) using upstream differencing and marching with the stream (the time-like) direction. Schemes with bad divergence should clearly be rejected (see an example in [B16, §4.2], the case of high Mach number). Those with fast convergence may also have some disadvantage (in case high-order corrections, as in Sec. 10.2, are desired; see for example [B17, §2.2]).

It is therefore advisable to add to the program of calculating $\bar{\mu}$ also a routine for checking the stability of the scheme examined, and to calculate, together with (3.1), also the value of $\max_{|\underline{\theta}| \leq \pi} |\mu(\underline{\theta})|$. It is also useful to calculate weighted mean squares of $\mu(\underline{\theta})$ for high-frequency $\underline{\theta}$'s . Such quantities predict the error decrease in a given number of multigridded relaxation sweeps for a given initial error [B19, §4.5]. Some of these measures are supplied by SMORATE [M3].

The value of local mode analysis becomes dubious at places of strong discontinuities, e.g., where the coefficients of the differential equation change their order of magnitude discontinuously (or within few meshsizes). This usually happens along manifolds of lower dimensionality, hence more computational work per gridpoint can there be afforded, hence an accurate measure of efficiency is not so needed. But some basic rules, outlined below, must still be followed.

3.3 Block relaxation rule. Semi smoothing.

The most basic rule in devising relaxation schemes is that a locally strongly coupled block of unknowns which is locally decoupled from (or weakly coupled with) the coarser-grid variables should be relaxed simultaneously. The reason is that a point-by-point relaxation smoothes only along the strong couplings.

This rule is of course important whether or not the equations are continuous. In case of persistent S_1 h-ellipticity (i.e., a difference operator with good S_1-h-ellipticity throughout a substantial subdomain, but without uniformly good S_2-h-ellipticity measure for any $S_2 \not\subset S_1$), the rule implies either the use of block relaxation in suitable directions (line relaxation, plane relaxation, etc.), or the use of suitable semi-coarsening, or both.

Semi coarsening, or more specifically S-coarsening, means that $H_j = 2h_j$ for $j \in S$ and $H_j = h_j$ otherwise, where H_j and h_j are the meshsizes of the coarse grid and the fine grid, respectively, in the x_j direction ($j = 1, \ldots, d$). In such a case we need to smooth the error only in directions S. The definition of the smoothing factor should accordingly be modified. We generalize (3.1) to any coarsening situation by defining

$$\bar{\mu} = \max \{\rho(\mu(\underline{\theta})) : |\underline{\theta}| \leq \pi, \max_{1 \leq j \leq d} |\theta_j| H_j/h_j \geq \pi\}. \tag{3.3}$$

Similarly we generalize (3.2) by defining $C_{jj} = I_q$ if $\max |\theta_j| H_j/h_j \geq \pi$, and $C_{jj} = 0$ otherwise. The S-smoothing factor is defined as (3.3), or the generalized (3.2), for S-coarsening.

If point (not block) relaxation is to be used, then S-h-ellipticity is a necessary and sufficient condition for the existence of relaxation schemes with good (i.e., bounded away from 1) S-smoothing factors. This is an easy generalization of a theorem proved in [B12, §4.2]. The more general situation, with block relaxation, is summarized by the following theorem:

THEOREM 3.1. Let S and S' be two sets of directions: $S, S' \subset \{1, \ldots, d\}$. A necessary and sufficient condition for the existence of an S-block relaxation scheme with good S'-smoothing rates is that the discrete operator L^h is uniformly coupled in all $S' - S$

<u>directions; that is,</u> $E^h_{S',S}(L^h) = O(1)$.

$E^h_{S',S}$ is the measure of uniform coupling in S' modulo S , defined by

$$E^h_{S',S}(L^h) = \min_{\substack{\pi/2 \le |\underline{\theta}|_{S'} \le \pi \\ \theta'_j = \theta_j \text{ for } j \in S}} |\tilde{L}^h(\underline{\theta})| / |\tilde{L}^h(\underline{\theta}')| . \qquad (3.4)$$

The theorem states, in other words, that the S'-smoothing factors, produced from L^h by a suitable S-block relaxation, are bounded away from 1 by a quantity which depends only on $E^h_{S',S}(L^h)$.

If the coefficients of L^h are not constant, the above block-relaxation rule suggests that the formal $\overline{\mu}$ (calculated at each point with frozen coefficients of L^h) may be allowed to be bad (close to 1) at some (but not many) grid points, without ruining the overall multigrid efficiency. Experience supports this. Thus, point relaxation schemes yield good full smoothing if and only if the operator has good h-ellipticity measure at <u>almost</u> all points; at some special points its measure of h-ellipticity may become small (or even vanish). This may happen for example when a flow, or any other (sub)characteristic direction, accidentally aligns itself with the grid at some isolated points. If the alignment does not persist, line relaxation is not needed, even if upstream differencing is used.

Only in cases of strong alignment (see Sec. 2.1) the corresponding block relaxation must be used. But it may be confined to the region of strong alingment. If you tried to have block relaxation wherever accidental alignment occurs, you would often need several sweeps with several block directions. This is seldom necessary. If, for example, the strong alignment is due to grid alignment of boundary layers, it is enough to perform line (or plane) relaxation only at the very lines adjacent to such boundaries (and sometimes not even there — see [B17, §3.3]), with just point relaxation elsewhere. If the alignment is strong because it occurs in a large subdomain, line (or plane) relaxation of only that special direction is needed there.

3.4 Distributive, box and collective Gauss-Seidel

To obtain efficient smoothing, a selection should be made from an abundance of available relaxation schemes. The choice depends on experience and on some physical insight, with $\bar{\mu}$ calculations serving for final quantitative judgment. We list here some important types of schemes Each of those can be operated pointwise or blockwise (see Sec. 3.3) and in different orderings (see Sec. 3.6 below). Some simple schemes are described in more detail in [S4]. We first describe the successive displacement species, then we mention their simultaneous-displacement counterparts (Sec. 3.5). In case of equations with many or complicated lower-order terms, it is recommended to apply any of these schemes with the scaled principal terms only (see Sec. 10.3).

The most basic scheme is the <u>Gauss-Seidel</u> (GS) scheme, in which all the discrete equations are scanned one by one in some prescribed order. Each of them in its turn is satisfied by changing the value of one corresponding discrete unknown. This is easy to do if the problem is linear and if there is a natural one-to-one correspondence between equations and unknowns (if, e.g., the matrix of coefficients is definite). In case the problem is <u>nonlinear</u>, each discrete equation may be a nonlinear equation in terms of the corresponding unknown. It is then usually best to make just one Newton step toward solving each equation in its turn. This <u>Gauss-Seide-Newton</u> (GSN) scheme is not related to any <u>global</u> linearization of the system of equations, it just linearizes one discrete equation in terms of one discrete unknown, yielding usually a very simple scheme which does not require any storage other than the storage of the (approximate) solution.

When relaxation is used as the prime solver, much may be gained by Successive Over Relaxation (SOR), in which the GS correction calculated for each unknown is multiplied by a <u>relaxation parameter</u> ω. The situation is different when relaxation is used only as a smoother in multigrid solvers. The best smoothing (lowest $\bar{\mu}$) is usually obtained for the natural value $\omega = 1$, so that GS is not only cheaper (per sweep), but also at least as effective (per sweep) as SOR. Lower $\bar{\mu}$ may be obtained by other parametrizations (e.g., the distributive GS described below), but for regular second-order elliptic equations this gain hardly justifies the extra work involved. Simple GS is probably the best known smoother (especially with red-black ordering -- see Sec. 3.6).

If block relaxation is required (cf. Sec. 3.3), <u>block GS</u> can be used. This means that the blocks are scanned one by one, the equations of each block are simultaneously satisfied by changing the corresponding block of unknowns. In the two-dimensional plane (x,y) if the blocks are lines parallel to x (lines with constant y), the relaxation is called x-line GS, or xLGS. Similarly yLGS is defined.

When there is no natural one-to-one correspondence between discrete equations and unknowns (the matrix is not approximately definite; e.g., non-elliptic and singular perturbation equations, or elliptic systems which are not strongly elliptic [B19, §3.6]), simple GS should be replaced either by Distributive Gauss-Seidel (DGS) or by Weighted Gauss-Seidel schemes. In DGS, with each discrete equation we associate a "ghost" unknown, with some prescription being selected for the dependence of regular unknowns on ghost unknowns. Usually, each regular unknown is written as a prescribed linear combination of neighboring ghost unknowns. Then, as in GS, the equations are scanned one by one, each being satisfied by changing the corresponding ghost unknown. This means in practice that a certain pattern of changes is distributed to several neighboring regular unknowns (hence the denomination "distributive GS); the ghost unknowns do not explicitly appear, nor stored in any way, they just serve for the description of DGS. (In fact their values are never known — only changes in their values are calculated to induce changes in the regular unknowns). In case of block (e.g., line) DGS relaxation, a block of ghost unknowns is simultaneously changed to simultaneously satisfy the corresponding block of equations. In two dimensions we thus have xLDGS and yLDGS schemes. The smoothing analysis of DGS schemes is best executed in terms of the ghost unknowns [B12, §4.1-4.2].

In Weighted GS (WGS) schemes, with each discrete unknown we associate a ghost equation, which is a preassigned linear combination of neighboring equations, and we perform GS in terms of the ghost equations. Taking work into account, WGS seems inferior to DGS, since each equation is calculated several times per sweep, unless the ghost equations explicitly replace the original equations -- which is just a preconditioning of the discrete system of equations. It pays to precondition the system (as against performing DGS) only if the resulting system is not more complicated than the original, which is seldom the case. Preconditioning that yields a simpler system could usually be done already in terms of the differential equations, giving a simpler differential system.

For systems of equations $(q > 1)$, simple GS is appropriate only in case the system is strongly elliptic [B19, §3.6]. Otherwise collective GS or DGS schemes should be employed. Collective Gauss-Seidel (CGS) is performed when the grid is not staggered, i.e., all the q grid equations and q unknown functions are defined on the same gridpoints: The grid points are scanned one by one, at each point we change simultaneously ("collectively") its q unknowns so as to simultaneously satisfy its q equations. In case of a staggered grid one can divide the domain into (usually overlapping) small boxes. The boxes are scanned, for each one we change simultaneously all unknowns interior to it so as to simultaneously satisfy all equations interior to it. This is called Box GS (BGS. See relevant remarks in Sec. 5.6).

DGS schemes are generally more efficient for staggered grids than BGS (except sometimes in very coarse grids; cf. Sec. 6.3): Instead of coupling the equations, we relax them individually. The method is to design the linear combinations (the dependence of regular unknowns on ghost unknowns) so that relaxing the discrete equations corresponding to one differential equations will not significantly alter the residuals corresponding to other differential equations; or at least that such significant alterations will not be circular. See examples of DGS for Cauchy-Riemann and compressible and incompressible Navier-Stokes and Euler systems in [B19, §5.3, 6.3, 7.3], [B14, §3.6] and [B16, §4].

For nonlinear systems, the same methods can be used, but instead of fully satisfying an equation (or a collective of q equations, or a box of equations), only one Newton step is made in terms of the corresponding (regular or ghost) unknown (or collective of q unknowns, or box of unknowns). For semi h-elliptic cases, block CGS (e.g. line CGS, meaning simultaneous solution of all equations on a line through changing all that line's unknowns), or block DGS, or block BGS, may be performed.

For some systems, the different relaxation passes for different interior equations may have different efficiencies, in which case the overall smoothing-per-work may be improved by making more passes on some of them, provided there is no serious feedback of errors from those equations to others. See an example in [B19, §7.3].

Higher-order equations are sometimes most efficiently relaxed by writing them as systems of lower order equations. For example, the biharmonic can be written as a pair of Poisson equations. Relaxing this system involves less work (per complete sweep) and yields better smoothing (per sweep) than relaxing the biharmonic. But special care should be taken in relaxing the boundary conditions for this system (see Sec.5.3).

3.5 Simultaneous displacement (Jacobi) schemes

The GS schemes described above are successive-displacement schemes: The new value of an unknown (or block of unknowns) replaces the old one as soon as it is calculated, and is immediately used in relaxing the next equations. In simultaneous displacement schemes new values replace old ones only at the end of the sweep, after all of them have been calculated; hence each of them is calculated explicitly in terms of old values only. Corresponding to each of the schemes above we have a simultaneous-displacement scheme, called : Jacobi relaxation, Jacobi-Newton, distributive Jacobi, weighted Jacobi, collective Jacobi, box Jacobi, line Jacobi, weighted line Jacobi etc. - corresponding to GS, GSN, DGS, WGS, CGS, BGS, line GS, line DGS, etc., respectively.

Unlike GS, Jacobi schemes often require under-relaxation ($\omega < 1$) in order to provide good smoothing. But with relaxation as a smoother (not an independent solver), good and optimal values of ω are independent of the domain, and can easily be calculated by local mode analysis.

Distributive and weighted Jacobi (under-)relaxation amounts actually to the same thing (equivalent also to the weighted Jacobi in [B7, §3.3]).

All experience so far shows that Jacobi schemes are inferior to the corresponding GS schemes. They not only require more work (for operating the relaxation parameter) and more storage (for storing the new values separately), but their smoothing factors are in fact worse. For the 5-point Poisson equation, for example, Jacobi under-relaxation ($\omega_{optimal} = .8$) yields $\bar{\mu} = .6$, while GS gives $\bar{\mu} = .5$ and .25 for lexicographic and red-black orderings, respectively. The situation is similar in all cases so far examined. The advantage of simultaneous-displacement schemes is in their being more amenable to rigorous analysis (but there seems to be no practical value to this — see Sec. 14) and their vectorizability and parallelizability (but red-black GS and similar schemes are also fully parallelizable — see Sec. 3.6).

3.6 Relaxation ordering. Vector and Parallel processing

For successive-displacement schemes, the order in which the equations (or blocks of equations) are relaxed has an important effect on the smoothing factors. The main orderings used are the usual lexicographic (LEX) order (in which the equation at grid point $(i_1, \ldots, i_d)$ is relaxed before $(j_1, \ldots, j_d)$ if $i_k = j_k$ for $1 \leq k < \ell$ and $i_\ell < j_\ell$), and related orders (LEX order for some permutation of the coordinates, some of them possibly reversed); symmetric relaxation (lexicographic sweep followed by a sweep in the reversed order); red-black (RB) ordering (in which all "red" gridpoints are relaxed before all "black" ones, where the coloring is similar to that of a checker board, namely a point $(i_1, \ldots, i_d)$ is red if $i_1 + \ldots + i_d$ is even, and black otherwise); and more general pattern relaxation (similar to RB, but with different coloring and possibly more colors). For difference equations involving more than nearest neighbors, RB schemes still depend on the ordering of points within each color. If all such points are relaxed simultaneously, the scheme is called Jacobi-RB; similarly LEX-RB, etc.

Each of these orderings has its block-relaxation versions. xLGS (or xLDGS) can be done lexicographically forward (increasing y) or backward (decreasing y), or symmetrically (forward alternating with backward). Or, corresponding to RB, we can first relax the even lines, then the odd lines. This is called zebra xLGS (or x-zebra) relaxation. Similarly, yLGS (or yLDGS) can be done upward, downward, symmetrically or zebra. Particularly robust schemes are the Alternating-Direction Zebra (ADZ = x-zebra alternating with y-zebra) and Alternating-Direction Symmetric LGS (ADS = symmetric xLGS alternating with symmetric yLGS). Many more block GS schemes are similarly defined in higher dimensions. The choice of blocks is governed by the rule in Sec. 3.3. Concerning the choice of ordering we have the following remarks.

It has been found that GS with RB ordering is the best for the 5-point Poisson equation [F2]. Similarly, DGS with RB ordering within each of its passes (called briefly Distributive RB, or DRB) is the best for many systems, such as Cauchy-Riemann and compressible and incompressible Navier-Stokes equations [B14, §3.4 - 3.6], [B16, §4.2]. For 5-point Poisson, RB-GS provides $\overline{\mu}_1 = .25$, $\overline{\mu}_2 = .25$ and $\overline{\mu}_3 = .32$ (cf. Eq. (3.2)), as against $\overline{\mu} = .5$ for LEX-GS. Moreover, RB-GS can be executed with only four operations per grid point, whereas lexicographic GS requires five. Similar comparisons hold for the more complicated elliptic systems.

In addition, the mentioned RB schemes (or Jacobi-RB, or more general pattern relaxation in cases based on larger stencils), are fully vectorizable and parallelizable: All the equations of the same color can be relaxed in parallel, thus taking full advantage of computers having vector or parallel processing capabilities. The zebra schemes are similarly parallelizable. (See more about parallelization of all multigrid processes in [B14].)

For non-elliptic equations or for elliptic equations with large non-isotropic lower-order terms (singular perturbation problems, in particular), the first approach ([B6], [S2], [B7], [B10]) was to employ "downstream" ordering, in which the equation at a point A is relaxed before (or simultaneously with) that at point B if the solution at B depends more heavily on the solution at A than vice-versa (e.g., if the fluid flows, or the convection transports, from A to B). This provides very good smoothing factors (better than those for regular elliptic problems). If different "downstream" directions exist at different parts of the domain, this may require a sequence of several relaxation sweeps in several directions. If for example line relaxation is also required, ADS relaxation may be needed, i.e., four passes over the domain. Each pass may be effective in only part of the domain, but the combined sweep will give excellent smoothing everywhere, for any combination of semi h-elliptic approximations in two dimensions (and also in three dimensions, if the grid is coarsened in only two directions. cf. Sec. 4.2.1). In some particular cases (when the reduced equation is hyperbolic in some time-like direction, and upstream differencing is employed) such schemes yield not only great smoothing but also great convergence, making coarse-grid corrections superfluous.

Since one or two (efficient) sweeps are usually all that is needed at each multigrid stage, the multi-direction procedure is sometimes not fully efficient. Also, it requires more complicated programs (several kinds of passes). Hence, ordering-free schemes were developed, in which good smoothing is obtained for any ordering, including RB and/or zebra (the block-relaxation rule should still be kept.) Such ordering-free schemes are obtained either by distributive relaxation [B10, §6], or by using slightly more artificial viscosity than that required for upstream-differencing [B12, §4.3], [B17, §5.7, 6.3, 7.2].

3.7 Other smoothers: ILU

The above list of relaxation schemes, although including the most efficient smoothers found so far, does not exhaust all possibilities.

Of special recent interest is the use of incomplete LU decomposition (ILU), and related schemes, as smoothers [W1], [K2]. These smoothers have been shown to be very robustly efficient for a wide range of simple 5-point and 9-point difference equations. A careful comparison, in which the total amount of operations in a full multigrid (FMG) algorithm is counted taking into account the ILU set-up operations, shows these schemes to be quite comparable to the GS schemes [T2]. They need however much larger storage and they become considerably less efficient for nonlinear problems, or for systems, or for more complicated stencils. Unlike other smoothers, they are not local. They are a "package deal", robust on one hand, but inflexible to special local needs on the other hand. For three-dimensional problems they become less robust.

4. INTERIOR TWO-LEVEL CYCLES

Having computed the smoothing factor $\bar{\mu}$, one should expect the asymptotic convergence factor per multigrid cycle to approach $\bar{\mu}^{\nu}$, where ν is the number of relaxation sweeps (on the fine grid h) per cycle. This ideal figure does not take into account the exact nature of the inter-grid transfers. The next task then is to design those transfers so as to approach the ideal figure. To separate their design from questions related to boundary conditions (which will be taken up at the next stage), we still think in terms of fully-periodic or infinite-space problems; we still restrict our attention to <u>interior</u> processes, because it is there that most of the computational work is invested. Furthermore, we simplify the multigrid situation at this stage by restricting our attention to two grids only, the finest grid $h = (h_1,\ldots,h_d)$ and the next coarser grid $H = (H_1,\ldots,H_d)$, where usually $H = 2h$. That is, we assume in our analysis that the grid-H equations are solved (exactly) each time the algorithm gets to that grid, without analyzing how that solution is obtained, hence without involving grids coarser than H in the analysis.

These assumptions indeed simplify our studies very much. First, the error can be expanded in a Fourier integral (or series) and the transformations of the amplitudes of different Fourier components by multigrid operations can be calculated. Indeed, for linear systems with constant coefficients only few Fourier components at a time are coupled by these two-level interior processes, hence transformations of Fourier amplitudes are expressed as small matrices (Sec. 4.1). In case of non-constant coefficients, we usually freeze them at some values (treated then as parameters of the analysis). In case of nonlinear equations, their Newton linearization is analyzed (although no such linearization is needed in the actual processes; see Sec. 8.3). The parameters of the analysis then depend on the solutions around which linearization is made.

This freezing of coefficients is reasonable as long as the real coefficients do not change too drastically over a mesh-size. Where they do, we can sometimes model them as changing periodically, again making mode analysis with small matrices possible [B19, §4.7]. Where mode analysis becomes too difficult or dubious, numerical experiments with periodic boundary conditions can be used instead. One should then restrict oneself, however, to simultaneous-displacement relaxation (Jacobi, or red-black or zebra Gauss-Seidel schemes), to avoid the special problems created along the starting line (or termination line) of the relaxation sweep, thus completely separating away issues related to boundaries. One should also make sure that at this stage (whether mode analysis or periodic numerical experiments are used) both grids are fine enough, and grid-H equations are solved accurately enough (without taking into account the work this accurate solution requires), in order to separate away questions related to coarser grids (see Sec. 6). Do not, however, forget in the process that this is a modeling for a multigrid solution, hence your model must be recursible: The H equations should have the same general form as the original h equations, with the same range of possible parameters.

In addition to the relaxation scheme, studied above, the main issues to be studied at this interior-two-level stage are when to switch (under what criteria, or after how many relaxation sweeps) from grid h to grid H; what should be the coarse-grid variables; and the type (in the interior) of three multigrid operators: The fine-to-coarse transfer of residuals I_h^H, the coarse grid operator L^H, and the coarse-to-fine interpolation of corrections I_H^h. These issues are one-by-one discussed in the subsections below. They are later reviewed again, from a more general perspective, in Sec. 11. Relevant to these issues are also Secs. 8.5 and 10.2 (nonlinear problems and higher-order techniques).

4.1 Two-level cycling analysis. Switching criteria

Details of the two-level mode analysis are described in [B19, §4.6 - 4.8] and in [S4, §3.3 - 3.5, §7,§8,§9]. The former also discusses modifications of the analysis to account for the fact that in practice the grid-H equations are only approximately solved, modification for the case of equations with highly oscillatory coefficients, and ways to make precise comparisons between mode analysis and numerical experiments (for debugging purposes).

On grid 2h the Fourier mode $\exp(i\underline{\theta}\cdot\underline{x}/\underline{h})$ aliases (coincides with) the mode $\exp(i\underline{\theta}'\cdot\underline{x}/\underline{h})$ whenever $|\theta_j - \theta'_j| = 0$ or π, $(1 \leq j \leq d)$. Hence each set of so aliasing components usually includes 2^d components $\{\underline{\theta}^1, \ldots, \underline{\theta}^{2^d}\}$, called harmonics of each other. They are coupled to each other by the two-level processes. (The special sets with less than 2^d different components do not require special analysis, since they are limits of regular sets.)

We define the two-level cycle as follows: Make ν_1 relaxation sweeps on grid h, then transfer the residual problem to grid H and solve it there exactly, then interpolate that grid-H solution to grid h and add it as a correction to the former grid-h solution, then make ν_2

more relaxation sweeps on grid h. It is easy to see that in the infinite space, if L^h, L^H, I_h^H and I_H^h are all constant operators, and if the error in the solution before such a cycle has the form $\Sigma_j A_j \exp(i\underline{\theta}^j \cdot \underline{x}/\underline{h})$, where the sum is over a set of 2^d harmonics, then the error after the cycle will have a similar form, and the new A_j's will be linear combinations of the old ones. If we deal with a system of q grid equations then each amplitude A_j is a q-vector, hence the overall transformation of the 2^d amplitudes by the two-level cycle is a $(2^d q)\times(2^d q)$ matrix M, which can be denoted $M(\underline{\theta})$, where $\underline{\theta}$ is the lowest harmonic $(|\theta| \leq \pi/2)$.

This matrix $M(\underline{\theta})$ is called <u>the two-level amplification matrix</u>. The easiest and most modular program for calculating it is to write a different routine for the general matrix-element of each of the five involved processes: relaxation, L^h, I_h^H, L^H, and I_H^h. Their respective matrices $\check{S}^h$, $\check{L}^h$, $\check{I}_h^H$, $\check{L}^H$ and $\check{I}_H^h$ have dimensions $(2^d q)\times(2^d q)$, $(2^d q)\times(2^d q)$, $q\times(2^d q)$, $q\times q$ and $(2^d q)\times q$, respectively, and each of their elements is a function of $\underline{\theta}$. Then program

$$M(\underline{\theta}) = (\check{S}^h)^{\nu_1}\,[I - \check{I}_H^h\,(\check{L}^H)^{-1}\,\check{I}_h^H\,\check{L}^h]\,(\check{S}^h)^{\nu_2} \tag{4.1}$$

The main performance measure of the two-level cycle is the <u>two-level asymptotic convergence factor</u> (per cycle)

$$\overline{\lambda} = \max_{|\underline{\theta}| \leq \pi/2} \rho(M(\underline{\theta})) . \tag{4.2}$$

Note that $\overline{\lambda}$ depends on the sum $\nu = \nu_1 + \nu_2$, but not on the separate values of ν_1 and ν_2. In fact, when many cycles are performed the separate values are immaterial. Various other performance measures can similarly be defined. (See [S4, §3.4 - 3.5], where the notation $\hat{M}_{h,n}^{2h}$ and ρ^* corresponds to our M and $\overline{\lambda}$, respectively. Additional two-level measures will be discussed in Sec. 7.4).

Using the two-level analysis we try to (roughly) optimize the involved processes; namely, the objective is to maximize $w^{-1}\log(1/\overline{\lambda})$, where $w = A(\nu w_0 + w_1 + w_2)$, w_0 is the work in one relaxation sweep, w_1 is the work of calculating and transferring the residuals, w_2 is the work of the I_H^h interpolation, and A is a factor through which the work on coarser grids is taken into account. For regular elliptic problems where V cycles are used (see Sec. 6.2) we can assume similar operations on each of the grids, hence $A = (1-\hat{\rho}_1 \ldots \hat{\rho}_d)^{-1}$, where $\hat{\rho}_j = h_j/H_j$ (usually $\hat{\rho}_j = .5$). In problems requiring W cycles (see Sec. 6.2), $A = (1-2\hat{\rho}_1 \ldots \hat{\rho}_d)^{-1}$. To avoid the laborious count of operations and the arbitrary assignment of proper weights to different arithmetic and non-arithmetic operations (which are really machine-dependent), one can use the work of a standard relaxation sweep as the work unit. In complicated problems, where calculating L^h outweighs interpolations, one can then neglect w_2 and take $w_1 = 1$ for full residual weighting and $w_1 = 2^{-d}$ for residual injection. The convergence factor per work unit is then denoted by $\mathring{\mu} = \overline{\lambda}^{1/w}$. As above (Sec. 3.2), in addition to the goal of minimizing $\mathring{\mu}$ we should take robustness and simplicity into account.

One can also partly separate the study of I_h^H, L^H and I_H^h from that of relaxation by the <u>Coarse-Grid Correction (CGC) mode analysis</u>, in [B7, §A.1]. But this is not simpler than the full two-level analysis, especially since relaxation schemes have already been selected in the previous stage. We use a CGC analysis in Sec. 4.3 below.

The ideal factor $\overline{\lambda} = \overline{\mu}^{\nu}$ is not always obtainable. If $\overline{\mu}^{\nu}$ is too small we will get $\overline{\lambda} > \overline{\mu}^{\nu}$, because of significant high-frequency amplitudes generated from low ones by interpolation or by RB-type relaxation. (see Sec. 4.3). Even when obtainable, too small values of $\overline{\lambda}$ will require too precise interpolations, hence too much investment in w_1 and w_2, and will at a later stage be frustrated by other interactions (boundaries and non-constant coefficients). Also, such small $\overline{\lambda}$ will not usually be needed in the final FMG algorithm (see Sec. 7.2 - 7.3). Hence, the optimal cycle always employs a small ν, typically $\nu \leqslant 3$.

In regular elliptic problems $\nu = 1$ is too small to be optimal (unless the sweep includes several passes, as in symmetric and alternating-direction schemes), since the overhead of w_1 and w_2 weighs too much against it. Hence usually the optimal number is $\nu = 2$ for very efficient smoothers ($\overline{\mu} \leqslant .3$ or so), and $\nu = 3$ otherwise. A small change in ν does not disturb the overall efficiency very much. Considerably larger ν are less efficient, because they bring the process into the range of larger feeding from low to high frequencies, while not much more is gained in reducing the overhead (already at $\nu = 3$, $w_1 + w_2$ is quite small compared with νw_0.

A possible approach is <u>accommodative</u>: do not fix ν in advance, but continue relaxation as long as it exhibits the fast convergence of high frequencies, e.g., as long as the convergence factor (some norm of the residuals divided by the same norm a sweep earlier) is smaller than the smoothing factor $\bar{\mu}$. For non-scalar $(q>1)$ systems, such a critorion can separately be applied to each equation, possibly resulting in more passes for part of the equations. Similarly it may separately be applied at different subdomains (since smoothing is a local process), possibly giving partial relaxation sweeps.

In case of non-elliptic and singular perturbation problems there are some particular smooth error components (smooth characteristic components of the differential operator or the reduced differential operator) for which L^H is a bad approximation to L^h , hence $\bar{\lambda}$ cannot be much smaller than .5 , no matter how small $\bar{\mu}^\nu$ is [B17, §5.1], [B3]. But for exactly the same components and the same reason, L^h itself is not a good approximation to the differential operator L . Hence, exactly for these components, we do not <u>need</u> much <u>algebraic</u> convergence (convergence to the discrete solution), since the discrete solution itself is far from the differential solution. Hence, for such cases the asymptotic convergence factor $\bar{\lambda}$ is not really the measure we need. The one we need is obtained by the two-level FMG analysis (see Sec. 7.4). Moreover, for non-elliptic or singular perturbation problems the usual assumption that high-frequency components are local does not hold. It is violated by high-frequency characteristic components in cases of strong alignment (Sec. 2.1). The interior mode analysis should then be supplemented with a half-space analysis (Sec. 7.5).

4.2 <u>Choice of coarse grid</u>

When the fine-grid, with meshsize $h=(h_1,\ldots,h_d)$, is given, the choice of a coarse-grid, with meshsize $H=(H_1,\ldots,H_d)$, is often straightforward: Take every other line (every other hyperplane, for $d>2$) of the fine grid in each direction. The coarsening ratio $H_j/h_j=2$ is usually optimal: it is the smallest recursively convenient number, and is already big enough to make the coarser-grids work quite small relative to the fine-grid work; larger H_j/h_j will not save significantly more work, but will significantly degrade the smoothing factors (see Eq. (3.3)). The smaller ratio $H/h=2^{\frac{1}{2}}$ may be as efficient (trading larger A for smaller ν) , and it is recursively convenient in some two-dimensional problems with rotatable operators; see [S4, §2.4] , [R1].

When the fine-grid discretizations are done in terms of "cells", with the discrete variables defined at certain cell positions (e.g., cell centers, or centers of vertical cell boundaries, etc.), and especially when the grid is <u>staggered</u> (different grid functions are defined at different cell positions), it is more convenient to coarsen in terms of the cells: Take every 2^d fine cell as a coarse cell, and then place coarse-grid variables at coarse-cell positions analogous to their positioning in the fine cells. The coarse grid points will then generally not be a subset of the fine grid points. (For another approach, see [D1, Fig. 1]).

In some cases the "fine-grid" is not a well-organized grid at all; e.g., a general finite-element triangulation, not based on any grid lines. Then one can still construct the coarse grid as a uniform grid, placed over the domain with no particular relation to the fine grid. Another approach is to base the choice of coarse-grid variables on purely algebraic considerations (Sec. 13.1). Mode analysis is of course not very suitable for analyzing such situations.

4.2.1 Semi-Coarsening. Semi-coarsening, or more specifically S-coarsening, is the technique of using grid H which is coarser than h in only a partial set S of coordinates; i.e., $H_j = 2h_j$ for $j \in S$ and $H_j = h_j$ for $j \notin S$. This means somewhat more work on coarse grids; but either this or block relaxation are needed in some cases — see the rule in Sec. 3.3. Semi coarsening is sometimes preferable to block relaxation. For example, in three-dimensional problems where there are two fixed coordinates with stronger couplings, full coarsening would require plane relaxation, which is inconvenient. (Solving these plane equations approximately by one multigrid cycle, if done simultaneously at all planes, will look very much like semi coarsening.) Also, exactly in those cases, semi-coarsening involves relatively small work on coarser grids, since two coordinates are still coarsened, hence the number of coarse grid points is at most one third the number of points on the finest grid.

Sometimes, a combination of block relaxation and semi coarsening may be the best. For example, the equation $aU_{xx} + bU_{yy} + cU_{zz}$ with $a \ll b \ll c$, discretized on a cubic grid ($h_x = h_y = h_z$), will best be solved by z-line relaxation and y-z semi-coarsening. Generally, rough calculations of S-smoothing factors (Sec. 3.3) immediately show what procedures can be taken.

In many semi elliptic cases block relaxation is of course preferable to semi-coarsening. For example, when the directions of semi ellipticity are different at different subdomains. To change accordingly the directions of semi-coarsening would be much messier than changing block directions (unless made in the framework of non-geometric coarsening, as in Sec. 13.1).

4.2.2 Modified Coarse-Grid Functions. When a difference operator L^h is given which has no good h-ellipticity or semi-h-ellipticity measure, then no relaxation can be efficient in reducing all high-frequency error components. To reduce all components efficiently we will then apply modified coarse-grid correction functions.

Suppose for example that the slow components (i.e., the components for which relaxation is inefficient) are all clustered around some known modes $\varphi_j(\underline{x})$, $(j = 1, \ldots, J)$. This means that the error $v^h = \tilde{u}^h - u^h$ can be written as $v^h(\underline{x}) = \Sigma\, v_j^h(\underline{x})\,\varphi_j(\underline{x})$, where v_j^h are smooth functions. It is then these smooth functions which we try to approximate by coarse-grid functions v_j^{2h}. See [B15, §3.2] and [B21].

4.3 Orders of interpolations and residual transfers

The most important aspect of the coarse-to-fine correction interpolation I_H^h and the residual transfer I_h^H is their orders, defined as follows: The order of I_H^h is m if the interpolation of the low-frequency Fourier component $\exp(i\underline{\theta}\cdot\underline{x}/h)$, with amplitude 1 on the coarse grid H, creates on the fine grid h high-frequency components (the harmonics of the low frequency) with amplitudes $O(|\underline{\theta}|^m)$. It also reproduces the $\underline{\theta}$ component itself on grid h with an amplitude $1 + O(|\underline{\theta}|^m)$. The order of the fine-to-coarse transfer I_h^H is said to be m, and its secondary order $\overline{m}$, if a high-frequency harmonic with amplitude 1 on grid h contributes $O(|\underline{\theta}|^m)$ to the amplitude of the corresponding low frequency when transferred to grid H, while a low frequency with amplitude 1 on grid h contributes $1 + O(|\theta|^{\overline{m}})$ to its grid-H amplitude. Thus, linear and bilinear interpolations have order 2, while cubic interpolation is fourth order. Residual transfer by injection $(I_h^H \equiv 1)$ has order 0 and infinite secondary order, whereas the usual full-weighting residual transfer (Eq. (4.6) below) is of order 2 and secondary order 2.

What orders should be used in the multigrid cycle? This depends on the orders of derivatives appearing in our equations. Suppose we have a system of q differential equations in q unknown functions, and let m_{ij} be the highest order of differentiation (or differencing) of the j-th unknown in the i-th equation, $(i,j = 1, \ldots, q)$. We assume, and this is usually the case, that the q unknown functions are interpolated independently of each other and that the residuals of each of the q grid equations are transferred separately from the others. Denote by m^j the order of I_H^h used in interpolating the j-th correction (correction to the j-th unknown function) and by m_i and $\overline{m}_i$ the order and secondary order, respectively, of the I_h^H used in transferring the i-th residual (residuals of the i-th equation).

What m^j, m_i and $\overline{m}_i$ $(i,j = 1, \ldots, q)$ should be used? Examining orders of magnitude in the CGC mode-analysis operator (the operator in brackets in (4.1)), under the asumption that all $m^j > 0$, we find the following basic rules and observations:

(A) The high-frequency harmonics of the lowest frequencies (those with $|\underline{\theta}| = O(h)$), are amplified by the CGC operator by a factor with a spectral radius $1 + O\left(\sum_{i,j} h^{m_i + m^j - m_{ij}}\right)$. Hence, to avoid large magnification of high-frequencies, we should have

$$m_i + m^j \geq m_{ij}, \qquad (4.3)$$

preferably even $m_i + m^j > m_{ij}$. On the other hand, larger values ($m_i + m^j > m_{ij} + 1$) would not significantly further reduce the spectral radius, hence they are <u>asymptotically</u> (when many cycles are made) not needed.

(B) Every high-frequency harmonic (before the CGC cycle) contributes to the corresponding low-frequency (after the cycle) through a $q \times q$ transformation matrix $(L^H)^{-1}B$, where $B_{ij} = O(h^{m_i - m_{ij}})$. This is usually not important asymptotically (for many cycles), but <u>if only one cycle is performed</u> (as in FMG algorithms) that transformation may cause large errors unless

$$m_i \geq m_{ij}. \qquad (4.4)$$

For relaxation schemes with interactions between high and low frequencies (e.g., RB schemes), this transformation may also cause <u>asymptotic</u> degradation unless $m_i > \Sigma_k (m_{ik} - r_{kj})$, where $O(h^{r_{kj}})$ is the size of the high-frequency errors in the k-th function generated by relaxation from an O(1) low-frequency error in the j-th function. RB and zebra schemes for $q = 1$ give $r_{11} = m_{11}$, hence the rule requires $m_1 > 0$, i.e., full weighting (see Sec. 4.4). This requirement can however be slackened by a more precise look at the nature of these particular schemes (allowing the use of simpler transfers such as the "half injection" $I_h^H \equiv .5$ or "half weighting"; see [F2, §2], [S4, §8.1]).

(C) The low-frequency error components themselves are reduced by a factor $O(h^{\tilde{m}})$, where $\tilde{m} = \min(\tilde{p}, \overline{m}_1, \ldots, \overline{m}_q, m^1, \ldots, m^q)$ and $\tilde{p}$ is the lowest of the approximation orders on levels h and H. Hence $\tilde{m}$ must be positive, which is indeed the case for any consistent differencing and interpolation schemes. Larger values of $\tilde{m}$ may of course give better cycle performance. Our experience indicates that $\tilde{m} = 2$ gives considerably better $\overline{\lambda}$ than $\tilde{m} = 1$. Since this is a low-frequency matter, hence non-local, higher $\tilde{m}$ may be effective only if they are carefully matched by corresponding high-order approximations and interpolations at boundaries. But one usually does not have to go into the trouble of $\tilde{m} > 2$. Rather, employ more cycles with $\nu \leq 3$ (see Sec. 4.1). As a result the factor $O(h^{\tilde{m}})$ will usually be dominated by $\bar{\mu}^{\nu}$ in determining $\bar{\lambda}$.

(D) We also note that every low-frequency error component (before the CGC cycle) contributes to every one of its harmonics (after the cycle) through a $q \times q$ transformation matrix D, where $D_{jj} = O(h^{m_j})$ and for $i \neq j$ D_{ij} has higher orders in h. This tells us something about the range where relaxation should be efficient (see Sec. 12).

4.4 Variable operators. Full weightings

The above mode-analysis rules are insufficient in case L^h is highly-varying, i.e., its coefficients substantially change between two neighboring gridpoints. For such L^h the residuals after relaxation are also highly varying, hence to represent them correctly on grid H full residual weighting should be used, i.e., I_h^H should satisfy, for any residual function R^h,

$$(H_1 \ldots H_d) \sum_{x^H} (I_h^H R^h)(x^H) = h_1 \ldots h_d \sum_{x^h} R^h(x^h) , \qquad (4.5)$$

where x^h are the fine-grid points and x^H are the coarse-grid points. In other words, full weighting "preserves integrals". (Throughout this discussion it is assumed that the difference equations on all grids are written in their divided form, analogous to the differential equations. If, however, they are multiplied through by factors which depend on the meshsize, then one should not forget to have those factors in (4.5), too.) One can regard full weighting as a scheme in which each residual $R^h(x^h)$ on the fine grid is being distributed to several coarse grid points, with weights whose sum is $\tilde{\rho} = h_1 \ldots h_d/(H_1 \ldots H_d)$. Hence each residual R^h is a weighted average of its transferred values on grid H, times $\tilde{\rho}$. This weighted average represents a certain interpolation, $\hat{I}_H^h$ say. Thus every full weighting I_h^H is the adjoint (or, in matrix terminology, the transpose) of an interpolation $\hat{I}_H^h$, times $\tilde{\rho}$. We denote this by $I_h^H = \tilde{\rho}\hat{I}_H^{h*}$. The normal (9-point symmetric) full weighting, defined by

$$\left(I_h^{2h} R^h\right)\left(x^{2h}\right) = \sum_{\max|\nu_j| \leq 1} 2^{-d-\Sigma|\nu_j|} R^h\left(x^{2h} + (\nu_1 h_1, \ldots, \nu_d h_d)\right), \qquad (4.6)$$

is for example the adjoint of bilinear interpolation, times 2^{-d}.

The requirement (4.5) is equivalent, in terms of the Fourier analysis, to the requirement that I_h^H has a positive order (see Sec. 4.3).

Such full weightings should perhaps be used in almost any case. Only in some particular cases non-full weightings happen to be asymptotically somewhat better. An example is injection in case of the standard 5-point Poisson operator, which yields lower $\overline{\lambda}$ as well as lower w_1 than the full weighting (4.6) (see [B19, §4.8]). But even in those cases, for the purpose of Full Multigrid (FMG) algorithms (see Sec. 7), full weightings may be preferable. (See rule (4.4) above and [S4, §3.6].)

4.5 The coarse-grid operator

The coarse grid operator L^H should be a proper homogenization of the fine-grid operator L^h. In smooth problems this is easily obtained by good discretizations of both L^h and L^H. In nonlinear problems this is effectively obtained by a suitable FAS averaging of the fine-grid solution (see $\hat{I}_h^H$ in Sec. 8.5), provided the coefficients (of the linearized equations) do not change drastically over a meshsize (see Secs. 4.6 and 11).

4.6 Strongly discontinuous, strongly asymmetric operators

As long as the fine-grid operator does not vary drastically, the above rules for I_h^H, L^H and I_H^h work fine. A more difficult case is that of a strong discontinuity in L^h, i.e., where its coefficients change their order-of-magnitude within a meshsize. Orders of interpolations are not so important then; rather, special forms should be used which take into account the particular nature of the discontinuity. The rule is first to analyze the behavior near the discontinuity of the error which is inefficiently reduced by relaxation. This error is approximately a solution to the homogeneous equations. (If it is not, then it has large residuals and therefore there locally exists a relaxation scheme for which it will be reduced efficiently). Hence its general behavior is like that of solutions to the homogeneous differential equations. The interpolation I_H^h of corrections should take this behavior into account. For example, if we have a diffusion problem $\nabla(a\nabla u) = F$, near a strong discontinuity of the diffusion coefficient $a(x)$ the derivatives of the solution to the homogeneous equation are not continuous; instead, $a\nabla u$ is continuous there, and this can be used to design good interpolation schemes [A1].

It is less clear how to generally design the residual transfers I_h^H and the coarse grid operators L^H near a strong discontinuity. In the symmetric case the variational rule

$$I_h^H = \tilde{\rho}\, I_H^{h*} \qquad (4.7)$$

$$L^H = I_h^H L^h I_H^h \qquad (4.8)$$

are most robust [A1]. For cases which are not essentially symmetric, the "Galerkin" rule (4.8) with "suitable" I_h^H may still be good, but not always with (4.7).

For non-elliptic and singular perturbation problems, the considerations and experiments in [B17] indicate that improved results are obtained by a full residual weighting in which residuals, on being transferred from a fine gridpoint to a different point (or points) on the coarse grid, are transferred roughly in the downstream direction. As for correction interpolation for such problems, however, it seems that the symmetric schemes are preferable to schemes with upstream bias. Coarse-grid operators identical with the fine-grid ones were used, with excellent FMG results (even where the asymptotic rates were slow).

A general perspective on these questions of coarsening a problem (designing I_h^H, L^H, I_H^h) is given in Sec. 11 below.

5. BOUNDARY CONDITIONS AND TWO-LEVEL CYCLING

The theoretical two-level mode analysis described above (Sec. 4.1) gives us the ideal convergence factor per cycle ($\bar{\lambda}$), or per work-unit ($\mathring{\mu}$). These are the interior convergence factors, obtained in the absence of boundary interference. The next stage is to construct an actual multigrid program for an actual, bounded domain, and in particular to decide on the special treatment the various processes should take at points near or on boundaries. The goal is to attain or approach the interior convergence factors. This is generally possible, since the boundary neighborhood is a lower-dimensional set of grid points, hence we can allow there more work (per point) than in the interior, without changing the total work by much. The comparison to the interior factors is a very important tool in debugging the program or finding conceptual mistakes, especially mistakes in treating boundary conditions or interior equations at points adjacent to boundaries.

In Secs. 5.2 – 5.5 below we mention some rules related to the multigrid processes near or on boundaries. The general remarks of Sec. 11 and the curved-boundary treatment in Sec. 9.3 are also relevant here.

In addition to boundary conditions, some problems have global conditions. These should also be incorporated at this stage. Their multigrid implementation is discussed in Sec. 5.6.

5.1 Simplifications and debugging

It is advisable to start with a program for rectangular domains whose boundaries coincide with grid lines at all levels. This will make the programming much easier (the two-page program in [B7, App. B] can serve as a model), and will separate away various difficulties related to the more general situation. Afterwards, in writing programs for non-rectangular domains, one would have to decide whether to write a general-domain or a specific-domain program. Experience shows general-domain multigrid programs to be considerably less efficient (typically requiring twice the CPU time). One can model one's general-domain program after MGO1 [S4, §10], or after MUGPACK, or actually use the MUGPACK or GRIDPACK software [M3]. But the efficiency of this software, too, is still considerably below the efficiency of specific-domain programs (where the efficiency of rectangular domains can be approached). The reason is the many checks that should be made to distinguish between various possible positions of gridpoints with regard to the boundary, especially in interpolation routines, where two grids are simultaneously involved.

It is also advisable to start programming cycling algorithms, before proceeding to the additional questions related to the Full Multi-Grid algorithm (taken up in Sec. 7). Cycling algorithms start with some arbitrary approximation on the finest grid and reduce its error by cycling between that grid and coarser grids. Types of cycles are defined in Sec. 6.1 below. At this stage, however, we avoid the question of what cycle is the best: For debugging purposes it is best to start with comparing the theoretical two-level asymptotic convergence factor ($\bar{\lambda}$) with the experimental one by an algorithm which simulates a two-level algorithm. This is done by returning from the next coarser grid H back to the finest grid h only when the H equations have been solved to a very good accuracy (e.g., by taking large γ or very small δ in the cycles of Sec. 6.2). In this way we still separate away questions particular to too-coarse grids or related to three or more levels (delaying them to Sec. 6). Also, to reach the asymptotic, i.e. the worst possible, experimental factor $\bar{\lambda}$ without spending too many cycles and without needing double precision, it is helpful to start with initial errors devised to contain a large amplitude of a worst component. Such a component is derivable from the mode analysis.

Debugging of multigrid programs can generally benefit from relations between the levels. Most bugs and conceptual errors immediately show as irregular behavior in the standard multigrid output (listing the history of the dynamic residual norms for every relaxation sweep on every level, as in [B7, App. B]). A preliminary error-detection table, based on that output, is provided in [B9, Lecture 18]. Troubles related to treatment of boundaries often show in the following way: The first couple of cycles exhibit the expected (interior) convergence factor, since the relative weight of errors near the boundaries is small. Later, however, the errors near the boundaries start to dominate and the convergence factor degrades. The coarser is the basic (finest) grid, the sooner this degradation appears.

5.2 Interpolation near boundaries and singularities

Interpolation should use the boundary conditions even when they are not explicitly shown on the grid (sometimes they are only implicit in the program). Exception is the case of discontinuity on the boundary, such as a boundary layer thinner than the meshsize.

Near boundary singularities, such as reentrant corners, the interpolation can be improved by using the asymptotic behavior, whenever known. That is, if the correction V^h to be interpolated from the coarser grid is expected to be of the form $V^h = W^h\psi$, where ψ is a known singular function and W^h is smooth, then polynomial interpolation should be used to interpolate W^h, not V^h.

Some experiments indicate, however, that such improvements may not be necessary in FMG programs (see Sec. 7), since, although the convergence rates are slowed down by the singularity, its presence also implies (much) larger truncation errors, hence the same amount of work is still enough to produce algebraic errors smaller than the truncation errors. In fact, FMG experiments with Poisson equation on singular domains or with singular right-hand sides [O1, App. Iv-9] show that, without using special interpolations, algebraic errors below the level of truncation errors are sometimes more easily obtained in such singular cases than in regular cases.

5.3 Relaxation of boundary conditions

Except for some simple Dirichlet problems, discrete boundary conditions should generally be relaxed and transferred to the coarser grid in the same way interior difference equations do. It is important to notice that the boundary relaxation may spoil very much the smoothness of interior residuals near the boundary. Indeed, for a smooth error function, the interior residuals formed near the boundary by relaxing the boundary conditions are $O(h^{\ell-m})$ times the typical magnitude of other interior residuals, where m is the order of the interior differential equation and ℓ is the order of the boundary condition (usually $\ell < m$).

One way around this difficulty is immediately realized by looking at the one-dimensional case. It is clear in that case that boundary conditions need not be relaxed at all. Their errors are not functions that can be smoothed out in any way; they are just isolated values, which can always very well be represented on the coarser grid. Analogously in higher dimensional cases, the role of relaxation should not be to impose the boundary conditions, but only to smooth their error <u>along</u> the boundary. Instead of Gauss-Seidel-type relaxation for the boundary condition $Bu = g$, say, one can make a Gauss-Seidel relaxation of the equation $\Delta_s Bu = \Delta_s g$, where Δ_s is an approximation to the Laplace operator along the boundary; e.g., in two-dimensional problems, $\Delta_s = \partial^2/\partial s^2$, where s is the boundary arclength. This increases the above ℓ by 2, making the perturbation to the interior smoothness negligible. In case the boundary smoothing factor is not as good as the interior one, a couple of boundary sweeps may be performed per each interior one.

Another way around the above difficulty is to ignore it and rely on more precise residual transfers (Sec. 5.4).

5.4 Residual Transfers near Boundaries

Relaxation seldom leaves quite smooth residuals near the boundaries, where the normal succession of relaxation steps breaks off (in lexicographic schemes, for example). And this is especially so when relaxation of boundary conditions is not done in the above (Sec. 5.3) manner. Thus, in many cases it is important that each individual fine-grid residual is correctly represented on the coarse grid. This is what we called <u>full</u> residual weighting. The full weighting near boundaries, and also near interfaces, is considerably more complicated than the interior full weighting (described in Sec. 4.4). This is because the influence of the residual on the solution depends on its distance from the boundary; e.g., in Dirichlet problems for m-order elliptic equations the influence is proportional to the $(m/2)$-th power of the distance. Thus the weight used in transferring a residual from a fine-grid point to a coarse-grid point depends on the distance of both points from the boundary. Near boundary corners the dependence is even more involved.

Hence, near boundaries the interior full-weighting rule (4.5) is modified to the requirement that

$$\sum_{x^H} (I_h^H R^h)(x^H)\, W^H(x^H)\, G(x^H) = \sum_{x^h} R^h(x^h)\, W^h(x^h)\, G(x^h) \qquad (5.1)$$

is satisfied for any given $R^h(x)$, where $\Sigma f(x^H) W^H(x^H)$ and $\Sigma f(x^h) W^h(x^h)$ are discrete approximations, on grids H and h respectively, to the integral $\int f dx$ for any function f, and where $G(\xi)$ has the behavior of the Green function near the boundary. That is, for two neighboring ξ_1 and ξ_2, the ratio $G(\xi_1)/G(\xi_2)$ roughly gives the ratio between the solutions of $Lu(x) = \delta_{\xi_1}(x)$ and $Lu(x) = \delta_{\xi_2}(x)$, with homogeneous boundary conditions. Usually one can take $G(\xi) = d_\xi^\alpha$, where d_ξ is the distance of the point ξ from the boundary, and $\alpha = m - \ell - 1$, where ℓ is the order of the highest normal derivative in the neighboring boundary condition.

Relation (5.1) need not of course be kept very precisely. Residual weightings I_h^H that deviate from it by 20% may easily still show the same convergence rates. Another way of deriving residual weighting near boundaries is by variational rules, like (4.7) in essentially-symmetric cases. And still other ways exist. It may all seem complicated, but, as explained in Sec. 11, it is in principle no more complicated than discretizing the original differential equations near the boundaries.

5.5 Transfer of boundary residuals

Residuals are defined and are transferred (with some averaging) to the coarser grid H, not only with respect to the interior equations, but also with respect to the boundary conditions. In order to do it in the right scale, the divided form of the boundary conditions (the form analogous to the differential conditions, without multiplying through by a power of h) should be used to calculate residuals, average them and transfer. For this purpose a clear conceptual separation should be made between boundary conditions and neighboring interior equations. Incorporating the former into the latter is often convenient to do: It is equivalent to assuming some ghost unknown which is automatically set to satisfy the boundary condition. It may easily lead to wrong transfers. To do it right, one should assume the given boundary condition is incorporated on the finest grid, while the corresponding homogeneous condition is incorporated on all coarser grids. This is equivalent to imposing the boundary condition at relaxation, which, as explained in

Sec. 5.3, will sometimes result in large neighboring residuals and hence slower convergence, unless more precise residual weighting is used.

In symmetric problems one can consistently use the variational relation (4.7), without ever distinguishing between interior equations and boundary conditions, provided good interpolation I_H^h is defined. For some classes of problems this interpolation may be based on the difference equations, interior and boundary alike [D1].

5.6 Treatment of global constraints

In addition to boundary conditions many problems also specify some global conditions, such as integral relations, etc. For example, the pressure $p(x)$ in incompressible Navier-Stokes equations is determined only up to an additive constant; for its unique determination one should add an integral condition like

$$\int p(\underline{x})\,d\underline{x} = 0 \tag{5.2}$$

(integrating over the entire flow field), or a pointwise condition such as

$$p(\underline{x}_o) = 0 \ . \tag{5.3}$$

Both conditions are of the "global" type which we like to consider here, even though the pointwise one does not look so global: We generally consider the type where a single discrete condition has a large global effect on the solution. Boundary conditions in one-dimensional problems are also of this type (cf. Sec. 5.3). The normalization condition $(u,u) = 1$ in eigenproblems is a nonlinear condition of this type. In continuation processes such a condition is often added to a problem to make it well-posed (converting into an unknown some global continuation parameter in terms of which the problem is ill-posed).

The way to treat such conditions in multigrid processes is quite natural and obvious, but is often overlooked. Misguided by the practice in relaxation solvers, one would tend to treat (5.3) at the relaxation phase. Imposing such a pointwise global condition just by changing p at x_o is really harmful to the multigrid solution, since it frustrates the error-smoothing processes near x_o.

Global conditions need not be treated at all on the fine grid. There can be no error-smoothing related to such single conditions. All one has to do is to transfer the residual of the condition to serve as the right-hand side for a similar condition on the next coarser grid. In case of a nonlinear condition, FAS should be used (Sec. 8.3). A condition like $(u^h, u^h) = b^h$, for example, will be transferred to the condition $(u^H, u^H) = b^H$, where

$$b^H = b^h + (I_h^H u^h, I_h^H u^h) - (u^h, u^h), \tag{5.4}$$

which is a special case of (8.5). The global nature of a condition like (5.3) becomes increasingly transparent as it is transferred to coarser grids by proper approximations.

The global condition must of course be operated in solving the coarsest-grid problem (cf. Sec. 6.3). Sometimes it should be operated on <u>several</u> of the coarsest grids. For example, approximations to a condition like $(w,u) = b$, where w is a given weight function which changes signs in the domain, must perhaps be operated on a grid fine enough to resolve these sign changes (or at least crudely simulate them). Similarly, the condition $(u^H, u^H) = b^H$ should be operated on a grid fine enough to crudely resolve the sign changes in the solution u.

When a global condition <u>is</u> treated in relaxation (on a coarse but not the coarsest grid, for example) this should be done in a global way. For example, the condition should be satisfied (at the end of each sweep, say) by adding a constant (or a smooth function) to the entire solution, or by multiplying the entire solution by a constant (or a smooth function), so that the error-smoothing process is not frustrated.

In some relaxation schemes, the global condition seems to be needed in the local relaxation. For example, in the BGS scheme (Sec. 3.4) one solves in small boxes little problems similar to the given boundary-value problem. For the solution in the box to be uniquely determined, a condition like the global condition is needed there. In solving discrete incompressible Navier-Stokes equations in a small local box, for example, a pressure condition similar to (5.2) or (5.3) is needed. The best then is to use in each box a "no change" kind of condition. That is, to require, for example, that some discrete approximation to $\int p(x)\,dx$ (integration being over the small box) retains its value from before the relaxation step.

6. <u>MANY-LEVEL CYCLES</u>

Having obtained satisfactorily performing two-level cycling algorithms, one needs next to turn on the complete sequence of grids, using now the two-level techniques in recursion. The new algorithmic questions which arise are discussed below. Some of them could theoretically be investigated by three-level mode analysis, but this trouble is neither needed nor normally taken.

6.1 Multigrid cycles. Initial and terminal relaxation

For any grid h, finest or intermediate, a multigrid h-cycle can recursively be defined as follows: Make ν_1 relaxation sweeps on grid h, then transfer the residual problem to the next coarser grid H (= 2h) and solve it there approximately, using γ H-cycles (unless H is the coarsest grid), then interpolate the grid-H solution and add it as a correction to the grid-h solution, and finally make ν_2 more sweeps on grid h. On the coarsest grid the problem is solved either directly or by ν_o relaxation sweeps (cf. Sec. 6.3).

In two-level cycles only the sum $\nu = \nu_1 + \nu_2$ matters of course. When h is an intermediate grid the separate values of ν_1 and ν_2 do make some difference, although not a big one. In regular elliptic solvers experience shows that $\nu_2 = [\nu/2]$ is probably the best prescription (see for example [S4, Tables 3.3]). In double-discretization schemes (Sec. 10.2) it is important to use $\nu_2 = 0$. In "accommodative" algorithms (see Sec. 4.1, 6.2) the values of ν_1 and ν_2 are determined internally.

Note also that the several passes of a complex relaxation sweep (such as ADZ) can be divided between the initial and the terminal stages of the cycle [S4, §7.3].

6.2 Switching criteria, types of cycles

The criteria when to switch from a fine grid h to the next coarser grid H = 2h were examined in a previous stage (Sec. 4.1). These same criteria can be used recursively, i.e., not only when h is the finest grid. We need in addition some criteria for switching from any grid H back to the next finer grid h. Two kinds of switches are used: Fixed and accommodative.

Fixed algorithms switch from H back to h after a preassigned number γ of H-cycles. The h-cycle is recursively defined to be of the type $C(\nu_1,\nu_2)^\gamma$, if all the H-cycles are of this same type. It is defined to be of type $F(\nu_1,\nu_2)$ if $\gamma = 2$ and the first H-cycle is an $F(\nu_1,\nu_2)$ cycle itself while the second H-cycle is a $C(\nu_1,\nu_2)^1$ cycle. See flowcharts and operation counts in [B17, §6.1]. The cycle $C(\nu_1,\nu_2)^1$ is also called a V cycle and denoted $V(\nu_1,\nu_2)$. The cycle $C(\nu_1,\nu_2)^2$ is also called a W cycle and denoted $W(\nu_1,\nu_2)$.

Cycles with $\gamma = 2$ (F or W cycles) are safe, because the H-grid problem is solved to a much better accuracy than the accuracy of the h-cycle, hence they perform practically as well as an exact solution of the H problem. Larger values of γ are not normally used, except for simulating two-level algorithms (see Sec. 5.1). F cycles are somewhat less expensive than W cycles in one-dimensional problems with many levels, and also in higher dimensions when semi coarsening is used, but otherwise they perform practically the same. V cycles may save considerable fraction (1/3, in two-dimensional problems) of the work. They are safe to use when the two-level convergence factor $\bar{\lambda}$ is small (e.g., $\bar{\lambda} \leq .15$, as in regular elliptic cases), in which case the convergence per V cycle will be close to $\bar{\lambda}$. If $\bar{\lambda} = .5$, on the other hand, a V cycle may not even attain that factor, because the coarse grid equations themselves will only be solved crudely, the error thus cascading through the levels. This situation arises in cases of severe singularities or in non-elliptic and singular perturbation problems. In the latter cases, for example, the artificial viscosity on grid kh is k times larger than on grid h, hence visiting grid kh only once per cycle would give an asymptotic convergence factor no better than $1 - 1/k$ [B17, §5.1]. Since on the coarsest grid $k = O(h^{-1})$, the asymptotic factor will be $1 - O(h)$, which is very poor indeed. In this situation W (or F) cycles are absolutely necessary for good <u>asymptotic</u> factors. V cycles may however still work quite satisfactorily in FMG algorithms (see Sec. 7.4 below and the numerical experiments in [B17, §7.1]).

<u>Accommodative algorithms</u> switch from grid H back to grid h when a certain norm of the residuals on grid H drops below some factor η times the latest value of the corresponding norm on grid h. The parameter η is not a sensitive one. A good general prescription seems to be $\eta = 1.1\,\bar{\lambda}$. If $\bar{\lambda}$ is not approximately known, take $\eta = 2^{-d}$, a value related to exchange-rate considerations (cf. Sec. 9.6).

Generally, accommodative algorithms may be troublesome at program development stages, since they may cause more complex interactions between the internal checks and the real questions one likes to examine. Their flexibility may prevent us from seeing some of the troubles, and they are not suitable for precise comparisons. In the production stages, accommodative algorithms have the disadvantage that they require the extra work of calculating the residual norms. On the other hand, accommodative algorithms are more robust. Also, in complicated problems (which is where this robustness is needed), the residual norm calculation is inexpensive relatively to other calculations, assuming <u>dynamic</u> residuals (calculated anyway in the relaxation process) are used.

6.3 Coarsest grids. Inhomogeneous operators

When the multigrid h-cycle performs considerably poorer than expected, it is first important to distinguish between fine-grid and coarse-grid troubles. This distinction is easy to make, by simulating two-level algorithms (taking large γ or small η) and examining whether this improves the convergence factor (per h-cycle), and how much this improvement depends on the size of h. Also examine whether reasonable convergence is obtained on your coarsest grid. If not, or if the trouble is confined to coarse h, the following remarks may be relevant.

Inhomogeneous operators are the main source for the special troubles appearing only on sufficiently coarse grids. On such grids, lower order terms of the operator start to affect, or even dominate, the smoothing and convergence factors. If we have neglected them in designing the fine-grid relaxation, we should now take them into account.

A typical example is the equation $-\Delta u + \sigma u = f$ with purely Neumann boundary conditions. If σ is positive but small, the smoothing factor of a GS relaxation is essentially the same as for Poisson equation, but the convergence factor is roughly $4/(4+h^2\sigma)$, which may be very slow even on the coarsest grid. Hence the coarsest-grid equations should be solved either directly (e.g., by elimination, which is inexpensive since the coarsest grid can contain just few points), or by relaxation, where after each sweep a suitable constant is subtracted from the approximate solution [A1, §4]. If $\sigma = 0$ everywhere except in some small subdomain, that constant subtraction should be employed on all grids which are not fine enough to resolve that small subdomain.

Indefinite case. If σ is negative, the situation is much worse, whatever the boundary conditions: For the coarse grid to approximate the slowly converging fine-grid component, its meshsize must be fine enough: For large $|\sigma|$, the coarsest meshsize must satisfy $H = O(R^{1/p}(-\sigma)^{-(p+1)/(2p)})$, where R is the radius of the domain and p is the approximation order. In many cases this H is smaller than the finest affordable meshsize. Hence, in any case, on sufficiently coarse levels completely different inter-grid transfers, such as those mentioned in Sec. 4.2.2, should be employed.

In designing the relaxation schemes for complex systems of equations, e.g. in fluid dynamics, we can take only subprincipal terms into account (Secs. 2.1, 3.1). On very coarse grids, however, this is no longer fully justified, and the smoothing factors may deteriorate. We may then have to use either more sweeps (by increasing ν and/or γ, or by using accommodative algorithms), or more sophisticated relaxation.

In solving Navier-Stokes equations, for example, improved results were obtained by using the high-speed DGS scheme ([B19, §7.3], [B13, §9], [B16, §4.2]) on all finer grids, while employing BGS (see Sec. 3.4) on the two coarsest grids.

Even for homogeneous operators, convergence of h-cycles can sometimes be slower on very coarse grids, because the convergence factor $\overline{\lambda}$ cannot be smaller than $O(h^{\tilde{m}})$; see (C) in Sec. 4.3. In such cases one can make more h-cycles, by increasing γ or switching accommodatively, which is inexpensive since h is coarse.

Sometimes troubles seen on coarse grids are only indications of bad procedures at some special, restricted regions, such as boundaries (see Sec. 5.1). Or they may signal the need to operate some global conditions, which are not enforced on finer grids (see Sec. 5.6).

Of special concern is the <u>coarsest grid</u> itself. Relaxation there should be converging, not just smoothing as on other grids. Various conditions not enforced on finer grids must be enforced on the coarsest one, calling for special procedures. If nothing better is known, one can always use either a direct solver or a slow but safe iterative process such as Kaczmarz relaxation; on the coarsest grid they are affordable. Finally note that the coarsest grid cannot efficiently contribute to convergence if all its points happen to lie too close to Dirichlet boundaries.

7. FULL MULTI-GRID (FMG) ALGORITHMS

The cycling algorithms developed in the previous stages are easily converted into full multigrid (FMG) programs. The main difference is that instead of starting with an arbitrary approximation (e.g., $u_0^h \equiv 0$) on the finest grid, the first approximation u_0^h is obtained by an interpolation Π_H^h from a coarse-grid (approximate) solution u^H. Namely, $u_0^h \equiv \Pi_H^h u^H$, where $H = 2h$ and where u^H has been calculated by a similar FMG process with H as its finest level. The full algorithm can be either "fixed" (e.g., the algorithm in [B11, Fig. 3], [B13, Fig.1], [B14, Fig. 1] and [B17, Fig. 4]), or "accommodative" (as in [B8, Sec. 1.3], [B12, Fig. 1], [B11, Sec. 3.6 and Fig. 1], [B19, Sec. 2.2]). Both versions are available in the model program FMG1 [M3].

FMG algorithms are in a sense <u>easier</u> to program than cycling algorithms. Their main driving routine is some lines longer, they may include an additional interpolation routine (Π_H^h), and they involve several more algorithmic questions (dealt with in the following subsections) -- but on the other hand they are much more <u>forgiving</u>. Their basic performance, which is to solve all problems to the level of truncation errors in just one or two cycles (see Sec. 7.3), is undisturbed by various little mistakes (conceptual mistakes or programming bugs, especially in treating boundaries) which may degrade very much the asymptotic convergence of cycling algorithms. These mistakes may however affect the FMG performance of other problems, hence it is safer to detect them by perfecting the multigrid cycling (as in Secs. 4, 5, and 6) before turning FMG on

7.1 Order of the FMG interpolation

The FMG interpolation operator Π_H^h is not necessarily the same as the correction interpolation operator I_H^h later used in the multigrid correction cycles. Often the order of Π_H^h should be higher than the order of I_H^h, since the first approximation is smoother than the corrections: In the right-hand side of the latter (i.e., in the residuals) the amplitude of high-frequency components is usually comparable to that of low-frequency components.

The optimal order of Π_H^h depends on the purpose of calculations. If one desires ultimately to get the algebraic error (i.e., the errors in solving the difference equations) to be very small (far below truncation errors), then Π_H^h should exploit all the smoothness of u^h in order not to produce unnecessary high-frequency algebraic errors. (High-frequency errors are the most expensive to liquidate in the multigrid cycling, since they are processed on the finest grid.) In fact, in such a case the first few correction interpolations I_H^h should also be of suitably high orders. The precise rules for scalar elliptic equations are given in [B7, App. A.2]. Note that these rules assume that the order of smoothness is known in advance.

Usually, however, the smoothness order is not known in advance. More importantly, we are not interested in solving to arbitrarily small algebraic errors; we like them only to be smaller than the truncation errors. The optimal order depends then on the norm by which we measure errors. Suppose we solve a $q \times q$ system of differential equations, and assume our error norm includes difference-quotients up to order ℓ_j in the j-th unknown function $(1 \leq j \leq q)$. Then the order $\hat{m}^j$ of the first interpolation of that function should not be less than $p + \ell_j$, where p is the order of approximation. Otherwise, the $O(h^{\hat{m}^j - \ell_j})$ high-frequency errors produced by interpolation would be much larger than the $O(h^p)$ (low-frequency) truncation errors.

In case of equations with strongly discontinuous coefficients, the higher-order interpolation Π_H^h should be of a different form, taking into account the different sense of smoothness in the solutions (cf.

Sec. 4.6. A higher-order interpolation of this sort is presented in [A1, Eq. (5.12)]). The remarks of Sec. 5.2 apply here as well.

In some programs, especially general-domain programs, the higher-order interpolation Π_H^h turned out to cost more CPU time than the rest of the algorithm [O1]. An interpolation of an order smaller than indicated above may then be more practical. In case of rotatable differential operators, simpler higher-order interpolations can be used, based on the equations themselves [H3], [F3, §3].

7.2 Optimal switching to a new grid

In designing the FMG algorithm one should decide how well the equations on level $H = 2h$ should be solved before the solution is interpolated for the first time to grid h and the h-cycles start. The optimal point to switch is when the work of h-cycles becomes as efficient as the work of H-cycles in reducing the differential error (the difference between the differential solution u and our current computed solutions, $\tilde{u}^H$ or $u_o^h = \Pi_H^h \tilde{u}^H$). This happens when the algebraic error on grid H, namely $e^H = \| u^H - \tilde{u}^H \|$, is about 2^{-d} times the algebraic error on grid h, $e^h = \| u^h - \tilde{u}^h \|$, where d is the dimension, u^H is the exact solution of the H-equations and u^h is the exact solution of the h-equations. This is because h-cycles are about 2^d times as expensive as H-cycles. The switching point $e^H \approx 2^{-d} e^h$ is roughly equivalent to

$$e^H \approx \beta E^H , \quad \beta = (1 - 2^{-p}) / (2^d - 1) , \tag{7.1}$$

where $E^H = \| u^H - u \|$ is the truncation error on grid H and p is the order of approximation (see [B15, §5.2] and also [F3, App. A]).

In practice the values of e^H and E^H are of course not known, but we can derive from (7.1) the algebraic reduction needed on level H before switching. Namely, denoting by e_o^H the value of e^H when H cycles are started and by e_*^H its value at the switching point (7.1), and assuming that the switching from the 2H cycles to the H cycles has been made when a relation similar to (7.1) was reached on level 2H, we find [B15, §5.2] that the algebraic reduction on grid H is roughly

$$e_*^H / e_o^H \approx 2^{-p-d} . \tag{7.2}$$

This can be obtained by about

$$(p + d) / \log_2 (1/\bar{\lambda}) \tag{7.3}$$

H cycles, where $\bar{\lambda}$ is the convergence factor per cycle (Sec. 4.1). The switch (7.2) can also be used in an accommodative algorithm, replacing e_*^H / e_o^H by the corresponding ratio of residuals, which of course can be measured. The number of H cycles usually turns out to be 1 or 2 .

7.3 Total computational work. Termination criteria

Suppose that on the finest grid h we wish to obtain an algebraic error smaller than a specified factor α times the truncation error: $e^h \leq \alpha E^h$. Suppose also that the switch from level $H = 2h$ is made roughly when (7.1) is met; i.e., when $e^H \approx 2^p \beta E^h$. Then the algebraic error reduction required on grid h is roughly $\alpha_1 \approx \alpha(1 - 2^{-d}) / (2^p - 1)$. The number of work units to obtain such a reduction is about $\log \frac{1}{\alpha_1} / \log \frac{1}{\mathring{\mu}}$, where $\mathring{\mu}$ is the interior convergence factor per work unit (see Sec. 4.1) and is usually just modestly larger than the interior smoothing factor $\bar{\mu}$. Counting also the work for the reduction (7.2) on coarser grids, we find that the total number of work units theoretically required by the Full Multi-Grid algorithm is about

$$\left\{ \log \frac{2^p - 1}{\alpha(1 - 2^{-d})} + \frac{p + d}{2^d - 1} \log 2 \right\} \Big/ \log \frac{1}{\bar{\mu}} . \qquad (7.4)$$

The actual total number of work units is usually larger than (7.4), because of the need to make integral numbers of relaxation sweeps and coarse grid corrections. Typically one V or W cycle for each level, with $\nu = 2$ or 3 , yields e^h considerably below E^h .

The observation that in FMG algorithms one cycle on each level is needed, and is also basically enough to reduce the algebraic errors to the level of truncation errors (even though sometimes two shorter cycles may be more efficient), can heuristically be understood as follows. The first approximation on grid h , obtained by interpolating the grid-2h solution, necessarily contains two types of errors: (A) High-frequency errors, i.e., errors having oscillations invisible and hence unapproximable on the coarser grid. (B) Aliasing errors, i.e., smooth errors introduced by high-frequency data because on the coarse grid high-frequency data is mistaken for smooth data. Relaxation on grid h can be efficient in removing the high-frequency errors (because of their local nature: At each point they are essentially determined by the neighboring residuals.) Having removed the high-frequency errors we have also removed the high-frequency data from the residual problem, hence we can then go back to grid 2h with the residual problem to remove the aliasing errors (which are smooth errors, hence not determined by neighboring residuals, hence inefficiently treated by relaxation).

The algorithm may indeed be terminated after a fixed number of cycles on the finest grid h. This number is roughly $\log\frac{1}{\alpha_1} / \log\frac{1}{\bar\lambda}$, and in practice it is one or two. Or else, especially if an estimate for $\bar\lambda$ is not known, termination can be done when a certain norm of the residuals on grid h becomes smaller than a corresponding norm of $\alpha\tau^h \approx \alpha(2^p - 1)^{-1} \tau_h^{2h}$ (see Sec. 8.4). One should of course check numerically, using a problem with a known solution or a solution computed on a finer grid, that, with these termination procedures $e^h \leq \alpha E^h$ is indeed obtained.

7.4 Two-level FMG Mode Analysis

Instead of developing Full Multi-Grid (FMG) programs from the cycling programs, including boundary conditions, one can first develop the FMG algorithm still within the framework of two-level mode analysis (immediately following the stage of Sec. 4). This may again serve to separate away questions related to boundary conditions (questions discussed in Sec. 5), and questions related to many levels and to very coarse grids (Sec. 6) from the particular questions of the FMG algorithm (Secs. 7.1 - 7.3). The latter can then be examined in the interior, without boundary interference (or also with ideal boundaries -- see Sec. 7.5), and the performance figures so calculated can serve as ideals against which the actual program can be developed and debugged. Such an analysis is particularly useful in cases the usual two-level analysis (that of Sec. 4.1 above) is too pessimistic because of the existence of different components with markedly different convergence properties. For example, in case of non-elliptic or singular perturbation problems there are smooth characteristic components which converge slower than others, since for such components L^H is not a very good approximation to L^h. But exactly for the same components and for the same reason, L^h itself is not a good approximation to L, hence these components do not need much algebraic convergence, and the fact that they have slow asymptotic rates does not matter [B17, §5.1, 5.2]. What we need then is an analysis which does not tell us the worst asymptotic rate, but tells us, separately in each mode, how well we solve the problem by an FMG algorithm with a given number of prescribed cycles.

To analyze the FMG solution of $Lu = f$, where L is a $q \times q$ system and has constant coefficients (or frozen local values, in case the original L was variable), we first analyze a single component $u(\underline{\theta}) = \exp(i\underline{\theta} \cdot \underline{x} / \underline{h})$. We calculate the corresponding f, and hence also the solution $u^H = u^H(\underline{\theta})$ to the coarse-grid equation $L^H u^H = f^H = I^H f$, where I^H is the local averaging used in our discretization for transfering a continuum function to grid H. The interpolation of u^H to the fine grid gives an approximation $u_o^h = \Pi_H^h u^H$ made up of 2^d Fourier components (the harmonics of $\underline{\theta}$, i.e. all components $\underline{\theta}'$ such that $\underline{\theta}' = \underline{\theta} + (\nu_1, \dots, \nu_d)\pi$, ν_j integer and $-\pi < \theta_j' \le \pi$). To the set of 2^d amplitudes we then apply the usual (cycling) two-level mode analysis (Sec. 4.1); i.e., using (4.1) we calculate the $2^d q \times 2^d q$ matrix $M(\underline{\theta})$ describing the transformation of these 2^d amplitudes by one cycle. The result of applying k such cycles on grid h we denote by $u_k^h(\underline{\theta}) = M(\underline{\theta})^k u_o^h(\underline{\theta})$. Having calculated $u_k^h(\underline{\theta})$ we can then examine its qualities by several measures.

One measure is <u>how well below truncation errors</u> u_k^h is. This is measured for example by

$$\max_{|\underline{\theta}| \le \pi} \frac{\| u_k^h(\underline{\theta}) - u(\underline{\theta}) \|}{\| u(\underline{\theta}) - u^h(\underline{\theta}) \|}, \qquad (7.5)$$

where $u^h(\underline{\theta})$ is the exact solution of the grid h equations, and $\| \cdot \|$ is any norm under which we want to guarantee convergence. Note that u_k^h is made of 2^d components, while u^h and u are made of only one of those; the norms can be taken anyway.

Another, perhaps more direct and important measure, is <u>how well we have solved the differential equations</u>. That is, we directly measure $\| u_k^h - u \|$, thus evaluating not only the performance of our fast solver, but also the quality of our discretization scheme itself. We evaluate the total quality of our procedures in solving the differential equations at a given amount of work. In measuring the error $\| u_k^h - u \|$ we should of course give smaller weights to high-frequency $\underline{\theta}$'s; we cannot and need not solve for them as accurately as for low frequencies. Thus, if we aim at an approximation order p, good measures of performance are

$$\max_{|\underline{\theta}| \le \pi} \| u_k^h(\underline{\theta}) - u(\underline{\theta}) \| / |\underline{\theta}|^p, \qquad (7.6)$$

or

$$\left\{ \int_{|\underline{\theta}| \le \pi} |\underline{\theta}|^{-2p} \| u_k^h(\underline{\theta}) - u(\underline{\theta}) \|^2 \, d\underline{\theta} \right\}^{\frac{1}{2}}, \qquad (7.7)$$

etc. Several such measures can easily be produced, approximately, by the program that calculates $u_k^h(\underline{\theta})$. The program is an easy extension of the usual (cycling) two-level mode-analysis program.

All the issues examined by the two-level cycling analysis (relaxation, the number $\nu = \nu_1 + \nu_2$ of sweeps per cycle, and the interior operators I_h^H, L^H and I_H^h — see Sec. 4) can further (more accurately) be optimized by the FMG mode analysis. In addition we can examine by this analysis the effect (in the interior) of various Π_H^h interpolation procedures, various values of ν_1, ν_2 and k, and, most importantly, various interior discretization procedures. An important advantage is that this analysis can be used even in cases where no algebraic convergence is desired or obtained (cf. Sec. 10.2).

Moreover, the predictions of the FMG-mode analysis are more robustly held than those of the cycling mode analysis when real boundaries are added. For example, take a problem with singularities in the boundary, such as reentrant corners. The effect of such singularities (similar to the effect of singular perturbations mentioned above) is that certain solution components (those that take on the singular behavior) are harder to approximate, in particular their L^H approximation is not so good, hence we cannot get convergence factors per cycle as good as predicted by the interior cycling mode analysis. On the other hand, exactly for those components, L^h is not a good approximation either, hence one easily gets below truncation error (sometimes even more easily than in regular cases; see [O1, §IV-9] and [S4, Table 10.2]), so that predictions like (7.5) are likely to be roughly held up.

A very simple example of two-level FMG mode analysis is given in [B17, §5.2] .

7.5 Half-space FMG mode analysis

Another advantage of the two-level FMG mode analysis is the possibility to make it also near a piece of boundary, modelled by a grid hyperplane, so that the entire domain is modelled by a half space. This is particularly important in non-elliptic or singular perturbation problems, where the high-frequencies far in the interior can still strongly be affected by the boundary. For a simple example — see [B17, §5.3].

Those examples in [B17] illustrate another technique which can be used whenever one wants to focus one's analysis on smooth components. One can then simplify the analysis very much by using the first-differential approximation (the first terms in a Taylor expansion) of the difference operators, instead of the difference operators themselves.

PART II. ADVANCED TECHNIQUES AND INSIGHTS

Having completed the above stages of developing the basic multigrid solver, one can start introducing various important improvements. Some possibilities are outlined in the following sections, followed by comments of more "philosophical" nature, which can however be readily useful to the practitioner. We then close with general remarks on multigrid applications to chains of problems and to evolution problems.

8. FULL APPROXIMATION SCHEME (FAS) AND APPLICATIONS

The Full Approximation Scheme (or Full Approximation Storage - FAS) is a widely used version of multigrid inter-grid transfers, explained for example in [B6, §4.3.1], [B7, §5], [B8, §1.2], [B10, §2.1], [B14, §2.3]. It has mainly been used in solving nonlinear problems, but it has so many other applications that it should perhaps be used in most advanced programs. The scheme, its programming, and several of its applications are sketched below. Another, perhaps the most important application of FAS is described in Sec. 9, and another one in Sec. 15.

8.1 From CS to FAS

Consider first a linear problem $L^h u^h = f^h$ on a certain grid h, with some approximate solution $\tilde{u}^h$ obtained, say, after several relaxation sweeps. Now we like to use coarser grids in order to supply fast approximations to smooth errors. Thus, it is the corrections $v^h = u^h - \tilde{u}^h$ which we try to approximate on the next coarser grid $H = 2h$. In the simpler multigrid programs, such as [B7, App. B], the coarse-grid unknown function is indeed v^H, intended to approximate the correction v^h. This multigrid version is therefore called the Correction Scheme (CS). Since $L^h v^h = R^h$, where

$$R^h = f^h - L^h \tilde{u}^h \qquad (8.1)$$

is the fine-grid residual, the CS coarse-grid equations are

$$L^H v^H = I_h^H R^h \,. \qquad (8.2)$$

where L^H approximates L^h on the coarse grid. Once (8.2) has been approximately solved, its approximate solution $\tilde{v}^H$ is interpolated to the fine grid and serves as a correction to the fine-grid solution:

$$\tilde{u}^h_{NEW} = \tilde{u}^h + I^h_H \tilde{v}^H . \tag{8.3}$$

In the Full Approximation Scheme (FAS) we perform exactly the same steps, but in terms of another coarse-grid variable. Instead of v^H we use

$$\hat{u}^H = \hat{I}^H_h \tilde{u}^h + v^H \tag{8.4}$$

as the coarse-grid unknown function, where $\hat{I}^H_h$ is some fine-to-coarse transfer which need not be similar to I^H_h in (8.2). (They are actually defined on different spaces.) This coarse-grid variable $\hat{u}^H$ approximates $\hat{I}^H_h u^h$, the full intended solution represented on the coarse grid, hence the name "Full Approximation Scheme". The FAS coarse-grid equations, derived from (8.2) and (8.4), are

$$L^H \hat{u}^H = \hat{f}^H \tag{8.5a}$$

where

$$\hat{f}^H = L^H(\hat{I}^H_h \tilde{u}^h) + I^H_h R^h . \tag{8.5b}$$

Having obtained an approximate solution $\tilde{u}^H$ to (8.5), the approximate coarse-grid correction is of course $\tilde{v}^H = \tilde{u}^H - \hat{I}^H_h \tilde{u}^h$, hence the FAS interpolation back to the fine grid, equivalent to (8.3), is

$$\tilde{u}^h_{NEW} = \tilde{u}^h + I^h_H (\tilde{u}^H - \hat{I}^H_h \tilde{u}^h) . \tag{8.6}$$

To use directly

$$\tilde{u}^h_{NEW} = I^h_H \tilde{u}^H \tag{8.6'}$$

would be worse, of course, since it would introduce the interpolation errors of the full solution u^H instead of the interpolation errors of only the correction v^H (but see end of Sec. 8.5). Notice that $\hat{I}^H_h \tilde{u}^h$ in (8.6) and in (8.5b) must be identically the same: A common programming mistake is to have a slight difference between the two. This difference may dominate the calculated correction function and hinder convergence. Also it is important to start with identically the same $\hat{I}^H_h \tilde{u}^h$ as the first approximation in solving (8.5).

Note: The FAS equations (8.5) are a model for each interior differential equation or boundary condition appearing in the problem: each of them is transferred, separately, by this prescription. Other side conditions, global constraints, etc., are transferred in exactly the same way. See for example Eq. (5.4). In case a double discretization is employed on all levels (see Sec. 10.2), two functions $\hat{f}^H$ should be calculated, one for each coarse-grid operator L^H [B17, §2.1].

For linear problems, the FAS steps (8.5-6) are fully equivalent to the CS steps (8.2-3). Indeed, the safest way to construct a correct FAS program is to start with a linear subcase, write first a CS program, then convert it to FAS and check that, at each iteration, the FAS results on the finest grid are identically the same as the CS results, except for round-off errors. Common mistakes in implementing FAS, especially in treating boundary conditions, are thus easily detected. The conversion of a CS program to FAS can be done by a trivial addition of three routines [B9, Lecture 12]. It can be done either for a cycling program or for an FMG program. The simplest examples are the cycling program FASCC and the program FMG1 [M3].

8.2 FAS: Dual point of view

To see why FAS is preferable to CS in many situations, we rewrite (8.5), using (8.1), in the form

$$L^H \hat{u}^H = f^H + \tau_h^H \tag{8.7}$$

where

$$\tau_h^H = L^H(\hat{I}_h^H \tilde{u}^h) - I_h^H(L^h \tilde{u}^h) . \tag{8.8}$$

and $f^H = I_h^H f^h$. Observe that (8.7) without the τ_h^H term is the original coarse-grid equation (with the particular discretization $f^H = I_h^H f^h$), and that $\hat{u}^H$ approximates $\hat{I}_h^H \tilde{u}^h_{NEW}$, and at convergence $\hat{u}^H = \hat{I}_h^H u^h$. Hence τ_h^H is the <u>fine-to-coarse defect correction</u>, a correction to the coarse-grid equation designed to make its solution coincide with the fine-grid solution.

We can now reverse our point of view of the entire multigrid process: Instead of regarding the coarse grid as a device for accelerating convergence on the fine grid, we can view the fine grid as a device for calculating the correction τ_h^H to the coarse-grid equations. Since this correction depends on the non-smooth components of the solution, we obtain it by interpolating the solution to the fine grid and correcting its non-smooth components by relaxation. Having obtained the correction τ_h^H by such a "<u>visit</u>" to grid h, we continue the solution process on grid H. Later we may "<u>revisit</u>" grid h, in order to update τ_h^H . In such a case the interpolation (8.6) should better be used if we do not want to lose the non-smooth components of the solution already obtained by relaxation in previous visits. This entire process will then yield a solution on grid h , which we can improve by inserting into it visits to grid $h/2$. Etc.

Since $\hat{u}^H$ is just an improvement to u^H , we can omit the $\hat{}$, and just understand that the meaning of u^H changes as soon as an approximation $\tilde{u}^h$ exists on the next finer grid h .

This point of view, and the fact that the full fine grid solution is represented on all coarser grids, open up many algorithmic possibilities, as we shall see below.

8.3 Nonlinear problems

The Correction Scheme is not applicable to nonlinear problems, since the correction equation $L^h v^h = R^h$ is valid only for linear L^h. In case L^h is nonlinear, the correction equation can instead be written in the form

$$L^h(\tilde{u}^h + v^h) - L^h(\tilde{u}^h) = R^h \quad . \tag{8.9}$$

Transferring this equation to the coarse grid (replacing L^h by L^H , $\tilde{u}^h$ by $\hat{I}_h^H \tilde{u}^h$, v^h by v^H and R^h by $I_h^H R^h$) we get the FAS equations (8.5) . Thus, one important advantage of FAS over CS is its direct applicability to nonlinear problems. (This is a general property of defect-correction schemes — see for example [L1], [S3]). The CS scheme can of course be applied to the linearization of L^h around the current approximation $\tilde{u}^h$. But FAS is usually preferable, because:

(i) No global linearization is needed. The only linearization is the local one used in relaxation (see the GSN relaxation in Sec. 3.4). Hence, no extra storage for coefficients of the linearized equation is required.

(ii) The programming is very convenient since the FAS equations (8.5) are exactly the same as the original discrete equations, except for a different right-hand side. Hence only one relaxation routine, one residual transfer routine, one boundary relaxation routine, etc., are ever needed in the program.

(iii) The multigrid rate of convergence is not constrained by the convergence rate of Newton iterations. It is still mainly determined by the interior smoothing rate. Solving the nonlinear problem is no more expensive than solving, just once, the corresponding linearized problem. (In many cases, though, Newton convergence rate is fast enough to impose no real constraint to an FMG algorithm).

(iv) FAS is useful in various other ways. Particularly important for nonlinear problems are its applications in continuation processes (Sec. 15), near discontinuities (Sec. 8.5) and in automatic local refinement procedures (Sec. 9).

Although not employed in the FAS multigrid processing, we still use Newton linearizations in the local mode analysis, to estimate smoothing and convergence factors (Secs. 3, 4).

Various nonlinear problems have been solved by FAS, including transonic flow problems [S2], [J1], [M4]; steady-state incompressible Navier-Stokes equations [B13], [B19], [D2]; and linear complementary problems arising from free boundary problems. A simple example of the latter is to calculate the nonnegative function u which minimizes the Dirichlet integral $\int(\underline{\nabla}u\cdot\underline{\nabla}u + 2fu)dx$. Without the $u \geq 0$ constraint the problem is of course equivalent to a problem with the Poisson equation $-\Delta u = f$. But the nonnegativity constraint introduces nonlinearity. Using a FAS - FMG algorithm this nonlinear problem is solved with essentially the same amount of work as Poisson problems [B18].

8.3.1 Eigenvalue problems can simply be regarded as nonlinear problems. They are nonlinear since the unknown eigenvalue λ_j multiplies the corresponding unknown eigenfunction u_j. Also, to fix the eigenfunctions nonlinear orthonormality conditions $(u_i, u_j) = \delta_{ij}$ are added as global constraints. The solution algorithm proceeds as a usual FAS - FMG multigrid, with the global constraints treated basically as in Sec. 5.6. The eigenvalues are updated once per cycle, together with a more precise determination of the individual eigenfunctions within the space spanned by them, by a Rayleigh-Ritz process. Experiments for model problems [B20] show that an FMG algorithm with one V(2,1) cycle on each level gives a discrete eigenfunction with algebraic errors much smaller than truncation errors. Similar work is needed for each additional eigenfunction.

8.3.2 Continuation (embedding) techniques. Nonlinear problems usually have many discrete solutions, and the desired one is obtained only if we start from a close enough first approximation. A common way to obtain a good first approximation, or generally to trace branches of solutions, is by continuation (also called "embedding" or "Davidenko method"): The problem, including its discretization and its approximate solution is written as depending on some parameter

$$\gamma_0 \leq \gamma \leq \gamma_*$$

such that for γ_0 the problem is easily solvable (e.g., it is linear), while for γ_* it is the problem we really need to solve. We advance γ from γ_0 to γ_* in steps $\delta\gamma$ small enough to ensure that the solution to the γ problem can serve as a good first approximation in solving the $\gamma+\delta\gamma$ problem. Sometimes γ is a physical parameter; sometimes the solutions are better defined in terms of a non-physical parameter, such as the arclength of the solutions path [K1].

As for the relation between multigrid and continuation, several situations arise. Sometimes, the FMG algorithm is a good continuation process by itself. In particular, in non-elliptic and singular perturbation problems where relaxation adds $O(h)$ artificial viscosity, the FMG algorithm starts from highly viscous solutions (since h is large) and gradually eliminates viscosity as it proceeds to finer grids. This is a natural continuation path since problems with large viscosity terms are well-defined and easier to solve. This continuation is carried even much further when the FMG algorithm is continued to include local refinements around thin viscous layers (see Sec. 9).

Sometimes, however, an explicit continuation over the problem path $\gamma_0 \leqslant \gamma \leqslant \gamma_*$ should be made, either because the intermediate problems are interesting themselves, or because they are necessary for reaching, or even defining, the desired γ_* solution. When the intermediate problems are not of interest, they can of course be solved to a lower accuracy, using coarser grids only. The grids cannot all be too coarse, however; the meshsize $\hat{h}$ must participate in the continuation process if components of wavelengths comparable to $\hat{h}$ are needed to keep the solutions in the "attraction region" of the desired solution path.

Even when components with $O(\hat{h})$ wavelengths are needed in the continuation process, in each $\delta\gamma$ step they do not usually change much. We can therefore employ the "frozen τ" techniques described below (Sec. 15), and perform most of the $\delta\gamma$ steps using, on most parts of the domain, very coarse grids, with only few "visits" to grid $\hat{h}$: Such a continuation process will often require less computational work than the final step of solving (to a better accuracy) the γ_* problem.

8.4 Estimating truncation errors. τ extrapolation

As with other defect-correction schemes, the defect can serve also as an error estimator. That is, τ_h^H is an approximation to the <u>local truncation error</u> τ^H (on the coarse grid H), defined by

$$\tau^H = L^H(\hat{I}^H u) - I^H(Lu) \qquad (8.10)$$

where u is the true differential solution and I^H and $\hat{I}^H$ are two continuum-to-H transfer operators, defined as follows. I^H is the operator used in our grid-H discretization for $f^H = I^H f$, and $\hat{I}^H$ represents the sense in which we want u^H to approximate u: We want u^H to actually approximate $\hat{I}^H u$. The injection $(\hat{I}^H u)(x^H) = u(x^H)$ is usually meant, but other local averagings are sensible too.

Note the analogy between (8.10) and (8.8). τ^H is the correction to grid-H right-hand side that would make the grid-H solution coincide with the true differential solution $\hat{I}^H u$, while τ_h^H is the correction that would make it, at convergence, coincide with the fine-grid solution $\hat{I}_h^H u^h$. It is hence clear that at convergence

$$\tau^H \approx \tau^h + \tau_h^H , \tag{8.11}$$

where τ^h is the fine-grid local truncation error, defined with $\hat{I}^h$ such that $\hat{I}^H = \hat{I}_h^H \hat{I}^h$. The sign $\approx$ means equality up to a higher order in h. Relation (8.11) means, more precisely, that if $\tau^h + \tau^H$ were used to correct the H-equations, then u^H would be a higher-order approximation to $\hat{I}^H u$; namely, $u^H - \hat{I}^H u$ would equal $w^H - \hat{I}_h^H w^h$, where $L^H w^H = I_h^H \tau^h$ and $L^h w^h = \tau^h$.

Relation (8.11) can be used to inexpensively raise the approximation order. If the local approximation order (order of consistency) at the point x is p, i.e., if $\tau^h(x) \approx c(x) h^p$ where $c(x)$ is independent of h, then $\tau^H(x) \approx 2^p c(x) h^p$, hence $\tau_h^H(x) \approx (2^p - 1) c(x) h^p$, and hence $\tau^H(x) \approx 2^p (2^p - 1)^{-1} \tau_h^H(x)$. To raise the approximation order all we have then to do is to change the grid-H equations (8.5) by writing them as in (8.7) and multiplying τ_h^H by the fixed factor $2^p (2^p - 1)^{-1}$. This operation is called <u>τ extrapolation</u>. It resembles Richardson extrapolation, but it can profitably be done even in cases the latter cannot (e.g., in cases $p = p(x)$ is not constant), because it extrapolates the equation, not the solution. The τ extrapolation can be shown to be a special case of the higher-order techniques of Sec. 10.2 below, but it is especially simple and inexpensive. It costs only one multiplication on the <u>coarser</u> grid. It is probably best to use it in an FMG algorithm with $W(\nu,0)$ cycles, since a terminal relaxation with the lower order discretization would impair the approximation (even though its order would remain higher). An option for τ extrapolation exists in the model program FMG1 [M3].

Because of the analogy to the local truncation errors, τ_h^H is also called the <u>relative local truncation error</u> — the local truncation error of grid H relative to grid h. It is a by-product of the FAS processing which can be used to estimate the true local truncation errors: $\tau^h \approx (2^p - 1)^{-1} \tau_h^H$. Hence it can be used in FMG stopping criteria (see Sec. 7.3) and in grid adaptation criteria (Secs. 9.5, 15).

8.5 FAS interpolations and transfers

The consideration in determining the residual transfer I_h^H, the correction interpolation I_H^h and the FMG interpolation Π_H^h in FAS are basically the same as in CS, but there are some additional possibilities and we should also specify now the FAS solution transfer $\hat{I}_h^H$.

For linear problems the choice of $\hat{I}_h^H$ does not of course matter; all choices will give identically the same results. The solution efficiency of many nonlinear problems is also insensitive to the exact choice of $\hat{I}_h^H$. The choice does matter only where the problem coefficients drastically vary over a meshsize. By the "problem coefficients" we mean those of the linearized problem. In practice, using FAS, we do not linearize the problem, but the transferred solution $\hat{I}_h^H u^h$ implicitly determines the problem coefficients on the coarse grid H — determining the coefficients may in fact be regarded as the purpose of this transfer (although, unlike the CS situation, the coefficients can change on grid H, with the changing approximation). When the problem coefficients are highly variable, it is important to have each coarse-grid coefficient a suitable average of neighboring values of the corresponding fine-grid coefficient. The coarse-grid problem, in other words, should be a proper "homogenization" of the fine-grid problem. Such homogenization is usually obtained by using full weighting for $\hat{I}_h^H$ (as for I_h^H in Eqs. (4.5) - (4.6)).

In some, very special situations the dominant solution-dependent term in the coefficients may have the form $g(u)$, where g is a sensitive function; large changes in g are caused by more-or-less normal changes in u over a meshsize. In such a case the weighting $\hat{I}_h^H$ should have the special form

$$\hat{I}_h^H \tilde{u}^h = g^{-1} \bar{I}_h^H (g(\tilde{u}^h)) , \qquad (8.12)$$

where $\bar{I}_h^H$ is a normal full weighting, such as (4.6), $g(\tilde{u}^h)$ is a grid-function such that $(g(\tilde{u}^h))(x^h) = g(\tilde{u}^h(x^h))$ for every fine-grid point x^h, and g^{-1} is the inverse function of g; that is, $g^{-1}(g(\alpha)) = \alpha$ for any value α. If several sensitive functions such as g appear in the coefficients, several $\hat{I}_h^H$ may correspondingly have to be used.

(So far we have not seen a practical problem where this was required.)

An important possibility offered by FAS is the interpolation near an interior discontinuity, such as a shock or an interface. The grid-H solution, introducing smooth changes to the grid-h solution, may change the location of such a discontinuity, moving it a meshsize or two. Near the discontinuity the correction $\tilde{v}^H = \tilde{u}^H - \hat{I}_h^H \tilde{u}^h$ will then be highly non-smooth; it will look like a pulse function. Interpolating it as a correction to the fine grid will introduce there unintended high oscillations. To avoid this, the FAS interpolation (8.6) should be replaced by (8.6') near the discontinuity. This is easy to implement, by adopting the following, more general rule.

Use (8.6) everywhere except near points where $\tilde{u}^H - \hat{I}_h^H \tilde{u}^h$ is comparable to $\tilde{u}^h$, where (8.6') should be used.

8.6 Application to integral equations

When the integral equation

$$\int K(x,y)\, u(y)\, dy = f(x,u(x)) \tag{8.13}$$

is discretized in a usual way on a grid with $n = O(h^{-d})$ points, the unknowns are all connected to each other; the matrix of the (linearized) discrete system is full. A solution by elimination would require $O(n^3)$ operations. An FMG solution would require $O(n^2)$ operations, since each relaxation sweep costs $O(n^2)$ operations. In case (8.13) is nonlinear in u, FAS-FMG would be used. Using the FAS structure, even for linear problems, we can often reduce this operation count very much, by exploiting smoothness properties of K.

In most physical applications $K(x,y)$ has a singularity at $y = x$. So usually $K(x,y)$ becomes either smaller or smoother as the distance $|y-x| = (\Sigma_1^d (y_j - x_j)^2)^{\frac{1}{2}}$ increases. "Smaller" and "smoother" are in fact related: The former is obtained from the latter by differentiations of (8.13) with respect to $x = (x_1, \ldots, x_d)$ (possibly replacing the integral equation by an integro-differential equation). In either case, one can obtain practically the same accuracy in the numerical integration using meshsizes that increase with $|y-x|$, cutting enormously the work involved in relaxation. Usually $u(y)$ is much less smooth than $K(x,y)$ for large $|y-x|$. The integration with increasing meshsizes cannot then use point values of $\tilde{u}^h$, but should use local averages of $\tilde{u}^h$, taken over boxes whose size increases with $|y-x|$. Exactly such averages are supplied by the sequence of coarser grids in the FAS structure. The FAS solution transfers $\hat{I}_h^H$ should of course represent full weighting. One can increase the accuracy of integration by using, in addition to the full-weighting averages, higher local moments, represented on additional coarser grids.

8.7 Small storage algorithms

Various effective methods for vastly reducing the storage requirement of the multigrid algorithm, without using external storage, can be based on the Full Approximation Scheme. One simple method [B10, §2.2] is to use the fact that a problem whose finest grid is h can satisfactorily be solved by an FMG algorithm with only one h-cycle (see Sec. 7.3). This means that only one visit is needed to grid h, including the FMG interpolation Π_H^h, a couple of relaxation sweeps, and the residuals and solution transfers I_h^H and $\hat{I}_h^H$, back to grid H. All these operations can be made "wave-like" by just one pass over grid h, requiring no more than few columns at a time kept in memory. (When the operations of one relaxation sweep have been completed up to a certain column, the operations of the next sweep can immediately be completed on the next column, etc.) This visit is enough to supply the corrected right-hand side $\hat{f}^H$ on grid H (cf. Sec. 8.2), hence enough to calculate $\tilde{u}^H$, without any storage allocated to grid h, except for the few mentioned continuously shifted columns.

$\tilde{u}^H$ is as precise as $\tilde{u}^h_{NEW}$. The usual terminal sweeps of the h-cycle are only done if we need the solution on grid h, their role is to <u>smooth</u> the interpolation errors, not to <u>reduce</u> the error. Moreover, suppose that what we really need from our calculations is some functional of the solution, $\Phi(u)$ say, so we would like to calculate $\Phi^h(\tilde{u}^h_{NEW})$. All we have to do is to calculate, incidentally to transferring the solution $\tilde{u}^h$ back to grid H, the values of both $\Phi^h(\tilde{u}^h)$ and $\Phi^H(I_h^H\tilde{u}^h)$. Then, having later obtained $\tilde{u}^H$, we can calculate

$$\hat{\Phi}^H(\tilde{u}^H) = \Phi^H(\tilde{u}^H) + \Phi^h(\tilde{u}^h) - \Phi^H(I_h^H\tilde{u}^h) , \qquad (8.14)$$

which is practically as accurate as $\Phi^h(\tilde{u}^h_{NEW})$, since $\tilde{u}^H - I_h^H\tilde{u}^h$ is small and smooth.

This simple procedure reduces the required storage typically by the factor 2^d, without increasing the computational work. Other procedures can reduce the storage much farther by avoiding the storage of coarser grids too, except for a certain $(n_k h_k)\times(n_k h_k)$ box on each grid $h_k = 2^k h$. The h_k box is shifted within the h_{k-1} box to supply the $\tau_{h_k}^{h_{k-1}}$ corrections. The amount of work increases since on "re-visiting" the h_k box we need to reproduce its own $\tau_{h_{k+1}}^{h_k}$ corrections. This can be done only in the interior of the box, distance $O(h_k|\log\varepsilon|)$ from the boundary of the box, where $\varepsilon = O(h^p)$ is the desired accuracy on the finest grid, because closer to that boundary the h_{k+1} high-frequency errors are larger than the desired accuracy. Hence we must have $n_k \geq O(|\log\varepsilon|)$. The overall required storage can therefore be reduced to $O(|\log\varepsilon|^d|\log h|)$ (not just $O(|\log h|)$ as mistakenly calculated in [B7, §7.5]). Such procedures are called segmental refinement techniques.

Another small-storage multigrid algorithm, not based on FAS, is described in [H1]. It is a region dissection procedure, particularly suited for elongated domains.

9. LOCAL REFINEMENTS AND GRID ADAPTATION

Non-uniform resolution is needed in many, perhaps most, practical problems. Increasingly finer grids are needed near singularities, near non-smooth boundaries, at boundary layers, around captured shocks, etc., etc. Increasingly coarser grids are needed for external problems on unbounded domains. The multi-level FAS approach gives a convenient way to create non-uniform adaptable structures which are very flexible, allowing fast local refinements and local coordinate transformation, and whose equations are still solved with the usual multigrid efficiency. Moreover, the grid adaptation can naturally be governed by quantities supplied by the FAS multigrid processing, and it can naturally be integrated with the FMG algorithm to give increasingly better approximations to the true differential solution, at a fast, nearly optimal rate. These techniques, outlined below, are described in more detail in [B7, §7, 8, 9], [B8, §2, 3, 4] or [B10, §3, 4]. An application to three-dimensional transonic flows is described in [B23].

Another highly flexible discretization using a multigrid solver is described in [B2]. It is based on a finite element approach, which makes the program simpler, especially for complicated structures, but the execution is less efficient. The techniques outlined below are also applicable to finite element formulations as in [B11], [B9, Lecture 4].

9.1 Non-uniformity organized by uniform grids

Our non-uniform discretization grows from the simple observation that the various grids (levels) used in usual multigrid algorithms need not all extend over the same domain. The domain covered by any grid may be only a proper part of the domains covered by coarser grids. Each grid h can be extended only over those subdomains where the desired meshsize is roughly less than 2h . In such a structure, the effective meshsize at each neighborhood will be that of the finest grid covering it (see Figs. 2, 3, and 4 in [B10]).

This structure is very flexible, since local grid refinement (or coarsening) is done in terms of extending (or contracting) uniform grids, which is relatively easy and inexpensive to implement. A scheme named GRIDPACK [M3] has been developed for constructing, extending and contracting general uniform grids, together with many service routines for such grids, including efficient sweeping aids, interpolations, displays, treatment of boundaries and boundary data, etc. It is partly described in [B8, §4]. One of its advantages is the efficient storage: The amount of logical information (pointers) describing a uniform grid is proportional to the number of <u>strings</u> of points (contiguous sets of gridpoints on the same gridline), and is therefore usually small compared with the number of points on the grid. Similarly, the amount of logical operations for sweeping over a grid is only proportional to the number of strings. Changing a grid is inexpensive too. One can easily add finer levels, or extend existing ones, thus effecting any desired refinements.

Moreover, this structure will at the same time provide a very efficient <u>solution process</u> to its difference equations, by using its levels also as in a multigrid solver. For this purpose the Full Approximation Scheme must be used, because in parts of the domain not covered by the finer grid h , the coarser grid $H = 2h$ must certainly show the full solution, not just a correction. Indeed, the FAS approach naturally fits here: We use on grid H the equations $L^H u^H = \hat{f}^H$, where $\hat{f}^H$ is given by (8.5b) wherever $L^h u^h$ is well defined (i.e., in the interior of grid h), and $\hat{f}^H = f^H$ otherwise. In other words (cf. Eqs. 8.7-8), the fine-to-coarse correction τ_h^H is simply cancelled wherever it is not defined. Applying an FMG algorithm with these structures and equations, we will get a solution that in each subdomain will satisfy its finest-grid equations, while at interior boundaries (of fine levels not covering the entire domain) the solution will automatically be as interpolated from coarser grid. Note that the coarse-grid solution is influenced by the finer grid even at regions not covered by the latter, since the coarse-grid <u>equations</u> are modified in the refined region.

In other words, a patch of the next finer level h can be thrown on any part of a given grid $H = 2h$, correcting there the latter's equations to the finer-grid accurracy. Moreover, several such patches may be thrown on the same grid. Some or all of the patches may later be discarded, but we can still retain their τ_h^H corrections in the grid H equations.

An important advantage is that difference equations are in this way defined on uniform grids only. Such difference equations on equidistant points are simple, inexpensive and standard, even for high-order approximations, whereas on general grids their weights would have to be calculated by lengthy calculations separately for each point. Relaxation sweeps are also made on uniform grids only. This simplifies the sweeping and is particularly essential for symmetric or line relaxation schemes.

9.2 Anisotropic refinements

It is sometimes desired to have a grid which resolves a certain thin layer, such as a boundary layer. Very fine meshsizes are then needed in one direction, namely, across the layer, to resolve its thin width. Even when the required meshsize is extremely small, not many gridpoints are needed, since the layer is comparably thin, provided, of course, that fine meshsizes are used only in that one direction. We need therefore a structure for meshsizes which get finer in one direction only.

In case the thin layer is along coordinate hyperplane $\{x_j = \text{const.}\}$ this is easily done by semi refinements: Some levels H are refined by the next level h only in the j coordinate, $h_j = H_j/2$ whereas $h_i = H_i$ for $i \neq j$. In fact, different patches may have different refinement directions, so that level h should be considered as a collection of grids. Thus, the set of all grids is arranged logically in a tree, each grid having a unique next-coarser grid, but possibly several next-finer grids, instead of the former linear ordering. All these grids are still uniform, and can still easily be handled by GRIDPACK.

Note that the next-finer grids of a given grid H may geometrically have some overlap. All that is needed in such cases is to set priority relations, to tell which correction τ_h^H applies at each point of grid H. Such priority relations are simply set by the order in which the corrections are transferred to grid H.

In case the thin layer is not along coordinate lines, the methods of the following sections could be used.

9.3 Local coordinate transformations

Another dimension of flexibility and versatility can be added to the above system by allowing each of the local patches to have its own set of local coordinates.

Near a boundary or an interface, for example, the most effective discretization is made in terms of coordinates in which the boundary (or interface) is a coordinate line. In such coordinates it is much easier to formulate high-order approximations near and on the boundary, and to introduce meshsizes which are much smaller across than along the boundary layer (Sec. 9.2); etc. In the interior, local patches of coordinates aligned with characteristic directions (along streamlines, for instance) can greatly reduce the cross-stream numerical viscosity (cf. Sec. 2.1).

Each set of coordinates will generally be used with more than one grid, so that (i) local refinements, isotropic or anisotropic, in the manner described above, can be made within each set of coordinates; and (ii) the multigrid processing retains its full efficiency by keeping the meshsize ratio between any grid and its next-coarser one properly bounded.

Since local refinement can be made within each set of coordinates, the <u>only</u> purpose of the coordinate transformation is to provide the grid with the desired orientation, i.e., to have a given manifold (such as a piece of the boundary) coincide with a grid hyperplane. Since, furthermore, this needs to be done only locally, it can be obtained by a simple and standard transformation. For example, in two-dimensional problems, let a curve be given in the general parametric form

$$x = x_0(s), \quad y = y_0(s), \quad (s_1 < s < s_2) \tag{9.1}$$

where s is the arclength, i.e.,

$$x_0'(s)^2 + y_0'(s)^2 = 1 . \tag{9.2}$$

To get a coordinate system (r,s) in which this curve coincides with the grid line $\{r=0\}$, we use the standard transformation

$$x(r,s) = x_0(s) - ry_0'(s), \quad y(r,s) = y_0(s) + rx_0'(s) . \tag{9.3}$$

Locally, near $r=0$, this transformation is isometric (simple rotation).

The main advantage of this transformation is that it is fully characterized by the single-variable functions $x_0(s)$, $y_0(s)$. These functions, together with $x_0'(s)$, $y_0'(s)$ and $q(s) = x_0''/y_0' = -y_0''/x_0'$ can be stored as one-dimensional arrays, in terms of which efficient interpolation routines from (x,y) grids to (r,s) grids, and vice versa, can be programmed once for all. The difference equations in (r,s) coordinates are also simple to write in terms of these stored arrays, since by (9.2-3),

$$\frac{\partial}{\partial x} = -y_0' \frac{\partial}{\partial r} + \frac{x_0'}{1+rq} \frac{\partial}{\partial s}, \quad \frac{\partial}{\partial y} = x_0' \frac{\partial}{\partial r} + \frac{y_0'}{1+rq} \frac{\partial}{\partial s}. \tag{9.4}$$

A different kind of multi-level procedure using a combination of cartesian grids and grids curved along boundaries is described in [S4, §11]. The main difference is that all levels, from coarsest to finest, are used there both for the cartesian and for the curved grids, and at each level the relaxation includes interpolations between the two types of grids, while the present approach is to regard the curved grids as a finer level which correct the finest cartesian grid near the boundary. The present approach is perhaps more economic and flexible, but it requires a (crude) approximation to the boundary conditions to be given on the cartesian grids, too.

9.4 Sets of rotated cartesian grids

Another variant of this procedure is required in case the location of the thin layer (interface, shock, etc.) is not fully defined. For this purpose, each level will be a set of rotated cartesian grids, possibly overlapping. The finer the level, the finer (richer) is also the set of rotations [B10, Fig. 4]. The self-adaptive criteria (see Sec. 9.5) can be employed to decide where to refine the set of rotations (together with refining the meshsize in one direction). Hence the scheme can capture discontinuities (thin layers), without defining them as such. The stronger the discontinuity, the better its resolution.

Since only rotated cartesian grids are needed in this scheme, the finite difference equations are as simple as ever. Hence this method is sometimes preferable even in cases where the location of the thin layer is known.

9.5 Self-adaptive techniques

The flexible organization and solution process, described above, facilitate the implementation of variable mesh-size $h(x)$ and the employment of high and variable approximation order $p(x)$. How, then, are mesh-sizes and approximation-orders to be chosen? Should boundary layers, for example, be resolved by the grid? What is their proper resolution? Should high-order approximation be used at such layers? How does one detect such layers automatically? In this section we survey a general multigrid framework for automatic selection of $h(x)$, $p(x)$ and other discretization parameters in a (nearly) optimal way. This system automatically resolves or avoids from resolving thin layers, depending on the goal of the computations, which can be stated through a simple function. (For more details see [B7, §8], [B8, §3]).

As our directive for sensible discretization we consider the problem of minimizing a certain error estimator E subject to a given amount of solution work W (or minimizing W for a given E. Actually, the control quantity will be neither E nor W, but their rate of exchange). This optimization problem should of course be taken quite loosely, since full optimization would require too much control work and would thus defeat its own purpose.

The error estimator E has generally the form

$$E = \int_\Omega G(x) \, \tau^h(x) \, dx , \tag{9.5}$$

where $\tau^h(x)$ is the local truncation error (cf. (8.10)) at x. $G(x) \geq 0$ is the error-weighting function. It should in principle be imposed by the user, thus defining his goal in solving the problem. In practice $G(x)$ serves as a convenient control. It is only the relative orders of magnitude of $G(x)$ at different points x that really matter, and therefore it can be chosen by some simple rules. For example, if it is desired to compute ℓ-order derivatives of the solution up to the boundary then $G(x) \approx d_x^{m-1-\ell}$, where d_x is the distance of x from the boundary, and m is the order of the differential equation.

The work functional W is roughly given by

$$W = \int_\Omega \frac{w(p(x))}{h(x)^d} \, dx , \tag{9.6}$$

where d is the dimension and h^{-d} is therefore the number of grid-points per unit volume. (Replace h^d by $h_1 \ldots h_d$ in case of anisotropic grids.) $w = w(p)$ is the solution work per grid-point. In multigrid processing, this work depends mainly on the approximation order (consistency order) $p(x)$. If the high-order techniques of Sec. 10 are used then usually $w(p) \approx w_0 p$, although sometimes $w(p) = O(p^3)$ for unusually high p (see Sec. 10.1).

Treating $h(x)$ as a continuous variable, the Euler equations of minimizing (9.5) subject to (9.6) can be written as

$$G \frac{\partial \tau}{\partial h} - \lambda d\, w(p)\, h^{-d-1} = 0, \qquad (9.7)$$

where λ is a constant (the Lagrange multiplier), representing the marginal rate of exchanging optimal accuracy for work: $\lambda = -dE/dW$.

In principle, once λ is specified, equation (9.7) determines, for each $x \in \Omega$, the local optimal values of $h(x)$, provided the truncation function $\tau^h(x)$ is fully known. In some problems the main behavior of $\tau^h(x)$ near singularities or in singular layers is known in advance by some asymptotic analysis, so that approximate formulae for $h(x)$ can apriori be derived from (9.7). More generally, however, equation (9.7) is coupled with, and should therefore be solved together with, the given differential equations. Except that (9.7) is solved to a cruder approximation. This is done in the following way:

In the FAS solution process we readily obtain the quantity τ_h^H. By (8.11) and (9.5), the quantity $-\Delta E(x) = G(x)\,\tau_h^H(x)$ can serve as an estimate for the decrease in E per unit volume owing to the refinement from H to h in the vicinity of x. By (9.6), this refinement requires the additional work (per unit volume) $\Delta W = w(p)\, h^{-d}\,(1-2^d)$. The local rate of exchanging accuracy for work is $Q = -\Delta E / \Delta W$. If Q is much larger than the control parameter λ, we say that the transition from H to h was highly profitable, and it pays to make a further such step, from h to $h/2$. So we will next establish grid $h/2$ in the neighborhood of x, as in any other neighborhood where there are points with $Q >> \lambda$.

The computer work invested in the test is negligible compared with the solution work itself, since Q is calculated by just a couple of operations per point on the coarser grid H once per cycle.

A similar test can be used to decide on changing the local approximation order $p(x)$, with τ_h^H being replaced by the p_1-to-p_0 defect-correction (10.1) and correspondingly $\Delta W = (w(p_1) - w(p_0))h^{-d}$. Or, if we treat p as a constant over the domain, but we like to optimize that constant, we can measure ΔE globally; i.e., measure directly the change in some quantity of interest (e.g., some functional of the solution we are most interested in), due to the transition from p_1 to p_0. Correspondingly, global ΔW will be used. Whether locally or globally, the order will be increased beyond p_1 if $-\Delta E/\Delta W >> \lambda$. Other discretization parameters, such as the computational boundaries (when the physical domain is unbounded), or refinements in grid orientations (see Sec. 9.4), can be decided by similar tests all based on comparing some $-\Delta E / \Delta W$ to the exchange-rate parameter λ. How to control λ and coordinate it with the solution algorithm is discussed in the next Section.

9.6 Exchange rate algorithms

Near a severe singularity many levels of increasingly finer grids on increasingly narrower subdomains may be needed, and formed by the above criteria. If the usual FMG algorithm were applied to these levels, too much work would be spent, since too many passes on coarser grids would be made. Only when all grids cover the same domain is it true that the coarse-grid work is small compared to the next-finer grid work, since the latter deals with about 2^d as many points. This is no longer so when local refinements are used: Finer grids may include less points than some coarser grids. The amount of work in a usual FMG algorithm would therefore be much greater than proportional to the total number of gridpoints; (9.6) would not hold.

A better procedure is to decrease the accuracy-to-work exchange rate λ in a gradual sequence $\lambda_1 > \lambda_2 > \ldots,$ and to use the solution obtained for λ_{j-1} as the first approximation to the solution on the grids formed for λ_j. Numerical experiments (in collaboration with Dov Bai) confirm that for some types of severe singularities (due to singular right-hand sides of the differential equations or the boundary conditions) such a first approximation followed by one V cycle leaves algebraic errors smaller than truncation errors. We have used $\lambda_j = \lambda_{j-1}/16$ for two-dimensional problems with second-order approximations; in the absence of singularities this ratio would reproduce the usual FMG algorithm.

For other types of singularities (due to singular boundary shapes or singular coefficients) exchange-rate controls should be used also in the correction cycles in order to restore (9.6). Typically, the algorithm continues relaxing on a given grid wherever $\hat{Q}(x) \equiv -G(x)\,\Delta R(x)/\Delta W > a_1\lambda$ where $-\Delta R^h(x)$ is the local decrease in the residual $|R(x)|$ per gridpoint per sweep and ΔW is the corresponding work. Wherever $\hat{Q}(x) < a_1\lambda$ but $-G(x)\,|R(x)|/\Delta W > a_2\lambda$, switch is made to the next coarser grid. Details will be given elsewhere.

Ultimately, such exchange-rate criteria unify all the multigrid switching and adaptation criteria, integrating them into one algorithm, in which λ is gradually decreased. The process can be continued indefinitely, with increasingly finer levels created, globally and locally, in an almost optimal way. It can be terminated when either E or W or λ reach preassigned limits. The above techniques of anisotropic grids, local transformations and rotations, and adaptation of approximation orders can all be integrated into such an exchange-rate algorithm.

10. HIGHER-ORDER TECHNIQUES

A sound way of constructing high-order approximations to a given differential problem $LU = F$, is first to construct a multigrid program with a low approximation order, and then convert it into a high-order program. The lower order is easier to develop and is also useful as a component in the higher-order program. Such programs are usually more efficient than programs which use high-order difference operators throughout. We mostly recommend the method of Sec. 10.2 below, especially for non-elliptic and singular perturbation problems.

10.1 Fine-grid defect corrections

Given a program for solving the (linear or nonlinear) low-order (order p_0) discrete system $L_0^h u^h = f^h$, an obvious multigrid approach for raising the approximation order is by high-order "deferred" (or "defect") corrections introduced once per cycle on the currently-finest grid [B10, §3.4]. That is, we add to f^h the correction

$$\sigma_{1,0}^h(x^h) = L_0^h \tilde{u}^h(x^h) - L_1^h \tilde{u}^h(x^h), \tag{10.1}$$

where L_1^h is the higher-order operator, its approximation order (consistency order) being $p_1 > p_0$, and $\tilde{u}^h$ is the current approximate solution. A similar correction is of course introduced to the discrete boundary conditions, too. To save h-cycles one should employ an FMG algorithm (Sec. 7), and use corrections like (10.1) at all the FMG stages (i.e., for every currently-finest grid). The total amount of work is then still basically given by (7.4). Note that that work is proportional to the approximation order p_1. However, this count does not take into account the calculation of (10.1) once per cycle. For lower p_1 this extra work may be less than the other work within the cycle (a couple of sweeps on each level), but for high p_1 it becomes dominant and makes the amount of work per cycle proportional to p_1 (assuming spectral-type methods cannot be used and the complexity of calculating L_1^h is thus proportional to p_1), hence the total work is in principle $O(p_1^2)$. Furthermore, for higher p_1 we have in principle to use higher computer precision, making the work of each arithmetic operation (in calculating (10.1)) again proportional to p_1, bringing the total work to $O(p_1^3)$. In the normally used range of p_1, however, the work is usually still dominated by relaxation and still doable in the basic computer precision, hence still proportional to p_1.

The deferred correction technique (suggested by L. Fox) is a special case of the concept of defect corrections (see [L1], [S3], [A2]). An important advantage of such a technique is that the higher-order operator L_1^h (and the corresponding higher order boundary conditions) need not be stable. This gives much freedom in the relatively difficult task of calculating L_1^h. This freedom is especially welcome in non-elliptic and singular perturbation cases, where convenient central approximations are unstable.

The reason that L_1^h need not be stable is that the defect correction iterations, even when converging to the solution of L_1^h, their convergence is fast only in the smooth solution components (for which L_0^h is a good approximation to L_1^h) and is very slow in the high-frequency components. Since instability is a property of high-frequencies, it can creep in only very slowly. The growth of unstable modes within the few cycles made is not too damaging.

The whole purpose of defect corrections is in fact to correct low-frequency components; only for such components higher-order approximation, such as L_1^h, are much better than lower-order approximations like L_0^h. Recognizing this and the fact that in multigrid processes low frequencies are converged via the coarse-grid corrections, we see that the main effect of the defect corrections can be obtained by applying them only at the stage of transferring residuals to the coarser grids. This would save about two work-units per cycle, and would give better approximations in case L_1^h is unstable. This idea, from a different point of view, is described in the next section.

10.2 Double discretization: High-order residual transfers

On any given grid participating in multigrid interactions, discrete approximations to the continuous operator L are used in two different processes: in relaxation sweeps, and in calculating residuals transferred to coarser grids. The two discretization schemes need not be the same [B19, §3.11]. The discretization L_0^h employed in the relaxation sweeps must be stable (see Sec. 12), but its accuracy may be lower than the one we wish to generate. The discretization L_1^h used in calculating the transferred residuals determines the accuracy of our numerical solution, but it need not be stable. This "double discretization" scheme is especially useful in dealing with non-elliptic and singular perturbation problems: One can use the most convenient (but sometimes unstable) central differencing for L_1^h, and add artificial viscosities (see Sec. 2.1) only to L_0^h. This will ensure stable solutions which still have the accuracy of the central differencing.

Note that such a multigrid process will not converge to zero residuals, since it uses two conflicting difference schemes. The very point is, indeed, that the solution produced may be a better approximation to the <u>differential</u> solution than can be produced by either scheme.

The lack of algebraic convergence makes the usual two-level mode analysis irrelevant for double discretization schemes. Instead they can be analyzed by the two-level FMG mode analysis (Sec. 7.4).

Double discretization schemes can of course similarly be applied to <u>boundary conditions</u>; e.g., to Neumann conditions: Simple first-order schemes can be used in relaxation, while second-order Neumann conditions (which are sometimes complicated and may sometimes be unstable) can be used to transfer boundary-condition residuals to coarser grids.

The double discretization scheme need not be confined to the currently finest level; it can <u>also be used on coarser levels</u>. This will give better coarse-grid corrections, and hence faster algebraic convergence. (In non-elliptic and singular perturbation cases the algebraic convergence is usually determined by the quality of the coarse-grid correction [B17, §5.1].) It is also more convenient to program, since the same residual transfer routine, based on L_1^h, is used on all levels. Moreover, if only L_0^h is used on coarser levels, the gain in approximation order per cycle cannot be more than p_0; hence the final approximation order cannot exceed $2p_0 + r_0$, where r_0 is the convergence order of relaxation [B17, §2.2]. Such a restriction does not exist if L_1^h is used for residual transfers on all levels. The approximation order p_1 can then be attained, perhaps even in one cycle, no matter how high p_1 is. In particular, pseudo spectral approximations can be used in L_1^h (if the boundary conditions are periodic, for example), yielding unbounded approximation orders in few cycles.

In order to obtain the high approximation orders several rules should be observed: Suitable interpolation orders and residual-transfer orders should be employed. The right orders can be derived by crude mode analysis, as in Sec. 4.3, but with particular attention to boundaries (see in particular rule (C) in Sec. 4.3). FMG algorithms with $W(\nu,0)$ cycles should be used (see Sec. 6.2), to ensure accurate enough solution of the coarse-grid equations and to avoid degradation of the approximation by terminal relaxation. Also, when double discretization is used on all levels together with the Full Approximation Scheme (see Sec. 8), notice that two different right-hand sides should be used on coarser grids, one for relaxation and a different one for residual transfers [B17, §2.1].

In case L_1^h is a better approximation than L_0^h not only for smooth components but also in the high-frequency range, the method of fine-grid defect corrections (Sec. 10.1) will eventually give smaller errors than the coarse-grid defect correction described here. But the gain will hardly justify the extra work involved in calculating (10.1) separately from the calculation of residuals. In problems where L_1^h is unstable, the present method is both faster and more accurate.

Double discretization schemes have already been used successfully in various cases, including fourth and sixth order approximations to Poisson equation [S1]; second-order approximations to simple singular perturbation problems [B17, §7], [B3, §7]; and second-order approximations to incompressible Navier-Stokes equations (in collaboration with D. Deswarte).

10.3 Relaxation with only subprincipal terms

A particularly useful application of the above techniques is to employ a simple relaxation operator L_0^h where non-principal terms are neglected; more precisely, to employ the simplest stable L_0^h which approximates the subprincipal terms of the differential operator (see Sec. 2.1). Other terms need to be approximated only in L_1^h. For some fluid-dynamics systems this procedure can save a substantial amount of work. The techniques of either Sec. 10.1 or 10.2 can be used with this relaxation, but more work is saved by the latter. On very coarse grids this type of relaxation may give worse performance. In such cases use more sweeps or reintroduce the neglected non-principal terms.

11. COARSENING GUIDED BY DISCRETIZATION

The term "coarsening" is used here for the entire process of transferring a residual problem $L^h v^h = R^h$ from a fine grid h to the next coarser grid H (=2h). This includes the formulation of the coarse-grid problem $L^H v^H = I_h^H R^h$, where the coarse-grid operator L^H and the fine-to-coarse transfer I_h^H should be determined both in the interior and near boundaries, and similar equations should be transferred for the boundary conditions themselves, and for any other side conditions the problem may have.

Any sufficiently general method of coarsening implies a discretization method, in the following sense: If the differential problem $Lu = f$ is discretized by any method, giving the problem $L^h u^h = f^h$ on grid h, and if this problem is then successively coarsened to $L^{2h} u^{2h} = f^{2h}$, $L^{4h} u^{4h} = f^{4h}$, etc., by successive applications of the <u>same</u> coarsening method, then in the limit (for a sufficiently coarse grid) we obtain a discretization of $Lu = F$ which does not depend on the original discretization $L^h u^h = F^h$, but only on the method of coarsening. The limit discretization, in this sense, is the <u>fixed point of the coarsening method</u>. (In practice the limit is almost fully established after just a couple of coarsening levels).

This quite trivial observation has important implications. It implies that coarsening is at least as rich and difficult as discretization. It implies that controversies and competing techniques will emerge concerning coarsening similar to the ones in the field of discretization. Indeed such competitions have already surfaced. For example, the competition between finite-difference and finite-element methods, a dispute which in fact consists of several separate issues: The variational derivation of discrete equations (or coarse-grid equations) vs. direct differencing; the interpolation issue (finite-elementers insist on using the same interpolation — the same "element" — as used in deriving the discrete equations, while finite-differencers allowing more freedom in interpolation, sometimes gaining higher accuracy in some error norms); the issue of general triangulation vs. uniform grids; and the issue of compactness of high-order approximations. These issues should not be confused with each other: variational derivation is possible and natural even without the use of elements [F4, §20.5]. Uniform grids can be used with finite-element solutions, too, changing the elements only near boundaries, a structure much more efficient computationally, especially in conjunction with multigrid methods [B11]. High-order compact operators arise quite naturally in the finite-element method, but such operators can also be derived by finite-difference approaches, such as the operator compact implicit method [C1] and also the Hodie method [L3].

All these and other issues arise as well with regard to coarsening, and the competing approaches are generally successful in coarsening wherever they are successful as discretization procedures — which is usually in problems where they are more natural. Variational approaches ([N2], [B7, App. A.5], are natural for self-adjoint problems, and provide the most robust and automatic coarsening procedures for such problems [A1], [D1], although they can be replaced by less expensive procedures (analogous to direct differencing) if the self-adjoint problem is not particularly complicated. In singular perturbation problems, such as those arising in fluid dynamics, discretization as well as coarsening are most successfully guided by physical understanding (artificial viscosity, upstream differencing, etc.). And so on.

The attempt to devise general fine-to-coarse transfers, good for all problems, is as hopeless as the attempt to have general, completely problem-independent discretization procedures.

But if this argument tells us how complicated coarsening can be, it also shows a general way to handle this difficulty. Namely: the coarsening method can always be guided by the discretization scheme.

Indeed, conversely to the statement above (that every coarsening implies discretization), one can say that every discretization scheme can be used to derive a coarsening procedure. This is done by imitation or analogy: Think about <u>discretizing</u> the problem $Lv = R$ on the coarse grid H; then replace the operations done on the continuous domain by analogous operations done on the fine-grid h; e.g., replace integrations by analogous summations (or by integrations by elements, in case v^h is given in terms of finite elements). Galerkin discretization schemes, for example, are easily translated in this way into analogous coarsening formulae [N2, §3].

A coarsening procedure analogous to the finest-grid discretization scheme is called <u>compatible coarsening</u>. It is not necessary to use compatible coarsening, but it almost always makes a good sense to do so. In case the discretization scheme is a bad one, this would give a bad coarsening and hence slow asymptotic convergence rates of the multigrid cycling. But experience with several such cases (e.g., boundary singularities improperly treated) show that, if compatible coarsening is used, this slowness does not matter, because the source of slowness (bad discretization) is also, and for the same components, a source for larger truncation errors, hence an FMG algorithm (with the same discretization scheme on all currently-finest levels) still solves <u>below truncation errors</u> in the usual number of cycles (one or two, depending on the interior processes). Moreover, the slower <u>asymptotic</u> rates can in this way serve as a detector for the bad discretization, which otherwise may be passed unnoticed.

Compatible coarsening makes sense also from the point of view of computer resources and programming effort. For example, if a great generality and simplicity of programming is obtained by a discretization scheme (e.g., finite elements) which on the other hand spends a lot of computer time and storage to assemble the discrete equations and store them, the coarsening procedure can do the same since the time and storage it spends will be smaller than those already spent on the fine grid.

There are some special cases in which compatible discretizations are not quite available. These are cases where the discretization scheme is not general enough, because it specifically uses features of the finest grid not present on coarser ones. It uses for example a finest grid exactly laid so that its lines coincide with special lines of the problem, such as boundaries or lines of strong discontinuities (as in [A1]). In such situations compatible discretization is not well defined. To define it we must think in terms of a more general discretization scheme. (Again, the coarsening process serves to detect a certain flaw in discretization: In this case the flaw is the lack of generality.)

When double discretization is used (Sec. 10.2), compatible coarsening means the use of such a double discretization on coarser levels, too (as indeed recommended in Sec. 10.2).

12. REAL ROLE OF RELAXATION

The role of relaxation in multigrid processes has often been stated: It is to smooth the error; i.e., to reduce that part of the error (the "high-frequency" part) which cannot be well approximated on the next coarser grid. Some elaboration and clarification of this statement is important.

What is the "error" we want to smooth? It is usually thought of as the algebraic error, i.e., the difference $u^h - \tilde{u}^h$ between our calculated solution $\tilde{u}^h$ and the discrete solution u^h (the exact solution to the discrete equations). However, in view of the double discretization scheme (Sec. 10.2), where u^h is not well-defined, it becomes clear that what relaxation should really do is to smooth the differential error, i.e., the difference $u - \tilde{u}^h$, where u is the solution to the given differential equations. In fact, this is the true role of relaxation even when double discretization is not used, if what we want to approximate is u, not u^h: It is the smoothness of $u - \tilde{u}^h$ which permits its efficient reduction via the coarser grid.

Thus, the important measure of relaxation efficiency is not the algebraic smoothing factor $\overline{\mu}$, but the differential smoothing factor, the factor by which the high-frequency part of $u - \tilde{u}^h$ is reduced per sweep. This is not usually recognized because the latter factor is not constant: It approximately equals $\bar{\mu}$ when the high-frequencies in $u - \tilde{u}^h$ are large compared with those in $u - u^h$ (where u^h is the local solution to the discrete equations employed in relaxation), but below this level $\bar{\mu}$ may mislead, and when $\tilde{u}^h$ is closer to u than to u^h in their non-smooth components, the differential factor may even be larger than 1. For example, in solving a singular perturbation problem with strong alignment (see Sec. 2.1), we can reduce the algebraic smoothing factors of point Gauss-Seidel relaxation by taking a larger artificial viscosity and, more importantly, by taking it isotropically, instead of anisotropically. This would not however improve the overall performance of our double-discretization FMG algorithm (see the experiments in [B17, §7]), since it would not reduce the differential smoothing factors.

The differential smoothing is the purpose of relaxation not only on the finest grid $h*$. On any grid h, its relaxation reduces its range of high frequencies in the error $u - \tilde{u}^{h*}$, where we interpret changes in $\tilde{u}^h$ as changes in $\tilde{u}^{h*}$ via the interpolation relations.

We can here also elaborate on what are those "high-frequency components" (of the differential error) that should be converged by relaxation on grid h. Generally speaking we say that these are the components "invisible" on the next coarser grid H, i.e., Fourier components $\exp(i\underline{\theta}\cdot\underline{x}/\underline{h})$ which on grid H coincide with lower components, that is to say components with $\pi < \max_j |\theta_j| H_j/h_j$, $\max|\theta_j| \leq \pi$ (cf. Eq. (3.3)). More precisely we should include in the "high-frequency" range all those components that are not efficiently reduced by relaxation on other grids, which can be any range of the form $\cup_{j=1}^{d} \{\alpha_j < \theta_j \leq \alpha_j H_j/h_j\}$ (assuming H_j/h_j is the same on all levels). An example is discussed in [B17, §5.7]. The main point of the example is to show that some further understanding is sometimes needed in interpreting mode-analysis results.

The range of frequencies to be reduced by relaxation can also be modified by modified coarse-grid functions of the type mentioned in Sec. 4.2.2. The corresponding smoothing factors were called "partial smoothing factors" in [B15, §2.1]. We can thus generally say that the role of relaxation is to reduce the information content of the error, so that it becomes approximable by a cruder (lower dimensional) approximation space.

Another important point to clarify is that relaxation should be efficient only as long as the high-frequency error components have relatively large amplitudes: When the high-frequency errors are too small compared with the low-frequency ones, relaxation cannot usually be efficient because of certain feeding from low to high components. Such feeding is caused by interaction with boundaries, and by non-constant coefficients, and by the high-frequency harmonics generated when the low-frequency error is corrected via the coarse-grid cycle (see observation (D) in Sec. 4.3). Sometimes such feeding is even caused by the interior relaxation itself; e.g., red-black relaxation of an order-m differential equation produces $O(h^m)$ high-frequency errors from $O(1)$ low-frequency errors. When the size of high-frequency amplitudes approaches the size fed from low frequencies, relaxation should be stopped; this is the point where the coarse-grid correction should be made. If relaxation is stopped in time, then the range of strong interactions with low-frequencies is not entered. It is exactly then that the multigrid convergence rates can accurately be deduced from the smoothing-rate analyses.

Finally, even though smoothing is the main role of relaxation, we should not forget its influence on other components. Some relaxation schemes with extremely good smoothing factors are either unstable or they cause large amplification of some low-frequency errors (see Sec. 3.2).

We can thus say in summary that the role of relaxation is to reduce large amplitudes of certain (usually high-frequency) components of the differential error, while avoiding from significantly amplifying its other components.

Stability of the difference equations used in relaxation is only a tool in performing this role, not an end by itself.

13. DEALGEBRAIZATION OF MULTIGRID

An interesting line in the development of multigrid can be viewed as a gradual "dealgebraization", a gradual liberation from algebraic concepts, and the development of methods that increasingly exploit the underlying differential nature of the problems. We'd like to briefly trace this line here, so as to bring out some concepts useful in practical implementations.

As the first step of dealgebraization we can regard the replacement of "acceleration" by "smoothing". The early two-grid and multi-grid approach viewed coarse-grid corrections mainly as a tool for accelerating the basic iterative process - the fine-grid relaxation. Only later it became clear that the only role of relaxation is to smooth the error. (cf. Sec. 12, where a further "dealgebraization" of the smoothing concept is described.) This slight shift in understanding revolutionized the multigrid practice: It made it clear that the fine-grid process is basically local, hence analyzable by local mode analysis. This understanding, together with that analysis, produced the truly efficient multigrid cycles, in which very few sweeps are made on each grid before switching to coarser ones, and in which the fine-to-coarse meshsize ratio assumes the (practically) optimal value of 1:2.

The next dealgebraization steps are related to the trivial understanding that we are not primarily interested in solving the algebraic equations (obtaining u^h), but we are interested in approximating the differential solution u. First, this implies that we have to solve the algebraic equations only "to the level of truncation errors", i.e., only to the point that our calculated solution $\tilde{u}^h$ satisfies $\|\tilde{u}^h - u^h\| \approx \|u - u^h\|$; further reduction of $\tilde{u}^h - u^h$ is meaningless.

This implies that the asymptotic convergence rate of the multigrid cycle is not important by itself. What counts is the amount of work we need in an FMG algorithm in order to reduce the error from its original value on grid 2h, which is approximately $\|\Pi_{2h}^{h} u^{2h} - u\| \approx \|u^{2h} - u\|$, to the desired level $\|u^h - u\|$. This is a reduction by a factor of 2^p only, which can usually be achieved in one cycle. (The fundamental reason for this is again non-algebraic: See Sec. 7.3.) Evidently it is then more relevant to think in terms of the FMG analysis (Sec. 7.4) than in terms of asymptotic rates.

Even the later viewpoint, that we want to reduce the errors to the level of truncation errors, is too algebraic-oriented. It is tied too much to one given discretization on one given grid. The optimal moment of switching from a certain currently-finest grid H to a new, finer grid $h = H/2$ is not necessarily when $\|\tilde{u}^H - u^H\| \approx \|u - u^H\|$. Rather, it is determined by comparing H-cycles to h-cycles in their efficiency at driving $\tilde{u}^h$ closer to u (see Sec. 7.2). What really counts is the behavior of the differential error $E = \|\tilde{u}^h - u\|$ as a function of the total accumulated computational work W.

<u>We want</u> $E(W)$ <u>to be as fast-decreasing as possible</u>.

From this as our objective we can derive correct switching criteria, i.e. decide when to establish a new finer grid. The next step is to realize that criteria based on $E(W)$ can be applied <u>locally</u>, to decide not only <u>when</u> to have a finer grid, but also <u>where</u> to have it. This naturally brings us to <u>grid adaptation</u> (Sec. 9.5). Indeed one can integrate the switching and self adaption criteria (discussed in Secs. 6.2, 7.2, 9.5) into a total <u>multi-level adaptive algorithm</u>, where switching between levels and creating new, or extending existing levels are all governed by the same exchange-rate criteria (see Sec. 9.6).

Another step away from fixed algebraic concepts is to allow variable discretization schemes, i.e., schemes which can be changed throughout the algorithm to promote faster decreasing E(W). This includes the use of higher-order, variable-order and adaptible order schemes, governed again by E(W) criteria (see Sec. 9.5 and [B10, §3.6, 4.3]). Using different discretization schemes in relaxation and in residual transfers (Sec. 10.2) is a further step in that direction.

By now we have gone quite far beyond the notion of multigrid as just a fast algebraic solver, toward viewing it as a total treatment of the original problem. This is proved to be a very beneficial general principle: Always think of multigrid in terms of as original a problem as possible: For example, instead of using Newton iterations, employing multigrid as a fast solver of the linearized problems, apply multigrid directly to the non-linear problem (Sec. 8.3). Instead of solving an eigenproblem by the inverse power method, with multigrid as the fast inverter, you can multigrid directly the original eigenvalue problem (Sec. 8.3.1). Instead of using multigrid for solving each step in some outer iterative process — be it a continuation process, a time-dependent evolution, a process of optimizing some parameters or solving an inverse problem, etc. — apply it directly to the originally given problem (cf. Secs. 8.3.2, 15, 16). Instead of a grid adaptation process where the discrete problem on each grid configuration is completely solved (by multigrid, say) and then used to decide on an improved grid configuration, the whole adaptation process can be integrated into a multigrid solver (Sec. 9.6). And so on.

An illustration to this approach is the solution of optimization problems, where the parameter to be optimized is some continuum function on which the solution u depends. This "parametric function" may for example be the shape of the boundary (e.g., the shape of an airplane section which we want to optimize in some sense), or a certain coefficient of the differential equations (e.g., in inverse problems, where one tries to determine this coefficient throughout the domain so that the solution will best fit some observational data), etc. Multigridding the original problem means that we solve it by some FMG algorithm, where already at the coarser FMG stages we treat the given optimization problem, by optimizing a coarser representation of the parametric function. On the finer grids, incidentally to relaxing the equations, we optimize that function locally (when this makes sense), and then we introduce smooth corrections to the function during the coarse-grid correction cycles. Instead of using the multigrid solver many times, we may end up doing work only modestly larger than just one application of that solver.

13.1 Reverse trend: Algebraic multigrid

Contrary to the above line of dealgebraization, there is a recent trend to develop purely Algebraic Multi-Grid (AMG) algorithms. By this we mean a multi-level algorithm without any geometry, without grids. An algebraic (linear or nonlinear) system of equations is given. To solve it fast, a sequence of increasingly "coarser" levels is created. A coarser level in this context is a related, but much smaller, algebraic system. The choice of the coarse-grid variables, and of the coarse-to-fine interpolation, is based not on geometric positions but on the algebraic equations themselves: They should be chosen so that each fine-grid variable strongly depends on one or more coarse-grid variables. The relaxation scheme is also guided by the algebraic connections, e.g. relaxing simultaneously every block of unknowns strongly coupled to each other but weakly coupled to all other unknowns.

Generally the efficiency that can be achieved by such algebraic algorithms is below that of algorithms built to exploit the geometric information, let alone the further efficiency obtainable by further de-algebraization. On the other hand these algebraic solvers may be used as black boxes for larger classes of problems. They may be especially useful in cases where the geometrical information is too complicated, such as finite-element equations based on arbitrary partitions, or various problems which are not differential in their origin but still lend themselves for fast multi-level solutions. Also, examples can be constructed, artificially perhaps, of finite-difference equations on a uniform grid, in which the usual geometric choice of coarse-grid variables is not good, since many fine-grid variables happen to depend too weakly on the coarse-grid variables (cf. Sec. 3.3). Algebraic multigrid may then perform better. Initial experiments with symmetric diagonally dominant systems gave excellent asymptotic convergence factors per relaxation sweep, better even than in geometric multigrid for Poisson equations, but with more expensive inter-grid processes. (The work is done in collaboration with Steve McCormick and John Ruge.)

14. PRACTICAL ROLE OF RIGOROUS ANALYSIS

A good deal of the literature on multigrid consists of articles with <u>rigorous</u> analyses of the algebraic convergence. For a growing class of problems the basic multigrid assertion is rigorously proven, namely, that an FMG algorithm will solve the algebraic system of n equations (n unknowns on the finest grid) to the level of truncation errors in less than Cn computer operations; or at least, that $Cn \log \frac{1}{\varepsilon}$ operations are enough to reduce the L^2 norm of the error by any desired factor ε. The emphasis is on C being independent of n; it may depend on various parameters of the given differential problem. This is clearly the best one can do in terms of the order of dependence on n, hence the result is very satisfying.

The question discussed below is what role such rigorous analyses can have in the practical development of multigrid techniques and programs. It is an important question for the practitioner, who may wonder how much of those proofs he should try to understand.

The main shortcoming of the rigorous results is that they are usually unrealistic in terms of the size of C. In most of the proofs C is not even determined. This does not change the important fact that the best the proof could do in terms of C is very unrealistic: In most cases the provable constant is many orders of magnitudes larger than the one obtainable in practice. In typical cases the rigorous bound is $C \approx 10^8$, while the practical one is $C \approx 10^2$. Only in the very simplest situation (equations with constant coefficients in a rectangle) one can obtain realistic values of C, by Fourier methods [F5], [B7, App. C], [S4, § 8]. Recent analyses with reasonable (although still several times larger than the practical) values of C for the five-point Poisson equation on a general convex domain is given in [B4], [V1].

What can then be the practical value of the Cn results, especially those where C is unreasonably large? Usually in complexity analyses results with undetermined constants are sought in cases where the size of the constants is indeed less important. A typical result would for example be that a solution to some problem, depending on some parameter n, is obtained in $C\,n!$ operations. Here C may be unimportant, since changing C by orders of magnitude will only slightly increase the range of n for which the problem is solvable. But this spirit of undetermined constants is clearly pushed way too far when the estimate is Cn, the typical constant is $C = 10^8$ and the typical value for n is 10^3 to 10^5. Here C becomes more important than n. In the practical range of n, the provable Cn result is vastly inferior to much simpler results obtained by simpler algorithms (such as banded elimination with typically $4n^2$ operations; not to mention drastic improvements obtainable by modern sparse-matrix packages [D3]). Thus, the values of n for which the unrealistic rigorous result can compete with much simpler solution methods is very far out in the range of overkilling the problem. In a sense, one proves efficiency of an algebraic solution process by taking an extremely unreasonable algebraic problem. The efficiency is not in terms of solving the given <u>differential</u> problem.

The existing rigorous theory, being too concerned on making C independent of n, is often careless about its dependence on various problem parameters. This dependence can be hair-raising indeed, something like exp(exp(...)), with as many compounded exponentials as there are stages in the proof. Hence, a very distorted picture is in fact supplied about the real complexity in solving the given differential problem.

The implied intention of "Cn" theorems with unspecified or unrealistic C is sometimes understood as follows: The rigorous analysis only tells us that a constant C <u>exists</u>, and its actual value we can then determine empirically. That is, if we have calculated with $n = 10^3$ and solved the problem in 10^5 operations, say, then the rigorous proof guarantees that for $n = 10^4$ we would solve the problem in 10^6 operations. This understanding is wrong: The nature of the rigorous proofs is such that the information for $n = 10^3$ does not help the estimates for $n = 10^4$. The only rigorous estimate is still $C10^4$ operations, with the same unrealistic C. The guess that the number of operations for $n = 10^4$ will be 10^6 is purely non-rigorous. Even <u>heuristically</u> it does not follow in any way from the "Cn" theorem. Nothing in that theorem excludes, even heuristically, an operation count such as $Cn/(1 + 10^4 C/n^2)$, for example, with an astronomically large C.

Thus, if one literally believed these rigorous bounds, one would not use the multigrid method in the first place. This indeed historically happened: The estimates in [F1] are so bad (although only the simplest problem is considered; cf. [B7, §10]), and those of [B1] so much worse (even though his constants are undetermined), that nobody was encouraged to use such methods. They were considered to be merely of asymptotic curiosity.

Several other cases from the multigrid history are known where wrong practical conclusions were derived from the asymptotic rigorous analysis. For example, non-smooth boundaries, reentrant corners in particular, gave troubles in the rigorous proofs. This led to the wrong conclusion that there are real troubles there. The practical fact is that such problems are solved to within truncation errors as easily as regular problems [O1, §IV-9], [S4, Table 10.2]. Even the asymptotic algebraic convergence rate in such cases can be made to attain the interior rates. The difficulties are purely difficulties of the proof, not of the computational process. The proof made us too pessimistic. In other cases similar proofs made people too optimistic, because their asymptotic relations did not show the real difficulties encountered in the real range. Some people did not realize, for example, the very real difficulties in solving degenerate and singular perturbation equations (in particular indefinite problems such as $\varepsilon\Delta + k^2 u = f$, where ε is positive but small), because these difficulties disappear for sufficiently small meshsizes. But such meshsizes are far too small to be used in practice. (The terrible growth of C as function of ε is not seen if all we are interested in is that C will not depend on n.) Fedorenko had a completely wrong idea about the practical meshsize ratios and the number of grids to be used. He writes: "The proposed method thus consists of a solution with the aid of an auxiliary net; if this latter is extremely large, the problem can also be solved on it by using a net of a particular type for the problem, and so on". And several similar historical examples could be given.

It is indeed not reasonable to expect unrealistic performance estimates to be of practical value. In practice we are interested in understanding the difference between one algorithm which solves the problem in few minutes CPU time and another algorithm which solves it in a few more minutes, or in hours. A rigorous result that tells us that the solution will surely be obtained within a few weeks (even years) of CPU time, cannot explain that difference. The factors important in the proof may only remotely and non-quantitatively be related to those operating in practice. Even in cases of much more reasonable C (such as [B4]), the relative values of C in two competing approaches (e.g., V cycles vs. W cycles) does not point to their relative efficiency in practice. The rigorous proof tells us more about the efficiency of the proof than about that of the actual algorithm. Hence, for all its pure-mathematical interest and intellectual challenge, the rigorous approach (to non-model problems) is not a practical tool. This is the present state of the art, at least. The main role of the rigorous analysis (at least in the few cases where more realistic constants are given) is generally to enhance our confidence in the method, a psychological role that should not be slighted.

The purpose of the analysis should be borne in mind. We are not trying here to prove any central mathematical idea. We are engaged in a very practical problem, namely, how to solve the equations fast. This is in its nature as practical a problem as, say, building an airplane or understanding nuclear fission. (In fact the only purpose of the fast solvers is to aid solving such engineering and scientific problems.) One would not postpone building airplanes until rigorous proofs of their flight capabilities are furnished. Clinging to rigorous mathematics, like clinging to any secure images, may have wrong contexts. Moreover, in this business of fast solvers what one tries to apriori estimate is nothing but the computer time (and other computer resources), which is after all exactly known in each particular case, even though aposteriori. The main practical aims of theoretical understanding should therefore be:

(i) To give us realistic and quantitative insights to the important factors affecting the overall efficiency. The insight should be simple enough and still precise enough so that one can use it to improve our algorithms, and perhaps even to debug our programs. The insight should of course be independent of the numerical experiments.

(ii) Even more important than quantitative performance prediction, one urgently needs to know whether the performance (predicted or found empirically) is as good as one could hope to get (see the situation described in Sec. 1.1). Hence the main theoretical task is to provide us with ideal performance figures, which the practical algorithm should then attempt to approach.

The local mode analysis is an example of a theory constructed with these aims in mind. This is amply emphasized throughout Part I of the present paper. At any rate, it is strongly recommended not to restrict oneself to numerical experiments only, without any supporting theory. The experiments can be, and have been, quite misleading: They may happen to show, for some particular cases, much better results than should generally be expected. More often, they show results much inferior to those that could be obtained, because of some conceptual mistakes and/or programming bugs. Experience has taught us that careful incorporation of (usually non-rigorous) theoretical studies is necessary for producing reliable programs which fully realize the potential of the multigrid method.

15. CHAINS OF PROBLEMS. FROZEN τ

We often need to solve not just one isolated problem but a sequence of similar problems depending on some parameter. For example, we may be studying the effect of changing some physical parameters on the "performance" of a system, where the performance is measured in terms of the solution u to a differential problem. We may want to find for what physical parameters the performance is optimal. Or, in "inverse problems", we may desire to find the physical parameters for which the solution best fits some physically observed behavior. Or we may need to solve a sequence of problems in a continuation process (see Sec. 8.3.2). Or, the most familiar case, the parameter may be the time t, and each problem in the sequence may represent the implicit equations of one time step.

The key to a highly efficient multi-level treatment of such a sequence of problems is to understand the behavior of high-frequency components. Most often, the change in one step (i.e., the change from one problem in the sequence to the next) is a global change, governed by global parameters. In some problems, the relative changes in high-frequency components are therefore comparable to the relative changes in low ones. Hence, for such problems, the absolute high-frequency changes in each step are negligible — they are small compared to the high frequencies themselves, and therefore small compared with the discretization errors. In such cases one need not use the finer grids at each step; the finer the level the more rarely it should be activated. Often this is the situation in most parts of the domain, but in some particular parts, such as near boundaries, significant high-frequency changes do take place in every step, hence more refinement levels should more often be activated in those parts only.

The Full Approximation Scheme (FAS) gives us a convenient structure in which to see smooth changes in the solution without (locally) activating finer grids. The way to neglect changes in wavelengths smaller than $O(h)$, without neglecting those components themselves, is to freeze τ_h^{2h} (see Sec. 8.2), i.e., to use on grid $2h$ the values of τ_h^{2h} calculated in a previous step and thus in the present step avoid any visit to grid h. Once in several steps visiting grid $2h$ a visit can be made to grid h, to update τ_h^{2h}. In visiting grid $2h$ changes in τ_{2h}^{4h} are made, their cumulative values since the last visit to grid h can serve to decide when a new visit to grid h is needed, using exchange-rate criteria (see Secs. 9.5-6 and [B11, §3.9]). Since these criteria can be applied locally, one can decide when and where to activate increasingly finer levels.

An obvious but important remark: Whether the above procedures are used or not, and whether FAS or CS are employed, in each step (i.e., for each problem in the chain) it is normally more economic to work on the correction problem, taking the previous-step solution as a first approximation. When FAS-FMG is used, this is easily done, even for non-linear problems, as follows. First, the old values of τ_h^{2h} should be used in the 2h-stage of the FMG algorithm (i.e., before grid h is ever visited in the present step). Secondly, the FMG interpolation (first interpolation to grid h in this step) should be a FAS-like interpolation, using the old values of $\tilde{u}^h$; i.e., like Eq. (8.6), but with possibly higher order Π_H^h replacing I_H^h (cf. Sec. 7.1).

Solving the chain of problems we usually need to monitor certain solution functionals. In order to calculate such a functional Φ with finest-grid accuracy even at steps not visiting the finest grid, transfers as in Eq. (8.14) can be used.

16. TIME DEPENDENT PROBLEMS

The experience with multigrid applications to evolution problems is quite sparse, but several possibilities can already be outlined, such as fast solvers to implicit equations, coarse-grid time steps, highly adaptible structures and high-order techniques.

One obvious application is to use fast multigrid solvers for solving the set of algebraic equations arising at each time step when implicit time differencing is employed. Such differencing is normally needed whenever the physical signal speed is considerably higher than the speed of substantial changes in the solution. The latter speed determines the size of time steps δt we need to approximate the solution accurately, but with such δt and explicit differencing the numerical signal speed will be slower than the physical one, causing numerical instability. Using implicit equations and solving them by multigrid can be viewed as a way to inexpensively obtain high signal speeds by propagating information on coarse grids. Indeed, with multigrid solvers the cost of an implicit time step is comparable to that of an explicit one.

In many cases one can even do much better using techniques as in Sec. 15 above. For second-order parabolic problems, for example, significant changes in high-frequency components, whose wavelength is $O(h)$, occur only in very particular places such as

(i) initially, for a short time interval $0 \leq t \leq O(h^2)$;

(ii) at distance h and time interval $O(h^2)$ from points where significant changes occur in boundary conditions or in forcing terms (source terms) of the equation.

At all other places significant high-frequency changes are induced by comparably significant low-frequency changes. Hence the frozen-τ technique, with a special control for time-dependent problems [B11, §3.9], can give us a solution with the fine-grid accuracy but where most of the time in most of the domain we use coarse grids only. The cost of an average step may then be far smaller than the cost of an explicit time-step. For the heat equation in the infinite space and steady forcing terms, for example, one can show by Fourier analysis that marching from initial state to 90% steady-state, following the solution throughout with close to finest-grid accuracy, can in this way cost computational work equivalent to just 10 explicit time steps. (The finest grid needs to be activated only in the first few time steps, and very rarely later.)

Notice that when we march (calculating smooth changes in the solution) on coarse grids we can also use large time steps with explicit differencing. When fully adapted grids are used there is no need for implicit differencing, because each range of components is in effect handled by a meshsize comparable to the wavelength and by a time-step corresponding to the propagation speed, so that no conflict arises between different characteristic speeds.

The multi-level techniques can also be applied to a parabolic part of the system, such as the implicit pressure equation in integrating Navier-Stokes equations [B23]. Here too, the techniques of Sec. 15 can further save a lot of fine-grid processing.

Whether the fast solver is used or not, the multi-level procedures can also give highly flexible discretization structures. Patches of finer grids with correspondingly finer time steps can be used in any part of the space-time domain, in a manner similar to Sec. 9.1. Anisotropic refinements, local coordinate transformations and rotated cartesian grids can be used as in Secs. 9.2, 9.3 and 9.4, all controlled by exchange-rate criteria (cf. Secs. 9.5, 9.6; but instead of criteria based on the τ_h^H of the approximate solution, criteria here will be based on recently accumulated changes in τ_h^H [B11, §3.9]).

Finally, independently of the above techniques, one could use a multigrid procedure similar to Sec. 10.2 above to efficiently increase the approximation order of a stable discretization

$$L_0^h u_0^h(x,t) = 0, \quad (t > 0) \tag{16.1}$$

of a time-dependent system, not necessarily linear, by using coarse-grid defect corrections. Typically L_0^h is a simple low-order implicit operator, allowing simple integration. One wants to use a simple (e.g., central in time) higher-order operator L_1^h, which may be unstable, to raise the approximation order. This can be done by integrating the defect equation

$$L_0^H v^H = -I_h^H L_1^h u_0^h, \tag{16.2}$$

and then correcting

$$u_1^h = u_0^h + I_H^h v^H , \tag{16.3}$$

where H may either be coarser than h (coarse-grid defect correction) or $H = h$ (defect correction on the same grid). If the order of consistency of L_j^h is p_j, $(j=0,1)$, then for low-frequency components

$$|u_1^h - u| = O(h^{p_1} t + h^{p_0} H^{p_0} t^2) , \tag{16.4}$$

where u is the differential solution and t is the time interval over which (16.1-3) is integrated from initial conditions $u_1^h(x,0) = u_0^h(x,0) = u(x,0)$.

The scheme (16.1-3) is always stable, since only the stable operators L_0^h and L_0^H are integrated. Notice that this is true only if u_0^h is integrated independently of u_1^h. The seemingly similar scheme, in which after each time step u_0^h is reinitialized by being replaced with the more accurate u_1^h, may well be unstable. On the other hand it does pay to reinitialize every $O(1)$ time interval, short enough to make the second term in (16.4) smaller than the first one.

The independent integration of u_0^h and v_0^h requires extra storage. By taking $H = 2h$ this extra storage becomes only a fraction of the basic storage (one time level of u_0^h), and the computational work is also just a fraction more than the work of integrating (16.1). These two-level schemes can be extended to more levels and more approximation orders. The multigrid exploits here the fact that the higher-order approximation L_1^h is desired only for sufficiently low frequencies; for the highest frequencies (where numerical instability occurs) L_0^h is in fact a better approximation than L_1^h.

REFERENCES

[A1] R.E. Alcouffe, A. Brandt, J.E. Dendy, Jr., and J.W. Painter: The multi-grid methods for the diffusion equation with strongly discontinuous coefficients. *SIAM J. Sci. Stat. Comput.* 2 (1981), 430-454.

[A2] W. Auzinger and H.J. Stetter: Defect corrections and multigrid iterations. This *Proceedings*.

[B1] N.S. Bakhvalov (Bahvalov): Convergence of a relaxation method with natural constraints on an elliptic operator. *Ž. Vyčisl. Mat. i Mat. Fiz.*, 6 (1966), 861-885. (Russian)

[B2] R.E. Bank: Multi-level Iterative Method for Nonlinear Elliptic Equations. *Elliptic Problem Solvers* (M. Schultz, ed.), Academic Press, New York 1981, 1-16.

[B3] C. Börgers: Mehrgitterverfahren für eine Mehrstellendiskretisierung der Poisson-Gleichung und für eine zweidimensionale singulär gestörte Aufgabe. Diplomarbeit, Universität Bonn, Bonn, 1981.

[B4] D. Braess: The convergence rate of a multigrid method with Gauss-Seidel relaxation for the Poisson equation. This *Proceedings*.

[B5] A. Brandt: Multi-level adaptive technique (MLAT) for fast numerical solution to boundary value problems. *Proc. 3rd Internat. Conf. on Numerical Methods in Fluid Mechanics* (Paris, 1972), Lecture Notes in Physics, 18, Springer-Verlag, Berlin and New York, 1973, 82-89.

[B6] A. Brandt: Multi-level adaptive techniques. IBM Research Report RC6026, 1976.

[B7] A. Brandt: Multi-level adaptive solutions to boundary-value problems. *Math. Comp.* 31 (1977), 333-390. ICASE Report 76-27.

[B8] A. Brandt: Multi-level adaptive techniques (MLAT) for partial differential equations: Ideas and software. *Mathematical Software* III (John R. Rice, ed.), Academic Press, New York 1977, 273-314. ICASE Report 77-20.

[B9] A. Brandt et al.: Lecture notes of the ICASE Workshop on Multigrid Methods. ICASE, NASA Langley Research Center, Hampton, VA., 1978.

[B10] A. Brandt: Multi-level adaptive solutions to singular-perturbation problems. *Numerical Analysis of Singular Perturbation Problems* (P.W. Hemker and J.J.H. Miller, eds.), Academic Press 1979, 53-142. ICASE Report 78-18.

[B11] A. Brandt: Multi-level adaptive finite-elements methods: I. Variational problems. *Special Topics of Applied Mathematics* (J. Frehse, D. Pallaschke, U. Trottenberg, eds.), North Holland 1980, 91-128. ICASE Report 79-8.

[B12] A. Brandt: Numerical stability and fast solutions of boundary-value problems. *Boundary and Interior Layers - Computational and Asymptotic Methods* (J.J.H. Miller, ed.), Boole Press, Dublin 1980, 29-49.

[B13] A. Brandt: Multi-level adaptive computations in fluid dynamics. *AIAA J.* 18 (1980), 1165-1172.

[B14] A. Brandt: Multigrid Solvers on Parallel Computers. *Elliptic Problem Solvers* (M. Schultz, ed.), Academic Press, New York 1981, 39-84.

[B15] A. Brandt: Stages in developing multigrid solutions. *Numerical Methods for Engineering* (E. Absi, R. Glowinski, P. Lascaux, H. Veysseyre, eds.), Dunod, Paris 1980, 23-44.

[B16] A. Brandt: Multigrid solutions to steady-state compressible Navier-Stokes equations. *Proc. Fifth International Symposium on Computing Methods in Applied Sciences and Engineering*, Versailles, France, December 14-18, 1981.

[B17] A. Brandt: Multigrid solvers for non-elliptic and singular-perturbation steady-state problems. Weizmann Institute of Science, Rehovot, Israel, 1981. *Computers and Fluids*, to appear.

[B18] A. Brandt and C.W. Cryer: Multigrid algorithms for the solution of linear complementary problems arising from free boundary problems. MRC Technical Summary Report #2131, University of Wisconsin, Madison, Oct. 1980. *SIAM J. Sci. Stat. Comp.*, to appear.

[B19] A. Brandt and N. Dinar: Multigrid solutions to elliptic flow problems. *Numerical Methods for Partial Differential Equations* (S. Parter, ed.), Academic Press 1979, 53-147. ICASE Report 79-15.

[B20] A. Brandt, S. McCormick and J. Ruge: Multigrid algorithms for differential eigenproblems. Submitted to *SIAM J. Sci. Stat. Comp.*, September 1981.

[B21] A. Brandt and S. Ta'asan: Multigrid methods for highly oscillatory problems. Research Report, September 1981.

[B22] A. Brandt, J.E. Dendy, Jr. and H. Ruppel: The multi-grid method for the pressure iteration in Eulerian and Lagrangian hydrodynamics. *J. Comp. Phys.* 34 (1980), 348-370.

[B23] J.J. Brown: A multigrid mesh-embedding technique for three-dimensional transonic potential flow analysis. Boeing Commercial Airplane Company, Seattle, Washington, April 1981.

[C1] M. Ciment, S.H. Leventhal and B.C. Weinberg: The operator compact implicit method for parabolic equations. *J. Comp. Phys.* 28 (1978), 135-166.

[D1] J.E. Dendy, Jr.: Black box multigrid. LA-UR-81-2337 Los Alamos National Laboratory, Los Alamos, New Mexico. *J. Comp. Phys.*, to appear.

[D2] N. Dinar: Fast methods for the numerical solution of boundary-value problems. Ph.D. Thesis, The Weizmann Institute of Science, Rehovot, Israel (1979).

[D3] I.S. Duff: Sparse matrix software for elliptic PDE's. This *Proceedings*.

[F1] R.P. Fedorenko: On the speed of convergence of an iteration process. *Ž. Vyčisl. Mat. i Mat. Fiz.* 4 (1964), 559-564. (Russian)

[F2] Ha. Foerster, K. Stueben and U. Trottenberg: Non-standard multigrid techniques using checkered relaxation and intermediate grids. *Elliptic Problem Solvers* (M. Schultz, ed.), Academic Press, New York, 1981, 285-300.

[F3] Ha. Foerster and K. Witsch: On efficient multigrid software for elliptic problems on rectangular domains. Preprint No. 458, SFB 72, Universität Bonn, Bonn, 1981.

[F4] G.E. Forsythe and W.R. Wasow: *Finite-Difference Methods for Partial Differential Equations*. Wiley, New York 1960.

[F5] P.O. Frederickson: Fast approximate inversion of large sparse linear systems. Math. Report 7-75, Lakehead University, Ontario, Canada, 1975.

[H1] W. Hackbusch: The fast numerical solution of very large elliptic differential schemes. *J. Inst. Maths. Applics.* 26 (1980), 119-132.

[H2] W. Hackbusch: Multi-grid convergence theory. This *Proceedings*.

[H3] J.M. Hyman: Mesh refinement and local inversion of elliptic partial differential equations. *J. Comp. Phys.* 23 (1977), 124-134.

[K1] H.B. Keller: Numerical solution of bifurcation and nonlinear eigenvalue problems. *Applications of Bifurcation Theory* (P. Rabinowitz, ed.), Academic Press, New York, 1977, 359-384.

[K2] R. Kettler: Analysis and comparison of relaxation schemes in robust multigrid and preconditioned conjugate gradient methods. This *Proceedings*.

[L1] B. Lindberg: Error estimation and iterative improvement for the numerical solution of operator equations. Report UIUCDCS-R-76-820, University of

Illinois, Urbana, Illinois, 1976.

[L2] J. Linden: Mehrgitterverfahren für die Poisson-Gleichung in Kreis und Ringgebiet unter Verwendung lokaler Koordinaten. Diplomarbeit, Bonn 1981.

[L3] R.E. Lynch and J.R. Rice: The Hodie method and its performance. *Recent Advances in Numerical Analysis* (C. de Boor, ed.), Academic Press, New York, 1978, 143-179.

[M1] J. Meinguet: Multivariate interpolation at arbitrary points made simple. *J. Applied Math. Phys.* (ZAMP) 30 (1979), 292-304.

[M2] P. Meissle: Apriori prediction of roundoff error accumulation in the solution of a super-large geodetic normal equation system. NOAA Professional Paper 12, National Oceanic and Atmospheric Administration, Rockville, Maryland, 1980.

[M3] MUGTAPE 82, A tape of multigrid software and programs, including GRIDPACK; MUGPACK; simple model programs (CYCLE C, FASCC, FMG1 and an eigenproblem solver); Stokes equations solver; SMORATE; BOXMG [D1]; MGOO and MGO1 [S4]. Available at the Department of Applied Mathematics, Weizmann Institute of Science, Rehovot, Israel, and at the GMD-IMA, Postfach 1240, Schloss Birlinghoven, D-5205, West Germany.

[M4] D.R. McCarthy, and T.A. Reyhner: A multi-grid code for three-dimensional transonic potential flow about axisymmetric inlets at angle of attack. AIAA Paper 80-1365, *Proceedings of AIAA 13th Fluid and Plasma Dynamics Conference*, Snowmass, Colo., July 1980.

[N1] K.A. Narayanan, D.P. O'Leary and A. Rosenfeld: Multi-resolution relaxation. TR-1070 MCS-79-23422, University of Maryland, College Park, Maryland, July 1981.

[N2] R.A. Nicolaides: On multiple grid and related techniques for solving discrete elliptic systems. *J. Comp. Phys.* 19 (1975), 418-431.

[N3] R.A. Nicolaides: On the ℓ^2 convergence of an algorithm for solving finite element equations. *Math. Comp.* 31 (1977), 892-906.

[O1] D. Ophir: Language for processes of numerical solutions to differential equations. Ph.D. Thesis (1979), The Weizmann Institute of Science, Rehovot, Israel.

[R1] M. Ries, U. Trottenberg and G. Winter: A note on MGR methods. Preprint 461, Universität Bonn, SFB 72, 1981. *Lin. Alg. and Appl.*, to appear.

[S1] S. Schaeffer: High-order multigrid methods. Ph.D. Thesis, Colorado State Univ., Fort Collins, Colorado, 1982.

[S2] J.C. South and A. Brandt: Application of multi-level grid method to transonic flow calculations. ICASE Report No. 76-8, NASA Langley Research Center, Hampton, Virginia, 1976.

[S3] H.J. Stetter: The defect correction principle and discretization methods. *Num. Math.* 29 (1978), 425-443.

[S4] K. Stueben and U. Trottenberg: Multigrid methods: fundamental algorithms, model problem analysis and applications. This *Proceedings*.

[T1] S.L. Tanimoto: Template matching in pyramids. *Computer Graphics and Image Processing* 16 (1981), 356-369.

[T2] C.A. Thole: Beiträge zur Fourieranalyse von Mehrgittermethoden: V-Cycle, ILU-Glättung, anisotrope Operatoren. Diplomarbeit, Universität Bonn. Bonn, to appear

[V1] R. Verfürth: The contraction number of a multigrid method with ratio 2 for solving Poisson's equation. Ruhr-Universität Bochum, West Germany, 1981

[W1] P. Wesseling: A robust and efficient multigrid method. This *Proceedings*.

THE MULTI GRID METHOD AND ARTIFICIAL VISCOSITY

E.J. van Asselt
Mathematical Centre
Amsterdam
The Netherlands

0. INTRODUCTION

We consider the convection diffusion equation in two dimensions:

$$(0.1) \qquad L_\varepsilon u = f,$$

with

$$L_\varepsilon = \varepsilon\Delta + b_1 \frac{\partial}{\partial x} + b_2 \frac{\partial}{\partial y},$$

$$b_1^2 + b_2^2 = 1, \quad u : \mathbb{R}^2 \to \mathbb{R}.$$

A well-known method to avoid unstable discretizations of (0.1) when the diffusion coefficient ε is small in comparison with the meshwidth h is the addition of artificial viscosity to ε.

Let $\beta = C_1 h$ be the amount of artificial viscosity added to ε on a fine grid, with mesh size $\bar{h} = (h,h)$, and $\bar{\beta}$ be the artificial viscosity added to ε on a corresponding coarse grid with mesh size $\bar{H} = (H,H) = (2h,2h)$ in a two level algorithm (TLA).

In this paper by local mode analysis we analyse two different choices of $\bar{\beta}$: $\bar{\beta} = \beta = C_1 h$ and $\bar{\beta} = 2\beta = C_1 H$. For this purpose in section 1 we give the definitions of amplification and convergence factors.

In section 2 we show that in the TLA, $\bar{\beta} = \beta$ gives a smaller convergence factor than $\bar{\beta} = 2\beta$. Further it is proved that the choice $\bar{\beta} = \beta$ corresponds with the Galerkin Approximation for the coarse grid operator up to order h^2.

In section 3, three variants of the choice of artificial viscosity on the coarse grids in a multi level algorithm (MLA) are examined.

1. THE AMPLIFICATION FACTOR AND THE CONVERGENCE FACTOR IN THE TWO LEVEL ALGORITHM AND THE COARSE GRID CORRECTION

In this section we give definitions of the amplification factor and the convergence factor of the TLA and the coarse grid correction (CGC).
Consider the linear partial differential equation:

$$(1.1) \qquad Lu(x) = f(x),\ x = (x_1,x_2) \in \mathbb{R}^2,\ u : \mathbb{R}^2 \to \mathbb{R}.$$

Let G_h and G_H be uniform grids with mesh size $\bar{h} = (h,h) \in \mathbb{R}^2$ and $\bar{H} = (H,H) = (2h,2h) \in \mathbb{R}^2$:

$$G_h = \{(jh,kh) \mid j,k \in \mathbb{Z}\},$$

$$G_H = \{(jH,kH) \mid j,k \in \mathbb{Z},\ H = 2h\}.$$

Let GF_h and GF_H be the spaces of grid functions:

(1.2) $$GF_h = \{u_h \mid u_h : G_h \to \mathbb{R}\}$$

$$GF_H = \{u_H \mid u_H : G_H \to \mathbb{R}\},$$

provided with norm

$$\|u_h\|_h = \sup_{j,k\in\mathbb{Z}} |u_h(jh,kh)|,$$

and

$$\|u_H\|_H = \sup_{j,k\in\mathbb{Z}} |u_H(jH,kH)|$$

respectively.

Let L_h and L_H be discretizations of L on G_h and G_H:

$$L_h u_h = f_h;\ L_H u_H = f_H,$$

with $u_h, f_h \in GF_h$ and $u_H, f_H \in GF_H$.

The amplification matrix M of one cycle of the TLA is given by:

(1.3) $$M = S^q C S^p,$$

where the number of pre- and post-relaxations is p and q respectively; S denotes the amplification matrix of the smoothing process; and C of the CGC.

With prolongation $P : GF_H \to GF_h$ and restriction $R : GF_h \to GF_H$, we have

(1.4) $$C = I - PL_H^{-1}RL_h.$$

In order to express the rate of convergence of the TLA in terms of local mode analysis, we use the following notations:

$$\hat{u}_h(\omega) = \frac{h^2}{2\pi} \sum_{j\in\mathbb{Z}^2} e^{-ijh\omega} u_h(jh),\ \omega \in LF_h \cup HF_h;$$

$$LF_h = \{(\omega_1,\omega_2) \mid \omega_1,\omega_2 \in [-\tfrac{\pi}{2h}, \tfrac{\pi}{2h}]\},$$ the range of low frequencies;

$$HF_h = \{(\omega_1,\omega_2) \mid \omega_1,\omega_2 \in [-\tfrac{\pi}{h}, \tfrac{\pi}{h}],\ (\omega_1,\omega_2) \notin LF_h\},$$ the range of high frequencies.

<u>REMARK</u>.

$\hat{u}_h : LF_h \cup HF_h \to \mathbb{C}$ is the Fourier transform of u_h.

The backtransformation formula reads

$$u_h(jh) = \frac{1}{2\pi} \int_{\omega\in LF_h \cup HF_h} e^{ijh\omega}\hat{u}_h(\omega)d\omega$$ (cf. HEMKER [4]).

$\hat{u}_h(\omega)$ is called the amplitude, and $e^{ijh\omega}$ the mode of frequency ω.

Let $\omega^{(1)} = (\omega_1^{(1)}, \omega_2^{(1)}) \in LF_h$, then we define its *harmonics* by

$$\omega^{(2)} = (\omega_1^{(1)}, \omega_2^{(1)} \pm \frac{\pi}{h}),$$
$$\omega^{(3)} = (\omega_1^{(1)} \pm \frac{\pi}{h}, \omega_2^{(1)}),$$
$$\omega^{(4)} = (\omega_1^{(1)} \pm \frac{\pi}{h}, \omega_2^{(1)} \pm \frac{\pi}{h}),$$

where the + or - sign are chosen such that $\omega^{(k)} \in HF_h$, k = 2,3,4. (see figure 1).

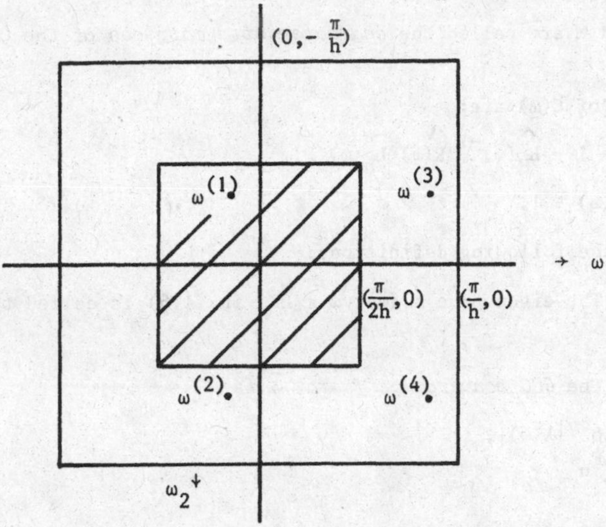

figure 1. $\omega^{(1)}$ and its harmonics.
LF_h is the shaded area.

Suppose that R and P are invariant for translations, then the four frequencies $\omega^{(k)}$, k = 1,2,3,4 are coupled by R and P.
For each $\omega^{(1)} \in LF_h$, we denote $\hat{u}_h(\omega^{(k)})$, k = 1,2,3,4 in vector notation simply by $\hat{\bar{u}}_h(\omega^{(1)})$.
For all $\omega \in LF_h$ we can define a matrix $\hat{R}(\omega): \mathbb{R}^4 \to \mathbb{R}$ such that

$$\widehat{Ru_h}(\omega) = \hat{R}(\omega)\hat{\bar{u}}_h(\omega).$$

Similarly, for the prolongation we can define a $\widehat{P(\omega)}: \mathbb{R} \to \mathbb{R}^4$ such that

$$\widehat{Pu_h}(\omega) = \hat{P}(\omega)\hat{\bar{u}}_h(\omega).$$

For all $\omega \in LF_h$ with harmonics $\omega^{(k)}$, k = 2,3,4 we can introduce a 4 × 4 matrix $\hat{M}(\omega): \mathbb{R}^4 \to \mathbb{R}^4$ which relates the error e_h before to the error Me_h after one cycle of the TLA.

This $\hat{M}(\omega)$ reads:

$$\hat{M}(\omega) = \tilde{S}(\omega)^q \, \hat{C}(\omega) \, \tilde{S}(\omega)^p,$$

with

$$\hat{C}(\omega) = I - \hat{P}(\omega)\ \widehat{L_H}(\omega)^{-1}\ \widehat{R(\omega)}\ \widetilde{L_h}(\omega),$$

$$\widetilde{L}_h(\omega) = \mathrm{diag}(\hat{L}_h(\omega),\ \hat{L}_h(\omega^{(2)}),\ \hat{L}_h(\omega^{(3)}),\ \hat{L}_h(\omega^{(4)}));$$

$$\widetilde{S}(\omega) = \mathrm{diag}(\hat{S}(\omega),\ \hat{S}(\omega^{(2)}),\ \hat{S}(\omega^{(3)}),\ \hat{S}(\omega^{(4)}));$$

where $\hat{L}_h$ and $\hat{S}$ are the characteristic forms (or symbols) of the operator L_h and the smoothing operator S.

The matrices $\hat{C}$ and $\hat{M}$ are called the *characteristic matrices* of the CGC and the TLA respectively.

The eigenvalues ρ of $\hat{C}(\omega)$ are:

(1.6) $$\rho_1(\omega) = 1 - \widehat{L_H}(\omega)^{-1}\ \widehat{R}(\omega)\ \widetilde{L_h}(\omega)\ \widehat{P}(\omega)$$

$$\rho_{2,3,4}(\omega) \equiv 1.$$

This leads us to the following definitions:

(1.7) DEFINITION. The eigenvalue $\rho_1(\omega)$, $\omega \in LF_h$ in (1.6) is called the *CGC amplification factor.*

(1.8) DEFINITION. The *CGC convergence factor* $\bar{\lambda}$ *is:*

$$\lambda = \sup_{\substack{\omega \in LF_h \\ \omega \neq 0}} |\lambda(\omega)|,$$

with $\lambda(\omega)$ the CGC amplification factor.

(1.9) DEFINITION. (cf. BRANDT, DINAR [1]). The *two level (TL) amplification factor* $\mu(\omega)$, $\omega \in LF_h$, is the eigenvalue of $\hat{M}(\omega)$ with largest modulus. $\bar{\mu} = \sup_{\substack{\omega \in LF_h \\ \omega \neq 0}} |\mu(\omega)|$ is called the *TL convergence factor.*

2. TWO LEVEL ANALYSIS OF THE CONVECTION DIFFUSION EQUATION

In section 2.1 we describe the addition of artificial viscosity to the diffusion coefficient ε when it is small in comparison with the meshwidth h.

In section 2.2 we express the CGC amplification factor in terms of the artificial viscosity on the fine ($\beta = C_1 h$) and on the coarse grid ($\bar{\beta}$).

In section 2.3 we show that the choice $\bar{\beta} = \beta$ gives a smaller CGC convergence factor than $\bar{\beta} = C_1 H$.

In section 2.4 by local mode analysis of a TLA with Symmetric Gauss Seidel (SGS) relaxation we obtain the same result for the TL convergence factor.

Finally in section 2.5 we show that the coarse grid discretization with $\bar{\beta} = \beta$ corresponds with the Galerkin Approximation of $L_{\varepsilon+\beta,h}$ up to terms of order h^2.

2.1. The convection diffusion equation

We study the convection diffusion equation (0.1) in two dimensions. Stability of the discretization is considered in the following sense:

(2.1.1) DEFINITION. Let Lu = f be a linear PDE with constant coefficients.

Let L_h be a discretization of L, with characteristic form $\hat{L}_h$.

The *stability of* L_h *with respect to the mode* $e^{ijh\omega}$ is the quantity $|\hat{L}_h(\omega)|$.

(2.1.2) DEFINITION. Let $L_{\varepsilon,h}$ be a discretization of (0.1) with characteristic form $\hat{L}_{\varepsilon,h}$.

The *asymptotic stability of* $L_{\varepsilon,h}$ *with respect to the mode* $e^{ijh\omega}$ is the quantity $\lim_{\varepsilon\downarrow 0} |\hat{L}_{\varepsilon,h}(\omega)|$.

Discretization of (0.1) by central differences gives the following scheme:

$$(2.1.3)\qquad (L_{\varepsilon,h}u_h)_{i,j} \equiv (\frac{\varepsilon}{h^2} - \frac{b_2}{2h})u^h_{i,j-1} + (\frac{\varepsilon}{h^2} + \frac{b_2}{2h})u^h_{i,j+1} +$$

$$+ (\frac{\varepsilon}{h^2} - \frac{b1}{2h})u^h_{i-1,j} + (\frac{\varepsilon}{h^2} + \frac{b1}{2h})u^h_{i+1,j} +$$

$$- \frac{4\varepsilon}{h^2}u^h_{i,j} = f^h_{i.j},$$

with $u_h = (\ldots, u^h_{i.j}, u^h_{i+1,j}, \ldots)$, $u^h_{i,j} = u(ih,jh)$, $f^h_{i,j} = f(ih,jh)$.

We consider $\varepsilon = O(h)$.

For all $\omega = (\omega_1,\omega_2)$ with $b_1 \sin\omega_1 h + b_2 \sin\omega_2 h = 0$ we find: $\lim_{\varepsilon\downarrow 0}|\hat{L}_{\varepsilon,h}(\omega)| = 0$. Hence the asymptotic stability of $L_{\varepsilon,h}$ with respect to the modes of these frequencies in zero.

The scheme is consistent of order 2, i.e. $\|J^1_h L_\varepsilon u - L_{\varepsilon,h} J^3_h u\|_h = O(h^2)$, with J^k_h the injection $C^k(\mathbb{R}^2) \to GF_h$, k = 1,3.

If we use artificial viscosity β for the discretization of (0.1), i.e. if we use $L_{\alpha,h}u_h = f_h$; $\alpha = \varepsilon + \beta = \varepsilon + h/2$, as a discretization of $L_\varepsilon u = f$, then this discretization has zero asymptotic stability: $\lim_{\varepsilon\downarrow 0} \hat{L}_{\alpha,h}(\omega) = 0$, only with respect to the mode of frequency $\omega = (\omega_1,\omega_2) = (0,0)$, and the consistency is of order 1.

2.2. The coarse grid correction amplification factor.

In this section we give an explicit expression for the CGC amplification factor.

For prolongation P we take linear interpolation and for restriction R we take transposed linear interpolation. (7 points restriction and prolongation, cf. HEMKER [4], WESSELING [6]).

The characteristic forms read:

$$\hat{P}(\omega) = \hat{R}(\omega) = \tfrac{1}{4}(1 + \cos\omega_1 h + \cos\omega_2 h + \cos(\omega_1 - \omega_2)h),$$

The characteristic form of $L_{\alpha,h}$ reads

$$\hat{L}_{\alpha,h}(\omega) = \frac{2\alpha}{h^2}(\cos\omega_1 h + \cos\omega_2 h) - \frac{4\alpha}{h^2}$$

$$+ \underline{i}\,\frac{1}{h}(b_1 \sin\omega_1 h + b_2 \sin\omega_2 h);$$

An analogous form exists for the coarse grid discretization $L_{\bar{\alpha},H}$.

Now we consider two choices for the amount of artificial viscosity $\bar{\beta}$ on the coarse grid:

$$\bar{\beta} = \beta = h/2, \qquad \text{i.e. } \bar{\alpha} = \alpha = \varepsilon + \beta = \varepsilon + h/2,$$

$$\bar{\beta} = H/2, \qquad \text{i.e. } \bar{\alpha} = \varepsilon + H/2 = \varepsilon + h.$$

We study the behaviour of the discretization in the limit for $\varepsilon \to 0$.

From (1.6) it follows that

$$|\lambda(\omega)| = |[\{p^2\bar{\alpha}(\alpha - \bar{\alpha}) - \frac{h^2}{2}qr(b_2 - b_1)\} + i\, hp\{\frac{\bar{\alpha}}{2} r(b_2 - b_1) + (\alpha - \bar{\alpha})q\}] / (p^2\bar{\alpha}^2 + h^2q^2)|,$$

with

$$p = S_1^{\,2} + S_2^{\,2},$$

$$q = b_1S_1C_1 + b_2S_2C_2,$$

$$r = S_1S_2S_{12},$$

and

$$S_i = \sin \omega_i h, \; C_i = \cos \omega_i h, \; i = 1,2;$$

$$S_{12} = \sin(\omega_1 - \omega_2)h, \; C_{12} = \cos(\omega_1 - \omega_2)h.$$

For the choice $\bar{\beta} = \beta = h/2$ in the limit for $\varepsilon \to 0$ we find:

(2.2.1) $$|\lambda_{\bar{\beta} = \beta}(\omega)| = |b_2 - b_1||r|/(p^2 + 4q^2)^{\frac{1}{2}}.$$

For the choice $\bar{\beta} = 2\beta = H/2$ in the limit for $\varepsilon \to 0$ we find

(2.2.2) $$|\lambda_{\bar{\beta} = 2\beta}(\omega)| = \tfrac{1}{2}((|b_1 - b_2|^2|r|^2 + p^2)/(p^2 + q^2))^{\frac{1}{2}}.$$

These two CGC amplification factors are compared with each other in the following section.

2.3. The choice of artificial viscosity on the coarse grid

Now we compare the two CGC amplification factors $\lambda_{\bar{\beta} = \beta}$ and $\lambda_{\bar{\beta} = 2\beta}$ for different values of the convection coefficients b_1 and b_2.

(2.3.1) LEMMA. *For all* b_1, b_2 *with* $b_1^{\,2} + b_2^{\,2} = 1$

a) $\lim_{\omega \to 0} |\lambda_{\bar{\beta} = \beta}(\omega)| = 0$, *and in particular*

b) $\lim_{\substack{\omega \to 0 \\ b_1\omega_1 + b_2\omega_2 = 0}} |\lambda_{\bar{\beta} = 2\beta}(\omega)| = \frac{1}{2}$

PROOF. Let $\theta_1 = \omega_1 h$, $\theta_2 = \omega_2 h$.

a) For $\theta_1 = 0$:

$$\lim_{\omega \to 0} |\lambda_{\bar{\beta} = \beta}(\omega)| = \lim_{\substack{\theta_2 \to 0 \\ \theta_1 = 0}} |\lambda_{\bar{\beta} = \beta}(\omega)| = 0, \qquad \text{(cf. (2.2.1))}.$$

For $\theta_2 = \xi\theta_1$:

$$\lim_{\omega\to 0} |\lambda_{\bar\beta = \beta}(\omega)| = \lim_{\theta_1\to 0} |b_2 - b_1|\theta_1^2\xi(1-\xi)/\{\theta_1^2(1+\xi^2)^2 + 4(b_1 + \xi b_2)^2\}^{\frac{1}{2}} = 0$$

independent of ξ, and a) is proved.

b) For $b_2 = 0$:

$$\lim_{\substack{\omega\to 0\\ b_1\omega_1+b_2\omega_2=0}} |\lambda_{\bar\beta = 2\beta}(\omega)| = \lim_{\substack{\theta_2\to 0\\ \theta_1=0}} |\lambda_{\beta = 2\beta}(\omega)| = \tfrac{1}{2}, \text{(cf 2.2.2)}.$$

For $b_2 \neq 0$, with $\xi = -\dfrac{b_1}{b_2}$:

$$\lim_{\substack{\omega\to 0\\ b_1\omega_1+b_2\omega_2=0}} |\lambda_{\bar\beta = 2\beta}(\omega)| = \lim_{\substack{\theta_1\to 0\\ \theta_2=\xi\theta_1}} |\lambda_{\bar\beta = 2\beta}(\omega)| = \tfrac{1}{2},$$

(cf. 2.2.2),

and b) is proved. Q.E.D.

Remark that (2.3.1) b) implies $\bar\lambda_{\bar\beta = 2\beta} \geq \frac{1}{2}$. For the following lemma we use the abbreviations p,q and r as in section 2.2.

(2.3.2) <u>LEMMA</u>. *Let* $b_1, b_2 \in \mathbb{R}$ *with* $b_1^2 + b_2^2 = 1$ *be such that for all* $\omega \in LF_h$:

$$3|b_1 - b_1|^2 r^2 \leq p^2 + 4q^2,$$

then

a) $|\lambda_{\bar\beta = \beta}(\omega)| \leq |\lambda_{\bar\beta = 2\beta}(\omega)|$ *for all* $\omega \in LF_h$, $\omega \neq 0$, *and*

b) $\lambda_{\bar\beta = \beta} \leq \lambda_{\bar\beta = 2\beta}$

<u>PROOF</u>.
a) For $\omega \neq 0$, $p^2 + 4q^2 \neq 0$, and $p^2 + q^2 \neq 0$, so $3|b_2 - b_1|^2 r^2 \leq p^2 + 4q^2$ implies

$$|b_2 - b_1|^2 r^2/(p^2+4q^2) \leq (|b_1 - b_2|^2 r^2 + p^2)/(4(p^2+q^2))$$

which proves a).

b) From a) and the continuity of $|\lambda_{\bar\beta}(\omega)|$ for $\bar\beta = \beta$ and $\bar\beta = 2\beta$ in the surrounding of the origin b) follows directly. Q.E.D.

From this lemma we can derive the following corollaries:

(2.3.3) <u>COROLLARY 1</u>. *For all* $b_1, b_2 \in \mathbb{R}$ *with* $b_1^2 + b_2^2 = 1$ *and* $|b_2 - b_1|^2 \leq 4/3$ $\bar\lambda_{\bar\beta=\beta} \leq \bar\lambda_{\bar\beta=2\beta}$.

<u>PROOF</u>. $|r|^2/|p^2| \leq \frac{1}{4}$, and $|b_2 - b_1|^2 \leq \frac{4}{3}$ imply $3r^2|b_2 - b_1|^2 \leq p^2 + 4q^2$.
Now apply lemma (2.3.2). Q.E.D.

(2.3.5) <u>COROLLARY 2</u>. *For all* $b_1 > 0$, $b_2 \geq 0$ (*or* $b_1 < 0$, $b_2 \leq 0$) $\bar\lambda_{\bar\beta=\beta} \leq \frac{1}{2}$.

PROOF. $\max_{b_1,b_2\geq 0} |b_2 - b_1| = 1$, and $|S_{12}| \leq 1$, hence $4|b_2 - b_1|^2 r^2 \leq 4S_1{}^2S_2{}^2 \leq p^2 + 4q^2$, and from this follows directly $|\lambda_{\bar{\beta}=\beta}(\omega)| \leq \frac{1}{2}$ for all $\omega \in LF_h$, $\omega \neq 0$. From the continuity of $|\lambda_{\bar{\beta}=\beta}|$ in the surrounding of the origin it follows that $\bar{\lambda}_{\bar{\beta}=\beta} \leq \frac{1}{2}$. Q.E.D.

In corollary 1 we proved that for all b_1 and b_2 with $|b_2 - b_1|^2 \leq 4/3$ the amount of the fine grid artificial viscosity on the coarse grid gives a smaller CGC convergence factor than the amount of artificial viscosity corresponding to the coarse grid mesh size. We were not able to prove or disprove this for all b_1 and b_2.
Numerical computations of the CGC convergence factors and the CGC amplification factors on the set of frequencies:

(2.3.6) $$FG_h = \{(\omega_1 h, \omega_2 h) | \omega_1 h = j.\pi/32,\ \omega_2 h = k.\pi/32;\ j,k \in \mathbb{Z},\ -16 \leq j,\ k \leq 16,\ (j,k) \neq (0,0)\},$$

suggest that it is true for all b_1 and b_2 indeed.

Table 1 shows the maxima of the CGC amplification factors on FG_h for different values of the convection coefficients b_1 and b_2, and $\varepsilon = 10^{-6}$. Because of the symmetry of $\bar{\lambda}$, we considered only (b_1,b_2) on a quarter of the unit circle.

(b_1,b_2)	$\lambda_{\bar{\beta}=\beta}$	$\lambda_{\bar{\beta}=2\beta}$
$(\sqrt{\frac{1}{2}},\sqrt{\frac{1}{2}})$	$1.5.10^{-11}$	0.50
$(\frac{1}{2}\sqrt{3},\frac{1}{2})$	0.17	0.51
(1,0)	0.40	0.53
$(\frac{1}{2}\sqrt{3},-\frac{1}{2})$	0.47	0.55
$(\sqrt{\frac{1}{2}},-\sqrt{\frac{1}{2}})$	0.48	0.54

Table 1. Maxima of the CGC Amplification factors on FG_h with $\varepsilon = 10^{-6}$.

Figure 2 shows the CGC amplification factors on FG_h, multiplied by 10, and rounded to the nearest integer.

Fig. 2a. $b_1 = \sqrt{\tfrac{1}{2}}$, $b_2 = \sqrt{\tfrac{1}{2}}$, $\bar{\beta} = \beta$

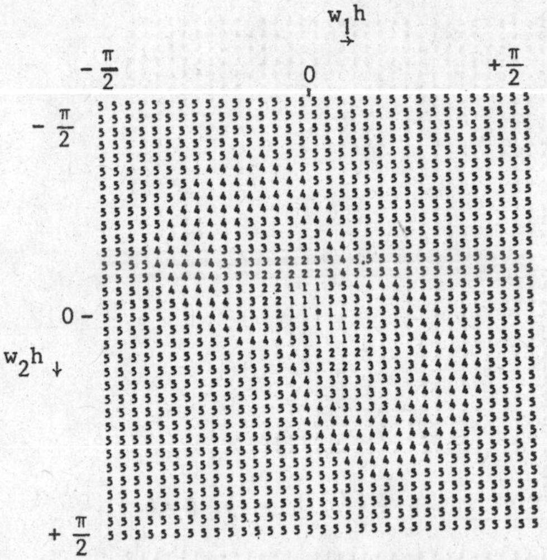

Fig. 2a'. $b_1 = \sqrt{\tfrac{1}{2}}$, $b_2 = \sqrt{\tfrac{1}{2}}$, $\bar{\beta} = 2\beta$

Fig. 2b. $b_1 = \tfrac{1}{2}\sqrt{3}$, $b_2 = \tfrac{1}{2}$, $\bar{\beta} = \beta$

Fig. 2b'. $b_1 = \tfrac{1}{2}\sqrt{3}$, $b_2 = \tfrac{1}{2}$, $\bar{\beta} = 2\beta$

Fig. 2c. $b_1 = 1$, $b_2 = 0$, $\bar{\beta} = \beta$

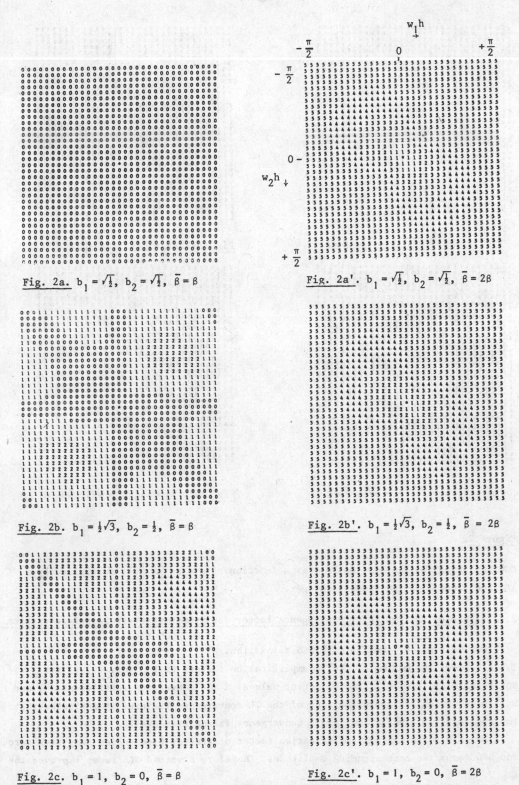

Fig. 2c'. $b_1 = 1$, $b_2 = 0$, $\bar{\beta} = 2\beta$

Fig. 2d. $b_1 = \frac{1}{2}\sqrt{3}$, $b_2 = -\frac{1}{2}$, $\bar{\beta} = \beta$

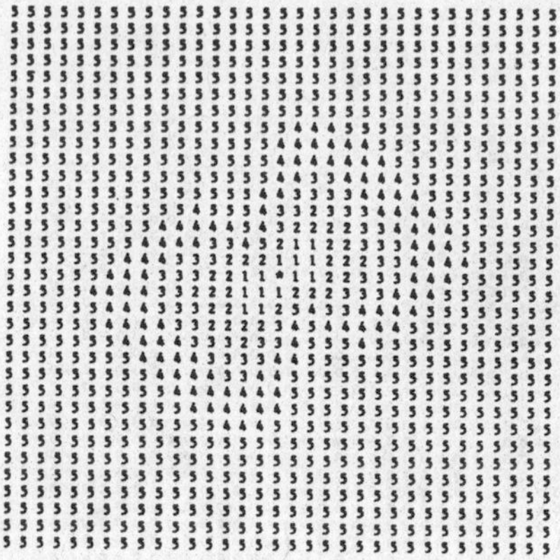

Fig. 2d'. $b_1 = \frac{1}{2}\sqrt{3}$, $b_2 = -\frac{1}{2}$, $\bar{\beta} = 2\beta$

Fig. 2e. $b_1 = \sqrt{\frac{1}{2}}$, $b_2 = -\sqrt{\frac{1}{2}}$, $\bar{\beta} = \beta$

Fig. 2e'. $b_1 = \sqrt{\frac{1}{2}}$, $b_2 = -\sqrt{\frac{1}{2}}$, $\bar{\beta} = 2\beta$

Figure 2.

CGC amplification factors, on FG_h, as a function of $\omega \in [-\frac{\pi}{h}, \frac{\pi}{h}]^2$, multiplied by 10, and rounded to the nearest integer.

2.4. Values of the two level convergence factor for Symmetric Gauss Seidel relaxation.

Now we consider the TLA with SGS relaxation.
Table 2 shows the maxima of the TL amplification factors on FG_h (cf. 2.3.6) for SGS relaxation, $\varepsilon = 10^{-6}$, and different values of the convection coefficients b_1 and b_2. These maxima are approximations of the TL convergence factors. It indicates that the choice $\bar{\beta} = \beta$ gives a smaller TL convergence factor than $\bar{\beta} = 2\beta$.

For $\bar{\beta} = \beta$ the maximum amplification factor occurs for frequencies away from zero, and SGS damps the corresponding amplitudes. Therefore a second SGS sweep improves the convergence factor. However for $\bar{\beta} = 2\beta$ the very low frequencies give high values of the CGC amplification factor (cf. Figure 2) whereas SGS has no influence on them.

	$\tilde{\lambda}_{\bar{\beta}} = \beta$		$\tilde{\lambda}_{\bar{\beta}} = 2\beta$	
number of SGS-sweeps (b_1,b_2)	1 SGS-sweep	2 SGS-sweeps	1 SGS-sweep	2 SGS-sweeps
$(\sqrt{\frac{1}{2}},\sqrt{\frac{1}{2}})$	0.15	0.06	0.50	0.50
$(\frac{1}{2}\sqrt{3},\frac{1}{2})$	0.17	0.08	0.48	0.47
$(1,0)$	0.23	0.10	0.50	0.50
$(\frac{1}{2}\sqrt{3},-\frac{1}{2})$	0.28	0.11	0.49	0.48
$(\sqrt{\frac{1}{2}},-\sqrt{\frac{1}{2}})$	0.24	0.11	0.50	0.50

Table 2. Maxima of TL amplification factors on FG_h as approximation to the TL convergence factor; $\varepsilon = 10^{-6}$.

2.5. Relation of $L_{\bar{\alpha},h}$ to the Galerkin Approximation of $L_{\alpha,h}$.

Here we show that the operator $L_{\bar{\alpha},H}$, $\bar{\alpha} = \varepsilon + \beta$ corresponds to the Galerkin Approximation of $L_{\alpha,h}$, $\alpha = \varepsilon + \beta$ (cf. FREDERICSON [2], HACKBUSCH [3], WESSELING [6]), up to terms of order h^2.

2.5.1. DEFINITION. $C^k(\mathbb{R}^2)$ is the space of real functions f with continuous partial derivatives $\frac{\partial^j f}{\partial^m x \partial^n y}$, $j = m+n = 0,\ldots,k$; $m,n \geq 0$, and with norm $\|f\|_k = \max_{j=0,\ldots,k} |f|_j$, where

$$|f|_j = \max_{j=m+n} \{ \sup_{(x,y)\in\mathbb{R}^2} |\frac{\partial f^j(x,y)}{\partial x^m \partial y^n}| \}.$$

2.5.2. DEFINITION. J_h^k is the injection $C^k(\mathbb{R}^2) \to GF_h$.

The mapping $\|\cdot\|_{3,h}$: $GF_h \to \mathbb{R}$ defined by

$$\|u_h\|_{3,h} = \max\{\sup_{i,j}|u^h_{i,j}|,\ \sup_{i,j}|\Delta_{xxx}\, u^h_{i,j}|,\ \sup_{i,j}|\Delta_{xxy}\, u^h_{i,j}|,\ \sup_{i,j}|\Delta_{xyy}\, u^h_{i,j}|,\ \sup_{i,j}|\Delta_{yyy}\, u^h_{i,j}|\},$$

where

Δ_{xxx}, Δ_{xxy},... are third order differences,

e.g.

$$\Delta_{xxx}\, u^h_{i,j} = \frac{1}{h^3}\,(u^h_{i+2,j} - 3u^h_{i+1,j} + 3u^h_{i,j} - u^h_{i-1,j}),$$

$$\Delta_{xxy}\, u^h_{i,j} = \frac{1}{h^3}\,(u^h_{i+1,j+1} - 2u^h_{i,j+1} + u^h_{i-1,j+1} - u^h_{i+1,j} + 2u^h_{i,j} - u^h_{i-1,j}),$$

is a norm on GF_h.

(2.5.3) THEOREM. *Let* P *be the 7-point prolongation* $P:GF_H \to GF_h$, *and* $\bar{R}$ *be the 7-point restriction*: $\bar{R}:GF_h \to GF_H$. *Let* R *be the injection*: $GF_h \to GF_H$. *Let* $G_{\bar{C}} = \{v \mid v \in GF_h,\ h \in \mathbb{R},\ \exists u \in C^4(\mathbb{R}^2),\ \|u\|_4 \le \bar{C},\ J^4_h u = v\}$, *then for all* $v_H \in GF_H$ *with* $Pv_H \in G_{\bar{C}}$

$$\|L_{\alpha,H}v_H - \bar{R}L_{\alpha,h}Pv_H\|_H \le h^2\tilde{C}\|Pv_H\|_{3,h},$$

where the constant $\tilde{C}$ *depends on* $\bar{C}$, b_1 *and* b_2.

PROOF. Let $v_H \in GF_H$ with $Pv_H \in G_{\bar{C}}$, hence there exists an $u \in C^4(\mathbb{R}^2)$ with $J^4_h u = Pv_H$. Application of Taylor expansion, and the mean value theorem, for $\bar{\alpha} = \alpha$ yields (cf. 1.2):

$$\|L_{\alpha,H}RJ^4_h u - \bar{R}L_{\alpha,h}J^4_h u\|_H \le C_1 h^2 |u|_3;$$

C_1 depends on b_1 and b_2.
Application of Taylor expansion and the mean value theorem yields:

$$\sup_{i,j} \left|\frac{\partial^3 u}{\partial x^3}(ih,jh) - \Delta_{xxx}(J^4_h u)_{i,j}\right| \le C.\bar{C}.h,$$

$$\sup_{i,j} \left|\frac{\partial^3 u}{\partial x^2 \partial y}(ih,jh) - \Delta_{xxy}(J^4_h u)_{i,j}\right| \le C.\bar{C}.h,$$

$$\sup_{i,j} \left|\frac{\partial^3 u}{\partial x \partial y^2}(ih,jh) - \Delta_{xyy}(J^4_h u)_{i,j}\right| \le C.\bar{C}.h,$$

$$\sup_{i,j} \left|\frac{\partial^3 u}{\partial y^3}(ih,jh) - \Delta_{yyy}(J^4_h u)_{i,j}\right| \le C.\bar{C}.h.$$

Hence

$$\|L_{\alpha,H}RPv_H - \bar{R}L_{\alpha,h}Pv_H\|_H \le \tilde{C}h^2\|Pv_H\|_{3,h},$$

where $\tilde{C}$ depends on $\bar{C}$, b_1 and b_2.

Since $RP = I$, the identity on GF_H the theorem is proved. Q.E.D.

REMARK. The operator $\bar{R}L_{\alpha,h}P$ is called the *Galerkin Approximation of* $L_{\alpha,h}$.

$$\|Pv_H\|_{3,h} \le \|P\|\,\|v_H\|_{3,H} \text{ with } \|P\| = \sup_{\|v_H\|_{3,H}=1} \|Pv_H\|_{3,h},$$

hence the right hand side of the inequality in theorem (2.5.4) can be replaced by $\hat{C}h^2\|v_H\|_{3,H}$ where $\hat{C}$ depends on $\bar{C}$, b_1, b_2 and P.

3. THE CHOICE OF ARTIFICIAL VISCOSITY IN THE MULTI LEVEL ALGORITHM

In this section we describe how the results of section 2 can be used in a MLA. We discuss three variants. Consider a MLA with n+1 levels: $\ell_0, \ell_1, \ldots, \ell_n$. To solve $L_{\alpha,h}\, u_h = f_h$ on level ℓ_n the MLA can be applied with different amounts of artificial viscosity on the levels $\ell_0, \ldots, \ell_{n-1}$. On each level the amount can be related either to the meshwidth of the finer or to that of the coarser grid.

Table 3 shows the three variants.

In variant 1 the artificial viscosity is the same on all levels.

In variant 2 the artificial viscosity on each level ℓ_k $(0 \le k < n+1)$ corresponds to the meshwidth h_{k+1} on the level ℓ_{k+1}.

In variant 3 the viscosity on each level ℓ_k corresponds to the meshwidth h_k.

level	variant 1	variant 2	variant 3
ℓ_n	β	β	β
ℓ_{n-1}	β	β	2β
ℓ_{n-2}	β	2β	4β
$\vdots$	$\vdots$	$\vdots$	$\vdots$
ℓ_k	β	$2^{n-k-1}\beta$	$2^{n-k}\beta$
$\vdots$	$\vdots$	$\vdots$	$\vdots$
ℓ_0	β	$2^{n-1}\beta$	$2^{n}\beta$

Table 3. Three variants for the choice of artificial viscosity on sublevels in a MLA.

Many authors applied the Galerkin Approximation with success (cf. MOL [5]). So according to section 2.5, this corresponds with variant 1. However if the number of levels becomes large, the asymptotic stability of the operators on the coarser grids descreases and numerical experiments show that (e.g. for SGS-relaxation) divergence may occur (cf. de ZEEUW, van ASSELT [7]).

With values of β even smaller than $h_n/2$ these experiments also show that the other variants still converge, and variant 2 has a better rate of convergence than variant 3.

Further, the asymptotic stability of the operators in variant 2 and in variant 3 are of the same order, and the amount of work for the three variants is the same.

4. CONCLUSIONS

For a two level algorithm the choice $\bar{\beta} = \beta$ is better than $\bar{\beta} = 2\beta$. The choice $\bar{\beta} = \beta$ corresponds with the Galerkin Approximation of $L_{\alpha,h}$ up to terms of order h^2.

For a MLA variant 2 is prefarable to variant 1 and variant 3.

ACKNOWLEDGEMENT

The author wishes to thank Dr. P.W. Hemker, and Drs. P.M. de Zeeuw for the stimulating discussions on the subject and the assistence in the computational work.

REFERENCES

[1] BRANDT, A.; DINAR, N., *Multi-grid solutions to elliptic flow problems*, Report Number 79-15, July 16, ICASE (1979).

[2] FREDERICKSON, P.O., *Fast approximate inversion of large sparse linear systems*, Mathematics report 7-75, Lakehead University, Ontario, Canada (1975).

[3] HACKBUSCH, W., *On the multigrid method applied to difference equations*, Computing 20 (1978), 291-306.

[4] HEMKER, P.W., *Fourier Analysis of gridfunctions, prolongations and restrictions*, Report NW 93/80, Mathematical Centre, Amsterdam (1980).

[5] MOL, W.J.A., *On the choice of suitable operators and parameters in multigrid methods*, Report NW 107/81, Mathematical Centre, Amsterdam (1981).

[6] WESSELING, P., *Theoretical and practical aspects of a multigrid method*, Report NA-37, Delft University of Technology (1980).

[7] ZEEUW, P. de; ASSELT. E.J. van, *Numerical computation of the convergence rate of the multi level algorithm applied to the convection diffusion equation*. To be published as Report NW, Mathematical Centre, Amsterdam (1982).

DEFECT CORRECTIONS AND MULTIGRID ITERATIONS

W. Auzinger, H.J. Stetter
Technical University of Vienna

Introduction

In several of his publications since 1978, Brandt describes a modification of the general multigrid algorithm which be calls "truncation extrapolation" or "τ-extrapolation" (see, e.g., [1]). In their analysis of this idea, the authors of this paper reinterpreted Brandt's τ-extrapolation as a defect correction step ([2]). Independently, W. Hackbusch had suggested a combination of the multigrid and defect correction approaches in [3], see also [4].

In the following, we will at first demonstrate how any defect correction iteration may be combined with an iterative procedure for its effective solution operator and derive a bound on the convergence factor for the combined iterative process. We will then consider the situation when the iterative solution procedure consists of multigrid cycles. Various implementation versions of a defect correction multigrid cycle will be discussed, particularly for the case where defect correction is to achieve a higher order approximation to the differential equation. We will also regard some algorithmic and quantitative details in the context of model problems. Finally we will present some numerical experiences.

Our approach permits the use of standard multigrid software in situations more general than those for which the software was designed. This will extend the applicability of existing and coming multigrid packages and thus reduce programming costs for many applications.

1. Defect Correction with an Iterative Solution Procedure

The principle of <u>defect correction</u> may be summarized thus (cf., e.g., [5]): We wish to solve the problem

$$F\,u = c \tag{1.1}$$

but we only have an efficient solution procedure for the related problem

$$\tilde{F} v = \tilde{c} ; \qquad (1.2)$$

however, we are able to evaluate the direct mapping F . Then the following two iterative processes are natural (see [5]):

$$\text{(A)} \quad u^{(i+1)} := u^{(i)} - [\tilde{F}^{-1} F u^{(i)} - \tilde{F}^{-1} c]$$

or (1.3)

$$\text{(B)} \quad \tilde{F} u^{(i+1)} := \tilde{F} u^{(i)} - [F u^{(i)} - c] .$$

If $\tilde{F}$ is a linear mapping (F may still be nonlinear), then (A) and (B) collapse into the familiar

$$u^{(i+1)} := u^{(i)} - \tilde{F}^{-1} [F u^{(i)} - c] . \qquad (1.4)$$

Convergence of these processes depends on the contractivity of the mappings

$$\text{(A)} \quad I - \tilde{F}^{-1} F \quad \text{or} \quad \text{(B)} \quad I - F \tilde{F}^{-1} ;$$

in the case of linear F these two mappings are related by the similarity transformation

$$I - \tilde{F}^{-1} F = \tilde{F}^{-1} (I - F \tilde{F}^{-1}) \tilde{F} .$$

(Here and in the following, we always assume that solutions of equations exist and are unique at least locally so that the notion of an inverse mapping makes sense.)

We will now assume that the efficient solution procedure for problems (1.2) is itself an iterative process and of the defect correction type; let us denote it by

$$v_{j+1} := v_j - K (\tilde{F} v_j - \tilde{c}) . \qquad (1.5)$$

In conjunction with the iterative process (1.4) or (1.3), one will not want to use more than one or a few iterations (1.5) - with an appropriate initial value v_o - in place of $\tilde{F}^{-1}$, even if (1.5) does not converge very rapidly towards the solution $\tilde{F}^{-1} \tilde{c}$ of (1.2).

In the case of (1.4), i.e. when $\tilde{F}$ is linear, we have to compute the correction $\Delta u^{(i)}$ for $u^{(i)}$ from

$$\tilde{F}\,\Delta u^{(i)} = F\,u^{(i)} - c =: d(u^{(i)}) ; \tag{1.6}$$

therefore our iterative solution procedure (1.5) will take the form

$$\begin{aligned} v_o^{(i)} &:= 0 , \\ v_{j+1}^{(i)} &:= v_j^{(i)} - K\,(\tilde{F}\,v_j^{(i)} - d(u^{(i)})), \quad j = 0(1)r-1 , \end{aligned} \tag{1.7}$$

and the next iterate in (1.4) will be

$$u^{(i+1)} := u^{(i)} - v_r^{(i)} . \tag{1.8}$$

For r = 1, (1.7) and (1.8) collapse into

$$u^{(i+1)} := u^{(i)} - K\,d(u^{(i)}) ; \tag{1.9}$$

this is (1.4), with $\tilde{F}^{-1}$ replaced by the operation K.

In the case of (1.3B), where we have to solve

$$\tilde{F}\,u^{(i+1)} = \tilde{F}\,u^{(i)} - d(u^{(i)}) ,$$

the corresponding iterative solution procedure is

$$\begin{aligned} u_o^{(i+1)} &:= u^{(i)} , \\ u_{j+1}^{(i+1)} &:= u_j^{(i+1)} - K\,[\tilde{F}\,u_j^{(i+1)} - \tilde{F}\,u^{(i)} + d(u^{(i)})], \quad j = 0(1)r-1, \end{aligned} \tag{1.10}$$

and $u^{(i+1)} := u_r^{(i+1)}$. For r = 1, this becomes (1.9) again.

The case of (1.3A) is slightly different through the occurence of $F^{-1}c =: \bar{u}$ which is normally used as $u^{(0)}$ in version A of the nonlinear defect correction procedure, see [5]. If we apply one iteration of (1.5), with starting value $\bar{u}$, in the computation of $\tilde{F}^{-1}F\,u^{(i)}$ in (1.3A), we arrive once more at (1.9). (If we start with $u^{(i)}$, which is more natural, we must replace both occurences of $\tilde{F}^{-1}$ in (1.3A) by one iteration step (1.5) in order to obtain (1.9).)

In any case, (1.9) appears as the natural combination of iterative defect correction and an iterative solution procedure for $\tilde{F}$ even if F is nonlinear. Because of its simple structure, the iterative procedure (1.9 admits also other natural interpretations; but in our context it seems most intuitive to regard it as a <u>composite iterative procedure</u> with the constituents

defect correction for (1.1) and

iterative solution for (1.2).

It is clear that (1.9) converges to the same limit as (1.4) resp. (1.3) if it converges. If we know the rates of convergence of (1.4) (or (1.3)) and (1.5), what can we tell about the rate of convergence of (1.9)?

<u>Proposition</u>: Assume that q and $\tilde{q}$ are the (local) convergence rates of (1.4) and (1.5) resp. Then the convergence rate $\bar{q}$ of (1.9) satisfies

$$\bar{q} \leq q + \tilde{q} + q\tilde{q}. \tag{1.11}$$

<u>Proof</u>: If all operations are linear we simply observe that

$$I - KF = (I - K\tilde{F}) + (I - \tilde{F}^{-1}F) - (I - K\tilde{F})(I - \tilde{F}^{-1}F) \tag{1.12}$$

which implies

$$\|I - KF\| \leq \|I - K\tilde{F}\| + \|I - \tilde{F}^{-1}F\| + \|I - K\tilde{F}\|\,\|I - \tilde{F}^{-1}F\| \quad .$$

If one or both of the operations are nonlinear, the convergence rates are local Lipschitz constants rather than norms. We have

$$\|u^{(i+1)} - u^*\| \leq q\, \|u^{(i)} - u^*\| \quad ,$$
$$\|\tilde{u}^{(i+1)} - u^{(i+1)}\| \leq \tilde{q}\, \|u^{(i)} - u^{(i+1)}\| \quad ,$$

where u* is the limit of (1.3), $u^{(i+1)}$ and $\tilde{u}^{(i+1)}$ are the results of (1.3) and (1.9) resp. ; the second inequality holds because the iterative processes (1.7) or (1.10) have $u^{(i+1)}$ as a limit and $u^{(i)} - u^{(i+1)}$ as initial error. Hence

$$\|\tilde{u}^{(i+1)} - u^*\| \leq \|\tilde{u}^{(i+1)} - u^{(i+1)}\| + \|u^{(i+1)} - u^*\|$$
$$\leq \tilde{q}\,\|u^{(i)} - u^{(i+1)}\| + q\,\|u^{(i)} - u^*\|$$
$$\leq \tilde{q}(1+q)\,\|u^{(i)} - u^*\| + q\,\|u^{(i)} - u^*\| . \qquad \square$$

Obviously, $q < 1$ and $\tilde{q} < 1$ are not sufficient for $\bar{q} < 1$. On the other hand, $q \ll 1$ and $\tilde{q} \ll 1$ imply $\bar{q} \approx q + \tilde{q} < 1$. If more than one iteration of (1.5) is employed $\tilde{q}$ is replaced by $\tilde{q}^r$; for $\tilde{q} < 1$, $\bar{q} \to q$ as $r \to \infty$.

2. Application to Multigrid Algorithms

We will now apply the composite iterative defect correction procedure introduced in section 1 to the following situation:

(1.2): a standard discretization of a standard differential problem on a given (fine) grid;

(1.1): a more sophisticated discretization of the same problem on the same grid, or

a discretization on the same grid of a weakly nonlinear version of the same problem, or

another related discrete problem more complicated than (1.2);

(1.5): one cycle of a multigrid algorithm for (1.2).

We assume that an efficient implementation of a multigrid algorithm for the standard problem (1.2) is available; thus our approach extends the use of such software to the wider range of problems (1.1). This extension of standard algorithms to wider classes of problems by means of defect correction may become an important tool in the design of easy-to-use multigrid software packages.

In accordance with the above interpretation, we will from now on assume that $\tilde{F}$ is a linear operation, i.e. we will consider (1.4) and (1.6) - (1.9). Note that the linearity of F has not been assumed in these equations.

In the more detailed "model problem analysis" of sections 4 and 5, (1.2) will be the standard five-point $O(h^2)$-discretization of a Helmholtz equation on a rectangle. (1.1) will either be the nine-point $O(h^4)$-discretization of the same differential equation problem or an $O(h^2)$-discretization of a nonlinear Helmholtz equation. It may be useful to have these interpretations in mind as a reference in this and the next section.

At first, we have to verify that a multigrid cycle for a linear problem (1.2) has indeed the structure (1.5). This is not immediately obvious from the common notation *) for an l-level multigrid cycle for $L_l v = c_l$ which is

$$v^{j+1} := M_l \; v^j + N_l \; c_l \tag{2.1}$$

where the l-grid iteration operator M_l is recursively defined by

$$\begin{aligned} M_1 &:= [I_1 - I_o^1 \, L_o^{-1} \, I_1^o \, L_1] \, S_1^\nu , \\ M_k &:= [I_k - I_{k-1}^k \, (I_{k-1} - M_{k-1}^\gamma) \, L_{k-1}^{-1} \, I_k^{k-1} \, L_k] \, S_k^\nu , \qquad k = 2(1)l. \end{aligned} \tag{2.2}$$

However, (2.2) does not display the fact that the smoothing operations S_k are relaxations w.r.t. the corrector equations $L_k v_k = c_k$ on the levels k, i.e.

$$\begin{aligned} v_k &\overset{S_k}{\rightarrow} \bar{v}_k := v_k - R_k \; (L_k v_k - c_k) \; , \\ v_k &\overset{S_k^\nu}{\rightarrow} \bar{v}_k := v_k - R_k^\nu \; (L_k v_k - c_k) \; , \end{aligned} \tag{2.3}$$

where R_k^ν, $\nu \geq 2$, is recursively defined by $R_k^\nu := R_k^{\nu-1} + R_k - R_k L R_k^{\nu-1}$. With (2.3) and the abbreviation $(L_l^{l-1})^{-1} := I_{l-1}^l \, L_{l-1}^{-1} \, I_l^{l-1}$, the two-grid algorithm for $L_l v = c_l$ takes the form

$$\begin{aligned} v_l^{j+1} &:= \bar{v}_l^j - (L_l^{l-1})^{-1} \; (L_l \bar{v}_l^j - c_l) \\ &= v_l^j - R_l^\nu \; (L_l v_l^j - c_l) - (L_l^{l-1})^{-1} \; (L_l (v_l^j - R_l^\nu (L_l v_l^j - c_l)) - c_l) \\ &= v_l^j - [R_l^\nu + (L_l^{l-1})^{-1} - (L_l^{l-1})^{-1} \; L_l \; R_l^\nu] \; (L_l v_l^j - c_l) \; , \end{aligned} \tag{2.4}$$

which has the defect correction structure of (1.5).

---

*) Cf. the introductory paper by U. Trottenberg in this volume. For simplicity, smoothing after correction is not considered here.

We now replace the operation L_{l-1}^{-1} in $(L_l^{l-1})^{-1}$ by another two-grid step to obtain a three-grid algorithm ($\gamma = 1$): The correction $e_{l-1} = L_{l-1}^{-1}\, d_{l-1}$ is approximatively computed by

$$e_{l-1}^{o} = 0 \xrightarrow{S_{l-1}^{\nu}} \bar{e}_{l-1} = e_{l-1}^{o} - R_{l-1}^{\nu}\,(L_{l-1}\, e_{l-1}^{o} - d_{l-1})$$

$$= R_{l-1}^{\nu}\, d_{l-1},$$

$$e_{l-1} = \bar{e}_{l-1} - (L_{l-1}^{l-2})^{-1}\,(L_{l-1}\, \bar{e}_{l-1} - d_{l-1})$$

$$= [R_{l-1}^{\nu} + (L_{l-1}^{l-2})^{-1} - (L_{l-1}^{l-2})^{-1}\, L_{l-1}\, R_{l-1}^{\nu}]\, d_{l-1}\,.$$

Hence, in (2.4), $(L_{l-1}^{l})^{-1}$ is simply replaced by

$$I_{l-1}^{l}\,[R_{l-1}^{\nu} + (L_{l-1}^{l-2})^{-1} - (L_{l-1}^{l-2})^{-1}\, L_{l-1}\, R_{l-1}^{\nu}]\, I_{l}^{l-1}$$

which leaves the structure of (2.4) unaltered.

It is now clear that this argument can be continued recursively and that $\gamma = 2$ (W-cycle) and smoothing after correction lead to the same structure (1.5) for (2.1). (The use of a notation displaying this structure - like the one employed in (2.3) and (2.4) - may also be helpful in the analysis of other aspects of multigrid algorithms.)

In the context described at the beginning of this section, one step of the composite iterative defect correction procedure (1.9)

$$u^{(i+1)} := u^{(i)} + K\, d(u^{(i)})$$

consists the following operations:

Computation of defect $d(u^{(i)}) := Fu^{(i)} - c$ on the given (fine) grid

Multigrid cycle for $\tilde{F}v = -d(u^{(i)})$:

- Smoothing of the initial correction $v_o^{(i)} = 0$ by relaxation w.r.t. $\tilde{F}v = -d(u^{(i)})$
- Transfer of the defect $F\bar{v}_o^{(i)} + d(u^{(i)})$ of the smoothed correction $\bar{v}_o^{(i)}$ to coarser grid
- . . .
- Generation of approximate solution $\tilde{v}^{(i+1)} = -K\, d(u^{(i)})$ of $\tilde{F}v = -d(u^{(i)})$ on given (fine) grid

Computation of $u^{(i+1)} := u^{(i)} - \tilde{v}^{(i+1)}$

The appearance of this <u>defect</u> <u>correction</u> <u>multigrid</u> <u>cycle</u> ("DCMG-cycle") becomes more familiar when we set $u := u^{(i)} - v$ and use the linearity of $\tilde{F}$ to obtain the right hand version of

one multigrid cycle for

$$\tilde{F}v = -d(u^{(i)}) \qquad\qquad \tilde{F}u = \tilde{F}u^{(i)} - d(u^{(i)}) \tag{2.5}$$

starting with

$$v_0^{(i)} := 0 \qquad\qquad u_0^{(i)} := u^{(i)}$$

yields

$$\tilde{v}^{(i+1)} \qquad\qquad \tilde{u}^{(i+1)}$$

which is used to form

$$u^{(i+1)} := u^{(i)} - \tilde{v}^{(i+1)} \qquad\qquad u^{(i+1)} := \tilde{u}^{(i+1)} .$$

This right hand version also corresponds to the iterative procedure (1.10) which must be used for nonlinear $\tilde{F}$. (2.5) may also be written in the forms

$$\begin{aligned} \tilde{F}u &= \tilde{c} - [(Fu^{(i)} - c) - (\tilde{F}u^{(i)} - \tilde{c})] \\ &= \tilde{c} - [d(u^{(i)}) - \tilde{d}(u^{(i)})] \\ &= c + \Delta F u^{(i)}, \quad \text{with} \quad \Delta F := \tilde{F} - F , \end{aligned} \tag{2.6}$$

which display various aspects of our defect correction approach.

In this version, the DCMG-cycle is distinct from the original multigrid cycle for $\tilde{F}u = \tilde{c}$ only by the replacement of $\tilde{c}$ by $\tilde{F}u^{(i)} - d(u^{(i)})$. The evaluation of $d(u^{(i)})$ and the replacement of $\tilde{c}$ can be implemented into a given multigrid program without much effort; in particular, none of the data structures and of the grid transfer operations (including the definition of $\tilde{F}$ on the coarser grids) are affected!

This DCMG-cycle may also be used like the original multigrid cycle: It may be repeated or it may be recursively embedded into a Full Multigrid Algorithm (FMG). No further changes in the respective standard algorithms are required.

In a FMG-algorithm sufficient care has to be exerted in the interpolation for an initial approximation on a next finer grid. Several authors have suggested that a higher order interpolation should be employed which makes use either of the difference equation or of the differential equation. In our context, the difference or differential equation connected with Fu = c must be used. Particularly in the case of nonlinear F, the following use of the differential equation leads to a simple procedure:

Assume that we wish to approximate the solution of

$$-\Delta u(x,y) = g(x,y,u(x,y)); \tag{2.7}$$

then the following two interpolation formulas with the substitution (2.7) permit an easy and accurate transfer of an approximate solution from a 2h-grid to an h-grid:

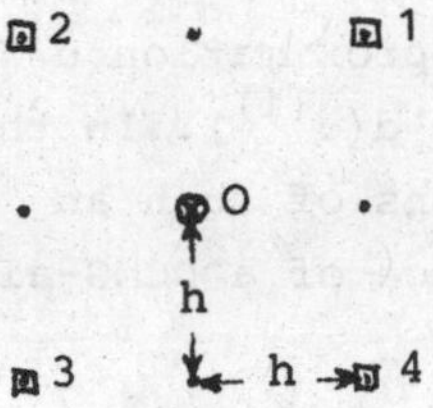

$$u_o = \frac{1}{4}\sum_{i=1}^{4} u_i - \frac{h^2}{8}\sum_{i=1}^{4} (\Delta u)_i + O(h^4)$$

$$\text{with } (\Delta u)_i := -g(x_i, y_i, u_i)$$

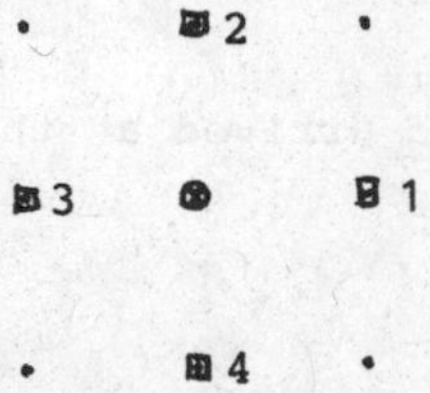

$$u_o = \frac{1}{4}\sum_{i=1}^{4} u_i - \frac{h^2}{16}\sum_{i=1}^{4} (\Delta u)_i + O(h^4)$$

So far we have taken the <u>convergence</u> of our DGMG-procedure for granted. It is clear that the contractivity of the multigrid cycle proper will remain unchanged, i.e. $\tilde{q}$ of (1.11) will be the convergence rate of the multigrid cycle. However, the convergence rate q of the defect correction component of the composite process has to be determined and checked:

With $\Delta\tilde{F} := \tilde{F} - F$, the iteration operator of iterative defect correction is (cf. section 1)

$$I - \tilde{F}^{-1} F = \tilde{F}^{-1} \Delta F .$$

In many typical applications (see the beginning of this section), $\tilde{F}^{-1}\Delta F u$ will be small for <u>smooth</u> gridfunctions u but $\|F^{-1}\Delta F\|$ will not be small. E.g., if F and $\tilde{F}$ are $O(h^4)$ and $O(h^2)$-discretizations of the same smooth,

2nd order differential equation problem, $\Delta F\, u = O(h^2)$ for smooth u because it is essentially the principal term of the local discretization error of $\tilde{F}$. But $\|\Delta F\| = O(h^{-2})$ and it seems at first that there is no uniform (in h) bound for $\|\tilde{F}^{-1}\Delta F\|$ at all.

Fortunately, a closer analysis reveals that - in this situation - ΔF is $h^{-2} \times$ (4th order difference operator) so that the application of the "summation operator" $\tilde{F}^{-1}$ compensates the h^{-2}; thus there exists a constant q independent of h such that $\|\tilde{F}^{-1}\Delta F\, u\| \leq q\|u\|$ or $\|\tilde{F}^{-1}\Delta F\| \leq q$. Yet q may not be sufficiently small in view of (1.11).

In these and similar cases the convergence rate q of the defect correction part of our procedure may be improved by an a-priori reduction of the non-smooth components in the current approximation $u^{(i)}$, i.e. by a smoothing of $u^{(i)}$ before the formation of $d(u^{(i)})$. In the next section we will consider various implementations of such an "a-priori smoothing" and their effect on the performance of a DCMG-algorithm.

3. Defect correction and smoothing

Through the considerations of section 1, we had arrived at the following form of a DCMG-cycle (see also (2.5)):

<u>Implementation (i)</u>

Form $d(u^{(i)}) := F\, u^{(i)} - c$

Perform one multigrid cycle for $\tilde{F}\, u = \tilde{F}\, u^{(i)} - d(u^{(i)})$ starting with $u^{(i)}$

This procedure which takes $u^{(i)}$ into $u^{(i+1)}$ has the structure (1.9); hence it is clear that its fixed point is the desired solution u^* of $F\, u = c$ and that - if it contracts - it contracts towards u^*.

An a-priori smoothing*) takes $u^{(i)}$ into $\bar{u}^{(i)}$, then we proceed as in implementation (i), with $u^{(i)}$ replaced by $\bar{u}^{(i)}$. Since our goal is an approximation of u^*, it is natural to use relaxation w.r.t. (1.1) for smoothing. Hence, with our notation (2.3),

---

*) This term is to indicate that the smoothing occurs outside the multigrid cycle proper and in addition to the smoothing inside the multigrid cycle.

$$\bar{u}^{(i)} := u^{(i)} - R\,(F\,u^{(i)} - c)\ , \tag{3.1}$$

where R is now the relaxation operator of the a-priori smoothing w.r.t. (1.1).

<u>Implementation (ii)</u>

Smooth $u^{(i)}$ by relaxation w.r.t. $F\,u = c$ to obtain $\bar{u}^{(i)}$

Form $d(\bar{u}^{(i)}) := F\,\bar{u}^{(i)} - c$

Perform one multigrid cycle for $\tilde{F}\,u = \tilde{F}\,\bar{u}^{(i)} - d(\bar{u}^{(i)})$ starting with $\bar{u}^{(i)}$

We will now consider the properties of this DCMG-cycle: Substitution of (3.1) into (1.9) yields

$$\begin{aligned} u^{(i+1)} &:= \bar{u}^{(i)} - K\,(F\,\bar{u}^{(i)} - c) \\ &= u^{(i)} - R\,(F\,u^{(i)} - c) - K\,(F(u^{(i)} - R\,(Fu^{(i)} - c)) - c)\ . \end{aligned} \tag{3.2}$$

If F is linear, (3.2) becomes

$$u^{(i+1)} := u^{(i)} - [R + K - K F R]\,(F\,u^{(i)} - c) \tag{3.3}$$

$$=: u^{(i)} - \bar{K}\,(F\,u^{(i)} - c) \tag{3.4}$$

which is again of type (1.9). The iteration operator is now

$$I - [R + K - K F R]F = (I - K F)\,(I - RF);$$

but if our smoothing has been successful we must have

$$\|I - \bar{K}\,F\| << \|I - K\,F\| \cdot \|I - R\,F\| \approx \|I - K\,F\| \tag{3.5}$$

because $I - R\,F$ reduces the unsmooth components and $I - K\,F$ the smooth ones. It is clear from (3.3) that the fixed point u* has remained unaltered.

Even when F is nonlinear, the invariance of u* follows immediately from (3.2) and $F\,u^* - c = 0$ because K and R are linear and map the origin into itself. The Frechet derivative of the right hand side of (3.2) at $u^{(i)}$ becomes $(I - K\,F'\,(u^{(i)}))(I - R\,F'\,(u^{(i)}))$ so that we may expect a similar effect as in (3.5) for the contraction rate of (3.2) which is now defined by (local) Lipschitz bounds which may be estimated by norms of Frechet derivatives.

Thus implementation (ii) is perfect except for the fact that it uses a smoother which is not the one contained in the program for a multigrid cycle for (1.2). Although the implementation of (3.1) should require little effort beyond the defect computation which we need anyway, it seems worthwile to consider how we would get away with a smoother for (1.2) as an a-priori smoother, i.e. with the standard smoother of our multigrid software:

Implementation (iii)

Smooth $u^{(i)}$ by relaxation w.r.t. $\tilde{F}\, u = \tilde{c}$ to obtain $\bar{u}^{(i)}$

Form $d(\bar{u}^{(i)}) := F\, \bar{u}^{(i)} - c$

Perform one multigrid cycle for $\tilde{F}\, u = \tilde{F}\, \bar{u}^{(i)} - d(\bar{u}^{(i)})$ starting with $\bar{u}^{(i)}$

Intuitively, the exchange of the a-priori smoother should not strongly affect the performance of the DCMG-cycle: The smooth components which would suffer from a contraction towards the "wrong" solution $\tilde{u}^*$ of $\tilde{F}\, u = \tilde{c}$ are hardly affected by a typical relaxation smoother and the unsmooth components need only be reduced. It is true that u^* will no longer be a fixed point of the DCMG-cycle (iii) because it is not reproduced by the a-priori smoother; but this will not hurt if the actual fixed point $\bar{u}^*$ differs from u^* by no more than the magnitude of the discretization error of (1.1).

With a relaxation smoother w.r.t. (1.2)

$$\bar{u}^{(i)} := u^{(i)} - \tilde{R}\, (\tilde{F}\, u^{(i)} - \tilde{c}) \tag{3.6}$$

we obtain for version (iii) of the DCMG-cycle in analogy to (3.2)

$$u^{(i+1)} := u^{(i)} - \tilde{R}\, (\tilde{F}\, u^{(i)} - \tilde{c}) - K\, (F(u^{(i)} - \tilde{R}(\tilde{F} u^{(i)} - \tilde{c})) - c) \tag{3.7}$$

Assuming F to be linear, we may subtract the identity

$$\begin{aligned} u^* = u^* - \tilde{R}\, (\tilde{F}\, u^* - \tilde{c}) - K\, (F(u^* - \tilde{R}(\tilde{F}\, u^* - \tilde{c})) - c) \\ + (I - K\, F)\, \tilde{R}\, (\tilde{F}\, u^* - \tilde{c}) \end{aligned}$$

from the fixed point equation for $\bar{u}^*$ (cf. (3.7)) to obtain, with $\bar{e} := \bar{u}^* - u^*$,

$$\bar{e} = (I - K\,F)(I - \tilde{R}\,\tilde{F})\;\bar{e} - (I - KF)\;\tilde{R}\;(\tilde{F}u^* - \tilde{c})\ .$$

If our DCMG-cycle is contractive, $\tilde{M} := (I - KF)(I - \tilde{R}\,\tilde{F})$ has a norm smaller than 1 and

$$\bar{e} = -(I - \tilde{M})^{-1}\;(I - K\,F)\;\tilde{R}\;(\tilde{F}\,u^* - \tilde{c})\ . \qquad (3.8)$$

(3.8) yields the desired characterization of the "fixed point shift" $\bar{e}$: E.g., if (1.1) and (1.2) are $O(h^4)$ and $O(h^2)$ discretizations of the same elliptic equation, we have

$$\tilde{F}\,u^* - \tilde{c} = O(h^2) \quad \text{(it is approx. the discr. error of (1.2))}$$

$$\|\tilde{R}\| = O(h^2) \quad \text{for standard relaxation smoothers}$$

while the first two operators are $O(1)$. Thus $\bar{e} = O(h^4)$ which is the order of the discretization error of (1.1). Thus the solution u^* of (1.1) may be approximated by our DCMG-cycle (iii) to within the magnitude of its discretization error. Actually, $\bar{e}$ will be a small multiple of h^4 because $I - K\,F$ reduces smooth components considerably.

For nonlinear F, a more complicated analysis leads to a result analogous to (3.8).

In its version (iii), the DCMG-cycle begins with a smoothing and a defect computation step which are immediately followed - within the multigrid cycle proper - by another smoothing and defect computation step, all of them on the finest grid level; both smoothers refer to (1.2). Thus it is natural to identify the defect correction part of the DCMG-cycle (iii) with the beginning of the multigrid cycle proper:

Smooth $u^{(i)}$ w.r.t. $\tilde{F}\,u = \tilde{c} \to \bar{u}^{(i)}$
Form $d(\bar{u}^{(i)}) := F\,\bar{u}^{(i)} - c$

- -	- - - - - - - - - - - - - - -
Smooth $\bar{u}^{(i)}$ w.r.t. $\tilde{F}u = \tilde{F}\bar{u}^{(i)} - d(\bar{u}^{(i)}) \to \bar{\bar{u}}^{(i)}$	Smooth $u^{(i)}$ w.r.t. $\tilde{F}u = \tilde{c} \to \bar{u}^{(i)}$
Form $d_1 := \tilde{F}\bar{\bar{u}}^{(i)} - \tilde{F}u^{(i)} + d(\bar{u}^{(i)})$	Form $d_1 := F\bar{u}^{(i)} - c$
. . .	. . .

(remainder of MG-cycle)

Implementation (iv)

Perform one multigrid cycle for $\tilde{F}u = \tilde{c}$ starting with $u^{(i)}$.

where the defect d_l on the finest grid level is computed w.r.t. $F u = c$

Obviously, this version requires the least change in a standard program for the multigrid cycle; only the defect computation on the finest grid has to be changed. This version has been suggested by Hackbusch ([3]); it is also essentially identical to Brandt and Dinar's τ-extrapolation approach.

From the point of view of our analysis, version (iv) is the least transparent: The two components of our composite iteration procedure have now been merged. If we continue to identify the initial smoothing and the computation of $d(\bar{u}^{(i)})$ as the defect correction phase then we no longer have a full l-level multigrid cycle but only an (l-1)-level one.

Actually, using the equivalence considered in connection with (2.5), we may relate implementations (iii) and (iv) in the following way (cf. implementation (iii)):

(iii)	(iv)
Smooth $u^{(i)}$ by relaxation w.r.t. $\tilde{F}u = \tilde{c}$ to obtain $\bar{u}^{(i)}$	
Form $d_l(u^{(i)}) := F_l \bar{u}^{(i)} - c$	
Perform one l-grid cycle for $\tilde{F}_l\ v = - d_l(\bar{u}^{(i)})$ starting with $v_l = 0$	Perform one (l-1)-grid cycle for $\tilde{F}_{l-1}\ v = - I_l^{l-1}\ d_l(\bar{u}^{(i)})$ starting with $v_{l-1} = 0$
Form $u^{(i+1)} := u^{(i)} - \tilde{v}_l^{(i+1)}$	Form $u^{(i+1)} := u^{(i)} - I_{l-1}^{l}\ \tilde{v}_{l-1}^{(i+1)}$

Thus version (iv) should behave rather like version (iii) and its fixed point should satisfy (3.8) with a slightly different multigrid operator $\tilde{M}$. Hackbusch ([3]) has shown that - under suitable assumptions - his defect correction process and hence our version (iv) of the DCMG-iteration converges to a limit $\bar{u}^*$ which satisfies $\bar{u}^* - u^* = O(h^4)$ in the case which we have considered for version (iii).

<u>Remark</u>*) :

Actually, the authors feel that version (iv) should not be regarded as a member of the DCMG-family but as a first step towards a more general multigrid algorithm concept which is just emerging in the work of Brandt (see his contributions in these Proceedings):

In this concept, the role of smoothing (of "relaxation") is seen like that of other stabilizing devices in classical discretizations of p.d.e.: They are necessary to control high frequency effects, and it is accepted that they may change the mathematical model slightly (like artificial viscosity). This means that - generally and throughout a multigrid cycle - the smoother need not be restricted by the request that it must have the solution of the corrector equation at the respective grid level as a fixed point.

In our context, this means that F takes the role of $\tilde{F}$ (in the form pertaining to the momentary grid level) throughout the multigrid cycle except in the smoothing steps. This implies that the direct solution on the coarsest grid level is also done with the respective version of F instead of $\tilde{F}$. This seems acceptable for two reasons:

a) At that stage, the size of the system is sufficiently small so that the more complicated F-system is cheaply solvable as well.

b) At the coarsest grid level, the stepsize parameter h is sufficiently large so that the F-system is stable.

Although the convergence analyses for multigrid algorithms (e.g. in Hackbusch's presentation in these Proceedings) are immediately extendable to such algorithms, it seem desirable to tailor their analysis more immediately to the new concept. In particular, the fixed point of such an MG-cycle has to be suitably characterized. Some progress in this direction will be reported by the authors in a separate paper.

4. Model problem analysis

We will now amplify some of the preceding discussion with a simple model problem

$$- \Delta u(x,y) + u(x,y) = 1 \quad \text{on} \quad G = [0,1] \times [0,1] \ , \qquad (4.1)$$
$$u(x,y) = 0 \quad \text{on} \quad \partial G \ .$$

*) This remark was not part of the presentation at the conference but arose as an afterthought on various remarks made by Achi B. during the conference.

We consider uniform grids $\mathbb{G}_h$, with $h = \frac{1}{n}$, $n = 2^{-m}$, $m \in \mathbb{N}$. The grid subscripts μ,ν vary over $0(1)n$ in $\mathbb{G}_h$ and over $1(1)n-1$ on the interior of $\mathbb{G}_h$.

At first we consider the computation of an $O(h^4)$-approximation to (4.1) by means of an $O(h^2)$ multigrid algorithm for (4.1). Thus (cf. sections 2 and 3), for $\mu,\nu = 1(1)n-1$,

$$(\tilde{F}u)_{\mu\nu} := \frac{1}{h^2}\begin{bmatrix} & -1 & \\ -1 & 4 & -1 \\ & -1 & \end{bmatrix} u_{\mu\nu} + \begin{bmatrix} & & \\ & 1 & \\ & & \end{bmatrix} u_{\mu\nu} ,$$

$$(Fu)_{\mu\nu} := \frac{1}{6h^2}\begin{bmatrix} -1 & -4 & -1 \\ -4 & 20 & -4 \\ -1 & -4 & -1 \end{bmatrix} u_{\mu\nu} + \frac{1}{12}\begin{bmatrix} & 1 & \\ 1 & 8 & 1 \\ & 1 & \end{bmatrix} u_{\mu\nu} ,$$

$$(\Delta Fu)_{\mu\nu} := \underbrace{\frac{1}{6h^2}\begin{bmatrix} 1 & -2 & 1 \\ -2 & 4 & -2 \\ 1 & -2 & 1 \end{bmatrix} u_{\mu\nu}}_{\approx\, h^4 (u_{xxyy})_{\mu\nu} \text{ for smooth } u} + \frac{1}{12}\begin{bmatrix} & -1 & \\ -1 & 4 & -1 \\ & -1 & \end{bmatrix} u_{\mu\nu} .$$

Because of their simple Fourier representations we consider damped Jacobi smoothers only ($\omega < 1$):

$$\tilde{S}_\omega : u \to u - \frac{\omega}{4} h^2 (\tilde{F} u - \tilde{c}) ,$$

$$S_\omega : u \to u - \frac{\omega}{10/3} h^2 (F u - c) ,$$

$$(I - \tilde{R}_\omega \tilde{F})u_{\mu\nu} = \frac{1}{4}\left(\begin{bmatrix} & \omega & \\ \omega & 4(1-\omega) & \omega \\ & \omega & \end{bmatrix} - h^2 \omega \begin{bmatrix} & & \\ & 1 & \\ & & \end{bmatrix}\right) u_{\mu\nu} ,$$

$$(I - R_\omega F)u_{\mu\nu} = \frac{1}{20}\left(\begin{bmatrix} \omega & 4\omega & \omega \\ 4\omega & 20(1-\omega) & 4\omega \\ \omega & 4\omega & \omega \end{bmatrix} - \frac{h^2}{2} \omega \begin{bmatrix} & 1 & \\ 1 & 8 & 1 \\ & 1 & \end{bmatrix}\right) u_{\mu\nu} .$$

With respect to the well-known orthogonal basis

$$(s_{kl})_{\mu\nu} := \sin k \frac{\mu}{n} \pi \cdot \sin l \frac{\nu}{n} \pi , \qquad k,l = 1(1)n-1 , \quad (4.2)$$

for the gridfunctions which vanish on ∂G, the representations of the above operations become, with

$$c_k := \cos \frac{k}{n}\pi, \quad k = 1(1)n-1, \tag{4.3}$$

$$\tilde{F}\, s_{kl} = \frac{2}{h^2}\,[(1-c_k) + (1-c_l) + \frac{h^2}{2}]\, s_{kl}$$

$$F\, s_{kl} = \frac{2}{h^2}\,[(1-c_k) + (1-c_l) - \frac{1}{3}(1-c_k)(1-c_l) + \frac{h^2}{6}\,(2 + \frac{c_k+c_l}{2})]\, s_{kl}$$

$$\Delta F\, s_{kl} = \frac{2}{3h^2}\,[(1-c_k)(1-c_l) + \frac{h^2}{4}\,(2-c_k-c_l)]\, s_{kl}$$

$$\tilde{F}^{-1}\,\Delta F\, s_{kl} = \frac{1}{3}\,\frac{(1-c_k)(1-c_l) + h^2(2-c_k-c_l)/4}{(1-c_k) + (1-c_l) + h^2/2}\, s_{kl} =: \rho_{kl}\, s_{kl}$$

$$(I - \tilde{R}_\omega\, \tilde{F})\, s_{kl} = [(1-\omega) + \frac{1}{2}\,\omega(c_k+c_l) - \frac{h^2}{4}\,\omega]\, s_{kl}$$

$$(I - R_\omega\, F)\, s_{kl} = [(1-\omega) + \frac{2}{5}\,\omega(c_k+c_l) + \frac{\omega}{5}\,c_k c_l - \frac{h^2}{10}\,\omega\,(2 + \frac{c_k+c_l}{2})]\, s_{kl}$$

It is easily checked that the individual contraction factors ρ_{kl} of the defect correction operator $\tilde{F}^{-1}\Delta F$ increase with k and l as we have assumed at the end of section 2. Although

$$\rho_{1,1} = \frac{h^2}{12}\,\frac{\pi^2(\pi^2+1)}{\pi^2 + 1/2} + O(h^4) \,\dot{<}\, h^2,$$

the convergence rate $q^{(i)}$ of the unsmoothed defect correction in version (i) of the DCMG-cycle is only

$$q^{(i)} = \|\tilde{F}^{-1}\Delta F\|_2 = \max_{k,l} \rho_{kl} = \rho_{n-1,n-1}$$

$$= \frac{1}{3}\,(1 - \frac{2\pi^2-1}{8}\,h^2 + O(h^4)) \approx \frac{1}{3}\,.$$

These numbers confirm the strong contraction of the defect correction for smooth components but also the poor overall contraction. Although $q = \frac{1}{3}$ would not destroy the convergence of the composite cycle for any reasonable multigrid algorithm for (4.1) it would still lead to an intolerably poor convergence rate, cf. (1.11).

It is true that - for FMG-algorithms - the poor convergence may not materialize because the starting approximations $u^{(o)}$ on a new finer grid level are formed by interpolation and may not contain a good deal of high frequency error components with a suitable interpolation procedure. But this smoothing effect of the interpolation may depend crucially on details of the algorithm and the problem.

We will now consider the effect of a-priori smoothing on the contraction rate of a DCMG-cycle. According to (3.3) we have the iteration operator $(I - KF)(I - RF)$ for version (ii) where $I - KF$ is the iteration operator for version (i). If we decompose $I - KF$ as in (1.12) we find

$$(I - KF)(I - RF) = (I - K\tilde{F})(I - RF) + (I - \tilde{F}^{-1}F)(I - RF) - (I - K\tilde{F})(I - \tilde{F}^{-1}F)(I - RF) .$$

Obviously, we must have

$$\|(I - \tilde{F}^{-1}F)(I - RF)\| << \|I - \tilde{F}^{-1}F\|$$

in order to achieve (3.5) because the contraction rate $\tilde{q} = \|I - K\tilde{F}\|$ of the multigrid cycle for (1.2) may be assumed to be small in any case and $\|I - RF\| \approx 1$.

From the above Fourier representations of $I - \tilde{F}^{-1}F = \tilde{F}^{-1}\Delta F$ and $I - R_\omega F$ we obtain

$$(I - \tilde{F}^{-1}F)(I - R_\omega F)\, s_{kl} = \frac{1}{3}\, \frac{(1-c_k)(1-c_l) + O(h^2)}{(1-c_k)+(1-c_l)+O(h^2)}\, [(1-\omega) + \frac{2}{5}\,\omega(c_k+c_l) + \frac{\omega}{5}\, c_k c_l + O(h^2)]\, s_{kl}$$

For $\omega = \frac{5}{8}$, where $[..] = O(h^2)$ for $c_k = c_l = -1$, we obtain

$$(I - \tilde{F}^{-1}F)(I - R_{\frac{5}{8}}F)\, s_{kl} = \left\{\frac{(1-c_k)(1-c_l)(3+2(c_k+c_l)+c_k c_l)}{24\,(2 - c_k - c_l)} + O(h^2)\right\} s_{kl}$$

and the principal term of $\{...\}$ remains below 0.065 for $(c_k, c_l) \in [-1,+1]^2$. Thus we have been able to reduce q in (1.11) approximately by a factor 5.

It is to be expected that other standard smoothers achieve comparable reductions in the effective contraction rate q of the defect correction component in the DCMG-cycle. For version (iii), we have to replace $I - R_\omega F$ by $I - \tilde{R}_\omega \tilde{F}$ and Jacobi relaxation with $\omega = \frac{1}{2}$ leads to

$$\|(I - \tilde{F}^{-1}F)(I - \tilde{R}_{\frac{1}{2}}\tilde{F})\|_2 = \frac{1}{12} + O(h^2).$$

Thus, from the contraction point of view, version (iii) with a suitable smoother should be equally effective.

Since version (iv) has been established as a simplified variant of version (iii) it should behave similarly. Here, the contraction rate $\tilde{q}$ of the multigrid component may be slightly smaller due to the truncated multigrid cycle.

5. Weakly nonlinear problems

To exhibit the possibilities of a DCMG algorithm for nonlinear problems, we consider a boundary value problem for the semilinear Helmholtz equation

$$- \Delta u(x,y) + p(x,y,u(x,y))\, u(x,y) = g(x,y) \;, \tag{5.1}$$

with $p(x,y,u) \geq 0$, and its standard $O(h^2)$-discretization. As indicated at the beginning of section 2, (1.1) will now signify this nonlinear discretization (= system of algebraic equations) while (1.2) will denote the analogous discretization of a linear equation

$$- \Delta u(x,y) + \tilde{p}(x,y)\, u(x,y) = g(x,y) \;, \tag{5.2}$$

where $\tilde{p}(x,y) \geq 0$ is a suitable approximation of $p(x,y,u(x,y))$ for the anticipated values of u. If the variation of p and/or u is small, $\tilde{p}$ may be constant (see below).

Our defect correction iteration operator $\tilde{F}^{-1}\Delta F$ is now nonlinear, with

$$(\Delta F u)_{\mu\nu} = [\tilde{p}(x_\mu,y_\nu) - p(x_\mu,y_\nu,u_{\mu\nu})]\, u_{\mu\nu} \;. \tag{5.3}$$

A Lipschitz bound q for $\tilde{F}^{-1}\Delta F$ is furnished by a bound on the Frechet derivative of $\tilde{F}^{-1}\Delta F$

$$q = \sup_u \|\tilde{F}^{-1}(\Delta F)'(u)\| \leq \|\tilde{F}^{-1}\| \cdot \sup_u \|(\Delta F)'(u)\|$$

where the sup is taken over a suitable convex domain of grid functions which contains the iterates and their limit. For (5.3),

$$[(\Delta F)'(u)v]_{\mu\nu} = [\tilde{p}(x_\mu,y_\nu) - \frac{\partial}{\partial u} p(x_\mu,y_\nu,u_{\mu\nu})u_{\mu\nu} - p(x_\mu,y_\nu,u_{\mu\nu})]\, v_{\mu\nu} \;; \tag{5.4}$$

$\|\tilde{F}^{-1}\|$ for (5.2) can be bounded by the well-known bound $(2\pi^2)^{-1} \approx 0.05$ for $\tilde{p} = 0$.

Due to this relatively low value for $\|\tilde{F}^{-1}\|$, the choice of $\tilde{p}(x,y)$ is not too crucial: If we can keep the Lipschitz bound for ΔF below 2, we have $q \approx 0.1$. Naturally, some a-priori estimate on the solution u will normally be needed for the selection of an appropriate $\tilde{p}$.

Since now ΔF is not a difference quotient there is no particular need for a-priori smoothing. Thus implementation (i) of our DCMG-cycle should be satisfactory. Due to the simple structure of ΔF (cf. (5.3)), it is advantageous to form the right hand side of the adjusted problem $\tilde{F}u = \tilde{F}u^{(i)} - d(u^{(i)})$ as $c + \Delta F\, u^{(i)}$ so that no explicit defect correction occurs:

Implementation (i)'

Form $\Delta F\, u^{(i)}$

Perform one multigrid cycle for $\tilde{F}u = c + \Delta F\, u^{(i)}$ starting with $u^{(i)}$.

Thus we have returned to our original approach in its most elementary form: Take the standard defect correction pattern (cf. (1.3B))

Form $\Delta F\, u^{(i)}$

Solve $\tilde{F}u = c + \Delta F\, u^{(i)}$ for $u^{(i+1)}$

and replace the direct solution by one multigrid cycle. Note that $\tilde{F}\, u = c + \Delta F\, u^{(i)}$ is normally not a local linearization of $Fu = c$ in the usual sense of the word, i.e. a Frechet derivative of the original problem; this would be more complicated in the present situation, cf. (5.4).

The following concrete example will show some details of the approach more clearly: Consider the boundary value problem

$$
\begin{aligned}
-\Delta u(x,y) + e^{u(x,y)} u(x,y) &= 1 \quad \text{on} \quad G = [0,1]^2 , \\
u(x,y) &= 0 \quad \text{on} \quad \partial G .
\end{aligned}
\tag{5.5}
$$

It is easily seen that $u \geq 0$ and $\Delta u \leq 0$ in G, hence $0 \leq e^u u < 1$ or $0 \leq u < 0.57$ must hold for the true solution of (5.5). Therefore

$$\tilde{p}(x,y) \equiv 1$$

is a reasonable approximation for $p(x,y,u) = e^u$. For it,

$$(\Delta F\, u)_{\mu\nu} = (1 - e^{u_{\mu\nu}})\, u_{\mu\nu}$$

and a Lipschitz bound for ΔF in a neighborhood of the true solution is (see (5.4))

$$\max_{u\in[0,0.6]} |1-e^u u - e^u| < 2 . \qquad (5.6)$$

This leads to a bound 0.1 for the contraction rate q of the defect correction as explained above.

But from (5.6) we also realize that a constant value $\tilde{p}$ at the upper end of the range of e^u would have been a better choice: For $\tilde{p} = 1.8$, we obtain a Lipschitz bound near 1 and $q \approx 0.05$. Note that the linear operator with $\tilde{p} = 1$ could have been constructed as $F'(u_0)u$ with $u_0 \equiv 0$ while the linear operator with $\tilde{p} = 1.8$, although it may formally be back-interpreted as $F'(u_0)u$ with $u_0 \equiv 0.314...$, has been obtained by a completely different reasoning.

All that remains to be done is to take a multigrid program which is suitable for $-\Delta u + 1.8\, u = 1$ and to insert a few lines of code which update the right hand side of the difference equation to

$$1 + (1.8 - e^{u_{\mu\nu}^{(i)}})\, u_{\mu\nu}^{(i)}$$

at the beginning of each multigrid cycle where $u^{(i)}$ is the approximation available at this time on the current finest grid. This modified program will compute an approximate solution for (5.5), with hardly any extra effort beyond that for a comparable linear problem.

It is obvious that the present approach may be combined with that of the previous sections to obtain a DCMG-cycle which computes an $O(h^4)$-approximation for (5.5), again with not much additional effort.

6. Numerical results

In a first preliminary set of test runs, we used the code MGOOD2 from the multigrid package MGOO of the GMD-IMA ([6]). It solves the standard $O(h^2)$-discretization of a Dirichlet problem for a Helmholtz equation on a rectangle in a fixed MG or an FMG mode, using checkered relaxation for smoothing. We used the V-cycle mode throughout. A subroutine for computing defects w.r.t. the $O(h^4)$-discretization of section 4 was added as well as various a priori smoothing procedures.

The following test problem was used:

$$-\Delta u(x,y) + (1+x^2+y^2)\, u(x,y) = g(x,y) \quad \text{on} \quad G = [0,1]^2$$
$$u(x,y) = 0 \quad \text{on} \quad \partial G \; ; \qquad (6.1)$$

g was chosen such that (6.1) yielded the true solution

$$U(x,y) = \sin \pi x \sin \pi y + 0.2 \sin 5\pi x \sin 5\pi y. \qquad (6.2)$$

To exhibit the efficiency of the various DCMG implementations of section 3 (see Table 1), we employ the accuracy of the exact $O(h^2)$-solution u_h^* has a reference, i.e. we define its maximal truncation error $\|u_h^* - U_h\|_\infty$ as 1. This accuracy level is reached by the FMG algorithm implemented in MGOOD2 with one MG-cycle per grid level; hence we count computing times in multiples of the time needed by this code to compute $\approx u_h^*$. Both references are used separately for finest grids with $h = \frac{1}{16}$ and $h = \frac{1}{32}$. All accuracies refer to the distance from the true solution (6.2); thus, in terms of the reference system for $h = \frac{1}{16}$, u_h^* for $h = \frac{1}{32}$ would have an accuracy of $\approx$ .25 and need an effort $\approx$ 4.

In Table 1, implementation (iia) employs a-priori smoothing w.r.t. the $O(h^4)$-discretization by means of one $\frac{5}{8}$ - damped Jacobi relaxation (cf. section 4), implementation (iib) uses one "4 color relaxation". In implementation (iii), one $\frac{1}{2}$ - damped Jacobi relaxation w.r.t. the $O(h^2)$-discretization is used. Implementation (iv) employs only the original smoothing of the MG cycle, cf. section 3.

The following facts are obvious from table 1:

- All versions lead to an $O(h^4)$ approximation of U, but only (i) and (ii) produce the solution determined by the defect operation.
- While the fixed point shift is harmless for version (iii), it is quite significant for version (iv) (which had been suggested by Hackbusch and Brandt/Dinar).
- The improvement in convergence with a-priori smoothing is significant. With a suitable smoother (e.g. (iib)), one V-cycle per grid level in an FMG-algorithm suffices for a close approximation of the $O(h^4)$-solution determined by the defect operation. Without a-priori smoothing, two V-cycles per grid level are necessary, which is more expensive.
- Defect correction achieves more accuracy than further grid refinement at less cost. Note that - at $h = \frac{1}{32}$ - the 50% time increase (over the original MGOOD2) of implementation (iib) buys a reduction of the max

error of the computed approximation of (6.2) by a factor of 150-200, from a poor $.5 \times 10^{-2}$ to a decent $.3 \times 10^{-4}$!

We also determined contraction rates by iterating the MG cycles on a fixed grid ($h = \frac{1}{32}$). Table 2 contains the reduction of the max error within the 5th cycle. The numbers have to be interpreted cautiously because in all versions the rates were much lower in the first cycles, probably due to the smooth error of our initial approximation $u_h \equiv 0$.

Finally, we tested the approach of section 5 on problem (5.5), again on the basis of the code MGOOD2. No a-priori smoothing was used (cf. section 5). Both $\tilde{p} = 1$ and $\tilde{p} = 1.8$ were tried.

The early contraction rate was $\approx .1$ for $p = 1$ and $\approx .05$ for $p = 1.8$, which confirms our considerations. In the FMG mode, with one V-cycle per grid level, however, defect correction with $\tilde{p} = 1$ produced a better approximation to the exact $O(h^2)$-solution than with $\tilde{p} = 1.8$.

In any case, the truncation error level was reached with one V-cycle per grid level. Although we have run no test, we believe that the time needed by our DCMG-version of a linear MG code remains below that of a nonlinear FAS-code for the same job.

Conclusions

We have considered some ways how defect correction may be combined with multigrid algorithms. We have indicated how effective algorithms may be designed and what is necessary to obtain the full power of the approach. Such algorithms need only minor modifications of standard multigrid software for linear problems.

Since several such modifications may be standarized they could be built into future multigrid software as modes or options. This would greatly increase the range and flexibility of such software at the cost of little extra code.

DCMG - Implementation

	NCYCL	(i) Accuracy	(i) Effort	(iia) Acc.	(iia) Eff.	(iib) Acc.	(iib) Eff.	(iii) Acc.	(iii) Eff.	(iv) Acc.	(iv) Eff.
$h = \frac{1}{16}$	1	.3459	1.35	.1001	1.53	.0308	1.49	.1218	1.45	.3518	1.35
	2	.0268	1.79	.0233	2.07	.0245	2.08	.0393	1.89	.3720	1.78
T.E. = $.1980_{10}{-1}$	3	.0248	2.14	.0243	2.63	.0244	2.62	.0376	2.47	.3825	2.14
$h = \frac{1}{32}$	1	.0980	1.33	.0078	1.52	.0050	1.52	.0097	1.47	.2664	1.42
	2	.0160	1.71	.0066	2.10	.0067	2.03	.0077	2.00	.1021	1.91
T.E. = $.4797_{10}{-2}$	3	.0070	2.10	.0067	2.78	.0067	2.66	.0076	2.43	.0984	2.35

Table 1

MGOOD2	(i)	(iia)	(iib)	(iii)	(iv)
.108	.299	.073	.106	.300	.068

Table 2: Contraction rates

Bibliography

[1] A. Brandt - N. Dinar: Multi-grid solutions to elliptic flow problems, ICASE Report 79-15, July 1979.

[2] W. Auzinger - H.J. Stetter: Extrapolation beim Multigrid-Verfahren, Vortrag, GAMM-Tagung 1981, Würzburg.

[3] W. Hackbusch: Bemerkungen zur iterierten Defektkorrektur und zu ihrer Kombination mit Mehrgitterverfahren, Report 79-13, Angew. Math. Univ. Köln, September 1979.

[4] W. Hackbusch: Introduction to multigrid methods for the numerical solution of boundary value problems, in: Computational Methods for Turbulent, Transonic, and Viscous Flows (J.A. Essers, ed.), Hemisphere Publ. Corp., to appear in 1982.

[5] H.J. Stetter: The defect correction principle and discretization methods, Num. Math. 29 (1978) 425-443.

[6] H. Foerster - K. Witsch: On efficient multigrid software for elliptic problems on rectangular domains, Math. and Computers in Simulation 28 (1981), no. 3.

ON MULTIGRID METHODS OF THE TWO-LEVEL TYPE.

O. Axelsson
Department of Mathematics
University of Nijmegen, The Netherlands

Abstract

Modified proofs of spectral equivalence in connection with methods of two-level grid type are presented. The upper bounds on the condition numbers are independent on the mesh parameter and on the smoothness of the solution. Some particular two-level methods are extended by recursion to multigrid methods.

1. Introduction

The numerical solution of elliptic boundary value problems by finite element methods is now a well established technique. Its great power is mainly due to its generality in coping with boundaries of various shapes and with various types of boundary conditions but is also due to the ease of getting higher order approximations.

For the solution of the resulting linear systems of algebraic equations, quite efficient methods of both iterative and direct types already exists if the problems are of not too large size. For very large problems and in particular for three dimensional problems, the iterative methods are usually more efficient. One type of such methods are based on certain preconditioning techniques coupled with a conjugate gradient method as an accelerating device (see for instance [1]). Another type of methods are based on multigrid techniques similar to those used for difference methods (see [8], [10], [4], and [5] and the references quoted therein).

These latter methods seem to base their success on utilizing the smoothness of the solution (in the interpolation process). The damping of the highly oscillatory iteration error components can be done by some simple smoothing process and the error reduction factor has been proven to be very small for model problems (see other talks in these proceedings). As is wellknown, the resulting computational complexity is then of optimal order or nearly of optimal order, O(N log N), where N is the number of unknowns.

However it is not clear how efficient the method is on more general problems, in particular those with many singularities due for instance to boundary and interior corners (the latter may result from discontinuous coefficients in the boundary value problem). Hence it is of interest to construct methods for which a small reduction factor can be proven utilizing purely algebraic means, i.e. not based on any smoothness of the solution.

A candidate for such a class of methods are methods based on the two-level grid method, originally considered by Bank-Dupont [6] (see also Braess [7] and Axelsson, Gustafsson [2]). Such methods may be implemented into a quite efficient numerical method in many ways. Some possibilities are discussed in [2] where an implementation based on incomplete factorization and a conjugate gradient method was used.

Here we shall mainly discuss an alternative approach, which is applicable when the domain consists of a union of axiparallell rectangles or when the given domain has been mapped, for instance by an isoparametric transformation, onto such a domain. In this method the problem is recursively reduced in size until such a small size is reached that the problem may be solved directly. This is not done by the usual substructuring techniques, for which the asymptotic computational complexity would be much larger, but with the help of nested iterations.

2. The bilinear form

Given a Hilbert space V and a coercive, symmetric and bounded bilinear form a(.,.) on V×V we let a(.,.) define an innerproduct. Let $|||\cdot|||$ be the associated norm, i.e.

$$|||v||| = a(v,v)^{\frac{1}{2}} \quad \forall v \in V.$$

Between the innerproduct and the norm we have the following relation, the extended C-B-S inequality,

$$|a(u,v)| \le |||u|||\ |||v||| \quad \forall u,v \in V.$$

Let V_1, V_2 be two nontrivial subspaces of V with only the trivial element in common, i.e. $V_1 \cap V_2 = \{0\}$. Then the stronger relation

(2.1) $$|a(u,v)| \le \gamma|||u|||\ |||v||| \quad \forall u \in V_1,\ v \in V_2,$$

where

(2.2) $$0 < \gamma = \sup_{u \in V_1, v \in V_2} \{a(u,v) \,/\, |||u|||\ |||v|||\} < 1,$$

is valid.
We note that

(2.3) $$\gamma \ge \tilde{\gamma},$$

where

(2.4) $$\tilde{\gamma} = \sup_{u \in V_1; v \in V_2} 2a(u,v) \,/\, [|||u|||^2 + |||v|||^2],$$

because

$$2|||u|||\ |||v||| \le |||u|||^2 + |||v|||^2.$$

However, due to homogeneity, the supremum in (2.2) is taken for some $\hat{u}$, $\hat{v}$, such that

$$|||\hat{u}||| = |||\hat{v}||| = 1. \text{ But}$$

$$\tilde{\gamma} \ge 2a(\hat{u},\hat{v}) \,/\, [|||\hat{u}|||^2 + |||\hat{v}|||^2] = a(\hat{u},\hat{v}) = \gamma.$$

Together with (2.3), this implies that

$$\gamma = \tilde{\gamma}.$$

In fact, due to symmetry of a(.,.), the supremum in (2.4) must be taken for $\hat{u},\hat{v}$ such that $|||\hat{u}||| = |||\hat{v}|||$. This implies immediately that $\gamma = \tilde{\gamma}$.

If a(.,.) is not coercive but only non negative, $a(u,u) \ge 0 \quad \forall u \in V$, then $|||\cdot|||$ is only a seminorm. If neither V_1 nor V_2 contains any non trivial element u_0 in the nullspace N_0 of a(u,v) (i.e. for which $a(u_0,v) = 0 \quad \forall v \in V$), then with $v = u_0 + u$ we get

$$a(v,v) = a(u_0 + u, u_0 + u) = a(u,u)$$

and

$$a(u,v) = a(u,u_0 + u) = a(u,u).$$

Hence $\gamma = 1$.

Therefore, in the following we assume that all such $u_0 \in N_0$ is in one of the two subspaces, say $u_0 \in V_1$, i.e. we assume that $N_0 \subset V_1$. It follows than that (2.1) and the analysis to come, remains valid even for non-coercive (but nonnegative) bilinear forms.

We shall now describe an application on a discretized boundary value problem for which we may choose the subspaces V_1, V_2 such that the corresponding γ does not depend on the mesh discretization parameter h. It will later be realized that this is a property of fundamental importance for the effectiveness of the solution method of the associated linear algebraic systems of equations.

Hence consider the boundary value problem

$$-\sum_{i,j=1}^{d} \frac{\partial}{\partial x_i}\left(a_{ij}\frac{\partial u}{\partial x_j}\right) + bu = f, \ \underset{\sim}{x} \in \Omega \subset \mathbb{R}^d,$$

$$u = 0, \ \underset{\sim}{x} \in \Gamma_1, \ \sum_{i,j} a_{ij}\frac{\partial u}{\partial x_j} n_i = g, \ \underset{\sim}{x} \in \Gamma_2 = \Gamma - \Gamma_1$$

where n_i are the components of the unit normal. Ω is a bounded (hyper-) polygonal domain. We assume that the symmetric matrix $[a_{ij}(\underset{\sim}{x})]_{i,j=1}^{d}$ is uniformly positive definite, that $b \geq 0$ and that $|a_{ij}|$ and b are uniformly bounded from above on $\overline{\Omega}$. To this boundary value problem we associate the bilinear form

$$(2.5) \qquad a(u,v) = \int_\Omega \Big[\sum_{i,j} a_{ij}\frac{\partial u}{\partial x_i}\frac{\partial v}{\partial x_j} + buv\Big]d\Omega, \ u,v \in V,$$

$$V = \{v \in H^1(\Omega), \ v = 0 \text{ on } \Gamma_1\}$$

and the variational formulation

$$a(u,v) = \int_\Omega fv d\Omega + \oint_{\Gamma_2} gv d\Gamma_2, \quad \forall v \in V.$$

$a(.,.)$ is symmetric and bounded. If meas $(\Gamma_1) \neq 0$, and/or if $b \geq b_0 > 0 \quad \forall \underset{\sim}{x} \in \overline{\Omega}$, then $a(.,.)$ is also coercive and then, as wellknown, to every $f \in L^2(\Omega)$, $g \in L^2(\Gamma_2)$, there exists a unique solution $\hat{u} \in V$. If $b \equiv 0$ and meas $(\Gamma_1) = 0$, then $a(.,.)$ is only nonnegative and the solution $\hat{u} \in V$ is unique up to a constant term. Note that even if $a(.,.)$ is coercive on Ω, its restriction to elements (see section 3) may not be coercive. Motivated by the discussion above about the nullspace N_0, we let $N_0 \subset V_1$.

3. Spectral equivalence

For ease of presentation we shall describe the method only for plane problems (d = 2). Let Ω_1 be a (coarse) subdivision of Ω into triangles, let Ω_h be the corresponding subdivision when each edge of Ω_1 has been divided into h^{-1} equal parts or intervals, and let T_h be the resulting number of triangles. At the endpoints of each interval we place a node and there may also be some interior nodes on each triangle. Let the total number of nodes on each triangle be q. To every node we associate a basisfunction with local support; those associated with vertex nodes span a space V_1 and the remaining ones a space V_2.

We assume that no function, whose restriction to an arbitrary triangle is a constant function, other than zero, is in V_2.

Let $e^{(\ell)}$ be an arbitrary triangle (or element), $\ell = 1,2,\ldots,T_h$. The restriction to $e^{(\ell)}$ of the basisfunctions with support on $e^{(\ell)}$ (i.e. those corresponding to nodes on $\bar{e}^{(\ell)}$ is denoted $\{\phi_i^{(\ell)}\}_{i=q-2}^{q}$ for the vertex nodes and $\{\phi_i^{(\ell)}\}_{i=1}^{q-3}$ for the remaining nodes (We do not always explicitly indicate the dependence on h.). To e_ℓ and its basisfunctions we associate the matrix

$$A^{(\ell)} = [a_\ell(\phi_j^{(\ell)},\phi_i^{(\ell)})]_{i,j=1}^{q},$$

where $a_\ell(.,.)$ is the restriction of a(.,.) in (2.5) to $e^{(\ell)}$. With the chosen ordering of the nodes, this matrix has the block structure

$$\text{(3.1a)} \qquad A^{(\ell)} = \begin{bmatrix} B^{(\ell)} & C^{(\ell)} \\ C^{(\ell)t} & A^{(\ell)} \end{bmatrix}$$

where

$$A^{(\ell)} = [a(\phi_j^{(\ell)},\phi_i^{(\ell)})]_{i,j=q-2}^{q}, \quad B^{(\ell)} = [a(\phi_j^{(\ell)},\phi_i^{(\ell)})]_{i,j=1}^{q-3}$$

and

$$C^{(\ell)} = [a(\phi_j^{(\ell)},\phi_i^{(\ell)})], \quad j = q-2,q-1,q, \quad i = 1,2,\ldots,q-3.$$

The global matrices are derived by assembly of the $A^{(\ell)}$'s in the usual way, i.e.

$$A = A^{(1)} \oplus A^{(2)} \oplus \ldots \oplus A^{(T_h)}$$

etc. We use a global ordering, where the vertex nodes appear first. Then A has the corresponding block matrix structure as $A^{(\ell)}$, i.e.

$$\text{(3.1b)} \qquad A = A_h = \begin{bmatrix} B_h & C_h \\ C_h^t & A_h \end{bmatrix}.$$

We use the following notations,

$$\underset{\sim}{A}^{(\ell)} = \begin{bmatrix} 0 & 0 \\ 0 & A^{(\ell)} \end{bmatrix}, \quad \underset{\sim}{B}^{(\ell)} = \begin{bmatrix} B^{(\ell)} & 0 \\ 0 & 0 \end{bmatrix}, \quad \ell = 1,2,\ldots,T_h,$$

both of order q, and

$$\tilde{A}_h = \begin{bmatrix} 0 & 0 \\ 0 & A_h \end{bmatrix}, \quad \tilde{B}_h = \begin{bmatrix} B_h & 0 \\ 0 & 0 \end{bmatrix}.$$

Lemma 3.1. The matrix B_h as defined above is positive definite and has a spectral condition number $O(1)$, $h \to 0$.

Proof. Let $v \in V_2$. Its restriction to $e^{(\ell)}$ satisfies $v = \sum_{i=1}^{q-3} \alpha_i^{(\ell)} \phi_i^{(\ell)}$. Let (in this proof) the vector $\tilde{\alpha}^{(\ell)}$ be defined by $\tilde{\alpha}^{(\ell)^t} = [\alpha_1^{(\ell)},\dots,\alpha_q^{(\ell)}]$, where $\alpha_i^{(\ell)} = 0$, $i = q-2,q-1,q$. We have

$$0 < \mu_\ell^{(1)} \tilde{\alpha}^{(\ell)^t} \tilde{\alpha} \le a_\ell(v,v) = \tilde{\alpha}^{(\ell)^t} A^{(\ell)} \tilde{\alpha}^{(\ell)} = \tilde{\alpha}^{(\ell)^t} \tilde{B}^{(\ell)} \tilde{\alpha}^{(\ell)}$$

$$\le \mu_\ell^{(0)} \tilde{\alpha}^{(\ell)^t} \tilde{\alpha} \quad \forall \tilde{\alpha}^{(\ell)} \ne \tilde{0}.$$

Here $\mu_\ell^{(1)}$, $\mu_\ell^{(0)}$ are the extreme eigenvalues of $B^{(\ell)}$. $\mu_\ell^{(1)}$ is positive, because functions which are constant on $e^{(\ell)}$ are excluded from the function space V_2. (Otherwise, if $b \equiv 0$ on $e^{(\ell)}$, then $a_\ell(v,v) = 0$ if $v|_{e^{(\ell)}} \equiv$ constant.) Further, since the entries of $A^{(\ell)}$ are $O(1)$, $h \to 0$, $\mu_\ell^{(0)}$ is bounded from above by a number independent on h. Since the derivative term in the integrand of a_ℓ does not vanish, $\mu_\ell^{(1)}$ is likewise bounded below by a positive number independent on h.

By summation over the triangles we get

$$0 < \min_\ell \mu_\ell^{(1)} p_1 \tilde{\alpha}^t \tilde{\alpha} \le a(v,v) = \tilde{\alpha}^t \tilde{A} \tilde{\alpha} = \tilde{\alpha}^t \tilde{B}_h \tilde{\alpha}$$

$$\le \max_\ell \mu_\ell^{(0)} p_2 \tilde{\alpha}^t \tilde{\alpha} \quad \forall \tilde{\alpha} \ne \tilde{0}.$$

Here $\tilde{\alpha}$ is the extension of the $\tilde{\alpha}^{(\ell)}$'s to Ω, i.e. the restriction of $\tilde{\alpha}$ to $e^{(\ell)}$ satisfies

$$\tilde{\alpha}|_{e^{(\ell)}} = \tilde{\alpha}^{(\ell)}, \quad \ell = 1,2,\dots,T_h.$$

Hence, in particular, the components of $\tilde{\alpha}$ corresponding to vertex nodes are all zero. p_1 and p_2 are the smallest and largest numbers, respectively of triangles having any of the nonvertex nodes in common. (If there exists any interior node, than $p_1 = 1$. Further $p_2 = 2$, except in trivial cases.) Hence the spectral condition number of B_h is bounded above by

$$H(B_h) \le \frac{p_2}{p_1} \max_\ell \mu_\ell^{(0)} \Big/ \min_\ell \mu_\ell^{(1)}$$

and this number is itself bounded above by a number independent on h. Hence the assertion of the Lemma follows. □

Definition 3.1. Let $N = N_h$ be the number of nodes in Ω_h and let $\{F_h\}$, $\{G_h\}$ be two sequences of matrices of order N_h. The sequences are said to be spectrally equivalent if there exists positive numbers δ_1, δ_0 independent on h, such that

$$\delta_1 \tilde{\alpha}^t F_h \tilde{\alpha} \le \tilde{\alpha}^t G_h \tilde{\alpha} \le \delta_0 \tilde{\alpha}^t F_h \tilde{\alpha} \quad \forall \tilde{\alpha} \in \mathbb{R}^{N_h}.$$

We call δ_0/δ_1 (an upperbound of) the condition number of the spectral equivalence.

<u>Lemma 3.2.</u> The matrices $(\underset{\sim}{A}_h + \underset{\sim}{B}_h)$ and A_h are spectrally equivalent with condition number $(1+\gamma)/(1-\gamma)$.

<u>Proof.</u> (This result is already found in Bank, Dupont [6], but is presented here for the sake of completeness.)
By the definition (2.1) of γ_ℓ corresponding to $e^{(\ell)}$ we have with $u \in V_1$, $v \in V_2$,

$$2a_\ell(u,v) \le 2\gamma_\ell\{a_\ell(u,u)\ a_\ell(v,v)\}^{\frac{1}{2}}$$
$$\le \gamma_\ell\{a_\ell(u,u) + a_\ell(v,v)\}.$$

Hence

$$(1-\gamma_\ell)[a_\ell(u,u) + a_\ell(v,v)] \le a_\ell(u,u) + a_\ell(v,v) + 2a_\ell(u,v) =$$
$$= a_\ell(u+v,\ u+v) \le (1+\gamma_\ell)[a_\ell(u,u) + a_\ell(v,v)].$$

By letting $\underset{\sim}{\alpha}^{(\ell)} = (\alpha_1^{(\ell)},\dots,\alpha_q^{(\ell)})$ be defined by

$$u = \sum_{i=q-2}^{q} \alpha_i^{(\ell)}\phi_i^{(\ell)}, \qquad v = \sum_{i=1}^{q-3} \alpha_i^{(\ell)}\ \phi_i^{(\ell)},$$

we have then

$$(1-\gamma_\ell)\underset{\sim}{\alpha}^{(\ell)t}(\underset{\sim}{A}^{(\ell)}+\underset{\sim}{B}^{(\ell)})\underset{\sim}{\alpha}^{(\ell)} \le \underset{\sim}{\alpha}^{(\ell)t}A^{(\ell)}\underset{\sim}{\alpha}^{(\ell)} \le (1+\gamma_\ell)\underset{\sim}{\alpha}^{(\ell)t}(\underset{\sim}{A}^{(\ell)}+\underset{\sim}{B}^{(\ell)})\underset{\sim}{\alpha}^{(\ell)}.$$

By summation over all triangles we get finally,

$$(3.2)\qquad (1-\gamma)\underset{\sim}{\alpha}^t(\underset{\sim}{A}_h+\underset{\sim}{B}_h)\underset{\sim}{\alpha} \le \underset{\sim}{\alpha}^t A_h\underset{\sim}{\alpha} \le (1+\gamma)\underset{\sim}{\alpha}^t(\underset{\sim}{A}_h+\underset{\sim}{B}_h)\underset{\sim}{\alpha}.$$

Here $\gamma = \max_\ell \gamma_\ell$ and $1-\gamma \ge \delta > 0$. By the uniform meshrefinement as described in section 2, it follows that all angles in the triangles are given by those in the original coarse mesh Ω_1. By a homogeneity argument it then follows that $\gamma \le 1-\delta$, for some $\delta > 0$, independent on h. Hence the spectral equivalence follows from (3.2). □

<u>Theorem 3.1.</u> A_h and $A_h - C_h^t\ B_h^{-1}\ C_h$ are spectrally equivalent with condition number $1/(1-\gamma^2)$.

<u>Proof.</u> By Lemma 3.1, B_h is invertible and positive definite. If A_h is nonsingular, then A_h is also positive definite and

$$A_h^{-\frac{1}{2}}(A_h - C_h^t\ B_h^{-1}\ C_h)A_h^{-\frac{1}{2}} = I - F_h^t\ F_h$$

where

$$F_h = B_h^{-\frac{1}{2}}\ C_h\ A_h^{-\frac{1}{2}}.$$

But

$$\begin{bmatrix} B_h^{-\frac{1}{2}} & 0 \\ 0 & A_h^{-\frac{1}{2}} \end{bmatrix} A \begin{bmatrix} B_h^{-\frac{1}{2}} & 0 \\ 0 & A_h^{-\frac{1}{2}} \end{bmatrix} = \begin{bmatrix} I & F_h \\ F_h^t & I \end{bmatrix}$$

whose spectrum by Lemma 3.2 is contained in the interval $[1-\gamma, 1+\gamma]$. Hence the spectrum of $\begin{bmatrix} 0 & F_h \\ F_h^t & 0 \end{bmatrix}$ is contained in $[-\gamma,\gamma]$, from which it follows that

$$||F_h||_2 = \rho(F_h^t F_h)^{\frac{1}{2}} \le \gamma.$$

Hence the spectrum of $I - F_h^t F_h$ is in $[1-\gamma^2,1]$ from which the assertion follows. If A_h is singular, then the nullspace $\mathrm{Ker}(A_h) \subset \mathrm{Ker}(C_h)$. ($A_h$ is singular iff $b \equiv 0$ and meas $(\Gamma_1) = 0$.) As above we realize that

$$(1-\gamma_\ell^2)\underset{\sim}{\alpha}^{(\ell)^t} A^{(\ell)} \underset{\sim}{\alpha}^{(\ell)} \le \underset{\sim}{\alpha}^{(\ell)^t} (A^{(\ell)} - C^{(\ell)^t} B^{(\ell)^{-1}} C^{(\ell)}) \underset{\sim}{\alpha}^{(\ell)} \le \underset{\sim}{\alpha}^{(\ell)^t} A^{(\ell)} \underset{\sim}{\alpha}^{(\ell)}$$

$$\forall \underset{\sim}{\alpha}^{(\ell)} = (\alpha_{q-2}^{(\ell)}, \alpha_{q-1}^{(\ell)}, \alpha_q^{(\ell)}), \underset{\sim}{\alpha}^{(\ell)} \notin \mathrm{Ker}(A^{(\ell)}).$$

Hence by summation with respect to ℓ, we get

$$(1-\gamma^2)\underset{\sim}{\alpha}^t A_h \underset{\sim}{\alpha} \le \underset{\sim}{\alpha}^t (A_h - C_h^t B_h^{-1} C_h) \underset{\sim}{\alpha} \le \underset{\sim}{\alpha}^t A_h \underset{\sim}{\alpha}$$

$\forall \underset{\sim}{\alpha}$, where $\underset{\sim}{\alpha}|_{e^{(\ell)}} = \underset{\sim}{\alpha}^{(\ell)} = (\alpha_{q-2}^{(\ell)}, \alpha_{q-1}^{(\ell)}, \alpha_q^{(\ell)})$, and the assertion follows. □

Finally we note that the actual calculation of $\gamma = \max_\ell \gamma_\ell$ may be done algebraically, utilizing the matrix F_h in the proof of Theorem 3.1. It follows namely, that on element levels,

$$\gamma_\ell = \sup_{\underset{\sim}{\alpha} \in \mathbb{R}^3} \frac{||F^{(\ell)^t} \underset{\sim}{\alpha}||}{||\underset{\sim}{\alpha}||}, \quad F^{(\ell)} = B^{(\ell)^{-\frac{1}{2}}} C^{(\ell)} A^{(\ell)^{-\frac{1}{2}}},$$

if A_ℓ is nonsingular (which is the case if $b|_{e^{(\ell)}} > 0$). But then

$$\gamma_\ell^2 = \sup_{\underset{\sim}{\alpha} \in \mathbb{R}^3} \frac{\underset{\sim}{\alpha}^t A^{(\ell)^{-\frac{1}{2}}} C^{(\ell)^t} B^{(\ell)^{-1}} C^{(\ell)} A^{(\ell)^{-\frac{1}{2}}} \underset{\sim}{\alpha}}{\underset{\sim}{\alpha}^t \underset{\sim}{\alpha}}$$

$$= \sup_{\underset{\sim}{\alpha} \in \mathbb{R}^3} \frac{\underset{\sim}{\alpha}^t C^{(\ell)^t} B^{(\ell)^{-1}} C^{(\ell)} \underset{\sim}{\alpha}}{\underset{\sim}{\alpha}^t A^{(\ell)} \underset{\sim}{\alpha}},$$

that is, γ_ℓ^2 is the largest eigenvalue of the generalized eigenvalue problem,

$$\lambda A^{(\ell)} \underset{\sim}{\alpha} = C^{(\ell)^t} B^{(\ell)^{-1}} C^{(\ell)} \underset{\sim}{\alpha}. \tag{3.3}$$

It is easily proven (for instance by perturbing $A^{(\ell)}$ slightly by a positive definite matrix, for instance by letting $b := b + \varepsilon$, $\varepsilon > 0$, in (2.5) and then letting $\varepsilon \to 0$) that (3.3) also determines γ_ℓ^2 when $A^{(\ell)}$ is singular.

This algebraic method of calculating γ_ℓ was used in [2] and in [9]. Another way of calculating it is by direct use of the defining equation (2.2),

$$\gamma_\ell = \sup_{u,v} a_\ell(u,v) \quad , \quad a_\ell(u,u) = a_\ell(v,v) = 1.$$

As an example, consider the case of a rectilinear triangle T with unit edge lengths and the bilinear form

$$a(u,v) = \int_T \underset{\sim}{\nabla} u \cdot \underset{\sim}{\nabla} v \, dx\, dy.$$

If we use quadratic basisfunctions on the midedge nodes, and linear on the vertex

nodes (see Figure 3.1),

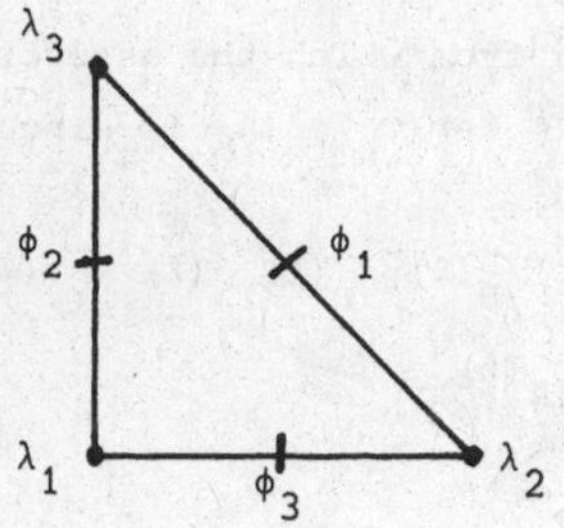

Figure 3.1. A rectilinear quadratic polynomial approximation element.

then any function $u \in V_1$ is on the form

$$u = \alpha_0 + \alpha_1 x + \alpha_2 y$$

and any function $v \in V_2$ is on the form

$$v = \beta_1\phi_1 + \beta_2\phi_2 + \beta_3\phi_3$$

where

$$\phi_1 = 4\lambda_2\lambda_3 \ , \ \phi_2 = 4\lambda_1\lambda_3 \ , \ \phi_3 = 4\lambda_1\lambda_2$$

and λ_i, $i = 1,2,3$ are the barycentric coordinates.
We find

$$(3.4) \qquad \begin{aligned} a(u,u) &= 1 \Rightarrow \tfrac{1}{2}(\alpha_1^2 + \alpha_2^2) = 1 \ , \\ a(v,v) &= 1 \Rightarrow \beta_1^2 + \beta_2^2 + \beta_3^2 - \beta_1\beta_2 - \beta_1\beta_3 = \frac{3}{8} \end{aligned}$$

and

$$a(u,v) = \iint_T \underset{\sim}{\nabla} u \, \underset{\sim}{\nabla} v \, dx\, dy = \frac{2}{3}\,[\alpha_1(\beta_1 - \beta_2) + \alpha_2(\beta_1 - \beta_3)].$$

An elementary calculation (for instance by use of Lagrange multiplicators) of maximizing $a(u,v)$ as a function of $\alpha_1,\alpha_2,\beta_1,\beta_2,\beta_3$, subject to the constraints (3.4), gives the solution

$$\alpha_1 = \alpha_2 = 1 \ , \ \beta_2 = \beta_3 = 0 \ , \ \beta_1 = \sqrt{3/8}$$

and

$$\gamma = a(u,v) = \sqrt{2/3}.$$

Similarly, for the case of so called serendipity quadratic approximation elements (see Figure 4.1), one finds

$$\gamma = \sqrt{5/11} \approx 0{,}67.$$

Note finally that if we already know that the functions u,v are interpolants of smooth functions, then much smaller condition numbers can be achieved. Such a case arises for instance if we work through a sequence of meshes $\{\Omega_h\}$, $h = 1,\frac{1}{2},\frac{1}{4},\ldots$ and let the interpolant of the approximate solution u_{2h} on the previous mesh be the initial approximation in the iterative process used for calculating the solution u_h on the mesh Ω_h. For the calculation of γ_ℓ we should then use the second of the two methods presented above.

4. Recursive use of two-level iterative methods

The results of section 3 may be applied in several ways in order to derive efficient methods for the iterative solution of

$$a(u_h,v_h) = (f,v_h) \quad \forall v_h \in V_1 \cup V_2$$

or, by (3.1b),

$$\mathcal{A}_h \begin{bmatrix} \underset{\sim}{\beta} \\ \underset{\sim}{\alpha} \end{bmatrix} = \begin{bmatrix} B_h & C_h \\ C_h^t & A_h \end{bmatrix} \begin{bmatrix} \underset{\sim}{\beta} \\ \underset{\sim}{\alpha} \end{bmatrix} = \begin{bmatrix} \underset{\sim}{f}_2 \\ \underset{\sim}{f}_1 \end{bmatrix} \tag{4.1}$$

(The definition of the blockvectors should be evident.)
Firstly we note that due to the spectral equivalence as outlined in Lemma 3.2, we may use

$$\mathcal{D}_h = \begin{bmatrix} B_h & 0 \\ 0 & A_h \end{bmatrix}$$

as a preconditioner of $\mathcal{A}_h$, i.e. we may use a basic iterative method on the form

$$\begin{bmatrix} B_h & 0 \\ 0 & A_h \end{bmatrix} \begin{bmatrix} \underset{\sim}{\beta}^{k+1} - \underset{\sim}{\beta}^{k} \\ \underset{\sim}{\alpha}^{k+1} - \underset{\sim}{\alpha}^{k} \end{bmatrix} = -\tau_k \left(\mathcal{A}_h \begin{bmatrix} \underset{\sim}{\beta}^{k} \\ \underset{\sim}{\alpha}^{k} \end{bmatrix} - \begin{bmatrix} \underset{\sim}{f}_2 \\ \underset{\sim}{f}_1 \end{bmatrix}\right), \quad k = 0,1,2,\ldots \tag{4.2}$$

where $\begin{bmatrix} \underset{\sim}{\beta}^{k} \\ \underset{\sim}{\alpha}^{k} \end{bmatrix}$ is the k'th iterate and $\{\tau_k\}$ is a sequence of iteration parameters. Since the corresponding condition number $(1+\gamma)/(1-\gamma)$ is independent on h, the number of iterations we have to perform is also independent on h. The linear systems with matrices A_h and B_h, respectively, may be solved in many ways. In the case A_h is non-singular, we may use a direct method for the first mentioned systems. This may be reasonable because the order of A_h in our applications will be much smaller than that of $\mathcal{A}_h$ (and of B_h as well). Typically, for triangular elements and quadratic polynomial basisfunctions, the order of A_h is about $\frac{1}{4}$ (i.e. $\frac{1}{4} N_h$) of that of $\mathcal{A}_h$. We may also use an incomplete factorization of A_h and solve the corresponding linear system by (preconditioned) iterations. Again, since the order of A_h is relatively small, it may not matter so much that the number of iterations now increase with $h \to 0$, as long as h is not too small.

For B_h we may use iteration which as well as (4.2) is a process of optimal computational complexity, because the spectral condition number of B_h is $O(1)$, $h \to 0$. The efficiency may be further improved upon by use of the diagonal part or some incomplete factorization of B_h as a preconditioner.

Such and similar techniques have already been discussed in [2]. Here I shall describe an alternative procedure which is particularly advantageous when B_h has such a special structure that linear systems with B_h are easily solved by a direct method. They may also be applied when B_h may be split into the sum of two matrices, $B_h = B_h^{(1)} + B_h^{(2)}$, where $B_h^{(1)}$ is of the above kind and where the spectral radius, $\rho(B_h^{(1)^{-1}} B_h^{(2)})$ is (much) smaller than 1. Such a case arises for instance when we use so called serendipity element approximations (no interior nodes) on rectangles as we shall now see.

We assume that $B_h^{(2)}$ is negative semidefinite. How to deal with the $B_h^{(2)}$ part shall be discussed in a forthcoming paper about nonlinear problems [3]. There at each outer iterative step we solve a simpler linear problem, in our case with the matrix $B_h^{(1)}$ taking the place of B_h in (3.1b).

Hence consider now the case where linear systems with matrix B_h are solved efficiently by a direct method.

Such a case arises for the bilinear form

$$a(u,v) = \int_\Omega p\underset{\sim}{\nabla}u \cdot \underset{\sim}{\nabla}v \, dx\, dy$$

when p is piecewise constant over each element and we use serendipity type elements with piecewise quadratic approximations, as shown in Figure 4.1.

Figure 4.1. Nodes for a serendipity element, (no interior node).

In the vertex nodes we use the usual bilinear basisfunctions. In the midedge points we use the biquadratic basisfunctions.

$$\phi_1(x,y) = 4\lambda_1\lambda_2\mu_1 \quad , \quad \phi_3(x,y) = 4\lambda_1\mu_1\mu_2$$

$$\phi_2(x,y) = 4\lambda_1\lambda_2\mu_2 \quad , \quad \phi_4(x,y) = 4\lambda_2\mu_1\mu_2$$

where λ_1, λ_2 and μ_1, μ_2 are linear functions in x and in y, respectively, and of Lagrange type, that is,

$$\lambda_1(N_5) = 1 \ , \ \lambda_1(N_i) = 0 \ , \ i = 6,7,8$$

etc. An easy calculation shows that the entries of the element stiffness matrix B satisfies

$$b_{1,3} = b_{1,4} = 0 \ , \ b_{2,3} = b_{2,4} = 0.$$

Hence the matrix in (3.1a) takes the form

$$(4.3)\qquad A = \left[\begin{array}{cc|cc|c} x & x & & & \\ x & x & \multicolumn{2}{c|}{0} & \\ \hline & & x & x & C \\ \multicolumn{2}{c|}{0} & x & x & \\ \hline \multicolumn{4}{c|}{C^t} & A \end{array}\right]$$

We use a global ordering as indicated in Figure 4.2, where edge nodes on horizontal edges are ordered first and in vertical order, then the edge nodes on vertical edges, which are ordered in horizontal order and finally the vertex nodes are ordered (in an arbitrary order). Then the global matrix (3.1b) takes the same structure as the matrix in (4.3).
Hence the corresponding B_h is a tridiagonal matrix, which is easily factored in bidiagonal triangular factors in the usual way.

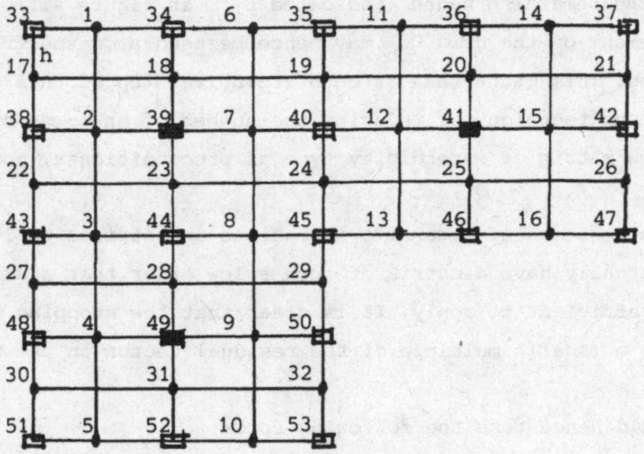

Figure 4.2. Global ordering on level h.

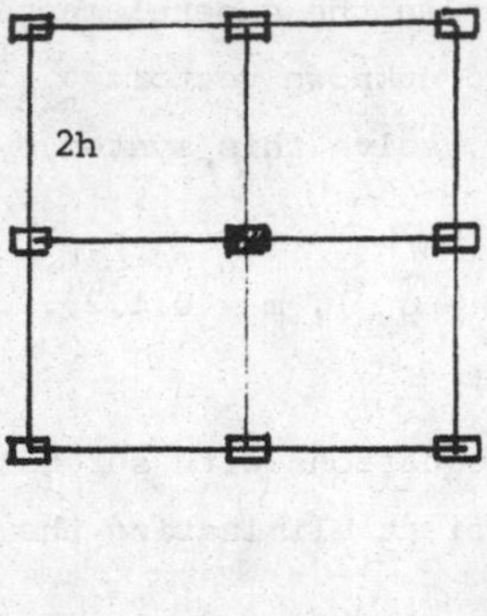

Figure 4.3. A super element on level 2h.
The interior node is eliminated.

In the corresponding block system (4.1) we now eliminate the block vector $\underset{\sim}{\beta}$. Hence we get

(4.4) $$(A_h - C_h^t B_h^{-1} C_h)\underset{\sim}{\alpha} = \underset{\sim}{f}_1 - C_h^t B_h^{-1} \underset{\sim}{f}_2.$$

Now we know from Theorem 3.1 that A_h is spectrally equivalent with $A_h - C_h^t B_h^{-1} C_h$. Hence we may solve (4.4) by iteration with A_h or another matrix which is similarly equivalent to A_h, as a preconditioner. The resulting number of iterations are then independent on h.

We note that A_h comes from the basisfunctions on the vertex nodes only. We group together four adjacent elements to a "superelement" as indicated in Figure 4.3. Here the interior node is only coupled to the nodes on the boundary of the super-element. We eliminate all such interior nodes and the resulting matrix has now the same nodes as a serendipity matrix on the mesh Ω_{2h}. In fact the corresponding global (assembled) matrix is even spectrally equivalent with the global serendipity element matrix corresponding to the remaining nodes (indicated by □ in figure 4.2).

This serendipity element on the grid Ω_{2h} may hence be used as a spectrally equivalent preconditioning. This means that at each iterative step of this process we may again apply the above technique of reducing the number of unknowns by elimination and construct a new matrix of serendipity type as preconditioner, now on level Ω_{4h}.

In this way we get a sequence of recursive iterations on matrices of lower and lower order until we eventually have a matrix of such a low order that a direct method would be the most efficient to apply. It is clear that the stopping test on each level should involve a (small) multiple of the residual vector on the nearest higher (finer grid) level.

The whole scheme would hence take the following form:
Solve (4.4) by iteration

$$A_h(\underset{\sim}{\alpha}_h^{k+1} - \underset{\sim}{\alpha}_h^k) = \underset{\sim}{g}_h := -\tau_k[A_h\underset{\sim}{\alpha}_h^k - \underset{\sim}{f}_1 - C_h^t B_h^{-1}(C_h\underset{\sim}{\alpha}_h^k - \underset{\sim}{f}_2)],\ k = 0,1,2,\ldots .$$

At each step of this, eliminate the interior nodes in the superelements, which results in a matrix $\tilde{A}_{2h}$, a new righthandside $\tilde{\underset{\sim}{g}}_{2h}$ and unknown vectors $\underset{\sim}{\delta}_{2h}^{k+1} = \underset{\sim}{\alpha}_h^{k+1} - \underset{\sim}{\alpha}_h^k$ in the remaining "exterior" vertex nodes Ω_{2h}. Now, solve this system by the basic iterative method

$$\mathcal{A}_{2h}(\underset{\sim}{\delta}_{2h}^{k+1,m+1} - \underset{\sim}{\delta}_{2h}^{k+1,m}) = -t_m(\tilde{A}_{2h}\,\underset{\sim}{\delta}_{2h}^{k+1,m} - \tilde{\underset{\sim}{g}}_{2h}),\ m = 0,1,2,\ldots$$

where $\underset{\sim}{\delta}_{2h}^{k+1,m} \to \underset{\sim}{\delta}_{2h}^{k+1}$, $m \to \infty$.

At each step of this, solve the system of linear equations with serendipity matrix $\mathcal{A}_{2h}$ on the mesh Ω_{2h} by the same process, i.e. by first eliminating the midedge node variables ($\underset{\sim}{\beta}_{2h}$) etc.

Another interesting possibility of deriving a sequence of nested iterations is the following. We assume that Ω has been mapped onto a square, $\Omega_0 = [0,1]^2$ and that a(.,.) is the resulting bilinear form on Ω_0. We successively subdivide Ω_0 into rectilinear triangles as shown in Figure 4.4. We introduce quadratic basisfunctions on all edge nodes and linear basisfunctions on the vertex nodes as before. The resulting matrix is denoted A_h. We let each group of four elements form a superelement as indicated in figure 4.5a. The interior edge nodes are eliminated (this is done locally on each superelement by so called "static condensation").

We now group together elements with orientation a rotation by 45 degrees. In the interior there are groups of four elements, at the boundaries there are only one or two such elements in each superelement. The resulting global matrix is denoted $\tilde{A}_{\sqrt{2}h}$ and the corresponding quadratic (or rather linear-quadratic) approximation matrix is denoted $A_{\sqrt{2}h}$.

By the same argument as for the serendipity elements (that is, by Theorem 3.1), $\tilde{A}_{\sqrt{2}h}$ and $A_{\sqrt{2}h}$ are spectrally equivalent.
In each superelement in the rotated grid we again eliminate the interior edge nodes of $A_{\sqrt{2}h}$ resulting in a matrix $\tilde{A}_{2h}$. The elements are now grouped together as indicated in Figure 4.5c. We form A_{2h} which by the same argument as before is spectrally equivalent to $\tilde{A}_{2h}$. A complete cycle,

$$\Omega_h \to \Omega_{\sqrt{2}h} \to \Omega_{2h}$$

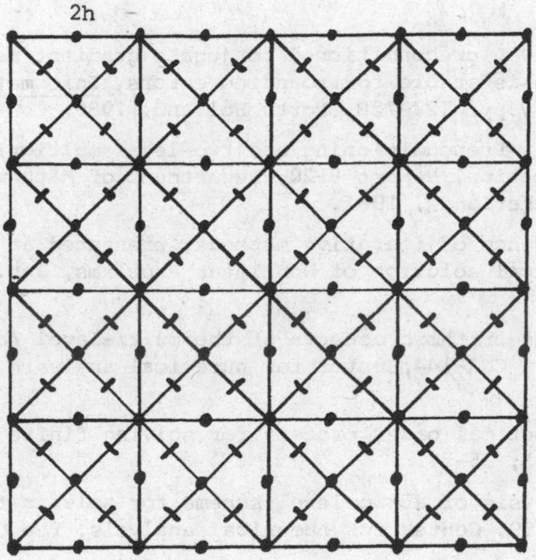

Figure 4.4. Nodes and elements in Ω_h; nodes to be eliminated denoted ✖.

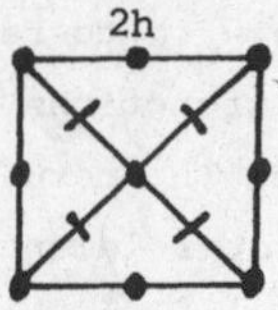

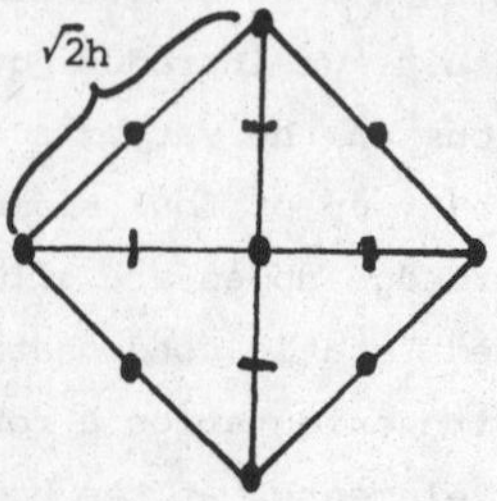

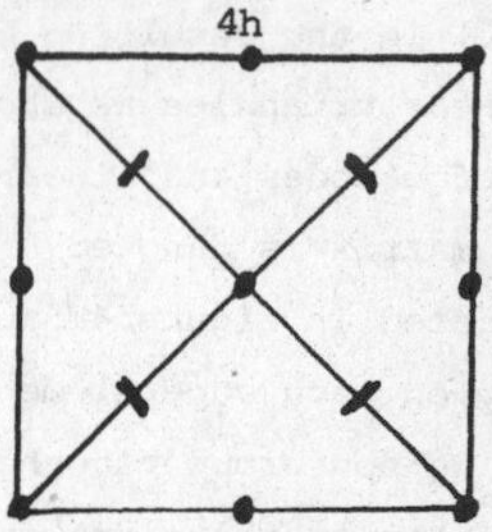

Figure 4.5a
Superelement on Ω_h.

Figure 4.5b
Superelement on $\Omega_{\sqrt{2}h}$.

Figure 4.5c
Superelement on Ω_{2h}.

is now completed and by recursion, the whole sequence of nested iterations is defined.

The last method has resemblances with the total reduction method of U. Trottenberg and his coworkers [11].

More details on nested iteration methods and their practical performance are currently under investigation.

References

1. Axelsson, O. and Gustafsson, I., A preconditioned conjugate gradient method for finite element equations, which is stable for rounding errors, Information Processing 80, Ed. Lavington, S.H., pp. 723-728, North Holland, 1980.

2. Axelsson, O. and Gustafsson, I., Preconditioning and two-level multigrid methods of arbitrary degree of approximation, Report 8120, Department of Mathematics, University of Nijmegen, The Netherlands, 1981.

3. Axelsson, O., On global convergence of iterative methods, presented at the Oberwolfach Symposium on Numerical solution of Nonlinear Problems, Jan. 31 - Febr. 5, 1982.

4. Bank, R.E. and Sherman, A.H., Algorithmic aspects of the multi-level solution of finite element equations, Report CNA-144,Center for numerical analysis, University of Texas at Austin, 1979.

5. Bank, R.E. and Dupont, T., An optimal order process for solving finite element equations, Math. Comp. 36 (1981), 35-51.

6. Bank, R.E. and Dupont, T., Analysis of a two-level scheme for solving finite element equations, Report CNA-159, Center for numerical analysis, The University of Texas at Austin, 1980.

7. Braess, D., The contraction number of a multigrid method for solving the Poisson equation, Numer. Math. 37 (1981), 387-404.

8. Hackbusch, W., On the convergence of a multi-grid iteration applied to finite element equations, Report 77-8, Mathematisch Institut, Köln. Univ., 1977.

9. Maitre, J.F. and Musy, F., The contraction number of a class of two-level

methods; an exact evaluation for some finite element subspaces and model problems, these proceedings.

10. Nicolaides, R., On finite-element multi-grid algorithms and their use, ICASE Report 78-8, NASA Langley Res. Center, Hampton, VA., 1978.

11. Stüben, K. and Trottenberg, U., these proceedings.

The Convergence Rate of a Multigrid Method with Gauss-Seidel Relaxation for the Poisson Equation

Dietrich Braess
Institut für Mathematik
Ruhr-Universität, D-463o Bochum, F.R. Germany

The numerical solution of the Poisson equation is treated by a multigrid method for a uniform grid. The convergence rate can be estimated even for the iteration with a V-cycle independently of the shape of the domain as long as it is convex and polygonal. The smoothing effect of the Gauß-Seidel relaxation is described by a discrete seminorm which is weaker than the energy norm.

1. Introduction

The linear equations which result from the discretization of elliptic boundary value problems typically have matrices with large condition numbers. This is due to the fact that the spectra of the associated differential operators are not bounded. Therefore the efficient numerical solution of the equations requires facilities for the adequate treatment of the smooth portions of the solution as well as for the non-smooth ones. Possibly one needs algorithms which combine two different devices for this aim. It has turned out during the last years that multigrid methods have the desired properties and can solve the equations very effectively [5,6,10].

For the same reason the numerical analysis of multigrid methods can probably not be done effectively if one restricts oneself to the use of one norm. (At a first glance [3] seems to be a counter-example.) One has to measure at least the amount and the smoothness of an approximate solution or of its error, respectively. This has been done in two different ways, and two different types of convergence proofs for multigrid methods are found in the literature.

A. *Fourier Method:* Exact bounds for the convergence rates are established by Fourier analysis. The results show the good performance of the method [5,8,11]. But the analysis is restricted to rectangular domains.

B. *Asymptotical Analysis:* A convergence rate bounded away from 1 is established for sufficiently many smoothing steps. The domains may be arbitrary as long as they are consistent with the regularity of the boundary value problem [1,6,9]. Usually norms of Sobolev type are used to control the process.

There is still the unanswered question whether and how the (good) performance depends on regularity and on the shape of the boundary. This question is important not only with the use of multigrid codes but also with respect to the reliability of the local mode analysis [5]. Though this heuristic method neglects the influence of boundaries and boundary conditions it often yields results which are close to the observed numbers.

Here we will consider a multigrid method for the discretization of the Poisson equation with Dirichlet boundary conditions for which the questions above can be partially answered (though we are still far from an answer in the general case).

$$\begin{aligned} -\Delta u &= f \quad \text{in } \Omega \subset \mathbb{R}^2 \\ u &= 0 \quad \text{on } \partial\Omega. \end{aligned} \tag{1.1}$$

Our method of analysis is in between the methods of Type A and B described above. We list some typical features.

1. The mesh is assumed to be uniform. The mesh ratio of subsequent meshes is $\sqrt{2}$. An analogous investigation with ratio 2 will be given in [13].

2. Ω may be an arbitrary convex, polygonal domain for which the boundary lies grid lines.

3. Explicit bounds for the convergence rates are evaluated, e.g., a convergence factor (=error damping factor per cycle) of

0.172 for the two grid cycle with 1 Gauß-Seidel relaxation

will be established. This result for convex domains is comparable to 0.125 for the same method if Ω is a rectangular domain [11].

4. The convergence rate for the complete multi-level iteration with V-cycles is shown to be better than $1/(1+r)$, if $r \geq 1$ Gauß-Seidel relaxation sweeps are performed on each level. We emphasize that usually convergence rates are established only for W-cycles.

5. To measure smoothness, two different norms are used. This is a feature one generally finds in the analysis of Type B.

6. The calculations break down, if the domains have reentrant corners (German: *einspringende Ecken*). Nevertheless, one is working with a net for the bad cases. The convergence factor (in the two-level case) cannot be worse than 1/2.

We mention that the reader will find more details of the calculations in [4]. On the other hand we will comment the analysis in this paper often from a different viewpoint.

2. The Multigrid Algorithm

Let $\Omega \subset \mathbb{R}^2$ be a bounded domain. Assume that there is a triangulation of Ω with rectangular isosceles triangles with sides of length h and $H=\sqrt{2}h$ (see Fig.1). The set of grid points $\{p_i\}$ which are contained in Ω is denoted by Ω_h, while Ω_H refers to the (rotated) coarser subgrid

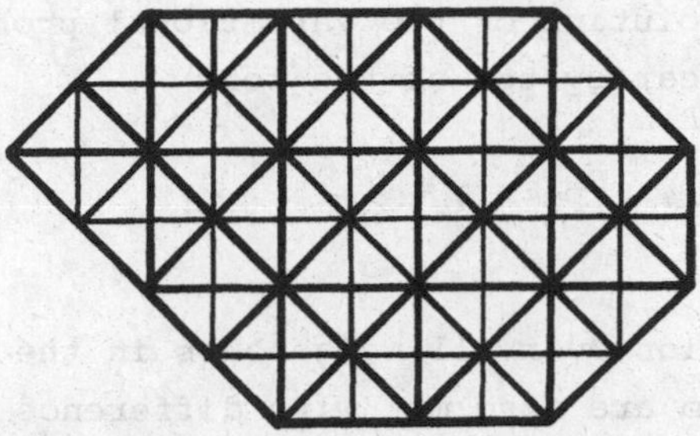

Fig. 1

formed by triangles with sides of length H and $\sqrt{2}H=2h$. Similarly $\Omega_{2h} \subset \Omega_H$ is defined. When we associate to each grid point a colour:

white to the points of Ω_H,
black to the points of $\Omega_h \setminus \Omega_H$,

then we get the structure of a checkered board.

As usual, let S_h denote the space of those functions from $C(\Omega)$ which are linear on each triangle of Ω_h and which vanish on the boundary $\partial\Omega$. The discretization of (1.1) for the grid Ω_h is understood to be the (solution of the) variational problem in S_h:

$$J(u):=a(u,u)-2(f,u)_0 \longrightarrow \min! \tag{2.1}$$

Here,

$$a(u,v)=\int_\Omega (u_\xi v_\xi + u_\eta v_\eta) \tag{2.2}$$

$$(u,v)_0=\int_\Omega u\, v.$$

When S_h is endowed with the energy norm $\|u\|=\sqrt{a(u,u)}$, the norm is easily expressed in terms of the values on the grid, $u_i=u(p_i)$:

$$\|u\|^2 = \sum_{\substack{i,j \\ d(i,j)=h}} (u_i - u_j)^2. \tag{2.3}$$

The summation runs over all pairs of grid points with Euclidean distance h, and the points of the boundary are included. Here and in subsequent symmetrical sums each pair is taken only once.

Because of (2.3) the solution of the variational problem (2.1) in S_h is determined by a linear system of the form

$$u_i = {\sum_j}'^{h} \frac{1}{4} u_j + b_i, \quad p_i \in \Omega_h, \tag{2.4}$$

where $\sum_h'$ refers to summation over all neighbours in the grid Ω_h. Linear equations of the same form are also met with difference approximation by the standard 5-point-star discretization. The numerical solution of these linear equations by a multigrid method is the central topic of this paper.

A classical tool for solving (2.4) is the Gauß-Seidel iteration which for convenience will be split into two parts.

$$(G_h^b u)_i = \begin{cases} {\sum_j}'^{h} \frac{1}{4} u_j + b_i, & \text{if } p_i \text{ is coloured black,} \\ u_i \quad , & \text{if } p_i \text{ is not black.} \end{cases}$$

Similarly G_h^w operates on the white coloured points. Then $G_h^b \circ G_h^w$ represents a point Gauß-Seidel relaxation, with the variables being enumerated such that the matrix of the linear system has Property A. For ease of notation we will put

$$G_h^\nu = \begin{cases} G_h^b, & \text{if } \nu \text{ is odd,} \\ G_h^w, & \text{if } \nu \text{ is even.} \end{cases}$$

In order to define a multigrid algorithm let h_q, $q=0,1,\ldots,q_{max}$ be a finite sequence of meshsizes with $h_{q-1} = \sqrt{2}\, h_q$, $q \geq 1$. The corresponding grids will be denoted by Ω^q instead of Ω_{h_q} and the associated finite element spaces of piecewise linear continuous functions by S^q.

The original grid is $\Omega^{q_{max}}$ and the other grids are auxiliary ones. When we consider the situation for a fixed q, we use some obvious abbreviations:

$$h = h_q, \quad \Omega_h = \Omega^q, \quad S_h = S^q,$$

$$H = h_{q-1}, \quad \Omega_H = \Omega^{q-1}, \quad S_H = S^{q-1}.$$

As usual, in the multigrid procedure the variables carry three superscripts: 1. the specification of the level, 2. a count of the loop, and 3. a count of the steps within a loop.

<u>Algorithm</u> (k-th loop at level $q \geq 1$ for the multigrid procedure with r smoothings)

0. *Start*. Given $u^{q,k,0}$, if k=0 replace it by $G_q^{r+1} u^{q,k,0}$.

1. *A priori smoothing*. For $\nu=1,2,\ldots,r-1$ compute

$$u^{q,k,\nu} = G_q^{r-1-\nu} \; u^{q,k,\nu-1}.$$

2. *Transfer step*. Put $u^{q,k,r} = G_q^b \, u^{q,k,r-1}$.

3. *Coarse grid approximation*. Let u^{q-1} denote the solution of the variational problem:

$$J(u^{q,k,r}+v) \longrightarrow \min_{v \in S^{q-1}}$$

If $q = 1$, compute $v_1 = u^{q-1}$.
If $q > 1$, compute an approximate solution v_1 to u^{q-1} by applying $\mu = 1$ or $\mu = 2$ iteration loops at level q-1 starting with $u^{q-1,0,0}=0$.
Put $u^{q,k,r+1} = u^{q,k,r}+v_1$.

4. *A posteriori smoothing*. For $\nu = 1,2,\ldots,r+1$ determine

$$u^{q,k,r+1+\nu} = G_q^{\nu} u^{q,k,r+\nu},$$

and put $u^{q,k+1,0} = u^{q,k,2r+2}$.

Specifically, r-1 half steps for the smoothing are performed before the correction and r+1 halves are done after it. The transfer is carried out in a way such that the right hand side for the linear equation on the lower level is easily obtained [3,11]. It corresponds to a half-weighting of the defects.

3. Decomposition of S_h and the strengthened Cauchy inequality

The central idea of our analysis is the decomposition of the finite element space S_h as a direct sum

$$S_h = S_H \oplus T_h, \tag{3.1}$$

where

$$T_h = \{w \in S_h;\ w = 0 \text{ on } \Omega_H\}.$$

Given $u \in S_h$ the optimization of the functional J in $u+S_H$ is a job which is (approximately) done in Step 3 of the multigrid algorithm. On the other hand the optimization of J in the other direction, i.e., in $u+T_h$, is actually done by the application of G_h^b.

A first bound for the convergence factor of the multigrid method is therefore obtained by analyzing the alternate optimization in S_H and T_h. To this end the following abstract lemma is given. Let P_V denote the orthogonal projection onto a linear subspace V of a Hilbert space. Note that $x-P_V x$ is the element in $x+V$ with least norm.

Lemma 3.1. Let the Hilbert space U be a direct sum of its linear subspaces V and W. Assume that $u=v+w$, $u \in U$, $v \in V$, $w \in W$ and that there is a $\gamma<1$ such that a *strengthened Cauchy inequality* holds:

$$|(v,w)| \leq \gamma \|v\| \|w\|. \tag{3.2}$$

If u is optimal in $u+W$, i.e., if $P_W u = 0$, then

$$\|u-P_V u\| \leq \gamma \|u\|. \tag{3.3}$$

A simple proof of this lemma is given here because we have changed a detail in contrast to [3]. Usually Inequality (3.2) is postulated for all $v \in V$ and all $w \in W$ [2,7,8].

First we consider the special case where dim V=dim W=1, as illustrated in Fig. 2. Then $\|u-P_V u\| = \|u\| \cos\alpha \leq \gamma \|u\|$, where α is an angle between vectors from V and W. In the general case we have $u \in V_1 \oplus W_1$, where V_1=span v and W_1 = span w. By assumption we have $(u,w)=0$. Since we know that the lemma is true for $V_1 \oplus W_1$, we conclude from $V_1 \subset V$:

$$\inf_{v \in V} \|u-v\| \leq \inf_{v \in V_1} \|u-v\| \leq \gamma \|u\|.$$

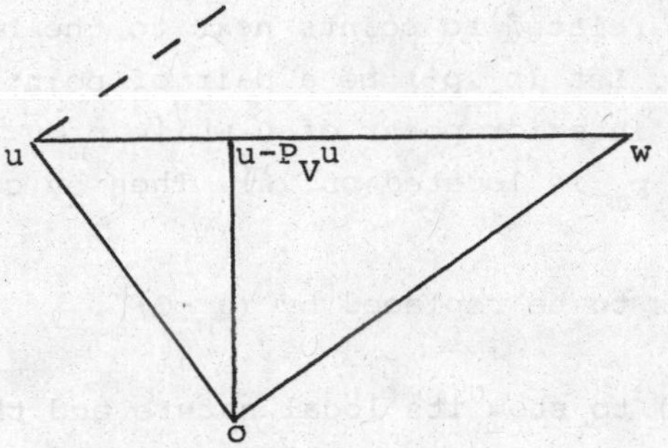

Fig. 2. Decomposition u=v+w after the optimization in W. Here cos $\alpha(v,w)/(\|v\|\ \|w\|)$.

This proves the lemma. □

Recall that $S_h \subset H_o^1(\Omega)$ is a Hilbert space when endowed with the inner product $(u,v):=a(u,v)$. Since

$$J(u) = \|u-u_h\|^2 + \text{const},$$

where u_h is the solution and const $= J(u_h)$, the variational problem is equivalent to the minimization of $\|.-u_h\|$. Obviously, Lemma 3.1 applies to the optimization problem with the shifted argument as well. We only have to compute the factors in the strengthened Cauchy inequality for the decomposition (3.1).

<u>Lemma 3.2.</u> If $v \in S_H$ and $w \in T_h$ then

$$|a(v,w)| \leq \frac{1}{\sqrt{2}} \|v\|\ \|w\|. \tag{3.4}$$

If, moreover v and w satisfy zero Dirichlet boundary conditions and if Ω_h has no reentrent corner at the points of $\Omega_h \setminus \Omega_H$ then

$$|a(v,w)| \leq \sqrt{\frac{1}{2}\left(1 - \frac{|v|^2}{\|v\|^2}\right)}\ \|v\|\ \|w\|. \tag{3.5}$$

Here $|.|$ is a seminorm on S_h:

$$|u|^2 = \sum_{\substack{n,m \\ p_n, p_m \in \Omega_H \\ d(n,m)=2h}} \frac{1}{2}(u_n - u_m)^2. \tag{3.6}$$

In this sum the terms related to points next to the boundary are to be understood as follows. Let (p_n,p_m) be a pair of points with distance 2h such that p_n is an interior point of Ω while $p_m \notin \bar{\Omega}$. (The point in the middle between p_n and p_m is located on $\partial\Omega$). Then by convention

$$\frac{1}{2}(u_n-u_m)^2 \quad \text{is to be replaced by } (u_n-0)^2.$$

We will prove (3.4) to show its local nature and the independency of the shape of the boundary and of the boundary conditions. At the same time we indicate how the improvement (3.5) is obtained.

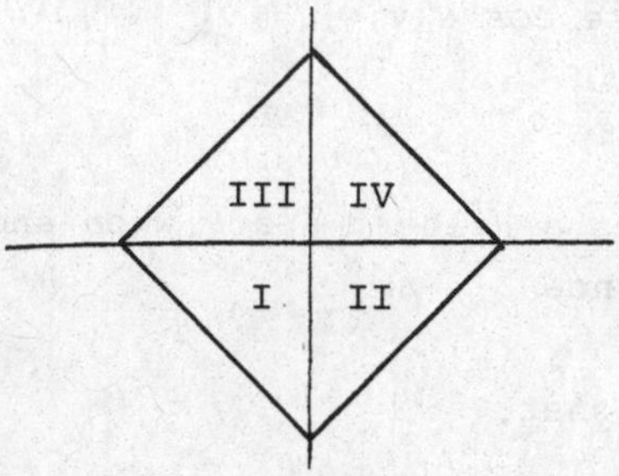

Fig. 3

Consider a triangle in the coarser grid (built from Triangles I and II in Fig. 3). Since $w \in T_h$ vanishes on the boundary of the square drawn in Fig. 3 and $v \in S_H$ it follows that

$$v_\xi, v_\eta \quad = \text{const},$$

$$|w_\xi| = |w_\eta| = \text{const}, \qquad w_\xi(I) = -w_\xi(II).$$

Consequently, when we understand that in the following expressions only the terms resulting from Triangles I and II are taken, then the ξ-portion of the inner products cancels and

$$\begin{aligned} a(v,w) &= \int (v_\xi w_\xi + v_\eta w_\eta) \\ &= \int v_\eta w_\eta \\ &= \frac{1}{2}\int v_\eta^{\,2} - \frac{1}{2}\int (v_\eta - w_\eta)^2 + \frac{1}{2}\int w_\eta^{\,2} \\ &= \frac{1}{2}\|v\|^2 - \frac{1}{2}\int v_\xi^{\,2} - \frac{1}{2}\int (v_\eta - w_\eta)^2 + \frac{1}{4}\int (w_\xi^{\,2} + w_\eta^{\,2}) \end{aligned}$$

$$= \frac{1}{2}\|v\|^2 - \frac{1}{2}\int v_\xi^{\,2} - \frac{1}{2}\int (v_\eta - w_\eta)^2 + \frac{1}{4}\|w\|^2. \tag{3.7}$$

If we abandon the negative terms in (3.7) and sum over all triangles in Ω, we get

$$a(v,w) \leq \frac{1}{2}\|v\|^2 + \frac{1}{4}\|w\|^2. \tag{3.8}$$

Hence, $a(v,w) \leq \sqrt{\frac{1}{2}}\ \|v\|\ \|w\|$ whenever $\|w\| = \sqrt{2}\ \|v\|$. From a simple homogeneity argument it follows that (3.4) is correct for any ratio $\|v\|/\|w\|$.

On the other hand v_ξ may be expressed here by a difference referring to points of distance 2h. Hence, $\int v_\xi^{\,2}$ is just the ξ-portion of $|v|^2$. Similarly, the third term in (3.7) corresponds to the η-portion of $|v|^2$, but to understand this Triangles I, II, III and IV have to be analyzed together. The details are found in [4], where it is also discussed that the boundary will not spoil the result as long as the additional assumptions hold. Then the second and the third term in (3.7) imply that we may replace $\|v\|^2$ in (3.8) by $\|v\|^2 - |v|^2$. This finally leads to (3.5). □

Now we are in a position to establish a rough estimate for the convergence rate. To make the basic idea more transparent, we restrict ourselves here to the two-grid algorithm ($q_{max}=1$). As it will become obvious in Section 5, the multilevel procedure is treated as a perturbed two-level procedure.

The solution of the considered equation is denoted by u_q.

First we recall that G_q^b and G_q^w correspond to the solution of (2.1) in the subspaces T_q and

$$\widehat{T}_q = \{u \in S_q;\ u = 0 \text{ on black points of } \Omega_q\},$$

resp. Therefore the distance to the solution does not increase with these operations and

$$\|u^{q,k,r} - u_q\| \leq \|u^{q,k,0} - u_q\|. \tag{3.9}$$

If we apply Lemma 3.1 with $V = S_{q-1}$ and $W = T_q$, we obtain with $\gamma = \sqrt{1/2}$:

$$\|u^{q,k,r+1} - u_q\| \leq \sqrt{1/2}\ \|u^{q,k,r} - u_q\|. \tag{3.10}$$

In the same way by putting $V = T_q$ and $W = S_{q-1}$ we get

$$\|u^{q,k,r+2} - u_q\| \leq \sqrt{1/2}\, \|u^{q,k,r+1} - u_q\|. \qquad (3.11)$$

Since the approximation is not spoiled during the remaining steps of the algorithm, the collection of the inequalities above yields

$$\|u^{q,k+1,0} - u_q\| \leq \frac{1}{2}\, \|u^{q,k,0} - u_q\|. \qquad (3.12)$$

Specifically in each cycle the error is reduced by a factor γ^2, when the decomposition allows a strengthened Cauchy inequality with the constant γ. However, (3.12) is a poor result, when compared with the actually observed efficiency of multigrid methods. Fortunately, the bound can be improved by combining the inequality (3.5) containing individual numbers with an investigation of the Gauß-Seidel relaxation.

Nevertheless, (3.12) is of its own interest, because no restriction on the shape of the boundary is necessary to derive (3.4). Consequently, the multigrid algorithms will not break down, even if there are reentrant corners. On the other hand a slow down of the iterations has been observed in examples with irregular regions.

In the framework of multigrid procedures the alternate method was probably considered first by Bank and Dupont [2]. They established that γ^2 with γ as above is the spectral radius of the iteration matrix. Therefore they restricted their attention to two-grid schemes. In this context we want to emphasize that results on two-grid methods in terms of spectral radii cannot be used to analyse iterations with more than two grids. This is at least true as long as multigrid procedures are treated as two-grid procedures with perturbations.

4. The Gauß-Seidel relaxation

In principle, the point Gauß-Seidel relaxation corresponds to the alternate optimization in the subspaces T_h and $\hat{T}_h$, which where defined is connection with (3.1) and (3.9). The decomposition $S_h=T_h \oplus \hat{T}_h$, however, cannot lead to a strengthened Cauchy inequality with a favourable constant, since the contraction number for the Gauß-Seidel relaxation is close to one if h is small. In Lemma 3.2 there is hidden the statement that both the smooth functions and the non smooth ones [more precisely, all the eigenfunctions of the matrix operator resulting from the bilinear form a(.,.)] can be reasonably approximated either in S_H or in T_h.

The advantage of the Gauß-Seidel relaxation is its smoothing effect. This is expressed by the following lemma which was proved in [4]. The solution of the underlying linear equation will be denoted as u_h.

<u>Lemma 4.1.</u> Assume that Ω is convex. If $u = G_h^b u$, then

$$\|G_h^w u - u_h\| \leq |u-u_h| \leq \|u-u_h\|. \tag{4.1}$$

To understand the meaning of the lemma, we compare the seminorm $|.|$ with the energy norm $\|.\|$. Actually $|.|$ is a norm on the subspace S_H but even there it is weaker than $\|.\|$. We have

$$|u| \leq \|u\| \quad \text{for any } u \in S_h. \tag{4.2}$$

Indeed, $|u|$ is evaluated by taking differences of values for arguments of distance 2h. These can be expressed by differences corresponding to points with distance h. Specifically, given a pair of points p_n, p_m of distance 2h, denote the grid point in between by p_{nm}. Then we have

$$\begin{aligned}(u_n-u_m)^2 &= (u_n-u_{nm}+u_{nm}-u_m)^2 \\ &\leq 2(u_n-u_{nm})^2+2(u_{nm}-u_m)^2.\end{aligned}$$

When treating each term on the right hand side of (3.6) in this way and summing up we get (4.2).

On the other hand $\|.\|$ cannot be estimated by $|.|$ in general. The ratio $|.|/\|.\|$ is a measure for the smoothness:

$|u| \ll \|u\|$, if u is not a smooth function, while
$|u| \approx \|u\|$, if u is a smooth function.

Therefore, Lemma 4.1 yields a quantitative statement about the smoothing effect of the Gauß-Seidel relaxation. The analysis of multigrid methods requires the measurement of two quantities: 1. the amount of the error, 2. its smoothness. Therefore we need (at least) two norms. In most investigations of Type B some Sobolev norms are chosen, and the transition from the discrete problem with meshsize h to the problem with the coarser grid is done via the continuous problem. The use of the (semi-)norm (3.6) is probably more direct, but possibly it yields reasonable results only for the Poisson equation. For the treatment of anisotropic elliptic problems, when line relaxations are applied, another appropriate seminorm has to be found.

Before we establish the final results we want to comment on another phenomenon: It is not only a coincidence that the constant in the strengthened Cauchy inequality for the decomposition (3.1) is small whenever the S_H-portion is a smooth function. Otherwise one would encounter the paradox described in [3, Section 8] that the Gauß-Seidel relaxation might enhance the non smooth portions. This will become obvious from the following arguments.

Let $u=v+w$, $v\in S_H$, $w\in T_h$. Let $w_o\in T_h$ be a vector such that $\cos\alpha := a(v,w_o)/(\|v\|\cdot\|w_o\|)$ is not small. Decompose $u' = G_h^b u$ in the same sense, $u'=v+w'$. It follows from $\|u'\|\leqslant\inf_{c\in\mathbb{R}}\|v-cw_o\|=\|v\|\cdot\sin\alpha$ that $\|w'\|\geqslant \|u'\|\cdot\cos\alpha/\sin\alpha$. Consequently, the constant in the Cauchy inequality must be small, when we expect that the Gauß-Seidel relaxation will produce only a small non-smooth portion $w'\in T_h$.

5. Final results

As we have already mentioned, we will treat the multigrid iteration as a perturbed two-grid iteration. In this framework Step 3 from the original algorithm will be replaced by the following (more general) step, which contains still a parameter δ to be specified later.

3'. Determine an approximation v_1 to u^{q-1}, where u^{q-1} is defined as in Step 3 such that

$$\|v^1-u^{q-1}\| \leqslant \delta \|u^{q-1}\|, \tag{5.1}$$

and put $u=u^{q,k,r}+v_1$.

For ease of notation we may restrict ourselves to the homogeneous equation and assume $u_q=0$ when estimating the convergence rate. Moreover, we will omit the superscripts q and k.

First we decompose the function u^r in the sense of (3.1), $u^r=v^r+w^r$, $v^r \in S_{q-1}$, $w^r \in T_q$ and put $\lambda=|v^r|/\|v^r\|$. Since u^r is optimal in T_q, from Fig.3 and Lemma 3.2 we have $\cos\alpha \leqslant \sqrt{(1-\lambda^2)/2}$ and

$$\begin{aligned} |u^r| &= |v^r| = \lambda \|v^r\| \\ &\leqslant \lambda \frac{\|u^r\|}{\sin\alpha} = \rho^{1/2} \|u^r\| \end{aligned} \tag{5.2}$$

where $\rho=2\lambda^2/(1+\lambda^2)$. From (5.2), Lemma 4.1 and $u^r=G_q^b u^{r-1}$ it follows that

$$\frac{\|G_q^w u^r\|}{\|u^r\|} \leqslant \frac{|u^r|}{\|u^r\|} \leqslant \rho^{1/2}. \tag{5.3}$$

Now we write $u^r = \tilde{G}u^o$, where

$$\tilde{G} = \underbrace{G_q^b \bullet G_q^w \bullet G_q^b \circ \ldots \circ G_q^{(r+1)}}_{r+1 \text{ alternating factors}}.$$

The last factor G_q^b is due to the transfer step, the next r-1 ones stem from the à priori smoothing, and the first one may be added, because it does not change $u^o=u^{k,q,o}$. (If k=0, look at the start, otherwise recall the computation of $u^{q,k,o}=u^{q,k-1,2r+2}$). Now (5.3) asserts that the next Gauß-Seidel halfstep would bring a reduction by a factor $\rho^{1/2}$. As was pointed out in [4, Section 5] then all the preceding halfsteps but the first one have gained a reduction of the error by the same factor at least. Hence,

$$\frac{\|u^r\|}{\|u^o\|} = \frac{\|\tilde{G}u^o\|}{\|u^o\|} \leq \rho^{r/2}. \tag{5.4}$$

Let $Q = 1-P_{S_{q-1}}$ denote the projector which is associated to the exact execution of the coarse grid correction $u'=Qu^r=u^r-u^{q-1}$. Since u^r is optimal in T_q, the lemmas 3.1 and 3.2 imply that

$$\begin{aligned} \|u'\| &= \|Q\tilde{G}u^o\| \leq \sqrt{(1-\lambda^2)/2}\ \|u^r\| \\ &\leq \rho^{r/2}\sqrt{(1-\lambda^2)/2}\ \|u^o\| = \rho^{r/2}\sqrt{\frac{1-\rho}{2-\rho}}\ \|u^o\|. \end{aligned} \tag{5.5}$$

By (5.1) the result of the perturbed correction step is expressed in the form

$$u^{r+1} = u'+\delta\cdot v_2, \tag{5.6}$$

with some $v_2 \in S_{q-1}$, $\|v_2\| \leq \|u^{q-1}\|$. From the well known characterization of closest points in subspaces of Hilbert spaces it follows that

$$\begin{aligned} \|u'+v_2\|^2 &= \|u'\|^2+\|v_2\|^2 \\ &\leq \|u'\|^2+\|u^{q-1}\|^2 = \|u^r\|^2. \end{aligned} \tag{5.7}$$

Finally, we have

$$u^{2r+2} = \tilde{G}^*u^{r+1},$$

where $\tilde{G}^*$ is the adjoint of $\tilde{G}$, because $\tilde{G}^*$ is the product of those projectors which enter into $\tilde{G}$, but in the opposite ordering. Therefore u^{2r+2} may be estimated by a duality argument: $\|u^{2r+2}\|=\sup_{\hat{u}}(\hat{u},u^{2r+2})/\|\hat{u}\|$. From (5.6) we obtain

$$\begin{aligned} (\hat{u},\ u^{2r+2}) &= (\hat{u},\ \tilde{G}^*u^{r+1}) \\ &= (\tilde{G}\hat{u},\ (1-\delta)u'+\delta[u'+v_2]) \\ &\leq (1-\delta)(\tilde{G}\hat{u},Q^2Gu^o)+\delta\|\tilde{G}\hat{u}\|\cdot\|u'+v_2\|. \end{aligned}$$

This and (5.7) imply that

$$(\hat{u},u^{2r+2}) \leqslant (1-\delta)(Q\tilde{G}\hat{u},\ Q\tilde{G}u^{o})+\delta\|\tilde{G}\hat{u}\|\cdot\|u^{r}\|$$

$$\leqslant (1-\delta)\|Q\tilde{G}\hat{u}\|\cdot\|Q\tilde{G}u^{o}\|+\delta\|\tilde{G}\hat{u}\|\cdot\|\tilde{G}u^{o}\|$$

$$\leqslant [\,(1-\delta)\|Q\tilde{G}\hat{u}\|^{2}+\delta\|\tilde{G}\hat{u}\|^{2}]^{1/2}$$

$$\cdot\ [\,(1-\delta)\|Q\tilde{G}u^{o}\|^{2}+\delta\|\tilde{G}u^{o}\|^{2}]^{1/2}.$$

Now from (5.4) and (5.5) we conclude that $(1-\delta)\|Q\tilde{G}u\|^2+\delta\|\tilde{G}u\|^2 \leqslant \delta_q\|u\|^2$, where

$$\delta_q = \max_{0\leqslant\rho\leqslant 1} \rho^r[\,(1-\delta)\,\frac{1-\rho}{2-\rho} + \delta]\,. \tag{5.8}$$

Consequently, the calculation above yields

$$\|u^{2r+2}\| \leqslant \delta_q\|u^o\|.$$

In particular, the two grid convergence rate δ_{TG} is obtained by evaluating the maximum on the right hand side of (5.8) for δ=0. On the other hand the factor for the complete multi level process is established by solving the equation

$$\delta_{MG,\mu} = \max_{0\leqslant\rho\leqslant 1} \rho^r\,\frac{1-\rho+\delta^{\mu}_{MG,\mu}}{2-\rho}\,.$$

The results are listed in Table 1 for different paramaters r and μ. Surprisingly, the convergence rate for the procedure with V-cycles, i.e. μ=1, turns out to be bounded away from 1, though the computed bounds are very conservative when compared with the observed good performance of multigrid methods. An empirical damping factor of ≈ 0.16 was reported [12]. The algorithm with a W-cycle, where μ=2 if $q_{max}-q$ is odd and μ=1 otherwise, is also of interest. Here the computing effort per cycle is roughly twice the effort for one V-cycle. If μ=1 and 2 are organized in the other order, then the damping factor increases but the computing effort is only 5o % higher than that for the V-cycle.

Table 1. Convergence rates (error damping per cycle)

r	0	1	2	3
δ_{TG}	0.5	0.172	0.114	0.086
W-cycles:				
$\delta_{MG,\mu=2}$		0.187	0.120	0.089
$\delta_{MG,\mu=2,1}$		0.205	0.127	0.093
$\delta_{MG,\mu=1,2}$		0.273	0.175	0.130
V-cycles:				
$\delta_{MG,\mu=1}$		$\frac{1}{2}$	$\frac{1}{3}$	$\frac{1}{4}$
if Ω is a square:				
δ_{TG}	0.5	0.125	0.053	0.042
ρ_{TG}	0.5	0.074	0.041	0.028
without intermediate grids:				
δ_{TG}		0.273	0.202	0.169
$\delta_{MG,2}$		0.291	0.219	0.183

The increase of the convergence factor with the number of levels when a V-cycle with r=1 is chosen becomes apparent from Table 2. It is probably more effective to change to a W-cycle after three or four levels have been passed because this improves the convergence factor while the computing effort is enhanced only by a small amount.

Table 2. Convergence rates for V-cycles with r=1 as a function of the number of levels

q	0	1	2	3	4	5	6	
δ_q	0	0.172	0.255	0.304	0.336	0.36o	0.377	$\rightarrow$ 0.5

The reader can get an impression from the seventh and eighth row in Table 1, how realistic our conservative bounds are. The damping factors in terms of the energy norm and the spectral radius for square domains have been determined in [11,12] by the Fourier method. The last two rows show the results from [18] for an algorithm with a mesh ratio 2. These numbers are given though here we compare algorithms, the computing efforts of which differ by ca. 2o %.

References

1. R.E.Bank and T.Dupont, An optimal order process for solving finite element equations. Math.Comp.36 (1981), 35-51.

2. R.E.Bank and T.Dupont, Analysis of a two-level scheme for solving finite element equations. Report CNA - 159, Austin 198o.

3. D.Braess, The contraction number of a multigrid method for solving the Poisson equation. Numer. Math. 37 (1981), 387-4o4.

4. D.Braess, The convergence rate of a multigrid method with Gauß-Seidel relaxation for the Poisson equation. (submitted)

5. A.Brandt, Multi-level adaptive solutions to boundary-value problems. Math. Comp. 31 (1977) 333-39o.

6. W.Hackbusch, On the convergence of multi-grid iterations. Beiträge zur Numer. Math. 9 (1981), 213-239.

7. J.F.Maitre and F.Musy, The contraction number of a class of two-level methods; an exact evaluation for some finite element subspaces and model problems. (These proceedings).

8. Th.Meis and H.-W.Branca, Schnelle Lösung von Randwertaufgaben. ZAMM 62.

9. R.A.Nicolaides, On the ℓ^2-convergence of an algorithm for solving finite element equations. Math. Comp. 31 (1977), 892-9o6.

1o. R.A.Nicolaides, On some theoretical and practical aspects of multi-grid methods. Math. Comp. 33 (1979), 933-952.

11. M.Ries, U.Trottenberg and G.Winter, A note on MGR methods, Preprint, Bonn 1981.

12. U.Trottenberg, private communication.

13. R.Verfürth, The contraction number of a multigrid method with mesh ratio 2 for solving Poisson's equation. (in preparation).

A MULTIGRID FINITE ELEMENT METHOD FOR THE TRANSONIC POTENTIAL EQUATION

Herman Deconinck
Research Assistant

Charles Hirsch
Professor

Vrije Universiteit Brussel, Department of Fluid Mechanics
Pleinlaan 2, 1050 Brussels, Belgium.

ABSTRACT

A non linear multigrid method has been applied to the transonic potential flow equation discretized with a classical Galerkin Finite Element approach on a curvilinear body fitted mesh.

The generality of the finite element method with regard to the treatment of non uniform arbitrarily generated grids is preserved by introducing non uniform interpolation in the coarse grid finite element space for the prolongation of coarse grid functions. The residual restriction is also constructed in a way fully consistent with the finite element approximation resulting in a non uniform weighting of fine grid residuals.

Substantial convergence acceleration is obtained with standard line relaxation as smoothing step and an implementation of the Kutta-Youkowski condition in cascade configurations has been achieved without adverse effect on the convergence speed. Computational results are shown for isolated airfoil, channel and cascade geometries.

1. INTRODUCTION.

The mixed elliptic-hyperbolic transonic potential equation is one of the first non elliptic and strongly non linear problems treated with the multigrid method. After the initial calculations by South and Brandt (1977) with the small perturbation equation various finite difference (F.D.) and finitevolume approaches were presented which mainly differ in the treatment of the artificial viscosity needed to guarantee a unique and stable solution and in the choice of the multigrid algorithm (linear or nonlinear) and smoothing method. Jameson (1979) used his well known artificial viscosity terms with the non linear multigrid scheme and a generalized ADI-method as smoothing step. Shortly after similar but three dimensional codes were developed by Mc Carthy and Reyhner (1980) and by Shmilovich and Caughey (1981) with line relaxation as smoothing method.

An alternative way was followed by Fuchs (1978) applying the linear multigrid scheme to the linearized equation in an external Newton iterative process and using a modified convective line relaxation as smoothing.

A similar finite difference multigrid scheme has been proposed recently by Boerstoel (1981), however using a flux vector splitting technique for the handling of shocks.

This paper reports on the progress made with a finite element (F.E.) multigrid method for the transonic potential equation, initiated in 1981 (Deconinck and Hirsch 1981b). The method takes advantage of the finite element interpolation theory to define systematic and natural restriction and prolongation operators for the transfer of functions and residuals between different grid levels, while conserving the usual flexibility of Finite Element methods to handle the equation in physical coordinates on arbitrarily shaped body fitted grids.

A Kutta-Youkowski condition has been incorporated in the basic non linear multigrid algorthm by simultaneously iterating on the boundary conditions until the correct circulation is obtained. In this process full advantage has been taken of the multigrid possibilities by updating the boundary conditions only on the coarsest grid allowing a minimal effort to eliminate the low frequency errors generated inside the domain due to the mismatching of the interior solution with the boundary conditions. In the subsequent transitions to fine meshes the boundary conditions are kept constant equal to the situation at the coarsest mesh. The results obtained with this technique in a simple V-cycle multigrid strategy are very satisfying and show an additional advantage to multigrid methods for this type of problems.

The smoothing operator used presently is one line relaxation sweep in the downstream direction over the whole flowfield.
No attempts have been undertaken until now to improve this technique although several possibilities exist, for instance the zebra relaxation suggested by Stüben and Trottenberg (1982).

2. ARTIFICIAL DENSITY FORM OF THE EQUATION AND GALERKIN FINITE ELEMENT APPROXIMATION

The potential equation in conservative form is given by

$$\partial_x(\rho\,\phi_x) + \partial_y(\rho\,\phi_y) = 0 \tag{1}$$

where x and y are the Cartesian coordinates in the physical plane and ϕ_x , ϕ_y the velocity components.

The density ρ is obtained from the isentropic relation :

$$\rho = \rho_t \left[1 - \frac{\gamma - 1}{\gamma\, r\, T_t} (\phi_x^2 + \phi_y^2) \right]^{1/\gamma-1} \tag{2}$$

where ρ_t and T_t are stagnation density and temperature and γ the ratio of specific heats.

In transonic flow regime equation (1) is mixed elliptic-hyperbolic and allows different weak solutions for a given set of boundary conditions. If proper viscosity terms are added to the equation a unique mathematical solution is again guaranteed which is equal to the physical solution except for a small region around shocks.

The artificial density form of the artificial viscosity terms, due to Hafez, Murman and South (1978) is particularly well suited for F.E. applications and works satisfactorily for flows with Machnumbers up to 1.5 (Deconinck & Hirsch, 1981 a). It is obtained by giving an upwind bias to the density which is replaced by :

$$\tilde{\rho} = \rho - \mu\, \rho_{\overleftarrow{s}}\, \Delta s \tag{3}$$

where $\rho_{\overleftarrow{s}}$ is the upwind derivative of ρ along the streamwise direction s, Δs the meshspacing and μ a switching function with cut-off Machnumber M_c which controls the amount of artificial viscosity

$$\mu = \max\left(0\, ,\, 1 - \frac{M_c^2}{M^2}\right) \tag{4}$$

In practical implementations the discretization of μ strongly influences the behaviour of the solution in the vicinity of a shock. A very reliable form is given by the following choice in which μ is taken constant over each element (fig.1).

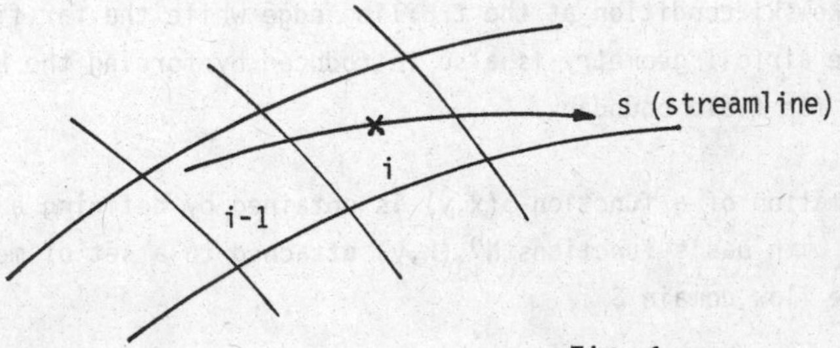

Fig. 1.

$$\mu_i = \max\left(0,\ 1 - \frac{M_c^2}{M_i^2},\ 1 - \frac{M_c^2}{M_{i-1}^2}\right) \tag{5}$$

where i - 1 denotes the upstream element of i, while M_c^2 is chosen between .9 and 1.

A finite Element weighted residual approach is based on the weak formulation of (1) given by :

$$R(\phi) = \int_S \tilde{\rho}\ \nabla W\ \nabla\phi\ dS - \oint_s W\ \tilde{\rho}\ \frac{\partial\phi}{\partial n}\ ds = 0 \tag{6}$$

for any continuous testfunction W, where S is the physical flowdomain with boundary s. The functional $R(\phi)$ is called residual. It can be shown that discontinuous solutions satisfying the weak form (6) will also satisfy the correct mass conserving jump relation across a shock (with normal $\bar{1}_n$) :

$$[\ \tilde{\rho}\ \ \nabla\phi\ .\bar{1}_n]_-^+ \ = \rho_+ (\frac{\partial\phi}{\partial n})_+ - \rho_-\ (\frac{\partial\phi}{\partial n})_- = 0$$

The integral over the boundary is the expression of the Neumann boundary conditions (B.C) which are part of the problem specification. Three types of geometry are considered each giving different specific Neumann B.C. : channel geometry, single airfoil and cascade geometry : Channel walls and blade or profile boundaries require the no flux condition :

$$\tilde{\rho}\ \frac{\partial\phi}{\partial n} = 0$$

Points belonging to periodic boundaries in cascade geometries are treated as interior points by letting coincide corresponding periodic boundaries.

At inlet and outlet boundaries either the solution is given (Dirichlet condition) or the mass flow rate $\tilde{\rho}$ $(\partial\phi/\partial n)$ is specified directly or in an iterative way by applying a Kutta-Youkowski condition at the trailing edge while the far field condition for the single airfoil geometry is also introduced by forcing the known mass-flow rate trough the far field boundary.

A F.E. approximation of a function $\phi(x,y)$ is obtained by defining a finite dimensional space S^h with basis functions $N_{ij}^h(x,y)$ attached to a set of meshpoints (i,j) spread over the flow domain S :

$$\phi^h(x,y) \ = \ \sum_{i,j}\ \phi_{ij}^h\ N_{ij}^h\ (x,y) \tag{7}$$

where ϕ^h_{ij} are the meshpoint values of ϕ^h and h the typical mesh size characteristic of the space S^h. It follows from (7) that :

$$N^h_{ij}\,(x^h_{kl}\,,\,y^h_{kl}) = \delta^{kl}_{ij} = \begin{cases} 1 \text{ for } (i,j) = (k,l) \\ 0 \text{ otherwise} \end{cases} \tag{8}$$

A discrete Galerkin approximation for the weak form (6) is obtained by taking the space of testfunctions W equal to the approximation space S^h, namely :

$$W^h = \sum_{i,j} \alpha^h_{ij}\, N^h_{ij}\,(x,y) \quad \epsilon \quad S^h \tag{9}$$

where α_{ij} are arbitrary reals. With these discretizations equation (6) reduces to the following non linear system for the meshpoint values of ϕ^h :

$$R^h_{ij} = \sum_{k,l} \phi^h_{kl}\; K^{kl}_{ij}\,(\phi^h) - f^h_{ij} = 0 \tag{10}$$

where $K(\phi^h)$ is the stiffness matrix and f^h the contribution of the Neumann B.C.

$$K^{kl}_{ij}\,(\phi^h) = \int_S \tilde{\rho}\,(\phi^h)\;\nabla N^h_{kl}\;\;\nabla N^h_{ij}\; dS \text{ and } f^h_{ij} = \oint_s \tilde{\rho}\;\frac{\partial\phi}{\partial n}\,N^h_{ij}\;ds \tag{11}$$

It is well known that exactly the same expression for the residual is found by solving the discrete minimization problem in S^h in cases where a minimum principle equivalent to the equation can be formulated (as in the fully elliptic subsonic case).

Expression (10) for the residual is developed in the physical plane and written in physical coordinates and can be evaluated for any trial function ϕ after a choice of the type of element has been made which determines the type of basisfunctions of the space S^h. In (Deconinck & Hirsch, 1980) bilinear and biquadratic Lagrange elements have been used, the latter allowing third order accuracy and parabolic approximation of the boundaries. With these elements the integrations over an element surface (eq.11) are usually carried out with Gauss quadrature after transformation of the arbitrarily shaped element to a unit square. In the standard F.E. treatment this transformation is the locally defined isoparametric mapping :

$$x^h(\xi^h,\,\eta^h) = \sum_{i,j} x^h_{ij}\; N^h_{ij}\;(\xi^h\,,\,\eta^h) \tag{12a}$$

$$y^h(\xi^h , \eta^h) = \sum_{i,j} y_{ij}^h N_{ij}^h (\xi^h , \eta^h) \qquad (12b)$$

which is completely determined by the mapping of the meshpoints of the space S^h causing arbitrarily located meshpoints of the grid $S^{h/2}$ not to be mapped uniformly in the (ξ^h , η^h) plane.

The discrete non linear system (eq.10) has been solved with the usual iterative methods such as successive line overrelaxation (SLOR) and approximate factorization (ADI) for which a F.E. version was developed (1981 a).
The simple SLOR method is reliable but extremely slow due to the fact that it eliminates effectively only the errors with wavelength comparable to the meshwidt h. Substantial convergence acceleration was achieved by solving the series of N+1 problems

$$R_{ij}^{2^n h} = 0 \qquad n = N , N-1 , \ldots, 1 , 0 \qquad (13)$$

defined in the space $S_{ij}^{2^n h}$ where the errors of wavelength $2^n h$ are eliminated effectively and the computational effort reduced.

In this grid refinement technique the influence of the coarse meshes is only sensible through the initial approximation for the next finer mesh, while in the full multigrid approach described subsequently the coarse grid equations are modified in order to represent meaningfull approximations of the fine grid corrections.

3. MULTIGRID ALGORITHM.

The nonlinear multigrid algorithm used in this paper is the Full Approximation scheme developed by Brandt (1977).

Considering the system of non linear equations (eq.10) constructed on the finest mesh with characteristic spacing h :

$$R^h(\phi^h)= K^h(\phi^h) - f^h = 0 \qquad (14)$$

If the high frequency errors have been eliminated effectively by means of a smoothing operation such as SLOR or ADI, the corrections $\delta\phi^h$ for an approximate solution ϕ_n^h can be approximated in a meaningful way on a coarse grid with typical spacing 2h obtained by dropping the odd numbered coordinate lines of the mesh S^h and an updated approximation ϕ_{n+1}^h can be calculated by interpolating the coarse grid approximation $\delta\phi^{2h}$ for $\delta\phi^{h}$ back to the original mesh

$$\phi^{h}_{n+1} = \phi^{h}_{n} + I^{h}_{2h}\ \delta\phi^{2h} \tag{15}$$

where I^{h}_{2h} is the coarse to fine grid function interpolation operator called "prolongation".

On the coarse grid, the correction $\delta\phi^{2h}$ defines a coarse grid solution ϕ^{2h} as the approximation of ϕ^{h} on the coarse grid :

$$\phi^{2h} = I^{2h}_{h}\ \phi^{h}_{n} + \delta\phi^{2h} \tag{16}$$

where I^{2h}_{h} is the function restriction. The equation to be solved on the coarse mesh takes again the usual form of eq. 14 :

$$K^{2h}\ (\phi^{2h}) = f^{2h} \tag{17}$$

where the right hand side is a known function of the fine grid approximate solution :

$$f^{2h} = -{}^{R}I^{2h}_{h}\ \ R^{h}(\phi^{h}_{n}) + K^{2h}(I^{2h}_{h}\ \phi^{h}_{n}) \tag{18}$$

After solving this equation the low frequency content of the fine grid approximation is updated according to (15) :

$$\phi^{h}_{n+1} = \phi^{h}_{n} + I^{h}_{2h}\ (\phi^{2h} - I^{2h}_{h}\ \phi^{h}_{n}) \tag{19}$$

The solution ϕ^{2h} in turn can be approximated on the mesh S^{4h} when it is sufficiently smooth i.e. the whole procedure can be applied in a recursive way to eq.(17). Brandt (1977) describes an adaptive strategy for the transition to a coarser or finer grid depending on the convergence level and speed on a particular grid. A more simple fixed strategy has been used in the present work, usually denoted as V-cycle.

Starting on the finest mesh with spacing h one line overrelaxation sweep is performed followed by the transition to the next coarser grid by means of eq.(17) and (18) until the coarsest grid is reached. On the coarsest grid some additional relaxation sweeps are performed and the solution of the next finer grid is updated by means of eq. (19) followed by one relaxation step until the second finest grid is reached.

The cycle terminates with the updating of the finest grid approximate solution with help of eq.(19).

As distinct from F.D. approaches the interpolation operators I_h^{2h}, I_{2h}^h and $^R I_h^{2h}$ are not arbitrary but based on the F.E. interpolation spaces S^h and S^{2h}. They are considered in some more detail in the following sections.

3.1. The coarse to fine grid function interpolation : operator I_{2h}^h.

The only natural choice for the interpolation of a coarse mesh function ϕ^{2h} to a fine mesh location (x_{ij}^h , y_{ij}^h) is to use the value of ϕ^{2h} in the location (x_{ij}^h , y_{ij}^h) given by the F.E. approximation in space S^{2h} :

$$[I_{2h}^h \phi^{2h}]_{ij} = \phi^{2h}(x_{ij}^h , y_{ij}^h) = \sum_{k,l} I_{kl}^{ij} \phi_{kl}^{2h} \tag{20}$$

where the matrix I_{kl}^{ij} is given by

$$I_{kl}^{ij} = N_{kl}^{2h} (x_{ij}^h , y_{ij}^h) \tag{21}$$

On an arbitrary mesh this results in non uniform interpolation coefficients I_{kl}^{ij} and for instance with bilinear elements (figure 2) uniform interpolation is only obtained if the fine grid meshpoints are situated in the middle of the coarse grid element sides and in the center giving only in this case the simple formula (figure 2a) :

$$|I_{2h}^h \phi^{2h}|_C = \frac{1}{4} (\phi_1^{2h} + \phi_2^{2h} + \phi_3^{2h} + \phi_4^{2h}) \quad \text{for the centernode}$$

$$|I_{2h}^h \phi^{2h}|_M = \frac{1}{2} (\phi_i^{2h} + \phi_j^{2h}) \qquad \text{for the midside node i-j} \tag{22}$$

$$|I_{2h}^h \phi^{2h}|_i = \phi_i^{2h} \qquad \text{for the corner nodes (identity)}$$

It follows that simple uniform interpolation is only possible for uniform refinements of the coarsest mesh which could be chosen arbitrarily.

In the general case with bilinear elements (figure 2b) four coefficients are needed for each fine grid meshpoint not coinciding with a coarse grid meshpoint. The computation of these general coefficients eq. (21) is not trivial since N_{kl}^{2h} (x,y) is not explicitly known for an arbitrarily shaped element and one has first to invert the isoparametric transformation eq.(12) to obtain ξ_{ij}^h from :

$$x_{ij}^h = \sum_{m,n} x_{m,n}^{2h} N_{m,n}^{2h} (\xi_{ij}^h , \eta_{ij}^h) \tag{23a}$$

$$y_{ij}^{h} = \sum_{m,n} y_{m,n}^{2h} \, N_{m,n}^{2h} \, (\xi_{ij}^{h} , \eta_{ij}^{h}) \tag{23b}$$

after which the computation is carried out in the ξ-η plane where the basis functions are simple polynomial expressions :

$$I_{kl}^{ij} = N_{kl}^{2h} \, (x_{ij}^{h} , y_{ij}^{h}) = N_{kl}^{2h} \, (\xi_{ij}^{h} , \eta_{ij}^{h}). \tag{24}$$

With the bilinear elements for instance N^{2h} (ξ,η) is of the form :

$$N^{2h}(\xi,\eta) = \frac{1}{4} \, (1 \pm \xi) \, (1 \pm \eta) \tag{25}$$

when the four corner points are situated at $(\xi,\eta) = (\pm 1 \, , \, \pm 1)$

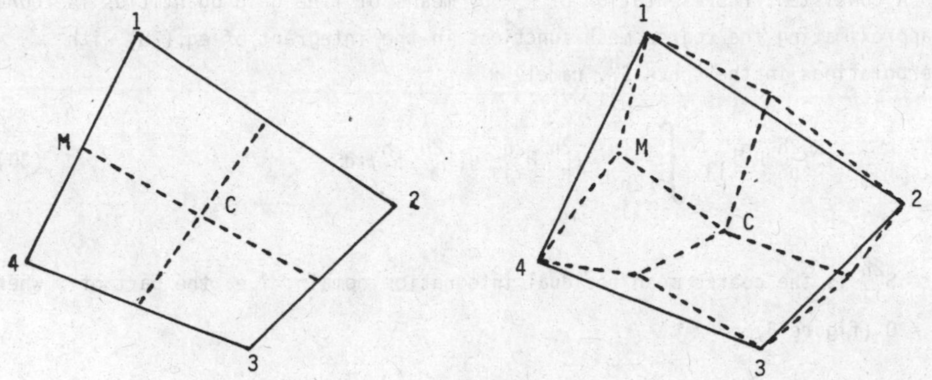

Fig. 2a : Uniform interpolation   Fig. 2b : Non uniform interpolation

3.2 Fine to coarse grid function interpolation (restriction) : operator I_h^{2h}.

The value of ϕ^h in the coarse mesh location calculated with the F.E. approximation in S^h leads to the identity since the coarse gridpoints belong also to the fine grid.

$$|I_h^{2h} \, \phi^h|_{ij} = \phi^h(x_{ij}^{2h} \, , \, y_{ij}^{2h}) = \sum_{k,l} \phi_{kl}^{h} \, N_{kl}^{h} \, (x_{ij}^{2h} \, , \, y_{ij}^{2h}) \tag{26}$$

which due to eq.(8) reduces to :

$$|I_h^{2h} \, \phi^h|_{ij} = \phi_{ij}^{h} \tag{27}$$

This type of restriction is sometimes called injection.

3.3 Fine to coarse grid integral interpolation : operator ${}^{R}I_h^{2h}$.

As distinct from the F.D. case the residual is an integral quantity which is scaled differently on different grids. It cannot be represented in the spaces S^h and S^{2h} and the previous interpolation rules are inapplicable. In the Galerkin approach the volume integrals are always of the form :

$$|R^{2h}(\phi^{2h})|_{ij} = \int_S N_{ij}^{2h}\, g(\phi^{2h})ds \tag{28}$$

For instance the residual in eq. (10) can be written in this form with :

$$g(\phi^{2h}) = \nabla(\rho\ (\phi^{2h})\ \ \nabla\phi^{2h}) \tag{29}$$

A consistent representation of R_{ij}^{2h} by means of fine grid quantities is found by approximating the coarse mesh functions in the integrant of eq.(28) with interpolations in the space S^h, namely :

$$|{}^{R}I_h^{2h}\ R^h|_{ij} = \int_{S_{ij}^{2h}} I_h^{2h}\ N_{ij}^{2h}\ \ g(I_h^{2h}\ \phi^h)\ dS \tag{30}$$

where S_{ij}^{2h} is the coarse mesh residual integration domain, i.e. the part of S where $N_{ij}^{2h} \neq 0$ (figure 3).

The interpolation of $\phi^h(x,y)$ to the coarse mesh leads again to the identity since the interpolation of meshpoint values is the identity by virtue of eq. (27):

$$I_h^{2h}\phi^h(x,y) = \sum_{k,l} N_{kl}^h(x,y)\ I_h^{2h}\ \phi_{kl}^h = \sum_{k,l} N_{kl}^h(x,y)\phi_{kl}^h = \phi^h(x,y) \tag{31}$$

In the same way the coarse mesh basisfunction $N_{ij}^{2h}(x,y)$ is approximated in the space S^h taking the following form after substitution of definition (21) :

$$I_h^{2h}\ N_{ij}^{2h}(x,y) = \sum_{k,l} N_{kl}^h(x,y)\ I_{ij}^{kl} \tag{32}$$

The final expression for the coarse mesh residual weighting by means of fine grid quantities is obtained from eq. (30) by inserting the expressions (31) and (32) :

$$|{}^{R}I_h^{2h}\ R^h|_{ij} = \sum_{k,l} I_{ij}^{kl} \int_{S_{ij}^{2h}} N_{kl}^h(x,y)\ g(\phi^h)dS \tag{33}$$

On a uniformly subdivided coarse mesh (figure 3a) it is clear that this general expression reduces to :

$$|{}^{R}I_h^{2h}\ R^h|_{ij} = \sum_{k,l} I_{ij}^{kl}\ R_{kl}^h \tag{34}$$

where the summation extends only over the 9 inner points in the domain S_{ij}^{2h} since the coefficients I_{ij}^{kl} are zero on and outside the boundaries of S_{ij}^{2h}.

Comparing eqs.(20) and (34) one concludes that the coarse to fine mesh interpolation I_{2h}^h is the adjoint of the residual weighting ${}^{R}I_h^{2h}$ since they have transposed coefficient matrices.

The following result known as full weighting is obtained for uniform subdivisions (figure 3a) , which corresponds to the uniform interpolation (22) :

$$|{}^{R}I_h^{2h}\ R^h|_{ij} = R_{ij}^h + \frac{1}{2}\,(R_{i,j+1}^h + R_{i,j-1}^h + R_{i-1,j}^h + R_{i+1,j}^h)$$

$$+ \frac{1}{4}(R_{i+1,j+1}^h + R_{i+1,j-1}^h + R_{i-1,j+1}^h + R_{i-1,j-1}^h) \tag{35}$$

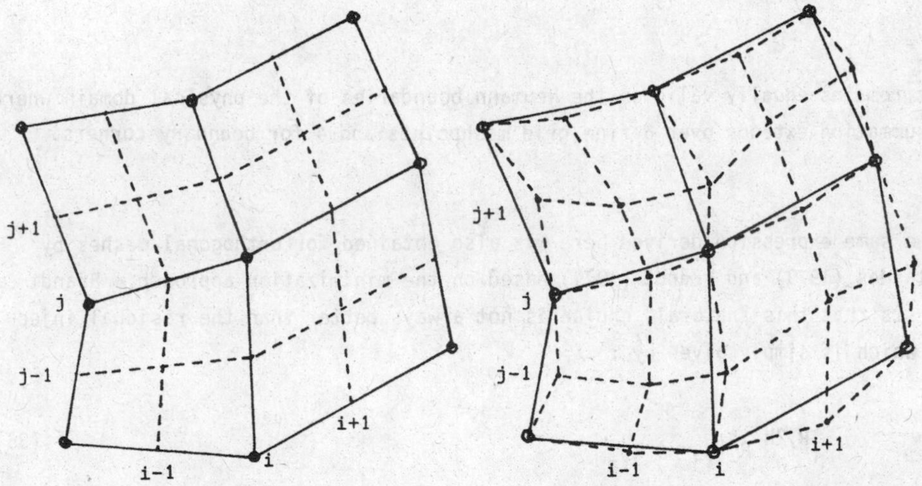

Fig. 3a : S_{ij}^{2h} uniformly subdivided mesh. Fig. 3b : Arbitrarily subdivided mesh.

On an arbitrarily subdivided mesh (figure 3b) the situation is different due to non overlapping integration domains for the coarse and fine mesh. If one is willing to apply the formula (34) with summation over the 9 innermost meshpoints in figue 3b and with the correct non uniform coefficients, two sources of errors are introduced with respect to the exact formula (33) :

. First the contributions of points $(i \pm 1 , j \pm 2)$ and $(i \pm 2 , j \pm 1)$ lying inside the coarse residual integration domain are omitted. For mildly distorted grids their contributions are negligible since the coefficients I_{ij}^{kl} are small near and zero on or outside the limits of S_{ij}^{2h} and also due to the fact that the integral in eq. (33) extends over only two fine mesh elements compared to 4 for the other points.

. Secondly the fine mesh integration domain for points $(i \pm 1, j \pm 1)$, $(i,j \pm 1)$ and $(i \pm 1,j)$ are not always completely contained in the coarse mesh domain S_{ij}^{2h} . Again the errors are small since the surface differences are small and more over since the integrants in eq. (33) approach zero near the limits of the fine mesh integration domain.

In conclusion, eq. (34) remains an extremely valuable approximation for eq.(33) in the arbitrary mesh case, of course only when used with the arbitrary mesh interpolation coefficient I_{ij}^{kl} already known from the non uniform interpolation.

It remains equally valid on the Neumann boundaries of the physical domain where the summation extends over 6 fine grid meshpoints and 4 for boundary corners.

The same expression derived here was also obtained for orthogonal meshes by Nicolaides (1979) and Brandt (1979) based on the minimization approach. Brandt suggests that this "natural" choice is not always better than the residual injection which is simply given by :

$$|{}^{R}I_{h}^{2h} \, R^{h}|_{ij} = 4 \, R_{ij}^{h} \tag{36}$$

in the uniform case.

On an arbitrary mesh residual injection could by constructed by supposing the function $g(\phi^{h})$ constant over the coarse mesh residual integration domain leading to the following general expression :

$$\left|R_{I_h}^{2h} R^h\right|_{ij} = R_{ij}^h \frac{\int N_{ij}^{2h}\, dS}{\int N_{ij}^h\, dS} \tag{37}$$

giving exactly expression (36) in the uniform case.

Computational experience with eq. (37) was highly unsatisfactory and showed that it is inapplicable, at least with the simple smoothing procedure used presently.

4. KUTTA-YOUKOWSKI CONDITION FOR CASCADE FLOWS.

For cascade flows the circulation around the blade is given by (fig. 4) :

$$\Gamma = \oint \bar{\nabla}\phi\, d\bar{\ell} = s(w_2 \sin \beta_2 - w_1 \sin \beta_1)$$

where s is the pitch, w_1 and w_2 the velocities at inlet and outlet, and β_1 , β_2 the in and outlet angles of the flow.

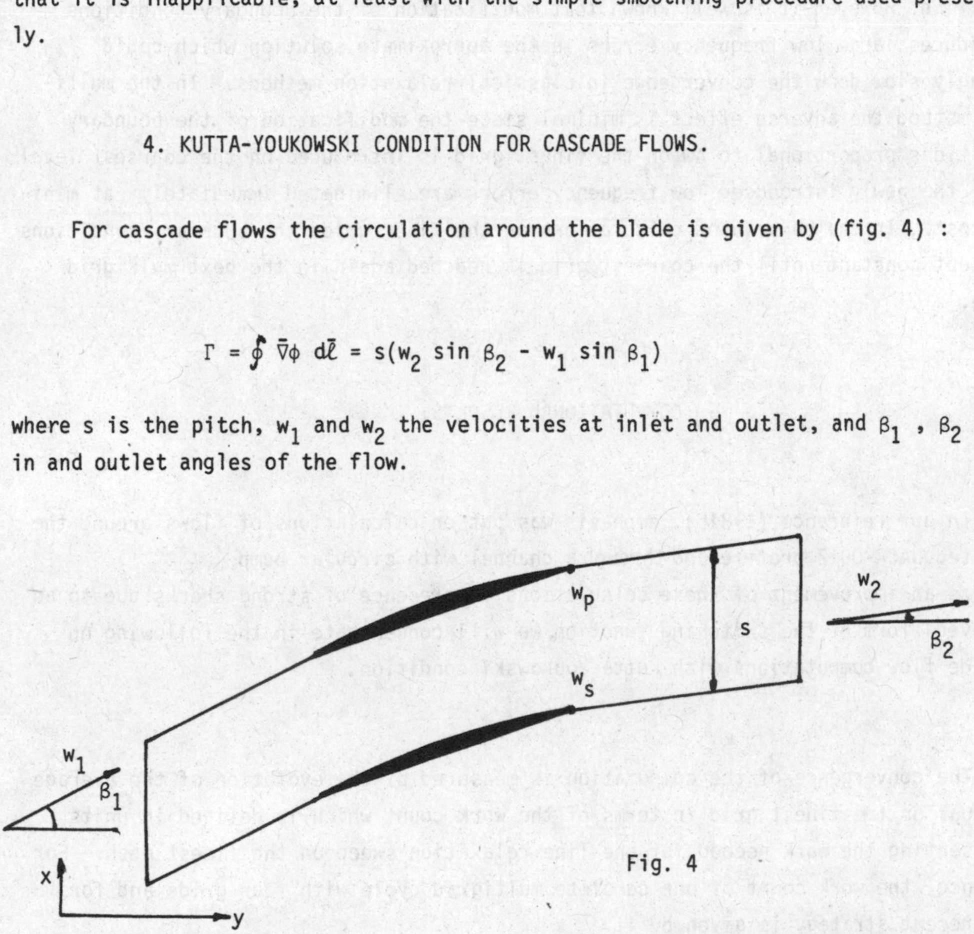

Fig. 4

The Kutta-Youkowski condition is satisfied for the value Γ^* of the circulation for which

$$\Delta w = w_p - w_S = 0 \tag{39}$$

with w_p and w_S the velocities at the trailing edge of the blade on the pressure and suction side.

A progressive change in outlet angle β_2 during the iterative convergence process :

$$\beta_2^{new} = \beta_2^{old} - \alpha . \Delta w^{old} \tag{40}$$

and corresponding adaptation of the inlet boundary conditions in order to conserve mass flow will lead to a converged solution which satisfies the Kutta-Youkowski condition. However it is well known that modification of the boundary conditions introduces large low frequency errors in the approximate solution which could strongly slow down the convergence in classical relaxation methods. In the multigrid method the adverse effect is minimal since the modification of the boundary conditions proportional to Δw on the finest grid is introduced on the coarsest level where the newly introduced low frequency errors are eliminated immediately at minimal cost. In the consequent calculations on the finer grids the boundary conditions are kept constant until the coarsest grid is reached again in the next multigrid cycle.

5. COMPUTATIONAL RESULTS.

In our reference (1981b), emphasis was put on calculations of flows around the isolated NACA-0012 profile and through a channel with circular bump.
Besides an improvement of these calculations in presence of strong shocks due to an improved form of the switching function, we will concentrate in the following on cascade flow computations with Kutta-Youkowski condition.

The convergence of the computation is measured by the evolution of the average residual on the finest grid in terms of the work count which is defined in units representing the work needed for one line relaxation sweep on the finest mesh. For instance, the work count of one complete multigrid cycle with four grids and for the present strategy is given by :

$$1 + 2(\frac{1}{4} + \frac{1}{16}) + \frac{10}{64} = 1,91 \text{ units}$$

plus the additional work for the residual weighting and other interpolations. The convergence rate as used below is defined as the mean reduction in the average residual per unit of work.

5.1 NACA-0012 airfoil at zero-angle of attack and $M_\infty = .85$.

This high Machnumber case was also considered in ref (1981b) and showed a decreased convergence rate of .957 compared to .90 for the flow at M = .80.
It is clear that the reason for this was to be sought in the increased shock intensity. A much better rate of .929 was obtained without changing the multigrid algorithm or smoothing operator but only by slightly modifying the artificial viscosity according to equation (5).
This modification affects only a very limited number of meshpoints in the shockregion. The new result is plotted on figure 5 and compared with the convergence histories with multigrid on two and three grids and with the nested iterations technique eq. (13), taken from ref. (1981b).

The pressure distribution is shown in figure 6 and compared with other conservative methods presented in a workshop held in Stockholm in 1979 (ref. Rizzi & Viviand, 1981). Our solution is converged within plottable accuracy after 40 work units or 15 multigrid cycles corresponding to an average residual drop of only $7\ 10^{-2}$.

5.2 Channel flow at M = .8435.

The same remarks hold for this testcase calculated with 73 x 25 meshpoints and four grids. Due to the modified viscosity the rate is improved from .963 to .935, figure 7.

The pressure distribution along the bump and opposite wall (figure 8) shows good agreement with the result obtained by Jameson, calculated on a 65 x 17 mesh and with the multigrid ADI scheme (MAD) for which a rate of .9742 was obtained (ref. Rizzi & Viviand, 1981).

5.3 Compressor cascade (DCA 9.5° camber blade).

The geometry of this and the next testcase is described in (ref. 1981a) and is generated by solving a system of elliptic partial differential equations for the coordinates. In the following a relatively coarse mesh was used with 481 meshpoints on the finest grid and two coarser grids containing 133 and 40 meshpoints.

The inlet and outlet Machnumbers were held fixed at a value of 1.05 and .761 while the inlet and outlet angles are adapted during the iterative process in order to satisfy the Youkowski condition. The Machdistribution showing a shock on the suction side is given in figure 9. The Machnumber distribution for pressure side and suction side are closed at the trailing edge due to the Youkowski condition.

The convergence history for mean residual (0) and maximum residual (X) is plotted in figure 10. The rate is .897 and is not affected by the iteration on the Youkowski condition.

The value of $\frac{\Delta w}{w_p}$ (see eq. (39))was -1.5 10^{-4} after 85 work units.

The structure of the shock extending from the suction side to the leading edge region where it is detached, can be seen figure 11.

5.4 Turbine Cascade (VKI - LS59 gasturbine blade).

Two cases are considered with an outlet Machnumber of 1.00 and 1.10 and with the same incidence of 30° which is kept constant instead of the inlet Machnumber. The Machdistributions and convergence histories are plotted in figures 12, 13, 14, 15.

Again the Machdistributions are closed at the trailing edge due to the Youkowski condition. The oscillatory behaviour of the convergence history is caused by an oscillatory behaviour of the circulation oscillating around the converged value with decreasing amplitude.

Convergence rates of .925 at M_2 = 1.00 and .948 at M_2 = 1.100 are obtained with final values of $\frac{\Delta w}{w_p}$ of 3.8 10^{-5} and 1.10^{-4}.

CONCLUSION.

A conceptual simple multigrid scheme has been developed consistent with the finite element method and applicable to general arbitrarily generated body fitted grids. Therefore non uniform interpolation and residual weighting operators had to be introduced. A fast and reliable method is obtained with the simple straight forward line relaxation scheme as smoothing step allowing the calculation of realistic transonic flows with about 10 to 20 multigrid cycles (30 to 50 work units).

A Kutta-Youkowski condition for cascade flows has been implemented with minimal cost and minimal effect on the convergence rate.

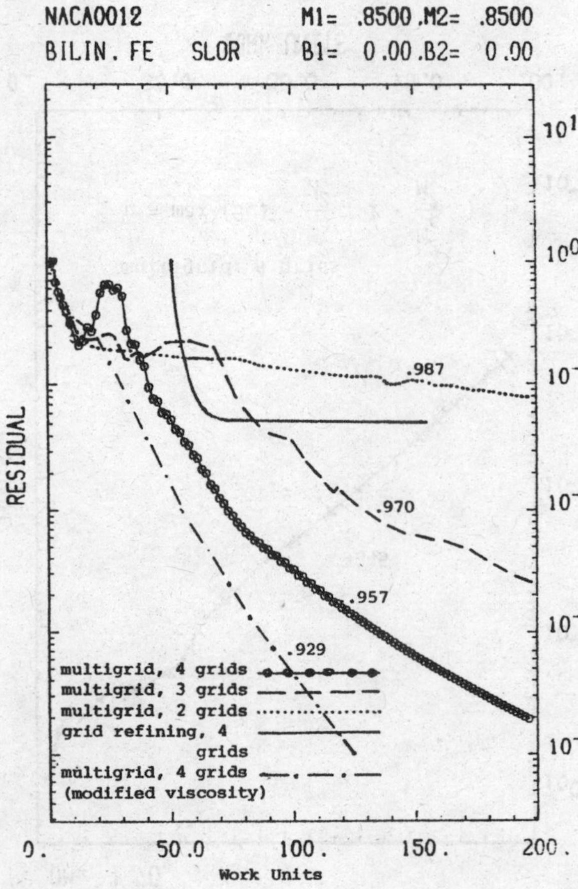
NACA0012
BILIN. FE SLOR
M1= .8500 .M2= .8500
B1= 0.00 .B2= 0.00
RESIDUAL
Work Units
.987
.970
.957
.929
multigrid, 4 grids
multigrid, 3 grids
multigrid, 2 grids
grid refining, 4 grids
multigrid, 4 grids (modified viscosity)

Fig. 5 : Convergence history

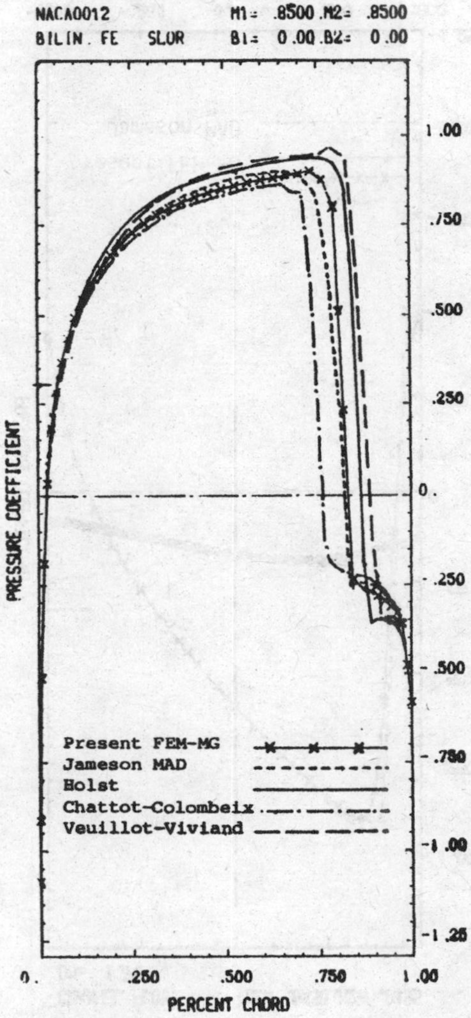
NACA0012
BILIN. FE SLOR
M1= .8500 .M2= .8500
B1= 0.00 .B2= 0.00
PRESSURE COEFFICIENT
PERCENT CHORD
Present FEM-MG
Jameson MAD
Holst
Chattot-Colombeix
Veuillot-Viviand

Fig.6

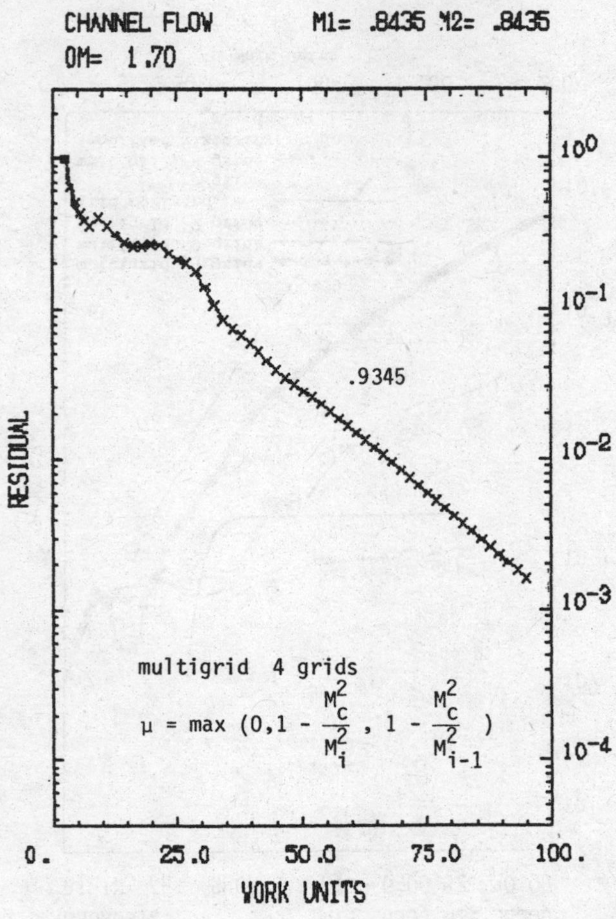

Fig. 7 : Convergence history

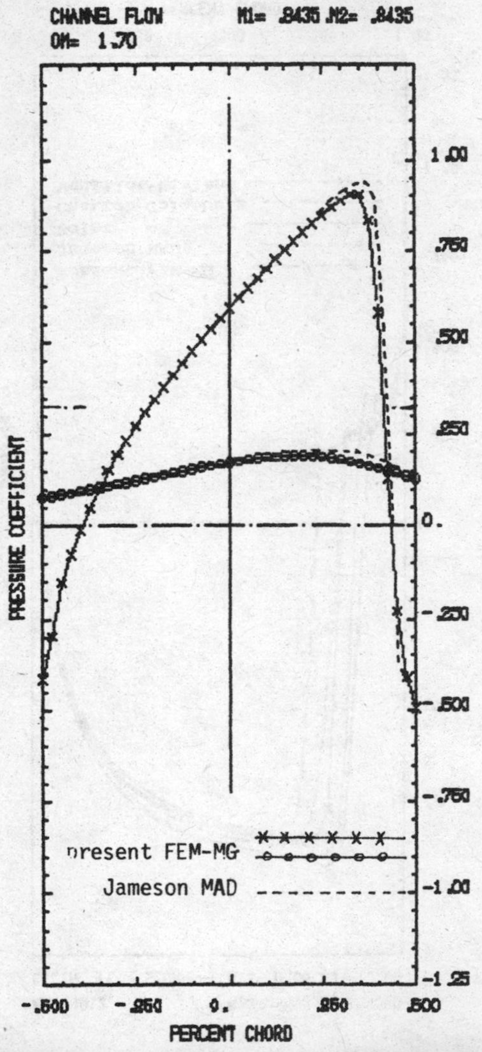

Fig. 8

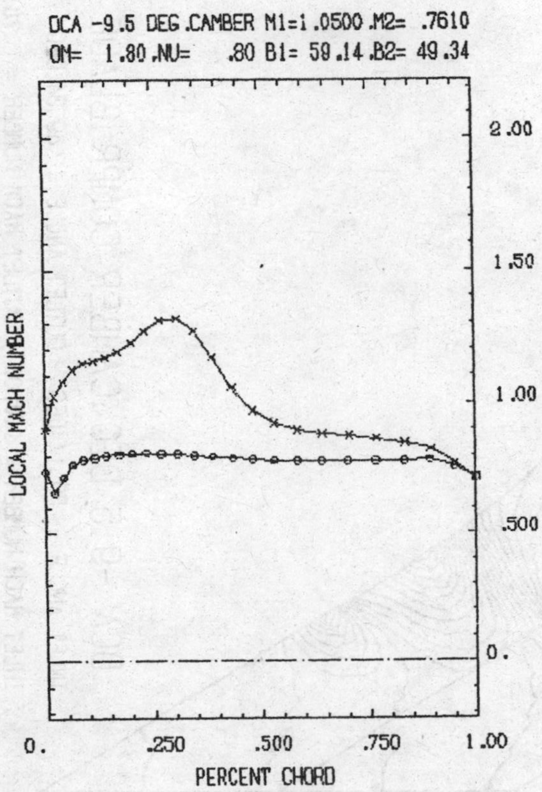

Figure 9.

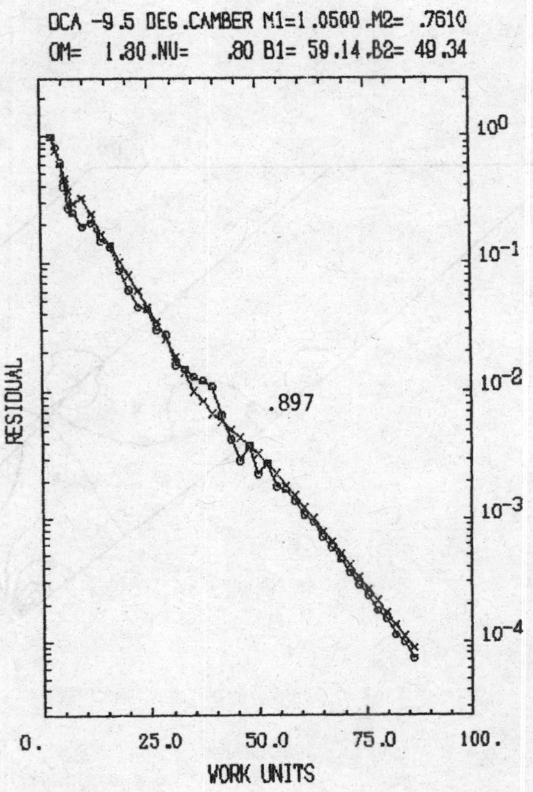

Fig. 10 : Convergence history

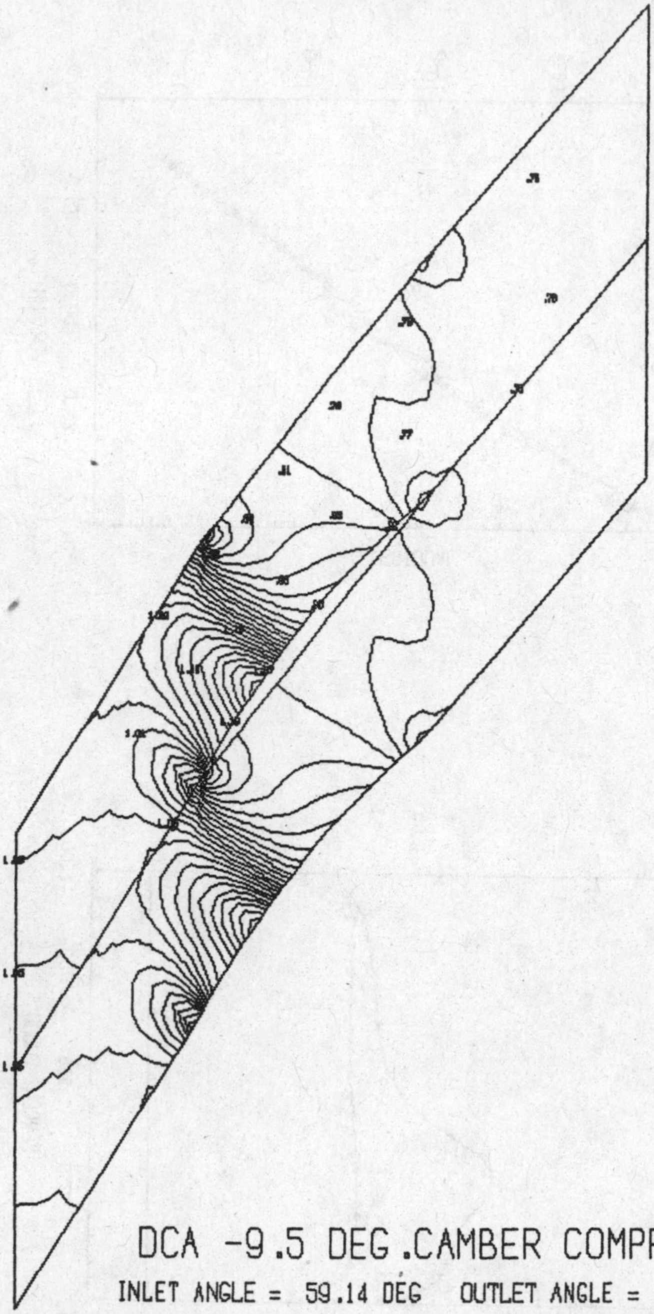

DCA -9.5 DEG.CAMBER COMPR.BLADE

INLET ANGLE = 59.14 DEG OUTLET ANGLE = 49.34 DEG

INLET MACH NUMBER = 1.0500 OUTLET MACH NUMBER = .7810

ISOMACH LINES

Figure 11.

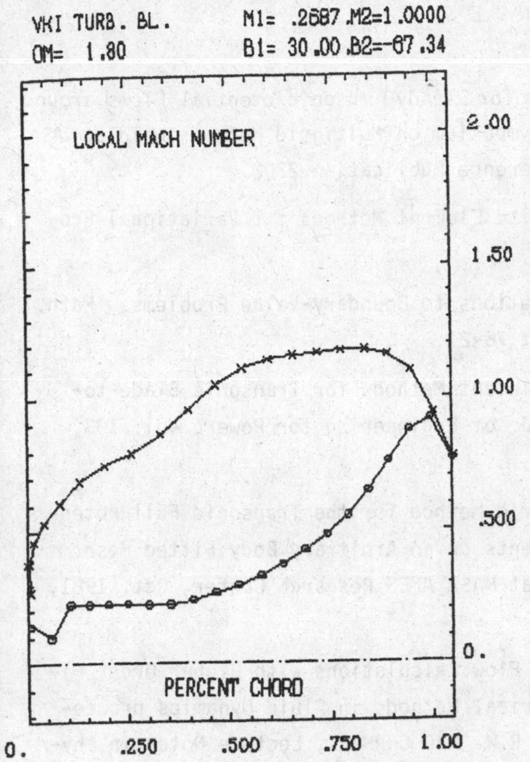

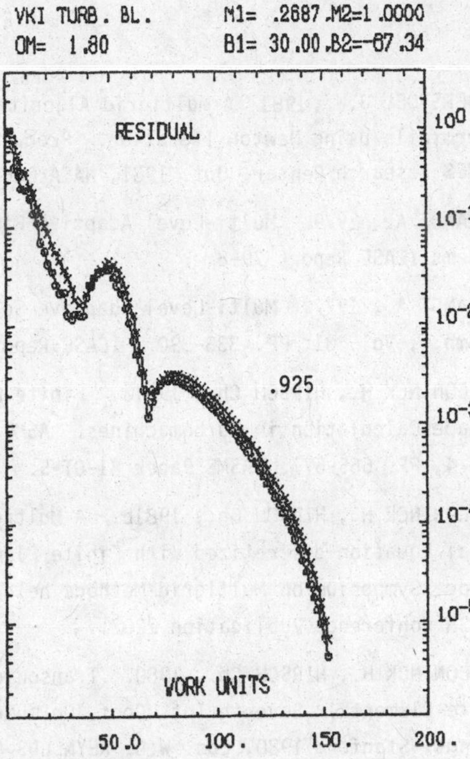

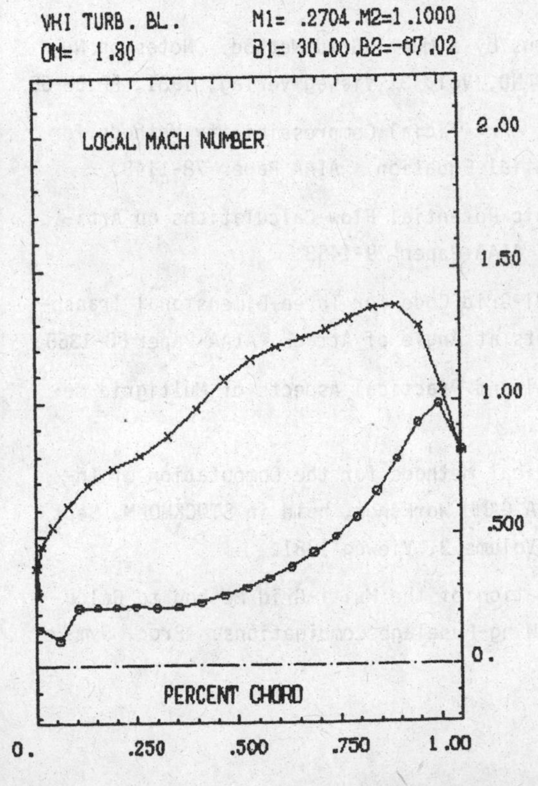

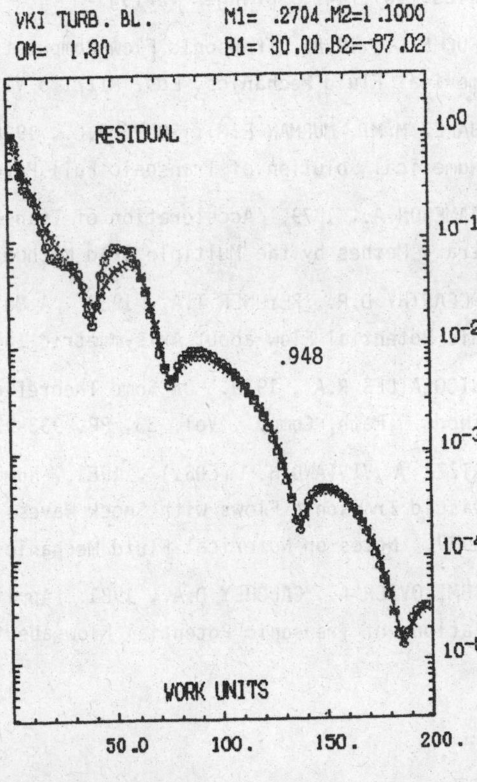

Figures 12, 13, 14, 15

R E F E R E N C E S.

BOERSTOEL J.W., 1981. A multigrid Algorithm for Steady Transonic Potential Flows around Aerofoils using Newton Iteration. Proc. Symposium on Multigrid Methods held at NASA AMES Research Center, Oct. 1981, NASA Conference Publication 2202.

BRANDT A., 1979. Multi-Level Adaptive Finite Element Methods : I Variational Problems ICASE Report 79-8.

BRANDT A., 1977. Multi-Level Adaptive Solutions to Boundary-Value Problems. Math. Comp. , Vol. 31, PP. 333-390. ICASE Report 76-27.

DECONINCK H., HIRSCH Ch., 1981a. Finite Element Methods for Transonic Blade-to-Blade Calculation in Turbomachines. ASME J. of Engineering for Power, Vol. 103, No 4, PP. 665-677. ASME Paper 81-GT-5.

DECONINCK H., HIRSCH Ch., 1981b. A Multigrid Method for the Transonic Full Potential Equation discretized with Finite Elements on an Arbitrary Body Fitted Mesh. Proc. Symposium on Multigrid Methods held at NASA AMES Research Center, Oct. 1981, NASA Conference Publication 2202.

DECONINCK H., HIRSCH Ch., 1980. Transonic Flow Calculations with Higher Order Finite Elements. Seventh Int. Conf. On Numerical Methods in Fluid Dynamics proceedings, Stanford 1980, Eds. W.C. REYNOLDS & R.W. MAC CORMACK, Lecture Notes in Physics, Vol. 141, Springer-Verlag.

FUCHS L., 1978. Transonic Flow Computations by a Multi-Grid Method. Notes on Numerical Fluid Mechanics, Eds. RIZZI & VIVIAND, Vol. 3, Vieweg Verlag, 1981, PP.58-65.

HAFEZ M.M., MURMAN E.M., SOUTH J.C., 1978. Artificial Compressibility Methods for Numerical Solution of Transonic Full Potential Equation. AIAA Paper 78-1148.

JAMESON A., 1979. Acceleration of Transonic Potential Flow Calculations on Arbitrary Meshes by the Multiple Grid Method. AIAA-Paper 79-1458.

MCCARTHY D.R., REYHNER T.A., 1980. A Multi-Grid Code for Three-Dimensional Transonic Potential Flow about Axisymmetric Inlets at Angle of Attack. AIAA-Paper 80-1365.

NICOLAIDES R.A., 1979. On Some Theoretical and Practical Aspects of Multigrid Methods. Math. Comp. , Vol. 33, PP. 933-952.

RIZZI A.,VIVIAND H. (EDS.) , 1981. Numerical Methods for the Computation of Inviscid Transonic Flows with Shock Waves. A GAMM Workshop, held in STOCKHOLM, Sept. 1979. Notes on Numerical Fluid Mechanics Volume 3, Vieweg 1981.

SHMILOVICH A., CAUGHEY D.A., 1981. Application of the Multi-Grid Method to Calculations of Transonic Potential Flow about Wing-Fuselage Combinations. Proc. Symposium on Multigrid Methods held at NASA AMES Research Center, Oct. 1981, NASA Conference Publication 2202.

SOUTH J.C., BRANDT A., 1977. Application of a Multi-Level Grid Method to Transonic Flow Calculations. Transonic Flow Problems in Turbomachinery, ED. by ADAMSON and PLATZER, Hemisphere, Washington 1977.

STÜBEN K., TROTTENBERG U., 1982. Multigrid methods: Fundamental algorithms, model problem analysis, and applications. These Proceedings.

SPARSE MATRIX SOFTWARE FOR ELLIPTIC PDE's

by

Iain S. Duff
Computer Science and Systems Division,
A.E.R.E. Harwell, Oxon.

Abstract

This paper surveys software for the solution of sparse sets of linear equations. In particular we examine codes which can be used to solve equations arising in the solution of elliptic partial differential equations.

In this paper, we are not concerned with algorithms per se but rather their embodiment in robust, portable, well-documented code. They may, to some extent, be considered as alternatives and competitors to multigrid techniques.

CONTENTS

1. Introduction

In this paper we describe mathematical software for solving sparse sets of linear equations. We pay particular attention to the solution of equations obtained from the discretization of elliptic partial differential equations. We restrict ourselves to discussing codes which are readily available, well documented and are easily transported over a wide range of computers. Thus we exclude uncoded algorithms, experimental versions of codes and software which is system dependent.

In addition to direct methods based on Gaussian elimination which we discuss in section 2, we consider software for iterative methods in section 3 and some of the recent advances in semi-direct methods in section 4. In section 5, we comment briefly on codes implementing fast methods based on Fourier analysis and cyclic reduction before ending with some concluding remarks in section 6.

Software based on multigrid algorithms is discussed in many of the other contributions to these proceedings (for example, Foerster et al (1982)). For this reason and because most multigrid software is still in the experimental stage, we have chosen not to include a discussion of any multigrid codes. We feel, however, justified in including this paper in a volume on multigrid techniques. The software we discuss is all of high quality and is currently used in the solution of partial differential equations in many application areas. Thus we are presenting market research on present brand leaders. This serves two purposes in the context of these proceedings. Firstly, it will provide multigrid protagonists with a guide to the strengths and weaknesses of currently used codes and will give an idea of present standards of high quality software in the area of partial differential equations. For multigrid techniques to be widely used and accepted the deficiencies in existing approaches must be answered while the level of software sophistication is maintained. Secondly, the established codes are essential for use as a benchmark in determining the viability of new work. Another reason for considering direct methods at a multigrid conference is that many multigrid approaches use a direct solver on the coarsest grid. Of course, for small problems it may well be that direct solution remains the method of choice. Throughout, we keep our discussion of the underlying algorithms and theory to a bare minimum and encourage readers wishing further details to read the books by Jennings (1977), George and Liu (1981), and Hageman and Young (1981) or the survey by Duff (1977a).

2. Direct methods

In this section we consider software based on Gaussian elimination for the solution of the sparse set of linear equations

$$A\underline{x} = \underline{b} \tag{2.1}$$

That is, we consider implicit or explicit factorizations of the form

$$PAQ = LU \tag{2.2}$$

where P and Q are permutation matrices and L and U are lower and upper triangular matrices respectively. When A is symmetric, we can use the decomposition

$$PAP^T = LDL^T \tag{2.3}$$

where P and L are as before and D is a diagonal matrix (for indefinite matrices A, D may need to be block diagonal with blocks of order 1 and 2).

An important feature of direct methods is that we can divide the solution of (2.1) into three stages viz.

(i) Selection of P and Q to preserve sparsity in the subsequent factorization. This is often followed by symbolic processing to create data structures for facilitating the factorization.

(ii) Computation of L,U,D.

(iii) Solution of (2.1) by forward and back substitution using the decomposition (2.2) or (2.3).

These three phases are often referred to as ANALYZE, FACTORIZE and SOLVE. The importance of this subdivision is that ANALYZE need only be performed once for a particular sparsity pattern and the FACTORIZE step is only required when the numerical values of the coefficient matrix changes. Since the SOLVE routine is usually very much faster than the first two stages (the ratio of FACTORIZE to SOLVE time is typically 10 to 1), direct methods can be very efficient when the solution of several problems with the same coefficient matrix is required. This would be the case, for example, when an elliptic partial differential equation is solved for varying boundary conditions.

When A is symmetric, the most common method of choosing the ordering in ANALYZE is that of minimum degree. This algorithm chooses as pivot, at each stage, the diagonal entry from the row with least number of non-zeros where fill-ins from previous elimination steps are included in the non-zero count. Remarkable advances have been made in recent years in the implementation of the minimum degree ordering. For example, on a matrix from a finite element problem of order 3466 with 13681 non-zeros, the best methods in use today run nearly 300 times faster than codes considered supreme twelve years ago.

Indeed, in recent years there have been significant advances in the implementation of all three phases when the matrix is symmetric (see, for example, George (1981)). For the matrix arising from the five-point discretization of the Laplace operator on an nxn grid, the first phase requires only $O(n^2)$ space and time, the second only $O(n^3)$ time and $O(n^2 \log n)$ space and the third $O(n^2 \log n)$ space and time. In the context of these proceedings, it is important to record that, in each case, the constant multiplying the dominant term is small (usually less than 10). Additionally, for simple geometries, the ANALYZE stage may be unnecessary or can be greatly reduced as indeed can FACTORIZE in cases where the coefficient matrix is of a standard form (for example, from the Laplacian operator). Thus, one can argue that the SOLVE times are those which should be compared with other methods of solution.

When A is unsymmetric it is common to combine the ANALYZE and FACTORIZE steps so that numerical stability can be ensured. This will increase the cost of processing a new pattern although for subsequent decompositions with changed numerical values a FACTORIZE by itself is usually sufficient. It is, however, common to monitor this FACTORIZE in case it is unstable.

We divide our subsequent discussion of software for the solution of (2.1) according to the properties of the matrix A. First we examine codes applicable when A is of a special structure (for example, banded). We then describe software applicable when A is symmetric positive definite or when no numerical pivoting is required. This is followed by a discussion of codes available when A is a general matrix and we conclude by considering cases when A is not formed explicitly but is generated a row at a time or as a sum of submatrices as is common in finite element applications. A critical analysis of several features available in direct codes and a comparison of some of the codes discussed here is given by Duff (1979).

Band matrices are a particularly simple example of sparse matrices and codes for solving both symmetric and unsymmetric band systems can be found in almost all mathematical software libraries (for example, HSL,IMSL,NAG,PORT) or collections (for example, LINPACK (Dongarra et al (1979)). The special and commonly occurring case of tridiagonal systems is often handled by a different code from that for matrices with a wider band.

A more general class of matrices which arise commonly in the solution of partial differential equations is that of variable band matrices (Jennings (1966)) where the first non-zero in each row and in each column need not be the same distance from the diagonal. Some of the standard orderings on grid based problems give matrices suitable for variable band storage and elimination schemes. Although the number of non-zeros in the LU factors for such schemes is usually higher than for the more general methods we will discuss later, variable band methods have the merit of being very simple and requiring much less storage overhead than the general schemes. Another benefit of both the band and variable band matrices is that only a relatively small part of the matrix need be in main storage at any one time. The Harwell code MA15 makes use of this observation in the solution of symmetric positive definite band systems. The same type of structures are also termed profile or skyline (for example, George et al (1980), Thompson and Shimazaki (1980), Hasbani and Engleman (1979) and Felippa (1975)).

There are a number of codes for solving general sparse symmetric positive definite systems. The SPARSPAK package of George et al (1980) has a particularly carefully designed user interface offering several options for the ordering P in (2.3) viz:

(i) Reverse Cuthill-McKee (profile elimination)

(ii) One-way dissection

(iii) Refined quotient tree

(iv) Nested dissection

(v) Minimum degree

using a storage, factorization and solution scheme appropriate to the particular ordering chosen. A good discussion of these orderings and their merits can be found in the book by George and Liu (1981). Great care has been taken by the authors of SPARSPAK to screen the user from the internal data structures. Although SPARSPAK has a commendably easy user interface, some efficiency has been sacrificed and the codes are difficult to modify for use in application packages. The Yale Sparse Matrix Package (Eisenstat et al (1977)) uses a minimum degree ordering although, in common with most other codes, a user specified ordering can be input. Both the Yale package and SPARSPAK have very efficient ANALYZE routines typically executing in $O(\tau)$ time and space where τ is the number of non-zeros in the factors. A similar approach has been used by Duff and Reid (1982) in the Harwell code MA27 which, in addition to having a very fast minimum degree ordering, has an option for obtaining a stable decomposition when the coefficient matrix is indefinite. Additionally, MA27 has been designed so that it will take advantage of vectorization on machines like the Cray-1.

Both SPARSPAK and the Yale package have an option for solving systems whose coefficient matrix is unsymmetric but has a symmetric structure. In both codes, pivots are chosen from the diagonal with no numerical restrictions and so the decomposition could be unstable. MA27 has been designed so that routines for obtaining a stable decomposition in this structurally symmetric case could be incorporated in the package but these modules have not yet been written.

The Yale package has a version which factorizes unsymmetric matrices using diagonal pivoting without numerical control employing an ordering supplied by the user or generated from applying the minimum degree ordering to the symmetric pattern whose upper triangle has the same structure as the upper triangle of the coefficient matrix. In order, however, to obtain a stable decomposition for general unsymmetric matrices, numerical pivoting must be used and, as we mentioned earlier, it is common to combine the symbolic processing and factorization phases. The NSPIV code of Sherman (1978) performs partial pivoting by columns within a given row ordering. His code has the merit of being short and produces a very stable decomposition but it is difficult to choose a good ordering for the rows. For many orderings there can be considerable fill-in and the computation time can be high. The code MA28 in the Harwell Subroutine Library (Duff (1977b)) selects its pivots by combining the Markowitz (1957) sparsity criterion with threshold pivoting where $a_{\ell k}$ can be used as a pivot only if

$$|a_{\ell k}| \geq u.\max_{i} |a_{ik}|$$

where u is a user set parameter in the range (0,1]. There is a corresponding code to MA28 in the Harwell Subroutine Library (ME28) for the solution of sparse sets of complex equations (Duff (1981b)). The current release of NSPIV also employs threshold pivoting to gain some sparsity over the original NSPIV code. Zlatev et al (1978) also use threshold pivoting with Markowitz' criterion in their code SSLEST. They also include options for restricting the pivot search to a specified number of columns and for dropping nonzeros below a specified value from the sparsity structure. The use of trop tolerances coupled with iterative refinement can greatly reduce the storage requirements for the factors L and U when solving (2.1). This approach is used in the codes SIRSM (Thomsen et al (1977)) and Y12M (Zlatev and Wasniewski (1978)) and work is currently under way to incorporate such a feature in the MA28 package. This use of drop tolerances is rather similar to the incomplete factorization semi-direct methods which we will discuss in section 4, although there the iterative scheme normally used is conjugate gradients rather than iterative refinement. The SLMATH code (IBM (1976)) based on work of Fred Gustavson uses full code when the reduced matrix becomes full and has a fast factorization entry (see, for example, Duff (1979)). It is planned to include such a switch in a future release of the MA28 package.

The last class of matrices which we consider are those which are not formed explicitly but are input row by row or element by element in a finite element calculation. As we mentioned earlier, the band and variable band matrices are well suited to row by row input but the first generally available code for element input was that of Irons (1970) for the frontal solution of positive definite element matrices. This was extended by Hood (1976) who allowed the assembled matrix to be unsymmetric and the MA32 package (Duff (1981a)) in the Harwell Subroutine Library combines a robust and stable version of this with a more general code allowing the input of a general unsymmetric matrix by rows. This mode of input is particularly designed for the direct solution of unsymmetric problems arising from finite difference discretizations of partial differential equations. Since these frontal codes use full matrix indexing in their innermost loops, it is possible to make them perform well on machines capable of vectorization. For example, the MA32 code running on the Cray-1 at Harwell has solved equations from discretizations of partial differential equationsat about 25 million floating point operations per second (Mflops). This time includes all the I/O overhead and is not far short of the best solution times for solving full systems on the Cray-1 using Fortran code (30 Mflops).

3. Iterative methods

Since the history and current development of iterative methods is tied strongly to the solution of partial differential equations, one might naturally assume that the section on iterative methods would form the major part of this paper. This is not the case, however, for two reasons. Firstly, mainly because of difficulties with user interface and parameter selection, there has been little production of high quality generally available software implementing iterative methods. Secondly, many recent developments and extensions of iterative methods use partial factorizations of the coefficient matrix to accelerate the convergence rate. We choose to discuss such methods, often called semi-direct methods, separately in section 4.

A major ongoing software project for iterative methods is ITPACK (Grimes et al (1980), Young and Kincaid (1980)). The ITPACK codes are written in a portable subset of Fortran and presently offer a unified interface to seven methods for solving linear equations with a sparse symmetric positive definite matrix.

The methods currently available are:

1. Jacobi iterative with Chebyshev acceleration
2. Jacobi iteration with conjugate gradient acceleration
3. Reduced system with Chebyshev acceleration
4. Reduced system with conjugate gradient acceleration
5. Successive overrelaxation
6. Symmetric successive overrelaxation with Chebyshev acceleration
7. Symmetric successive overrelaxation with conjugate gradient acceleration

where the reduced system corresponds to implementing Richardson's method on the system obtained after eliminating all the red nodes of a red black ordering. A principal feature of the package is the automatic selection of acceleration parameters and stopping criteria. The package is, however, oriented towards teaching and research with the emphasis on flexibility rather than maximum efficiency and so may not be suitable for use in a production environment. Although the routines are primarily designed to solve symmetric positive definite systems and are only guaranteed theoretically to converge on such problems, the interface allows matrices which are not symmetric and the experience of the developers of the package has been that the algorithms work well when the coefficient matrix is nearly symmetric. These routines have been incorporated into the ELLPACK software package (Rice (1978)), for the solution of elliptic partial differential equations in a teaching and research environment. The authors plan to extend the ITPACK package by including robust algorithms for the solution of unsymmetric problems.

The conjugate gradient algorithm for the solution of positive definite systems has a long history (Hestenes and Stiefel (1952)) but the method has not been greatly favoured on systems arising from discretizations of partial differential equations because it uses more storage and is not significantly faster than traditional methods such as SOR. Some of the attempts to improve the convergence have led to semi-direct methods which we discuss in the next section.

The strongly implicit procedure of Stone (1968) is particularly geared to the structures obtained from finite difference discretizations of elliptic equations and Jacobs (1980) has developed a series of routines DI205,DI209,DI213 and DI307 implementing this method for 5,9,13 diagonal systems derived from 2 dimensional models and 7 diagonal systems from 3 dimensional models, respectively. These subroutines are also available from NAG as subroutines D03EB,D03EC,D03UA and D03UB respectively. Jacobs has also developed SIP codes for use when, in addition to the diagonal structure, a few extra non-zeros or a number of small non-zeros are present. However, these algorithms exhibit very slow convergence on large problems (see, for example, Jacobs (1981a)). Their author now favours techniques based on preconditioned conjugate gradients which we will discuss in the following section.

4. Semi-direct methods

The development of methods which combine direct and iterative methods (some multigrid work could be considered in this class) is currently one of the most active areas of sparse matrix research. Quite recently some good software has been produced implementing algorithms based on this semi-direct approach. The main impetus for research on semi-direct methods has come from the desire to solve more complicated partial differential equations in three dimensions and on finer grids in two dimensions.

As we mentioned in section 3, a major class of semi-direct methods is geared to improving the convergence of the conjugate gradient method. Preconditioning is employed on the coefficient matrix to yield a matrix whose spectrum is more favourable. The normal effect of preconditioning is to bunch the eigenvalues. A common preconditioning technique is to perform a partial decomposition

$$A = LL^T + C \tag{3.1}$$

where the matrix C is not at round-off level and we have omitted permutations for the sake of clarity. We then use the preconditioned matrix in the conjugate gradient iteration. It is because of the partial factorization (3.1) that such methods are often termed semi-direct.

When using such techniques to solve partial differential equations whose discrete operator has a regular pattern it is normal to take advantage of the structure by constraining L so that its sparsity pattern is either identical to A or differs from it in some regular way. We illustrate such a possibility in Figure 4.1 where the original matrix arises from the five-point discretization of the Laplacian and L is allowed extra non-zeros in bands immediately outside and inside the central and outer bands of A respectively.

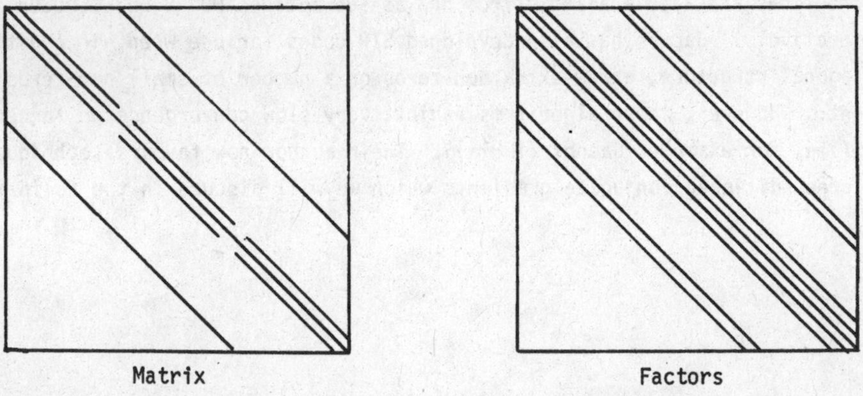

Figure 4.1 The structure of coefficient and partially factorized matrices on a model problem.

van Kats et al (1980) at ACCU Utrecht have developed software based on research by Meijerink and van der Vorst (1977). Their software include codes for conjugate gradients with or without a user supplied preconditioning together with several subroutines for use when the matrix is obtained from the five-point formula (the ICCG methods) with possible periodic boundary conditions. Their codes are in portable Fortran. Kuo-Petravic and Petravic (1981) have a similar ICCG code in the Computer Physics Communications Library and have a version (Kuo-Petravic and Petravic (1979)) implementing an unsymmetric preconditioning Jacobs (1981b) has also recently written preconditioned conjugate gradient codes particularly geared to the solution of partial differential equations.

When the structure of the matrix is more general or when diagonal pivoting is inadvisable, a preconditioning based on a partial factorization of the form (3.1) can still be applied but some form of stabilization will be required and the codes will normally be much more complicated. The codes by Ajiz and Jennings (1981) and Munksgaard's (1980) code MA31 in the Harwell Subroutine Library stabilize the partial factorization by modifying diagonal entries during the decomposition.

In the symmetric indefinite case, the conjugate gradient method (with or without preconditioning) can break down and biconjugate gradients (Fletcher (1976)) or a variant of the Lanczos algorithm (Parlett (1978) or Paige and Saunders (1975)) are commonly used. Code implementing the SYMMLQ algorithm of Paige and Saunders (1975) with optional preconditioning is available from Saunders at Stanford.

Although many of the methods used in the symmetric case can be extended to solve unsymmetric systems, the theory and practice of preconditioning is not as well understood. For unsymmetric systems it is possible to form the normal equations (usually implicitly) and solve the resulting symmetric positive definite system but this is generally not satisfactory for two reasons. The first is that the amount of computation for each iteration is doubled and the second is that the convergence of the iterative methods on the normal equations is often slow.

Software utilising semi-direct methods for solving unsymmetric systems is based on two main approaches; extensions of conjugate gradients is one avenue and the use of Chebyshev iterations another. Many unsymmetric matrices can be split

$$A = M - N \tag{3.3}$$

where the matrix M is symmetric positive definite and N is skew symmetric. A version of the conjugate gradient algorithm (Concus and Golub (1975)) can then be applied to solve

$$M\underline{x}^{(k+1)} = N\underline{x}^{(k)} + \underline{b} \tag{3.4}$$

but we do not know of any easily available software which implements this approach.

Another way in which conjugate gradients can be extended is to ensure local conjugacy between recent search directions by explicit orthogonalization. This approach, usually coupled with an incomplete LU factorization as a preconditioning, is used in several large oil reservoir codes (for example, Vinsome (1976)) but we know of no readily available software which implements it. The main problems lie in the lack of good theoretical results, the difficulties in choosing a good preconditioning, and the amount of reorthogonalization required on realistic problems. The biconjugate gradient algorithm can be extended to general systems (including those with complex coefficient matrices) and Jacobs (1981b) has produced software based on this approach. Paige and Saunders (1982) have written a code based on the bidiagonalization procedure of Golub and Kahan (1965) for solving either unsymmetric equations or the least squares problem. Their code, LSQR, is equivalent to the unsymmetric conjugate gradient method of Hestenes and Steifel but is numerically more reliable on ill-conditioned systems. It is possible to use any son-singular preconditioning matrix, C say, with LSQR (as long as the equations $C\underline{x}=\underline{b}$ and $C^T\underline{x}=\underline{b}$ can be solved) but again the main problem is in calculating a good preconditioning without which convergence can be slow.

Manteuffel (1977) has written five codes which solve both symmetric and unsymmetric systems using Chebyshev iteration. His codes are

TCHSYM: Matrix symmetric positive definite

TCHSYP: Matrix symmetric. Symmetric preconditioning matrix

TCHFPS: Matrix unsymmetric. Symmetric part used as preconditioner

TCHEB: Matrix unsymmetric

TCHEBP: Matrix unsymmetric with preconditioning

where, in each case, if C is the preconditioning matrix (C=I when there is no preconditioning), the eigenvalues of $C^{-1}A$ must lie in the right half plane.

A feature of these routines, which are written in portable Fortran and will shortly appear in the SLATEC library (a mathematical library from a subcommittee of the Sandia, Los Alamos, Air Force Weapons Laboratories, Technical Exchange Committee), is that the optimal iteration parameters are determined automatically. The user must, however, supply the preconditioning matrix. The ADACHE code of van der Vorst and van Kats (1979) with an optional stabilized partial LU preconditioning for 5 diagonal systems also implements Manteuffel's algorithms in the case where all the eigenvalues of A lie in the right half plane.

5. Fast methods

The very regular structure of Laplace, Poisson or Helmholtz equations give rise to a number of methods termed fast or fast direct methods. They are based on several techniques including marching methods, cyclic reduction, and the use of the fast Fourier transform (Swarztrauber (1977)). A recent survey of Detyna (1981) discusses many of the current approaches. They are termed fast since their computation time varies almost linearly with the number of grid points and an idea of their actual speed on model problems can be gleaned from the GAMM report (Schumann (1978)) on a contest between such solvers. This is an area in which fine tuning has resulted in many experimental codes which are unattractive pieces of software and are difficult to use. However, we can recommend the FISHPAK package (Swarztrauber and Sweet (1979)) as well designed and easy to use. It is written in portable Fortran and solves Laplace, Poisson and Helmholtz equations in Cartesian and polar coordinates using a combination of cyclic reduction and fast Fourier transforms. Another package by Swarztrauber (1981), FFTPACK, contains routines for computing periodic as well as symmetric discrete Fourier transforms and can be used to solve Poisson's equation subject to the standard boundary conditions including Dirichlet and Neumann conditions on opposite boundaries. van Kats and van der Vorst (1979) also have a code (TBPSXY) implementing cyclic reduction with Buneman stabilization (Buzbee et al (1970)) which can be used to solve Poisson's or Helmholtz' equation on a rectangular domain with a uniform grid and with Dirchlet boundary conditions on two opposite sides and Dirichlet or Neumann on the other sides. Although these fast methods are quite limited in the geometries to which they are applicable, it is sometimes possible to divide the region so that fast methods will work on each part and then use some correction technique to obtain the solution for the whole region. The capacitance matrix method is one way of doing this and O'Leary and Widlund (1981) have written a code using this method to solve the Helmholtz equation with Dirichlet boundary conditions on general bounded three-dimensional regions.

6. Concluding remarks

Although the use of direct methods is not practical for really large systems on general geometries, nearly all large production codes (particularly in the oil industry) have a direct solver option for use as a benchmarking tool, because of its robustness, and as the preferred method for small to medium two dimensional problems. The out-of-core frontal (e.g. Harwell's MA32) and variable band methods (e.g. Harwell's MA15) can be used even when the system is quite large particularly when the equation is being solved on a narrow pipe geometry. For example, problems in fluid flow on 31x200 grids with five variables per mesh point have been solved by MA32. Direct methods are particularly useful when the coefficient matrix is not symmetric or diagonally dominant, an uncommon occurrence in model problems but very common in many practical situations for example in fluid flow calculations. Another benefit of direct solvers is that they provide an accurate solution to the discrete equations. When solving partial differential equations in isolation it is reasonable to solve the discrete equations only to the accuracy of the discretization error (a philosophy often adopted by researchers in multigrid techniques). However, when the solution to the differential equations is required within the loop of a parametric study, for example in the generation of function values for a minimization routine or for use by data fitting packages, then an accuracy and smoothness far greater than that of the discretization is usually required. Iterative or semi-iterative methods will generally need considerably more work to achieve a satisfactory accuracy while this is obtained automatically by direct methods.

The iterative and semi-direct methods are the subject of much current research and high quality software is beginning to appear. These techniques show much promise in the solution of three dimensional problems with spherical (or cubic) geometries where even out-of-core direct methods are limited by storage. The smoother in the multigrid approach is often an iterative or semi-direct method and it is this class of methods which is in most direct competition with multigrids.

The main limitation of fast solvers lies in the difficulty in applying them to all but the simplest equations and geometries. It should be stressed, however, that the best of such methods, the FACR(ℓ) approach (Swartztrauber (1977)), executes in $O(n^2 \log \log n)$ time (on an nxn grid) and so is asymptotically very competitive with multigrids particularly since the constant multiplying the dominant term is 3.

In this paper, we have not given details of how one can obtain the codes described. We recommend that the reader contact the author(s) referenced or consult Duff (1982) where more details on code availability are given.

The ELLPACK project (Rice (1978)) represents a major vehicle for the comparison of equation solving codes in the context of the solution of partial differential equations. Its modular structure allows an easy interface for different methods of solving the discrete equations and one of the ELLPACK design criteria was to facilitate such comparisons. Presently there are direct and iterative solvers in the package and some codes based on semi-direct methods are likely to be incorporated in the near future. Perhaps it might be possible to examine the performance of multigrid codes within the framework of this project.

In this survey, we have quite intentionally avoided actual comparisons of codes. In many areas only very few comparisons of software have been conducted and in others (for example, Duff (1979), Duff and Reid (1979)) it is a highly specialised enterprise with the difficulties only just being appreciated. We have, however, some familiarity with the codes described or their authors and believe the software we have discussed to be of high quality.

Acknowledgements

During the writing of this paper I consulted many of the researchers to whose work I have referred. I would like to record my thanks to them for sending me information on their work and for commenting on early drafts of the manuscript.

I would like to thank John Reid and Chris Thompson at Harwell for reading a draft of this paper and making several helpful comments.

References

Ajiz, M.A. and Jennings, A. (1981). A robust ICCG algorithm. Report Civil Eng. Dept., Queen's University, Belfast.

Buzbee, B.L., Golub, G.H. and Nielson, C.W. (1970). On direct methods for solving Poisson's equations. SIAM J. Numer. Anal. 7, pp.627-656.

Concus, P. and Golub, G.H. (1975). A generalized conjugate gradient method for nonsymmetric systems of linear equations. Comput. Sci. Dept. Stanford Report STAN-CS-75-535. Presented at IRIA Meeting, Paris, December 1975.

Detyna, E. (1981). Rapid elliptic solvers. In Sparse Matrices and their Uses. I.S. Duff (Ed.), Academic Press pp.245-264.

Dongarra, J.J., Bunch, J.R., Moler, C.B. and Stewart, G.W. (1979). LINPACK Users' Guide. SIAM Press.

Duff, I.S. (1977a). A survey of sparse matrix research. Proc. IEEE 65, pp.500-535.

Duff, I.S. (1977b). MA28 - a set of Fortran subroutines for sparse unsymmetric linear equations. AERE Report R.8730, HMSO, London.

Duff, I.S. (1979). Practical comparisons of codes for the solution of sparse linear systems. In Sparse Matrix Proceedings 1978. I.S. Duff and G.W. Stewart (Eds.). SIAM Press pp.107-134.

Duff, I.S. (1981a). MA32 - A package for solving sparse unsymmetric systems using the frontal method. AERE Report R.10079, HMSO, London.

Duff, I.S. (1981b). ME28 - A sparse unsymmetric linear equation solver for complex equations. ACM Trans. Math. Softw. 7, pp.505-511.

Duff, I.S. (1982). A survey of sparse matrix software. AERE Report CSS 121. To appear in Sources and Development of Mathematical Software. W.R. Cowell (Ed.). Prentice-Hall.

Duff, I.S. and Reid, J.K. (1979). Performance evaluation of codes for sparse matrix problems. In Performance Evaluation of Numerical Software. L. Fosdick (Ed.), North Holland, pp.121-135.

Duff, I.S. and Reid, J.K. (1982). The multi-frontal solution of indefinite sparse symmetric linear systems. AERE Report CSS 122.

Eisenstat, C.S., Gursky, M.C., Schultz, M.H. and Sherman, A.H. (1977). Yale sparse matrix package I. The symmetric codes. II. The nonsymmetric codes. Reports 112 and 114. Dept. Computer Science, Yale University.

Felippa, C.A. (1975). Solution of linear equations with skyline-stored symmetric matrix. Computers and Structures 5, pp.13-29.

Fletcher, R. (1976). Conjugate gradient methods for indefinite systems. In Numerical Analysis, Dundee (1975). Lecture note 506. G.A. Watson (Ed.), pp.73-89.

Foerster, H. and Witsch, K. (1982). Multigrid software for the solution of elliptic problems on rectangular domains: MGOO (Release 1). These proceedings.

George, A. (1981). Direct solution of sparse positive definite systems: some basic ideas and open problems. In Sparse Matrices and their Uses. I.S. Duff (Ed.). Academic Press, pp.283-306.

George, A. and Liu, J.W.-H. (1981). Computer Solution of Large Sparse Positive Definite Systems. Prentice Hall.

George, A., Liu, J. and Ng, E. (1980). User guide for SPARSPAK: Waterloo Sparse Linear Equations Package. Research Report CS-78-30 (revised Jan.1980). Department of Computer Science. University of Waterloo.

Golub, G. and Kahan, W. (1965). Calculating the singular values and pseudo-inverse of a matrix. J. SIAM Numer. Anal. 2, pp.205-224.

Grimes, R.G., Kincaid, D.R. and Young, D.M. (1980). ITPACK 2A: A Fortran implementation of adaptive accelerated iterative methods for solving large sparse linear systems. Report CNA-164. Center for Numerical Analysis. University of Texas at Austin.

Hageman, L.A. and Young, D.M. (1981). Applied Iterative Methods. Academic Press.

Hasbani, Y. and Engelman, M. (1979). Out-of-core solution of linear equations with non-symmetric coefficient matrix. Computers and Fluids 7, pp.13-31.

Hestenes, M.R. and Stiefel, E. (1952). Methods of conjugate gradients for solving linear systems. J. Res. Nat. Bur. Standards 49, pp.409-436.

Hood, P. (1976). Frontal solution program for unsymmetric matrices. Int. J. Numer. Meth. Engng. 10, pp.379-399.

IBM (1976). IBM System/360 and System/370 IBM 1130 and IBM 1800. Subroutine Library - Mathematics. User's Guide. Program Product 5736-XM7. IBM Catalogue SH12-5300-1.

Irons, B.M. (1970). A frontal solution program for finite element analysis. Int. J. Numer. Meth. Engng. 2, pp.5-32.

Jacobs, D.A.H. (1980). A summary of subroutines and packages (employing the strongly implicit procedure) for solving elliptic and parabolic partial differentia equations. Report RD/L/N 55/80, Central Electricity Research Laboratories.

Jacobs, D.A.H. (1981a). The exploitation of sparsity by iterative methods. In Sparse Matrices and their Uses. I.S. Duff (Ed.). Academic Press, pp.191-222.

Jacobs, D.A.H. (1981b). Preconditioned conjugate gradient methods for solving systems of algebraic equations. Report RD/L/N 193/80. Central Electricity Research Laboratories.

Jennings, A. (1966). A compact storage scheme for the solution of symmetric linear simultaneous equations. Comput. J. 9, pp.281-285.

Jennings, A. (1977). Matrix Computation for Engineers and Scientists. Wiley.

Kuo-Petravic, G. and Petravic, M. (1979). A program generator for the incomplete LU decomposition-conjugate gradient (ILUCG) method. Computer Physics Comm. 18, pp.13-25.

Kuo-Petravic, G. and Petravic, M. (1981). A program generator for the Incomplete Cholesky Conjugate Gradient (ICCG) method with a symmetrizing preprocessor. Comput. Phys. Comm. 22, pp.33-48.

Manteuffel, T. (1977). The Tchebychev iteration for nonsymmetric linear systems. Numer. Math. 28, pp.307-327.

Markowitz, H.M. (1957). The elimination form of the inverse and its application to linear programming. Management Science 3, pp.255-269.

Meijerink, J.A. and van der Vorst, H.A. (1977). An iterative solution method for linear systems of which the coefficient matrix is a symmetrix M-matrix. Math. Comp. 31, pp.148-162.

Munksgaard, N. (1980). Solving sparse symmetric sets of linear equations by pre-conditioned conjugate gradients. TOMS 6, pp.206-219.

O'Leary, D.P. and Widlund, O. (1981). Algorithm 572: Solution of the Helmholtz equation for the Dirichlet problem on general bounded three-dimensional regions. TOMS 7, pp.239-246.

Paige, C.C. and Saunders, M.A. (1975). Solution of sparse indefinite systems of linear equations. SIAM J. Numer. Anal. 12, pp.617-629.

Paige, C.C. and Saunders, M.A. (1982). LSQR: An algorithm for sparse linear equations and sparse least squares. ACM Trans. Math. Softw. To appear.

Parlett, B.N. (1978). A new look at the Lanczos algorithm for solving symmetric systems of linear equations. Lin. Alg. and its Applics. 29, pp.323-346.

Rice, J.R. (1978). ELLPACK 77. User's Guide. Report CSD-TR-289. Computer Science Department, Purdue University.

Schumann, U. (Ed.) (1978). Computers, Fast Elliptic Solvers, and Applications. Advance Publications.

Sherman, A.H. (1978). Algorithm 533. NSPIV, a Fortran subroutine for sparse Gaussian elimination with partial pivoting. TOMS 4, pp.391-398.

Stone, H.L. (1968). Iterative solution of implicit approximations of multi-dimensional partial differential equations. SIAM J. Numer. Anal. 5, pp.530-558.

Swarztrauber, P.N. (1977). The methods of cyclic reduction, Fourier analysis and the FACR algorithm for the discrete solution of Poisson's equation on a rectangle. SIAM Rev. 19, pp.490-501.

Swarztrauber, P.N. (1981). Vectorizing the FFT's. To appear in Methods in Computational Physics. Volume on Parallel Algorithms. Lawrence Livermore Series.

Swarztrauber, P.N. and Sweet, R.A. (1979). Algorithm 541. Efficient Fortran subprograms for the solution of separable elliptic partial differential equations. TOMS 5, pp.352-364.

Thompson, E. and Shimazaki, Y. (1980). A frontal procedure using skyline storage. Int. J. Num. Meth. Engng. 15, pp.889-910.

Thomsen, et al (1977). SIRSM: Fortran package for the solution of sparse systems by iterative refinement. Report NI-77.13. Numerisk Institut, Lyngby, Denmark.

van der Vorst, H.A. and van Kats, J.M. (1979). Manteuffel's algorithm with pre-conditioning for the iterative solution of certain sparse linear systems with a non-symmetric matrix. Report TR.11, ACCU, Utrecht.

van Kats, J.M. and van der Vorst, H.A. (1979). Software for the discretization and solution of second order self-adjoint elliptic partial differential equations in two dimensions. Technical Report TR10, ACCU, Utrecht.

van Kats, J.M., Rusman, C.J. and van der Vorst, H.A. (1980). ACCULIB documentation. Minimanual. Report No.9, ACCU, Utrecht.

Vinsome, P.K.W. (1976). Orthomin, an iterative method for solving sparse sets of simultaneous linear equations. Paper number SPE 5729. 4th SPE Symposium on Numerical Simulation of Reservoir Performance.

Young, D.M. and Kincaid, D.R. (1980). The ITPACK package for large sparse linear systems. Report CNA.160. Center for Numerical Analysis. University of Texas at Austin.

Zlatev, Z., Barker, V.A. and Thomsen, P.G. (1978). SSLEST: A Fortran IV subroutine for solving sparse systems of linear equations. User's Guide. Report NI-78-01, Numerisk Inst., Lyngby, Denmark.

Zlatev, Z. and Wasniewski, J. (1978). Package Y12M - Solution of large and sparse systems of linear algebraic equations. Report No.24. Mathematisk Institut, Copenhagen, Denmark.

Multigrid Software for the Solution of Elliptic Problems on Rectangular Domains: MGOO (Release 1)

Hartmut Foerster *

Kristian Witsch ** †

* GMD, St. Augustin, Federal Republic of Germany

** Universität Düsseldorf, Düsseldorf, Federal Republic of Germany

Abstract

MGOO is a modular structured collection of Fortran subprograms which implement multigrid algorithms to solve second-order elliptic problems on rectangular domains subject to general boundary conditions. MGOO is designed for efficiency and storage economy. In particular, fast Poisson and Helmholtz solvers are included.

Table of contents

---

† Supported in part by the Minister für Wissenschaft und Forschung des Landes Nordrhein-Westfalen under Project "Mehrgittermethoden zur Lösung partieller Differentialgleichungen" (supervisor: U. Trottenberg)

1. Introduction

MGOO (Release 1) is a collection of multigrid solution modules for the standard five-point finite-difference approximation to a second order elliptic differential equation of the form

$$Lu := - a(x,y)u_{xx} - b(x,y)u_{yy} + c(x,y)u = f(x,y) \qquad ((x,y)\in R)$$

which is defined on a rectangle R with boundary conditions

$$\alpha(x,y)u + \beta(x,y)u_n = g(x,y) \qquad ((x,y)\in\partial R)$$

on the boundary ∂R of R. The subscript n denotes the derivative in the outward-pointing normal direction. In particular, the following cases are included,

Dirichlet boundary condition	$(\alpha = 1,\ \beta = 0)$
Neumann boundary condition	$(\alpha = 0,\ \beta = 1)$
Robbins boundary condition	$(\alpha\beta > 0)$.

In the general case the type of boundary condition may change along the sides of R.

The boundary-value problem is discretized by central difference approximations of order two on a uniformly spaced grid with grid lines matching the boundary ∂R. Corresponding discretizations are used on a sequence of subgrids which are generated by successively doubling the grid spacings. This hierarchy of uniform grids participates in the solution process. Multigrid solution methods are well known to have several important benefits:

Cyclic multigrid iterations, for example, reduce the error by a factor which is significantly less than one and independent of the grid size. Full Multigrid algorithms achieve discretization error accuracy within a number of operations which is a small multiple of the number of unknowns. For more details, see [24], [5], e.g.

In this paper we report on the usage, the structure and the performance (Sec. 2 - 4, 6) of the MGOO package, release 1. Moreover, the algorithmic components are described in detail (Sec. 5) and quantitative convergence factors as well as operation counts are provided for the solution schemes employed. Different multigrid solution schemes are automatically invoked for different problems to allow a maximum of efficiency and robustness. In particular, isotropic ($a \equiv b$) and anisotropic operators are distinguished. As a consequence, MGOO meets the benchmark provided by fast direct solution methods for solving Laplace, Poisson or Helmholtz equations. A comparison of some fast solvers' run-time on a model problem is given in Sec. 6. From MGOO performance evaluations, which are compiled in extracts in Tables 10 - 15 below, it is obvious that MGOO provides a convenient tool to solve a variety of real problems. Moreover, the expensive components of the employed solution schemes allow a high degree of vectorization.

Release 1 of MGOO excludes a multigrid solution scheme which applies also to first-order terms, since singular perturbation problems ought to be treated with the same efficiency as well. Any forthcoming implementation (cf. Sec. 7) must also take possible instabilities into account which might be introduced by coarse-grid central difference approximations to first-order derivatives.

For the frequently arising pure Neumann problem for Poisson's equation, MGOO will calculate a solution iff the data satisfies the discrete compatibility condition. This condition may be automatically checked and/or enforced by means of the trapezoidal rule for the Euclidean inner product (cf. Sec. 2.4).

MGOO is not a black-box program which returns a meaningful solution of guaranteed accuracy for every boundary value problem that is accepted. For indefinite problems it may even happen that the solution process diverges. MGOO issues warning messages to indicate to the user when performance is expected to be less efficient than predicted, e.g., as for positive-type discretizations (see Sec. 5). At run-time information is provided on request for purposes of analysis and debugging.

From this point of view MGOO is also a research oriented tool for the design and evaluation of multigrid software. The modular program structure is built upon a systematic separation of subtasks to facilitate maintenance as well as the implementation of modifications and extensions. The codes themselves are extensively documented.

MGOO is a user-oriented system, written in portable Fortran. The user interface relies on the specification of interfaces found valuable in the ELLPACK system design. ELLPACK itself is a cooperative effort started in 1976 to provide a tool for research in the evaluation and development of numerical methods for solving elliptic partial differential equations. One result of this project is the ELLPACK system which consists of a high-level language tailored for solving elliptic problems, plus a large collection of software modules to use. The final ELLPACK system will also include MGOO as a module. Distribution is made from the Computer Science Department at Purdue University. For details about ELLPACK see [20] and other papers of Rice. - The original stand-alone version of MGOO is available from GMD.

2. User interface and parameters of MGOO

The user interface of MGOO consists of a set of global interface (common) variables, the algebraic equation solution vector, four user-supplied subprograms and five multigrid module parameters. It relies on the specification of interfaces which is used in the ELLPACK system design [20]. Access to interface variables is gained through common blocks or through a MGOO module's calling sequence.

Global variables are arranged in common blocks according to their type, an arrangement which is essential for portability. The use of multigrid module parameters especially allows flexibility and robustness. To mention also efficiency, MGOO implements a variety of fixed multigrid solution schemes (cf. [11] and Sec. 5). According to the properties of the specific problem to be solved the most efficient scheme is automatically assigned at run-time.

User-supplied subprograms to define the partial differential equation and the boundary conditions, MGOO module parameters as well as arrays whose lengths depend upon the problem size (vector for discrete solution, workspace areas) are to be included as arguments in subroutine calls for MGOO preprocessing and solution modules (cf. Sec. 3).

To avoid name conflicts when users construct programs which use the MGOO modules a systematic naming convention has been applied. All internal MGOO common blocks and subprograms have six-character names with first character M, second G, and third and fourth O. Thus, to avoid name conflicts, MGOO users must not use six-character names of the form MGOOxy.

MGOO users specify their boundary-value problem due to the following conventions.

For pde coefficients there is a canonical ordering that MGOO modules use. It is

1. coefficient of u_{xx}
2. coefficient of u_{xy}
3. coefficient of u_{yy}
4. coefficient of u_x
5. coefficient of u_y
6. coefficient of u

Note that the release at hand only accepts coefficients of u_{xx}, u_{yy}, and u. The corresponding ordering for boundary condition coefficients is

1. coefficient of u
2. coefficient of u_n

The subscript n denotes the derivative in the outward-pointing normal direction.

The sides of the rectangular domain are assumed to be numbered in a standard way. For the rectangle $(x_a,x_b) \times (y_a,y_b)$ this is

1. southern side $(y = y_a)$
2. eastern side $(x = x_b)$
3. northern side $(y = y_b)$
4. western side $(x = x_a)$

Two subprograms are used to define the partial differential equation. They are

```
SUBROUTINE PDECOE(X,Y,CPDE)
REAL X,Y,CPDE(6)
```

and

```
REAL FUNCTION PDERHS(X,Y)
REAL X,Y
```

On return CPDE(I) contains the value of the ith coefficient of the differential operator at (x,y). PDERHS returns the right-hand side f evaluated at (x,y).

Similarly two additional subprograms are used to define the boundary conditions, namely

```
SUBROUTINE BCCOE(K,X,Y,CBC)
INTEGER K
REAL X,Y,CBC(2)
```

and

```
REAL FUNCTION BCRHS(K,X,Y)
INTEGER K
REAL X,Y
```

On return CBC(I) contains the value of the ith boundary condition coefficient on side k at (x,y) while BCRHS returns the corresponding boundary data g(x,y).

In order to reduce storage requirement and to save operational work the linear algebraic equations are transformed iff ICFST(3) > 1, IBCST(2,K) > 1 for any K, 1 ≤ K ≤ 4, or if the grid is rectangular (i.e., different grid spacings in x and y). Since MGOO is designed for speed respective specifications should be avoided whenever possible.

Next, we describe the interface variables and multigrid module parameters which have to be defined by the user.

2.1 Problem definition

Two integer arrays, ICFST and IBCST, are used to determine the status of the differential operator and boundary condition coefficients. ICFST(I) gives the status of the ith operator coefficient. IBCST(I,K) contains the status of the ith term in the boundary condition on side k. Possible values for coefficient status are

0 if the coefficient is identically zero,

1 if the coefficient is identically one,

2 if the coefficient is constant,

3 if the coefficient depends upon x and y.

Constant values of pde coefficients are provided by the global common variables CA, CB, and CC. Two logical variables afford further information upon the specified problem: HEQ indicates a homogeneous equation and NUNQ is the non-unique solution switch. In case the solution of the specified boundary-value problem is defined only up to a constant the value QU at some fixed grid point (QX,QY) may be specified by the user. Finally, another integer array, called IBCT, is used to specify boundary condition types. IBCT(K) contains the type of boundary condition specified along side k. The following values are possible choices.

1 if the solution is specified (Dirichlet boundary condition)

2 if the normal derivative is specified (Neumann boundary condition)

3 if a Robbins boundary condition is specified (nonzero product of boundary condition coefficients)

4 if a boundary condition is of mixed type (partly Dirichlet and/or Neumann and/or Robbins)

Note that the release of MGOO which we are considering is not intended for treating periodic boundary conditions.

2.2 Discrete domain

Six variables provide information about the rectangular grid on which a solution is to be calculated. XA, XB and YA, YB specify the ranges of the x and y grid, respectively. NXP gives the number of grid lines in x and NYP the corresponding number in y. Since a uniformly spaced grid is assumed, the grid spacings are calculated simply by HX = (XB-XA)/(NXP-1) and HY = (YB-YA)/(NYP-1). The locations of grid lines in x and y are then defined by

```
GRDX(I) = XA + (I-1) * HX,          I = 1,..,NXP,
GRDY(J) = YA + (J-1) * HY,          J = 1,..,NYP.
```

Recall that different grid spacings in x and y turn finite-difference schemes for the Laplacian and the Helmholtz operator into discrete analogues for anisotropic differential operators.

2.3 Algebraic equation solution

The vector UH is of length NXP * NYP and may contain a user-supplied initial approximation to the exact discrete solution (cf. MGOO module parameter ITYPE below). MGOO solution modules store the multigrid solution to the linear algebraic system at (GRDX(I),GRDY(J)) into UH(I,J) which corresponds to UH(I+(J-1)*NXP) where I = 1,..,NXP and J = 1,..,NYP.

2.4 MGOO module parameters

ITYPE, given in decimal expansion of the form ITYPE = MN, determines the type of the MG algorithm and which initial approximation shall be used. The Correction Scheme applies to linear problems. The Full Approximation Scheme for non-linear problems is not implemented in the release at hand, see however Sec. 7. If M = 1, then cyclic multigrid iterations (MGI) are carried out. If M = 0, the Full Multigrid (FMG) algorithm solves the problem to discretization error accuracy in one cycle per level. N = 0 means that zero should be the initial guess, while there is a user-supplied initial approximation to the exact discrete solution in case N = 1.

Additional information for the MGOO preprocessing modules is provided by NMIN and IDCC. NMIN gives the number of grid spacings the coarsest grid must have in both x and y. Usually we have NMIN = 2. Since MGOO attempts to find a solution even for non-definite problems (cf. the error handling below) relaxation on coarser grids may magnify, instead of reduce, the error. Then, increasing the value of NMIN to allow the coarsest grid to be finer is a remedy to handle at least some of these problems appropriately. See, however, also [5], Sec. 6.3, and [4], Sec. 3.2.

For the pure Neumann problem for Poisson's equation, MGOO will calculate a solution iff the data satisfies the discrete compatibility condition. This condition may be checked and/or enforced by modifying the right-hand side f. IDCC determines whatever action is to be taken.

Two parameters specify to what extent MGI cycles are to be performed during solution: ITER and IGAMMA. ITER, more precisely max(ITER,1), prescribes the number of multigrid iterations. In case ITER = 0 one MGI cycle is performed with less smoothing sweeps per grid than ITER = 1 and, correspondingly, ITER > 1 would imply. This alternative also applies to the FMG algorithm, i.e. when M = 0 for ITYPE < 2. Such schemes save about 25% of the operational work (up to nearly 40% in case of anisotropic operators) but are less accurate (cf. Secs. 5, 6). The number of MGI cycles on each FMG level, i.e. if ITYPE < 2, is fixed since one cycle per grid yields already discretization error accuracy when combined with a FMG interpolation that is based on the discrete equations (cf. Sec. 5 and [11]). If ITER > 1, however, then ITER-1 MGI steps follow the FMG algorithm to further reduce the algebraic error, i.e. the error in solving the difference equations.

IGAMMA determines the type of MGI cycles. V-type (IGAMMA = 1) and W-type cycles (IGAMMA = 2) are distinguished. W-cycles require 50% more work than V-cycles (cf. Sec. 6) but have proved to be especially robust.

2.5 <u>Global control information</u>

Two integer interface variables are concerned with printed output. Printed output at different output levels is available on request providing the user with information for purposes of analysis and debugging. LUOUT is the logical unit number for all printed output. LPO specifies the user's output level request. Only fatal error messages are output if LPO = 0. LPO = 3, 4 are debug modes. - EPS is a machine-dependent constant establishing the smallest positive magnitude. EPS is used to decide uncertainties during calculation.

The logical variable FESW is the fatal error switch. Fatal errors are detected until multigrid solution has begun and corresponding error messages will get printed. Through FESW MGOO requests error termination. - We do <u>not</u> guarantee that MGOO efficiently calculates a meaningful solution <u>for every</u> boundary value problem that is accepted. For indefinite problems it may even happen that the solution process diverges. To indicate to the user when both CP time and returned multigrid solution should be analyzed carefully, MGOO issues warning messages. So are the following

COARSEST GRID IS TOO FINE FOR EFFICIENT MULTIGRID SOLUTION.

For efficient multigrid iterations NXP and NYP should have the form:

(small integer times some power of two) plus one.

COEFFICIENT OF U IN THE DIFFERENTIAL EQUATION IS NEGATIVE AT LEAST AT ONE GRID POINT.

COEFFICIENT IN BOUNDARY CONDITION IS NEGATIVE AT LEAST AT ONE GRID POINT.

Other sources that may possibly cause a degradation of MGOO convergence rates, like discontinuities or singularities, are not automatically detected during preprocessing.

2.6 Workspace access

MGOO modules have access to reusable workspace which is used for temporary storage. On the one hand, uninitialized workspace is passed along to store grid function values during solution. On the other hand, during preprocessing arrays of variable dimensions are placed into workspace to store problem-dependent information for later use by MGOO solution modules. In addition, integer workspace is required to contain grid-dependent information necessary to process a hierarchy of grids.

Enough space is to be allocated to satisfy the largest workspace requirement for the set of modules in the package. Therefore, we give a bound for the proper length of reusable workspace. It is - apart from the usually negligible space for the direct solution on the coarsest grid -

$$\frac{13}{3} * NXP * NYP + 15 * (NXP+NYP)$$

which reduces to

$$3 * NXP * NYP + 10.5 * (NXP+NYP)$$

and

$$\frac{5}{3} * NXP * NYP + 13 * (NXP+NYP)$$

for the generalized Helmholtz equation where c is a function and for the Poisson equation, respectively. Additional integer workspace of

$$4 * (NXP+NYP) + 102$$

words is sufficient to satisfy the modules' requirements. However, if the boundary condition coefficients α, β satisfy $\alpha\beta = 0$ along the boundary, more precisely, if Dirichlet or Neumann boundary conditions are specified at each single side, there are only 83 words required. The actual size of allocated workspace is available at the user interface through two arguments in the subroutine call for MGOO preprocessing modules.

2.7 Restrictions

MGOO will attempt to find a solution even if $c \geq 0$ does not hold and/or the coefficients of the boundary condition do not satisfy $\alpha\beta \geq 0$ in which case a solution may not exist. In any such case, convergence has to be controlled by the user, the convergence behavior predicted by local Fourier analysis can no longer be guaranteed. - For rapidly changing coefficients, MGOO solution modules are not well suited. It is preferable to use Galerkin-type approximations to the coarse-grid differential operators instead (cf. Sec. 5.1). - For highly oscillatory solutions Full Multigrid algorithms should better rely on ITER = 1 or = 2 instead of using more modest cycles with ITER = 0 (cf. Sec. 2.4). - To solve the pure Neumann problem for Poisson's equation on machines with short word length, due to round-off orthogonalization of the data is required also on coarse grids. One extra subroutine call has to be inserted after each residual transfer to satisfy the discrete compatibility condition on each level.

This list is far from being complete. Users are welcome to report on any problems found with MGOO.

3. Program structure

MGOO is a modular structured collection of subprograms which implement multigrid algorithms to solve elliptic boundary value problems on rectangular domains. The package consists of two driver subroutines, several auxiliary subroutines and three sets of problem-dependent modules. The version at hand has approximately 7,500 lines of code, including comments, and is written in portable Fortran (PFORT subset [21] of ANSI Fortran 66).

The first driver, called MGOOSU, performs preprocessing to transform the user interface into one natural for the multigrid solution process. For this, there is a variety of subtasks to be performed. First, MGOOSU determines whether the input is admissible to call the second driver which invokes the multigrid solution modules. Then, the routine sets up the discrete problem and incorporates the boundary conditions. To meet the design goals of efficiency and storage economy different operators are distinguished during set-up and storage allocation. For the Laplacian, the Helmholtz operator $-\Delta + cI$ with c a constant or a function and the anisotropic operator $-a\partial_{xx} - \partial_{yy} + cI$ with $a \neq 1$, Robbins and mixed-type boundary operators are distinguished from those whose coefficients satisfy either $\alpha = 0$ or $\beta = 0$ at each single side of the boundary. For the Full Multigrid algorithm corresponding problems on coarser grids are also defined (cf. [24], Sec. 6). And finally, a LDU-decomposed system is generated for direct solution on the coarsest grid.

The second driver, called MGOOMN, invokes different multigrid solution modules for different problems to allow a maximum of efficiency and robustness. Assuming a discretization on a square grid boundary-value problems for the Helmholtz equation

$$-\Delta u + cu = f \quad \text{(c constant)}$$

and for the generalized Helmholtz equation, where c is a function, are treated separately. It should be noticed that because of the problem-dependent preprocessing there is no operational overhead incured for special cases such as the Dirichlet or the Neumann problem for Poisson's equation ($c \equiv 0$). (Cf. Tables 8 and 10 - 11). Also, anisotropic operators, namely with $a \neq 1$, are distinguished and different smoothing procedures and grid transfer operators are applied (cf. Sec. 5).

Typically, each (Correction Scheme) solution module is composed in the following way.

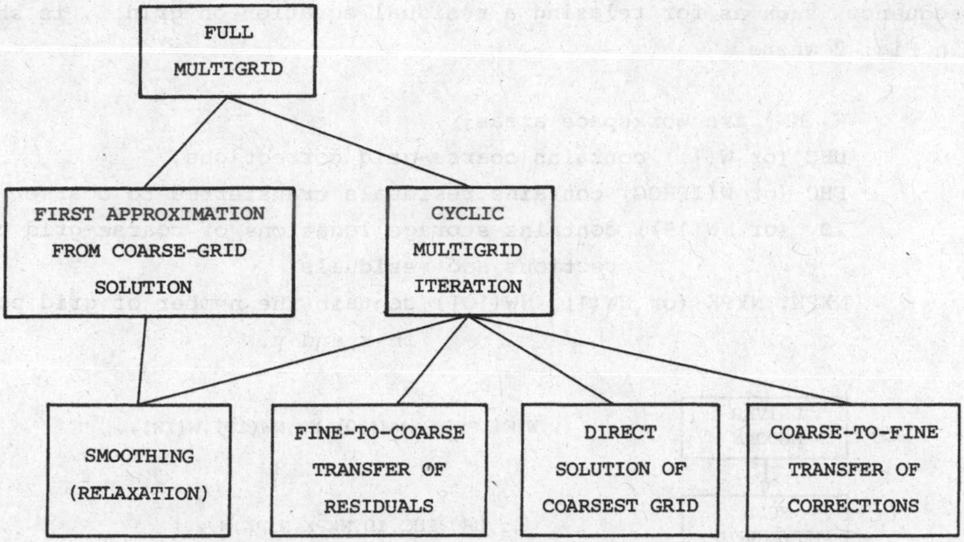

Fig. 1. Composition of a Correction Scheme solution module

For a description of the algorithmic components see Sec. 5. Additional problem-dependent routines are provided to relax the initial guess zero for coarse-grid corrections. So routines exist to calculate the (discrete L^2-) norm of the residual (used on request).

4. Data structure

Let a uniformly spaced grid be given. Its origin is defined to be the lower left corner point of the rectangular domain. Then, as long as grid lines match the boundary, subgrids are generated by successively doubling the grid spacings. (Note that staggered grids are not accepted by the modules at hand.) Therefore, the principle structure information which describes a hierarchy of grids for use by MGOO is reduced to the number of grid points in the x and y coordinate of which each grid consists (when intersected with the rectangular domain). Four additional arrays, also placed in integer workspace during preprocessing, are used to allow access to corresponding storage locations of grid function values. Pointers associated with solution values and sources, corrections and residuals as well as with boundary data on both horizontal and vertical grid lines are organized due to a separation in storage. Fine grid values are separated from values on the coarser levels 2 to m. This concept makes storage handling more flexible. A typical calling sequence, such as for relaxing a residual equation on grid ℓ, is shown in Fig. 2 where

W, NW are workspace areas,
UHC (or W(1)) contains coarse-grid corrections,
FHC (or W(IFHCG) contains residuals transferred to coarser grids,
ID (or NW(19)) contains storage locations of coarse-grid corrections and residuals,
NXPK, NYPK (or NW(1), NW(10)) contain the number of grid points in x and y.

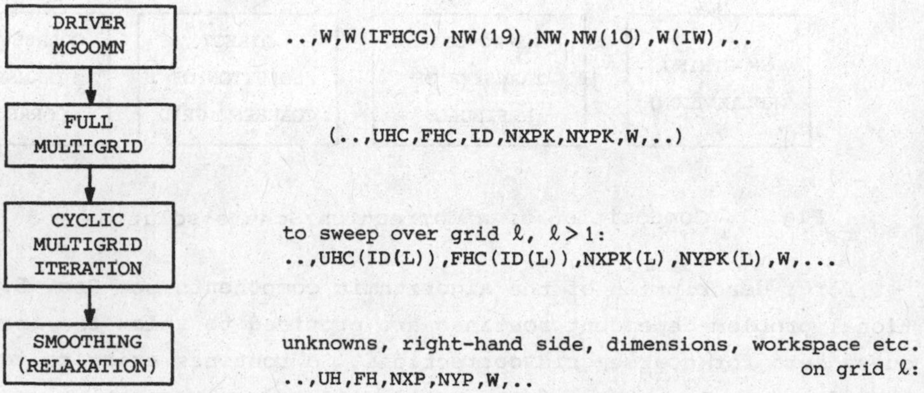

Fig. 2. Calling sequence during multigrid solution for relaxation of a residual equation

The fourth pointer array IDCE contains storage locations of center elements of difference stars that apply to grid ℓ, $\ell = 1,..,m$. To save operational work during solution center elements of difference stars are calculated once for all in advance and placed into workspace. This part of workspace is symbolically addressed by CEDS. In the constant coefficient case one center element per level ℓ is stored at CEDS(IDCE(L)). In case neither Dirichlet nor Neumann boundary conditions are specified at any one side of the rectangular domain center elements of boundary point difference stars are stored consecutively for the southern, eastern, northern, and western side at CEDS(IDCE(L)+1). Differential operators with variable coefficients mean variable center elements, not required only at boundary points where the solution is known. Such elements are stored gridwise.

5. Multigrid algorithms

In this chapter we describe the algorithmic components of the multigrid solution modules as they are implemented in MGOO, release 1. There are three different solution modules M1, M2, and M3 to allow a maximum of efficiency and robustness. Module selection is automatic and depends on the type of the differential operator.

5.1 Discretization and multigrid iterations

The Correction Scheme is used for multigrid iterations. Cyclic multigrid iterations start with some arbitrary (zero or user-supplied) approximation on the given grid and reduce the error by cycling between that grid and coarser grids. V-type and W-type cycles are distinguished (cf. Section 2.4). For both schemes the total computational work to achieve a fixed accuracy is proportional to the number of grid points on the finest (= the given) grid (cf. Sec. 5.3 for precise counts). Depending on the MGOO module parameter ITER (cf. Sec. 2.4) MGI spectral radii for V-cycles are approximately 1/20 (0 < ITER = number of iterations) or about 1/6 - 1/10 (ITER = 0: one iteration) independent of the grid size. W-cycles require 50% more work than V-cycles but have proved to be especially robust. Corresponding spectral radii are 1/30 and 1/8 - 1/14, respectively. Tables 1 - 4 below provide the user with all necessary information to select MGI cycles with an asymptotic convergence rate most convenient for his problem. The corresponding operational work can be derived from Table 6 in Sec. 5.3.

Following some remarks on the kind of discretization that is used in MGOO, we first give a description of those algorithmic components which are common to all three the solution modules. Then, we describe the different smoothing (relaxation) procedures, namely red-black (RB) or checkered (CH in [11], [10]), zebra line (ZL) and alternating zebra (AZ) Gauss-Seidel relaxation, as well as the fine-to-coarse residual transfer operators.

The elliptic operator

$$Lu := - au_{xx} - u_{yy} + cu$$

is discretized by central difference approximations of order 2 on a uniformly spaced grid with grid lines matching the boundary. For non-Dirichlet boundary conditions the normal derivatives are approximated by central differences. Function values at grid points beyond the boundary are eliminated yielding asymmetric four-point difference operators which apply to boundary grid function values. The standard five-point operator is also applied on coarser grids. We have decided not to use a Galerkin-type approximation to the coarse-grid differential operators because of the following reasons. First, there is an obvious storage penalty for operators of the form

$$L_\ell = I_{\ell-1}^{\ell} \, L_{\ell-1} \, I_{\ell}^{\ell-1} \qquad (\ell = 2,..,m)$$

with some appropriate interpolation operator $I_\ell^{\ell-1}$ and a nine-point fixed weighting like constant times $(I_\ell^{\ell-1})^T$ for $I_{\ell-1}^{\ell}$. Then, an implementation really pays only in special situations, e.g. if discontinuities of orders of magnitude exist in the coefficients. And finally, we know of existing multigrid software like Dendy's BOXMG [8] and Wesseling's MGD1 [27] which is based on Galerkin-type discretizations.

In all three solution modules bilinear interpolation is used for the coarse-to-fine transfer of corrections. However, due to the subsequent point or line relaxation pattern only half the fine-grid values have to be corrected. For solution on the coarsest grid a LDU decomposition of the coarse-grid matrix is used by the solution modules. The decomposed system is generated in advance by MGOO preprocessing modules.

As already indicated in Sec. 2 coefficients of central differences in y are normalized to 1 when setting up the discrete equations. Therefore, the coefficients a_{ij} of central differences in x determine the strength with which grid function values are coupled. This coupling must be taken into account by any satisfactory smoothing procedure. (If there is no transformation of the equations during set-up we have $a_{ij} = a(x_i,y_j)$ for grid points (x_i,y_j).) In MGOO, the following cases are distinguished.

RB relaxation if all coefficients a_{ij} are identically one,

ZL relaxation by lines in x if all coefficients a_{ij} are greater than one,

ZL relaxation by lines in y if all coefficients a_{ij} are less than one,

AZ relaxation otherwise.

The MGOO solution modules M1, M2 which implement RB relaxation as well as half injection (HI) for fine-to-coarse residual transfers are described in [11]. According to the specified value of the multigrid module parameter ITER (cf. Sec. 2.4) either two or three relaxation sweeps per cycle are performed on each level of discretization. The two-grid convergence factors contained in Table 1 are obtained by local Fourier analysis (see [24]) and indicate how MGOO RB-HI schemes are supposed to perform for Poisson's equation. (For Helmholtz' equation with c > 0 performance is even better.)

	ρ^*	σ_S^*	σ_E^*
ITER = 0	0.125	0.141	0.125
ITER > 0	0.034	0.046	0.045

Table 1: Two-grid convergence factors for MGOO implementations of RB-HI schemes

In Tables 1 - 4 ρ^*, σ_S^* and σ_E^* denote the suprema of ρ, σ_S and σ_E with respect to grid spacings h. For a given two-grid iteration operator ρ denotes the spectral radius, σ_S the spectral norm, and σ_E the energy norm which involves the discrete differential operator. While ρ^* illustrates the asymptotic convergence behavior, the norms σ_S^* and σ_E^* measure the error and defect reduction per iteration step. For details see [24]. From two-grid convergence factors Stüben and Trottenberg also derive h-independent convergence estimates for multigrid cycles ([24], Sec. 4.3).

For anisotropic operators line relaxation schemes provide satisfactory smoothing rates when strongly coupled blocks of unknowns are relaxed simultaneously (cf. [5], Sec. 3.3). The MGOO solution module M3 implements ZL relaxation by lines in x, ZL relaxation by lines in y as well as AZ relaxation (ZL by lines in x alternating with ZL by lines in y).

In ZL relaxation we change all the unknowns along a grid line so as to simultaneously satisfy all the corresponding equations. For this purpose we have to solve a tridiagonal system of equations. There are alternatives to the used LU decomposition which are faster, like cyclic reduction, but which incur more coding, considerable storage overhead or both. In particular, we have decided not to decompose the system in advance because of the storage penalty involved. - Every other grid line is processed in this way. Likewise we simultaneously displace the unknowns along the remaining lines in a second stage.

Thole [26] has investigated the efficiency of various ZL relaxation schemes by local Fourier analysis. His results clearly recommend the way relaxation sweeps on each discretization level should be arranged. In case ITER = 0 two sweeps per cycle are performed according to

$$o\ e\ o\ \bullet\ e\ ,$$

while for ITER > 0 three sweeps are most efficient when arranged to

$$o\ e\ o\ \bullet\ e\ o\ e\ .$$

e means line relaxation by (even) lines in x or y which are also coarse grid lines. Correspondingly, o stands for line relaxation by those (odd) lines in x or y which are not found on coarser grids. Finally, • denotes coarse-grid correction. The multigrid solution module MGOOM3, which we are considering, implements <u>full weighting</u> (FW) for fine-to-coarse residual transfers. The FW operator degenerates to a column or row weighting, i.e.

$$I_{\ell-1}^{\ell} = \frac{1}{16}\begin{bmatrix}2\\4\\2\end{bmatrix} = \frac{1}{8}\begin{bmatrix}1\\2\\1\end{bmatrix} \quad \text{or} \quad I_{\ell-1}^{\ell} = \frac{1}{16}[2 \quad 4 \quad 2] = \frac{1}{8}[1 \quad 2 \quad 1]\ ,$$

due to the preceding "o" relaxation which causes zero residuals along the respective grid lines. Table 2 contains corresponding two-grid convergence factors derived by local Fourier analysis for differential operators $L = -a\partial_{xx} - \partial_{yy}$ with constant $a \geq 1$ ([26]). For $0 < a \leq 1$ similar results are obtained for ZL relaxation by lines in y.

The MGOOM3 implementations of <u>AZ relaxation</u> schemes start with one oe ZL sweep by lines in y. Then, one eo ZL sweep by lines in x completes one AZ sweep. In any case, one such sweep is performed before the coarse-grid correction. Note that twice the work of one ZL sweep is required! Local two-grid analysis prove that • followed by "o" columnwise is more efficient than just ending up with • since the norms are considerably improved ([26]). Also, the operational overhead involved is reduced by

	a	ρ^*	σ_S^*	σ_E^*
ITER = 0	1	0.063	0.140	0.171
	2	0.028	0.060	0.118
	10	0.047	0.079	0.080
	10^2	0.052	0.089	0.075
	10^5	0.053	0.090	0.074
ITER > 0	1	0.017	0.105	0.032
	2	0.016	0.034	0.019
	10	0.030	0.049	0.036
	10^2	0.033	0.052	0.041
	10^5	0.033	0.053	0.042

Table 2: Two-grid convergence factors for MGOO implementations of ZL-FW schemes.

half the work required for coarse-to-fine correction transfers. Instead, for ITER > 0 one additional AZ sweep is performed after the coarse-grid correction. After "o" rowwise we again have fine-to-coarse residual transfers of the form

$$I_{\ell-1}^{\ell} = \frac{1}{8} [1 \quad 2 \quad 1] .$$

Corresponding two-grid convergence factors for elliptic operators $L = - a\partial_{xx} - \partial_{yy}$ with constant a are contained in Tables 3, 4. The entries indicate that MGOO will supply efficient solutions also for varying anisotropies. For more details see Thole [26].

a	ρ^*	σ_S^*	σ_E^*
10^{-5}	0.124	0.141	0.125
10^{-2}	0.119	0.135	0.121
10^{-1}	0.082	0.099	0.095
1	0.023	0.043	0.118
10	0.082	0.171	0.236
10^2	0.119	0.239	0.268
10^5	0.124	0.249	0.272

Table 3: Two-grid convergence factors for the MGOO AZ-FW scheme in case ITER = 0.

a	ρ^*	σ_S^*	σ_E^*
10^{-5}	0.053	0.087	0.074
10^{-2}	0.051	0.085	0.074
10^{-1}	0.038	0.062	0.049
1	0.009	0.031	0.009
10	0.038	0.075	0.049
10^2	0.051	0.119	0.071
10^5	0.053	0.127	0.074

Table 4: Two-grid convergence factors for the MGOO AZ-FW scheme in case ITER > O.

The most expensive component of MGI cycles is smoothing the error. MGOO smoothing procedures achieve high rates of vectorization since RB and zebra patterns yield decoupling relaxation schemes in case of five-point-star equations. This means that all the equations or blocks of equations corresponding to the same colour can be relaxed in parallel. Decoupling pattern relaxation improves the performance of the inter-grid transfers as well. A significant fraction of the operational work is saved and exploiting vector or parallel processing capabilities is simplified. See [3] for more details about parallelization of multigrid algorithms.

5.2 Full Multigrid

FMG solution processes (sometimes called "nested iterations") are described and analysed in various papers (cf. [5], [24], [11] et al.). In MGOO, the algorithm flowcharted as Fig. 1 in [3] is implemented with RB-HI, ZL-FW, and AZ-FW MGI schemes, respectively. One V-type or W-type MGI cycle is performed on each FMG level. The size of the coarsest grid, the starting level, is implicitly defined by the interface variables NXP, NYP and the MGOO module parameter NMIN. The FMG interpolation uses the discrete differential equations to obtain initial approximations from coarse-grid approximate solutions. This process incorporates an inherent partial smoothing. The interpolation is of order four, diminished to three at boundary points in case of non-Dirichlet boundary conditions.

To apply a FMG interpolation based on the equations themselves red-black coarsening or semi coarsening ([24], [5]) is required. In this context, coarsening means the characterization of fine subgrids on which the discrete equations can be used to obtain initial approximations from coarse-grid solution values. A useful coarsening is determined by the very first relaxation pattern of the subsequent MGI step. For RB, ZL, and AZ relaxation initial values are required on nearly half the fine-grid points only. Fine subgrids are defined accordingly.

For the Poisson equation FMG interpolation uses the skewed Laplacian together with red-black coarsening. Details are given in [11], see also [16]. For the discrete Helmholtz operator, however, the same procedure requires skewed operators (difference stars) with different center elements to be used on the red-black coarsened grids. If CEDS(IJ) contains the center element at some fine-grid point which is also found on the corresponding subgrid, then

$$2 * (\mathrm{CEDS(IJ)} - 2)$$

must be used instead when the skewed operator is applied. This transformation is determined by the red-black-coarsened-grid spacing which is $\sqrt{2}$ times the fine-grid spacing.

In case of anisotropic differential operators rotational symmetry is lost and, therefore, red-black coarsening is not useful. Due to the subsequent MGI step only fine-grid values on coarse-grid lines in either x (marked by $+$ and $\oplus$ in Fig. 3) or y are required. In detail, the initial solution values on a y-semi coarsened grid are derived as follows: Values on coarse-grid points which are also fine-grid points ($+$) can simply be prolonged to fine-grid values. Then, the remaining solution values on coarse-grid lines in x ($\oplus$) are determined by relaxing, in the notation of the preceding paragraph, o-columnwise the equations defined on the y-semi coarsened grid. Solution values at non-Dirichlet boundary points have to be precalculated by cubic interpolation though.

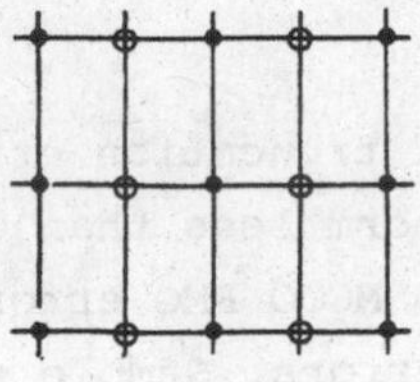

Fig. 3. y-semi coarsening

Note that the latter FMG interpolation also applies to nine-point difference stars. There are multigrid algorithms, e.g. [8] for problems with interfaces, and related methods ([10], [22]), too, which use the discrete differential equations in the coarse-to-fine correction transfer as well. Dendy's approach [8] is different, however: semi coarsening is not involved.

Estimates for the FMG error are derived in various papers, see [11], [13], [5], e.g. In the notation used in [11], App. A, we have with $\gamma^* = 1 + \beta^*$

$$\|\tilde{u}_\ell - u\| \le (\gamma^* + o(1)) \|u_\ell - u\| \quad \text{for all } \ell$$

where $\tilde{u}_\ell$ denotes the FMG solution for the discrete ℓ-level problem $L_\ell u_\ell = f_\ell$ and $u_\ell - u$ is the discretization error on grid (level) ℓ with grid spacing h_ℓ. The estimates for γ^* (with respect to the discrete L^2-norm) in Table 5 indicate the basic performance of FMG algorithms,

	RB-HI	ZL-FW	AZ-FW
γ^* ITER = 0	2.6	1.4 - 1.7	1.2 - ∞
γ^* ITER > 0	1.3	1.2 - 1.4	1.2 - 2.2

Table 5: Bounds for MGOO FMG errors over discretization errors

namely to solve a given problem to the level of truncation error in one cycle per level if a MGI scheme with spectral norm less than 0.2 is provided. Numerical performance is even better, MGOO FMG errors are often observed to be less than discretization errors. Sample examples are given in Sec. 6.

5.3 Operational work

To measure efficiency of MGOO solution modules the number of arithmetic operations remains to be specified. Having in mind the asymptotic two-grid convergence factors given in Sec. 5.1, we introduce quantities $W_\ell^{\ell+1}$ which fill this gap appropriately. This means that $w^{-1}\log(1/\rho^*)$ with $w = W_\ell^{\ell+1}$, for example, is a reasonable measure which we want to maximize within some appropriate range of robustness and simplicity. Here,

$$W_\ell^{\ell+1} \doteq w_0 + w_1 + w_2$$

is the operational work per fine-grid point performed by corresponding two-grid methods. "$\doteq$" indicates equality up to lower order terms. w_0 is the work of relaxing the algebraic equations (before and after the coarse-grid correction), w_1 is the work of residual calculation and fine-to-coarse transfer, and w_2 is the work involved in the coarse-to-fine correction transfer including addition to the previous fine-grid solution. Proper weights to different arithmetic operations are machine-dependent. Therefore, Tables 6 and 7 contain separate counts for divisions, multiplications as well as additions. Coarser grids are taken into account through the factor $r = (1-\gamma/4)^{-1}$ with $\gamma = 1$ for V-cycles and $\gamma = 2$ in case of W-cycles. Disregarding lower order terms the total work of one cyclic multigrid iteration step on level ℓ is

$$W_\ell^{MGI} \doteq r\, W_\ell^{\ell+1}\, N$$

with N = NXP * NYP the number of grid points on level ℓ.

	ITER = O			ITER > O		
	÷	*	+/-	÷	*	+/-
M1 RB-HI	O	$2\frac{7}{8}$	$8\frac{1}{2}$	O	$3\frac{7}{8}$	$11\frac{1}{2}$
M2 RB-HI	$2\frac{1}{8}$	$1\frac{1}{2}$	$8\frac{1}{2}$	$3\frac{1}{8}$	$1\frac{1}{2}$	$11\frac{1}{2}$
M3 yZL-FW	4	$9\frac{5}{8}$	$13\frac{3}{4}$	6	$13\frac{5}{8}$	$18\frac{3}{4}$
M3 AZ-FW	5	$10\frac{3}{8}$	16	8	$15\frac{3}{8}$	$23\frac{1}{2}$

Table 6: Two-grid operation counts

The two-grid multiplication counts for ZL-FW schemes are diminished to $7\frac{3}{4}$ and $10\frac{3}{4}$, respectively, in case of ZL relaxation by lines in x.

The total work W_ℓ^{FMG} of FMG algorithms is determined by W_ℓ^{MGI} and w_3, the work of one FMG interpolation related to a fine-grid point. Consequently, for MGOO FMG algorithms we have

$$W_\ell^{FMG} \doteq r\,(W_\ell^{MGI} + w_3\,N)$$

where w_3 is composed of the respective operation counts contained in Table 7.

	÷	*	+/-
M1	O	$\frac{3}{4}$	$2\frac{1}{2}$
M2	$\frac{3}{4}$	$\frac{1}{4}$	$2\frac{3}{4}$
M3	$\frac{1}{2}$	$1\frac{1}{4}$	$1\frac{1}{2}$

Table 7: Operational work of MGOO implementations of FMG interpolation

For example, a FMG solution for the Poisson or Helmholtz equation costs about 6 multiplications and $18\frac{1}{2}$ additions per fine grid point.

6. Performance

In Sec. 5 the algorithm performance of MGOO solution modules is evaluated in more theoretical ways. Algorithmic complexity, function evaluations, IF tests and index calculations in inner loops etc. are not taken into account, for example. The user, however, is particularly interested in the performance of his program measured in terms of CP time and storage requirement when solving a practical problem. Processing times have to be interpreted carefully, though. In a multiple user environment CP times can only be measured approximately. To some extent, they depend also on the system architecture and the compiler involved ([19]). In this paragraph we provide performance measurements of MGOO, release 1, running on a variety of boundary-value problems. Calculations were all carried out on the GMD IBM/370-158 computer with the Fortran H Extended compiler (OPT=2). The numerical performance strongly confirms the efficiency of MGOO solution modules as predicted by local Fourier analysis and asymptotic operation counts.

First, we compare some elliptic solvers efficiency when treating the discrete Dirichlet boundary-value problem for Poisson's equation on the unit square. Tables 8 and 9 below contain computing times and relative errors of different direct/semi-direct/approximate and iterative solution methods, respectively. The sample boundary-value problems with which we deal here have the following exact continuous solutions

1. $u(x,y) = 1$
2. $u(x,y) = x(1-x) + y(1-y)$
3. $u(x,y) = \sin 7\pi x \sin \pi y$
4. $u(x,y) = \exp[-100((x-0.5)^2+(y-0.5)^2)](x^2-x)(y^2-y)$.

Column 2 of Tables 8 and 9 provides the CP times t in seconds, while columns 3 - 6 contain the relative errors d_h to the exact discrete solution u_h,

$$d_h = \frac{\|u_h - \tilde{u}_h\|}{\|u_h\|}$$

with the approximate solution $\tilde{u}_h$ obtained with the method in question. $\|\cdot\|$ is the discrete maximum norm. The entries in each box refer to grid spacings $h = {}^1/64$ (upper) and $h = {}^1/256$ (lower entries), respectively. a(b) stands for $a \cdot 10^b$. In addition, in the last line of Table 8 δ_h indicates the relative discretization error

$$\delta_h = \frac{\|u - u_h\|}{\|u_h\|} .$$

Tables 8 and 9 are extended versions of Table 1 in [11].

The algorithms listed in Table 8 are characterized below.

Buneman: Buneman algorithm (stabilized cyclic reduction) [7]. Buneman's program XYPOIS ([6]) was used.

FACR(ℓ): Hockney's FACR method [15]. Optimal values for ℓ (steps of cyclic reduction) are ℓ=2 and ℓ=3 in case $h={}^1/64$ and $h={}^1/256$, respectively. Temperton's program PSOLVE ([25]) was used.

KRFFT: marching algorithm of Bank and Rose [2] using FFT. The march-

algorithm	t	$d_h(1)$	$d_h(2)$	$d_h(3)$	$d_h(4)$
Buneman	0.80 15.50	0.46(-4) 0.96(-3)	0.39(-4) 0.80(-3)	0.11(-4) 0.63(-4)	0.11(-4) 0.11(-3)
FACR(ℓ) (ℓ=opt.)	0.55 8.06	0.42(-4) 0.62(-3)	0.40(-4) 0.40(-3)	0.54(-5) 0.20(-4)	0.49(-5) 0.16(-3)
KRFFT (d. p.)	0.40 6.99	0.19(-4) 0.57(-4)	0.28(-4) 0.17(-4)	0.22(-4) 0.20(-5)	0.74(-5) 0.12(-5)
TR2DOO	0.62 9.99	0.58(-4) 0.73(-4)	0.39(-4) 0.66(-4)	0.24(-4) 0.44(-4)	0.16(-4) 0.18(-4)
NMICCG	5.30 63-115	-	-	0.75(-4) 0.76(-4)	0.20(-3) 0.12(-3)
MGOO (FMG)	0.49 7.46	0.0 0.0	0.0 0.0	0.22(-2) 0.27(-3)	0.87(-3) 0.50(-4)
δ_h	-	0.0 0.0	0.0 0.0	0.97(-2) 0.60(-3)	0.64(-2) 0.40(-3)

Table 8: Computing times and relative errors of direct, semi-direct and approximate Poisson solvers.

ing parameter is K = 5. Marching algorithms are known to be unstable, about 10 digits are destroyed during solution for K = 5. Therefore, KRFFT has been run in double precision.

<u>TR2DOO:</u> method of Total Reduction (TR) [23]. A special TR version ([9]) for the discrete Poisson equation on a square grid was used.

<u>NMICCG:</u> nested MICCG. Here MICCG (see below) is applied on different levels of discretization comparable to FMG. For each level the number of iterations is determined a posteriori such that the predescribed accuracy is achieved within a minimum amount of work. By this procedure we find a lower bound for the CP time of a MICCG nested-iteration algorithm. Use was made of the MICCG(1,3) algorithm derived from [17] (see [14]). It is not reasonable to treat sample problems 1 and 2 by this method.

<u>MGOO:</u> The FMG algorithm described in [11] was used with ITER = 0, ITYPE = 0 and IGAMMA = 1 (cf. Sec. 2.4).

Table 8 shows that MGOO establishes a very fast approximate Poisson solver. In contrast to fast direct and semi-direct (TR) methods MGOO is much more generally applicable, as is NMICCG.

A similar comparison of some iterative solution methods is summarized in Table 9. For convenience the iteration numbers are determined so as to achieve a relative error of 10^{-4} using zero as initial guess. The algorithms are characterized as follows.

ADI: alternating direction implicit iterative method by Peaceman-Rachford with optimal Wachspress parameters. 7 and 9 iterations were carried out in case $h={}^1/64$ and $h={}^1/256$, respectively, to achieve a relative error of about 10^{-4}.

ICCG: conjugate gradient (CG) method with preconditioning using an incomplete Cholesky factorization (IC) [18]. The ICCG(1,3) version of a program of van Kats and van der Vorst [17] was used which exploits the special matrix structure corresponding to a 5-point difference star. 20 and 50 iterations were carried out in case of $h={}^1/64$ and $h={}^1/256$, respectively.

MICCG: similar to ICCG but with a modified Cholesky factorization to improve preconditioning [12]. MICCG as derived from ICCG requires only half the number of iterations to solve Poisson's equation (see [14]). Use was made of MICCG(1,3).

algorithm	t	$d_h(1)$	$d_h(2)$	$d_h(3)$	$d_h(4)$
ADI	6.1 126.5	0.16(-2) 0.27(-2)	0.31(-4) 0.48(-3)	0.17(-4) 0.14(-3)	0.55(-5) 0.73(-3)
ICCG	14.5 629.	0.66(-4) 0.69(-3)	0.25(-4) 0.20(-3)	0.56(-5) 0.78(-5)	0.48(-5) 0.25(-4)
MICCG	7.3 315.	-	0.12(-3) 0.12(-4)	0.60(-4) 0.14(-4)	0.15(-3) 0.15(-4)
MGOO (V-type MGI)	1.14 18.05	0.18(-3) 0.62(-4)	0.22(-4) 0.23(-3)	0.11(-3) 0.10(-3)	0.30(-4) 0.26(-4)
MGOO (W-type MGI)	1.17 18.25	0.84(-3) 0.85(-3)	0.41(-3) 0.43(-3)	0.15(-3) 0.17(-3)	0.70(-4) 0.94(-5)

Table 9: Computing times and relative errors of iterative Poisson solvers.

MGOO: MGI cycles as described in Sec. 5.1 (see also Sec. 2.4) were used with ITER = 3 and ITER = 2 in case of IGAMMA = 1 (V-cycles) and IGAMMA = 2 (W-cycles), respectively.

MGOO MGI schemes perform very favourable in comparison with other iterative methods. However, it should be realized that quite often it is preferable to rely on a multigrid approach, like Full Multigrid, which is orientated to the continuous solution u rather than to the algebraic solution u_h (see Brandt [5], Part II, for details).

Up to here, performance evaluations refer only to Dirichlet boundary-value problems. Next, we emphasize, therefore, the important feature of MGOO that general boundary conditions are treated with the same efficiency. Corresponding performance evaluations for a variety of different boundary-value problems are summarized in Tables 10 - 15 below. Three differential equations of the form $Lu = f$ on $R = (0,1)^2$ are distinguished to take all three MGOO solution modules into consideration which make up the release at hand (cf. Sec. 3). For each equation we consider several boundary conditions which are characterized by the corresponding boundary condition number IBC = senw (integer with decimal expansion). The decimal digits specify the type of boundary condition on the southern, eastern, northern, and western side of ∂R, respectively (cf. Sec. 2.1). In all cases, $u(x,y) = \sin 10x \sin 10y$ is the solution of the boundary-value problem in question, the right-hand side f is chosen accordingly. Each digit of IBC has either one of the following meanings,

1 $\quad u = g$,

2 $\quad \frac{\partial u}{\partial n} = g$,

3 $\quad u + \beta\frac{\partial u}{\partial n} = g$ with $\beta = \beta(x,y) = 1 + x$ along the southern and and northern side of ∂R and $= 1 + y$ otherwise,

4 $\quad u + \beta\frac{\partial u}{\partial n} = g$ with $\beta = \beta(x,y) = \max\left((x^2-x+\frac{1}{8})(y^2-y+\frac{1}{8}),0\right)$ (ie., Dirichlet boundary conditions if $|x-\frac{1}{2}| \le \frac{1}{4}\sqrt{2}$ or $|y-\frac{1}{2}| \le \frac{1}{4}\sqrt{2}$).

g is chosen accordingly. In Tables 10 - 15 the entries in each box refer again to grid spacings $h={}^1/64$ and $h={}^1/256$, respectively.

The following information is relevant to measure the performance of MGOO solution modules.

t_F CP time (in seconds) of the FMG solution with ITER = 0 ,
rt_F $10^6\, t_F$ over number of unknowns,
t_P preprocessing time (in seconds) in case of FMG,
$\tilde{\delta}_h$ FMG error $\| u - \tilde{u}_h \|_\infty$ (cf. Sec. 5.2),
$\tilde{\gamma}_h$ $\tilde{\delta}_h$ over the discretization error $\| u - u_h \|_\infty$,
w actual workspace requirement (in thousand words),
t_I CP time (in seconds) of one V-type MGI cycle (ITER=1),
rt_I $10^6\, t_I$ over number of unknowns,
$\tilde{\rho}$ experimental spectral radius of the multigrid iteration employed.

IBC	t_F	rt_F	t_P	$\tilde{\delta}_h$	$\tilde{\gamma}_h$
1111	0.51 7.43	128 114	0.02 0.21	2.33(-3) 1.43(-3)	0.98 0.84
1212	0.56 7.60	137 116	0.02 0.23	2.64(-3) 1.80(-4)	0.94 0.98
2222	0.59 7.68	140 116	0.08 1.07	4.00(-3) 2.50(-4)	1.01 0.97
1233	0.59 7.71	142 117	0.03 0.26	5.79(-3) 3.89(-4)	0.98 0.97
1244	0.54 7.60	130 116	0.03 0.25	2.69(-3) 1.84(-4)	0.95 0.97

Table 10: Performance of MGOO solution module M1, FMG with RB-HI cycles to solve $-\Delta u = f$.

First, consider Poisson's equation once again. The ratios of FMG error over discretization error are usually less than 1.0, the theoretical estimates given in Table 5 are quite pessimistic. The solution times per unknown are remarkably constant with respect to non-Dirichlet boundary conditions (for h=1/256 the deviation is less than 3%). For the pure Neumann problem (IBC=2222) orthogonalization of the data to satisfy the discrete compatibility condition results in four to five times larger preprocessing times. For Poisson's equation performance is worse if more smoothing steps are involved, ITER = 1 yields somewhat larger values for $\tilde{\gamma}_h$ and relative solution times of 150 - 185.

IBC	w	t_I	rt_I	$\tilde{\rho}$
1111	7.4 111.1	0.41 6.03	103 93	0.038 0.042
1212	7.9 113.2	0.45 6.13	110 94	0.054 0.047
2222	7.9 113.2	0.47 6.21	111 94	0.051 0.051
1233	8.4 115.0	0.47 6.21	111 94	0.052 0.049
1244	8.4 115.0	0.42 6.13	102 93	0.050 0.049

Table 11: Performance of MGOO solution module M1, one RB-HI MGI cycle for $-\Delta u = f$.

Table 11 confirms the efficiency of RB-HI cycles also for non-Dirichlet boundary conditions, in particular see rt_I and $\tilde{\rho}$. Preprocessing times for multigrid iterations are negligible, except when coefficients of the differential operator have to be evaluated or the discrete compatibility condition is to be enforced. In any case, the preprocessing time is slightly less than t_P.

IBC	t_F	rt_F	t_P	$\tilde{\delta}_h$	$\tilde{\gamma}_h$
1111	0.62 9.84	155 151	0.26 4.01	2.31(-3) 5.25(-4)	0.98 1.08
1212	0.67 10.05	163 153	0.26 4.02	2.64(-3) 5.78(-4)	0.94 1.08
2222	0.71 10.20	168 154	0.27 4.04	1.76(-2) 2.11(-3)	0.90 1.12
1233	0.70 10.26	168 156	0.28 4.02	5.08(-3) 7.17(-4)	0.97 1.12
1244	0.67 10.05	163 153	0.29 4.06	2.67(-3) 5.13(-4)	0.95 1.00

Table 12: Performance of MGOO solution module M2, FMG with RB-HI cycles to solve $-\Delta u + (x+y)u = f$.

IBC	w	t_I	rt_I	$\tilde{\rho}$
1111	13.1 199.5	0.50 8.10	126 125	0.037 0.042
1212	13.9 202.6	0.53 8.23	130 126	0.053 0.047
2222	13.9 202.6	0.56 8.33	133 126	0.050 0.051
1233	13.9 202.6	0.56 8.41	134 128	0.050 0.049
1244	13.8 202.4	0.53 8.22	129 126	0.050 0.049

Table 13: Performance of MGOO solution module M2, one RB-HI MGI cycle for $-\Delta u + (x+y)u = f$.

Tables 12 and 13 show similar performance for the generalized Helmholtz equation where c is a function. Larger values for the relative solution times rt_F and rt_I are obtained since Gauss-Seidel relaxation is more expensive than for constant c (cf. Table 6), the variable coefficient requires a division instead of a multiplication at each grid point.

IBC	t_F	rt_F	t_P	$\tilde{\delta}_h$	$\tilde{\gamma}_h$
1111	1.50 24.92	378 380	0.29 4.28	3.72(-3) 2.41(-4)	1.40 1.46
1212	1.54 24.90	376 380	0.29 4.31	1.16(-2) 7.77(-4)	1.33 1.42
2222	1.63 25.31	386 383	0.30 4.33	3.59(-2) 2.31(-3)	0.91 0.93
1233	1.59 25.15	376 381	0.30 4.31	1.44(-2) 5.29(-3)	1.20 0.98
1244	1.55 24.99	376 381	0.30 4.30	1.18(-2) 3.56(-3)	1.41 0.93

Table 14: Performance of MGOO solution module M3, FMG with AZ-FW cycles to solve $-a(x,y)u_{xx} - u_{yy} + (x+y)u = f$ with $a(x,y) = 100^{x+y-1}$.

IBC	w	t_I	rt_I	$\tilde{\rho}$
1111	18.9 288.4	1.57 25.97	396 399	0.054 0.077
1212	19.5 290.5	1.63 26.30	398 401	0.069 0.077
2222	19.5 290.5	1.70 26.60	402 403	0.064 0.073
1233	19.5 290.8	1.67 26.50	395 401	0.064 0.081
1244	19.5 290.8	1.64 26.38	398 402	0.056 0.078

Table 15: Performance of MGOO solution module M3, one AZ-FW MGI cycle for $-a(x,y)u_{xx} - u_{yy} + (x+y)u = f$ with $a(x,y) = 100^{x+y-1}$.

In Tables 14 and 15 we deal with an anisotropic operator for which AZ relaxation is best. AZ relaxation is rather costly, the solution times are correspondingly large. Performance evaluations when MGOOM3 is applied to Poisson's equation yield values less than 1.22 for $\tilde{\gamma}_h$ and 0.011 to 0.0125 for the experimental spectral radii. This improvement for $a \simeq 1$ is expected according to Table 3.

The actual workspace requirement can hardly be reduced further if we do not want to evaluate function and coefficient values more than once per grid point. The performance evaluations summarized in Tables 10 - 15 above show the efficiency and storage economy of MGOO in solving large scale problems with general boundary conditions.

7. Extensions, further developments

Several extensions of the MGOO subroutine package are possible. One is to more general equations. Two and three are to the Full Approximation Scheme (FAS) and to higher-order techniques. Staggered grids are of interest in many applications, number four. The fifth extension, in this list, allows local mesh refinement.

Efficient multigrid solution schemes for more general equations are under investigation. A forthcoming release of MGOO will also cover equations in conservative form as well as with first-order terms including singular perturbation problems. - The conversion of Correction Scheme (CS) programs to FAS programs is technically simple. Although another coarse-grid variable is used, the algorithmic composition remains unchanged. One advantage of FAS over CS is its direct application (no global linearization needed) to non-linear problems (cf. [5], Sec. 8.3). - Auzinger and Stetter [1] have adapted a defect correction scheme to multigrid solution modules. A related approach is Brandt's τ extrapolation ([5], Sec. 8.4) which is especially simple and inexpensive. τ extrapolation uses the fine-to-coarse defect correction to raise the local approximation order. - For staggered grids it is not clear at this time how much of the MGOO structure can be retained. - Fixed as well as adaptive mesh refinement is a story by itself. However, much effort is under way in this direction, especially in the U.S., and it is expected that MGOO will get extended to local mesh refinement at some time.

Acknowledgements

We would like to thank K. Solchenbach for carrying out the performance measurements summarized in Tables 8 and 9. R. Hempel investigated nested iterations in the framework of ICCG and MICCG. Finally, we are especially grateful to Ch. Tillmann and C.A. Thole who tested the code and programmed numerical experiments.

References

[1] Auzinger, W., and Stetter, H.J.: "Defect correction and multigrid iterations", these Proceedings.

[2] Bank, R.E., and Rose, D.J.: "Design and implementation of an elliptic equation solver for rectangular regions", Computers, Fast Elliptic Solvers and Applications (U. Schumann, ed.), 112 - 124. Advance Publications, London, 1978.

[3] Brandt, A.: "Multigrid solvers on parallel computers", Elliptic Problem Solvers (M. Schultz, ed.), 39 - 83. Academic Press, New York, 1981.

[4] Brandt, A.: "Stages in developing multigrid solutions", Numerical Methods for Engineering GAMNI 2 (E. Absi, R. Glowinski, P. Lascaux, H. Veysseyre, eds.), vol. 1, 23 - 45. Dunod, Paris, 1980.

[5] Brandt, A.: "Guide to multigrid development", these Proceedings.

[6] Buneman, O.: "A compact non-iterative Poisson solver", Institute for Plasma Research Report 294, Stanford University, 1969.

[7] Buzbee, B.L., Golub, G.H., and Nielson, C.W.: "On direct methods for

solving Poisson's equation", SIAM J. Numer. Anal. 7, 627 - 656, 1970

[8] Dendy, J.E.: "Black box multigrid", Los Alamos Report LA-UR-81-2337, Los Alamos Scientific Laboratory, 1981.

[9] Foerster, H., Förster, H., and Trottenberg, U.: "Modulare Programme zur schnellen Lösung elliptischer Randwertaufgaben mit Reduktionsverfahren", Sonderforschungsbereich 72 Preprints 216/420, Bonn University, 1978/1980.

[10] Foerster, H., Stüben, K., and Trottenberg, U.: "Non-standard multigrid techniques using checkered relaxation and intermediate grids", Elliptic Problem Solvers (M. Schultz, ed.), 285 - 300. Academic Press, New York, 1981.

[11] Foerster, H., and Witsch, K.: "On efficient multigrid software for elliptic problems on rectangular domains", Math. Comput. Simulation XXIII, 293 - 298, 1981.

[12] Gustafsson, I.: "A class of first-order factorization methods", BIT 18, 142 - 156, 1978.

[13] Hackbusch, W.: "Multi-grid convergence theory", these Proceedings.

[14] Hempel, R.: "Lösung elliptischer Randwertaufgaben mit dem ICCG-Verfahren und seinen Varianten", Thesis submitted for a diploma, Institute for Applied Mathematics, Bonn University, 1982.

[15] Hockney, R.W.: "The potential calculation and some applications", Meth. Comp. Phys. 9, 135 - 211, 1970.

[16] Hyman, J.M.: "Mesh refinement and local inversion of elliptic partial differential equations", J. Comp. Phys. 23, 124 - 134, 1977.

[17] van Kats, J.M., and van der Vorst, H.A.: "Software for the discretization and solution of second-order self-adjoint elliptic partial differential equations in two dimensions", ACCU Technical Report 10, Academic Computer Centre Utrecht, 1979.

[18] Meijerink, J.A., and van der Vorst, H.A.: "An iterative solution method for linear systems of which the coefficient matrix is a symmetric M-matrix", Math. Comp. 31, 148 - 162, 1977.

[19] Rice, J.R.: "Machine and compiler effects on the performance of elliptic pde software", 10th IMACS World Congress Proceedings. IMACS, New Brunswick, 1982.

[20] Rice, J.R., and Boisvert, R.F.: Solving Elliptic Problems Using ELLPACK, Springer-Verlag, New York, to appear.

[21] Ryder, B.G.: "The PFORT verifier", Software Practice and Experience 4, 359 - 377, 1977.

[22] Ries, M., Trottenberg, U., and Winter, G.: "A note on MGR methods", Lin. Alg. Appl., to appear.

[23] Schröder, J., Trottenberg, U., and Witsch, K.: "On fast Poisson solvers and applications", Numerical Treatment of Differential Equations (R. Bulirsch, R.D. Grigorieff, J. Schröder, eds.), 153 - 187. Springer-Verlag, Berlin, 1978.

[24] Stüben, K., and Trottenberg, U.: "Multigrid methods: fundamental algorithms, model problem analysis and applications", these Proceedings.

[25] Temperton, C.: "On the FACR(ℓ) algorithm for the discrete Poisson equation", Research Report 14, ECMWF, Bracknell, 1977.

[26] Thole, C.A.: "Beiträge zur Fourier Analyse von Mehrgittermethoden: V-cycle, ILU-Glättung, anisotrope Operatoren", Thesis to be submitted.

[27] Wesseling, P.: "A robust and efficient multigrid method", these Proceedings.

ON MULTI-GRID ITERATIONS WITH DEFECT CORRECTION

W. Hackbusch
Mathematisches Institut, Ruhr-Universität Bochum
Postfach 10 21 48, D - 4630 Bochum 1, Germany

Abstract. Defect correction methods produce approximations of higher order without solving complicated equations. The defect correction iteration requires a repeated (exact) solution of a basic discretization with varying right-hand sides. The defect correction method can be combined with the multi-grid iteration so that the algorithm converges to a result of higher order. We prove the convergence of the modified multi-grid iteration and give error estimates.

1. Introduction

In the last years many papers on iterated defect corrections have appeared. Applications are mentioned for ordinary differential equations (e.g. by Frank and Ueberhuber [4], Hairer [9]), for partial differential equations (Frank and Hertling [3], Böhmer [1], Hertling (in [15])), and for eigenvalue problems (Lin Qun [11]). The paper of Stetter [12] describes the general principle and contains historical references. The results of this contribution are mainly those from Hackbusch [5], but here we give a more detailed analysis for the special case of a second order Dirichlet boundary value problem.

The defect correction method can be applied to nonlinear problems. In this case there are two variants of the method (version A and B of [12]). The situation is simpler in the linear case since then both versions coincide. The analysis of the nonlinear defect correction does not differ very much from the analysis of the linear one. Therefore, for this survey it suffices to treat the linear case only.

In the sequel we give a brief description of the general defect correction method. Let

(1.1) $$Lu = f$$

be the (continuous) linear problem. A discretization of (1.1) is given by means of L_h and R_h^F:

(1.2a) $$L_h u_h = f_h, \quad \text{where } f_h = R_h^F f.$$

This discretization will be assumed to be invertible. Let κ be the consistency order of (1.2a). In order to obtain an approximation with error better than $O(h^\kappa)$ we define a second discretization by L_h' and $R_h'^F$:

(1.2b) $$L_h' u_h' = f_h', \quad \text{where } f_h' = R_h'^F f,$$

assuming a higher consistency order κ' for (1.2b). Stability of Eq. (1.2b) is not required; L_h' may fail to be invertible.

The iterated defect correction is defined by

(1.3) $$u_h^o = L_h^{-1} f_h, \quad u_h^{i+1} = u_h^i - L_h^{-1}(L_h' u_h^i - f_h') \qquad (i=o,1,2,\ldots).$$

Usually, $\lim_{i\to\infty} u_h^i$ will not exist, but for fixed i we expect that the discretization error of u_h^i is of order $O(h^{\min(\kappa',(i+1)\kappa)})$.

In §2 error estimates are proved in the case of a Dirichlet boundary value problem. We recall this proof to prepare the basic assumptions of stability (regularity) and consistency that we need in the last chapter, too. The multi-grid algorithm and its combination with the defect correction principle is defined in §3. In §4 we prove the convergence of the iteration to a discrete function $\hat{u}_h$. We show that the Euclidean norm of the discretization error of $\hat{u}_h$ is of order $O(h^{\min(\kappa',2+\kappa)})$. Possible generalizations are discussed in §5. An example in §6 shows that our estimates are almost optimal.

2. Defect Correction Analysis for Second Order Dirichlet Boundary Value Problems

In order to avoid general notations and general assumptions on the kind of the problem (1.1), we restrict our considerations to the case of a Dirichlet boundary value problem of second order in a domain $\Omega\subset\mathbb{R}^d$,

$$L u = f \text{ in } \Omega \;, \qquad u = o \text{ on } \Gamma = \partial\Omega,$$

where L is an elliptic differential operator of second order. Assuming the coefficients of L and the boundary to be smooth enough, we know that $f\in L^2(\Omega)$ implies $u\in H^2(\Omega)\cap H_o^1(\Omega)$, or more generally, $f\in H^s(\Omega)$ $(s\geq o)$ implies $u\in H^{s+2}(\Omega)\cap H_o^1(\Omega)$[1]. This shows that there is a scale of Sobolev spaces $U_H^s\subset F_H^s$ defined by

$$U_H^s=\begin{cases} H^s(\Omega)\cap H_o^1(\Omega) & (s>1)\\ H_o^s(\Omega) & (o\leq s\leq 1)\\ (H^{-s}(\Omega))'=\text{dual space of } H^{-s}(\Omega) & (s<o)\end{cases},\quad F_H^s=\begin{cases} H^s(\Omega) & (s\geq -1)\\ (H^{-s}(\Omega)\cap H_o^1(\Omega))' & (s<-1)\end{cases}$$

such that

$$L:\ U_H^s\to F_H^{s-2} \qquad \text{and} \qquad L^{-1}:\ F_H^{s-2}\to U_H^s$$

are bounded mappings. Here, s may be bounded, $s_o\leq s\leq s_1$, according to the smoothness of the coefficients of L and of Γ; furthermore, s=1/2 (mod 1) has to be excluded or the definitions of U_H^s and F_H^s have to be modified for these values.

Usually, U_H^s and F_H^s are not identical. Only if $\Omega=\mathbb{R}^d$, we have $U_H^s=F_H^s=H^s(\mathbb{R}^d)$ for all s. But we mention

NOTE 1 Let $\Omega\neq\mathbb{R}^d$. U_H^s and F_H^s coincide (and have equivalent norms) iff $|s|<1/2$.

We denote the norms of U_H^s and F_H^s by

$$\|\cdot\|_{U,s} \quad\text{and}\quad \|\cdot\|_{F,s}\,,\quad \text{resp.}$$

For the analysis of the discrete problems (1.2a,b) we have to introduce discrete ana-

[1] Here we use the common notation of Sobolev spaces; $H^s(\Omega)$: space of differentiability order s, $H_o^s(\Omega)$: completion in $H^s(\Omega)$ of functions vanishing in a neighbourhood of Γ, $L^2(\Omega)=H^o(\Omega)$. $H^{-s}(\Omega)$ = dual space of $H_o^s(\Omega)$ for $s\geq o$.

logues of the spaces ${}^{U}H^{s}$, ${}^{F}H^{s}$, and their norms:

$${}^{U}H^{s}_{h} \text{ and } {}^{F}H^{s}_{h} \text{ with norms } \|\cdot\|_{U,h,s} \text{ and } \|\cdot\|_{F,h,s}\text{, resp.}$$

The simplest case is s=o: ${}^{U}H^{o}_{h}={}^{F}H^{o}_{h}$ is the space of grid functions equipped with the Eulicean norm. The norm of ${}^{U}H^{1}_{h}$ can be defined by means of first differences. In part I of [8] the norms of ${}^{U}H^{s}_{h}$ $(-1/2<s<3/2)$ and ${}^{F}H^{s}_{h}$ $(-3/2<s<1/2)$ are described. For the norm of ${}^{U}H^{2}_{h}$ compare [8, part II].

For an operator $A: {}^{U}H^{s} \to {}^{F}H^{t}_{h}$ we use the operator norm

$$\|A\|_{t\leftarrow s} = \|A\|_{F,t\leftarrow U,s} = \sup\{\|Au\|_{F,h,t}/\|u\|_{U,s} : o \neq u \in {}^{U}H^{s}\}.$$

Similarly, $\|\cdot\|_{U,t\leftarrow U,s}$, $\|\cdot\|_{U,t\leftarrow F,s}$, etc. are defined. If no confusion is possible, the shorter notation $\|\cdot\|_{t\leftarrow s}$ is used.

The discrete counterpart of $L^{-1}: {}^{F}H^{\sigma-2} \to {}^{U}H^{\sigma}$ (H^{σ}-regularity) is the *discrete regularity*

(2.1) $$\|L_h^{-1}\|_{\sigma\leftarrow\sigma-2} \leq C.$$

C is a generic constant independent of h. For the analysis of the iterated defect correction the condition (2.1) can be weakened (cf. [5]). However, the strong form (2.1) is needed for the multi-grid analysis, too (cf. [7]). The discrete regularity (2.1) is proved in [8, part I] for $1/2<\sigma<3/2$ and in [8, part II] for $\sigma=2$, if L_h is a suitable difference scheme. For finite element discretizations compare Hackbusch (in [15]).

The *consistency* condition describes the error $L_h R^U_h u^* - f_h$ (u^* solution of $Lu=f$), where

$$R^U_h : {}^{U}H^{\sigma} \to {}^{U}H^{\sigma}_{h}$$

is a suitable mapping of (continuous) functions into discrete grid functions. Let

$$u^*_h = R^U_h u^* \qquad (u^* = L^{-1} f).$$

Since $L_h u^*_h - f_h = (L_h R^U_h - R^F_h L)u^*$, the consistency condition can be written as

(2.2a) $$\|R^F_h L - L_h R^U_h\|_{\sigma-2\leftarrow\tau} \leq Ch^{\min(\kappa,\tau-\sigma)}.$$

κ is the consistency order of (L_h, R^F_h). The analogous consistency condition for (L'_h, R'^F_h) is

(2.2b) $$\|R'^F_h L - L'_h R^U_h\|_{\sigma-2\leftarrow\tau} \leq Ch^{\min(\kappa',\tau-\sigma)}.$$

Always, we assume $\kappa'>\kappa$; otherwise, the defect correction makes no sense. Under some additional assumptions it is possible to obtain a direct comparison of L_h and L'_h:

(2.2c) $$\|L'_h - L_h\|_{\sigma-2\leftarrow\tau} \leq Ch^{\min(\kappa,\tau-\sigma)}.$$

In the sequel we need only the estimates (2.1), (2.2a-c) for *special* values of σ and τ. Note that, e.g., (2.2a) holds for all $\tau\leq\tau_o$ if it is valid for $\tau=\tau_o$ and if the inverse estimate

$$(2.3)\qquad \|\cdot\|_{U,h,t} \le Ch^{s-t}\|\cdot\|_{U,h,s}, \qquad \|\cdot\|_{F,h,t} \le Ch^{s-t}\|\cdot\|_{F,h,s} \qquad (t\ge s)$$

holds for $t=\tau_o$ and $s=\sigma$.

Using the notation $u^*=L^{-1}f$ and $u_h^*=R_h^U u^*$ as before, we can estimate the error $u_h^i-u_h^*$ by the following

THEOREM 1 Let $s\in\mathbb{R}$ and $i\ge o$ be fixed and suppose $u^*\in {}^U H^{s+(i+1)\kappa}$. Assume

- (2.1) for $\sigma = s, s+\kappa, \dots, s+i\kappa$;
- (2.2a) for $\sigma = s+i\kappa$ and $\tau = s+(i+1)\kappa$;
- (2.2b) for $\sigma = s, s+\kappa, \dots, s+(i-1)\kappa$ and $\tau = s+(i+1)\kappa$;
- (2.2c) for $\sigma = s, s+\kappa, \dots, s+(i-1)\kappa$ and $\tau = s+\kappa$.

Then the ith iterate u_h^i of the defect correction (1.3) satisfies

$$(1.4)\qquad \|u_h^i - u_h^*\|_{U,h,s} \le C_i h^{\min(\kappa',(i+1)\kappa)} \|u^*\|_{U,s+(i+1)\kappa}.$$

The optimal order κ' is obtained for $i = \lceil \kappa'/\kappa - 1 \rceil$.

Proof. Use the representations

$$u_h^o-u_h^*=L_h^{-1}(R_h^F L-L_h R_h^U)u^*,$$

$$u_h^k-u_h^*=u_h^{k-1}-u_h^*-L_h^{-1}(L_h'u_h^{k-1}-f_h')=L_h^{-1}(L_h-L_h')(u_h^{k-1}-u_h^*)+L_h^{-1}(R_h'^F L-L_h'R_h^U)u^* \qquad (k\ge 1)$$

and prove by induction. ∎

3. Multi-Grid Iteration Combined with Defect Correction

3.1 Notation

Instead of only one discretization parameter h we consider a sequence

$$h_o > h_1 > \dots > h_{l-1} > h_l.$$

We change the notation by replacing any subscript h_k with the level number k:

$$L_k = L_{h_k},\ f_k = f_{h_k},\ {}^U H_k^s = {}^U H_{h_k}^s,\ \|\cdot\|_{U,k,s} = \|\cdot\|_{U,h_k,s},\ \text{etc}.$$

The restriction (fine to coarse grid) and the prolongation (coarse to fine grid) are denoted by

$$r:\ {}^F H_k^s \to {}^F H_{k-1}^s, \qquad p:\ {}^U H_{k-1}^s \to {}^U H_k^s \qquad (k=1,2,\dots,l).$$

The smoothing procedure is some iteration $u_k^i \to u_k^{i+1}$ defined by $\mathcal{S}_k$:

$$(3.1)\qquad u_k^{i+1} = \mathcal{S}_k(u_k^i,f_k) = S_k u_k^i + T_k f_k.$$

S_k is the iteration matrix of $\mathcal{S}_k$. We remark that T_k is determined by

$$(3.2)\qquad T_k = (I - S_k)L_k^{-1}.$$

An example for $\mathcal{S}_k$ is the modified Jacobi iteration

$$(3.3)\qquad \mathcal{S}_k(u_k,f_k) = u_k - \omega h_k^2(L_k u_k - f_k) \qquad \text{(for } \omega \text{ compare [6]).}$$

3.2 Multi-Grid Algorithm

The usual multi-grid iteration (compare e.g. [6]) consists of a smoothing step and a coarse grid correction. The latter step uses the restricted defect $r(L_k\bar{u}_k-f_k)$. In order to imitate the defect correction (1.3) we can replace $r(L_k\bar{u}_k-f_k)$ by $r(L'_k\bar{u}_k-f_k)$ at least for the finest level $k=l$ (cf. Brandt [2], Hemker [1o]). The resulting iteration is defined by the following recursive procedure *mgdc:*

procedure *mgdc*(k,u,f);
if k=o then $u:=L_0^{-1}*f$ else
begin integer j; array d,v;
for $j:=1$ step 1 until ν do $u:=\mathcal{S}_k(u,f)$;
(3.4) $d:=$if $k=l$ then $r*(f'_l-L'_l*u)$ else $r*(f_k-L_k*u)$,
$v:=$o; for $j:=1$ step 1 until γ do *mgdc*$(k-1,v,d)$;
$u:=u+p*v$
end;

One step of the multi-grid iteration with defect correction at level l is performed by calling *mgdc* with $k=l$ and $f=f_l$ (not f'_l). For $k<l$ this iteration is identical to procedure *MGM* from [6]. The number ν of smoothing steps and the number γ of coarse-grid iterations is discussed in [6].

There are two reasons for applying the modified algorithm (3.4). (i) The matrix L_l is simpler than L'_l, e.g., L_l has not so many entries as L'_l. Then, smoothing by $\mathcal{S}_k$ ($o\le k\le l$) corresponding to L_k is cheaper. (ii) Because of stability, L_l has lower order of consistency (cf. [1o]). Simple discretizations L'_l of higher order are possible but unstable. In this case the usual iteration with $L_l:=L'_l$ cannot be used, whereas the modification (3.4) still works well.

3.3 Two-Grid Iteration

In [6] we demonstrated that two-grid convergence is almost sufficient for multi-grid convergence. Therefore, we can reduce the analysis of the multi-grid iteration *mgdc* to the analysis of the corresponding two-grid case. The usual two-grid iteration of level l is

$$(3.5)\qquad \begin{aligned} &u_l:=u_l^j;\quad \text{for } \mu:=1\ (1)\ \nu \text{ do } u_l:=\mathcal{S}_l(u_l,f_l);\\ &u_l^{j+1}:=u_l+p\,L_{l-1}^{-1}\,r(f_l-L_lu_l); \end{aligned}$$

The two-grid defect correction iteration reads as follows:

$$(3.6)\qquad \begin{aligned} &u_l:=u_l^j;\quad \text{for } \mu:=1\ (1)\ \nu \text{ do } u_l:=\mathcal{S}_l(u_l,f_l);\\ &u_l^{j+1}:=u_l+p\,L_{l-1}^{-1}\,r(f'_l-L'_lu_l); \end{aligned}$$

In §4.1 we shall recall the convergence proof of the usual multi-grid iteration (3.5). The modified iteration (3.6) will be studied in §4.2 and §4.3.

4. Convergence and Error Estimates

4.1 Convergence of the Usual Two-Grid Iteration

The two-grid iteration (3.5) can be written as

(4.1) $$u_l^{j+1} = M_l u_l^j + g_l \quad \text{with} \quad M_l = M_l(\nu) = (L_l^{-1} - p L_{l-1}^{-1} r) L_l S_l^\nu,$$

where S_l is defined in (3.1). The iteration (4.1) converges if $\|M_l\|_{s\leftarrow s} < 1$ holds for some s. Sufficient conditions for this estimate are the *approximation property*

(4.2a) $$\|L_l^{-1} - p L_{l-1}^{-1} r\|_{s\leftarrow t-2} \le C h_l^{t-s}$$

for a suitable t and the *smoothing property*

(4.2b) $$\|L_l S_l^\nu\|_{t-2\leftarrow s} \le \eta(\nu) h_l^{s-t} \quad \text{for} \quad 1 \le \nu < \nu_{max}(h_l) \quad \text{with}$$

$$\eta(\nu) \to o \quad (\nu\to\infty), \qquad \nu_{max}(h) = \infty \quad \text{or} \quad \nu_{max}(h) \to \infty \quad (h\to o)$$

as discussed in [6][2]. Usually, one tries to prove (4.2a) for s=o, t=2, since then $\|\cdot\|_{s\leftarrow t-2} = \|\cdot\|_{t-2\leftarrow s}$ = spectral norm (cf. §3.2.1 of [6]). Sometimes one obtains the estimate (4.2a) for $s \in (o,1)$, $t=2-s$ (cf. §3.2.2 of [6], [7]). If L_{l-1} is not constructed in a special manner (e.g. by the Galerkin approach $L_{l-1} = rL_l p$), the exponent t-s of (4.2a) does not exceed κ: $t-s \le \kappa$. Usually, the function $\eta(\nu)$ behaves like $C\nu^{(s-t)/2}$. Therefore, $\eta(\nu) \to o$ requires $t > s$.

The two-grid convergence for iteration (3.5) follows from the following note (cf. [6]):

NOTE 2 *(two-grid convergence)* Assume the approximation and smoothing properties (4.2a,b). For given $\rho \in (o,1)$ the estimate

(4.3) $$\|M_l(\nu)\|_{s\leftarrow s} \le \rho < 1$$

holds for $\nu \in [\nu_{min}, \nu_{max}(h_l)]$, $h_l < h_{max}$ (ν_{min}, h_{max} depending on ρ only).

The estimate (4.3) implies

$$\|u_l^j - u_l\|_{U,l,s} \le \rho^j \|u_l^o - u_l\|_{U,l,s} \qquad (u_l = L_l^{-1} f_l).$$

4.2 Convergence of the Two-Grid Iteration with Defect Correction

We recalled the sufficient conditions (4.2a,b) since we want to prove the convergence of the modified algorithm under similar assumptions. The two-grid iteration (3.6) with defect correction leads to

(4.4) $$u_l^{j+1} = M_l' u_l^j + g_l'$$

with

$$M_l' = M_l'(\nu) = (I - p L_{l-1}^{-1} r L_l') S_l^\nu,$$

$$g_l' = g_l'(f_l, f_l', \nu) = (I - p L_{l-1}^{-1} r L_l') (I - S_l^\nu) L_l^{-1} f_l + p L_{l-1}^{-1} r f_l'.$$

The following theorem shows that we get the same convergence result for iteration (3.6) as for (3.5):

[2] The norms in [6] are to be defined by $\|\cdot\|_U := \|\cdot\|_{U,h,s}$, $\|\cdot\|_F := \|\cdot\|_{F,h,t-2}$.

THEOREM 2 Assume $t-s \leq \kappa$ and
- (2.1): $\|L_l^{-1}\|_{s \leftarrow s-2} \leq C$, $\|L_l^{-1}\|_{t \leftarrow t-2} \leq C$,
- (2.2c): $\|L_l'-L_l\|_{s-2 \leftarrow t} + h_l^{t-s}\|L_l'-L_l\|_{t-2 \leftarrow t} \leq Ch_l^{t-s}$,
- (4.2a,b): approximation and smoothing property.

Then for any $\rho' \in (o,1)$ there are ν_{min} and h_{max} such that

(4.5) $$\|M_l'(\nu)\|_{s \leftarrow s} \leq C' \eta(\nu) \leq \rho' < 1$$

provided that $\nu \in [\nu_{min}, \nu_{max}(h_l)]$, $h_l \leq h_{max}$.

Proof. Use the decomposition

$$M_l'(\nu) = \{(L_l^{-1} - p\, L_{l-1}^{-1} r)[I + (L_l'-L_l)L_l^{-1}] - L_l^{-1}(L_l'-L_l)L_l^{-1}\}(L_l S_l^{\nu})$$

and estimate each single factor or round bracket separately. ■

The assumptions of Theorem 2 are known from Theorem 1 [(2.1), (2.2c)] and Note 2 [(4.2a,b)]. For the choice of s and t we add

REMARK 1 The numbers s and t should satisfy $o \leq s \leq t \leq 2$. Otherwise, it might be difficult to prove (2.1) and (4.2b).

Thanks to (4.5) the iterates u_l^j of the two-grid iteration with defect correction converge to some limit

(4.6) $$\hat{u}_l = \hat{u}_l(\nu) := \lim_{j\to\infty} u_l^j \qquad (u_l^j \text{ from (3.6)}).$$

Note that $\hat{u}_l$ is *not* the solution of $L_l'u_l'=f_l'$ and no fixed point of the smoothing iteration $\mathcal{S}_l$.

4.3 Error Estimates for the Limit $\hat{u}_l$

Inequality (4.5) ensures that $I-M_l'$ is regular. Since $\hat{u}_l$ is a fixed point of iteration (4.4) we have

LEMMA 1 $$\hat{u}_l = [I - M_l'(\nu)]^{-1}\{[I-pL_{l-1}^{-1}rL_l'][I-S_l^{\nu}]L_l^{-1}f_l + pL_{l-1}^{-1}rf_l'\}.$$

Let $u_l^*=R_l^U u^*$, $u^*=L^{-1}f$, be the restricted solution. We want to estimate the error $\hat{u}_l-u_l^*$. By

$$[I - M_l']u_l^* = [I - pL_{l-1}^{-1}rL_l'][I-S_l^{\nu}]u_l^* + pL_{l-1}^{-1}rL_l'u_l^*$$

we obtain

LEMMA 2 The error $\hat{u}_l - u_l^*$ can be decomposed into

(4.7a) $$\hat{u}_l - u_l^* = [I - M_l'(\nu)]^{-1}(v_l + w_l)$$

with

(4.7b) $$v_l = [I - pL_{l-1}^{-1}rL_l'][I-S_l^{\nu}]L_l^{-1}(f_l-L_l u_l^*),$$

(4.7c) $$w_l = p\, L_{l-1}^{-1} r\, (f_l' - L_l'u_l^*).$$

It suffices to study v_l and w_l because of

LEMMA 3 Let $\|M_l'(\nu)\|_{s\leftarrow s} \le \rho' < 1$ (cf. (4.5)). Then we have

(4.8) $$\|\hat{u}_l - u_l^*\|_{U,l,s} \le (\|v_l\|_{U,l,s} + \|w_l\|_{U,l,s}) / (1-\rho').$$

The decomposition v_l+w_l separates the influences of the defects $f_l-L_lu_l^*$ and $f_l'-L_l'u_l^*$. w_l can be expected to be of order $O(h_l^{\kappa'})$:

LEMMA 4 Let $t \le \tau \le s+\kappa'$ and assume

- (2.1): $\|L_l^{-1}\|_{s\leftarrow s-2} \le C$,
- (2.2b): $\|L_l'u_l^*-f_l'\|_{F,l,s-2} + h_l^{t-s}\|L_l'u_l^*-f_l'\|_{F,l,t-2} \le Ch_l^{\tau-s}\|u^*\|_{U,\tau}$,
- (4.2a): approximation property.

Then (4.9) holds:

(4.9) $$\|w_l\|_{U,l,s} \le Ch_l^{\tau-s}\|u^*\|_{U,\tau}.$$

Note that $\tau-s$ does not exceed κ' and that $\tau-s=\kappa'$ for the possible choice $\tau:=s+\kappa'$ ($\ge t$).

Proof. Estimate $w_l = \{L_l^{-1} - (L_l^{-1} - pL_{l-1}^{-1}r)\}(f_l'-L_l'u_l^*)$ by

$$\|w_l\|_{U,l,s} \le \|L_l^{-1}\|_{s\leftarrow s-2}\|f_l'-L_l'u_l^*\|_{F,l,s-2} + \|L_l^{-1}-pL_{l-1}^{-1}r\|_{s\leftarrow t-2}\|f_l'-L_l'u_l^*\|_{F,l,t-2}.$$ ∎

Thanks to (4.8) it remains to estimate v_l. But the analysis of v_l is more complicated. We start with the factorization

(4.10) $$v_l = AB(f_l-L_lu_l^*), \quad A:= I - pL_{l-1}^{-1}rL_l, \quad B:=(I - S_l^{\nu})L_l^{-1}.$$

One expects that $f_l-L_lu_l^*$ yields a factor h_l^{κ}. But what are the contributions of A and B ? An obvious estimate of v_l is

(4.11) $$\|v_l\|_{U,l,s} \le \|A\|_{s\leftarrow p}\,\|B\|_{p\leftarrow q}\,\|f_l-L_lu_l^*\|_{F,l,q},$$

where p and q are chosen suitably.

The nest lemma shows that at the most A gives a factor h_l^{κ}:

LEMMA 5 Let $s \le p \le t \le s+\kappa$ and assume

- (2.1): $\|L_l^{-1}\|_{s\leftarrow s-2} \le C$,
- (2.2c): $\|L_l'-L_l\|_{s-2\leftarrow p} \le Ch_l^{p-s}$,
- (4.2a): approximation property

and $\|L_l'\|_{t-2\leftarrow p} \le Ch_l^{p-t}$ (t from (4.2a)).

Then (4.12) holds:

(4.12) $$\|A\|_{s\leftarrow p} = \|I - pL_{l-1}^{-1}rL_l'\|_{s\leftarrow p} \le Ch_l^{p-s}.$$

For the choice $p=t=s+\kappa$ we would gain a factor h_l^{κ}. Again, Remark 1 applies: s and t should be chosen from the interval [o,2].

Proof. Use $A = I - pL_{l-1}^{-1}rL_l' = (L_l^{-1} - pL_{l-1}^{-1}r)L_l' - L_l^{-1}(L_l'-L_l)$. ∎

It will turn out that B plays a decisive rôle. By virtue of (3.2) B can be rewritten as

(4.13) $$B = (\sum_{\mu=o}^{\nu-1} S_l^{\mu}) T_l .$$

To approach the problem we consider the simplest smoothing procedure (3.3) with $S_l = I - \omega h_l^2 L_l$ and $T_l = \omega h_l^2 I$. The conjectures $\|S_l\|_{p\leftarrow p} \leq C$ and $\|T_l\|_{p\leftarrow p} \leq C$ would yield the estimate $\|B\|_{p\leftarrow p} \leq Ch_l^2$. Hence, the choice $p=q=s+\kappa$ would imply

(4.14) $$\|v_l\|_{U,l,s} \leq Ch_l^{2+2\kappa}\|u^*\|_{U,s+2+2\kappa}$$

(cf. Lemma 5, (2.2a)). But the estimate $\|B\|_{p\leftarrow p} \leq Ch_l^2$ cannot be expected for all p, since the norm $\|\cdot\|_{p\leftarrow p}$ is $\|\cdot\|_{U,p\leftarrow F,p}$ and not $\|\cdot\|_{U,p\leftarrow U,p}$. $\|T_l\|_{U,p\leftarrow F,p} \leq Ch_l^2$ is equivalent to

(4.15) $$\|\cdot\|_{U,l,p} \leq C\|\cdot\|_{F,l,p}.$$

Obviously, (4.15) holds for p=o since $\|\cdot\|_{U,l,o} = \|\cdot\|_{F,l,o}$ =Euclidean norm. It is not valid for p=1 because of the additional boundary condition. By Note 1, $\|\cdot\|_{U,p} \leq C\|\cdot\|_{F,p}$ holds iff $|p|<1/2$. As a consequence (4.15) is valid for $|p|<1/2$ with C=C(p). Using the inverse estimate (2.3), $\|\cdot\|_{U,l,p} \leq Ch_l^{q-p}|\cdot|_{U,l,q}$ $(q\leq p)$, we obtain

$$\|I\|_{U,l,p\leftarrow F,l,q} \leq Ch_l^{q-p} \qquad \text{iff } q\leq p,\ |q|<1/2$$

and therefore

$$\|T_l\|_{p\leftarrow q} \leq Ch_l^{2+q-p} \qquad \text{iff } q\leq p,\ |q|<1/2$$

for T_l from (3.3). A further analysis shows

$$\|S_l\|_{p\leftarrow p} \leq C \qquad \text{iff } -1/2<p<5/2.$$

These inequalities lead us to

NOTE 3 The estimate $\|B\|_{p\leftarrow q} \leq Ch_l^{2+p-q}$ can be expected only for $|q|<1/2$, $q\leq p$.

REMARK 2 The optimal order $O(h_l^{2+2\kappa})$ can be obtained for $\|v_l\|_{U,l,s}$ (cf. (4.14)), if we choose $p=q\in(-1/2,1/2)$ and $s=p-\kappa$ (cf. Lemma 5, Note 3, (2.2a)).

But Remark 1 and Remark 2 conflict. By Remark 2 one should choose s with $s=p-\kappa<\frac{1}{2}-\kappa\leq o$. By Remark 1, it might be difficult to satisfy the assumptions of Theorem 2 and of Lemmata 4,5 for negative s. Moreover, one is usually more interested in the Euclidean norm (s=o) of the error than in some negative norm $\|\hat{u}_l-u_l^*\|_{U,l,s}$ $(s<o)$. For these practical reasons we choose s=o in the following. In order to get simpler conditions we state first the result for the non-optimal choice p=q=o.

THEOREM 4 Set $\tau:=\min(\kappa',2+\kappa)$. Assume for some $t\leq\kappa$:

- (2.1): $\|L_l^{-1}\|_{o\leftarrow -2} \leq C$, $\|L_l^{-1}\|_{t\leftarrow t-2} \leq C$,
- (2.2a): $\|f_l-L_lu_l^*\|_{F,l,o} \leq Ch_l^{\tau-2}\|u^*\|_{U,\tau}$,
- (2.2b): $\|f_l'-L_l'u_l^*\|_{F,l,-2} + h_l^t\|f_l'-L_l'u_l^*\|_{F,l,t-2} \leq Ch_l^\tau\|u^*\|_{U,\tau}$,
- (2.2c): $\|L_l'-L_l\|_{-2\leftarrow t} \leq Ch_l^t$, $\|L_l'-L_l\|_{t-2\leftarrow t} \leq C$, $\|L_l'-L_l\|_{-2\leftarrow o} \leq C$,

(4.16) $\|L_l'\|_{t-2\leftarrow o} \leq Ch_l^{-t}$, $\|S_l\|_{o\leftarrow o} \leq C$, $\|T_l\|_{o\leftarrow o} \leq Ch_l^2$,

- (4.2a,b): approximation and smoothing property.

Then for suitable ν the two-grid iteration (3.6) converges to $\hat{u}_l$ with

(4.17) $$\|\hat{u}_l - u_l^*\|_{U,l,o} \leq Ch_l^\tau\|u^*\|_{U,\tau} \qquad (\tau:=\min(\kappa',2+\kappa)).$$

Proof. Convergence follows from Theorem 2, while the error estimate is derived from Lemmata 3,4,5, from (4.13) and (4.16). ■

Theorem 4 gives the optimal result $\|\hat{u}_l-u^*_l\|_{U,l,o}=O(h_l^{\kappa'})$ for the usual cases $\kappa=1$, $\kappa'=2$ and $\kappa=2$, $\kappa'=4$. If $\kappa'>2+\kappa$, one can improve the estimate (4.17) by choosing $q=p=1/2-\varepsilon$ $(\varepsilon>o)$ instead of $p=q=o$:

THEOREM 5 Let $\varepsilon>o$ be arbitrary and set $\tau_\varepsilon:=\min(\kappa',2.5-\varepsilon+\kappa)$. Assume for some $t\in[\frac{1}{2},\kappa]$ the following conditions:

- (2.1) with $\sigma=o,t$;
- (2.2a) with $\sigma=2.5-\varepsilon$, $\tau=\tau_\varepsilon$;
- (2.2b) with $\sigma=s,t$, $\tau=\tau_\varepsilon$;
- (2.2c) with $(\sigma,\tau)=(s,t),(t,t),(s,1/2-\varepsilon)$;
- (4.2a,b) and $|L_l'\|_{t-2\leftarrow p}\leq Ch_l^{p-t}$, $\|S_l\|_{p\leftarrow p}\leq C$, $|T_l\|_{p\leftarrow p}\leq Ch_l^2$ for $p:=1/2-\varepsilon$.

Then the assertion of Theorem 4 holds with (4.18) instead of (4.17):

(4.18) $$\|\hat{u}_l-u^*_l\|_{U,l,o}\leq Ch_l^{\tau_\varepsilon}\|u^*|_{U,\tau_\varepsilon}.$$

The example of §6 shows that, in general, estimate (4.18) is optimal modulo ε.

The restriction $|q|<1/2$ of Note 3 was a consequence of Note 1: ${}^UH^q={}^FH^q$ iff $|q|<1/2$. Another situation arises if $\Omega=\mathbb{R}^d$. Then ${}^UH^q={}^FH^q$ holds for all q, and also the discrete norms $\|\cdot\|_{U,l,q}$ and $\|\cdot\|_{F,l,q}$ coincide for all q. Hence, the restriction of Note 3 does not apply for $\Omega=\mathbb{R}^d$ (case of no boundary conditions). A similar case holds for function spaces with *periodic* boundary conditions and difference operators with constant coefficients. Since the local mode analysis (cf. Stüben and Trottenberg [14]) makes use of an infinite grid $(\Omega=\mathbb{R}^d)$ or periodic boundary conditions we are led to

NOTE 4 The local mode analysis indicates an error $\|\hat{u}_l-u^*_l|_{U,l,o}$ of order $O(h_l^{\min(\kappa',2+2\kappa)})$ (cf. (4.14)).

Indeed, the example of §6 shows that the worse estimate (4.18) is caused by the presence of the boundary that is neglected by the local mode analysis.

5. Generalizations

We consider some variants of iteration (3.6). Instead of one defect correction one may perform i corrections:

(5.1) $$\begin{aligned}&u_l:=u_l^j;\\&\text{for } \mu:=1\ (1)\ \nu \text{ do } u_l:=\mathcal{S}_l(u_l,f_l);\\&\text{for } \mu:=1\ (1)\ i \text{ do } u_l:=u_l-pL_{l-1}^{-1}r(L_l'u_l-f_l');\\&u_l^{j+1}:=u_l.\end{aligned}$$

Asuming convergence $u_l^j\to\hat{u}_l$ one obtains the representation (4.7a) of the error $\hat{u}_l-u^*_l$ with

$$v_l=A^iB(f_l-L_lu^*_l) \qquad \text{(cf. (4.10))}$$

instead of (4.7b). Again, B forms a threshold. The possibly attainable estimates

$\|A^i\|_{s\leftarrow s+i\kappa} \le Ch_l^i$ can be exhausted only with $s\in(-1/2-i\kappa, 1/2+i\kappa)$ contradicting Remark 1. For s=o the order of $\|\hat{u}_l-u_l^*\|_{U,l,o}$ is the same as for i=1 [i.e. for (3.6)], although the local mode analysis would indicate $O(h_l^{\min(\kappa',2+(i+1)\kappa)})$.

If one uses a multi-grid iteration (3.4) but with defect $r*(f_k'-L_k'*u)$ for all $k\le l$, the corresponding two-grid iteration would be

$$(5.2)\qquad \begin{aligned} &u_l := u_l^j;\quad \text{for } \mu:=1\ (1)\ \nu \text{ do } u_l := \mathcal{S}_l(u_l,f_l);\\ &u_l^{j+1} := u_l - p\,L_{l-1}'^{-1}\, r\,(L_l'u_l - f_l'); \end{aligned}$$

where L_{l-1}' is assumed to be stable and of consistency order κ'. Again, the error $\hat{u}_l-u_l^*$ is given by (4.7a) with $v_l=A'B(f_l-L_lu_l^*)$, where $A'=I-pL_{l-1}'^{-1}rL_l'$. Under suitable assumptions (also on p or r) one could expect $\|A'\|_{s\leftarrow s+\kappa'}\le Ch_l^{\kappa'}$. As in case of (5.1) the estimate (4.18) cannot be improved, whereas the local mode analysis indicates $O(h_l^{\kappa'})$.

Another variant results if smoothing step and coarse-grid correction are interchanged:

$$(5.3)\qquad \begin{aligned} &u_l := u_l^j - pL_{l-1}^{-1}r(L_l'u_l^j - f_l');\\ &\text{for } \mu:=1\ (1)\ \nu \text{ do } u_l := \mathcal{S}_l(u_l,f_l);\\ &u_l^{j+1} := u_l; \end{aligned}$$

Now, the representation $\hat{u}_l-u_l^* = (I-M_l')^{-1}(v_l+w_l)$ holds with

$$v_l=B(f_l-L_lu_l^*),\quad w_l=S_l^{\nu}pL_{l-1}^{-1}r(f_l'-L_l'u_l^*)\qquad \text{(B from (4.10) or (4.13))}.$$

Therefore, it is possible to obtain

$$\|\hat{u}_l - u_l^*\|_{U,l,o} \le C\,h_l^{\min(\kappa',2+\kappa)}$$

as in (4.17), but (4.18) does not hold.

Note 3 holds only for Dirichlet boundary value problems of second order. For a Neumann problem of second order we may expect $\|\hat{u}_l-u_l^*\|_{U,l,o}=O(h_l^{\min(\kappa',3.5+\kappa-\varepsilon)})$, while for a Dirichlet problem of order 2m the estimate may be or order $O(h_l^{\min(\kappa',m+1.5+\kappa-\varepsilon)})$.

The restrictions of Note 3 can be overcome by introducing inhomogeneous boundary conditions. Instead of $U_{H^s}=H^s(\Omega)\cap H_0^1(\Omega)$ $(s\ge 1)$ use $U_{H^s}=H^s(\Omega)\times H^{s-1/2}(\Gamma)$. This is not only another theoretical framework. The discretizations (1.2a,b) are to be changed so that the difference equations corresponding to $Lu=f$ (in Ω) are separated from discrete equations corresponding to $u=g$ (on Γ).

6. Example

Let $\Omega=[o,1]$, $Lu=-u''+u'$, $f=-1-2x$. Choose L_l and L_l' by

$$(L_lu_l)(x) = h_l^{-2}[-(1+h_l)\,u_l(x-h_l) + (2+h_l)\,u_l(x) - u_l(x+h_l)],\qquad \kappa=1,$$

$$(L_l'u_l)(x) = h_l^{-2}[-(1+\tfrac{h_l}{2})\,u_l(x-h_l) + 2\,u_l(x) - (1-\tfrac{h_l}{2})\,u_l(x+h_l)],\qquad \kappa'=2.$$

$R_l^F = R_l'^F$ is defined by $(R_l^F f)(x) = h_l^{-1} \cdot \int_{x-h_l/2}^{x+h_l/2} f(\xi)d\xi$, whereas R_l^U may be chosen as trivial injection: $(R_l^U u)(x) = u(x)$ for $x=h_l, 2h_l, \ldots, 1-h_l$.

Since $u^*(x) = x(1-x)$, the defects are

$$f_l' - L_l' u_l^* = 0 \quad \text{and} \quad f_l - L_l u_l^* = -h_l \mathbb{1} \quad \text{with } \mathbb{1}(x)=1.$$

Therefore, w_l from (4.7c) vanishes. The error $\hat{u}_l - u_l^* = (I-M_l')^{-1} v_l$ depends only on v_l:

$$\|\hat{u}_l - u_l^*\|_{U,l,o} \geq \frac{1}{1+\rho'} \|v_l\|_{U,\ ,o} \geq \frac{1}{2} \|v_l\|_{U,l,o}.$$

The *explicit* calculation of v_l for the two-grid iteration (3.6) with $\nu=2$ and $\mathcal{S}_k$ defined by (3.3) with $\omega = 1/4$ shows

$$(6.1) \qquad \|v_l\|_{U,l,o} = \frac{\sqrt{18}}{16} h_l^{3.5} + O(h_l^4)$$

confirming that the leading error is $O(h_l^{2.5-\varepsilon+\kappa})$ and not $O(h_l^{2+2\kappa}) = O(h_l^4)$. For weaker norms better estimates result:

$$\|v_l\|_{U,l,-1/2} = O(h_l^4) \qquad \text{(cf. Remark 2).}$$

The norm $\|\cdot\|_{U,l,o}$ coincides with $\|\cdot\|_{l_2}$, where

$$\|w_l\|_{l_p} = [h_l \Sigma_{\nu=1}^{h_l^{-1}-1} |w_l(\nu h_l)|^p]^{1/p}, \quad \|w_l\|_{l_\infty} = \max_\nu |w_l(\nu h_l)|.$$

For the l_p norms the results are

$$(6.2) \qquad \|v_l\|_{l_p} = (3 \cdot 2^{1/p}/16)\, h_l^{3-1/p} + O(h_l^4), \qquad \|v_l\|_{l_\infty} = \frac{3}{16} h_l^3 + O(h_l^4).$$

The variant (5.2) yields the estimates (6.1) and (6.2) with the *same* leading terms, but other remainders $O(h_l^4)$.

References

[1] BÖHMER, K.: *Discrete Newton methods and iterated defect corrections.* Numer. Math. *37* (1981) 167 - 192.

[2] BRANDT, A.: *Numerical stability and fast solutions to boundary value problems.* In: Boundary and Interior Layers (J.J.H.Miller, ed.), Boule Press, Dublin 198o

[3] FRANK, R. and J. HERTLING: *Die Anwendung der Iterierten Defektkorrektur auf das Dirichletproblem.* Beiträge zur Numerischen Mathematik *7* (1979) 19 - 31.

[4] FRANK, R. and C. UEBERHUBER: *Iterated defect correction for differential equations - part I: theoretical results.* Computing *2o* (1978) 2o7 - 228.

[5] HACKBUSCH, W.: *Bemerkungen zur iterierten Defektkorrektur und zu ihrer Kombination mit Mehrgitterverfahren.* Rev. Roumaine Math. Pures Appl. *26* (1981) 1319 - 1329.

[6] HACKBUSCH, W.: *Multi-grid convergence theory.* This volume.

[7] HACKBUSCH, W.: *Convergence of multi-grid iterations applied to difference equations.* Math. Comp. *34* (198o) 425 - 44o.

[8] HACKBUSCH, W.: *On the regularity of difference schemes.* Arkiv för Matematik *19* (1981) 71 - 95. - *Part II* to appear

[9] HAIRER, E.: *On the order of iterated defect correction.* Numer. Math. *29* (1978) 409 - 424.

[10] HEMKER, P.: *Mixed defect correction iteration for the accurate solution of the convection diffusion equation.* This volume.

[11] LIN QUN: *Iterative corrections for nonlinear eigenvalue problem of operator equations.* Research Report IMS-1, Academia Sinica, Beijing, 1981.

[12] STETTER, H.J.: *The defect correction principle and discretization methods.* Numer. Math. *29* (1978) 425 - 443.

[13] AUZINGER, W. and H. J. STETTER: *Defect corrections and multi-grid iterations.* This volume.

[14] STÜBEN, K. and U. TROTTENBERG: *Multigrid method: fundamental algorithm, model problem analysis, and applications.* This volume.

[15] WHITEMAN, J. R. (ed.): *The Mathematics of Finite Elements and Applications IV, MAFELAP 1981.* To be published by Academic Press, London.

Adaptive-grid methods for time-dependent partial differential equations

G. W. Hedstrom and G. H. Rodrique
Lawrence Livermore National Laboratory, University of California
Livermore, California 94550

Abstract

This paper contains a survey of recent developments of adaptive-grid algorithms for time-dependent partial differential equations. Two lines of research are discussed. One involves the automatic selection of moving grids to follow propagating waves. The other is based on stationary grids but uses local mesh refinement in both space and time. Advantages and disadvantages of both approaches are discussed. The development of adaptive-grid schemes shows promise of greatly increasing our ability to solve problems in several spacial dimensions.

1. Introduction

It has been common practice for many years to use adaptive grids and methods of varying orders of accuracy in the integration of ordinary differential equations [11]. Until recently, this has not been the case for time-dependent partial differential equations, but it seems likely that multidimensional computations will be very costly or even impossible without the development of adaptive-grid methods. In this paper we survey briefly two lines of current research, and we suggest some promising future developments.

Hyperbolic partial differential equations are known to exhibit propagating waves [22], of both characteristic and subcharacteristic nature. Consequently, numerical computation on a fixed grid tends to be costly - a fine grid is required everywhere. Admittedly, it is not always necessary to resolve the behavior in the wave front, and special methods have been developed for such cases. We omit these special problems from our discussion, and hence, we exclude shock tracking as done, for example, by deNeef [8], Zhu [28], and Salas [17]. We also exclude special algorithms for computing shocks and contact discontinuities on fixed grids, such as the methods of Glimm [7], Boris and Book [4,5], and Woodward and Colella [25]. We concentrate on problems for which the solution must be computed accurately in the frontal region, such as chemical combustion [9] and the accretion of matter to form a star [21, 24].

One approach to adaptive grids is to use a fixed number of grid points and to let them move with whatever fronts are present. We discuss work in this direction in Section 2, including the work of Dwyer and his coworkers [9] on combustion and the work of Winkler [21, 24] on star and galaxy formation. We also describe the general moving-grid algorithm of Miller and his coworkers [12, 14, 15]. The other approach is the use of local mesh refinement, and in Section 3 we describe the work of Bolstad [3] and Berger and Gropp [1, 2, 13]. In Section 4 we point out the respective advantages of the moving-grid and local-mesh-refinement approaches, and we point out some directions for future development.

2. Moving-grid schemes

The most direct way to set up a moving grid is to do it via a coordinate transformation. Thus, with the notation $\partial_x = \partial/\partial_x$, a hyperbolic system

$$\partial_t u + \partial_x f(u) = F(u) \tag{2.1}$$

may be transformed into

$$\partial_t u + \partial_\xi u \, \partial_t g + \partial_\xi f(u) \, \partial_x g = F(u) \tag{2.2}$$

by means of the transformation $\xi = g(x,t)$. Extensions to higher-order equations and to several dimensions are obvious. The numerical method for solving (2.1) then consists of a construction somehow of the mapping function g and a finite-difference or finite-element approximation to (2.2) on a fixed, uniform grid. In typical applications of this method the (ξ,t)-coordinate system follows a traveling wave, so that $\partial_t u$ in (2.2) is small. Consequently, it is desirable to use an implicit method for (2.2) with long time steps and to couple the construction of g directly with the solution of (2.2), and all of the methods described in this section do so.

Implementations of the moving-grid algorithm differ in their choice of a difference scheme for (2.2), but more importantly, they differ in the algorithm used to define g. One possible choice is to use Lagrangian coordinates, and this is useful if we want to follow the motion of the material. The criteria for g are problem dependent, and so far, no robust algorithm for its construction exists. It is possible to make good choices of g in special problems, however, and in this section we present two such special problems. We begin with Dwyer's method for combustion [9], and then we discuss Winkler's method for computing the formation of stars [21, 24]. We close this section with a description of Miller's method [12, 14, 15], which is more general than the others.

Let us mention here that several authors, including Rai and Anderson [16] and Yanenko and his coworkers [26, 27] have used similar ideas in fluid-dynamical computations. We do not expand on this application because there is some validity to the claim that in such problems there is no need to put many grid points in the wave fronts, and that special difference schemes on fixed grids are adequate.

In one dimension the equations of a combusting gas mixture are a modification of (2.1) in that second-derivative terms representing viscosity and heat diffusion must be added [23, pp. 2-4]. From a qualitative point of view it is clear that it is a good idea to select a grid which is stationary with respect to the flame front. Thus, we need to know the flame speed accurately. Now, the flame speed depends strongly on the behavior of the solution in the flame region. Consequently, we need a fine grid in the flame region. Combustion problems have the special property that the rate of chemical rection depends very strongly on the temperature [23, p.4]. In fact, a flame front may be identified numerically as a region of large temperature gradient. In the work of Dwyer [9] the basic idea is to select the mapping $\xi = g(x,t)$ so as to make $|\partial_\xi T|$ nearly constant, where T denotes the temperature. Some variation in $|\partial_\xi T|$ must be permitted, for otherwise, there would be no grid points in regions of constant temperature, such as might be present in large unburned regions. It is also noted in [9] that in flame-initation regions, it may happen that u is changing rapidly, even though $|\partial_x T|$ is small. Therefore, [9] recommends monitoring of a linear combination of $|\partial_\xi T|$ and $|\partial_\xi^2 T|$ in such flames.

Some two-dimensional computations with this method are also reported in [9], but only in some special, essentially one-dimensional geometries. A set of fixed grid lines is drawn so that they will be approximately orthogonal to the flame front throughout its course, and the one-dimensional moving-grid algorithm is used on each of these grid lines. Examples of geometries where this approach is inadequate are also given in [9].

Another application of the moving-grid method (2.2) is the computation of the formation of a star or galaxy [21, 24]. We begin with a description of the physical processes leading to the formation of a star, which is assumed to be spherically symmetric. We have an extended, diffuse cloud of matter which is being pulled by the force of gravity onto a protostar at the center. As the protostar grows, it becomes hotter and denser, until thermonuclear fusion begins. Since we are interested only in the growth of the protostar, we want to use a grid which moves with the surface of the protostar and not with the falling particles [24]. The speed of the moving surface depends on the rate of accumulation of falling matter, on the gravitational forces acting the protostar, and on the outward pressure of the light radiated by the protostar.

Because several processes are important, it is not sufficient to select the mapping g by monitoring changes in a single variable. In fact, Winkler [24] chooses g so as to make

$$\sum \log^2 (|\partial_\xi u_i|/|u_i|)$$

nearly constant as a function of ξ. The need for the relative rates $|\partial_\xi u_i|/|u_i|$ is dictated by the extremely large changes in such things as particle velocity, density, and temperature across the surface of the protostar. The choice of $\log^2$ came about through numerical experience [24]. It is clear that Winkler's mapping algorithm is a generalization of Dwyers [9].

In [21] Tscharnuter and Winkler report on a three-dimensional computation using a modification of this method to study the evolution of a galaxy. A spherical coordinate system is used with a fixed, uniform grid for the angles. Along the radial lines moving one-dimensional grids are chosen as in the protostar computation. Thus, the method is basically one-dimensional, and spiral galaxies are not treated.

Let us comment on the choice of criterion for the construction of the mapping function in [9] and [21, 24]. From the point of view of a numerical analyst, the grid should be fine at points where the local truncation error is large, not at points where $|\partial_x u|$ or $\sum|\partial_x u_i|/|u_i|$ is large. Because of the high degree of nonlinearity in the problems treated in [9, 21, 24], $|\partial_x u|$ is probably a pretty good estimator of the local truncation error. It is easy, though, to find examples of problems in which $|\partial_x u|$ is a poor estimate of the local truncation error. One such example is any second-order accurate scheme for a linear, hyperbolic equation. Similarly, a glance at the figures from Sod's gas-dynamical computations [18] shows that in the middle of a rarefaction wave $|\partial_x u|$ may be large, but the error is small. This point should be kept in mind when the methods of [9, 21, 24] are applied in other settings.

We turn now to the moving grid method of Keith Miller and his coworkers [12, 14, 15], which does use the local truncation error as a consideration in the construction of the grid. In this method an approximate solution to (2.1) is written as a piecewise linear spline in x over a grid which varies in time. The spline u and the grid are chosen so as to minimize

$$||\partial_t u + \partial_u f(u) - F(u)||^2 + \text{penalty} \quad . \tag{2.3}$$

The Euler equation for minimizing (2.3) is a discretization of (2.2), coupled to a discretization of a partial differential equation

$$L(u, \partial_t, \partial_\xi)G = 0 \tag{2.4}$$

for the function G such that $x = G(\xi,t)$ is the inverse mapping to $\xi = g(x,t)$.

As is explained in [14], a penalty function is needed in (2.3) for two reasons. One reason is that the method is ill posed without a penalty. In fact, without the penalty, the line $t = t_0$ is tangent to a characteristic curve for (2.2), (2.4) at every point (x_0, t_0) at which $\partial_x^2 u = 0$. The paper [14] gives examples of nonuniqueness of solutions. The other reason for the penalty

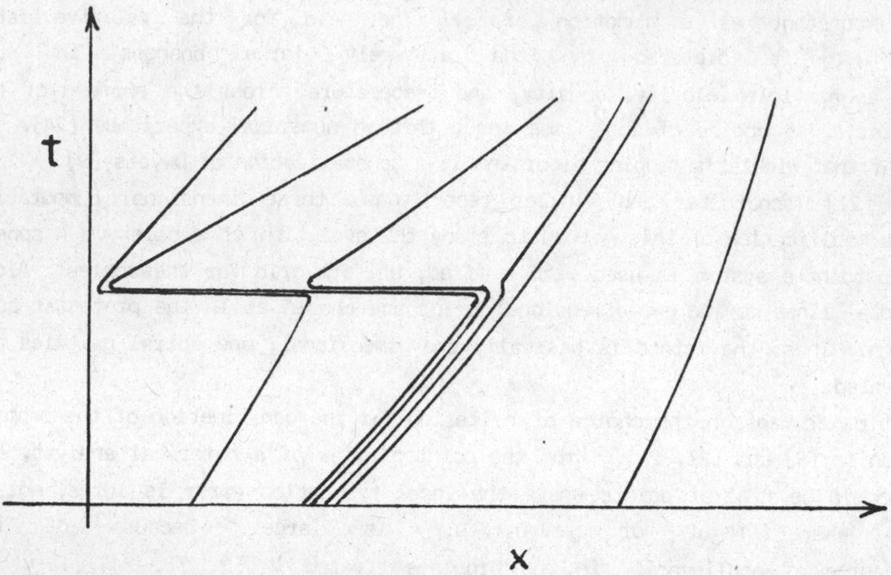

Figure 1. Moving grids

function in (2.3) is that without it, for a scalar equation (2.1) with F = 0 the (ξ,t)-coordinates follow the characteristic curves of (2.1). Such behavior has some advantages, of course, but it also means that grid points accumulate at a shock. Thus, the penalty function also serves to keep the grid points sufficiently well separated. Examples of penalty functions for specific problems are given in [12, 14, 15].

One advantage of Miller's method over that of Dwyer and Winkler is that Miller's method may be extended directly to higher dimensions. Much work remains to be done on the choice of penalty functions in higher dimensions, however. Another advantage of Miller's method is that the transformation of the equation from (2.1) to (2.2) is not done explicitly, but comes about automatically. This can be very useful for higher-order equations in several dimensions.

The primary disadvantage of the methods discussed in this section is that the number of grid points is fixed throughout the entire course of the computation. Thus, if the grid is following one wave front and another one arises somewhere, no new grid is created for the new wave, but rather the old grid has to adjust itself abruptly. A situation of this sort would occur in a combustion problem if a rarefaction wave were to enter from a boundary after the flame had started. A representative grid for this problem is shown in Figure 1.

In fact, this rigidity in the number of grid points is not inherent in the method but is primarily a question of computer science, namely, facility in the manipulation of data structures. The difficulty arises from the fact that, so far,

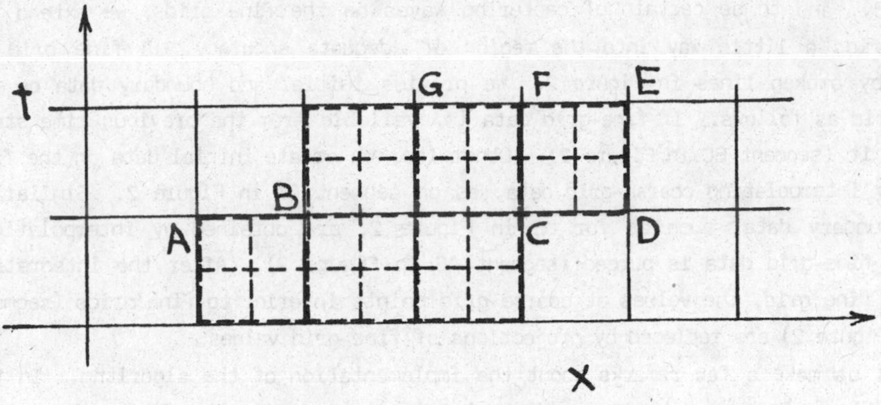

Figure 2. Local mesh refinement

the computer programs for the method are all written in FORTRAN, a language in which the manipulation of data structures is extremely awkward. It would be far easier to add and delete grids if the program were written in a language such as PASCAL, SIMULA, or possibly VAL.

3. Local mesh refinement

In this section we describe the local-mesh-refinement algorithm of Oliger and his students, Berger, Bolstad, and Gropp. This method does use a flexible data structure. Descriptions of the method in the open literature may be found in the works of Gropp [13] and of Berger, Gropp, and Oliger [2]. The theses of Berger [1] and Bolstad [3] will contain further discussion of the method, but they are not yet finished. We should point out that the publication dates for these papers do not correspond exactly with the chronological order in which the work was done. In fact, Bolstad's work [3] predates the others. It is for one-dimensional problems, and its data structures cannot be used in two dimensions. Gropp's paper [13] is a test of whether the method is reasonable at all in two dimensions. In the later papers [1, 2] the test for refinement is improved and the fine grids are permitted to have arbitrary orientation.

Let us briefly describe one time step of a one-dimensional version of the algorithm. Suppose that an approximate solution u is given at time $t = t_0$. We first compute the solution at time $t = t_0 + \Delta t$ by using a difference scheme on a coarse grid over the whole domain. This grid is denoted by solid lines in Figure 2. We then locate regions of low accuracy, using estimates of local truncation error, for example. We create fine grids on these regions of low accuracy, and to be certain of capturing waves on the fine grids, we extend the fine grids a little way into the region of adequate accuracy. A fine grid is shown by broken lines in Figure 2. We provide initial and boundary data on the fine grid as follows. If fine-grid data is available from the previous time step, we use it (segment BC in Figure 2). Otherwise, we create initial data on the fine grid by interpolating coarse-grid data, as on segment CD in Figure 2. Similarly, the boundary data, such as for DE in Figure 2, are obtained by interpolation. Unused fine-grid data is purged (segment AB in Figure 2). After the integration on the fine grid, the values at coarse grid points interior to fine grids (segment GF in Figure 2) are replaced by projections of fine-grid values.

Let us make a few remarks about the implementation of the algorithm. In the computations done to date the ratio $\Delta t/\Delta x$ is the same on the fine grids as on the coarse grid. This is because the test problems have all involved explicit methods for hyperbolic equations. Note that the algorithm is recursive, so that it is possible to have many levels of refinement. Experience indicates [1, 3] that three levels is a good number with Δx on each fine grid taken to be 1/4 of the Δx for the next coarser grid. We also point out that at any time level there may be any number of fine grids. (Only a certain number will fit, of

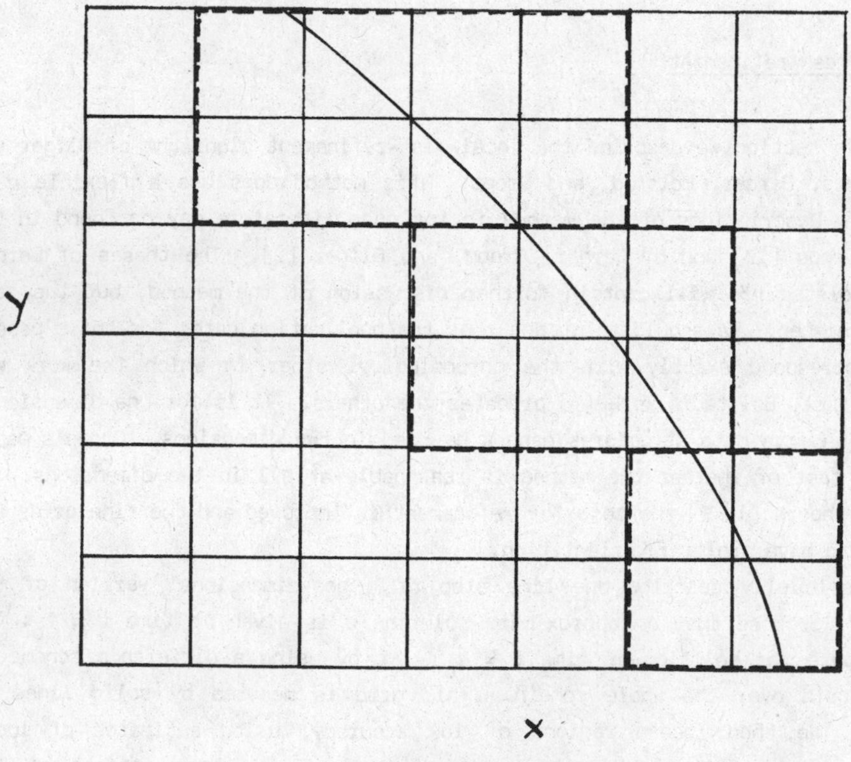

Figure 3. Parallel fine grids

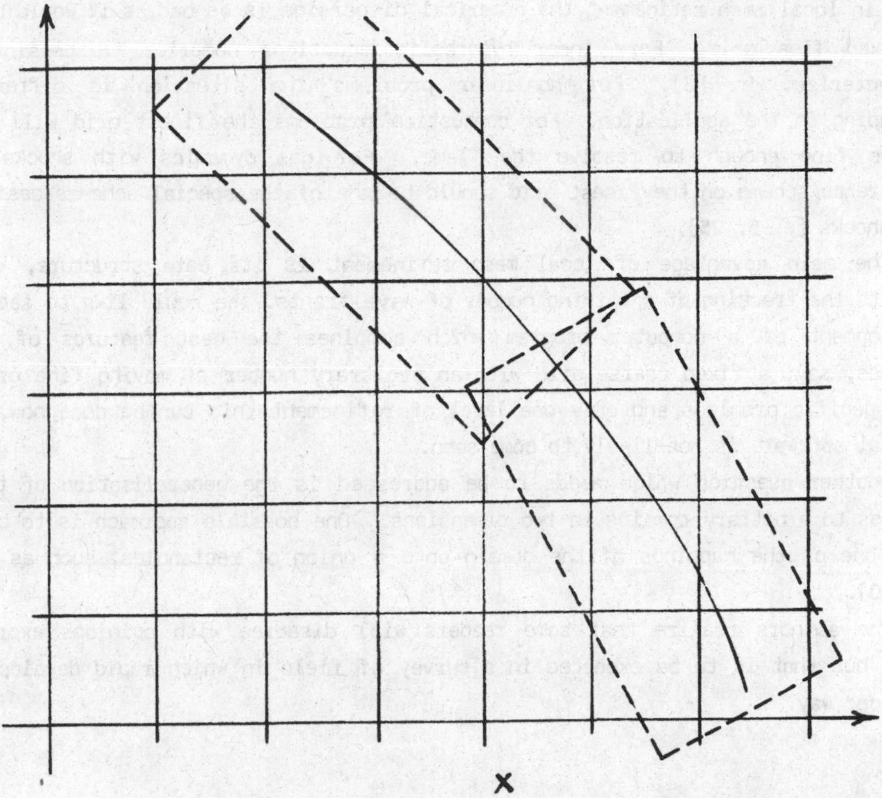

Figure 4. Fine grids with arbitrary orientation

course.) Consequently, the data structure is a tree, and it is appropriate to organize it via a linked list. Finally, we remark that on a multiprocessor the integrations on the different fine grids may be carried out simultaneously on different components.

The method has been extended to two dimensions in two different ways. Gropp [13] aligned the fine grids with the coarse grid, and Berger [1] permits fine grids with arbitrary orientation. Gropp's approach requires a larger number of fine grids to cover a wave front moving skew to the grid. See Figure 3. Berger's approach requires fewer fine grids, but the interpolation is much more complicated, and grid overlap causes some difficulty. See Figure 4. It is not yet clear which approach is better, although some people have strong opinions on the matter.

4. Conclusions

The principal advantage of moving-grid methods over local mesh refinement is that in local mesh refinement the numerical dispersion is as bad as it would be on a fixed fine grid. For linear hyperbolic equations numerical dispersion is characterized in [6]. For nonlinear problems the situation is different, depending on the application. For combustion problems the finest grid will have to be fine enough to resolve the flame. For gas dynamics with shocks the difference scheme on the finest grid should be one of the special schemes designed for shocks [4, 5, 25].

The main advantage of local mesh refinement is its data structure, which permits the tracking of a varying number of wave fronts. We would like to see the development of a computer program which combines the best features of both methods, say, a fixed coarse grid with an arbitrary number of moving fine grids. For specific problems and only one level of refinement this can be done now, but general software is not likely to come soon.

Another question which needs to be addressed is the generalization of these methods to arbitrary domains in two dimensions. One possible approach is to begin with one of the happings of the domain onto a union of rectangles, such as [10, 19, 20].

The authors realize that some readers will disagree with opinions expresed here, but that is to be expected in a survey of field in which rapid development is under way.

Acknowledgment

The authors thank the U. S. Department of Energy Office of Basic Energy Sciences, Applied Mathematics and Statistics Division. Work performed under the auspices of the U.S. Department of Energy by the Lawrence Livermore National Laboratory under contract No. W-7405-ENG-48.

References

1. Marsha Berger, Ph.D. Thesis, Stanford University, in prepartion.

2. Marsha Berger, W. D. Gropp, and Joseph Oliger, Grid Generation for time-dependent problems: Criteria and methods, Numerical grid generation techniques, NASA Conference Publication 2166, October 1981.

3. John Boltstad, Ph.D. Thesis, Stanford University, in preparation.

4. J. P. Boris, Flux-corrected transport modules for solving generalized continuity equations, NRL Memorandum Report 3237, March 1976.

5. J. P. Boris and D. L. Book, Solution of continuity equations by the method of flux-corrected transport, in Methods in Computational Physics, Vol. 16, Ed. by J. Killeen (Academic Press, Inc., New York, 1976).

6. R.C.Y. Chin and G. W. Hedstrom, A dispersion analysis for difference schemes: Tables of generalized Airy functions, Math. Comput. 32 (1978), pp. 1163-1170.

7. Alexandre Joel Chorin, Random choice solution of hyperbolic systems, J. Comput. Phys. 22 (1976), pp. 517-533.

8. Tom DeNeef and Charles Hechtman, Numerical study of flow due to a spherical implosion, Comput. Fluids 6 (1978), pp. 185-202.

9. H. A. Dwyer, F. Raiszadeh, and G. Otey, A study of reactive diffusion problems with stiff integrators and adaptive grids, Lecture Notes in Physics, Vol. 141, pp. 170-175, Springer-Verlag, New York, 1981.

10. Peter Eiseman, A multi-surface method of coordinate generation, J. Comput. Phys. 33 (1979), pp. 118-150.

11. C. William Gear, Numerical initial value problems in ordinary differential equations, Prentice-Hall, Englemwood Cliffs, New Jersey 1971.

12. R. J. Gelinas, S. K. Doss, and Keith Miller, The moving finite element method: Application to general partial differential equations with multiple large gradients, J. Comput. Phys. 40 (1981), pp. 202-249.

13. William D. Gropp, A test of moving mesh refinement for 2-D scalar hyperbolic problems, SIAM J. Sci. Stat. Comput. 1 (1980), pp. 191-197.

14. Keith Miller and Robert N. Miller, Moving finite elements. I, SIAM J. Numer. Anal. 18 (1981), pp. 1019-1032.

15. Keith Miller, Moving finite elements. II, SIAM J. Numer. Anal. 18 (1981), pp. 1033-1057.

16. Man Mohan Rai and Dale A. Anderson, The use of adaptive grids in conjunction with shock capturing methods, AIAA Paper 81-1012, Proceedings of Computational Fluid Dynamics Conference, 1981.

17. M. D. Salas, Shock fitting method for complicated two-dimensional supersonic flows, AIAA J. 14 (1976), pp. 583-588.

18. Gary A. Sod, A survey of several finite difference methods for systems of nonlinear hyperbolic conservation laws, J. Comput. Phys. 27 (1978), pp. 1-31.

19. Göran Starius, Composite mesh difference methods for elliptic boundary value problems, Numer. Math. 28 (1977), pp. 243-258.

20. Joe F. Thompson, Frank C. Thames, and C. Wayne Mastin, TOMCAT - a code for numerical generation of boundary-fitted curvilinear coordinate systems on fields containing any number of arbitrary two-dimensional bodies, J. Comput. Phys. 24 (1977), pp. 274-302.

21. W. M. Tscharnuter and K.-H. Winkler, A method for computing selfgravitating gas flows with radiation, Compt. Phys. Comm. 18 (1979), pp. 171-199.

22. G. B. Whitham, Linear and nonlinear waves, Wiley, New York, 1974.

23. Forman Williams, Combustion Theory, Addison-Wesley, Rading, Massachusetts, 1965.

24. K.-H. Winkler, Über ein numerisches Verfahren zur Berechnung instationärer sphärischer Stossfronten mit Strahlung, Max Planck Institute for Physics and Astrophysics, Report MPI-PAE/ASTRO 90, Munich, 1976.

25. Paul Woodward and Phillip Colella, High resolution difference schemes for compressible gas dynamics, Lecture Notes in Physics, Vol. 141, pp. 434-441, Springer-Verlag, Berlin-Heidelberg-New York, 1981.

26. N. N. Yanenko, E. A. Kroshko, V. V. Liseikin, V. M. Fomin, V. P. Shapeev, and Yu. A. Shitov, Methods for the construction of moving grids for problems of fluid dynamics with big deformations, Lecture Notes in Physics, Vol. 59, pp. 454-459, Springer-Verlag, New York, 1976.

27. N. N. Yanenko, V. M. Kovenya, V. D. Lisejkin, V. M. Formin, and E. V. Vorozhtsov, On some methods for the numerical simulation of flows with complex structure, Comput. Meth. Appl. Mech. Eng. 17/18 (1979), pp. 659-671.

28. Zhu Youlan, Chen Bingmu, Zhang Zuomin, Zhong Xichang, Qin Boliang, and Zhang Guanquan, difference schemes for initial-boundary value problems of hyperbolic systems and examples of application, Sci. Sinica, Special issue II (1979), pp. 261-278.

MIXED DEFECT CORRECTION ITERATION FOR THE ACCURATE SOLUTION OF THE CONVECTION DIFFUSION EQUATION

P.W. Hemker
Mathematical Centre
Amsterdam, The Netherlands

1. INTRODUCTION

In this paper we study properties of an iteration scheme for the numerical solution of the convection-diffusion equation

$$(1.1) \qquad L_\varepsilon u \equiv -\varepsilon\Delta u + \vec{a}.\nabla u = f,$$

in two dimensions, in particular in the case of a small diffusion coefficient ε. For small ε the equation is dominated by the convection term and boundary- or interior layers may appear along or downstream the convection direction $\vec{a}$. If the meshwidth of a discretization h is small compared with the thickness of these layers, standard methods for the numerical solution of (1.1) may fail to be efficient. Accurate methods such as finite element discretizations with piecewise polynomials become unstable for small ε/h, whereas stable methods, such as obtained by upwinding or application of artificial diffusion, are only 1st order accurate.

A proper mesh-refinement in the boundary layers may solve many of these problems in practice, but it requires a priori knowledge about the location and the shape of the boundary layer. For the automatic solution of (1.1) we seek a method (i) which does not make use of a-priori knowledge about the solution, (ii) of which the discretization is independent of a convection direction $\vec{a}$, (iii) which is accurate $\mathcal{O}(h^2)$ in the smooth parts of the solution and (iv) which locates the boundary and interior layers properly. Further we want that the width of the numerical boundary layer is at most $\mathcal{O}(h)$ for small values of ε/h. Such a method is suited for application in an algorithm which resolves the boundary layers by adaptive mesh-refinement.

We propose an iteration scheme that satisfies the above requirements and that can readily be incorporated in an iteration scheme for the solution of the discretized system. The method is based on an idea by BRANDT (1980) to increase the order of accuracy of a discrete solution in a multilevel algorithm by computing the residual - for transfer to a coarser grid - relative to a more accurately discretized operator than the operator that is used for the relaxation. This idea is very much related to the defect correction principle (see e.g. STETTER (1978), HACKBUSCH (1979), HEMKER (1981, 1982)).

Although we apply our method also to problems (1.1) with variable coefficients $\vec{a} = \vec{a}(x,y)$ in section 5, we restrict the analysis to the problem with constant coefficients. In order to show some details of a computation we also resort to the simple one-dimensional problem

$$(1.2)\qquad \varepsilon y'' + 2y' = f.$$

As was indicated by BRANDT (1980), the one-dimensional problem has properties that cannot be generalized to more dimensions. However, some basic techniques that are used in the 2-D case are more easily shown with the 1-D example.

In this paper the same method is used for the solution of (1.1) as was used in HEMKER (1982). In the present paper we extend previous results with a treatment of the convergence properties of the iteration and of the behaviour of the solution in the neighbourhood of the boundary. Also, some additional numerical results are presented.

2. THE ELEMENTARY DISCRETIZATIONS

For the discretization of the equation (1.1) we essentially use only a simple finite element or finite difference discretization. The analysis is made for the discretization on a regular square grid. In this case the difference stars are given by

$$(2.1)\qquad L_{h,\varepsilon} = \frac{-\varepsilon}{h^2}\begin{bmatrix} & 1 & \\ 1 & -4 & 1 \\ & 1 & \end{bmatrix} + \frac{a_1}{(4+2p)h}\begin{bmatrix} -p & +p & \\ -2 & 0 & 2 \\ -p & p & \end{bmatrix} + \frac{a_2}{(4+2p)h}\begin{bmatrix} & +2 & p \\ p & 0 & -p \\ -p & -2 & \end{bmatrix}.$$

With $p = 0$ it corresponds to the central difference discretization; with $p = 1$ to the finite element discretization with piecewise linear test and trial functions. The discretization operator is used either with the given diffusion coefficient ε or with this coefficient replaced by an artificially enlarged diffusion coefficient $\alpha = \varepsilon + Ch$, where C is independent of ε and h.

Analogous to (2.1), for the one-dimensional problem we have

$$L_{h,\varepsilon} = \frac{+\varepsilon}{h^2}[1,-2,1] + \frac{2}{h}[-\tfrac{1}{2},0,\tfrac{1}{2}]$$

For the 1-D problem the discretization with artificial diffusion $\alpha = \varepsilon + h$ is equivalent with the usual upwind discretization.

3. MIXED DEFECT CORRECTION ITERATION

The iteration scheme we propose is a special case of the following "mixed defect correction iteration" - scheme (MDCP):

$$\begin{cases} \tilde{L}_h^1 u_h^{(i+\frac{1}{2})} = \tilde{L}_h^1 u_h^{(i)} - L_h^1 u_h^{(i)} + f_h , & (3.1a) \\ \tilde{L}_h^2 u_h^{(i+1)} = \tilde{L}_h^2 u_h^{(i+\frac{1}{2})} - L_h^2 u_h^{(i+\frac{1}{2})} + f_h. & (3.1b) \end{cases}$$

In this process the operators L_h^1, L_h^2, $\tilde{L}_h^1$ and $\tilde{L}_h^2$ are discretizations of the operator L in the continuous problem

$$Lu = f. \qquad (3.2)$$

For the process (3.1) the following theorem is easily proved:

(3.3) <u>THEOREM</u>. *Let $\tilde{L}_h^1$ and $\tilde{L}_h^2$ satisfy the stability condition:*

$$\| \tilde{L}_h^{-1} \| < C, \text{ uniform in } h,$$

and let $L_h^k u_h = f_h$ and $\tilde{L}_h^k u_h = f_h$ be discretizations of order p_k and $q_k \le p_k$ respectively, $k = 1,2$. If for (3.1) a stationary solution

$$u_h^A = \lim_{i\to\infty} u_h^{(i)}$$

exists, then

$$\| R_h u - u_h^A \| \le C\, h^{\min(p_1+q_2,p_2)} ,$$

where $R_h u$ is the restriction of the solution u of (3.2) to the grid. □

For the convection diffusion equation (1.1) we make the following choice of operators in (3.1):

$$\begin{aligned} L_h^1 &= L_{h,\varepsilon} , & \tilde{L}_h^1 &= L_{h,\alpha}, \\ L_h^2 &= L_{h,\alpha} , & \tilde{L}_h^2 &= D_{h,\alpha} = 2 \text{ diag } (L_{h,\alpha}). \end{aligned} \qquad (3.4)$$

By these choices, (3.1a) is a defect correction step towards the 2nd order accurate solution of $L_{h,\varepsilon}\, u_h = f_h$, by means of the operator $L_{h,\alpha}$ as an approximation to $L_{h,\varepsilon}$. The second step (3.1b) is only a damped Jacobi-relaxation step towards the solution of the problem

$$L_{h,\alpha}\, u_h = f_h.$$

In the process (3.1)-(3.4) only linear systems for the discrete operator $L_{h,\alpha}$ have to be solved explicitly. For the treatment of these equations the multiple grid method can be used. For details about this solution method we refer to VAN ASSELT (these proceedings). In this paper we shall only be concerned with the convergence of the iteration process (3.1)-(3.4) and with the properties of its fixed points (the "*stationary solutions*" of (3.1)).

After substitution of the operators (3.4) in the process (3.1), its two stationary solutions - if they exist -

$$u_h^A = \lim_{i\to\infty} u_h^{(i)} \quad \text{and} \quad u_h^B = \lim_{i\to\infty} u_h^{(i+\frac{1}{2})},$$

can be characterized as solutions of the linear equations

$$(3.5) \qquad [L_{h,\varepsilon} + (\alpha-\varepsilon)L_{h,\alpha}D_{h,\alpha}^{-1}\Delta_h]u_h^A = f_h,$$

$$(3.6) \qquad [L_{h,\varepsilon} + (\alpha-\varepsilon)\Delta_h D_{h,\alpha}^{-1} L_{h,\alpha}]u_h^B = [I + (\alpha-\varepsilon)\Delta_h D_{h,\alpha}^{-1}]f_h.$$

For a brief notation, we denote (3.5) also as

$$(3.7) \qquad M_{h,\varepsilon}\, u_h^A = f_h.$$

By means of theorem (3.2), it is easily shown that, for a fixed ε and $h \to 0$, the solution u_h^A is 1st order accurate, whereas u_h^B is 2nd order. With the aid of local mode analysis the behaviour of the solution can be analyzed more precisely.

4. LOCAL MODE ANALYSIS

To analyze the process (3.1)-(3.4) and its stationary solutions u_h^A and u_h^B, we use local mode analysis. For the one-dimensional model problem (1.2) the characteristic forms of the operators (3.4) are given by

$$(4.1) \qquad \hat{L}_{h,\varepsilon}(\omega) = \frac{-4\varepsilon}{h^2} S^2 + \frac{4i}{h} SC \text{ and } D_{h,\alpha}(\omega) = \frac{-4\alpha}{h^2},$$

with $S = \sin(\omega h/2)$ and $C = \cos(\omega h/2)$. For $M_{h,\varepsilon}$ this results in the characteristic form

$$\hat{M}_{h,\varepsilon}(\omega) = \frac{-4\varepsilon}{h^2} S^2[1 + \frac{\alpha-\varepsilon}{\varepsilon} S^2] + \frac{4i}{h} SC[1 + \frac{\alpha-\varepsilon}{\alpha} S^2],$$

or, with the upwinding amount of artificial diffusion, $\alpha = \varepsilon + h$,

$$(4.2) \qquad \hat{M}_{h,\varepsilon}(\omega) = \frac{-4\varepsilon}{h^2} S^2[1 + \frac{h}{\varepsilon} S^2] + \frac{4i}{h} SC\,[1 + \frac{h}{h+\varepsilon} S^2].$$

For the two-dimensional problem (1.1) the characteristic forms are

(4.3) $\hat{L}_{h,\varepsilon}(\omega) = \frac{-4\varepsilon}{h^2} S^2 + \frac{4i}{h} T, \qquad \hat{D}_{h,\alpha}(\omega) = \frac{-8\alpha}{h^2}$ and

and

$$\hat{M}_{h,\varepsilon}(\omega) = \frac{-4\varepsilon}{h^2} S^2[1 + \frac{\alpha-\varepsilon}{2\varepsilon} S^2] + \frac{4i}{h} T[1 + \frac{\alpha-\varepsilon}{2\alpha} S^2], \tag{4.4}$$

where

$$T = -[a_1 S_\phi (2C_\phi + pC_{\phi+2\theta}) + a_2 S_\theta (2C_\theta + pC_{\theta+2\phi})]/(4+2p),$$

$$S^2 = S_\phi^2 + S_\theta^2, \; S_\phi = \sin\phi, \; C_\phi = \cos\phi,$$

$$\phi = \omega_1 h/2 \text{ and } \theta = \omega_2 h/2.$$

Local consistency and stability of the operators is studied in HEMKER (1982). We recollect the following remarks (cf. also BRANDT, 1980). In the limit for $\varepsilon \to 0$, the continuous operator L_ε is unstable for the modes $u_\omega = e^{i\omega x}$ with frequencies $\omega = (\omega_1, \omega_2)$ that are perpendicular to $a = (a_1, a_2)$, i.e. for all modes with $\omega \perp a$ we have $a.\nabla u_\omega = 0$.

In the limit for $\varepsilon \to 0$, the discrete operator $L_{h,\varepsilon}$ is unstable for the modes $u_{h,\omega} = e^{i\omega h}$; for which ω satisfies $T(\omega) = 0$. In the finite difference discretization (p=0), these modes $\omega = (\omega_1, \omega_2)$ are simply characterized by

$$a_1 \sin(\omega_1 h) + a_2 \sin(\omega_2 h) = 0.$$

The operator $L_{h,\alpha}$ has no unstable modes for $\varepsilon \to 0$ and it is consistent (of order one) with $L_{\varepsilon,h}$ if and only if $|\alpha - \varepsilon| = \mathcal{O}(h)$ as $h \to 0$.

Both in the 1-D and in the 2-D case (n = 1,2), the discretization (3.7) is consistent of order 2 and the operator $M_{h,\varepsilon}$ is *asymptotically stable*. Here we use the following

DEFINITION. The operator $M_{h,\varepsilon}$, a discretization of L_ε, is *asymptotically stable* if

$$\forall \rho > 0 \;\; \exists \eta > 0 \;\; \forall \omega \in [-\pi/h, \pi/h]^n \;\; \lim_{\varepsilon\to 0} |\hat{L}_\varepsilon(\omega)| > \rho \Rightarrow \lim_{\varepsilon\to 0} \frac{|\hat{M}_{h,\varepsilon}(\omega)|}{|\hat{L}_\varepsilon(\omega)|} > \eta, \tag{4.5}$$

(cf. HEMKER, (1982). In the one dimensional case, n = 1, we even find that $M_{h,\varepsilon}$ is ε-uniformly stable, i.e. (4.5) is satisfied with η independent of ρ.

5. THE CONVERGENCE OF MDCP ITERATION

In this section we consider the rate of convergence for the MDCP iteration

(3.1)-(3.4). By local mode analysis we show at what rate the different frequencies in the error are damped in the process, when the stationary solutions u_h^A and u_h^B are computed. The transition matrices of the defect correction step (3.1,a) and the damped Jacobi step (3.1,b) have the following characteristic forms

$$\hat{R}^{DCP}(\omega) = \widehat{(\tilde{L}_h^1 - L_h^1)}(\omega)/\hat{\tilde{L}}_h^1(\omega) = (\alpha-\varepsilon)\hat{\Delta}_h(\omega)/\hat{L}_{h,\alpha}(\omega),$$

$$\hat{R}^{JAC}(\omega) = \widehat{(\tilde{L}_h^2 - L_h^2)}(\omega)/\hat{\tilde{L}}_h^2(\omega) = (\hat{L}_{h\alpha}(\omega)-\hat{D}_{h,\alpha})/\hat{D}_{h,\alpha}.$$

The MDCP transition operator reads

$$\hat{R}^{MDCP}(\omega) = \hat{R}^{DCP}(\omega).\hat{R}^{JAC}(\omega).$$

For the one-dimensional model problem we find

$$\hat{R}^{MDCP}(\omega) = \frac{(\alpha-\varepsilon)SC}{-\alpha}\;\frac{[SC(\alpha^2-h^2)+ih\alpha]}{\alpha^2S^2+h^2C^2}, \tag{5.1}$$

and

$$|\hat{R}^{MDCP}(\omega)| \le \frac{\alpha-\varepsilon}{\alpha}\,\tfrac{1}{2}\sqrt{\frac{1}{4}\left(\frac{\alpha}{h}-\frac{h}{\alpha}\right)^2 + \max{}^2\left(\frac{h}{\alpha},\frac{\alpha}{h}\right)}. \tag{5.2}$$

With the upwinding amount of artificial viscosity, $\alpha = \varepsilon + h$, we find

$$|\hat{R}^{MDCP}(\omega)| \le \frac{1}{2}\,\frac{1}{\varepsilon+h}\sqrt{(\varepsilon+h)^2+\varepsilon^2} \le \tfrac{1}{2}\sqrt{2} < 1.$$

i.e. the process converges with a finite rate for all frequencies. In the limit for $\varepsilon \to 0$ we find

$$|\hat{R}^{MDCP}(\omega)| \le \frac{1}{2}.$$

The slowest convergence rate is shown for the intermediate frequencies ($\omega = \pm\,\pi/2h$); both the low and the high frequencies damp much faster.

For the 2-D problem the situation is somewhat more complex; we find

$$|\hat{R}^{MDCP}(\omega)| = \frac{(\alpha-\varepsilon)S^2\sqrt{\left(\frac{\alpha}{h}C^2S^2-\frac{h}{\alpha}T^2\right)^2+4T^2}}{2h\qquad\left[\left(\frac{\alpha}{h}S^2\right)^2+T^2\right]}, \tag{5.3}$$

where $C^2 = C_\phi^2 + C_\theta^2$. It is easy to show that $|\hat{R}^{MDCP}(\omega)| \le 1$ for all ω. However, for some frequencies convergence is slow. E.G. for the unstable modes ω of $L_{h,\varepsilon}$, for which $T(\omega) = 0$, we find

$$|\hat{R}^{MDCP}(\omega)| = \frac{\alpha-\varepsilon}{2\alpha}\,(C_\phi^2+C_\theta^2),$$

i.e. for small ε, along $T(\omega) = 0$, in the neighbourhood of $\omega = 0$, convergence is slow. If we set $\alpha = \varepsilon + \gamma h$, then, considering the limit for $\varepsilon \to 0$, we obtain

$$(5.4) \qquad |\hat{R}^{MDCP}(\omega)| = \left|\frac{\gamma S^2}{T}\right| \frac{\sqrt{1 + \frac{1}{4}\left(\frac{T}{\gamma} - C^2\left(\frac{\gamma S^2}{T}\right)\right)^2}}{1 + \left(\frac{\gamma S^2}{T}\right)^2}$$

To understand this expression, we introduce a new coordinate system in the frequency space. We define lines with constant $y = \gamma S^2(\omega)/T(\omega)$ and lines with constant $t = T(\omega)/\gamma$. Then

$$(5.5) \qquad |\hat{R}^{MDCP}(\omega)| = \frac{y}{1+y^2} \sqrt{1 + \frac{1}{4}(t-(2-yt)y)^2} \; .$$

In the neighbourhood of the origin lines of constant y are approximately circles, tangent in the origin to the line $T(\omega) = 0$, the value of y is proportional to the radius. Lines of constant t are lines approximately parallel to the line $a_1\omega_1 + a_2\omega_2 = 0$, t is proportional to the distance to this line.

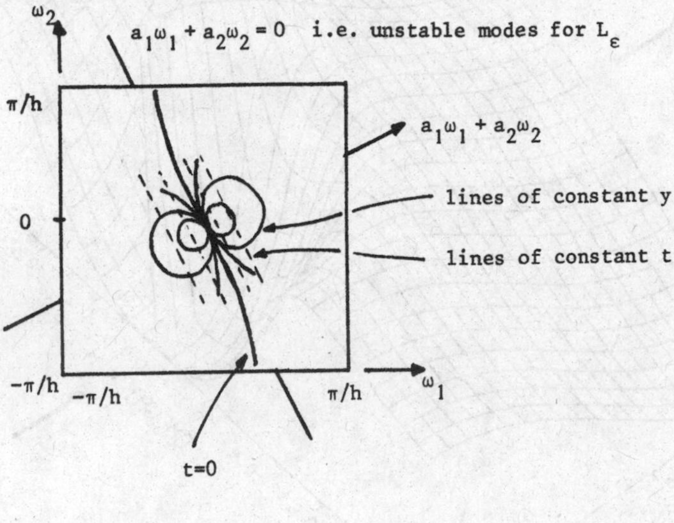

Figure 1. Lines of constant $y = \gamma S^2/T$ and constant $t = T/\gamma$ in the frequency space.

From (5.5) we see

(i) for small y

$$|\hat{R}^{MDCP}(\omega)| \approx y\sqrt{1 + \frac{1}{4}t^2} = O(y);$$

(ii) for large y, i.e. in the neighbourhood of t = 0,

$$|\hat{R}^{MDCP}| \approx \sqrt{y^{-2} + \frac{1}{4}C^2} \approx \frac{1}{2}\,(C_\phi^2 + C_\theta^2).$$

We see that low frequencies converge fast along the convection direction $\vec{a}$ and that convergence is slow (only!) in the direction perpendicular to the convection direction (i.e. for those ω with T(ω) = 0). In figure 2 we give an impression of the MDCP convergence rate as a function of ω. In this figure the rate is shown for the finite difference discretization (e.g. (2.1) with p = 0). The same behaviour is seen for the finite element discretization (p = 1). For other convection directions $\vec{a}$ the sharp ridge at the origin turns around the origin correspondingly.

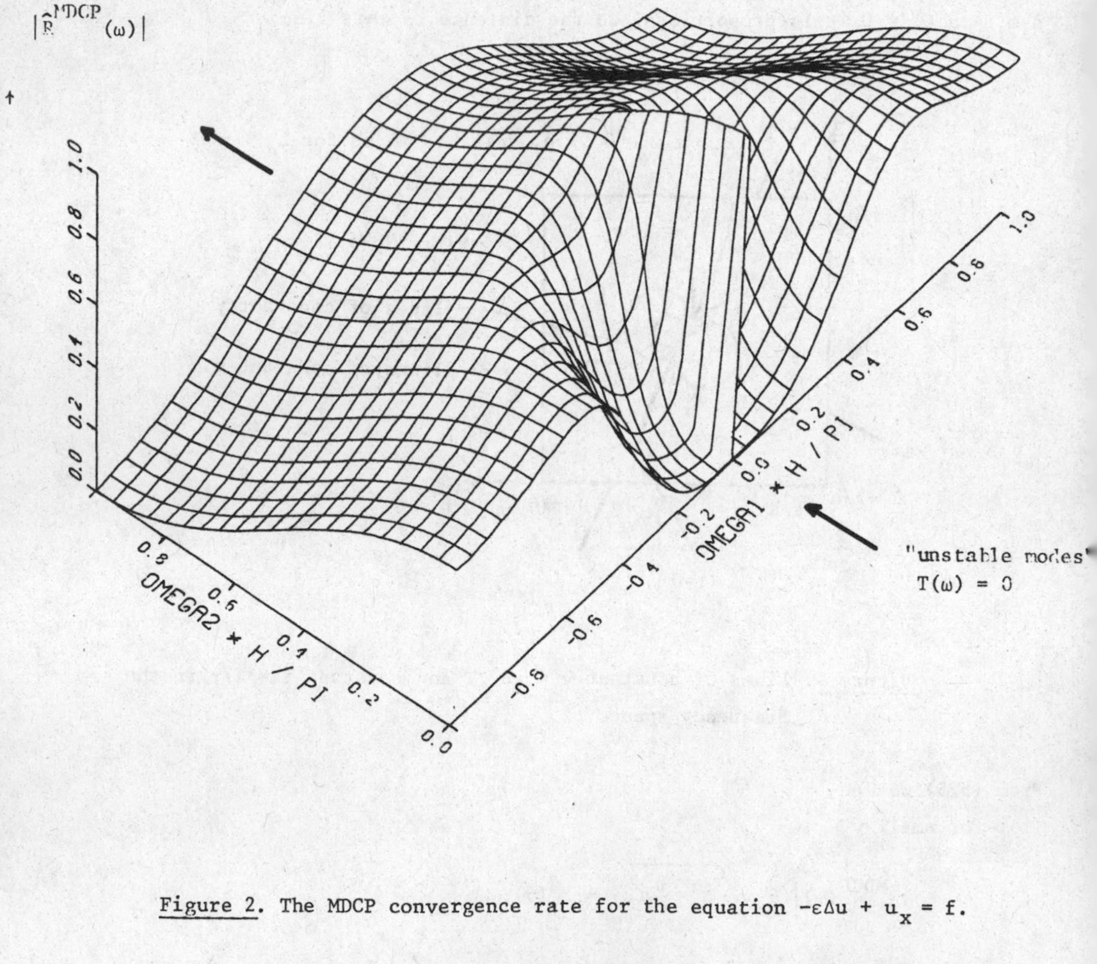

Figure 2. The MDCP convergence rate for the equation $-\varepsilon\Delta u + u_x = f$.

6. BOUNDARY ANALYSIS OF THE MDCP SOLUTION

Away from the boundary we already know that the MDCP discretization is asymptotically stable with respect to the right-hand side f_h. To analyze the effects of the boundary data, we consider the homogeneous problem

$$Lu_\omega = 0$$

in a discretization of the right half-space $(x \geq 0)$, the boundary $x = 0$ being a gridline. Dirichlet boundary data are given on this boundary and we consider solutions that are bounded at infinity. This situation is again studied by mode analysis, cf. BRANDT (1980). Now we use complex modes, $\omega = (\omega_1, \omega_2) \in \mathbb{C}^2$; $\omega_2 \in \mathbb{R}$ is given by the boundary data and $\hat{L}_h(\omega) = 0$ is solved for $\omega_1 \in \mathbb{C}$. Those solutions ω_1 for which $\operatorname{Im} \omega_1 > 0$ determine the behaviour of the discretization near the boundary at $x = 0$.

In this way, we first treat the one-dimensional model problem (1.2) with $\alpha = \varepsilon + h$. For this problem the only possible inhomogeneous boundary data are $u_h(0) = 1$. The modes

$$u_{h,\omega}(jh) = e^{i\omega hj} = \lambda^j$$

for which $M_{h,\varepsilon}\, u_{h,\omega} = 0$ are determined by

$$\hat{M}_{h,\varepsilon}(\omega) = 0, \tag{6.1}$$

which is a 4th degree polynomial in λ. In the limit for $\varepsilon \to 0$ we find for (6.1) the solutions $\lambda = 1$, $\lambda = 0$, $\lambda = 2 \pm \sqrt{5}$. From (4.2) it is clear that for all $\varepsilon/h > 0$, $\lambda = 1$ is a solution and no other solutions with $|\lambda| = 1$ exist. Since all λ are continuous functions of ε, we have for all $\varepsilon \geq 0$ two λ's with $|\lambda| < 1$ and two λ's with $|\lambda| \geq 1$. The two λ's with $|\lambda| < 1$ determine the behaviour of the solution near the boundary at $x = 0$. For small values of ε/h we find

$$\lambda_1 = \frac{3}{2}\frac{\varepsilon}{h} + O\left(\left(\frac{\varepsilon}{h}\right)^2\right)$$

and

$$\lambda_2 = 2 - \sqrt{5} - \left(2 - \frac{2}{5}\sqrt{5}\right)\left(\frac{\varepsilon}{h}\right) + O\left(\left(\frac{\varepsilon}{h}\right)^2\right)$$

$$\approx -0.236.$$

These values show that in the numerical boundary layer, for small ε/h, the influence of the boundary data decreases with a fixed rate per meshpoint. I.e. the width of the numerical boundary layer is only $O(h)$.

Of course, λ_1 and λ_2 only determine what modes appear in the solution of

$$M_{h,\varepsilon}\, u_h = 0 \quad u_h(0) = 1;$$

their relative amount is determined by the difference operator in the mesh-point next to the boundary. A closed analysis shows

(6.2)
$$u_h^A(jh) = -(2+2\sqrt{5})\,\frac{\varepsilon}{h}\,\lambda_1^j + [1+(2+2\sqrt{5})\frac{\varepsilon}{h}]\lambda_2^j + O((\frac{\varepsilon}{h})^2),$$
$$u_h^B(jh) = -(\tfrac{1}{2}+\tfrac{1}{2}\sqrt{5})\,\lambda_1^j + (\tfrac{3}{2}+\tfrac{1}{2}\sqrt{5})\lambda_2^j + O((\tfrac{\varepsilon}{h})).$$

This describes completely the behaviour of the 1-D numerical boundary layer solution.

For the 2-D model problem we proceed similarly. For given boundary data

$$u_{h,\omega}(jh) = e^{i\bar{\omega}h_2 j_2} \qquad \text{for } j_1 = 0,$$

we compute the modes

$$u_{h,\omega}(jh) = e^{i\omega h j} = e^{i\omega_1 h_1 j_1}\, e^{i\omega_2 h_2 j_2} = \lambda^{j_1}\, e^{i\bar{\omega}h_2 j_2},$$

that satisfy

$$M_{h,\varepsilon}\, u_{h,\omega} = 0 \qquad \text{for } j_1 > 0,$$

and we determine the corresponding $|\lambda|$.

To simplify the computation, we restrict ourselves to the finite difference star $(p=0)$ and artificial diffusion $\alpha = \varepsilon + h\,|a_1|/2$, $a_1 \neq 0$. First we consider boundary data with $\bar{\omega} = 0$ fixed. In the limit for $\varepsilon/h \to 0$ we determine λ from

$$\hat{M}_{h,0}(\omega) = 0\,.$$

We find the solutions $\lambda_0 = 0$, $\lambda_1 = 1$, $\lambda_{2,3} = 3 \pm \sqrt{12}$. Next we consider $\bar{\omega} \neq 0$. From (4.4) with $\varepsilon = 0$,

(6.3)
$$\hat{M}_{h,0}(\omega) = \frac{-2\alpha}{h^2}\, S^4 + \frac{2i}{h}\, T[2+S^2],$$

it follows that no real $\omega \in [-\pi/h, \pi/h]^2$, $\omega \neq (0,0)$, exists such that $\hat{M}_{h,0}(\omega) = 0$. Hence, except for $\bar{\omega} = 0$, no λ exists with $|\lambda| = 1$. All λ's are continuous functions of $\bar{\omega}$ and for small $\bar{\omega}$ we know

$$\lambda_1 = 1 - i\,\frac{a_2}{a_1}\,\bar{\omega}h + O((\bar{\omega}h)^4).$$

Hence $|\lambda_1| \geq 1$ and for all $\bar{\omega} \in [-\pi/h,\pi/h]$ there are two λ's with $|\lambda| < 1$ and two λ's with $|\lambda| \geq 1$. The two small λ's, considered as functions of $\bar{\omega} \in [-\pi/h,\pi/h]$ describe a curve inside the unit circle in C. These curves are closed subsets of $\mathbb{C}$ and have no point in common with the unit-circle. Thus, we see that $C = \max_{\bar{\omega}} |\lambda_{\bar{\omega}}|$ exists and

$$|\lambda| \leq C < 1.$$

If, instead of (6.3), we take (4.4) with $\varepsilon \neq 0$ as a starting point, the same continuity argument yields the same result for all ε which satisfy $0 \leq \varepsilon \leq Ch$. We conclude that, also in the two-dimensional case for small ε/h, the influence of the boundary data decreases with a fixed rate per meshpoint, i.e. the width of the numerical boundary layer is $O(h)$.

The above, non-constructive proof for the existence of $\max_{\bar{\omega}}|\lambda_{\bar{\omega}}| < 1$, allows the possibility of a large $|\lambda_{\omega}| < 1$, such that the existence may be of little practical use. In the numerical examples we show that the numerical boundary layer extends only over a few meshlines in the neighbourhood of the boundary indeed.

7. NUMERICAL RESULTS

In this section we show two numerical examples. The first example is to demonstrate the $O(h)$-width of the numerical boundary layer in the downstream direction. In the 2nd example we show that MDCP-iteration can also be applied to problems with interior layers and with variable coefficients. In both problems we discretize by the finite element method with piecewise linear test- and trial functions on a regular triangularization. The solution of the linear systems is obtained by a multiple grid method that is fully consistent with the finite element discretization. It used the 7-point Incomplete LU-relaxation, the 7-point prolongation and the 7-point restriction of which the weights are properly adapted to the Dirichlet or Neumann boundary conditions; no additional artificial diffusion is applied on coarser levels.

EXAMPLE 1.

We solve the constant coefficient problem

$$\varepsilon\Delta u + u_x = f(x,y),$$

on the unit square with Dirichlet boundary conditions. The boundary data and the data $f(x,y)$ are chosen such that

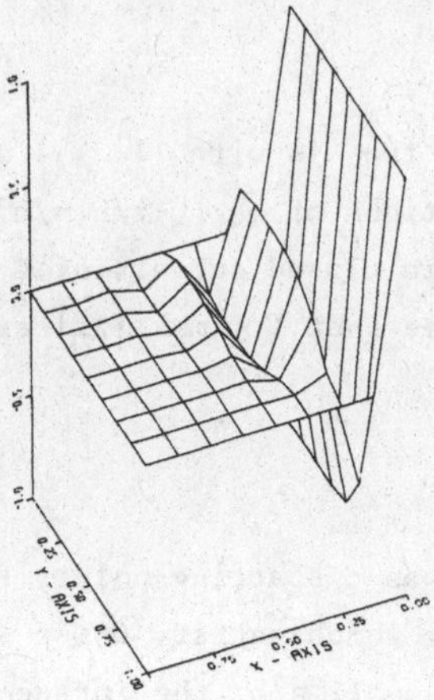

Fig. 3a

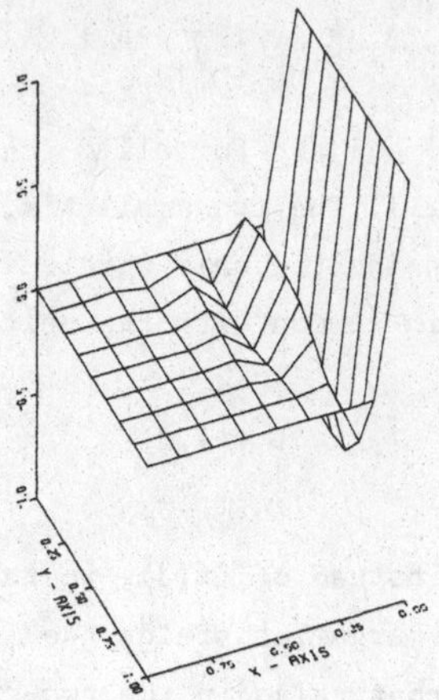

Fig. 3a'

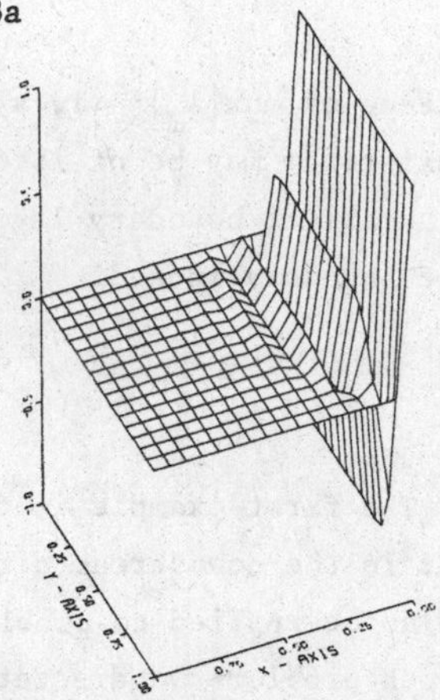

Fig. 3b

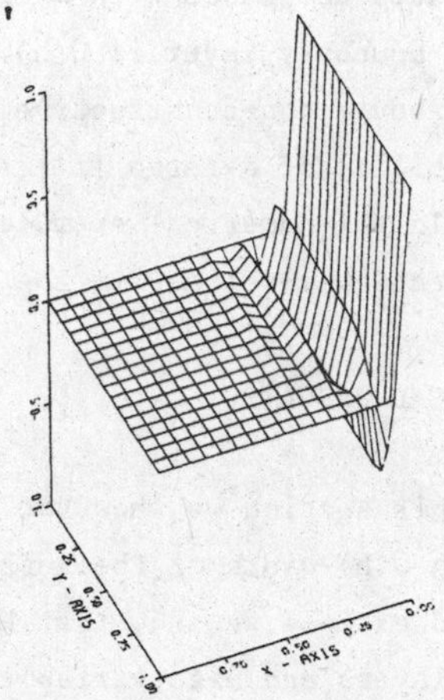

Fig. 3b'

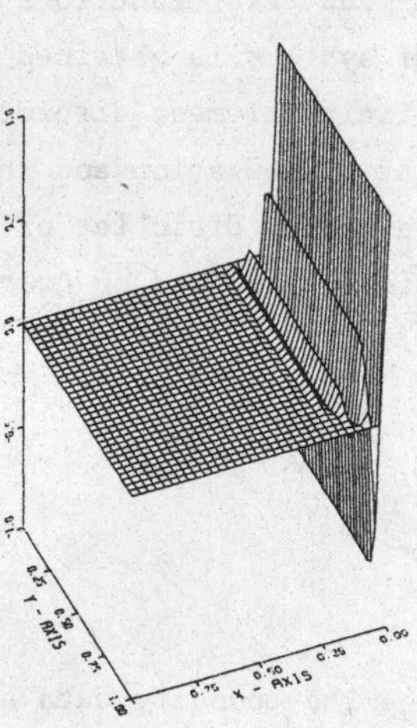

Fig. 3c

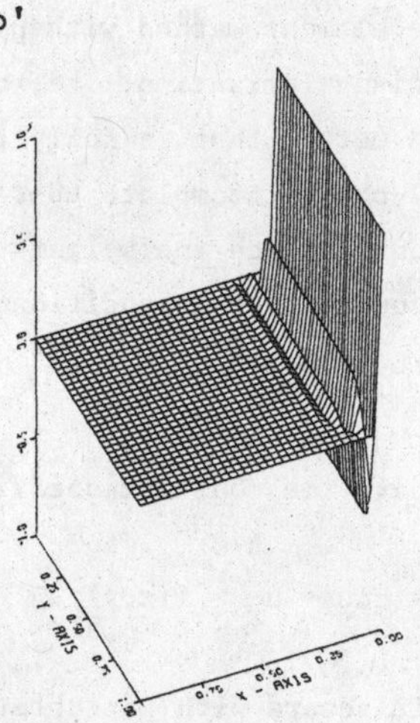

Fig. 3c'

Fig. 3 The numerical boundary layer in example 1.

$$u(x,y) = (\exp(-x/\varepsilon) - \exp(-1/\varepsilon))/(1-\exp(-1/\varepsilon)).$$

This solution only shows the exponential boundary-layer near the down-stream boundary. No smooth components in the solution are present. (In HEMKER (1982) we showed that smooth components - if present - are represented with an accuracy of order $O(h^2)$).

In the figures 3a, 3b, 3c, 3a', 3b', 3c' the solutions u_h^A and u_h^B as obtained by the MDCP-iteration are shown for mesh-widths h = 1/8, h = 1/16, h = 1/32. We see that the numerical boundary layer extends over a fixed number of mesh-lines away from the down-stream boundary, regardless of the meshwidth used.

EXAMPLE 2.

This second example describes a circular flow around a point. The equation reads

$$-\varepsilon\Delta u + \vec{b}.\nabla u = 0 \qquad \text{on } [-1,+1] \times [0,1]$$

$$\vec{b} = (2y(1-x^2),\ -2x(1-y^2))^T.$$

The boundary data are

(i) a Dirichlet boundary condition on $-1 \le x \le 0$, $y = 0$

$$u(x) = D(x) = 1 + \tanh(\gamma(1+2x)),$$

$$\gamma = 10,$$

and (ii) homogeneous Neumann boundary conditions on the remaining part of the boundary.

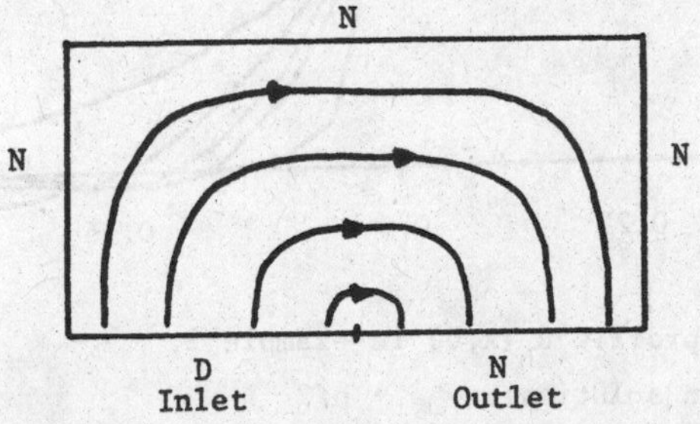

The convection direction is clockwise around the origin; the Dirichlet boundary condition is given at the upstream boundary, whereas the other boundary conditions are of Neumann type. The Dirichlet boundary data are such that s sharp interior layer is created in the interior of the domain. In the limit for $\varepsilon \to 0$, the exact solution at the outflow boundary is

$$u(x,0) = D(-x) \qquad 0 \le x \le 1.$$

The discretization is made on a regular grid with $h = 1/16$ and with $\varepsilon = 10^{-6}$.

In Figure 4 we compare the shapes of the outflow profile for 4 numerical solutions. One solution is obtained by straightforward application of artificial diffusion, $\alpha = \varepsilon + h/2$. (For a smaller artificial diffusion $\alpha = \varepsilon + h/4$ the MG-method for the solution of the system

$$L_{h,\alpha}\, u_h = f_h$$

diverges.) The other 3 solutions are obtained by MDCP-iteration. (At the outflow boundary both solutions u_h^A and u_h^D coincide!). These solutions are computed for different values of the artificial diffusion viz. $\alpha = \varepsilon + 2h$, $\alpha = \varepsilon + h$ and $\alpha =$ $= \varepsilon + h/2$.

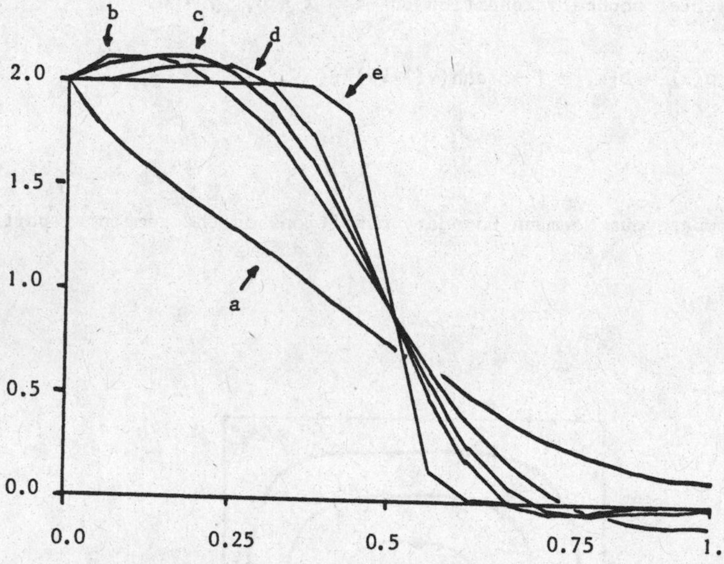

Figure 4. The outflow profile $u_h(x,0)$ in example 2.

a) artificial diffusion solution $\alpha = \varepsilon + h/2$
b) MDCP-iteration $\alpha = \varepsilon + 2h$,
c) $\alpha = \varepsilon + h$,
d) $\alpha = \varepsilon + h/2$,
e) pointwise exact limit-solution for $\varepsilon \to 0$.

In figure 4 we see that the interior layer is very well transported through the interior domain. Smaller values of the artificial diffusion in MDCP-iteration yield more accurate solutions.

EXAMPLE 3.

This example uses the same equation as example 2. Here the boundary conditions are

(i) Dirichlet boundary conditions

$$u(x) = D(x) \text{ on } -1 \le x \le 0 \quad , \quad y = 0,$$
$$u(x) = 0 \quad \text{ on } -1 \le x \le +1 \quad , \quad y = 1,$$
$$u(x) = 0 \quad \text{ on } 0 \le y \le 1 \quad \quad x = \pm 1;$$

and

(ii) homogeneous Neumann boundary conditions at the outlet boundary

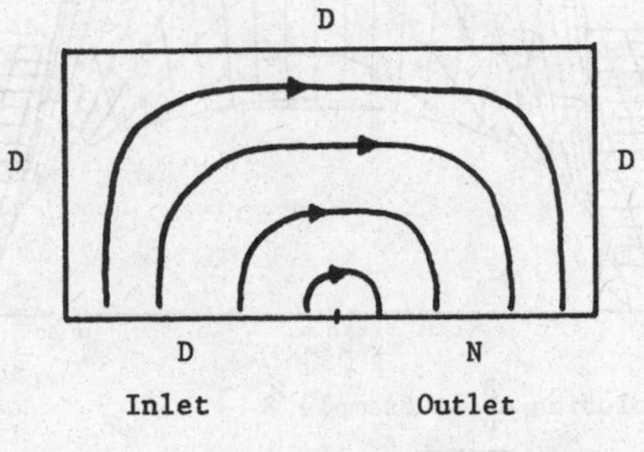

In figure 5a and 5b we give an impression of the numerical solution u_h^A and u_h^B obtained by MDCP-iteration and in figure 5c of the numerical solution obtained by artificial diffusion only. In figure 5d we show the computed solution that is obtained by application of one single defect correction step (3.1a) to correct the artificial diffusion solution as shown in fig. 5c. For all figures $\alpha = \varepsilon + h/2$ was used. From the figures we see that MDCP-iteration yields a better solution than is found by simple defect correction only, although both are theoretically 2nd order approximations.

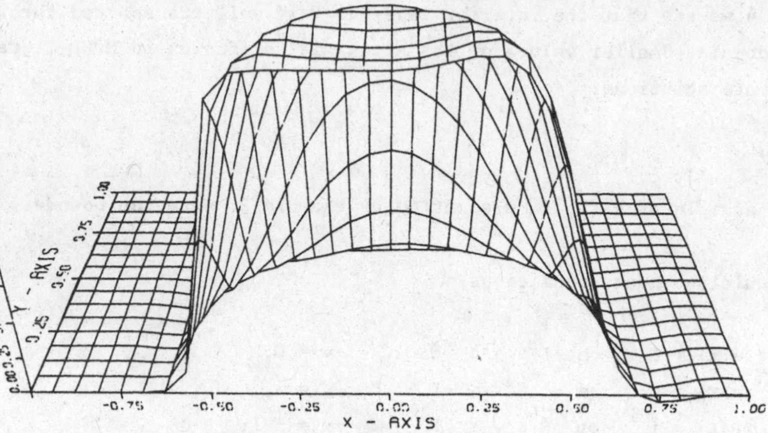

Fig. 5a The solution u_h^A in example 3

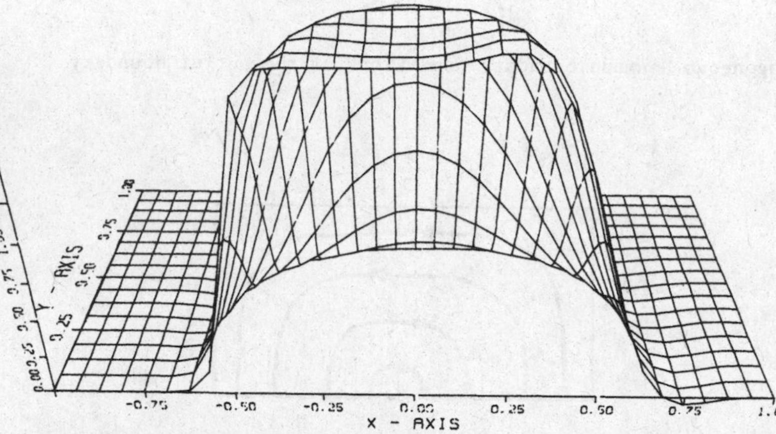

Fig. 5b The solution u_h^B in example 3

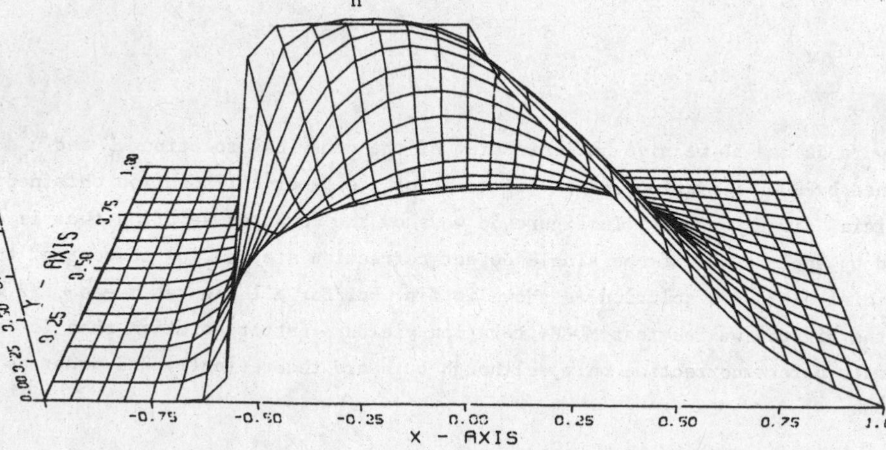

Fig. 5c The artificial diffusion solution in example 3

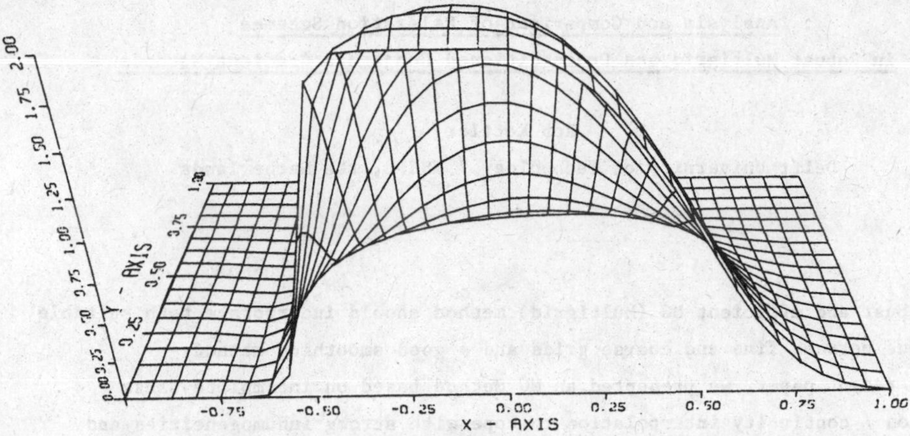

Fig. 5d The numerical solution of example 3 obtained by a single defect correction iteration step applied to the solution in figure 5c.

REFERENCES

BRANDT, A. [1980] *Numerical stability and fast solutions to boundary value problems.* in: Boundary and Interior Layers - Computational and asymptotic methods (J.J.H. Miller ed.) Boole Press, Dublin, 1980.

HACKBUSCH, W. [1979], *Bemerkungen zur iterierten Defektkorrektur und zu ihrer Kombination mit Mehrgitterverfahren* . Report 79-13, Math. Inst. Univ. Köln, 1979.

HEMKER, P.W. [1981], *Introduction to multi-grid methods*, Nw. Arch. Wisk. 29 (1981) 71-101.

HEMKER, P.W. [1982], *An accurate method without directional bias for the numerical solution of a 2-D elliptic singular perturbation-problem.* Procs. Oberwolfach meeting on Singular Perturbation Problems, to appear in Springer LNM, 1982.

STETTER, H. [1978], *The defect correction principle and discretization methods.* Num. Math. 29 (1978) 425-443.

Analysis and Comparison of Relaxation Schemes in Robust Multigrid and Preconditioned Conjugate Gradient Methods

Rob Kettler
Delft University of Technology / KSEPL, the Netherlands

Abstract

A robust and efficient MG (Multigrid) method should incorporate both suitable transitions between fine and coarse grids and a good smoothing method.

In a recent paper, we presented an MG method based on incomplete Gauss elimination / continuity interpolation to cope with strong inhomogeneities and general domains. Further acceleration has been achieved by using the MG method as preconditioning for CG, the result being called MGCG. In addition, a robust smoothing method is required for problems with strong anisotropies.

In this paper, several Gauss-Seidel and incomplete LU relaxation schemes are more extensively compared with respect to their smoothing factors and efficiencies. Besides, the practical behaviour of some of these schemes is compared for three iterative methods (MG, MGCG and ICCG) and two difficult test problems (according to Stone and Kershaw).

Contents

1. INTRODUCTION

Let us consider the convection-diffusion equation

$$-\nabla.D(x)\nabla u(x) + v(x).\nabla u(x) + c(x)u(x) = f(x) \ , \quad x\in\Omega\subset\mathbb{R}^2 \qquad (1.1)$$

with Dirichlet, Neumann or mixed conditions on $\partial\Omega$. D is a positive diagonal 2×2 matrix function and c is a non-negative function. Equation (1.1) may have the following characteristics:

(a) a general (polygonal) geometry of the domain Ω,
(b) strong inhomogeneities, resulting in strong discontinuities (i.e. discontinuities of several orders of magnitude) in D, and
(c) strong anisotropies, so that $D_{11} \ll D_{22}$ and/or $D_{22} \ll D_{11}$, locally or globally.

A computational grid Ω^0 of mesh size h^0 and a corresponding set of grid functions U^0 are defined by

$$\Omega^0 \equiv \{(x_1,x_2)\in\Omega \mid x_i=m_ih^0, \ m_i\in\mathbb{Z} \ , \ i=1,2\}; \quad U^0=\{u^0:\Omega^0\to\mathbb{R}\}. \qquad (1.2)$$

Then, finite difference approximations of equation (1.1) and the boundary conditions give rise to a (generally large) system of linear equations, which is denoted by

$$A^0u^0 = f^0 \qquad (u^0,f^0\in U^0; \ A^0:U^0\to U^0). \qquad (1.3)$$

A^0 represents a five- (or nine-) point difference molecule. Furthermore, we assume that:

(d) discontinuities in D occur only at grid lines; boundaries are situated along portions of grid lines.

The MG (Multigrid) method is an optimal-order iterative method for large sparse systems that arise from discretisations of partial differential equations. In this method, the process of relaxation alternates with a process of coarse-grid approximation, relaxation being such that it possesses certain smoothing properties (cf. [2]).

A robust MG method should be able to handle problems with characteristics (a), (b), (c) and (d), without loosing efficiency. In its conventional form (e.g. with simple Gauss-Seidel relaxation and linear interpolation), however, the performance of the MG method usually deteriorates drastically when applied to problems more difficult than a Poisson-type equation, as pointed out in Section 2.1. An improved MG method, based on incomplete Gauss elimination / continuity interpolation (cf. [1]), is briefly described in Section 2.2. For further acceleration, the MG method is combined with conjugate gradients in Section 2.3, this combination being called MGCG (Multigrid & Conjugate Gradients). Results in Chapter 2 have been presented in [7].

The main topic of this paper is the comparison of relaxation schemes for a variety of problems. In Chapter 3, several Gauss-Seidel and incomplete LU relaxation schemes are compared by means of local mode analysis with respect to their smoothing factors and efficiencies. The comparisons include the recent incomplete line LU method (see [10]); its application in MG and ICCG is new. Model problems are anisotropic-diffusion and singularly-perturbed convection-diffusion equations.

Chapter 4 covers experimental results for two well-known difficult test problems: Stone's problem [11], with strong anisotropies and a no-flow region, and Kershaw's problem [6], with a strongly inhomogeneous and non-rectangular domain. To these problems we applied the MG method described in Section 2.2, the MGCG method mentioned in Section 2.3 and the ICCG method (cf. [8]), using various relaxation schemes.

Chapter 5 contains concluding remarks.

2. PRELIMINARIES

2.1. Limitations of conventional MG methods

For k=0(1)M-1 (M+1 is the number of levels), let us introduce:

- $\Omega^{k+1} = \Omega^k_{0,0}$ — grid of mesh size $h^{k+1} = 2^{k+1}h^0$, where (2.1)

  $\Omega^k_{i,j} = \{(m_1h^k, m_2h^k) \mid m_1 \equiv i \pmod 2 \wedge m_2 \equiv j \pmod 2\}$; i,j=0,1 (2.2)
- $U^{k+1} = \{\Omega^{k+1} \to \mathbb{R}\}$ — set of real valued functions on Ω^{k+1}
- $P^k : U^{k+1} \to U^k$ — prolongation (interpolation) operator
- $R^k : U^k \to U^{k+1}$ — restriction operator
- $A^{k+1} : U^{k+1} \to U^{k+1}$ — Ω^{k+1} approximation of A^0
- $NREL^k_i$ (i=1,2) — number of relaxations
- $u^k := RELAX^k_i(A^k = f^k)u^k$ — application of relaxation

Then, we define the following (M+1)-grid algorithm for equation (1.3):

<u>Algorithm 2.1</u>

<u>Preprocessing</u>: <u>initialise</u> A^0, f^0, u^0

<u>for</u> k:=0(1)M-1 <u>do</u> <u>compute</u> $RELAX^k_1, RELAX^k_2, P^k, R^k, A^{k+1}$

Iterations: **for** it:=1(1)$NREL_1^0$ **do** $u^0:=RELAX_1^0(A^0.=f^0)u^0$

compute r^0

if termination criterion **then** terminate algorithm

$f^1:=R^0r^0$

for k:=1(1)M-1 **do**

| **for** it:=1(1)$NREL_1^k$ **do** $u^k:=RELAX_1^k(A^k.=f^k)0$

| **compute** r^k

| $f^{k+1}:=R^kr^k$

$u^M:=(A^M)^{-1}f^M$

for k:=M-1(-1)0 **do**

| $\delta u^k:=P^ku^{k+1}$

| $u^k:=u^k+\delta u^k$

| **for** it:=1(1)$NREL_2^k$ **do** $u^k:=RELAX_2^k(A^k.=f^k)u^k$

go to Iterations

$(r^k=f^k-A^ku^k)$.

Here, we have selected a fixed V-cycle MG strategy. The strategy, however, has no influence on the limitations described below.

P^k, R^k, A^{k+1}, $NREL_i^k$ and $RELAX_i^k$ (i=1,2 ; k=0(1)M-1) still have to be chosen. In the MG literature (e.g. [2], [3], [5], [12], [13]), common choices for P^k, R^k, A^{k+1} and $RELAX_i^k$ are bilinear interpolation for P^k, injection or transposed bilinear interpolation for R^k, a finite difference or Galerkin approximation for A^{k+1} and a Gauss-Seidel or incomplete point LU scheme for $RELAX_i^k$, whereas $(NREL_1^k, NREL_2^k) \in \{(1,0),(0,1),(1,1)\}$.

We analyse the reasons for the failure of the conventional MG methods when applied to equation (1.3) and therefore consider the characteristics (a), (b) and (c) mentioned in Chapter 1 (cf. [7]):

re (a)

If a boundary is situated inside a coarse-grid block, bilinear interpolation cannot be applied (in the case of an external boundary) or bilinear interpolation makes no sense (in the case of an internal boundary). During the last few years, various approaches to general domains have been suggested: most of them rather complicated and none of them robust, i.e. independent of type of boundary conditions. In the literature, general domains have only been tackled for the case of Dirichlet boundary conditions (cf. [4], [5]).

re (b)

If a discontinuity in D occurs inside a coarse-grid block, the error function on the finer grid may not be smooth, so that it cannot be approximated adequately by means of bilinear interpolation. Even in the case of simple problems with discontinuous D (e.g. a diffusion equation with D=1 in part of a square and D=1000 in the remainder), the performance of MG with bilinear interpolation degenerates severely (see [1], [7]).

re (c)

Fourier analysis and experiments show, that several relaxation schemes behave badly in the MG context (i.e. they are bad smoothers) when $D_{11} \ll D_{22}$ or $D_{22} \ll D_{11}$. Besides, some Gauss-Seidel schemes are bad smoothers for the case of simple singular-perturbation problems with a small diffusion parameter.

With regard to (a) and (b), we propose in Section 2.2 the use of incomplete Gauss elimination / continuity interpolation, instead of bilinear interpolation (details are given in [7]; a similar proposal has been made previously in [1]). Discussion with respect to (c) is the main topic of this paper.

2.2. The modified MG method

Using algorithm 2.1 and definitions (2.1) and (2.2), we make the following choices for P^k, R^k, A^{k+1} and $NREL_i^k$ $(i=1,2\ ;\ k=0(1)M-1)$:

- In the MG algorithm, $\delta u^k = P^k u^{k+1}$ is the approximation of the error $e^k (e^k \in U^k)$, which satisfies $A^k e^k = r^k$. Suppose u^{k+1} is given and exact, then

$$e_x^k = \delta u_x^k = u_x^{k+1}, \qquad x \in \Omega_{0,0}^k \qquad (2.3)$$

and, in the remaining points of Ω^k, an approximation better than bilinear interpolation is given by taking

$$\delta u_x^k = \frac{A_{x,x-\delta x}^k \delta u_{x-\delta x}^k + A_{x,x+\delta x}^k \delta u_{x+\delta x}^k}{A_{x,x-\delta x}^k + A_{x,x+\delta x}^k}, \qquad x \in \Omega_{i,j}^k \ (i+j=1), \qquad (2.4)$$

where $\delta x = (ih^k, jh^k)$, and then taking

$$\delta u_x^k = (r_x^k - \sum_{\delta x} A_{x,x+\delta x}^k \delta u_{x+\delta x}^k)/A_{x,x}^k, \qquad x \in \Omega_{1,1}^k, \qquad (2.5)$$

where $\delta x \in \{(\pm h^k,0),(0,\pm h^k),(\pm h^k,\pm h^k)\}$.
Equation (2.4) denotes continuity of $D\nabla u$ along the grid lines (this formula is derived in [1]); equation (2.5) denotes Gauss substitution in $\Omega_{1,1}^k$.

- Analogously, the restriction $f^{k+1} = R^k r^k$ consists in the application of incomplete Gauss elimination and subsequent transposed continuity interpolation.

- The coarse-grid matrix A^{k+1} is constructed by

$$A^{k+1} = R^k A^k P^k \quad \text{(Galerkin).} \tag{2.6}$$

- $NREL_1^k=0$, $NREL_2^k=1$ (called "a sawtooth cycle" in [12])

N.B. Choices for $RELAX_1^k$ have also been made in [7], but these are reconsidered in this paper.

Advantages of the presented choices for A^{k+1} and $NREL_i^k$ are given in [12]. The main advantage of this method, however, is the use of the adaptive incomplete Gauss elimination / continuity interpolation, instead of fixed bilinear interpolation. The following is noted:

(i) the new scheme takes care of discontinuities in D [if these are represented in (2.3)],

(ii) it automatically performs constant extrapolation at the boundary of a general domain (because of zero coefficients in A^k),

(iii) for the case of constant coefficients, the new method differs from the methods given in Section 2.1 (with bilinear interpolation) only in the use of r^k in equation (2.5); in our experiments omission of r^k from (2.5) yields slower convergence,

(iv) the amount of additional computational work and storage for the modified MG method is not substantial,

(v) for various problems with characteristics (a), (b), (c) and (d), the new method performs well.

Further details and numerical results of the modified MG method can be found in [7].

2.3. The MGCG method

If one iteration of the modified MG algorithm 2.1 is denoted by $u^0:=MG(A^0.=f^0)u^0$, then an MGCG algorithm is the CG algorithm 2 of [7], in which the statements <u>compute</u> K^0 (i.e. preprocessing) and $q^0:=(K^0)^{-1}r^0$ (i.e. preconditioning) are, respectively:

<u>for</u> k:=0(1)M-1 <u>do</u> <u>compute</u> $RELAX_1^k, RELAX_2^k, P^k, R^k, A^{k+1}$

and

$$q^0 := MG(A^0.=r^0)0.$$

For reasons of symmetry we choose: $NREL_1^k = 1$ and $RELAX_1^k$ is the adjoint of $RELAX_2^k$.

It should be noted that CG improves the convergence of MG, whereas implementation of CG is easy to perform. In the case of non-self-adjoint problems or an asymmetric MG scheme, one should apply, for instance, IDR (Induced Dimension Reduction, see [13]) instead of CG. Details and experiments of the MGCG method have been given in [7].

3. LOCAL MODE ANALYSIS AND SMOOTHING COMPARISON OF SEVERAL RELAXATION SCHEMES

3.1. Local mode analysis in general and model problems

To gain an idea of the behaviour of relaxation schemes in iterative methods, local mode (Fourier) analysis is a helpful tool (cf. [2] for the case of MG).

We assume a domain $\Omega=(0,1)\times(0,1)$, periodic boundary conditions, a uniform difference molecule and row-wise upward point-numbering, and we denote the system of linear equations at level k by

$$A^k u^k = f^k. \tag{3.1}$$

In the remaining part of Chapter 3 the superscript k is omitted.

Relaxation on system (3.1) is given by

$$u_{n+1} = u_n + K^{-1} r_n \quad (r_n = f - Au_n,\ n \in \mathbb{N}), \tag{3.2}$$

where

$$K = A + N \tag{3.3}$$

is an approximation of A such that $K^{-1} r_n$ is easy to compute. Then, the error after n+1 relaxations $e_{n+1} = u - u_{n+1}$ can be calculated from

$$e_{n+1} = K^{-1} N\, e_n. \tag{3.4}$$

We now expand the error in a Fourier series (of eigenfunctions of $K^{-1}N$) as follows

$$e_n(x_1,x_2) = \sum_{\theta=(\theta_1,\theta_2)} \varepsilon_\theta^n\, e^{i(\theta_1 x_1 + \theta_2 x_2)} \tag{3.5}$$

$[\theta_1,\theta_2 = -\pi(2\pi h)\pi]$, and we define the reduction factor of the θ Fourier component (which is only a function of θ and independent of n) by

$$\rho(\theta) = \frac{\left|\varepsilon_\theta^{n+1}\right|}{\left|\varepsilon_\theta^{n}\right|}. \tag{3.6}$$

K and N are periodic Toeplitz matrices. We denote their coefficients by κ_j and ν_j, respectively ($j=(j_1,j_2)\in\mathbb{Z}\times\mathbb{Z}$). $\rho(\theta)$ can then be calculated from

$$\rho(\theta) = \frac{\left|\sum\limits_{j\in J_\nu} \nu_j e^{i(j.\theta)}\right|}{\left|\sum\limits_{j\in J_\kappa} \kappa_j e^{i(j.\theta)}\right|}, \tag{3.7}$$

where J_κ is the set j for which $\kappa_j \neq 0$ and J_ν is the set j for which $\nu_j \neq 0$ (i.e. the molecular structures). Now, assuming A to be a periodic five-point Toeplitz matrix with coefficients σ_j, examples of (3.7) are given for the relaxation schemes (in point-numbering order) Jacobi, Line Jacobi and Point Gauss-Seidel (PGSF):

- Jacobi : A=L+D+U (L is strictly lower triangular, U is strictly upper triangular, D is diagonal), K=D, N=-L-U, so that

$$\rho(\theta_1,\theta_2) = \frac{\left|\sigma_{0,-1}e^{-i\theta_2}+\sigma_{-1,0}e^{-i\theta_1}+\sigma_{1,0}e^{i\theta_1}+\sigma_{0,1}e^{i\theta_2}\right|}{|\sigma_{0,0}|} \tag{3.8}$$

- Line Jacobi : A=L+D+U (L is strictly lower block-triangular, U is strictly upper block-triangular, D is block-diagonal), K=D, N=-L-U, so that

$$\rho(\theta_1,\theta_2) = \frac{\left|\sigma_{0,-1}e^{-i\theta_2}+\sigma_{0,1}e^{i\theta_2}\right|}{\left|\sigma_{-1,0}e^{-i\theta_1}+\sigma_{0,0}+\sigma_{1,0}e^{i\theta_1}\right|} \tag{3.9}$$

- PGSF : A=L+D+U (L is strictly lower triangular, U is strictly upper triangular, D is diagonal), K=L+D, N=-U, so that

$$\rho(\theta_1,\theta_2) = \frac{\left|\sigma_{1,0}\,e^{i\theta_1}+\sigma_{0,1}e^{i\theta_2}\right|}{\left|\sigma_{0,-1}e^{-i\theta_2}+\sigma_{-1,0}e^{-i\theta_1}+\sigma_{0,0}\right|} \tag{3.10}$$

For the case of the Poisson equation, contour plots of the reduction factors $\rho(\theta_1,\theta_2)$ are given in Fig. 3.1. Lines with $\rho(\theta_1,\theta_2)=0$ have been omitted, but they can easily be constructed.

The factor of interest in MG (and MGCG) is the smoothing factor

$$\bar{\rho} = \sup_{\theta\in H} \rho(\theta) \tag{3.11}$$

(cf. [2]), where $H = \{(\theta_1,\theta_2)\,|\,-\pi < \theta_1,\theta_2 < \pi,\ |\theta_1| > \frac{\pi}{2} \vee |\theta_2| > \frac{\pi}{2}\}$, i.e. the region of components that have a high frequency relative to h. From Fig. 3.1 it can be seen that $\bar{\rho}_{JACOBI}$(Poisson) = $\bar{\rho}_{LINEJACOBI}$(Poisson) = 1, whereas $\bar{\rho}_{PGSF}$(Poisson)=0.5. Hence, PGSF is a good smoothing scheme for this problem.

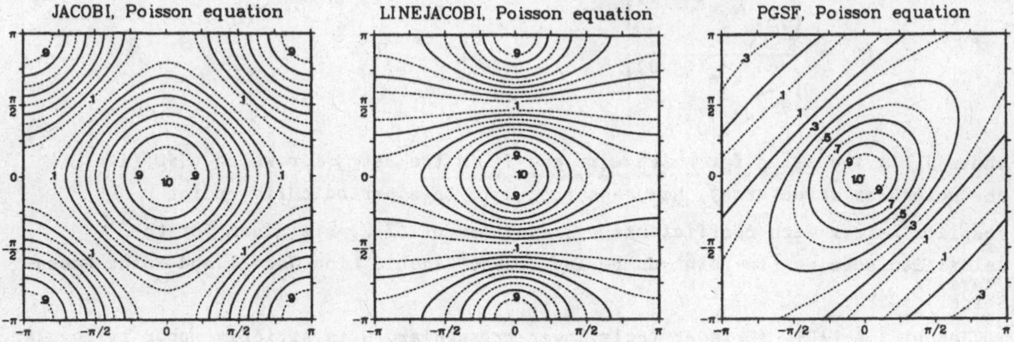

Figure 3.1 Reduction factors of Fourier components for Jacobi, Line Jacobi and PGSF

In the subsequent sections, we examine the behaviour of several relaxation schemes for the following set of model problems:

Problem No.	Equation	Symbol	Type
0	$\Delta u=f$	⊡	Poisson equation
1	$u_{xx}+\varepsilon u_{yy}=f$	▤	anisotropic-diffusion equations
2	$\varepsilon u_{xx}+u_{yy}=f$	▥	
3	$\varepsilon\Delta u-u_x=f$	→	singularly-perturbed convection-diffusion equations
4	$\varepsilon\Delta u+u_x=f$	←	
5	$\varepsilon\Delta u-u_y=f$	↑	
6	$\varepsilon\Delta u+u_y=f$	↓	
7	$\varepsilon\Delta u-u_x-u_y=f$	↗	
8	$\varepsilon\Delta u+u_x+u_y=f$	↙	
9	$\varepsilon\Delta u-u_x+u_y=f$	↘	
10	$\varepsilon\Delta u+u_x-u_y=f$	↖	

where $\varepsilon=10^{-p}$; $p=0(1)5$. Second-order derivatives are discretised by central differences and first-order derivatives are discretised by upward differences (Il'in schemes show analogous results).

3.2. Local mode analysis of relaxation schemes

3.2.1. Gauss-Seidel schemes

The various PGS (Point Gauss-Seidel) and LGS (Line Gauss-Seidel) relaxation schemes are denoted as follows:

PGSF - PGS in point-numbering order;

PGSFB - PGS in point-numbering order, followed by PGS in anti-point-numbering order;

LGSFi - LGS with lines in the x_i-direction marching forwards;

LGSBi - LGS with lines in the x_i-direction marching backwards;

LGSFiBj - LGSFi, followed by LGSBj;

(other GS methods are denoted in the usual manner).

We apply local mode analysis as described in Section 3.1 for five basically different GS schemes:

.PGSF: $\rho(\theta_1,\theta_2)$ is given by equation (3.10)

.PGSFB:
$$\rho(\theta_1,\theta_2) = \frac{\left|\sigma_{1,0}e^{i\theta_1}+\sigma_{0,1}e^{i\theta_2}\right|}{\left|\sigma_{0,-1}e^{-i\theta_2}+\sigma_{-1,0}e^{-i\theta_1}+\sigma_{0,0}\right|}\cdot\frac{\left|\sigma_{0,-1}e^{-i\theta_2}+\sigma_{1,0}e^{-i\theta_1}\right|}{\left|\sigma_{0,0}+\sigma_{1,0}e^{i\theta_1}+\sigma_{0,1}e^{i\theta_2}\right|} \tag{3.12}$$

.LGSF1:
$$\rho(\theta_1,\theta_2) = \frac{\left|\sigma_{0,1}e^{i\theta_2}\right|}{\left|\sigma_{0,-1}e^{-i\theta_2}+\sigma_{-1,0}e^{-i\theta_1}+\sigma_{0,0}+\sigma_{1,0}e^{i\theta_1}\right|} \tag{3.13}$$

.LGSF1F2:
$$\rho(\theta_1,\theta_2) = \frac{\left|\sigma_{0,1}e^{i\theta_2}\right|}{\left|\sigma_{0,-1}e^{-i\theta_2}+\sigma_{-1,0}e^{-i\theta_1}+\sigma_{0,0}+\sigma_{1,0}e^{i\theta_1}\right|}\cdot\frac{\left|\sigma_{1,0}e^{i\theta_1}\right|}{\left|\sigma_{0,-1}e^{-i\theta_2}+\sigma_{-1,0}e^{-i\theta_1}+\sigma_{0,0}+\sigma_{0,1}e^{i\theta_2}\right|} \tag{3.14}$$

.LGSF1B1F2B2: $\rho(\theta)$ is equal to the right-hand side of (3.14) multiplied by two similar terms.

Appendix A contains contour plots of $\rho(\theta)$ for the schemes PGSF, PGSFB, LGSF1 and LGSF1F2 (because of resemblance to LGSF1F2, plots of LGSF1B1F2B2 are omitted), and for the set of model problems; ε=0.1. We have also depicted PGSF; ε=0.01, to gain an impression of the behaviour of the figures as a function of ε (other plots will be included in a subsequent report). Contours ρ=0.1(0.2)0.9,1.0 are labelled, if possible; if no labels are provided, the exterior contour is ρ=0.1.

The figures contain information on the behaviour of a relaxation scheme:

(i) as iterative method. The relaxation scheme PGSF, for instance, is a very good iterative method for problem 7, but its performance degenerates severely for problems 4 and 6. Furthermore, we notice that (disregarding reflexion) the behaviour of PGSF is similar for the following pairs of problems: 1 and 2, 3 and 5, 4 and 6, and 9 and 10.

(ii) as preconditioning for CG. Considering the spectrum of eigenvalues, the scheme will in any case be a good preconditioning for CG if it is a (moderately) good iterative method. (Nevertheless, bad iterative methods may also be good preconditionings.)

(iii) in MG (or MGCG). From the figures, one can estimate smoothing factors for various coarsening strategies, i.e. coarsening by a factor other than 2 or coarsening in one direction only. It can be seen that PGSF is a good smoother for problem 7: in that case the factor of coarsening might be larger than 2; for problems 1, 2, 3, 5, 9 and 10, coarsening in one direction is more adequate than coarsening in two directions; for problems 4, 6 and 8, however, PGSF always fails as a smoother.

(Similar remarks can be made for schemes other than PGSF.)

For all schemes and model problems, the smoothing factor $\bar{\rho}$ of definition (3.11) is given in Appendix B. Moreover, we have determined the limit factors $\lim_{\varepsilon\downarrow 0}\bar{\rho}$ either theoretically or experimentally. A proof for the value of a limit factor is given below.

Theorem 3.1 $\lim_{\varepsilon\downarrow 0}\bar{\rho}_{PGSF}$ (problem 3) $= \frac{1}{3}\sqrt{2}$

Proof 3.1 Discretisation of $\varepsilon\Delta u - u_x = f$ yields $\sigma_{0,-1}=\sigma_{1,0}=\sigma_{0,1}=\varepsilon$, $\sigma_{-1,0}=1+\varepsilon$ and $\sigma_{0,0}=-1-4\varepsilon$. Substitution of (3.10) into (3.11) gives

$$\bar{\rho} = \sup_{\theta\in H} \frac{\left|\varepsilon e^{i\theta_1} + \varepsilon e^{i\theta_2}\right|}{\left|\varepsilon e^{-i\theta_2} + (1+\varepsilon)e^{-i\theta_1} - 1 - 4\varepsilon\right|}. \tag{3.15}$$

The supremum is achieved at $|\theta_2| = \frac{\pi}{2}$ and $\theta_1 = O(\varepsilon)$. By substitution of $\theta_2 = \frac{\pi}{2}$ and $\theta_1 = c\varepsilon$ into (3.15) and by expanding $e^{ic\varepsilon}$ into $e^{ic\varepsilon} = 1 + ic\varepsilon + O(\varepsilon^2)$, we obtain

$$\bar{\rho} = \sup_{c} \frac{|\varepsilon + i\varepsilon + O(\varepsilon^2)|}{|-3\varepsilon + i(c-1)\varepsilon + O(\varepsilon^2)|}. \tag{3.16}$$

For ε small, the supremum is achieved in the point $(\varepsilon, \frac{\pi}{2})$, so that $\lim_{\varepsilon\downarrow 0}\bar{\rho}_{PGSF} = \frac{1}{3}\sqrt{2}$.

(N.B. $\bar{\rho} = 5^{-\frac{1}{2}}$ in $(0, \frac{\pi}{2})$.) □

Similar proofs can be given for other GS limit factors.

3.2.2. Incomplete point LU schemes

An incomplete point (as opposed to line) LU factorisation of A can be formulated by

$$(\bar{L}+\bar{D})(\bar{D})^{-1}(\bar{D}+\bar{U}) = A + N , \tag{3.17}$$

where $\bar{L}$ is strictly lower triangular, $\bar{U}$ is strictly upper triangular and $\bar{D}$ is diagonal. Equation (3.17) is rewritten as

$$\bar{L}(\bar{D})^{-1}\bar{U} + \bar{L} + \bar{D} + \bar{U} = A + N. \tag{3.18}$$

We introduce

$$S = \bar{L} + \bar{D} + \bar{U}, \tag{3.19}$$

so that $\bar{L}_i^j = S_i^j$ $(j<0)$, $\bar{D}_i^0 = S_i^0$ and $\bar{U}_i^j = S_i^j$ $(j>0)$ [M_i^j denotes the element $(i,i+j)$ of matrix M]. For an ILU(ncd_1,ncd_2) method (cf. [9]; ncd = number of co-diagonals), S has the prescribed non-zero pattern

$$P^* = \{j \in \mathbb{Z} \,|\, 0<|j|<ncd_1 \vee n_1+1-ncd_2<|j|<n_1+1\}, \tag{3.20}$$

where we have assumed a nine-point difference molecule and n_m points in the x_m-direction. Furthermore, we require: $N_i^j=0$ if $j \in P^*$.

Then, an algorithm for obtaining the factorisation (3.18) is the following:

Algorithm 3.1

for $i:=1(1)|\Omega|$ **do**
  for $j:=1-i(1)|\Omega|-i$ **do**
    if $j \in P^*$ **then** $S_i^j := A_i^j - \sum_{\substack{\ell<0;\, j-\ell>0 \\ \ell,\, j-\ell \in P^*}} S_i^\ell (S_{i+\ell}^0)^{-1} S_{i+\ell}^{j-\ell}$

Relaxation on system (3.1) is given by equation (3.2) with $K=(\bar{L}+\bar{D})(\bar{D})^{-1}(\bar{D}+\bar{U})$, and requires the solution of two triangular systems and a residual calculation. If N is sparser than A, the residuals can be most efficiently calculated by

$$r_n = N(u_n - u_{n-1}) \tag{3.21}$$

Further details can be found in [9] and [7].

In order to calculate the reduction factors of the Fourier components, we have to compute the periodic ILU factorisation (3.17) for a uniform difference molecule. Uniformity and periodicity imply that in algorithm 3.1 the subscripts can be omitted.

For the solution of the resulting system of linear equations, we propose the following analogon of algorithm 3.1 (which is new in the literature):

Algorithm 3.2

S:=I
for i:=1,i+1 **while** **not** convergence **do**
  for j:=$-n_1-1(1)n_1+1$ **do**
    if $j\in P^*$ **then** $S^j := A^j - \sum_{\substack{\ell<0;\, j-\ell>0\\ \ell,\, j-\ell\in P^*}} S^\ell (S^0)^{-1} S^{j-\ell}$

This is a stable Gauss-elimination process converging to a real solution S^j ($j\in P^*$). (For high accuracy, i may run up to, say, 10.000 iterations, depending on the test problem.) Other initial iterands, such as S=A, yield the same solution.

$K=(\bar{L}+\bar{D})(\bar{D})^{-1}(\bar{D}+\bar{U})$ and N=K-A can now be computed and the reduction factors can be calculated from equation (3.7), as usual. We consider the following set of ILU schemes: ILU(1,1) (the so-called "five-point ILU"), ILU(1,2) (the so-called "seven-point ILU"), ILU(1,3), ILU(2,4) and ILU(7,9) (cf. [9]; the set is assumed to be representative).

In Appendix A, reduction-factor plots are given for the schemes ILU(1,1) and ILU(1,2), and for the set of model problems; ε=0.1. The contours for the other ILU schemes bear close resemblance to those for ILU(1,2), but are smaller. The limit behaviour as $\varepsilon\downarrow 0$ will generally be clear from the figures; for problems 1 and 2, however, we additionally present Fig. 3.2.

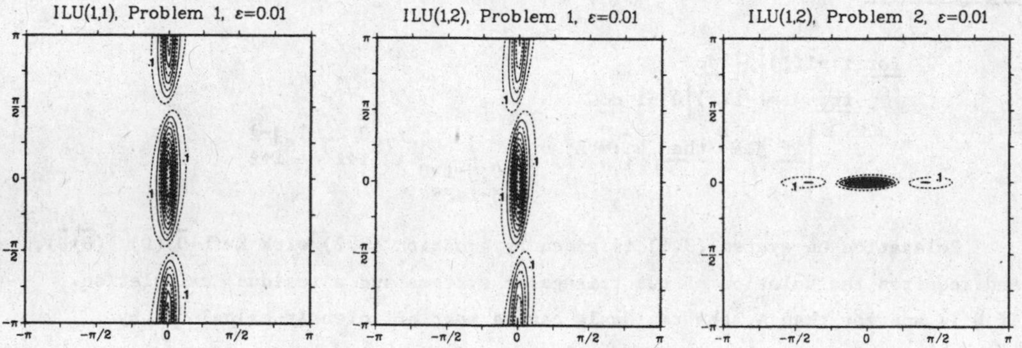

Figure 3.2 Reduction factors of Fourier components for ILU(1,1) and ILU(1,2); ε=0.01

For ILU(1,1), the plot of problem 2 is obtained by reflexion of the plot of problem 1.

Smoothing factors for all ILU schemes and model problems are contained in Appendix B, together with some limit factors. A proof for the value of the limit factor $\lim_{\varepsilon\downarrow 0} \bar{\rho}_{ILU(1,2)}$ (problem 2) is given below, since it is, unexpectedly, not equal to zero.

Theorem 3.2 $\lim_{\varepsilon\downarrow 0} \bar{\rho}_{ILU(1,2)}$ (problem 2) = $(3 + 2\sqrt{2})^{-1}$

Proof 3.2 Discretisation of $\varepsilon u_{xx}+u_{yy}=f$ yields $A^{-1}=A^{1}=\varepsilon$, $A^{-m}=A^{m}=1$ and $A^{0}=-2-2\varepsilon$. Because of symmetry: $S^{j}=S^{-j}$ in algorithm 3.2.

Let $d=(S^{0})^{-1}$, $b=S^{1}$, $c=S^{n_1}$, $e=S^{n_1-1}$ and $r=N^{n_1-2}$ (other elements of N are zero; n_1 is finite and large), then we obtain (cf. [9])

$$1/d = -2-2\varepsilon-(1+b^2+e^2)d \qquad (3.22a)$$

$$b = \varepsilon-de \qquad (3.22b)$$

$$e = -db \quad ; \quad c = 1 \qquad (3.22c)$$

$$r = edb \qquad (3.22d)$$

($d,b,e,r\in\mathbb{R}$; $b,e>0$). Multiplication of (3.22a) by d gives $d=-1-\varepsilon+\sqrt{2\varepsilon+\varepsilon^2-b^2-e^2}$ (the other root is not achieved). Substitution of (3.22c) into (3.22b) yields

$$b = \varepsilon/(1-d^2). \qquad (3.23)$$

We now assume $b=O(\varepsilon^p)$. Then, p must be equal to $\frac{1}{2}$, since:

(*) if $p<\frac{1}{2}$, then d will no longer be real;

(**) if $p>\frac{1}{2}$, then $d=-1+O(\varepsilon^{\frac{1}{2}})$, so that (3.23) gives $b=O(\varepsilon^{\frac{1}{2}})$, which contradicts $b=O(\varepsilon^p);p>\frac{1}{2}$.

We neglect higher-order terms of ε and denote $b=k\varepsilon^{\frac{1}{2}}$ ($k>0$), so that $e=-db=k\varepsilon^{\frac{1}{2}}$ and $d=-1+\sqrt{2-2k^2}.\varepsilon^{\frac{1}{2}}$. From (3.23) it follows that $k=\frac{1}{2}(2-2k^2)^{-\frac{1}{2}}$, so that $k=\frac{1}{2}(2-\sqrt{2})$ (the other root is not achieved). Now $k<1$, so that $d\in\mathbb{R}$. Substitution of e, d and b in (3.22d) gives $r=-k^2\varepsilon=\frac{1}{4}(\sqrt{2}-2)\varepsilon$ [thus, r is $O(\varepsilon)$!!]. By substitution of the reduction factors (3.7) and a supremum $(\theta_1,\theta_2)=(\frac{\pi}{2},0)$ into (3.11), we then obtain $\bar{\rho} = |r|/|r+\varepsilon| = (3+2\sqrt{2})^{-1}$. □

Similar proofs can be given for other ILU limit factors.

3.2.3. An incomplete line LU scheme

An incomplete block LU factorisation (cf. [10]) of A is presented by (3.18), where $\bar{L}$ is strictly lower block-triangular, $\bar{U}$ is strictly upper block-triangular and $\bar{D}$ is block-diagonal. We assume each block to represent a line and call the scheme ILLU (Incomplete Line LU).

Equations (3.18) and (3.19) still hold for ILLU, and factorisation can then be performed by algorithm 3.1 with n_2 instead of $|\Omega|$, $P^*=\{-1,0,1\}$ and M_i^j denoting the $n_1\times n_1$ block $(i,i+j)$ of matrix M. It easily follows that $\bar{L}=L$ and $\bar{U}=U$. Then, assuming $N_i^j=0$ on the discretisation molecule, the algorithm simplifies to:

Algorithm 3.3

$\bar{D}_1 = D_1$

for $i:=2(1)n_2$ *do* $\bar{D}_i := D_i - \text{tridiag}(L_i(\bar{D}_{i-1})^{-1}U_{i-1})$,

where $\bar{D}_i = S_i^0$, $L_i = A_i^{-1}$, $D_i = A_i^0$ and $U_i = A_i^1$. For relaxation, we store the tridiagonal LDU-decomposition of $\bar{D}$.

The relaxation statement $u:=u+[(L+\bar{D})(\bar{D})^{-1}(\bar{D}+U)]^{-1}(f-Au)$ is composed of:

Algorithm 3.4

$r:=f-Au$

$r_1:=(\bar{D}_1)^{-1}r_1$

for $i:=2(1)n_2$ *do* $r_i:=(\bar{D}_i)^{-1}(r_i-L_i r_{i-1})$

for $i:=n_2-1(-1)1$ *do* $r_i:=r_i-(\bar{D}_i)^{-1}U_i r_{i+1}$

$u:=u+r$,

where v_i denotes the i'th $n_1 \times 1$ block of vector v. We notice that one iteration of ILLU basically requires the solution of two tridiagonal systems and a residual calculation, whereas additional storage (compared with schemes such as PGSF) is 4 vectors. For extensive treatment of incomplete block LU methods, we refer to [10].

In order to calculate the reduction factors $\rho(\theta)$, we apply the ILLU analogon of algorithm 3.2, i.e. algorithm 3.3 without subscripts and with n_2 large, so that K and N can be computed and substituted in equation (3.7). We observe that reduction factors (and, hence, smoothing factors) depend on the block size n_1. In our experience, however, the rate of smoothing decreases as n_1 tends to infinity. We therefore consider the worst case $n_1 = \infty$ only; for this case reduction-factor plots and smoothing factors can be found in Appendices A and B, respectively.

3.3. Comparison of smoothing efficiencies

For efficiency comparison, one has to take into account aspects of computational work. Hence, we have estimated W for each relaxation scheme; W = the number of operations +, -, *, / per grid point necessary for one iteration, assuming variable matrix coefficients. The smoothing efficiency is then defined by

$$\bar{\tau} = W / |\ln \bar{\rho}| , \tag{3.24}$$

i.e. the number of operations per grid point to achieve a reduction by a factor of e in the components $\theta \in H$ with the largest reduction factor. For all schemes considered in Section 3.2 and for all model problems of Section 3.1, smoothing efficiencies are given in Appendix C.

From this appendix we notice that "robust" smoothers are LGSF1F2, LGSF1B1F2B2 and ILLU, the last scheme being somewhat more efficient than the other two. The remaining schemes fail as a smoother for anisotropic-diffusion problems, and sometimes for convection-diffusion problems. It should be noted that, as ε tends to zero for the anisotropic case, the performance of the ILU schemes deteriorates less rapidly than that of schemes such as PGSF. (For $\varepsilon=0.01$, ILU efficiency factors are still acceptable). Moreover, it can be seen that ILU(1,2) is more robust as a smoother than ILU(1,1) [see problem 2], whereas it is more efficient than ILU(1,3), ILU(2,4) and ILU(7,9) [see problems 3 and 4; these ILU schemes also require more memory storage than ILU(1,2)]. Further analysis of Appendix C is omitted and left to the reader.

A set of methods not considered in this chapter is that of pattern-relaxation schemes such as ZLGS (Zebra Line Gauss-Seidel, see [5]; these schemes can be efficiently implemented in MGR, cf. [4]). Local mode analysis for these schemes is contained in [3]; smoothing efficiencies appear to be better than those of corresponding non-pattern-relaxation schemes. In our experiments, however, significant differences in rates of convergence were never observed (see Chapter 4).

Of course, local mode analysis gives only a rough indication of the behaviour of relaxation schemes in MG, MGCG and ICCG (for MG, two-level analysis is more involved than smoothing analysis). Hence, the following chapter is restricted to the practical behaviour of schemes.

4. PRACTICAL BEHAVIOUR OF RELAXATION SCHEMES IN THE MODIFIED MG, THE MGCG AND THE ICCG METHOD: RESULTS AND COMPARISON FOR TEST PROBLEMS OF STONE AND KERSHAW

The modified MG method outlined in Section 2.2, the MGCG method mentioned in Section 2.3 and the ICCG method (cf. [8]) have been applied to the following test problems:

Stone's test problem (cf. [11])

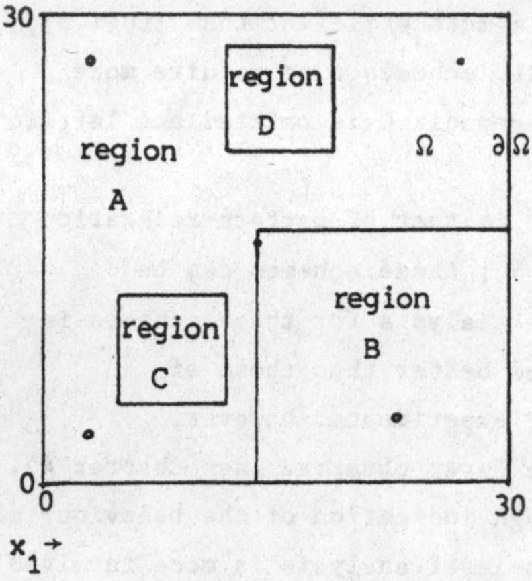

$$-\nabla . D\nabla u = f \text{ on } \Omega \quad (4.1a)$$

$$n.\nabla u = 0 \quad \text{on } \partial\Omega \text{ (Neumann)} \quad (4.1b)$$

$$f(3,3)=1.0,\ f(3,27)=0.5,\ f(23,4)=0.6,$$
$$f(14,15)=-1.83,\ f(27,27)=-0.27 \quad (4.1c)$$

region:	A	B	C	D
D_{11}=	1	1	10^p	0
D_{22}=	1	10^p	1	0

where we selected $p \in \{0,1,2,3,4,5\}$ (p=0 yields a Poisson equation).

Kershaw's test problem (cf. [6])

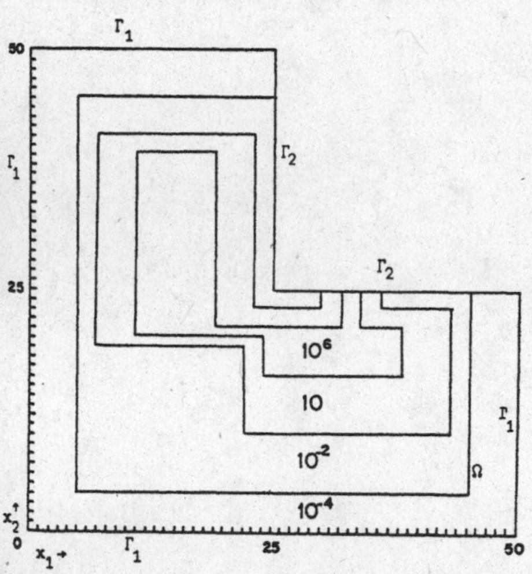

$$-\nabla . D\nabla u + u = f \quad \text{on } \Omega \quad (4.2a)$$

$$\left.\begin{array}{ll} n.\nabla u = 0 & \text{on } \Gamma_1 \text{ (Neumann)} \\ u = 0 & \text{on } \Gamma_2 \text{ (Dirichlet)} \\ \Gamma_1 \cup \Gamma_2 = \partial\Omega & \end{array}\right\} \quad (4.2b)$$

$$f(x_1,x_2) = 10^{(x_1 - x_2 + 49.)/24.5} \quad (4.2c)$$

D is strongly discontinuous on Ω; the orders of magnitude of D are marked in the figure (exact values of D are given in [6]).

The following relaxation schemes have been chosen:

- in MG/MGCG: PGSF, PGSFB, LGSF1, ZLGSF1, LGSF1F2, ZLGSF1F2, LGSF1B1F2B2, ILU(1,1), ILU(1,2) and ILLU;
- in ICCG: ILU(1,1), ILU(1,2) and ILLU.

In the zebra schemes ZLGSF1 and ZLGSF1F2, even lines are solved first. (Similar results have been obtained for the case of odd lines first.)

In the algorithms, we selected $u^0=0$, M=3, LU solution or 100 symmetric relaxations on Ω^3, and the termination criterion

$$||r_\eta^0|| < 10^{-10}||r_0^0|| \quad \vee \quad \textbf{MAXIT iterations,} \tag{4.3}$$

where η is the number of iterations, $||.||$ is the discrete ℓ^2 norm, MAXIT=50 for MG/MGCG and MAXIT=∞ for ICCG. The rate of reduction is defined by

$$\rho = (||r_\eta^0||/||r_0^0||)^{\frac{1}{\eta}} . \tag{4.4}$$

Furthermore, we have estimated the iteration work W for each method, and define: efficiency $\tau=W/|\ln \rho|$. Rates of reduction and efficiencies are given in Appendix D. (Encircled figures for MGCG indicate a much better limit convergence.)

From this appendix, it can be seen that, for anisotropic problems, a suitable relaxation scheme is indispensable for rapid convergence of MG. (It is obvious, however, that but for the modifications outlined in Section 2.2, MG would have failed for both test problems, because coefficients are discontinuous or domains are non-rectangular.) PGSF, PGSFB and LGSF1 demonstrate a similar behaviour (they are not robust relaxation schemes in MG/MGCG), and zebra relaxation schemes yield results analogous to those of corresponding non-zebra methods.

For further comparison, we make use of the graphical results depicted in Appendix E, where we consider Stone's problem with p=5 and Kershaw's problem. Markers are denoted every five iterations (and sometimes at the end of a curve). Preprocessing work has been incorporated.

<u>Results of modified MG</u> are plotted in Fig. E-1. We consider the following basically different relaxation schemes: LGSF1, LGSF1F2, LGSF1B1F2B2, ILU(1,1), ILU(1,2) and ILLU. For comparing schemes, the results for Stone's problem appear to be more significant than those for Kershaw's problem. It can be seen that LGSF1 is not a robust smoother, as predicted in Chapter 3. Besides, for Stone's anisotropic problem, ILU schemes are better relaxation methods in MG than might be expected from smoothing analysis. Applying MG to $u_{xx}+\varepsilon u_{yy}=f$ with $\Omega=(0,30)\times(0,30)$, Neumann boundary conditions and Stone's right-hand side, we obtain the results expected from smoothing analysis:

ρ	ILU(1,1)	ILU(1,2)
$\varepsilon=10^{-5}$	0.9000	0.8416
$\varepsilon=10^{5}$	0.9142	0.0014

Apparently, ILU only fails as a smoother in MG if the anisotropies extend from boundary to boundary (with Neumann boundary conditions; results for the case of Dirichlet boundary conditions are better, see [12]). Of the "robust" schemes proposed in Chapter 3, LGSF1B1F2B2 is less efficient than LGSF1F2 and ILLU.

Results of MGCG are plotted in Fig. E-2. The same relaxation schemes are considered as those in Fig. E-1. With respect to the schemes, Fig. E-2 is (almost) similar to Fig. E-1. However, for Stone's problem, the inefficiency of a scheme is definitely compensated by the use of CG; even the curve of LGSF1 drops (outside the figure): the average reduction factor for MGCG with LGSF1 decreases from 0.9561 (after 50 iterations) to 0.8723 (after 177 iterations), whereas the same factor for MG with LGSF1 increases from 0.9534 (after 50 iterations) to 0.9869 (after 177 iterations). Also, for the anisotropic diffusion equation mentioned above, the performance of ILU schemes in MG partly recovers as a result of the combination with CG:

ρ	ILU(1,1)	ILU(1,2)
$\varepsilon=10^{-5}$	0.3270	0.2545
$\varepsilon=10^{5}$	0.3354	0.00007

In addition, for MG the use of CG has a significant positive effect on Kershaw's problem.

Results of ICCG are plotted in Fig. E-3. As preconditioning for CG, ILLU is obviously better than ILU(1,1) and ILU(1,2).

Finally, results of MG, MGCG and ICCG have been gathered in Fig. E-4. For each iterative method, we have selected ILLU as the relaxation scheme, because of its robustness. We also depict MGCG with ILU(1,2) for Kershaw's problem. The methods considered appear to be equally efficient for Stone's problem. For Kershaw's problem, however, we notice that MG is significantly slower than MGCG and ICCG. Besides, ICCG suffers from a slow start and is rather inefficient for Poisson-type equations (see Appendix D). Hence, MGCG is to be preferred.

5. CONCLUDING REMARKS

In view of robustness of the MG method, we propose:

(i) the use of Gauss elimination / continuity interpolation to cope with strong inhomogeneities and general domains, and

(ii) the use of a robust smoother to cope with strong anisotropies and singularly-perturbed problems.

Results with respect to (i) are given in [1], [7]; investigations on (ii) have been presented in this paper.

Local mode analysis shows that simple PGS, LGS and ILU relaxation schemes are not robust smoothers (they fail for anisotropic-diffusion equations, and sometimes for singularly-perturbed convection-diffusion equations), whereas LGS in alternating directions and ILLU are efficient smoothers for all model problems considered.

Results of smoothing analysis have been confirmed by several experiments; for test problems containing a variety of difficulties, LGSF1F2 and ILLU appear to be robust and efficient smoothers in MG. Hence, these schemes (or, for the case of parallel computers, their zebra analogons, cf. [3], [10]) are to be preferred in a robust MG solver.

Combination of MG with CG (i.e. MGCG, cf. [7]) further improves the robustness. In MGCG, for instance ILU methods are acceptable relaxation schemes.

Finally, it should be noted that CG with ILLU as preconditioning (i.e. a special case of ICCG) is in strong competition with MGCG for difficult problems, if high accuracy is required. In any case, one should not bet on one horse only.

Acknowledgements

The author would like to thank P. Wesseling (Delft University of Technology) and J.A. Meijerink (KSEPL) for reviewing this paper and for their guidance.

References

[1] Alcouffe, R.E., Brandt, A., Dendy Jr., J.E. and Painter, J.W., The multi-grid method for the diffusion equation with strongly discontinuous coefficients, SIAM J. Sci. Stat. Comput. 2, pp. 430-454, 1981.

[2] Brandt, A., Multi-level adaptive solutions to boundary-value problems, Math. Comp. 31, pp. 333-390, 1977.

[3] Brandt, A., Multigrid solvers on parallel computers, Elliptic Problem Solvers, Proceedings, M.H. Schultz, ed., Academic Press, New York, pp. 39-83, 1981.

[4] Foerster, H., Stüben, K. and Trottenberg, U., Non-standard multigrid techniques using checkered relaxation and intermediate grids, Elliptic Problem Solvers, Proceedings, M.H. Schultz, ed., Academic Press, New York, pp. 285-300, 1981.

[5] Hackbusch, W., On the multi-grid method applied to difference equations, Computing 20, pp. 291-306, 1978.

[6] Kershaw, D.S., The incomplete Cholesky-conjugate gradient method for the iterative solution of systems of linear equations, J. Comp. Phys. 26, pp. 43-65, 1978.

[7] Kettler, R. and Meijerink, J.A., A multigrid method and a combined multigrid-conjugate gradient method for elliptic problems with strongly discontinuous coefficients in general domains, Shell Publication 604, KSEPL, Rijswijk, The Netherlands, 1981. (submitted)

[8] Meijerink, J.A. and Van der Vorst, H.A., An iterative solution method for linear systems of which the coefficient matrix is a symmetric M-matrix, Math. Comp. 31, pp. 148-162, 1977.

[9] Meijerink, J.A. and Van der Vorst, H.A., Guidelines for the usage of incomplete decompositions in solving sets of linear equations as they occur in practical problems, J. Comp. Phys. 44, pp. 134-155, 1981.

[10] Meijerink, J.A., Incomplete block LU factorisations, publication in preparation.

[11] Stone, H.L., Iterative solution of implicit approximations of multidimensional partial differential equations, SIAM J. Numer. Anal. 5, pp. 530-558, 1968.

[12] Wesseling, P., A robust and efficient multigrid method, in this volume.

[13] Wesseling, P. and Sonneveld, P., Numerical experiments with a multiple grid and a preconditioned Lanczos method, Approximation Methods for Navier-Stokes Problems, Proceedings, Paderborn 1979, R. Rautmann, ed., Lecture Notes in Math. 771, Springer-Verlag, Berlin, pp. 534-562, 1980.

Appendix A:

Plots of reduction factors of Fourier components

PGSF, Problem 3, ε=0.1

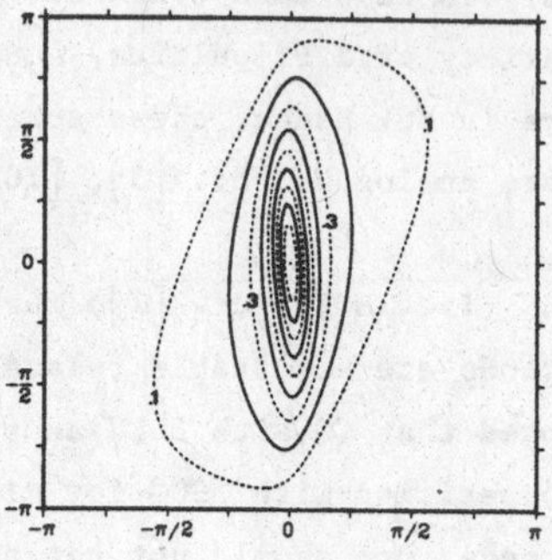

PGSF, Problem 7, ε=0.1

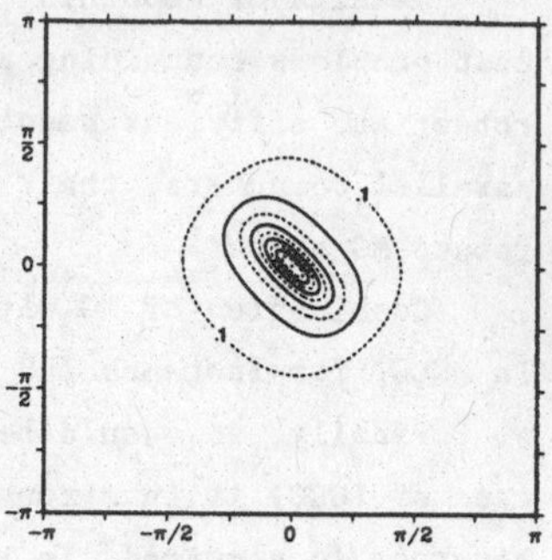

PGSF, Poisson equation

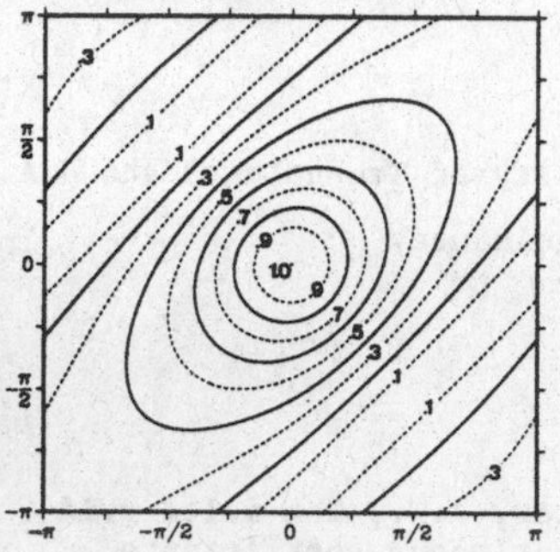

PGSF, Problem 4, ε=0.1

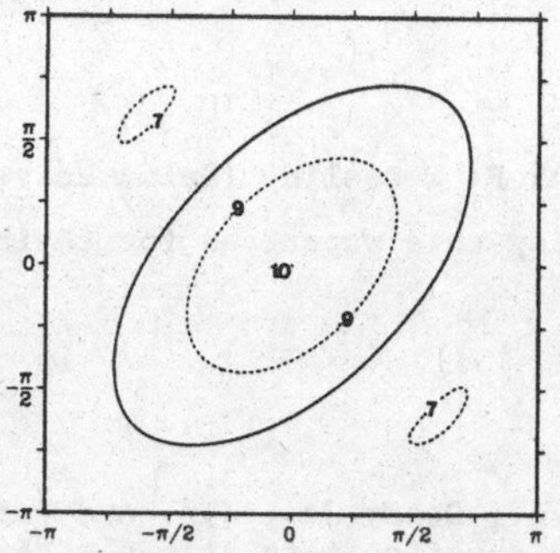

PGSF, Problem 8, ε=0.1

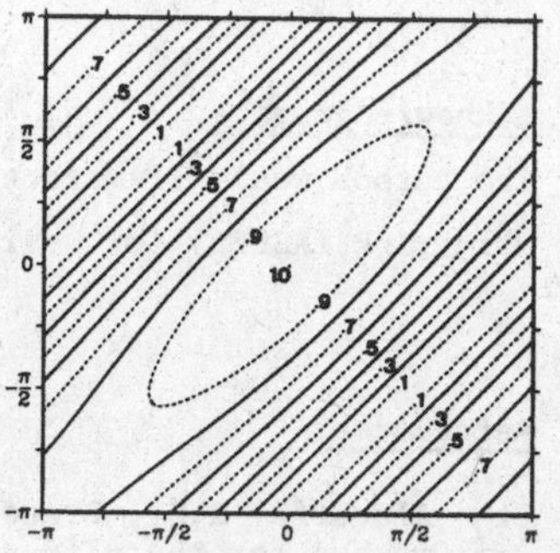

PGSF, Problem 1, ε=0.1

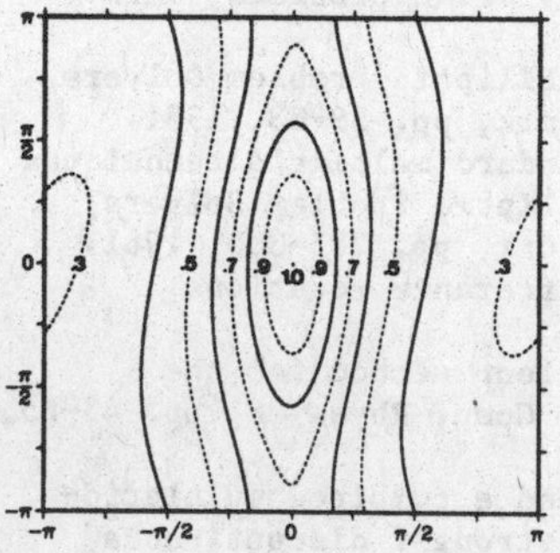

PGSF, Problem 5, ε=0.1

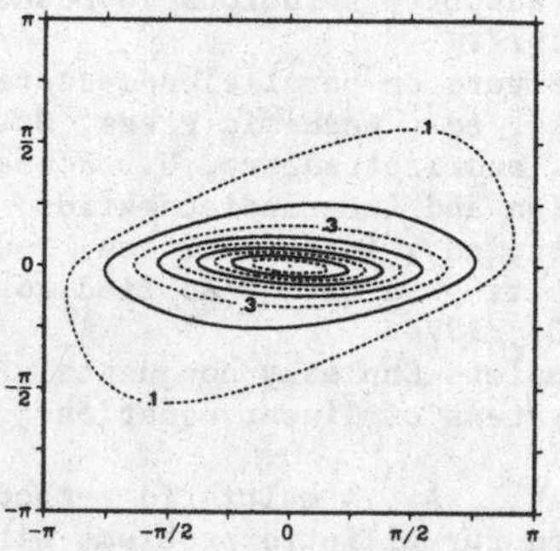

PGSF, Problem 9, ε=0.1

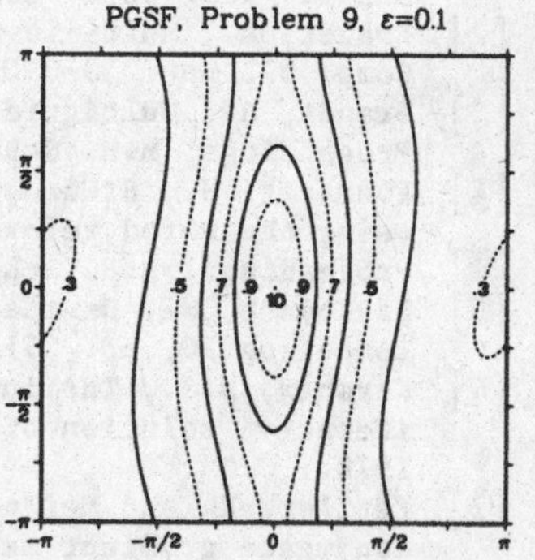

PGSF, Problem 2, ε=0.1

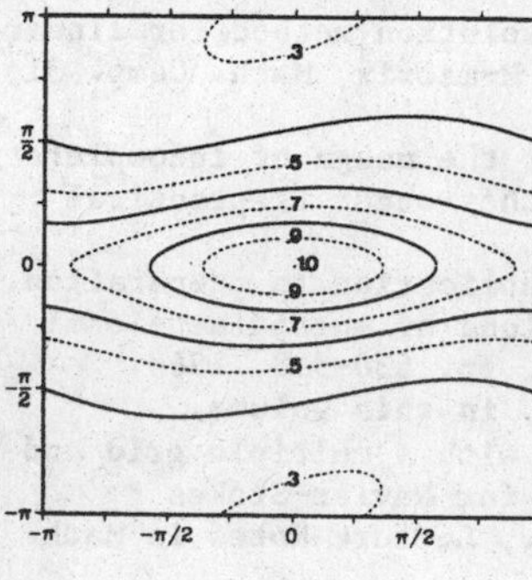

PGSF, Problem 6, ε=0.1

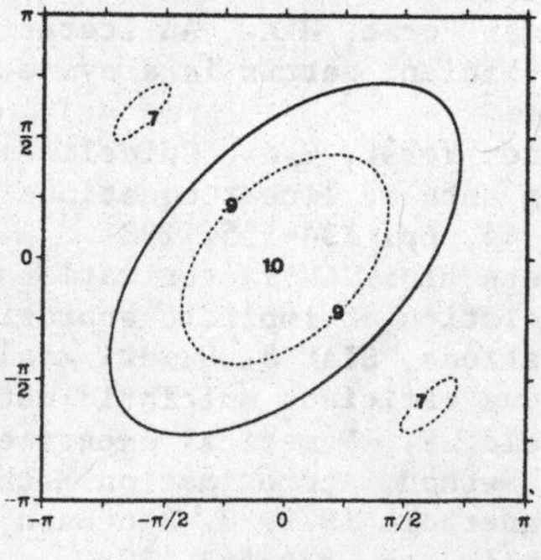

PGSF, Problem 10, ε=0.1

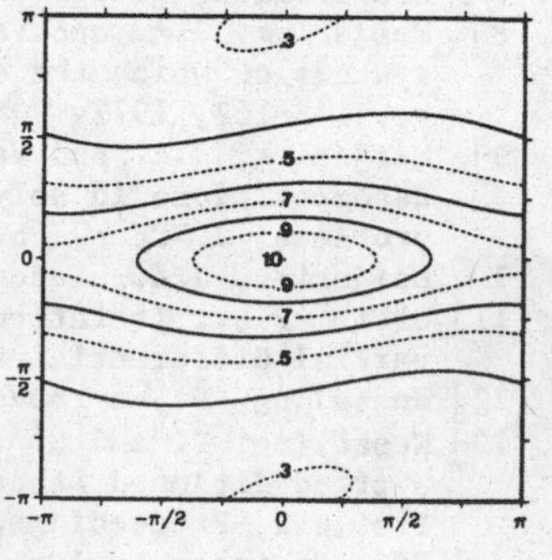

PGSF, Problem 3, ε=0.01

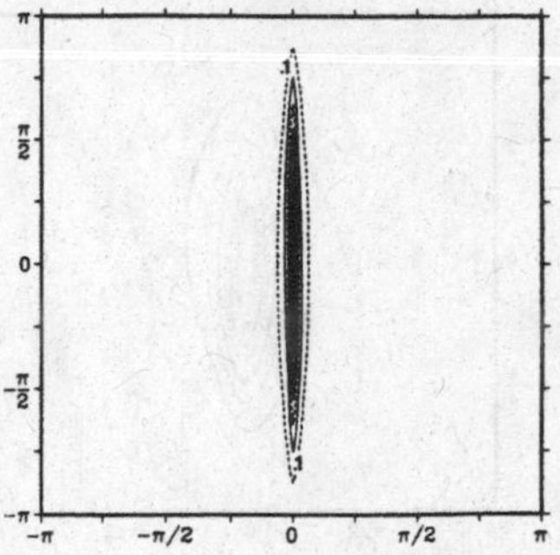

PGSF, Problem 7, ε=0.01

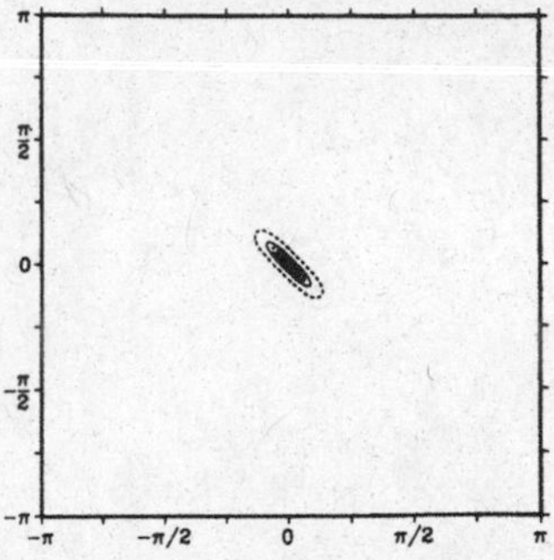

PGSF, Problem 4, ε=0.01

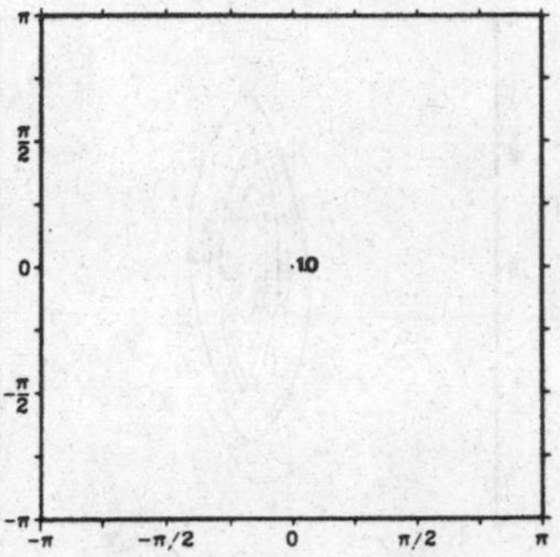

PGSF, Problem 8, ε=0.01

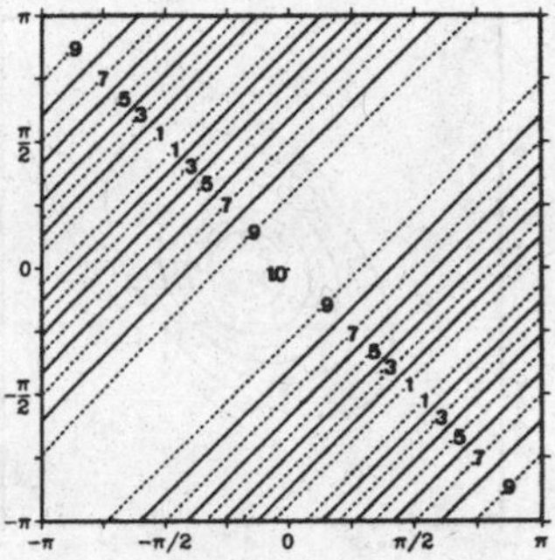

PGSF, Problem 1, ε=0.01

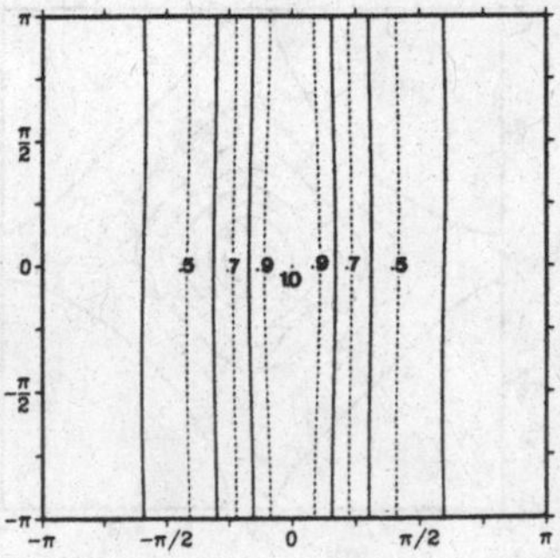

PGSF, Problem 5, ε=0.01

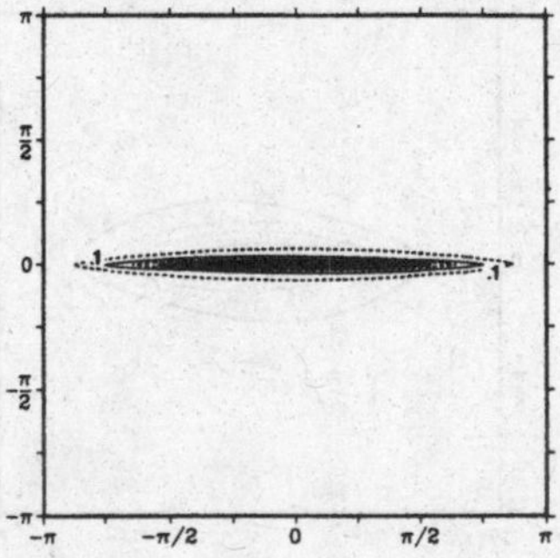

PGSF, Problem 9, ε=0.01

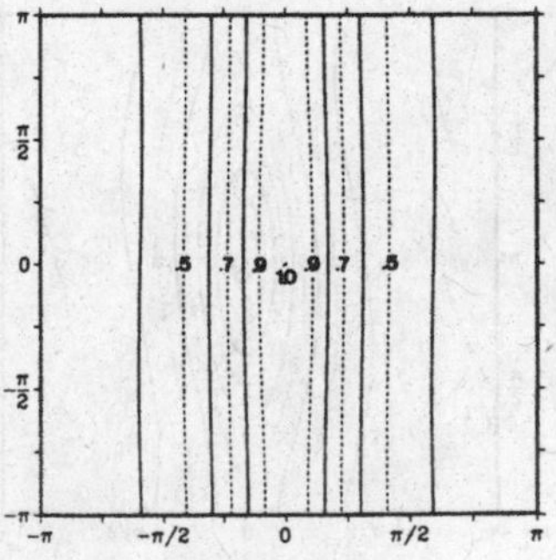

PGSF, Problem 2, ε=0.01

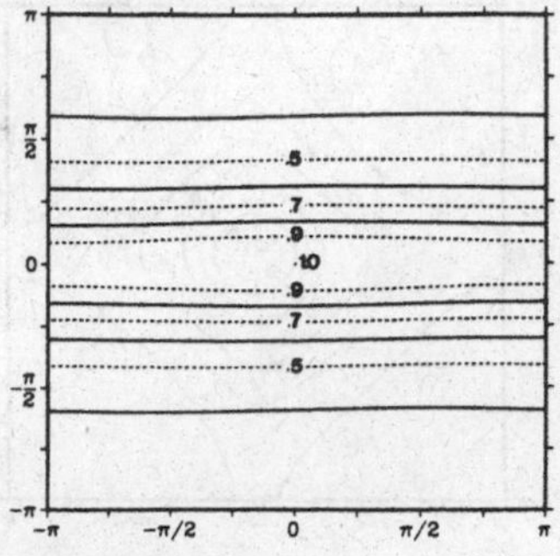

PGSF, Problem 6, ε=0.01

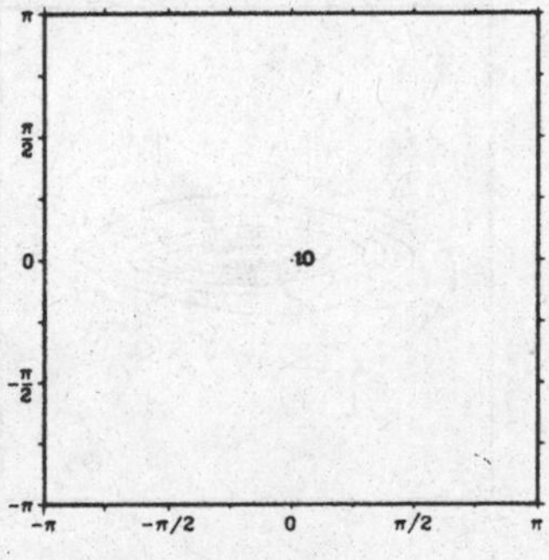

PGSF, Problem 10, ε=0.01

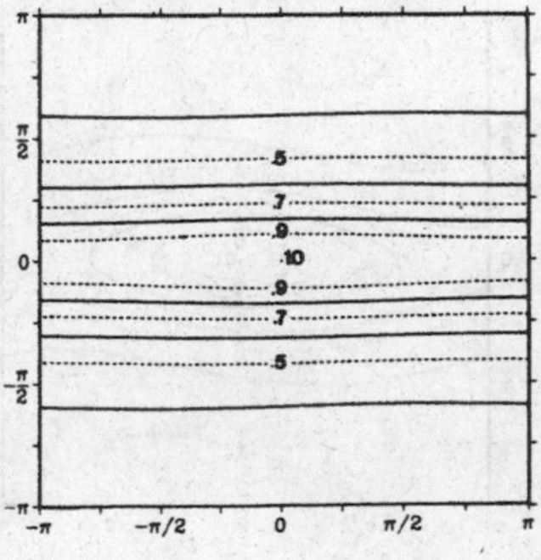

PGSFB, Problem 3, ε=0.1

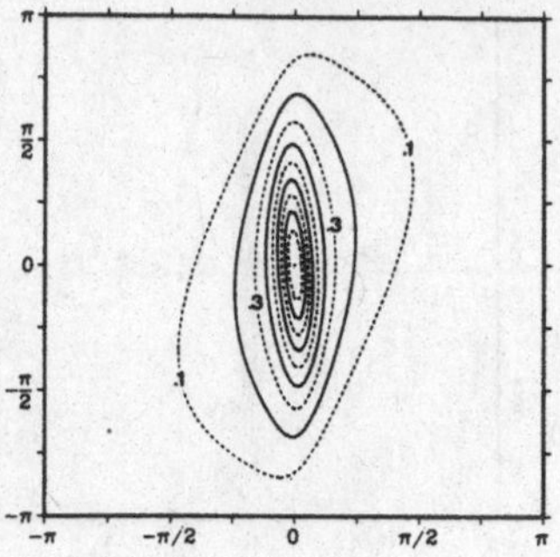

PGSFB, Problem 7, ε=0.1

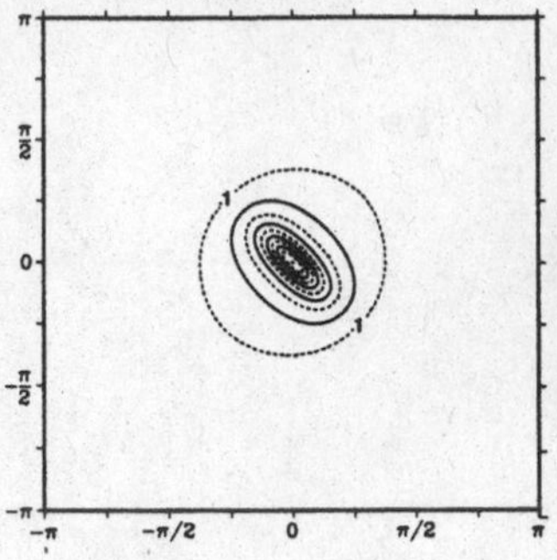

PGSFB, Poisson equation

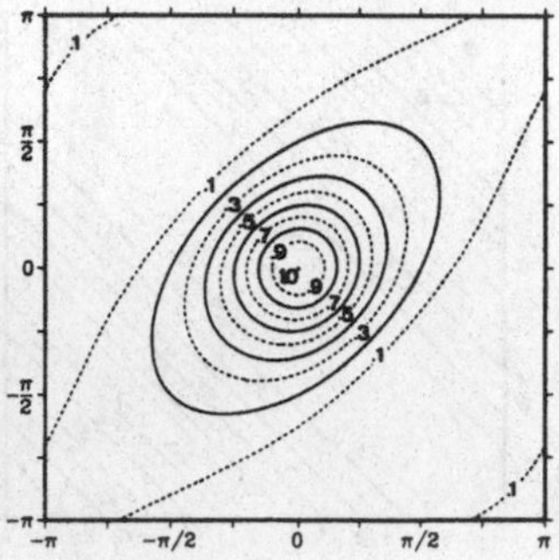

PGSFB, Problem 4, ε=0.1

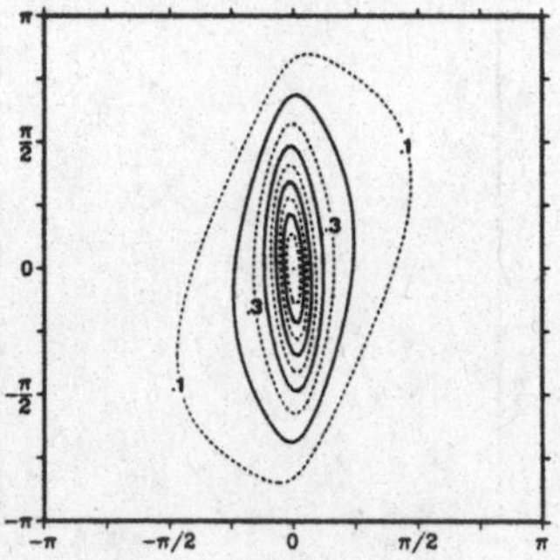

PGSFB, Problem 8, ε=0.1

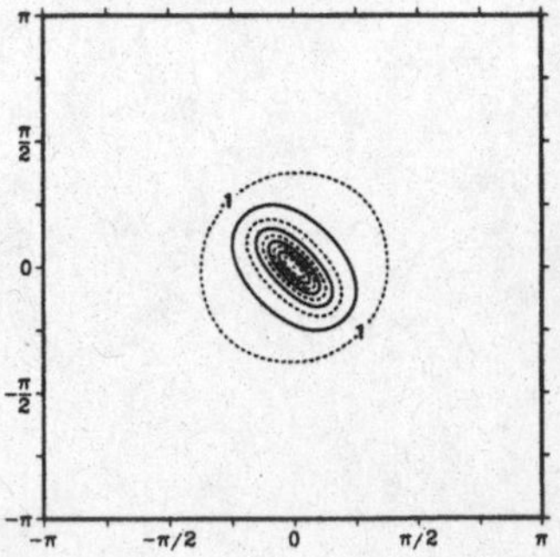

PGSFB, Problem 1, ε=0.1

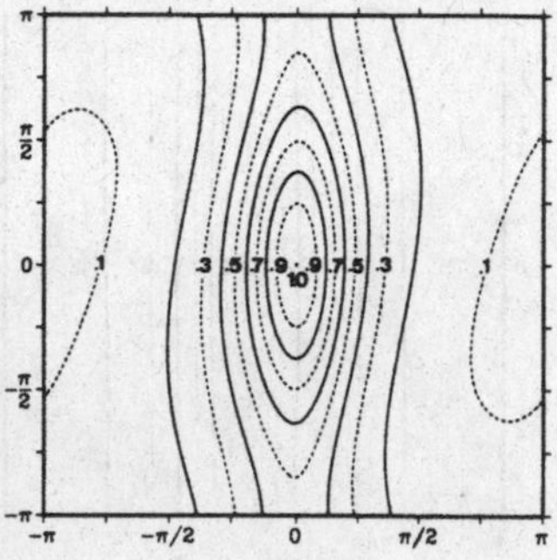

PGSFB, Problem 5, ε=0.1

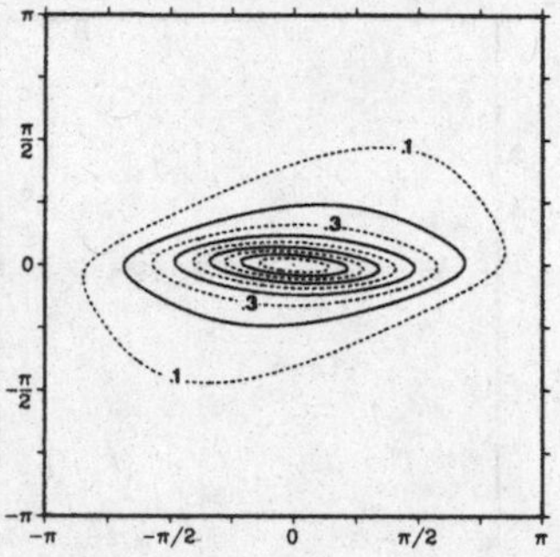

PGSFB, Problem 9, ε=0.1

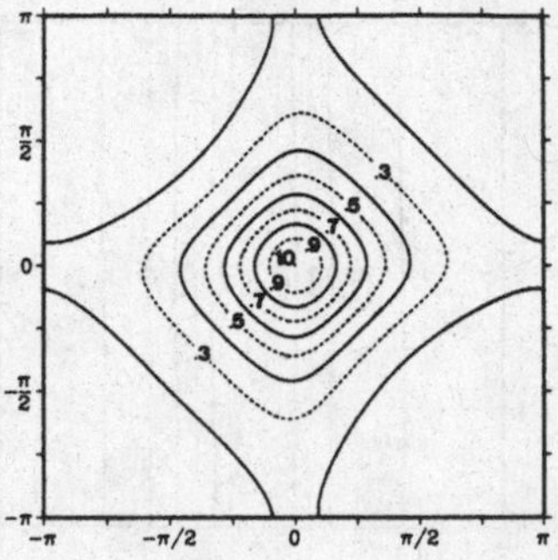

PGSFB, Problem 2, ε=0.1

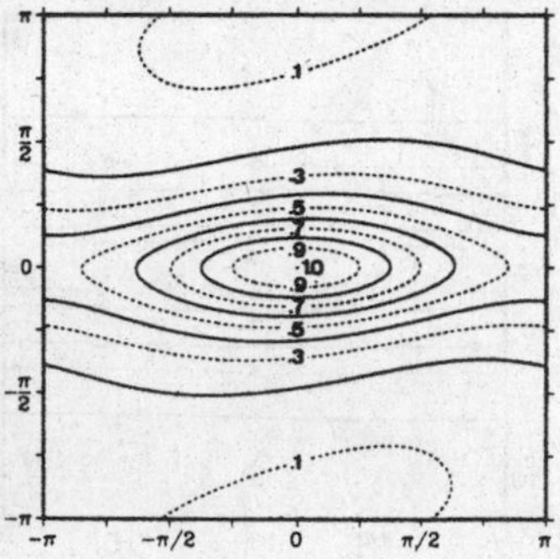

PGSFB, Problem 6, ε=0.1

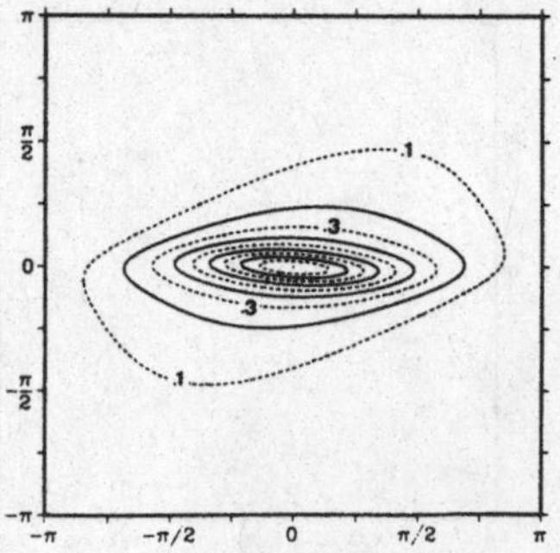

PGSFB, Problem 10, ε=0.1

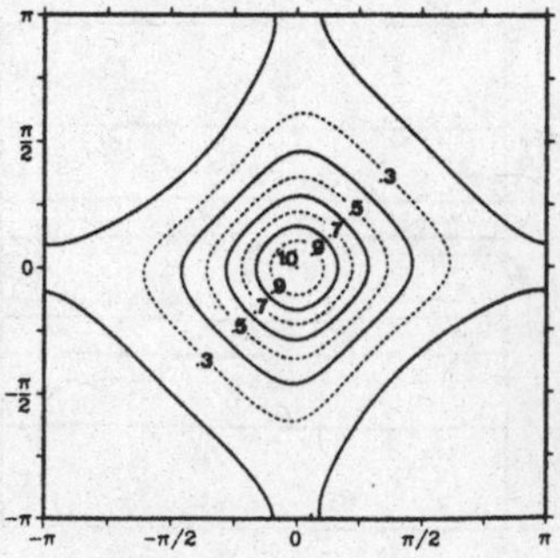

LGSF1, Problem 3, ε=0.1

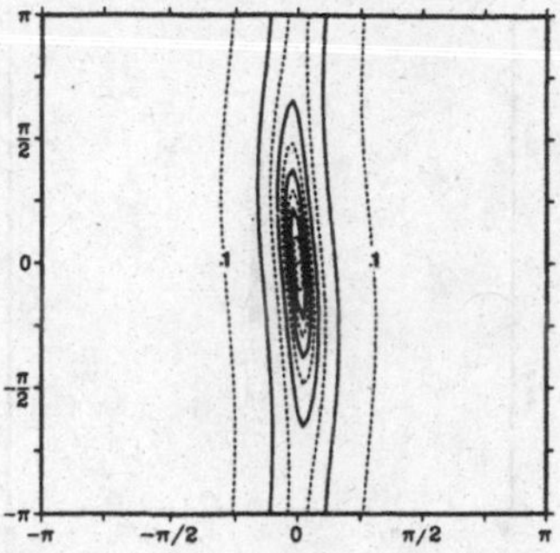

LGSF1, Problem 7, ε=0.1

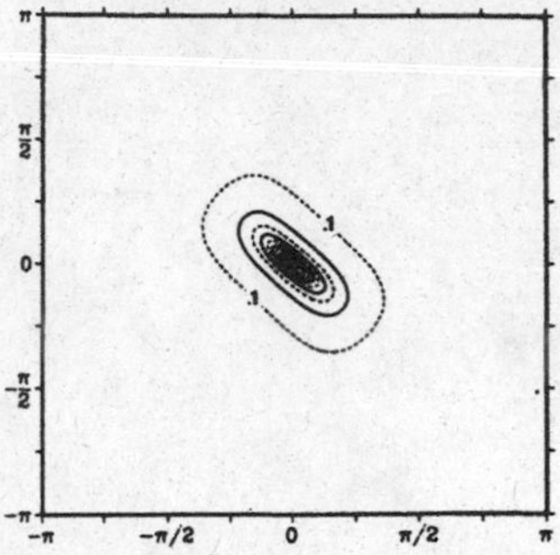

LGSF1, Poisson equation

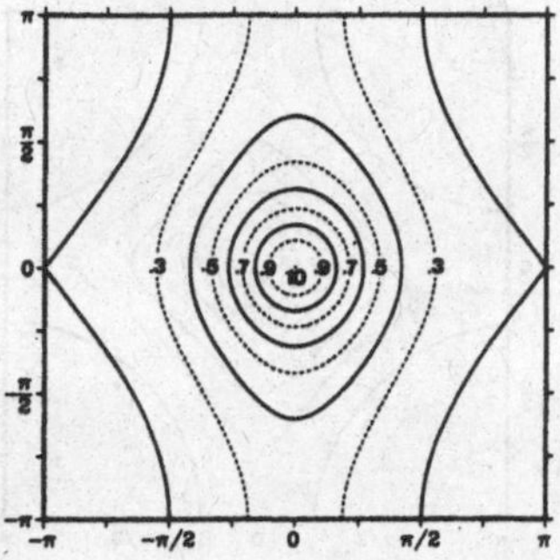

LGSF1, Problem 4, ε=0.1

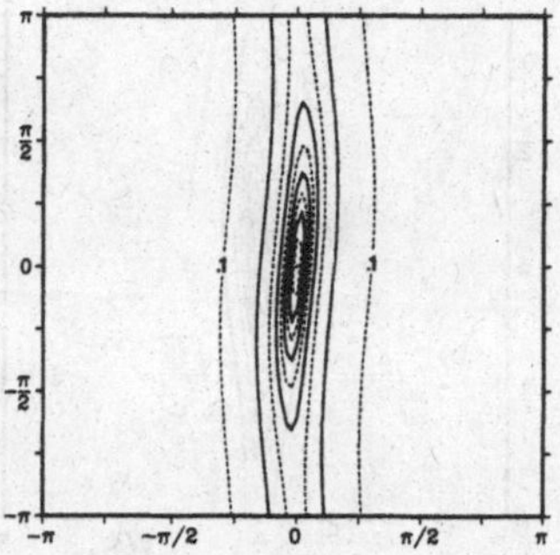

LGSF1, Problem 8, ε=0.1

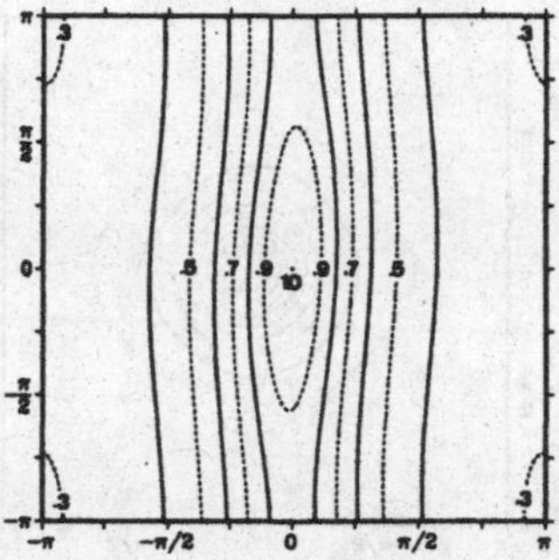

LGSF1, Problem 1, ε=0.1

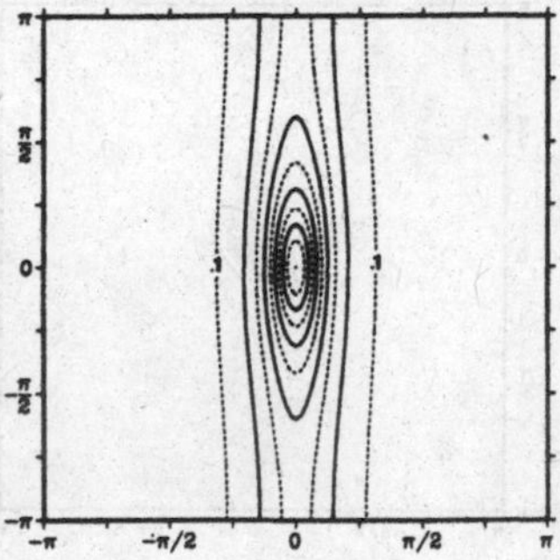

LGSF1, Problem 5, ε=0.1

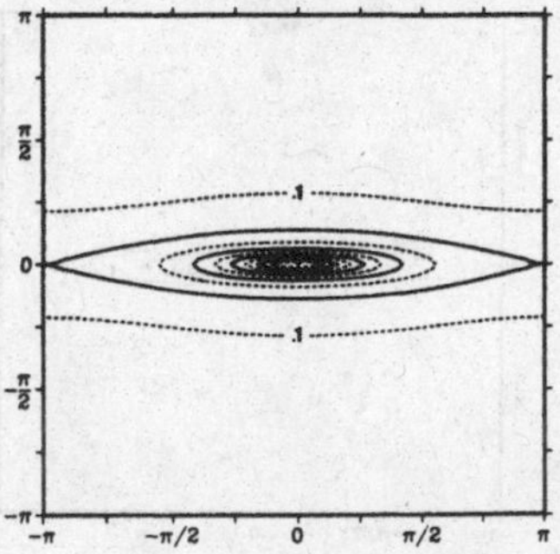

LGSF1, Problem 9, ε=0.1

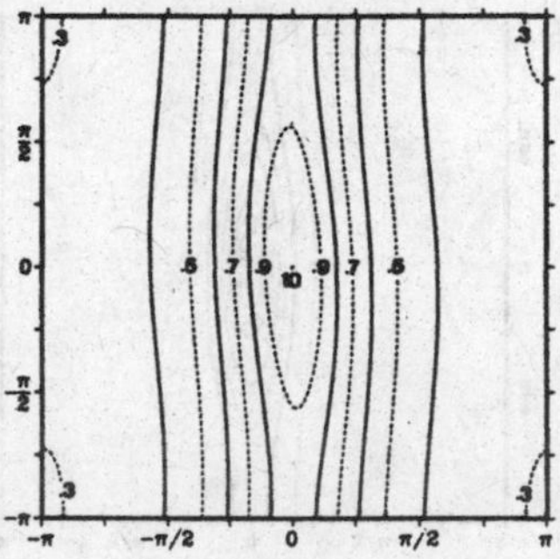

LGSF1, Problem 2, ε=0.1

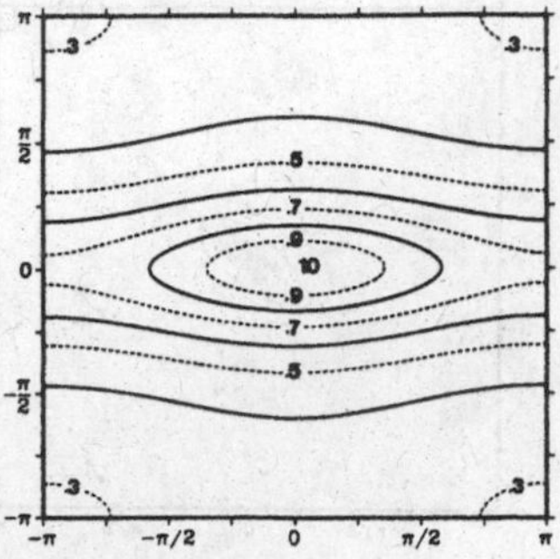

LGSF1, Problem 6, ε=0.1

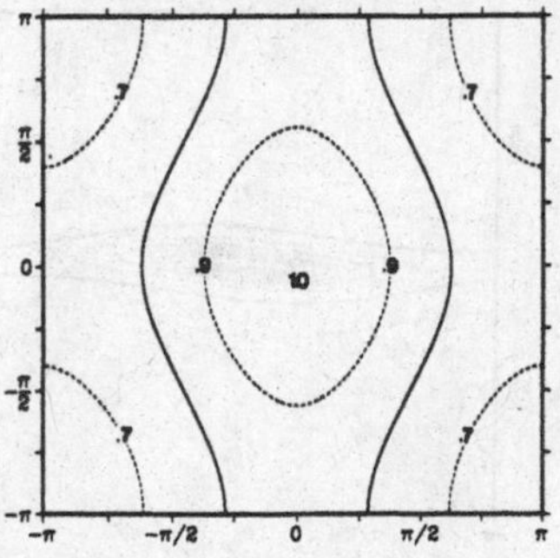

LGSF1, Problem 10, ε=0.1

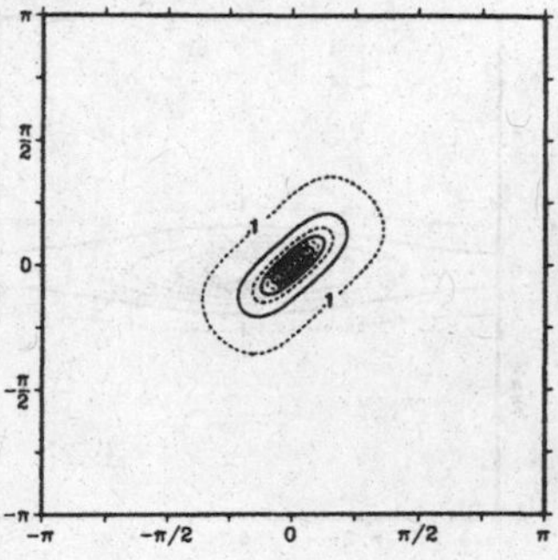

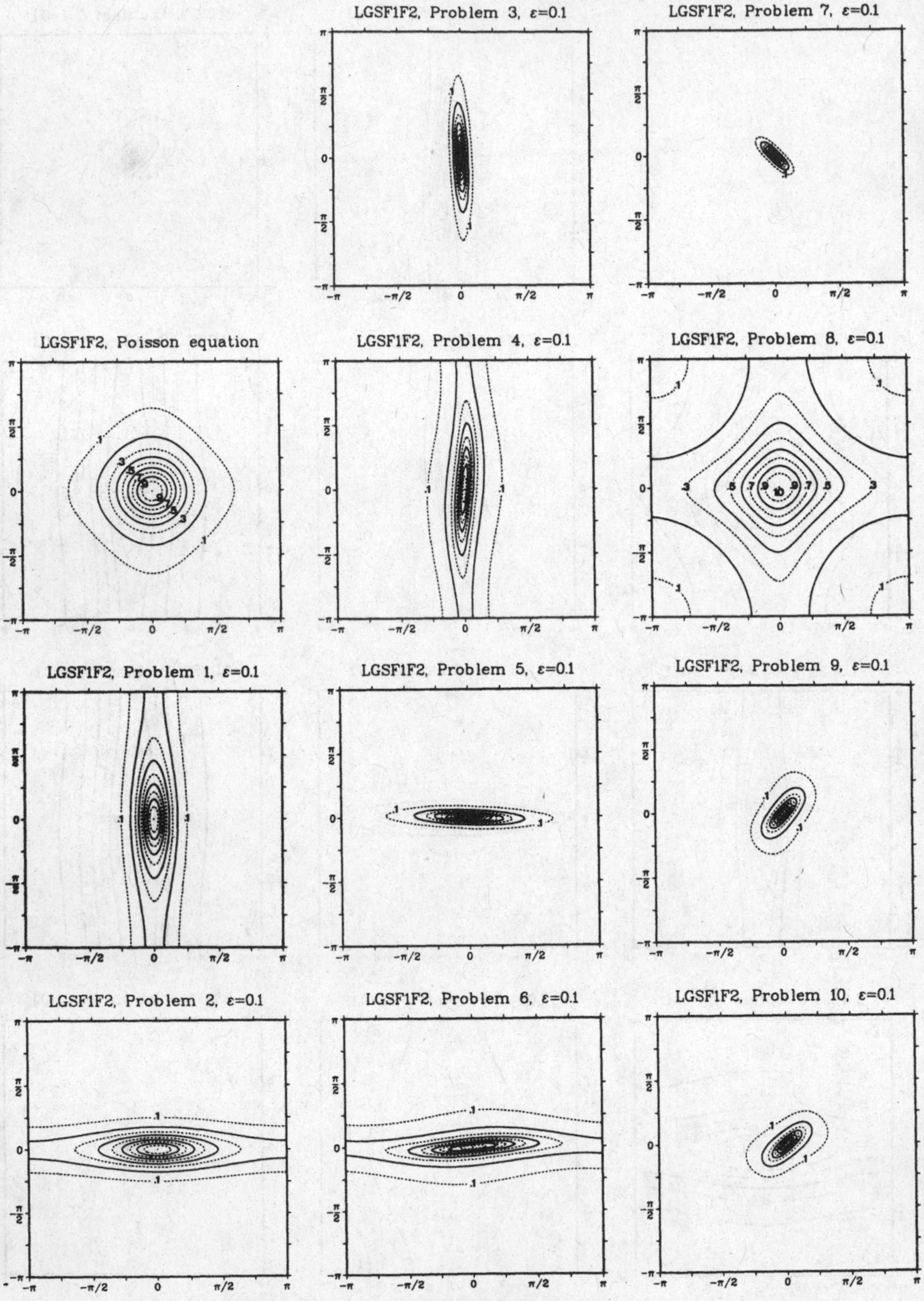
LGSF1F2, Problem 3, ε=0.1
LGSF1F2, Problem 7, ε=0.1
LGSF1F2, Poisson equation
LGSF1F2, Problem 4, ε=0.1
LGSF1F2, Problem 8, ε=0.1
LGSF1F2, Problem 1, ε=0.1
LGSF1F2, Problem 5, ε=0.1
LGSF1F2, Problem 9, ε=0.1
LGSF1F2, Problem 2, ε=0.1
LGSF1F2, Problem 6, ε=0.1
LGSF1F2, Problem 10, ε=0.1
−π
−π/2
0
π/2
π

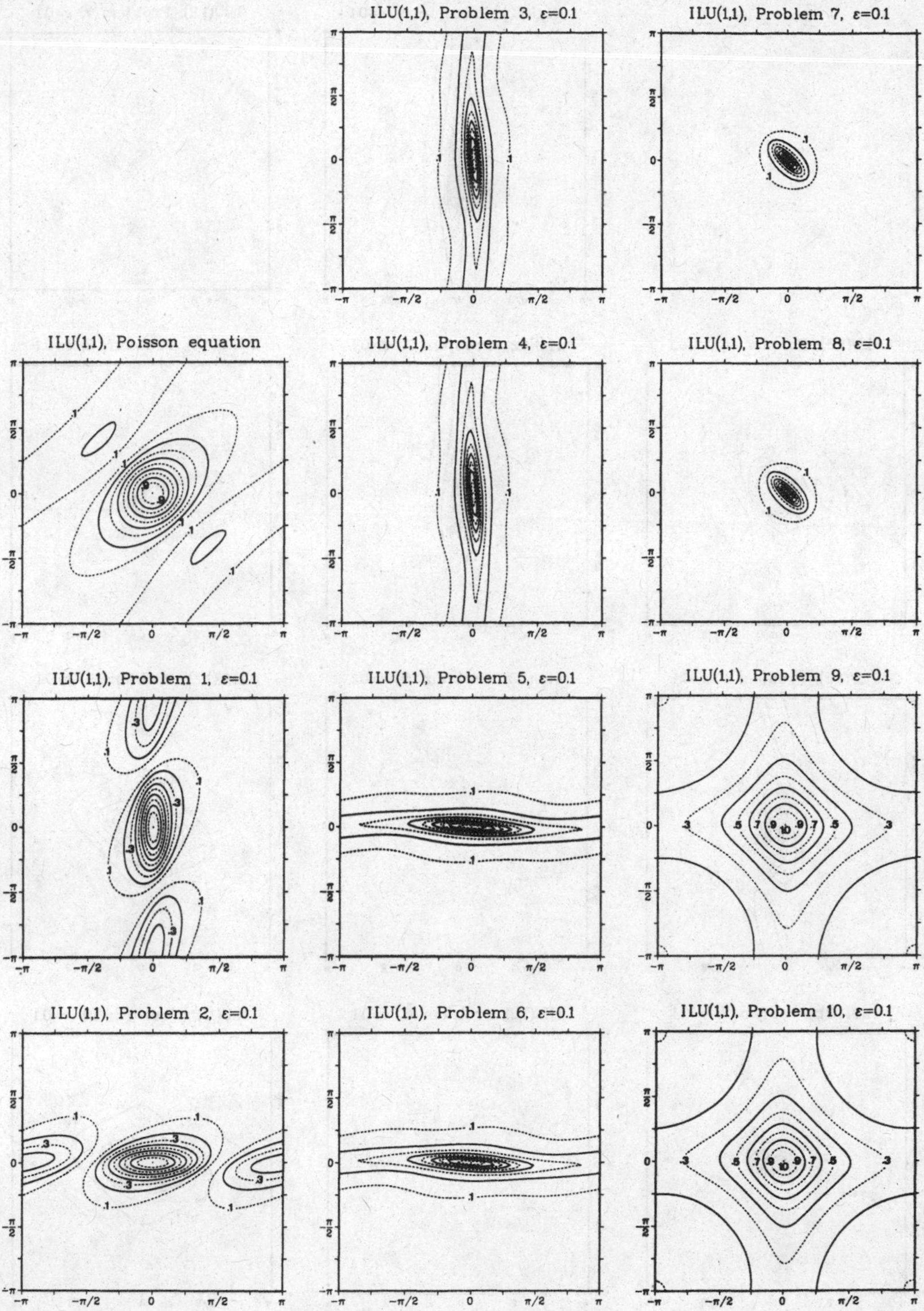
ILU(1,1), Problem 3, ε=0.1
ILU(1,1), Problem 7, ε=0.1
ILU(1,1), Poisson equation
ILU(1,1), Problem 4, ε=0.1
ILU(1,1), Problem 8, ε=0.1
ILU(1,1), Problem 1, ε=0.1
ILU(1,1), Problem 5, ε=0.1
ILU(1,1), Problem 9, ε=0.1
ILU(1,1), Problem 2, ε=0.1
ILU(1,1), Problem 6, ε=0.1
ILU(1,1), Problem 10, ε=0.1
−π
−π/2
0
π/2
π

ILU(1,2), Poisson equation

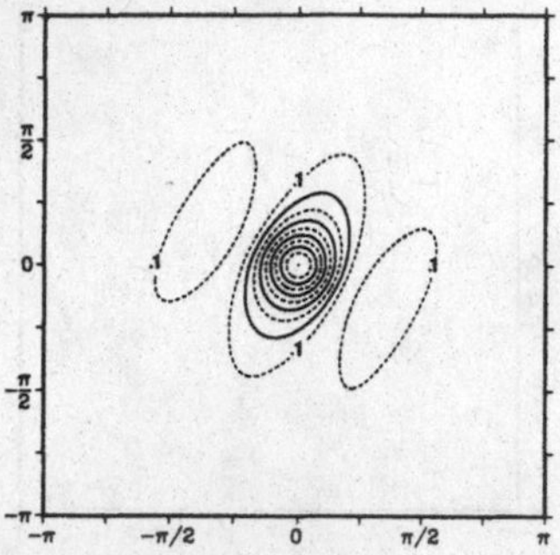

ILU(1,2), Problem 1, ε=0.1

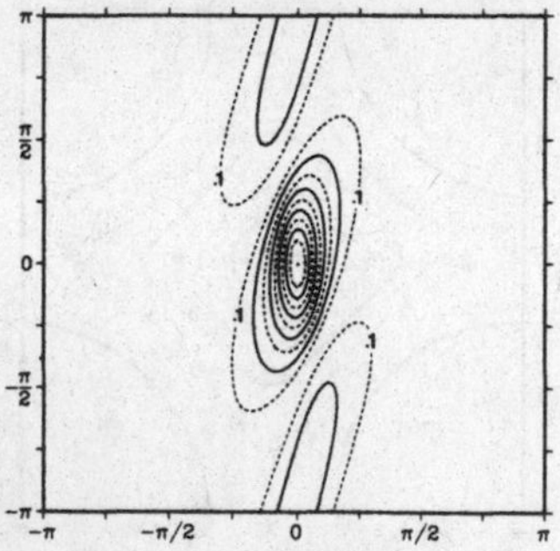

ILU(1,2), Problem 2, ε=0.1

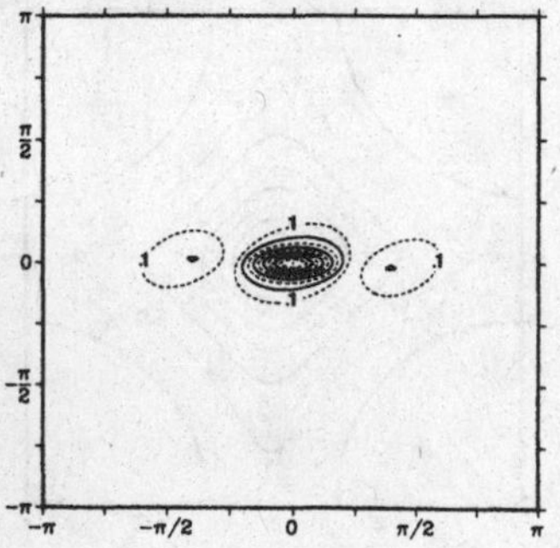

ILU(1,2), Problem 3, ε=0.1

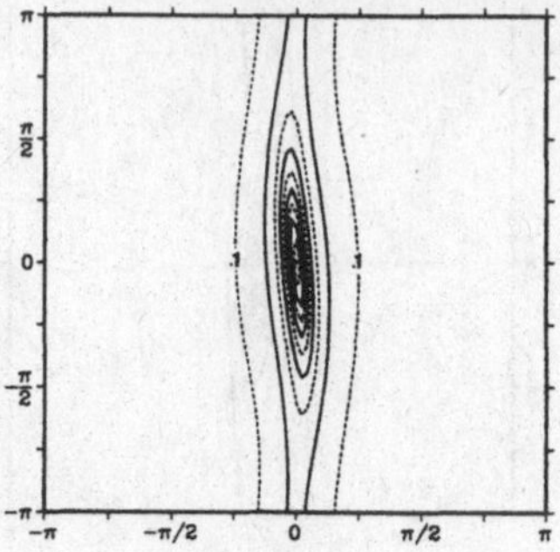

ILU(1,2), Problem 4, ε=0.1

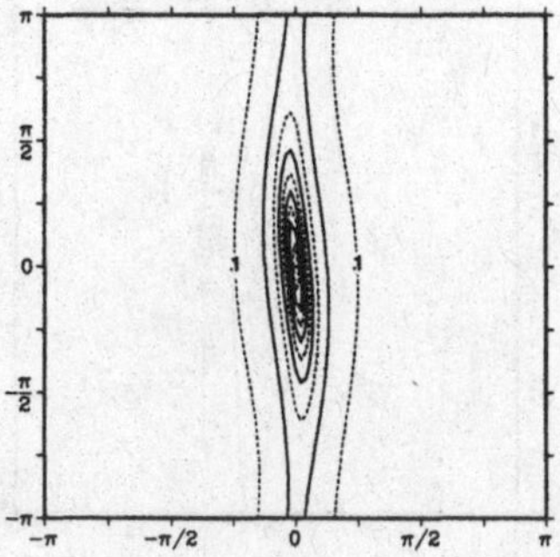

ILU(1,2), Problem 5, ε=0.1

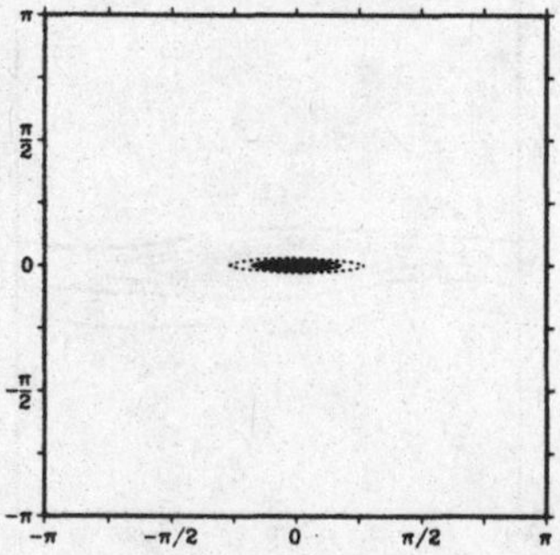

ILU(1,2), Problem 6, ε=0.1

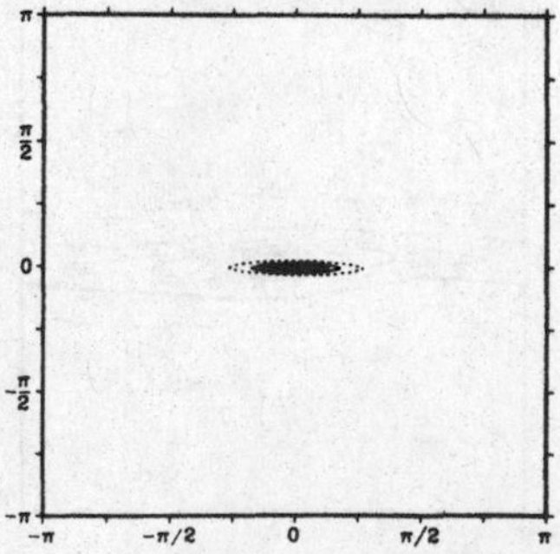

ILU(1,2), Problem 7, ε=0.1

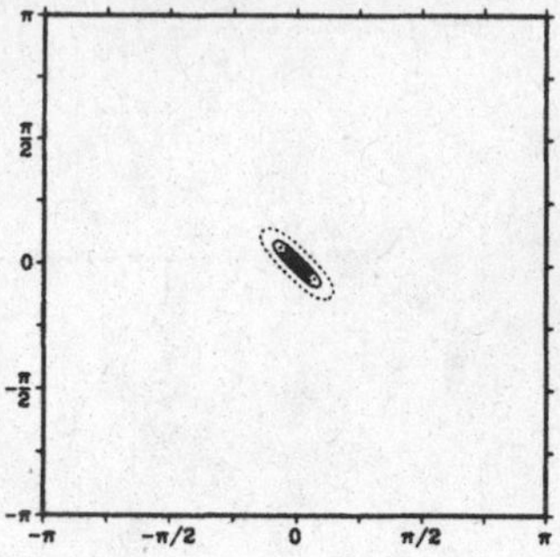

ILU(1,2), Problem 8, ε=0.1

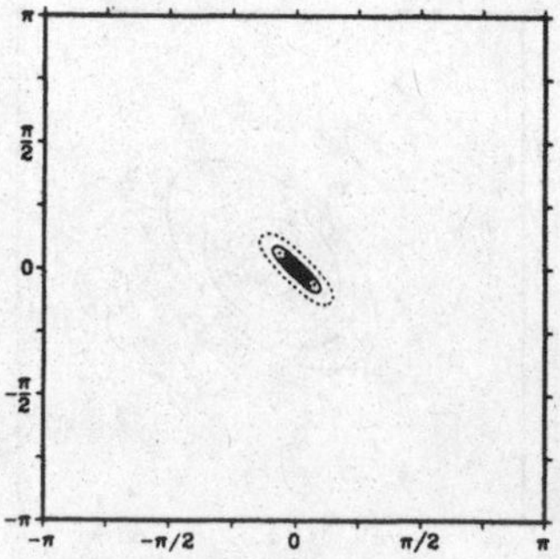

ILU(1,2), Problem 9, ε=0.1

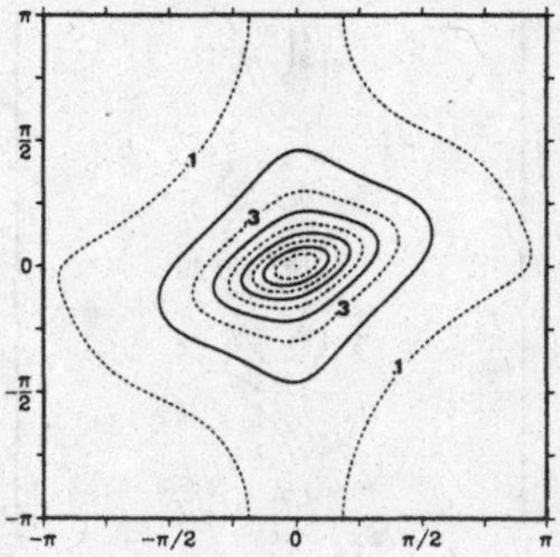

ILU(1,2), Problem 10, ε=0.1

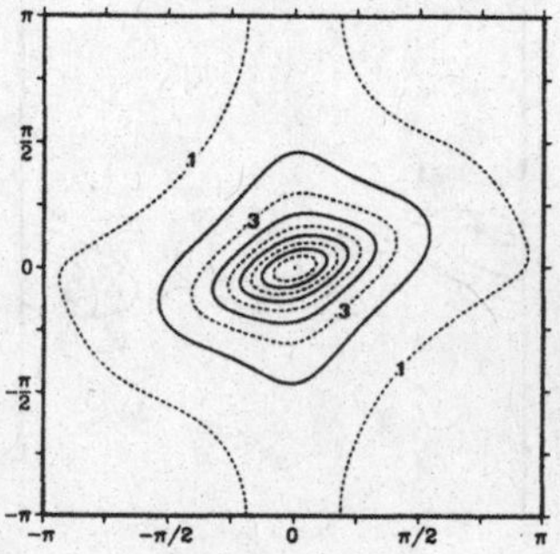

ILLU, Problem 3, ε=0.1

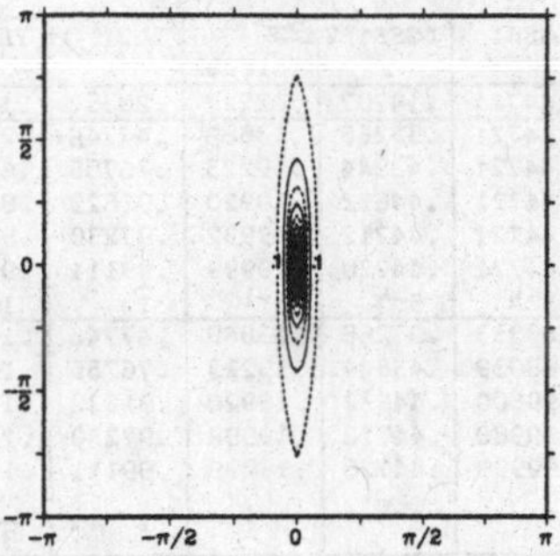

ILLU, Problem 7, ε=0.1

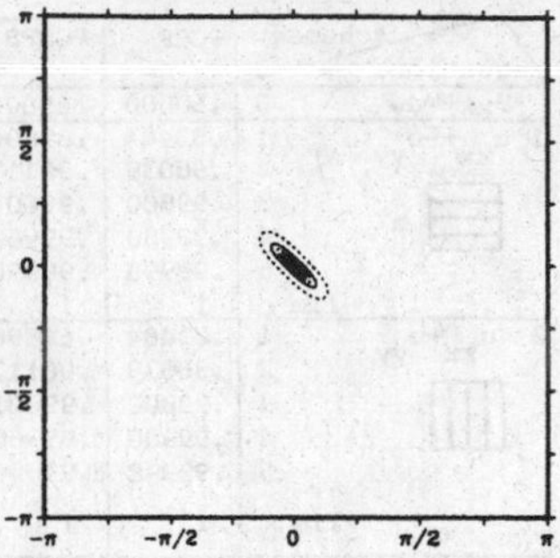

ILLU, Poisson equation

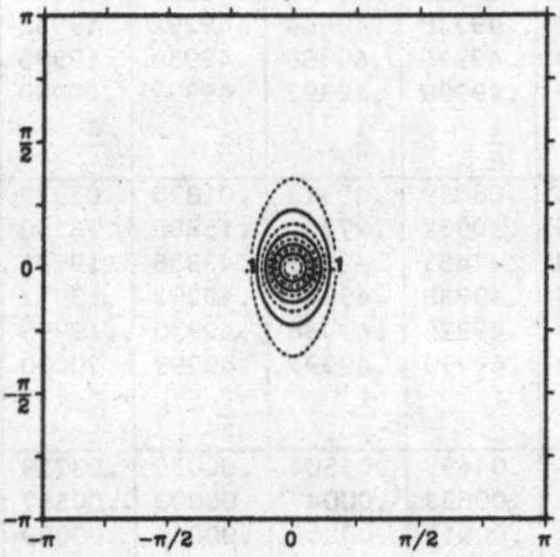

ILLU, Problem 4, ε=0.1

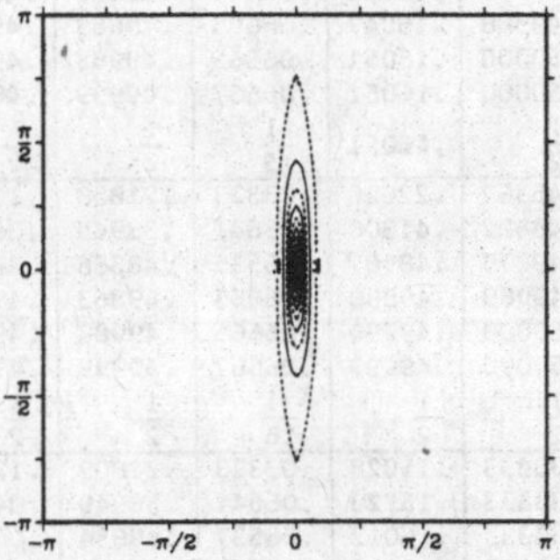

ILLU, Problem 8, ε=0.1

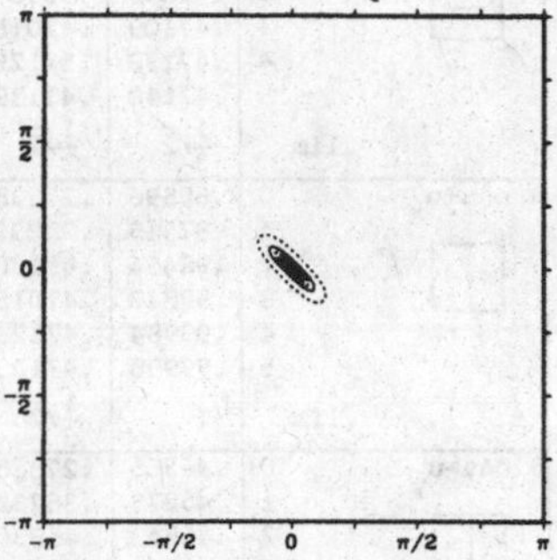

ILLU, Problem 1, ε=0.1

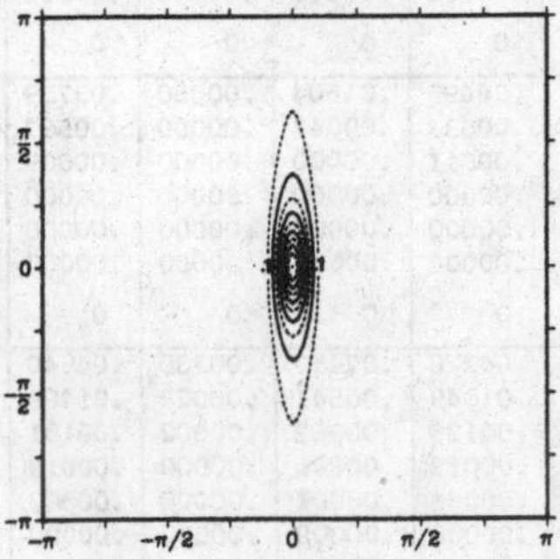

ILLU, Problem 5, ε=0.1

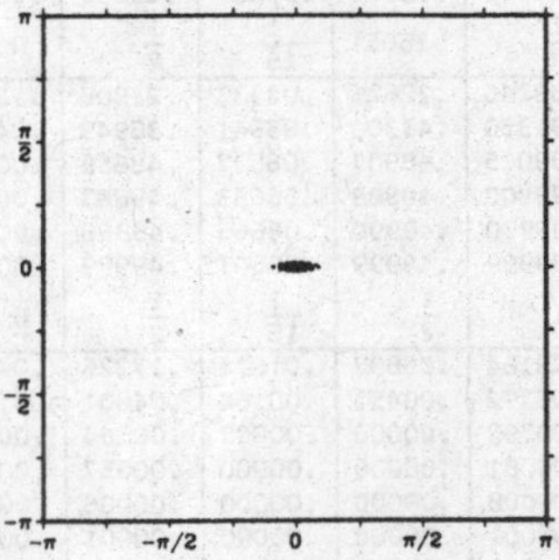

ILLU, Problem 9, ε=0.1

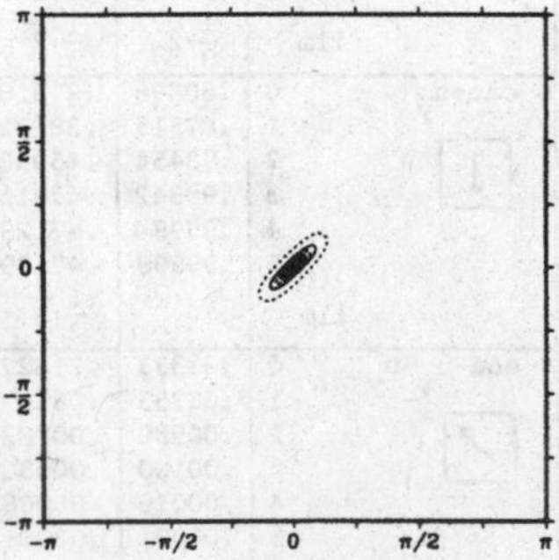

ILLU, Problem 2, ε=0.1

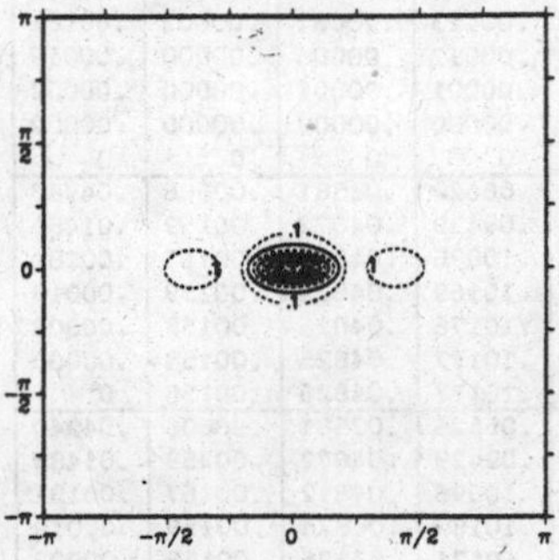

ILLU, Problem 6, ε=0.1

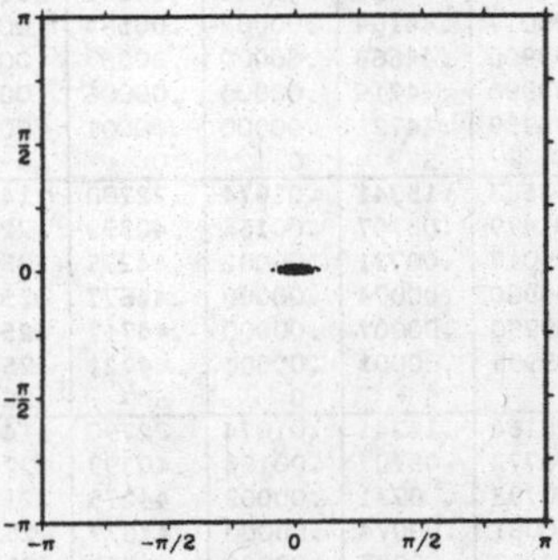

ILLU, Problem 10, ε=0.1

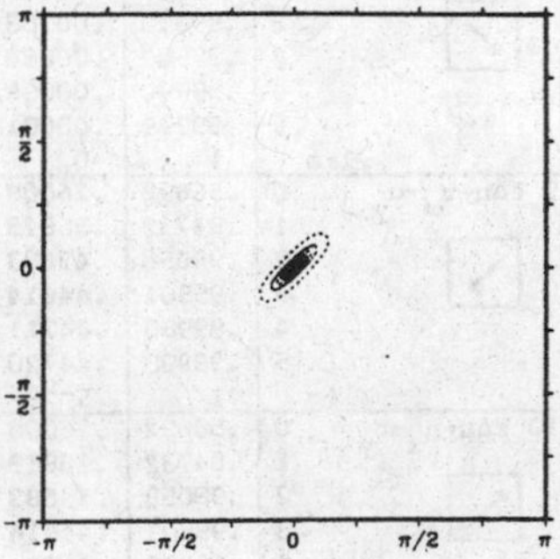

Appendix B: Smoothing factors

equation	p	PGSF	PGSFB	LGSF1	LGSF1F2	LGS F1B1F2B2	ILU(1,1)	ILU(1,2)	ILU(1,3)	ILU(2,4)	ILU(7,9)	ILLU
$u_{xx}+u_{yy}$	0	.50000	.25000	.44721	.14907	.02222	.20352	.12589	.05474	.02924	.00501	.05788
1 $u_{xx}+\varepsilon u_{yy}$	1	.83484	.69696	.44721	.37268	.13889	.47746	.27339	.16475	.13871	.04770	.13272
	2	.98039	.96117	.44721	.43844	.19223	.76755	.60700	.48279	.39794	.19292	.17590
	3	.99800	.99601	.44721	.44632	.19920	.91622	.84251	.77582	.71794	.50567	.19209
	4	.99980	.99960	.44721	.44712	.19992	.97230	.94574	.92007	.89560	.78667	.19747
	5	.99998	.99996	.44721	.44720	.19999	.99111	.98235	.97368	.96514	.92409	.19920
	lim	1	1	$5^{-1/2}$	$5^{-1/2}$	5^{-1}	1	1	1	1	1	$1/5$
2 $\varepsilon u_{xx}+u_{yy}$	1	.83484	.69696	.83333	.37268	.13889	.47746	.16510	.10781	.02460	.00294	.10769
	2	.98039	.96117	.98039	.43844	.19223	.76755	.17090	.14901	.03063	.00381	.14901
	3	.99800	.99601	.99800	.44632	.19920	.91622	.17150	.16422	.03317	.00415	.16422
	4	.99980	.99960	.99980	.44712	.19992	.97230	.17157	.16923	.03402	.00426	.16923
	5	.99998	.99996	.99998	.44720	.19999	.99111	.17157	.17083	.03430	.00430	.17083
	lim	1	1	1	$5^{-1/2}$	5^{-1}	1	$\frac{1}{3+2\sqrt{2}}$	$\frac{1}{3+2\sqrt{2}}$	.03441	.00431	$\frac{1}{3+2\sqrt{2}}$
3 $\varepsilon\Delta u-u_x$ (→)	0	.46355	.27338	.45367	.15028	.03333	.21808	.11164	.08059	.05313	.01870	.07977
	1	.45275	.38232	.48822	.15723	.05641	.38949	.36946	.30937	.27220	.15886	.16150
	2	.46843	.45930	.49875	.16012	.06537	.48656	.48408	.47453	.46745	.43338	.19531
	3	.47109	.47015	.49988	.16047	.06653	.49863	.49838	.49738	.49664	.49292	.19952
	4	.47138	.47129	.50000	.16051	.06665	.49988	.49985	.49975	.49968	.49930	.19995
	5	.47140	.47139	.50000	.16051	.06667	.49999	.49999	.49998	.49997	.49993	.20000
	lim	$\frac{1}{3}\sqrt{2}$	$\frac{1}{3}\sqrt{2}$	$\frac{1}{2}$	.16051	$\frac{1}{15}$	$\frac{1}{2}$	$\frac{1}{2}$	$\frac{1}{2}$	$\frac{1}{2}$	$\frac{1}{2}$	$\frac{1}{5}$
4 $\varepsilon\Delta u+u_x$ (←)	0	.60596	.27338	.45367	.22626	.03333	.21808	.11164	.08059	.05313	.01870	.07977
	1	.87515	.38232	.48822	.41300	.05641	.38949	.36946	.30937	.27220	.15886	.16150
	2	.98454	.45930	.49875	.48907	.06537	.48656	.48408	.47453	.46745	.43338	.19531
	3	.99842	.47015	.49988	.49888	.06653	.49863	.49838	.49738	.49664	.49292	.19952
	4	.99984	.47129	.50000	.49990	.06665	.49988	.49985	.49975	.49968	.49930	.19995
	5	.99998	.47139	.50000	.49999	.06667	.49999	.49999	.49998	.49997	.49993	.20000
	lim	1	$\frac{1}{3}\sqrt{2}$	$\frac{1}{2}$	$\frac{1}{2}$	$\frac{1}{15}$	$\frac{1}{2}$	$\frac{1}{2}$	$\frac{1}{2}$	$\frac{1}{2}$	$\frac{1}{2}$	$\frac{1}{5}$
5 $\varepsilon\Delta u-u_y$ (↑)	0	.46355	.27338	.33333	.15028	.03333	.21808	.12277	.04495	.01504	.00050	.03759
	1	.45275	.38232	.33333	.15723	.05641	.38949	.04184	.00633	.00047	.00000	.00567
	2	.46843	.45930	.33333	.16012	.06537	.48656	.00492	.00011	.00000	.00000	.00009
	3	.47109	.47015	.33333	.16047	.06653	.49863	.00050	.00000	.00000	.00000	.00000
	4	.47138	.47129	.33333	.16051	.06665	.49988	.00005	.00000	.00000	.00000	.00000
	5	.47140	.47139	.33333	.16051	.06667	.49999	.00001	.00000	.00000	.00000	.00000
	lim	$\frac{1}{3}\sqrt{2}$	$\frac{1}{3}\sqrt{2}$	$\frac{1}{3}$	.16051	$\frac{1}{15}$	$\frac{1}{2}$	0	0	0	0	0
6 $\varepsilon\Delta u+u_y$ (↓)	0	.60596	.27338	.63246	.22626	.03333	.21808	.12277	.04495	.01504	.00050	.03759
	1	.87515	.38232	.91350	.41300	.05641	.38949	.04184	.00633	.00047	.00000	.00567
	2	.98454	.45930	.99015	.48907	.06537	.48656	.00492	.00011	.00000	.00000	.00009
	3	.99842	.47015	.99900	.49888	.06653	.49863	.00050	.00000	.00000	.00000	.00000
	4	.99984	.47129	.99990	.49990	.06665	.49988	.00005	.00000	.00000	.00000	.00000
	5	.99998	.47139	.99999	.49999	.06667	.49999	.00001	.00000	.00000	.00000	.00000
	lim	1	$\frac{1}{3}\sqrt{2}$	1	$\frac{1}{2}$	$\frac{1}{15}$	$\frac{1}{2}$	0	0	0	0	0
7 $\varepsilon\Delta u-u_x-u_y$ (↗)	0	.33333	.21622	.28164	.06897	.01674	.17226	.09267	.04278	.02257	.00330	.04940
	1	.08333	.06509	.06772	.00451	.00164	.04801	.01807	.01049	.00541	.00028	.01489
	2	.00980	.00783	.00793	.00006	.00002	.00564	.00200	.00123	.00062	.00002	.00181
	3	.00100	.00080	.00081	.00000	.00000	.00057	.00020	.00012	.00006	.00000	.00019
	4	.00010	.00008	.00008	.00000	.00000	.00006	.00002	.00001	.00001	.00000	.00002
	5	.00001	.00001	.00001	.00000	.00000	.00001	.00000	.00000	.00000	.00000	.00000
	lim	0	0	0	0	0	0	0	0	0	0	0
8 $\varepsilon\Delta u+u_x+u_y$ (↙)	0	.66667	.21622	.63563	.25001	.01674	.17226	.09267	.04278	.02257	.00330	.04940
	1	.91667	.06509	.91479	.40150	.00164	.04801	.01807	.01049	.00541	.00028	.01489
	2	.99020	.00783	.99017	.44194	.00002	.00564	.00200	.00123	.00062	.00002	.00181
	3	.99900	.00080	.99900	.44668	.00000	.00057	.00020	.00012	.00006	.00000	.00019
	4	.99990	.00008	.99990	.44716	.00000	.00006	.00002	.00001	.00001	.00000	.00002
	5	.99999	.00001	.99999	.44721	.00000	.00001	.00000	.00000	.00000	.00000	.00000
	lim	1	0	1	$5^{-1/2}$	0	0	0	0	0	0	0
9 $\varepsilon\Delta u-u_x+u_y$ (↘)	0	.56692	.26008	.63563	.15341	.01674	.22790	.14608	.06529	.03581	.00606	.04940
	1	.84732	.36813	.91479	.05707	.00164	.40399	.22422	.09439	.04672	.00259	.01489
	2	.98058	.43683	.99017	.00721	.00002	.44275	.25010	.10096	.04812	.00167	.00181
	3	.99801	.44614	.99900	.00074	.00000	.44677	.25318	.10169	.04824	.00159	.00019
	4	.99980	.44711	.99990	.00007	.00000	.44717	.25350	.10176	.04825	.00158	.00002
	5	.99998	.44720	.99999	.00001	.00000	.44721	.25353	.10177	.04825	.00158	.00000
	lim	1	$5^{-1/2}$	1	0	0	$5^{-1/2}$	.25353	.10177	.04826	.00158	0
10 $\varepsilon\Delta u+u_x-u_y$ (↖)	0	.56692	.26008	.28164	.15341	.01674	.22790	.14608	.06529	.03581	.00606	.04940
	1	.84732	.36813	.06772	.05707	.00164	.40399	.22422	.09439	.04672	.00259	.01489
	2	.98058	.43683	.00793	.00721	.00002	.44275	.25010	.10096	.04812	.00167	.00181
	3	.99801	.44614	.00081	.00074	.00000	.44677	.25318	.10169	.04824	.00159	.00019
	4	.99980	.44711	.00008	.00007	.00000	.44717	.25350	.10176	.04825	.00158	.00002
	5	.99998	.44720	.00001	.00001	.00000	.44721	.25353	.10177	.04825	.00158	.00000
	lim	1	$5^{-1/2}$	0	0	0	$5^{-1/2}$	.25353	.10177	.04826	.00158	0

Appendix C: Smoothing efficiencies

equation	ρ	PGSF	PGSFB	LGSF1	LGSF1F2	LGS F1B1F2B2	ILU(1,1)	ILU(1,2)	ILU(1,3)	ILU(2,4)	ILU(7,9)	ILLU
		W=9	W=12	W=9	W=17	W=34	W=12	W=16	W=24	W=34	W=74	W=25
$u_{xx}+u_{yy}$	0	13.0	8.7	11.2	8.9	8.9	7.5	7.7	8.3	9.6	14.0	8.8
1 $u_{xx}+\varepsilon u_{yy}$	1	49.9	33.2	11.2	17.2	17.2	16.2	12.3	13.3	17.2	24.3	12.4
≡	2	454.5	303.0	11.2	20.6	20.6	45.4	32.0	33.0	36.9	45.0	14.4
	3	4504.5	3003.0	11.2	21.1	21.1	137.1	93.4	94.5	102.6	108.5	15.2
	4	******	******	11.2	21.1	21.1	427.3	286.8	288.1	308.4	308.4	15.4
	5	******	******	11.2	21.1	21.1	1344.6	898.4	899.7	958.3	937.4	15.5
	lim	******	******	11.2	21.1	21.1	******	******	******	******	******	15.5
2 $\varepsilon u_{xx}+u_{yy}$	1	49.9	33.2	49.4	17.2	17.2	16.2	8.9	10.8	9.2	12.7	11.2
|||	2	454.5	303.0	454.5	20.6	20.6	45.4	9.1	12.6	9.8	13.3	13.1
	3	4504.5	3003.0	4504.5	21.1	21.1	137.1	9.1	13.3	10.0	13.5	13.8
	4	******	******	******	21.1	21.1	427.3	9.1	13.5	10.1	13.6	14.1
	5	******	******	******	21.1	21.1	1344.6	9.1	13.6	10.1	13.6	14.1
	lim	******	******	******	21.1	21.1	******	9.1	13.6	10.1	13.6	14.2
3 $\varepsilon\Delta u-u_x$	0	11.7	9.3	11.4	9.0	10.0	7.9	7.3	9.5	11.6	18.6	9.9
→	1	11.4	12.5	12.6	9.2	11.8	12.7	16.1	20.5	26.1	40.2	13.7
	2	11.9	15.4	12.9	9.3	12.5	16.7	22.1	32.2	44.7	88.5	15.3
	3	12.0	15.9	13.0	9.3	12.5	17.2	23.0	34.4	48.6	104.6	15.5
	4	12.0	16.0	13.0	9.3	12.6	17.3	23.1	34.6	49.0	106.5	15.5
	5	12.0	16.0	13.0	9.3	12.6	17.3	23.1	34.6	49.0	106.7	15.5
	lim	12.0	16.0	13.0	9.3	12.6	17.3	23.1	34.6	49.1	106.8	15.5
4 $\varepsilon\Delta u+u_x$	0	18.0	9.3	11.4	11.4	10.0	7.9	7.3	9.5	11.6	18.6	9.9
←	1	67.5	12.5	12.6	19.2	11.8	12.7	16.1	20.5	26.1	40.2	13.7
	2	577.5	15.4	12.9	23.8	12.5	16.7	22.1	32.2	44.7	88.5	15.3
	3	5685.3	15.9	13.0	24.4	12.5	17.2	23.0	34.4	48.6	104.6	15.5
	4	******	16.0	13.0	24.5	12.6	17.3	23.1	34.6	49.0	106.5	15.5
	5	******	16.0	13.0	24.5	12.6	17.3	23.1	34.6	49.0	106.7	15.5
	lim	******	16.0	13.0	24.5	12.6	17.3	23.1	34.6	49.1	106.8	15.5
5 $\varepsilon\Delta u-u_y$	0	11.7	9.3	8.2	9.0	10.0	7.9	7.6	7.7	8.1	9.7	7.6
↑	1	11.4	12.5	8.2	9.2	11.8	12.7	5.0	4.7	4.4	4.0	4.8
	2	11.9	15.4	8.2	9.3	12.5	16.7	3.0	2.6	2.4	2.1	2.7
	3	12.0	15.9	8.2	9.3	12.5	17.2	2.1	1.8	1.6		1.8
	4	12.0	16.0	8.2	9.3	12.6	17.3	1.6	1.3	1.3		1.4
	5	12.0	16.0	8.2	9.3	12.6	17.3	1.4	1.0	1.1		1.3
	lim	12.0	16.0	8.2	9.3	12.6	17.3	0	0	0	0	0
6 $\varepsilon\Delta u+u_y$	0	18.0	9.3	19.6	11.4	10.0	7.9	7.6	7.7	8.1	9.7	7.6
↓	1	67.5	12.5	99.5	19.2	11.8	12.7	5.0	4.7	4.4	4.0	4.8
	2	577.5	15.4	909.1	23.8	12.5	16.7	3.0	2.6	2.4	2.1	2.7
	3	5685.3	15.9	9009.0	24.4	12.5	17.2	2.1	1.8	1.6		1.8
	4	******	16.0	******	24.5	12.6	17.3	1.6	1.3	1.3		1.4
	5	******	16.0	******	24.5	12.6	17.3	1.4	1.0	1.1		1.3
	lim	******	16.0	******	24.5	12.6	17.3	0	0	0	0	0
7 $\varepsilon\Delta u-u_x-u_y$	0	8.2	7.8	7.1	6.4	8.3	6.8	6.7	7.6	9.0	13.0	8.3
↗	1	3.6	4.4	3.3	3.1	5.3	4.0	4.0	5.3	6.5	9.0	5.9
	2	1.9	2.5	1.9	1.8	3.2	2.3	2.6	3.6	4.6	6.9	4.0
	3	1.3	1.7	1.3	1.2	2.2	1.6	1.9	2.7	3.5	5.6	2.9
	4	1.0	1.3	1.0	.9	1.7	1.2	1.5	2.1	2.8	4.8	2.3
	5	.8	1.0	.8	.7	1.4	1.0	1.2	1.8	2.4	4.2	1.9
	lim	0	0	0	0	0	0	0	0	0	0	0
8 $\varepsilon\Delta u+u_x+u_y$	0	22.2	7.8	19.9	12.3	8.3	6.8	6.7	7.6	9.0	13.0	8.3
↙	1	103.4	4.4	101.1	18.6	5.3	4.0	4.0	5.3	6.5	9.0	5.9
	2	913.5	2.5	911.2	20.8	3.2	2.3	2.6	3.6	4.6	6.9	4.0
	3	9013.2	1.7	9011.1	21.1	2.2	1.6	1.9	2.7	3.5	5.6	2.9
	4	******	1.3	******	21.1	1.7	1.2	1.5	2.1	2.8	4.8	2.3
	5	******	1.0	******	21.1	1.4	1.0	1.2	1.8	2.4	4.2	1.9
	lim	******	0	******	21.1	0	0	0	0	0	0	0
9 $\varepsilon\Delta u-u_x+u_y$	0	15.9	8.9	19.9	9.1	8.3	8.1	8.3	8.8	10.2	14.5	8.3
↘	1	54.3	12.0	101.1	5.9	5.3	13.2	10.7	10.2	11.1	12.4	5.9
	2	459.0	14.5	911.2	3.4	3.2	14.7	11.5	10.5	11.2	11.6	4.0
	3	4509.0	14.9	9011.1	2.4	2.2	14.9	11.6	10.5	11.2	11.5	2.9
	4	******	14.9	******	1.8	1.7	14.9	11.7	10.5	11.2	11.5	2.3
	5	******	14.9	******	1.4	1.4	14.9	11.7	10.5	11.2	11.5	1.9
	lim	******	14.9	******	0	0	14.9	11.7	10.5	11.2	11.5	0
10 $\varepsilon\Delta u+u_x-u_y$	0	15.9	8.9	7.1	9.1	8.3	8.1	8.3	8.8	10.2	14.5	8.3
↖	1	54.3	12.0	3.3	5.9	5.3	13.2	10.7	10.2	11.1	12.4	5.9
	2	459.0	14.5	1.9	3.4	3.2	14.7	11.5	10.5	11.2	11.6	4.0
	3	4509.0	14.9	1.3	2.4	2.2	14.9	11.6	10.5	11.2	11.5	2.9
	4	******	14.9	1.0	1.8	1.7	14.9	11.7	10.5	11.2	11.5	2.3
	5	******	14.9	.8	1.4	1.4	14.9	11.7	10.5	11.2	11.5	1.9
	lim	******	14.9	0	0	0	14.9	11.7	10.5	11.2	11.5	0

Appendix D: Numerical results

1. Average reduction factor ρ

MG		PGSF	PGSFB	LGSF1	ZLGSF1	LGSF1F2	ZLGSF1F2	LGS F1B1F2B2	ILU(1,1)	ILU(1,2)	ILLU
Stone	0	.2036	.1765	.2211	.2323	.0557	.0725	.0347	.1518	.0630	.0387
	1	.7365	.6457	.7245	.7845	.3103	.3592	.2517	.3909	.2148	.1644
	2	.9200	.9019	.9194	.9326	.2067	.2214	.2013	.6731	.3312	.1140
	3	.9524	.9483	.9511	.9584	.2011	.2447	.2204	.7794	.4043	.0912
	4	.9581	.9573	.9559	.9620	.2018	.2481	.2221	.7988	.4120	.0946
	5	.9579	.9577	.9543	.9605	.2026	.2485	.2223	.8010	.4165	.0942
Kershaw		.6068	.5982	.5930	.5781	.5550	.5295	.5031	.5805	.4591	.3978

MGCG		PGSB PGSF	PGSFB PGSFB	LGSB1 LGSF1	ZLGSB1 ZLGSF1	LGSB2B1 LGSF1F2	ZLGSB2B1 ZLGSF1F2	LGSF2B2F1B1 LGSF1B1F2B2	ILU(1,1) ILU(1,1)	ILU(1,2) ILU(1,2)	ILLU ILLU
Stone	0	.0647	.0255	.0562	.0740	.0174	.0214	.0088	.0202	.0088	.0060
	1	.2739	.1661	.2569	.3871	.0543	.0413	.0243	.0747	.0273	.0153
	2	.6232	.5259	.6228	.7273	.0571	.0401	.0252	.1459	.0499	.0152
	3	.8329	.7757	.8250	.8766	.0619	.0405	.0242	.2197	.0565	.0146
	4	.9259	.8834	.9050	.8939	.0623	.0401	.0250	.2480	.0585	.0146
	5	.9961	.9450	.9561	.9294	.0624	.0401	.0251	.2535	.0594	.0133
Kershaw		.1208	.0579	.0889	.1280	.0456	.0431	.0223	.0510	.0261	.0150

ICCG		ILU(1,1)	ILU(1,2)	ILLU
Stone	0	.6618	.5031	.3988
	1	.6716	.4838	.3926
	2	.6806	.5114	.3980
	3	.6850	.4942	.3595
	4	.6928	.4965	.3402
	5	.6856	.4873	.3503
Kershaw		.6767	.4965	.3172

2. Average efficiency factor τ

MG		PGSF W=32	PGSFB W=37	LGSF1 W=30	ZLGSF1 W=30	LGSF1F2 W=44	ZLGSF1F2 W=44	LGSF1B1F2B2 W=71	ILU(1,1) W=35	ILU(1,2) W=40	ILLU W=61
Stone	0	20.1	21.3	19.9	20.6	15.2	16.8	21.1	18.6	14.5	18.8
	1	104.6	84.6	93.1	123.6	37.6	43.0	51.5	37.3	26.0	33.8
	2	383.8	358.3	357.0	429.9	27.9	29.2	44.3	88.1	36.2	28.1
	3	656.1	697.0	598.4	706.0	27.4	32.5	46.9	140.4	44.2	25.5
	4	747.6	847.9	665.2	774.4	27.5	31.6	47.2	155.8	45.1	25.9
	5	744.0	856.1	641.3	744.4	27.6	31.6	47.2	157.7	45.7	25.8
Kershaw		64.1	72.0	57.4	54.7	74.7	69.2	103.4	64.4	51.4	66.0

MGCG		PGSB PGSF W=66	PGSFB PGSFB W=75	LGSB1 LGSF1 W=64	ZLGSB1 ZLGSF1 W=64	LGSB2B1 LGSF1F2 W=90	ZLGSB2B1 ZLGSF1F2 W=90	LGSF2B2F1B1 LGSF1B1F2B2 W=144	ILU(1,1) ILU(1,1) W=71	ILU(1,2) ILU(1,2) W=82	ILLU ILLU W=118
Stone	0	24.1	20.4	22.2	24.6	22.2	23.4	30.4	18.2	17.3	23.1
	1	51.0	41.8	47.1	67.4	30.9	28.2	38.7	27.4	22.8	28.2
	2	139.6	116.7	135.2	201.0	31.4	28.0	39.1	36.9	27.4	28.2
	3	361.0	295.3	332.7	485.9	32.3	28.1	38.7	46.6	28.5	27.9
	4	857.3	605.0	641.2	570.6	32.4	28.0	39.0	50.9	28.9	27.9
	5	16890.1	1325.8	1425.6	874.1	32.4	28.0	39.1	51.7	29.0	27.3
Kershaw		31.2	26.3	26.4	31.1	29.1	28.6	37.9	23.9	22.5	28.1

ICCG		ILU(1,1) W=27	ILU(1,2) W=31	ILLU W=34
Stone	0	65.4	45.1	37.0
	1	67.8	42.7	36.4
	2	70.2	46.2	36.9
	3	71.4	44.0	33.2
	4	73.6	44.3	31.5
	5	71.5	43.1	32.4
Kershaw		69.1	44.3	29.6

Appendix E: Graphical results

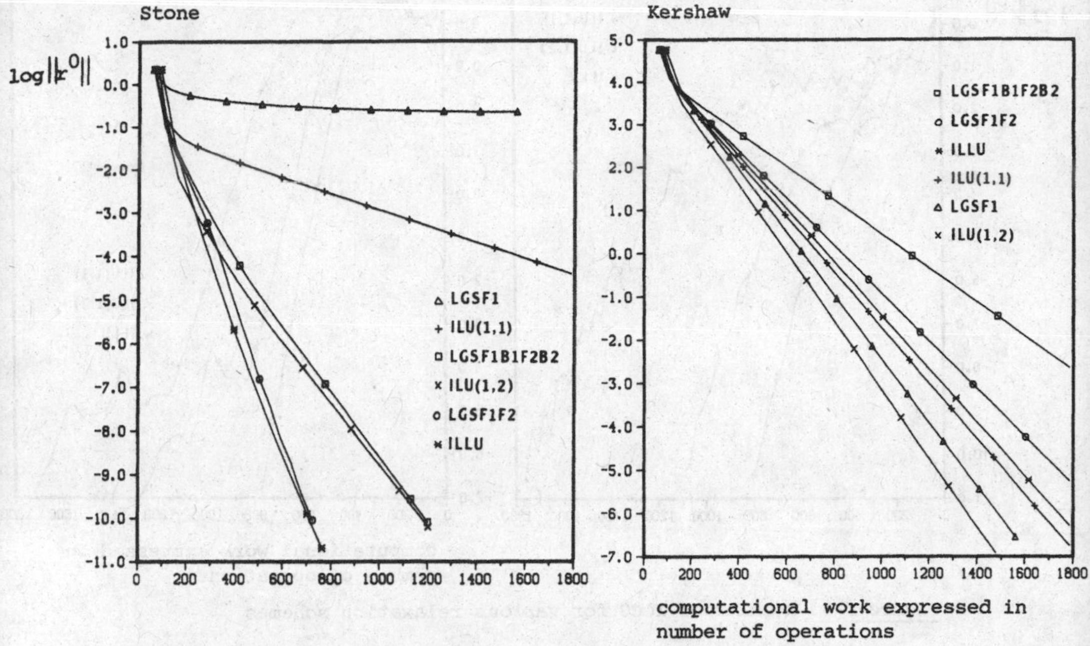

Figure E-1 Results of MG for various relaxation schemes

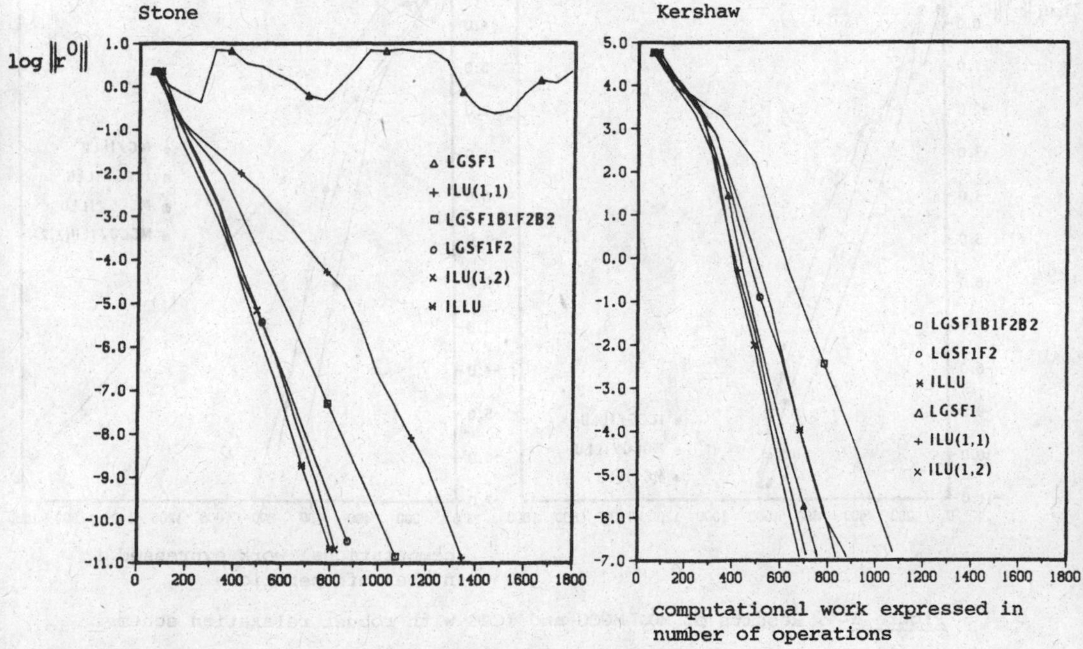

Figure E-2 Results of MGCG for various relaxation schemes

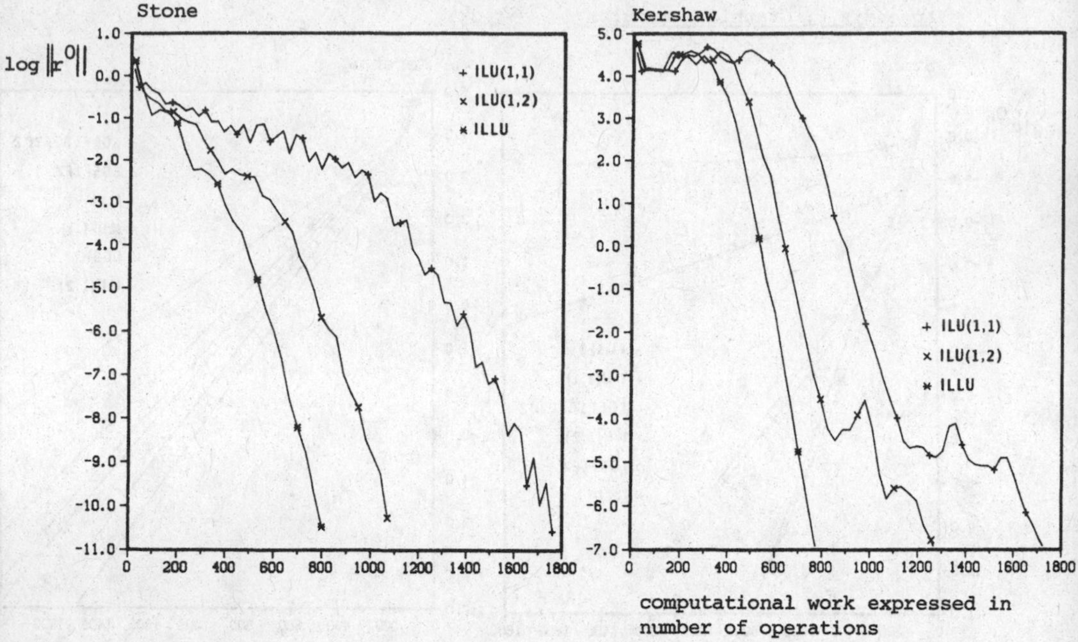

Figure E-3 Results of ICCG for various relaxation schemes

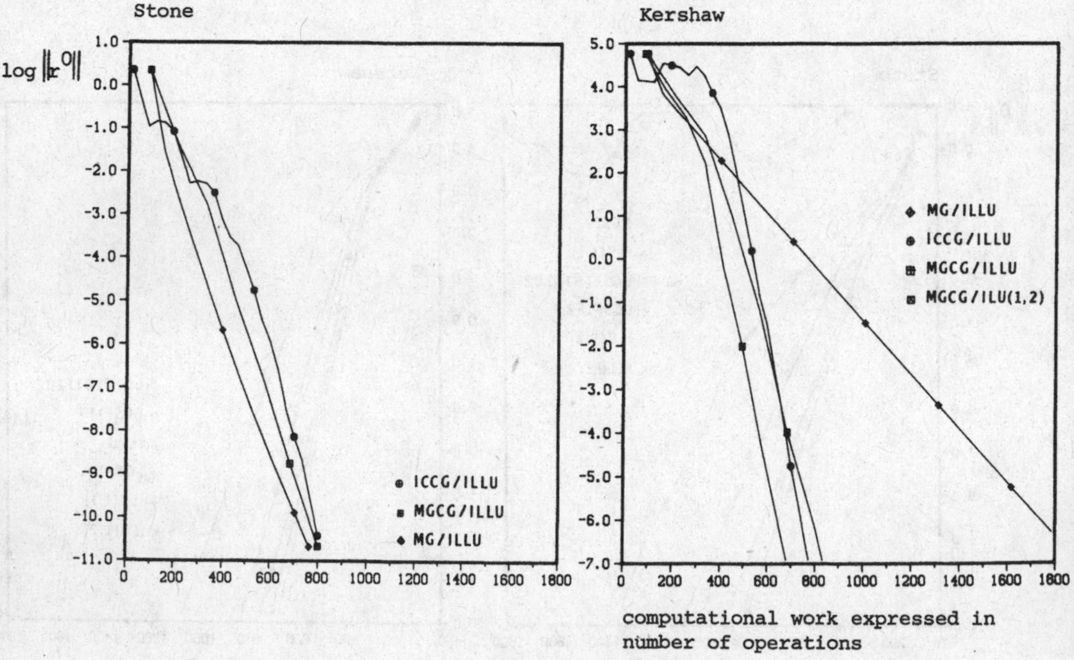

Figure E-4 Results of MG, MGCG and ICCG with robust relaxation schemes

THE CONTRACTION NUMBER OF A CLASS OF TWO-LEVEL METHODS ; AN EXACT EVALUATION FOR SOME FINITE ELEMENT SUBSPACES AND MODEL PROBLEMS

J.F. MAITRE and F. MUSY
Mathématiques-Informatique-Systèmes
Ecole Centrale de Lyon
69130 Ecully (France)

1. INTRODUCTION

For evaluating the convergence rate of some two-level methods, some authors (D. AXELSSON-I. GUSTAFSSON [1] ; R.E. BANK-T. DUPONT [2] ; D. BRAESS [3]) consider a decomposition of the Hilbert space H, $H = U \oplus V$, and relatively to a symmetric, positive, bounded bilinear form a (.,.) on H, the number

(1.1) $$\gamma = \sup \{a(u,v) \; ; \; u \in U, \; v \in V, \; a(u,u) = a(v,v) = 1\}.$$

For instance, for solving the problem

$$(P) \begin{cases} u^* \in H \\ a(u^*,v) = L(v), \; \forall \, v \in H \end{cases}$$

where L is a linear form on H, we may consider as in [2] the following two-level iteration :

(1.2) $$\begin{cases} u^o \text{ given} \\ \text{(i)} \quad u^{n+1/2} \in u^n + U \;\; \text{s.t.} \; a(u^{n+1/2},u) = L(u), \;\; \forall \, u \in U \\ \text{(ii)} \; u^{n+1} \in u^{n+1/2} + V \;\; \text{s.t.} \; a(u^{n+1}, v) = L(v), \;\; \forall \, v \in V \end{cases}$$

With the norm associated with a(.,.), the convergence rate is given by γ^2 :

(1.3) $$||u^{n+1} - u^*|| \leqslant \gamma^2 \; ||u^n - u^*||, \; n \geqslant 1.$$

If U and V are finite dimensional spaces, (1.2) corresponds to a two-block Gauss-Seidel iteration. In the particular case of a finite element space H associated with a mesh T_{h_1} and a "similar" space U associated with a mesh T_{h_2} with $h_2 > h_1$, (1.2) may be viewed as a two-grid method. But quite different decompositions are possible.

In this paper we evaluate γ for two cases of space decomposition, relatively to different a(.,.). In the first case H is a space generated by triangular quadratic elements, U the space of piecewise linear polynomials corresponding to the vertices of the mesh, and V is generated by the basis functions of H corresponding to the edge nodes. In the second case U is the space of piecewise linear polynomials on a mesh, H is similar but on the finer mesh obtained by dividing each triangle in 4 smaller ones as usually, and V is generated by the basis functions of H corresponding to the nodes not belonging to the coarse mesh.

Let U and V respectively be generated by the basis $\{u_i\}_{i=1,m}$, $\{v_j\}_{j=1,p}$. The iteration (1.2) is a block Gauss-Seidel method, the matrix of the system being partitioned as follows :

$$(1.4)\qquad \begin{bmatrix} A & B \\ B^t & C \end{bmatrix} \quad , \quad \begin{aligned} A_{ij} &= a(u_i, u_j),\ i,j = 1 \text{ to } m \\ B_{ij} &= a(u_i, v_j),\ i = 1 \text{ to } m,\ j = 1 \text{ to } p \\ C_{ij} &= a(v_i, v_j),\ i,j = 1 \text{ to } p. \end{aligned}$$

If A and C are invertible (a(.,.) coercive on H) :

$$(1.5)\qquad \gamma^2 = \rho\,(A^{-1}\ B\ C^{-1}\ B^t) = \rho\,(C^{-1}\ B^t\ A^{-1}\ B).$$

If C is invertible, and A singular (a(.,.) coercive on V but not on U), which will be the case in the sequel with $a(u,v) = \int \nabla u.\nabla v$ and U containing the constant functions, then :

$$(1.6)\qquad \gamma^2 = \max\ \{\ \lambda\ ;\ \{0\} \neq \text{Ker}\ (B\ C^{-1}\ B^t - \lambda\ A) \subset \text{Ker}^{\perp} A\ \}.$$

For finite element spaces, the calculation can be made on each element of the mesh corresponding to H. We have

$$a(u,v) = \sum_e\ a_e(u,v) \leqslant \sum_e \gamma_e\ a_e^{1/2}\ (u,u)\ a_e^{1/2}(v,v) \leqslant (\max_e \gamma_e)\ a^{1/2}(u,u)\ a^{1/2}(v,v),$$

where γ_e is defined on the element of number e :

$$(1.7)\qquad \gamma_e = \sup\ \{a_e(u,v)\ ;\ u \in U,\ v \in V,\ a_e(u,u) = a_e(v,v) = 1\}.$$

Then

$$(1.8)\qquad \gamma \leqslant \max_e \gamma_e$$

and each γ_e can be obtained by the eigenvalue problem (1.6) with the elementary matrices A_e, B_e, C_e.

In the subsequent sections, we exactly evaluate γ_e for different particular cases of the general bilinear form :

$$a(u,v) = \int a\ \nabla u.\nabla v + a_o \int u\ v$$

on a triangle for the two decompositions just described ("first case" and "second case").

2. THE ELEMENTARY MATRICES

For all the subsequent calculations, we need the elementary matrices A, B, C corresponding to the bilinear form

(2.1) $$a(u,v) = \int_K \nabla u.\nabla v \, dK + a_o \int_K uv \, dK$$

where K is a general triangle, and to the two announced space decompositions of the form $U \oplus V$.

According to the figure 2.1, the first decomposition is as follows

(2.2) $$P_2(K) = P_1(K) \oplus \{p \in P_2(K) \;;\; p(a_i) = 0, \; i=1 \text{ to } 3\}$$

where $P_k(K)$ denotes the space of degree k polynomials on K.

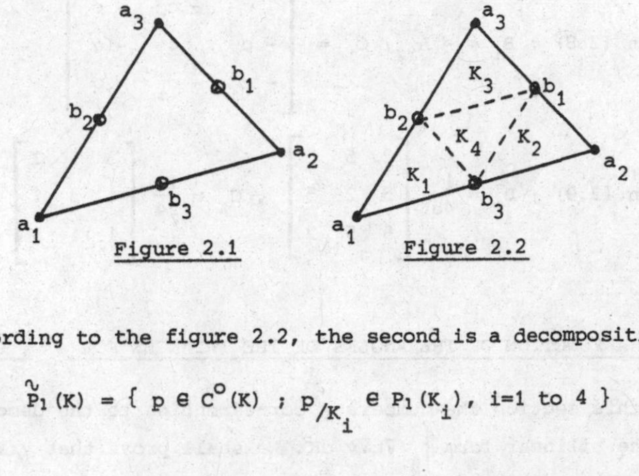

Figure 2.1 Figure 2.2

According to the figure 2.2, the second is a decomposition of the space

$$\tilde{P}_1(K) = \{ p \in C^o(K) \;;\; p_{/K_i} \in P_1(K_i), \; i=1 \text{ to } 4 \}$$

in

(2.3) $$\tilde{P}_1(K) = P_1(K) \oplus \{ p \in \tilde{P}_1(K) \;;\; p\,(a_i) = 0, \; i=1 \text{ to } 3 \}.$$

In the two cases, we choose for $U = P_1(K)$ the basis of the three area coordinates ($u_i = \lambda_i$, i=1 to 3), and for V the basis of the three shape functions of the entire space respectively associated to the nodes b_i, i=1 to 3.

We write the different matrices in the form

(2.4) $$A = A_1 + t\,A_2, \quad B = B_1 + t\,B_2, \quad C = C_1 + t\,C_2$$

where the parameter t is

(2.5) $$t = a_o.S, \quad S : \text{area of the triangle.}$$

If θ_i denote the angle of the triangle at the vertex a_i, we introduce the notations

(2.6) $$c_i = \text{cotg}\,(\theta_i)\ ,\quad i=1 \text{ to } 3\ ;\quad c = \sum_{i=1}^{3} c_i$$

(2.7) $$d = \sum_{i=1}^{3} \cos^2(\theta_i)$$

For the decomposition (2.2), the corresponding matrices are :

(2.8) $$A_1 = \frac{1}{2}\begin{bmatrix} c_2+c_3 & -c_3 & -c_2 \\ -c_3 & c_3+c_1 & -c_1 \\ -c_2 & -c_1 & c_1+c_2 \end{bmatrix} ;\ B_1 = -\frac{4}{3}A_1\ ;\ C_1 = \frac{4}{3}\begin{bmatrix} c & -c_3 & -c_2 \\ -c_3 & c & -c_1 \\ -c_2 & -c_1 & c \end{bmatrix}$$

(2.9) $$A_2 = \frac{1}{12}\begin{bmatrix} 2 & 1 & 1 \\ 1 & 2 & 1 \\ 1 & 1 & 2 \end{bmatrix} ,\ B_2 = \frac{1}{15}\begin{bmatrix} 1 & 2 & 2 \\ 2 & 1 & 2 \\ 2 & 2 & 1 \end{bmatrix} ,\ C_2 = \frac{8}{90}\begin{bmatrix} 2 & 1 & 1 \\ 1 & 2 & 1 \\ 1 & 1 & 2 \end{bmatrix}$$

For the decomposition (2.3), the matrices are :

(2.10) $$A_1 = \text{as in (2.8)}\ ;\ B_1 = -A_1\ ;\ C_1 = \begin{bmatrix} c & -c_3 & -c_2 \\ -c_3 & c & -c_1 \\ -c_2 & -c_1 & c \end{bmatrix}$$

(2.11) $$A_2 = \text{as in (2.9)}\ ;\ B_2 = \frac{1}{48}\begin{bmatrix} 2 & 5 & 5 \\ 5 & 2 & 5 \\ 5 & 5 & 2 \end{bmatrix} ,\ C_2 = \frac{1}{24}\begin{bmatrix} 3 & 1 & 1 \\ 1 & 3 & 1 \\ 1 & 1 & 3 \end{bmatrix}$$

3. EXPRESSION OF γ AS A FUNCTION OF THE ANGLES OF THE TRIANGLE FOR $a_o = 0$

We evaluate in this section the numbers γ corresponding to the decompositions (2.2) and (2.3), for the bilinear form $\int_K \nabla u \nabla v \, dK$. We shall prove that γ is a function of the form parameter d (2.7).

3.1 The quadratic case

Here the matrices are given by (2.8), and γ^2 is the largest eigenvalue of

(3.1) $$\frac{16}{9} A_1 C_1^{-1} A_1 x = \lambda A_1 X,$$

or of $$\frac{16}{9} A_1 C_1^{-1} y = \lambda y.$$

By evident arguments we obtain

$$\gamma^2 = \frac{2}{3}\ (1 - \lambda_{min}\ (D_c\, C_1'^{-1}))$$

where $C_1' = 3/4\ C_1$, $D_c = \text{diag}\ (c_1, c_2, c_3)$.

and λ_{min} (.) is the smallest eigenvalue of (.).

Using trigonometric relations in the triangle, we can write

$$\det\,(D_C - \lambda\, C_1') = 2\,(c_1c_2c_3 - c)\,\lambda^3 + 3\,(c - c_1c_2c_3)\,\lambda^2 - \lambda c + c_1c_2c_3$$

and the corresponding eigenvalues

$$\lambda_1 = 1,\quad \lambda_2 = \frac{1}{4} + \frac{1}{4}\sqrt{4d-3},\quad \lambda_3 = \frac{1}{4} - \frac{1}{4}\sqrt{4d-3}\ ;$$

the introduction of the parameter d (2.7) resulting of the identity

$$\frac{c_1c_2c_3}{c} = \frac{1-d}{3-d}\,.$$

We obtain then :

Result 3.1. For the quadratic case (2.2) and the bilinear form $\int \nabla u.\nabla v$, the number γ is given by

(3.2) $$\gamma^2 = \frac{1}{2} + \frac{1}{3}\sqrt{d - \frac{3}{4}}\,.$$

3.2 The piecewise linear case

Replacing in (3.1) the matrices of (2.8) by those of (2.10), it appears that here γ^2 is equal to $\frac{3}{4}$ of γ^2 given by (3.2).

Result 3.2. For the piecewise linear case (2.3) and the bilinear form $\int \nabla u.\nabla v$, the number γ differs from that of the quadratic case by the factor $\frac{\sqrt{3}}{2}$; its expression is given by

(3.3) $$\gamma^2 = \frac{3}{8} + \frac{1}{4}\sqrt{d - \frac{3}{4}}\,.$$

Remarks • We must emphasize that in the two case γ^2 depends only on the form parameter $\sum_{i=1}^{3} \cos^2\theta_i$ of the triangle. The optimal case corresponds to the equilateral triangle (d = 3/4) and the worst to the degenerate triangle with an angle equal to Π (d = 3). In this latter case γ is equal to 1 for the quadratic, but there is no catastrophe for the piecewise linear case since γ equals $\sqrt{3}/2$.

• We summarize the results of this section by drawing the curves of the functions $d \to \gamma^2$ for the two cases :

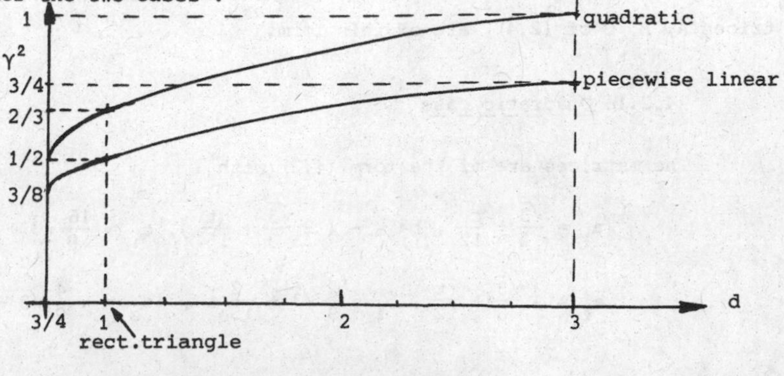

Figure 3.1

4. EXPRESSION OF γ AS A FUNCTION OF a_o FOR AN EQUILATERAL TRIANGLE

In this section, we give exact evaluation of the numbers γ in function of a_o for the bilinear form (2.1). We consider only the particular case of an equilateral triangle, which is the only situation in which the results can be explicited by hand without difficulty. We begin with a simple calculation of γ for the bilinear form $\int uv$, corresponding to the limit as a_o tends to ∞ ; this number is a constant (independant of the triangle).

4.1 The γ's for $\int uv$

The eigenvalue problems correspond respectively to the matrices of (2.9), (2.11) for the quadratic and bilinear case. In both cases the three matrices A, B, C belong to the multiplicative group of the matrices of the form $\alpha I + \beta ee^t$, where I is the identity and $e^t = (1, 1, 1)$, which possess only 2 eigenvalues λ_1, λ_2 with respective eigenspaces e and $\{ x ; e^t x = 0 \}$. In such a situation we have immediatly

$$(4.1) \quad \begin{cases} \gamma^2 = \max \{ \lambda_1, \lambda_2 \}, \text{ with} \\ \lambda_1 = \dfrac{(b_1 + 3b_2)^2}{(a_1 + 3a_2)(c_1 + 3c_2)}, \quad \lambda_2 = \dfrac{b_1^2}{a_1 c_1} \end{cases}$$

if the 3 matrices are :

$$(4.2) \qquad A = a_1 I + a_2 ee^t, \quad B = b_1 I + b_2 ee^t, \quad C = c_1 I + c_2 ee^t.$$

For the matrices of (2.9), (2.11) we obtain successively :

$$(4.3) \qquad \gamma^2 = \max \left\{ \frac{15}{16}, \frac{3}{5} \right\} = \frac{15}{16} \quad \text{(quadratic)}$$

$$(4.4) \qquad \gamma^2 = \max \left\{ \frac{9}{10}, \frac{9}{16} \right\} = \frac{9}{10} \quad \text{(piecewise bilinear)}$$

4.2 The γ's for $\int \nabla u . \nabla v + a_o \int uv$ and an equilateral triangle

The matrices A_2, B_2, C_2 are of the form $\alpha I + \beta ee^t$ for any triangle, but it is only for the equilateral triangle that the matrices A_1, B_1, C_1, and then the matrices A, B, C of (2.4), are of this form.

4.2.1. Quadratic case

The matrices are of the form (4.2) with :

$$a_1 = \frac{\sqrt{3}}{2} + \frac{t}{12}, \quad b_1 = - \left(2 \frac{\sqrt{3}}{3} + \frac{t}{15} \right), \quad c_1 = \frac{16}{9} \sqrt{3} + \frac{4}{45} t$$

$$a_2 = - \frac{\sqrt{3}}{6} + \frac{t}{12}, \quad b_2 = \frac{2}{9} \sqrt{3} + \frac{2}{15} t, \quad c_2 = - \frac{4}{9} \sqrt{3} + \frac{4}{45} t$$

and γ^2 is given as in (4.1) :

$$(4.5)\quad \begin{cases} \gamma^2 = \max\{\lambda_1, \lambda_2\}, \text{ with} \\ \lambda_1 = \dfrac{15}{16}\dfrac{t}{t+\frac{5}{4}\sqrt{3}}, \quad \lambda_2 = \dfrac{3}{5}\dfrac{(t+10\sqrt{3})^2}{(t+6\sqrt{3})(t+20\sqrt{3})} \end{cases}$$

4.2.2. Piecewise linear case

In the same way we have here :

$$a_1 = \frac{\sqrt{3}}{2} + \frac{t}{12}, \quad b_1 = -\left(\frac{\sqrt{3}}{2} + \frac{t}{16}\right), \quad c_1 = \frac{4}{3}\sqrt{3} + \frac{t}{12}$$

$$a_2 = -\frac{\sqrt{3}}{6} + \frac{t}{12}, \quad b_2 = \frac{\sqrt{3}}{6} + \frac{5}{48}t, \quad c_2 = -\frac{\sqrt{3}}{3} + \frac{t}{24}$$

and

$$(4.6)\quad \begin{cases} \gamma^2 = \max\{\lambda_1, \lambda_2\}, \text{ with} \\ \lambda_1 = \dfrac{9}{10}\dfrac{t}{t+\frac{8}{15}\sqrt{3}}, \quad \lambda_2 = \dfrac{9}{16}\dfrac{(t+\frac{8}{3}\sqrt{3})^2}{(t+2\sqrt{3})(t+\frac{16}{3}\sqrt{3})} \end{cases}$$

Result 4.1. For the bilinear form $\int \nabla u.\nabla v + a_o \int uv$ on an equilateral triangle, the expressions of γ in function of $t = a_o S$ are given respectively by (4.5), (4.6) for the quadratic and piecewise linear cases.

The figure 4.1 shows the curves of $a_o h^2 \to \gamma^2$ (h : side length of the equilateral triangle).

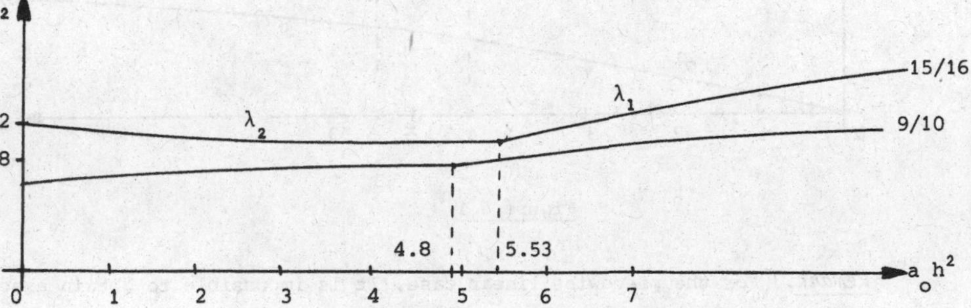

Figure 4.1

Remarks 1) Usually $a_o h^2$ is small, and γ^2 can be considered as approximatively constant in the two cases, slowly decreasing for (1) and slowly increasing for (2). It is only in exceptional situations that large values of $a_o h^2$ could happen (very coarse mesh, or singular perturbation problems of the type $-\varepsilon \Delta u + u = f$ ...).

2) We studied the effect of numerical integration on these γ's. With the following formulas

$$\int \phi \sim \frac{S}{3} \sum_{i=1}^{3} \phi(b_i) \quad \text{(quadratic)}$$

$$\int \sigma \sim \frac{S}{12} \left(\sum_{i=1}^{3} \phi(a_i) + 3 \sum_{i=1}^{3} \phi(b_i) \right) \quad \text{(piecewise linear)} ;$$

the corresponding γ is for all a_o, greater than the exact γ in the quadratic case and smaller in the piecewise linear. For small $a_o h^2$ the approximate γ's are very near of the exact ones.

5. COMPARISON OF THE $\gamma(a_o)$ FOR ONE -, TWO - AND THREE-DIMENSIONAL ELEMENTS

Proceeding in the same way as in 4.2.1. it is not difficult to evaluate γ in function of a_o for the one-and the three-dimensional problem $- \Delta u + a_o u$, respectively on a segment of length h (2 vertices, 1 middle node) or a regular tetrahedron of edge length h (4 vertices, 6 middle edge nodes).

The comparison of the 3 variable spaces, for the quadratic case, is summarized on the figure 5.1.

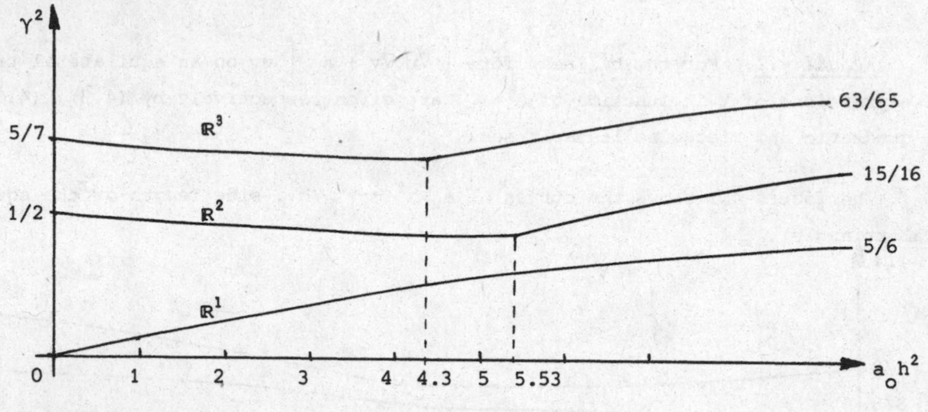

Figure 5.1

Remark. For the piecewise linear case, it is impossible to obtain exactly the same situation in $\mathbb{R}^3$ as in $\mathbb{R}^2$. In fact, it is possible to divide a regular tetrahedron into 8 tetrahedrons with vertices at the 10 nodes, but only 4 of these tetrahedrons are regular!

6. EVALUATION OF γ FOR $\int a(x)\, \nabla u . \nabla v$

In order to evaluate the effect of variable coefficients, we studied in the quadratic case for the model operator $- \nabla\, (a\, \nabla u)$, $a > 0$. We present here the results for an equilateral triangle and the numerical integration formula

$$\int_K \phi \, dK \sim \frac{S}{3} \sum_{i=1}^{3} \phi\,(b_i) \quad \text{(cf. figure 2.1)} \tag{6.1}$$

Whith the notations

$$\alpha_i = a(b_i),\ i=1 \text{ to } 3\ ;\ m = \frac{1}{3}\left(\sum_{i=1}^{3} \alpha_i\right), \tag{6.2}$$

the bilinear form is here :

$$a(u,v) = \frac{S}{3} \sum_{i=1}^{3} \alpha_i \quad (\nabla u.\nabla v)\ (b_i). \tag{6.3}$$

The 3 corresponding matrices A, B, C can be expressed, for a general triangle, using the matrix A_1 of (2.8) and the matrix

$$T = \begin{bmatrix} -\alpha_1 & \alpha_1 & \alpha_1 \\ \alpha_2 & -\alpha_2 & \alpha_2 \\ \alpha_3 & \alpha_3 & -\alpha_3 \end{bmatrix}$$

We have :

$$A = m\,A_1,\quad B = \frac{2}{3} A_1\,T,\quad C = \frac{4}{3} D\,T \tag{6.4}$$

with $D = \begin{bmatrix} -a_{11} & a_{22} & a_{33} \\ a_{11} & -a_{22} & a_{33} \\ a_{11} & a_{22} & -a_{33} \end{bmatrix}$, where $a_{ii} = (A_1)_{ii}$.

For an equilateral triangle, the eigenvalue problem can be solved by hand and we obtain :

Result 6.1. For the bilinear form $\int a(x)\ \nabla u.\nabla v$ on an equilateral triangle with the numerical integration formula (6.1), the number γ for the quadratic case is given by

$$\gamma^2 = \frac{1}{2} + \frac{\sqrt{2}}{4}\frac{\sigma}{m} \tag{6.5}$$

where $\sigma^2 = \frac{1}{3} \sum_{i=1}^{3} (\alpha_i - m)^2$.

Remark. The parameter is here the relative dispersion of the 3 values $a(b_i)$. It is possible to give the simple upper bound :

$$\gamma^2 \leqslant 1 - \frac{1}{2}\frac{\underline{a}}{\bar{a}}. \tag{6.6}$$

if $0 < \underline{a} \leqslant a(x) \leqslant \bar{a}$ on the triangle.

7. CONCLUDING REMARKS

The aim of this paper was to study the effect on the number γ of some parameters (element's geometry, coefficients of the bilinear form, dimension of the problem).

We have limited ourselves to evaluating γ for one triangle and two typical finite element spaces decompositions. Similar local results can be obtained for other spaces.

For a general mesh, γ can be bounded by $\max_e \gamma_e$.

For an equilateral, or other regular mesh, all the γ_e are equals to the same value γ which is the exact γ for the whole Neumann problem. With Dirichlet conditions, the γ is not greater than that for Neumann, and we are studying more precisely the effect of boundary conditions.

REFERENCES

AXELSSON O., and I. GUSTAFSSON, Preconditionning and two-level multigrid methods of arbitrary degree of approximation. Report 8120, Mathematisch Institut Katholieke Universiteit, Nijmegen (July 1981).

BANK R.E., and T. DUPONT, An optimal order process for solving finite element equations. Preprint 1977, and Math. Comp. 36 (1981), 35-51.

BRAESS D., The contraction number of a multigrid method for solving the Poisson equation. Numer. Math. 37 (1981), 387-404.

APPLICATION OF THE MULTIGRID METHOD TO A NONLINEAR INDEFINITE PROBLEM

Th. Meis, H. Lehmann, H. Michael
Mathematisches Institut
der Universität zu Köln

In this paper we want to demonstrate the efficiency of the multigrid method by means of a nonlinear example. Therefore we consider the boundary value problem

$$\begin{aligned} -\Delta u(x,y) &= \lambda \exp(u(x,y)) + q(x,y) && \text{on } Q = (0,1) \times (0,1) \\ u(x,y) &= 0 && \text{on } \partial Q. \end{aligned}$$

As multigrid method we intentionally choose a simple and quite rigid iteration scheme. That is the best way to prove the effectiveness of the basic principle.

1. The discrete problem

Let $n \in \mathbb{N}$, $N = 2^n$, $h = 1/N$ and $d = (N-1)^2$. The grid G_h is given by

$$\{(x,y) \in \mathbb{R}^2 \mid x = ih,\ y = jh \text{ for } i,j = 1(1)N-1\}.$$

Its "discrete closure" is

$$\overline{G}_h = \{(x,y) \in \mathbb{R}^2 \mid x = ih,\ y = jh \text{ for } i,j = 0(1)N\}.$$

The vector space of all functions $G_h \to \mathbb{R}$ is denoted by V_h with $\dim V_h = d$. Moreover, we need an interpolation $J_h : V_{2h} \to V_h$ and a projection $P_h : V_h \to V_{2h}$. We always use the same maps:

$$J_h v_{2h}(x,y) = \begin{cases} v_{2h}(x,y) & \text{for } (x,y) \in G_{2h} \\ \frac{1}{2}[v_{2h}(x-h,y) + v_{2h}(x+h,y)] & \text{for } (x-h,y) \in \overline{G}_{2h} \\ \frac{1}{2}[v_{2h}(x,y-h) + v_{2h}(x,y+h)] & \text{for } (x,y-h) \in \overline{G}_{2h} \\ \frac{1}{4}[v_{2h}(x-h,y-h) + v_{2h}(x-h,y+h) & \\ \quad + v_{2h}(x+h,y-h) + v_{2h}(x+h,y+h)] & \text{for } (x-h,y-h) \in \overline{G}_{2h} \end{cases}$$

$$P_h v_h(x,y) = \frac{1}{4}v_h(x,y) + \frac{1}{8}[v_h(x-h,y) + v_h(x+h,y) + v_h(x,y-h) + v_h(x,y+h)] + \frac{1}{16}[v_h(x-h,y-h) + v_h(x+h,y-h) + v_h(x-h,y+h) + v_h(x+h,y+h)].$$

For simplicity, we suppress the index h in the following, but not the index 2h. For example, we write v and P instead of v_h and P_h, but v_{2h} and P_{2h}. The meshsize of the coarsest grid, we use in the multigrid method, is denoted by h_{max}. In most cases, we take $h_{max} = 1/8$.

The usual five-point formula is used to discretize the Laplacian operator. v represents the discrete approximation to the solution u of the boundary value problem. For the finest grid, we set $r(x,y) = q(x,y)$ and for the coarser grids we recursively define $r_{2h} = P_h r_h$. On each grid d difference equations with d unknowns are obtained:

$$\frac{4}{h^2}v(x,y) - \frac{1}{h^2}[v(x-h,y) + v(x+h,y) + v(x,y-h) + v(x,y+h)] - \lambda\exp(v(x,y)) - r(x,y) = 0$$

for all $(x,y) \in G$.

We simply write $F_h(v_h) = F(v) = 0$ for the whole system of equations.

2. *The full scheme*

For $\lambda = 0$ the boundary value problem and the discrete systems are linear. This means $F(v) = Av - r$ with a symmetric, positive definite $d \times d$-matrix. The discrete solutions can be computed in a well-known manner (cf. e.g. [4]) by the multigrid method. The iteration scheme is given by

$$v^{(j+1)} = Sv^{(j)} + g \quad , \; j = 0(1)\infty ,$$

where

$$S = K(M_m \cdots M_1)$$

$$K = \begin{cases} I - \omega J(I_{2h} - S_{2h}^k)A_{2h}^{-1}PA & \text{for } 2h < h_{max} \\ I - \omega J A_{2h}^{-1}PA & \text{for } 2h = h_{max} . \end{cases}$$

The maps

$$I : V \to V \qquad , I_{2h} : V_{2h} \to V_{2h}$$

are the identities, M_1 to M_m are the matrices of the smoothing procedure and $\omega \in \mathbb{R}$, $\omega \approx 1$. In each iteration step on G_h, k steps on G_{2h} are performed. On the coarsest grid, the system is exactly solved. $A^{-1}r$ is as well a fixed point of the smoothing procedure as of the multigrid method itself. Replacing the m smoothing sweeps by one operator $H(v,r)$, yields

$$Sv + g = K(M_m \cdots M_1)v + g = KH(v,r) + \tilde{g}$$

$$A^{-1}r = SA^{-1}r + g = KA^{-1}r + \tilde{g}$$

$$\tilde{g} = (I - K)A^{-1}r$$

$$Sv + g = K(H(v,r) - A^{-1}r) + A^{-1}r \; .$$

Setting

$$w = H(v,r) \; ,$$

we obtain

$$Sv + g = w - \omega J(I_{2h} - S_{2h}^{k})A_{2h}^{-1}PF(w) \qquad \text{for } 2h < h_{max}$$

$$Sv + g = w - \omega JA_{2h}^{-1}PF(w) \qquad \text{for } 2h = h_{max} \; .$$

Since for each $z_{2h} \in V_{2h}$

$$A_{2h}^{-1}PF(w) = [F_{2h}^{-1}(z_{2h} + cPF(w)) - F_{2h}^{-1}(z_{2h})]/c \; ,$$

the iteration scheme for the linear case can be written as

$$w = H(v^{(j)},r) \qquad , v^{(j+1)} = w - \omega J(a_{2h} - b_{2h})/c \; .$$

For $2h < h_{max}$ we have

$$a_{2h} = (I_{2h} - S_{2h}^{k})F_{2h}^{-1}(z_{2h} + cPF(w))$$

$$b_{2h} = (I_{2h} - S_{2h}^{k})F_{2h}^{-1}(z_{2h}) \; .$$

These grid functions are the k-th multigrid approximations to the solutions $\tilde{v}$, $\hat{v}$ of the equations

$$F_{2h}(\tilde{v}) = z_{2h} + cPF(w)$$

$$F_{2h}(\hat{v}) = z_{2h} \; .$$

For $2h = h_{max}$, a_{2h} and b_{2h} are the exact solutions of the same equations.

In the nonlinear case $\lambda \neq 0$, we recursively define the multigrid method for the different values of h

$$v^{(j+1)} = w - \omega J(a_{2h} - b_{2h})/c \ , \ j = 0(1)\infty \ ,$$

where

$$w = H(v^{(j)}, r)$$

$$a_{2h} = \begin{cases} \text{k-th multigrid approximation} & \\ \text{to } F_{2h}^{-1}(z_{2h} + cPF(w)) & \text{for } 2h < h_{max} \\ F_{2h}^{-1}(z_{2h} + cPF(w)) & \text{for } 2h = h_{max} \end{cases}$$

$$b_{2h} = \begin{cases} \text{k-th multigrid approximation} & \\ \text{to } F_{2h}^{-1}(z_{2h}) & \text{for } 2h < h_{max} \\ F_{2h}^{-1}(z_{2h}) & \text{for } 2h = h_{max} \ . \end{cases}$$

When computing a_{2h} and b_{2h}, the two iterations must be started with the same function on G_{2h}.

The algorithm depends on the choice of the following parameters:

H	smoothing procedure
ω	correction parameter (here always $\omega = 1$)
c	differentiation parameter (here always $c = 1$)
k	number of multigrid steps on G_{2h} for one step on G_h (here always $k = 2$ or $k = 3$)
$v_h^{(0)}$	initial values on all grids
z_{2h}	auxiliary functions on all grids but the finest one.

The smoothing procedure and the choice of the initial values will be explained in detail in the next two chapters.
There are several ways, how to choose z_{2h}. Basically, one has to take two conditions into account:

(1) The term $(a_{2h} - b_{2h})/c$ corresponds to a numerical differentiation of F_{2h}^{-1} at z_{2h}. $\| z_{2h} \|$ has to be small, because $Pv^{(j)}$ lies in a neighbourhood of $F_{2h}^{-1}(0)$. Moreover, the function F_{2h}^{-1} is often only defined close to zero.

(2) It is not desirable to compute both a_{2h} and b_{2h} in each multigrid step. Therefore, it is convenient to choose z_{2h} so that b_{2h} is known.

The following strategies are evident:

(1) Set $z_{2h} = 0$. b_{2h} has to be computed only once at the beginning.

(2) Use the information of the previous step and set

$$(z_{2h})_{new} = (z_{2h} + cPF(w))_{old}$$

$$(b_{2h})_{new} = (a_{2h})_{old} .$$

In most cases, the choice of the initial values immediately results from the calculation of $v^{(0)}$.

(3) Compute

$$b_{2h} = Pv^{(j)} \qquad , z_{2h} = F_{2h}(b_{2h}) .$$

Hence $b_{2h} = F_{2h}^{-1}(z_{2h})$. In order to compute a_{2h}, b_{2h} is taken as initial guess for the multigrid method. A slight modification of this strategy is given by

$$b_{2h} = Pw \qquad , z_{2h} = F_{2h}(b_{2h}) .$$

This yields the FAS method of A. Brandt [1].

From a programmers point of view, strategy (2) seems to be easy and economical. In some critical cases we obtained better convergence by applying one single step of strategy (3) after several steps in the sense of (2). Therefore, we prefer this "mixed" method.

3. *The smoothing procedure*

The choice of the smoothing procedure is of great importance for the convergence of the multigrid method. In this paper, we only consider the pointwise Gauß-Seidel-Newton relaxation as classified by Ortega-Rheinboldt [5]. The sequence of the grid points of G_h is defined by red-black ordering (checkerboard). At first we treat the points $x = ih$, $y = jh$ for $i + j$ odd (red points), then those for $i + j$ even (black points). To solve the equation

$$v(x,y) - \frac{1}{4}[v(x-h,y) + v(x+h,y) + v(x,y-h) + v(x,y+h)]$$

$$- \frac{1}{4}h^2\lambda\exp(v(x,y)) - \frac{1}{4}h^2r(x,y) = 0$$

we perform two Newton steps during the first multigrid iterations and one step afterwards. This change is connected with the single application of the strategy (3) (see above). The value of m is always equal to 1, i.e. there is only one Gauß-Seidel step per level. After the relaxation, the defect nearly vanishes for all black points. For this reason, we have to calculate the defect only for the red points, which simplifies the projection as follows

$$PF(w)(x,y) = \frac{1}{8}[F(w)(x-h,y) + F(w)(x+h,y) + F(w)(x,y-h) + F(w)(x,y+h)] \quad , (x,y) \in G_{2h}.$$

Using the Gauß-Seidel method in connection with an indefinite system of equations, involves some difficulties, because the pure Gauß-Seidel iteration diverges. The above problem is indefinite for large values of v ("upper" solution). After linearization the diagonal elements of the matrix are

$$\delta(x,y,h) = 1 - \frac{1}{4}h^2\lambda\exp(v(x,y)).$$

In the worst case, they become negative. The numerical results show that the multigrid method converges if $\delta(x,y,h) \geq 1/2$ for all grids G_h on which pointwise relaxation is used. Going far below that limit causes divergence of the iteration. For $h = h_{max}$, a smoothing step is not necessary. Hence, the condition $\delta(x,y,h_{max}/2) \geq 1/2$ is sufficient. It follows

$$h^2_{max} \leq \frac{8}{\lambda} \min \exp(-v(x,y))$$

or

$$\max \exp(v(x,y)) \leq \frac{8}{\lambda h^2_{max}} \quad .$$

This restriction characterizes the limitation of our simple algorithm. There is only a small range of parameters, in which the iteration changes from good convergence to divergence. Using linewise Gauß-Seidel relaxation or Kaczmarz's method [3] can improve this multigrid method.

4. The initial values

In the following sections, we restrict ourselves to $q(x,y) \equiv 0$. The boundary value problem now reads as follows

$$-\Delta u(x,y) = \lambda\exp(u(x,y)) \quad \text{on } Q$$
$$u(x,y) = 0 \quad \text{on } \partial Q.$$

The discrete problems have at least two solutions for $\lambda \in (0,\lambda^*)$. These solutions coincide at λ^*, the so-called turning point. The choice of the initial values $v^{(0)}$ determines which solution is calculated. The attractive region of the upper solution is comparatively small. Our starting procedure is given by the following rules:

(1) Take a constant grid function on G_h for $h = h_{max}$. To approximate the lower solution, the constant can be chosen as 0. With respect to the upper solution it should be rather a large positive number, for instance 2.6 for $\lambda = 5.4$. Then improve this approximation by a damped Newton's method until the defect nearly vanishes. In critical cases it is recommendable to choose the damping coefficient equal to 1/16 for the first step.

(2) Compute successively approximations on G_h for $h = h_{max}/2, h_{max}/4, \ldots$ On each level one has to interpolate the function on the coarser grid and to perform k multigrid steps. Thus, the lower solution can be calculated as long as it is defined on the coarsest grid. The upper solution causes more complications, but we succeeded with $\lambda \in (5,6)$.

The starting procedure requires rather a great effort. However, it is sufficient to get one point on the lower (upper) solution in this way and to proceed with a continuation method.

Our continuation method is given by:

(1) Set $\lambda_2 = \lambda_1 + \Delta\lambda$. On G_h, for $h = h_{min}$, take the best approximation to v depending on λ_1 as initial value $v^{(0)}$ for the parameter λ_2.

(2) Compute recursively

$$v_{2h}^{(0)} = Pv^{(0)} \quad , z_{2h} = F_{2h}(v_{2h}^{(0)})$$

for all coarse grids with $h < h_{max}$.

(3) Perform one multigrid iteration considering the following modifications: on each grid omit the first smoothing step, because the solution is still smooth. When the solution on the coarsest grid has to be evaluated for the first time, apply a damped Newton's iteration.

Each continuation step is followed by as many multigrid steps as required.

5. The change of parameters at the turning point

Near the turning point the Jacobians of the equations $F(v) = 0$ are nearly singular. There one may not expect convergence. Prescribing $\eta := v(1/2,1/2)$ on all grids and calculating λ overcomes the difficulties in that region. In particular: given η and λ, the multigrid iteration is performed on all grids $\overset{\circ}{G}_h = G_h - \{(1/2,1/2)\}$. After a sufficient number of steps the defect nearly vanishes on G_h.

At (1/2,1/2) the defect on the finest grid becomes

$$f(\lambda,\eta) = \frac{4}{h^2}\eta - \frac{1}{h^2}[v(1/2-h,1/2) + v(1/2+h,1/2) + v(1/2,1/2-h) + v(1/2,1/2+h)] - \lambda\exp(\eta) .$$

Solving the equation $f(\lambda,\eta) = 0$ yields λ. To do this, we use a modified Steffensen method. The Steffensen method acts as an outer, the multigrid method as an inner iteration. All interesting cases work with only one up to three outer iteration steps. The procedure can be initialized with η and λ near the turning point. To take the constant η as a first approximation to v for $h = h_{max}$ proved to be a good choice to start the inner iteration. One can also use a continuation method as mentioned in section 4.

We want to explain the modified Steffensen method in more detail. Therefor we omit the argument η of f. The procedure is given by

$$\lambda^{(\nu+1)} = \lambda^{(\nu)} - \beta \frac{f^2(\lambda^{(\nu)})}{f(\lambda^{(\nu)} + \beta f(\lambda^{(\nu)})) - f(\lambda^{(\nu)})} , \quad \beta \in \mathbb{R}, \beta \neq 0.$$

Taking $\beta = -1$ leads to the usual Steffensen method. If λ satisfies the equation $f(\lambda) = 0$, $\beta = -1/f'(\lambda)$ is an optimal choice. Then the convergence is of order 3. In our problem $f'(\lambda)$ always remains negative. The derivative can be estimated by the last values of $f(\lambda^{(\nu)})$, which leads to a nearly optimal β. The whole procedure can be interpreted as two steps of the regula falsi method, as we were told by W. Werner (Mainz).

6. The results

It is well-known (cf. e.g. [2]), that the boundary value problem has two non negative solutions for $0 < \lambda < \lambda^*$. Figure 1 shows the course of both solutions along the bisector for two substantially different parameters λ. The characteristic shape of the solution depending on λ is depicted in Figure 2.

During the numerical treatment, difficulties only arise from the turning point λ^* and from large values of $\|v\|_\infty$. The lower solution $\underline{v}$ can be evaluated without any problems by the simple multigrid-continuation procedure close up to λ^*. For $\Delta\lambda = 0.1$ and $h_{max} \leq 1/4$ one

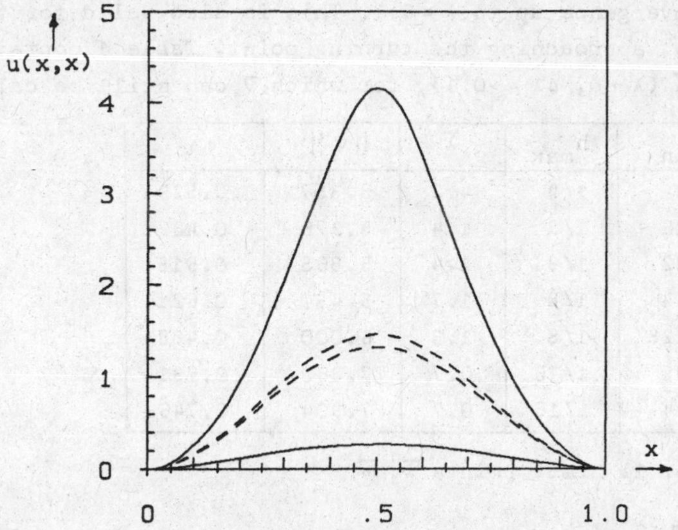

Fig. 1: u(x,x) for λ = 6.8 (---) and λ = 3.0 (——)

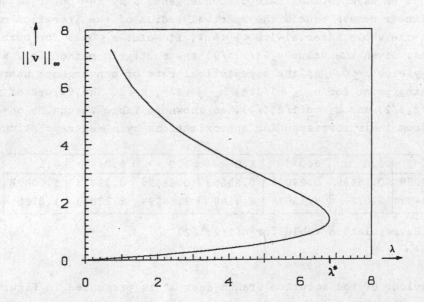

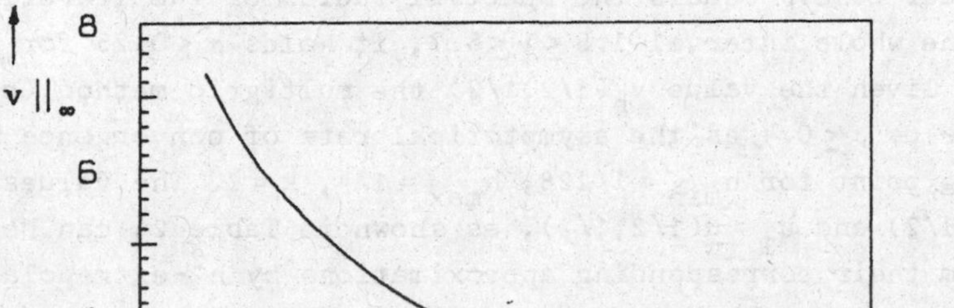

Fig. 2: $\|v\|_\infty$ for h = 1/64

can obtain convergence up to $\lambda = 6.7$. This is also valid for the upper solution $\overline{v}$, approaching the turning point. Table 1 contains the limit points $\overline{\lambda}$ ($\lambda \to 0$, $\Delta\lambda = -0.1$), for which $\overline{v}$ can still be calculated.

h_{min}	h_{max}	$\overline{\lambda}$	$\|\|v\|\|_\infty$	γ
1/8	1/4	4.2	3.357	0.529
1/16	1/8	1.4	5.976	0.462
1/32	1/8	1.4	5.865	0.518
1/64	1/8	1.7	5.430	0.621
1/128	1/8	1.3	6.000	0.488
1/32	1/16	0.7	7.363	0.731
1/64	1/16	0.7	7.304	0.746

Tab. 1: Limit points $\overline{\lambda}$, $\overline{v}$.

Here $\gamma = 1 - \frac{1}{16}h^2_{max}\overline{\lambda}\exp(\|v\|_\infty)$ (cf. section 2).

As well in the neighbourhood of the turning point as for large values of $\|v\|_\infty$ these results can be improved by choosing smaller stepsizes $\Delta\lambda$.

For all combinations of h_{min} and h_{max}, as mentioned in Table 1, we determined an asymptotical rate of convergence ρ by the power method. In the linear case ρ equals the spectral radius of the iteration matrix. In the whole interval $1.5 \leq \lambda \leq 6.7$, it holds $\rho \leq 0.25$ for both solutions. Given the value $v_h(1/2,1/2)$ the multigrid method (cf. section 5) yields $\rho \leq 0.4$ as the asymptotical rate of convergence near the turning point for $h_{min} = 1/128$, $h_{max} = 1/4$, $k = 2$. The values of $u_2 = \overline{u}(1/2,1/2)$ and $u_1 = \underline{u}(1/2,1/2)$, as shown in Table 2, can be obtained from their corresponding approximations by h^2-extrapolation.

	1.5	2.0	3.0	4.0	5.0	6.0	6.5
u_1	0.12096	0.16689	0.27037	0.39553	0.55696	0.79711	1.00428
u_2	5.694	5.072	4.1567	3.4549	2.84594	2.23991	1.8720

Tab. 2: Extrapolation table for u(1/2,1/2)

The behaviour of the solution branch near λ^* is presented in Figure 3 for a few discretizations.

We get the values of λ^* and $u^* = u(1/2,1/2)$ by interpolation with respect to $\|v_h\|_\infty = v_h(1/2,1/2)$ (cf. section 5) and by subsequent extrapolation. So λ^*_o results from λ by quadratic interpolation;

h	$\|\| v_h \|\|_\infty$	λ	λ_0^*	λ_1^*	λ_2^*
1/4	1.342 1.343 1.344	6.69050668 6.69050812 6.69050623	6.690508		
				6.814252	
1/8	1.379 1.380 1.381	6.78331296 6.78331621 6.78331610	6.783316		6.808073
				6.808459	
1/16	1.388 1.389 1.390	6.80217285 6.80217406 6.80217189	6.802174		6.808124
				6.808145	
1/32	1.390 1.391 1.392	6.80665117 6.80665273 6.80665090	6.806652		6.808124
				6.808125	
1/64	1.390 1.391 1.392	6.80775377 6.80775710 6.80775705	6.807757		6.808124
				6.808124	
1/128	1.391 1.392 1.393	6.80803211 6.80803251 6.80802953	6.808032		

Tab. 3: Extrapolation table for λ

h	$\|\| v_h \|\|_\infty$	u_0^*	u_1^*	u_2^*
1/4	1.342 1.343 1.344	1.34293		
			1.39297	
1/8	1.379 1.380 1.381	1.38046		1.39156
			1.39165	
1/16	1.388 1.389 1.390	1.38885		1.39166
			1.39166	
1/32	1.390 1.391 1.392	1.39096		1.39166
			1.39166	
1/64	1.390 1.391 1.392	1.39148		1.39166
			1.39166	
1/128	1.391 1.392 1.393	1.39161		

Tab. 4: Extrapolation table for u(1/2,1/2)

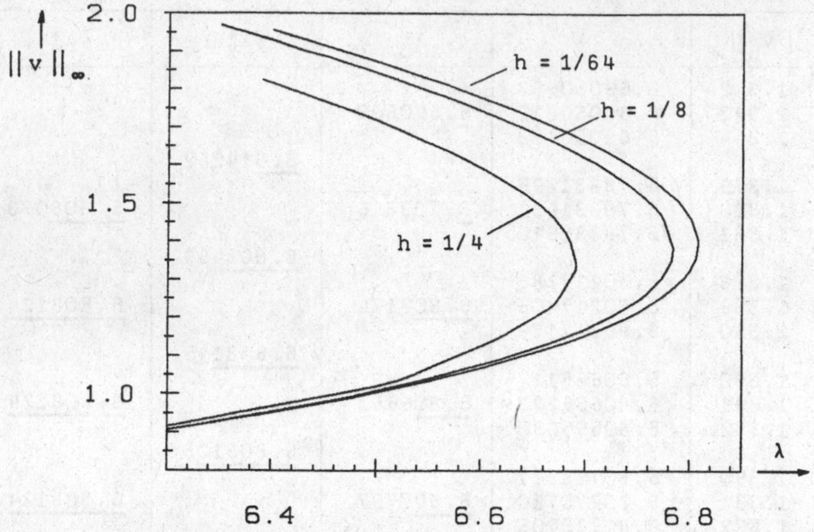

Fig. 3: Solutions near λ*

λ_1^* and λ_2^* are extrapolated from λ_o^* in the sense of a h^2-expansion (Tab. 3). Table 4 contains the corresponding values for u . One should notice the monotone behaviour of the columns in both tables. Our best results for the turning point are

$$\lambda^* \approx 6.808124$$
$$u^* \approx 1.39166 \ .$$

Paumier [6] computes $\lambda^* \approx 6.8082$ and $u^* \approx 1.392$.

References

[1] Brandt, A., Multi-level adaptive solution to boundary-value problems, Math. Comp., 31 (1977), pp. 333-390.

[2] Crandall, M.G. and Rabinowitz, P.H., Some continuation and variational methods for positive solutions of nonlinear elliptic eigenvalue problems, Arch. Rat. Mech. Anal., 58 (1975), pp. 207-218.

[3] Kaczmarz, S., Angenäherte Auflösung von Systemen linearer Gleichungen, Bulletin de l'academie polonaise des sciences et lettres A, 1937, pp. 355-357.

[4] Meis, Th. and Branca, H.W., Schnelle Lösung von Randwertaufgaben, ZAMM 62, 4/5 (1982), to appear.

[5] Ortega, J.M. and Rheinboldt, W.C., Iterative solution of nonlinear equations in several variables, Academic Press, New York/ London, 1970.

[6] Paumier, J.C., Une méthode numérique pour le calcul des points de retournement. Application à un probléme aux limites nonlinéaire, Num. Math., 37 (1981), pp. 433-452.

MULTI-GRID METHODS FOR SIMPLE BIFURCATION PROBLEMS

H. D. Mittelmann
Abteilung Mathematik
Universität Dortmund
Postfach 50 05 00
D-4600 Dortmund 50/FRG

Abstract A multi-grid method is developed for solving a class of nonlinear eigenvalue problems. The method makes essential use of the generalized inverse iteration given in [13]. Hence it does neither encounter any difficulties at turning points of the solution curve nor does it need any special modifications. A convergence proof is given and numerical results are presented for a typical problem from the theory of chemical reactions.

0. Introduction

Recently the analysis of numerical methods for bifurcation problems has undergone a rapid development, see, for example, [14]. On the other hand many linear and nonlinear problems in ordinary and partial differential equations have been treated successfully by very efficient numerical methods such as preconditioned conjugate gradient algorithms or multi-grid methods. So far, however, only very few applications of these latter techniques to bifurcation or eigenvalue problems are known, see, for example, [3, 6, 7, 10].

In this paper we consider simple bifurcation problems in the following sense. The solution curves in the usual bifurcation diagrams may exhibit bifurcation from the trivial solution at simple eigenvalues of the linearized problem or they may have one or more turning (limit) points. We include, however, problems which may be described by variational inequalities. Bifurcation for such problems has been investigated theoretically, for example, in [11] and numerically in [12, 13].

Instead of trying to describe the underlying class of problems in an abstract setting we give two typical examples one of which will be solved by a multi-grid method in sections 3 and 4. The first example is taken from structural mechanics. Consider a beam (a plate) which, for simplicity, is clamped at its ends (its edges) and is initially flat occupying a certain region Ω in $\mathbb{R}$ ($\mathbb{R}^2$) . The vertical displacement will be denoted by u. We assume that a compressing axial force acts on the beam (the plate) but that the displacement is restricted by in-plane obstacles, i. e.

there are sets C, $D \subset \Omega$ and u has to be nonnegative (nonpositive) for points in C(D). These problems have been formulated as variational problems respectively variational inequalities in [11] while the corresponding free boundary problem for the beam was derived, for example, in [12] It was shown in [11] that bifurcation from the trivial solution occurs at the first eigenvalue of the linearized problem. To determine this bifurcation point, however, is a nonlinear problem because of the obstacles, respectively the unknown free boundary.

The second example is taken from the theory of chemical reactions. If a material filling the volume $\Omega \subset \mathbb{R}^n$, $n \leq 3$, undergoes an exothermic reaction, the steady-state equation may in dimensionless variables be written as

$$(0.1) \qquad -\Delta u = \mu F(u) \quad \text{in } \Omega.$$

For simplicity we assume Dirichlet boundary conditions

$$(0.2) \qquad u = 0 \quad \text{on } \partial\Omega.$$

Here u is proportional to the difference of the temperatures of the reacting and the surrounding material and μ is proportional to the exothermicity of the reactant. The correct form of the nonlinearity is the Arrhenius Function

$$(0.3) \qquad F(u) = \exp(u/(1+\varepsilon u))$$

where $\varepsilon \geq 0$ is proportional to the temperature of the surrounding material. The Frank-Kamenetskii approximation is obtained from (0.3) by setting $\varepsilon = 0$. For a detailed derivation we refer, for example, to [4] . It is of interest to find the positive solutions of this problem and the first turning point of the curve of solutions since that corresponds to the point of self-ignition of the reactant.

In the first two sections we shall describe an iterative algorithm of Newton type for the computation of solutions to variational equalities respectively inequalities. For the first case we present in section 3 a two-grid method and in 4. a very simple multi-grid method. A convergence proof is given in 5. and numerical results for problem (0.1) - (0.3) in the last section show the effectivity of the proposed algorithms.

1. The Generalized Inverse Iteration

In the following we consider finite-dimensional problems obtained by discretization of nonlinear eigenvalue problems of the form (0.1) - (0.3) which we assume to be varirationally posed. Let two functionals f, g on $\mathbb{R}^n$ be given, where g has the special form

$$g(x) = 1/2(Bx,x),\ x\in\mathbb{R}^n, \tag{1.1}$$

B a symmetric and positive-definite matrix, while f is twice Fréchet-differentiable and its Hesse-matrix F(x) is positive-definite. In analogy to (0.1) we look at problems of the form

$$\nabla f(x) - \lambda Bx = 0 \tag{1.2}$$

where λ corresponds to μ^{-1}. If we add the constraint

$$\rho^2/2 - g(x) = 0 \tag{1.3}$$

then (1.2), (1.3) represent the Kuhn - Tucker conditions satisfied by critical points x_o of f with respect to the level sets

$$\partial S_\rho = \{x\in\mathbb{R}^n,\ g(x) = \rho^2/2\}$$

of the functional g. The parameter λ may be interpreted as Lagrange multiplier.

For the special case that f is also a quadratic functional, say,

$$f(x) = 1/2\ (Ax,x) \tag{1.4}$$

(1.2) is the usual generalized matrix eigenvalue problem and ρ may without restriction of generality be chosen as 1. Two algorithms are well known for this problem: The (simple) inverse iteration

$$\begin{aligned} A\tilde{x}_{k+1} &= Bx_k \\ x_{k+1} &= \tilde{x}_{k+1}/\|\tilde{x}_{k+1}\|_B \end{aligned} \tag{1.5}$$

for k = 1,2,..., where $x_1\in\partial S_1$ and the inverse iteration with Rayleigh-quotient shift

$$\begin{aligned} (A-\lambda_k B)\tilde{x}_{k+1} &= Bx_k \\ x_{k+1} &= \tilde{x}_{k+1}/\|\tilde{x}_{k+1}\|_B \end{aligned} \tag{1.6}$$

where $\lambda_k = (Ax_k,x_k)$ and again $x_1\in\partial S_1$. Here we have used $\|.\|_B$ for the norm on $\mathbb{R}^n$ introduced by B. For theoretical results on these methods see, for example, [18, 20].

The generalization

$$\begin{aligned} B\tilde{x}_{k+1} &= \nabla f(x_k) \\ x_{k+1} &= \rho\tilde{x}_{k+1}/\|\tilde{x}_{k+1}\|_B \end{aligned} \tag{1.7}$$

of (1.5) to (1.2), (1.3) with eigenvalues μ instead of λ was analysed in [5] while the following generalization of (1.6) was considered in [13]

$$\tilde{x}_{k+1} = x_k + p_k,\ p_k = -H_k \nabla f(x_k)$$

(1.8)
$$H_k = \begin{bmatrix} F_k - \lambda_k B & -Bx_k \\ -x_k^T B & 0 \end{bmatrix}^{-1}_{nxn}$$

$$x_{k+1} = \rho\, \tilde{x}_{k+1} / \|\tilde{x}_{k+1}\|_B$$

where $\lambda_k = (\nabla f(x_k), x_k)/\rho^2$, $F_k = F(x_k)$ and $x_1 \in \partial S_\rho$ has to be given in both cases. Under the assumption that $F_o - \lambda_o B$ is negative-definite on $\{x_o\}^\perp = \{x \in \mathbb{R}^n,\ (Bx, x_o) = 0\}$ for a solution (x_o, λ_o) of (1.2), (1.3), i. e. x_o is a local maximum of f with respect to ∂S_ρ, it was shown that (1.7) ((1.8)) converges linearly (quadratically) against x_o.

We note that H_k is the nxn principal submatrix of a matrix which is invertible in a neighbourhood of x_o. The singularity of $A - \lambda_o B$ does not influence crucially the performance of (1.6) (cf. [17]). For the nonlinear problem (1.2), however, the singularity of $F_o - \lambda_o B$ at the turning point did not allow to pass this point with a multi-grid method without modifying the algorithm considerably (cf. [3]).

In the sequel we shall see that a multi-grid method making essential use of the generalized inverse iteration may be defined which does not encounter similar difficulties and solves problem (1.2) faster than algorithm (1.8). The generalized inverse iteration compared very favourably with several other algorithms (cf. [13]) and the numerical results presented in section 4.2 of [3] show that the steplength for the pseudo-arclength continuation procedure has to be taken smaller near the limit point than for (1.8) (cf. 6. and [13]) thus reducing the effectivity.

2. An Algorithm for Variational Inequalities

The algorithm of the last section is in general only locally convergent. This means that in order to compute a point on the solution curve it could be necessary to use the method in the fashion of a continuation by following the path in several steps taking the latest result as starting value for the next point. If then the convergence is quadratic from the beginning this will still result in a very efficient algorithm, especially if the steps may be chosen relatively large. This will be seen in the last section. Here we describe an alternative technique using damping and regularization but essentially we now generalize the method to variational inequalities. We do this by introducing a suitable active-set strategy. There is no analogous technique in function spaces. That is the reason why we consider discretizations of the given problems instead of applying the proposed algorithms for the continuous problem and to discretize afterwards. This latter way, however, is possible for (1.2) (cf. Remark 2.6 in [13]).

We assume that a problem of the type of the axially compressed beam or plate, as described in the introduction, is discretized yielding a variational inequality

$$\begin{aligned}
&\lambda_o(\nabla g(x_o), x-x_o) \geq (\nabla f(x_o), x-x_o) \ \forall x \in K, \\
(2.1)\qquad &K = \{x \in \mathbb{R}^n, x_i \geq o, i \in J_1, x_i \leq o, i \in J_2\}, \\
&J_1, J_2 \subset \{1,\dots,n\}, J_1 = \{i_1,\dots,i_{n_1}\}, \\
&J_2 = \{j_1,\dots,j_{n_2}\}.
\end{aligned}$$

Now we assume

$$(2.2)\qquad f(o) = o, \nabla f(o) = o,$$

so that (2.1) always has the trivial solution. Under further natural assumptions it may be shown that solution branches exist bifurcating from the eigenvalues of the linearized problem (cf. [12]).

In order to be able to define the analogue of the generalized inverse iteration we need some further notation. Let $G = (g_1,\dots,g_{n_1+n_2})$, where $g_k = e_{i_k}$, $k = 1,\dots,n_1$, $g_{n_1+k} = e_{j_k}$, $k = 1,\dots,n_2$, $e_i \in \mathbb{R}^n$ the i-th unit vector. Then K in (2.1) may be re-written as

$$(2.3)\qquad K = \{x \in \mathbb{R}^n, G^T x \geq o\}.$$

For any $x \in \mathbb{R}^n$ let $I(x) = \{i \in \{1,\dots,n\}, g_i^T x = o\}$ and define $G_I = (g_i)_{i \in I}$, $Q_I = E_n - G_I G_I^T$, E_n the nxn identity matrix. For $x = x_k$ denote $I_k = I(x_k)$, $G_k = G_{I_k}$ and Q_k analogously.

<u>The Generalized Inverse Iteration for Variational Inequalities</u>

(2.4) Let $x_1 \in K \cap \partial S_\rho$ be arbitrary. Set k = 1 and $\mu_k = o$, $\mu_k \in \{o,1\}$.

1. Determine I_k and $u_k = r_k - \lambda_k Bx_k$, $\lambda_k = r_k^T x_k/\rho^2$, $r_k = \nabla f(x_k)$. Terminate the iteration if $G_k^T u_k \leq o$ and $\|Q_k u_k\| = o$.

2. Compute $|u_{kj}| = \max \{|u_{ki}|, (G_k^T u_k)_i > o\}$. If $\{(Q_k u_k, r_k) < |u_{kj}| \cdot \|Q_k u_k\|$ and $\mu_k = o\}$ or $\|Q_k u_k\| = o$ then set $\tilde{I}_k = I_k - \{j\}$ and determine $\tilde{Q}_k$. Otherwise let $\tilde{I}_k = I_k$, $\tilde{Q}_k = Q_k$.

3. Replace $F_k - \lambda_k B$ in (1.8) by $F_k - \lambda_k B - \tau_k E_n$, where $\tau_k = \max\{o, \delta+\sigma_k\}$ and τ_k is the largest eigenvalue of $F_k - \lambda_k B$ on $\{x_k\}^{\perp} \cap \{x \in \mathbb{R}^n, \tilde{Q}_k x = x\}$, $\delta > o$ a given constant. Compute the direction vector p_k as in (1.8) but in the variables x_{ki} with $(\tilde{Q}_k)_{ii} = 1$ (the free variables) only, fixing the others.

4. Determine the maximal admissible steplength $\bar{\alpha}_k$ in direction p_k and $\tilde{\alpha}_k = 2^{-j}$ where

$$j = \min\{i \in \mathbb{N} \cup \{o\}, f(\rho(x_k + 2^{-i} p_k)/\|x_k + 2^{-i} p_k\|_B) \geq 2^{-i-2} |p_k^T r_k|\}$$

and set

$$x_{k+1} = \rho(x_k + \alpha_k p_k)/\|x_k + \alpha_k p_k\|_B.$$

where $\alpha_k = \min\{\bar{\alpha}_k, \tilde{\alpha}_k\}$. If $\alpha_k = \bar{\alpha}_k$ then set $\mu_{k+1} = 1$ otherwise set $\mu_{k+1} = 0$. Set $k = k+1$ and goto 1.

The following convergence result was proved in [13].

<u>Theorem 2.5</u> Let there exist a constant $M = M(\rho) > o$ such that

$$o < (\nabla f(x+y) - \nabla f(x), y) \leq M\|y\|^2, \ \forall x \in S_\rho, \forall y \in S_{2\rho}, \rho > o$$

and, for simplicity, let M be chosen such that also the following inequality holds

$$(\nabla f(y), y) \leq M\|y\|^2, \ \forall y \in \partial S_\rho.$$

Assume that g is as in (1.1) and that the set

$$\Gamma = \{x^* \in K \cap \partial S_\rho, G^{*T} x^* \leq o, \|Q^* x^*\| = o\}$$

is finite and that $G^{*T} x^* < o$ for all $x^* \in \Gamma$ and $o < \delta < -\sigma^*$. Then the sequence $\{x_k\}$, $k = 1,2,\ldots$ generated by algorithm (2.4) converges to a point $x^* \in \Gamma$. If $f \in C^3(U(x^*))$ then the asymptotic (Q-)order of convergence is two.

The above algorithm was used for the numerical solution of a conforming finite element discretization with numerical integration of the beam-buckling problem. It proved to be considerably more efficient than a suitable analogue of the simple inverse iteration (1.7), of λ-continuation with Euler predictor step and a general purpose optimization routine.

3. A Two-Grid Method for Nonlinear Eigenvalue Problems

We now state a two-grid method for problems of the form (1.2)

(3.1) $\nabla f(x) - \lambda Bx = o.$

This algorithm makes essential use of the generalized inverse iteration (1.8). We do not generalize the method to variational inequalities (2.1). This could be done combining the ideas used in the sequel and those for the numerical solution of variational inequalities by multi-grid methods (cf.[8]).

We assume that (3.1) is the discretization of a corresponding continuous problem and that we have a sequence of grids $G^{(o)}, G^{(1)},\ldots$ with grid constants $h^{(o)} > h^{(1)} > \ldots > o$ yielding a sequence of discretizations of the form (3.1). For simplicity we assume that the original problem is posed on a square Ω and that each $G^{(i)}$ represents the grid points of a square mesh covering $\bar{\Omega}$ and that $h^{(i)} = 2h^{(i+1)}$, $i = o,1,\ldots$. $x^{(i)}$ will denote a mesh function defined on the grid $G^{(i)}$ with $n^{(i)}$ interior grid points and I_i^{i+1} (I_{i+1}^i) are interpolation (restriction) operators mapping the functions on $G^{(i)}$ $(G^{(i+1)})$ onto those on $G^{(i+1)}$ $(G^{(i)})$.

For the computation of a solution of

(3.2) $$\nabla f^{(i)}(x^{(i)}) - \lambda^{(i)} B^{(i)} x^{(i)} = o,$$

$i > o$, we propose the following algorithm. For simplicity we formulate it for the case $i = 1$. We denote $\|\cdot\|_i = \|\cdot\|_{B^{(i)}}$.

The Two-Grid Method

(3.3) Let $\rho^{(o)} > o$ be given. Set $k = 1$.

1. Compute $x^{(o)}$ with $\|x^{(o)}\|_o = \rho^{(o)}$ and $\lambda^{(o)}$ using algorithm (1.8) on $G^{(o)}$.

2. Interpolate $x^{(o)}$ to $\tilde{x}_k^{(1)} = I_o^1 x^{(o)}$ and smooth that by performing $\nu^{(1)}$ SOR-steps for the solution of the linear system

$$B^{(1)} x = \nabla f^{(1)}(\tilde{x}_k^{(1)})/\lambda^{(o)} .$$

For the result $x_k^{(1)}$ compute $\lambda_k^{(1)} = \nabla f^{(1)}(x_k^{(1)})^T x_k^{(1)} / \|x_k^{(1)}\|_1$.

3. Compute $r_k^{(1)} := - \nabla f^{(1)}(x_k^{(1)}) + \lambda_k^{(1)} B^{(1)} x_k^{(1)}$ and solve $(B = B^{(o)})$

$$\begin{bmatrix} F^{(o)}(x^{(o)}) - \lambda^{(o)} B & -Bx^{(o)} \\ -x^{(o)T} B & o \end{bmatrix} \begin{bmatrix} \delta^{(o)} \\ \bar{\delta} \end{bmatrix} = \begin{bmatrix} I_1^o r_k^{(1)} \\ o \end{bmatrix}$$

(The value of $\bar{\delta}$ will not be needed.)

4. Set $\tilde{\tilde{x}}_k^{(1)} = x_k^{(1)} + I_o^1 \delta^{(o)}$ and smooth it as in 2. using the system

$$B^{(1)} x = \nabla f^{(1)}(x_k^{(1)})/\lambda_k^{(1)}$$

yielding $x_{k+1}^{(1)}$. Compute $\lambda_{k+1}^{(1)}$. If a suitable termination criterion is satisfied, stop, otherwise set k=k+1 and go to 3.

In the first step of (3.3) we compute to, say, working accuracy a point on the solution curve for the coarse grid $G^{(o)}$. The philosophy behind the next step is that we find a point close to the solution curve for $G^{(1)}$ and let then a two-grid cycle find a point on this curve which somehow corresponds to $x^{(o)}$. This algorithm is only one possible multi-grid approach using ideas from the generalized inverse iteration and others will be investigated, too.

The following lemma will be proved for algorithm (3.3).

<u>Lemma 3.4</u> Any solution $x^{(1)}$ of (3.1) with $\lambda^{(1)} \neq o$ is a fixed point of the algorithm (3.3). If $x^{(1)}$ is a fixed point of (3.3) with $\lambda^{(1)} \neq o$ and sufficiently many SOR-steps are taken in (3.3) then $(x^{(1)}, \lambda^{(1)})$ is a solution of (3.1).

<u>Proof</u> We write the cycle given by the steps 3. and 4. of (3.3) as

(3.5) $$x_{k+1}^{(1)} = \phi(x_k^{(1)}),$$

$$\phi(x) = L_\omega^\nu x + L_\omega^\nu I_o^1 H^{(o)} I_1^o(-\nabla f^{(1)}(x) + \lambda^{(1)}(x) B^{(1)} x) + (E_n - L_\omega^\nu) B^{-1} \nabla f^{(1)}(x)/\lambda^{(1)}(x),$$

where $\nu = \nu^{(1)}$, $n = n^{(1)}$, $B = B^{(1)}$, $o < \omega < 2$, $H^{(o)}$ as in (1.8),

$$L_\omega = (D-\omega L)^{-1}[(1-\omega)D + \omega L^T]$$

and $B = D-L-L^T$ is the usual decomposition into lower triangular, diagonal and upper triangular part.

For a solution $(x^{(1)}, \lambda^{(1)})$ of (3.1) we have

$$\phi(x^{(1)}) = L_\omega^\nu x^{(1)} + (E_n - L_\omega^\nu) x^{(1)} = x^{(1)}.$$

If $x^{(1)}$ is a fixed point of (3.5) for the parameter $\lambda^{(1)} \neq o$ then a simple computation yields

$$[B^{-1} - L_\omega^\nu (B^{-1} + \lambda^{(1)} I_o^1 H^{(o)} I_1^o)]\,(-\nabla f^{(1)}(x^{(1)}) + \lambda^{(1)} B x^{(1)}) = o.$$

The matrix in brackets is regular for all sufficiently large ν since the spectral radius of L_ω is less than one (see, for example, [19]) and B is positive-definite. Thus $(x^{(1)}, \lambda^{(1)})$ is a solution of (1.2) and this proves the lemma.

If we want to compute a certain point on the branch for $G^{(1)}$, for example the turning point, it would be necessary to know the proper $\rho^{(o)}$ in step 1. Hence it would be desirable that the above algorithm for the $\rho^{(o)}$ corresponding to, say, the turning point for $G^{(o)}$ yields a point close to the turning point for $G^{(1)}$. The numerical results will show that for suitable choices of the interpolation and restriction operators there is in fact such a correspondence.

4. A simple Multi-Grid Method

We note that multi-grid methods often do not work satisfactorily if applied to indefinite problems as, for example, the Helmholtz equation, Stokes problem, mixed finite element approximations to the biharmonic equation etc. In the case of the Helmholtz equation the usual multi-grid methods without major modifications do not converge for wide ranges of the Helmholtz parameter even for a relatively fine coarsest mesh. If they converge they are often not faster than suitable conjugate gradient methods and, in fact, much slower than those for very indefinite problems. The latter is also true if the multi-grid method is made convergent by using the normal equations for the smoothing step, i. e. by smoothing with the Kaczmarz algorithm (cf. [9,18]).

The inherent indefiniteness in problems of type (3.1) may also cause difficulties unless special care is taken in generalizing a two-grid method as (3.3) to several grids. We present next a very simple multi-grid method and postpone the investigation of other multi-grid approaches. We assume that $\ell + 1$ grids are to be used, $\ell > 1$.

The Multi-Grid Method

(4.1) Let $\rho^{(o)} > o$ be given. Set $k = 1$.

1. Compute $x^{(o)}$ with $\|x^{(o)}\|_o = \rho^{(o)}$ and $\lambda^{(o)}$ using algorithm (1.8). With the two-grid method (3.3) compute the points $x^{(i)}$, $i = 1,\ldots,\ell-1$ with corresponding $\lambda^{(i)}$ using the grids $G^{(i-1)}$, $G^{(i)}$ and $\nu^{(i)}$ smoothing steps.

2. Interpolate $x^{(\ell-1)}$ to $\tilde{x}_k^{(\ell)} = I_{\ell-1}^{\ell} x^{(\ell-1)}$ and smooth $\tilde{x}_k^{(\ell)}$ by performing $\nu^{(\ell)}$ SOR-steps for the system

$$B^{(\ell)} x = \nabla f^{(\ell)}(\tilde{x}_k^{(\ell)})/\lambda^{(\ell-1)} .$$

   For the result $x_k^{(\ell)}$ compute $\lambda_k^{(\ell)}$.

3. Compute $r_k^{(\ell)} := -\nabla f^{(\ell)}(x_k^{(\ell)}) + \lambda_k^{(\ell)} B^{(\ell)} x_k^{(\ell)}$ and solve

$$\begin{bmatrix} F^{(o)}(x^{(o)}) - \lambda^{(o)} B^{(o)} & -B^{(o)} x^{(o)} \\ -x^{(o)T} B^{(o)} & o \end{bmatrix} \begin{bmatrix} \delta^{(o)} \\ \bar{\delta} \end{bmatrix} = \begin{bmatrix} I_1^{o} \cdots I_{\ell}^{\ell-1} r_k^{(\ell)} \\ o \end{bmatrix}$$

4. Set $\tilde{\tilde{x}}_k^{(\ell)} = x_k^{(\ell)} + I_{\ell-1}^{\ell} \ldots I_o^1 \delta^{(o)}$ and smooth it as in 2. using the system

$$B^{(\ell)} x = \nabla f^{(\ell)} (x_k^{(\ell)})/\lambda_k^{(\ell)}$$

yielding $x_{k+1}^{(\ell)}$. Compute $\lambda_{k+1}^{(\ell)}$. If a suitable termination criterion is satisfied, stop, otherwise set $k = k+1$ and goto 3.

It is obvious that an analogue of Lemma 3.4 holds for this algorithm. $I_o^1 H^{(o)} I_1^o$ has to be replaced in (3.5) by $I_{\ell-1}^{\ell} \ldots I_o^1 H^{(o)} I_1^o \ldots I_{\ell}^{\ell-1}$. (4.1) is a full multi-grid algorithm of repeated nested iteration-type with smoothing on the finest grid only.

5. Convergence Proof

For the two-grid method of 3. as well as for the multi-grid method of the last section formally the same convergence theory may be applied. Local convergence for the algorithms (1.7) and (1.8) was proved in [5,13] by writing the algorithms as fixed-point iterations

(5.1) $$x_{k+1} = \phi(x_k)$$

and by showing that under suitable assumptions $\phi'(x_o)P_{x_o}$, x_o a solution of (1.2), has a spectral radius less than one respectively equal to zero. P_z denotes the orthogonal projector on $\{z\}^{\perp}$. In [13] global convergence was proved for the algorithm (2.4).

In the following we shall outline a corresponding analysis for a slight modification of the algorithms (3.3) and (4.1). We assume that after a finite number k_o, say, of cycles the vector obtained is normalized in each of the following cycles to have the norm $\rho_o = \|x_{k_o}^{(\ell)}\|_{\ell}$. This modification did not strongly influence the performance of the algorithms but it was omitted in the computations in order to save the additional work. From (3.5) we conclude that the iteration function finally may be written as

(5.2) $$\psi(x) = \rho_o \phi(x)/\|\phi(x)\|_{\ell} = \rho_o \tilde{\phi}(x)/\|\tilde{\phi}(x)\|_{\ell}$$

where $\tilde{\phi}(x) = \lambda(x)\phi(x)$, $\phi(x)$ as in (3.5).
The derivative of $\tilde{\phi}$ at a solution x_o of (1.2) with parameter $\lambda_o > o$ is easily computed as

(5.3) $$\tilde{\phi}'(x_o) = \lambda_o L_{\omega}^{\nu} + L_{\omega}^{\nu} x_o \lambda'(x_o)^T + \lambda_o L_{\omega}^{\nu} \tilde{H}_o(-F_o + \lambda_o B + Bx_o\lambda'(x_o)^T) + (E_n - L_{\omega}^{\nu})B^{-1}F_o$$

where $\tilde{H}_o = I_{\ell-1}^{\ell} \ldots I_o^1 H^{(o)} I_1^o \ldots I_{\ell}^{\ell-1}$, $\ell = 1$ respectively $\ell > 1$.

We have

$$\psi'(x_o)P_{x_o} = P_{x_o}\tilde{\phi}'(x_o)P_{x_o}/\lambda_o . \tag{5.4}$$

Hence from (5.3) it may already be seen that for all sufficiently large ν the spectral radius of $\psi'(x_o)P_{x_o}$ is less than one. We recall that $L_\omega^\nu \to o$ for $\nu \to \infty, o<\omega<2$, and that it was shown in [5] that $P_{x_o}B^{-1}F_oP_{x_o}/\lambda_o$ has a spectral radius less than one if x_o is a local maximum of f(x) with respect to ∂S_ρ and if λ_o is not in the spectrum of $P_{x_o}B^{-1}F_oP_{x_o}$. Then x_o was called a nondegenerate local maximum. We have thus proved

Theorem 5.4 Let x_o be a nondegenerate local maximum of the functional f in (1.2) with respect to the level set ∂S_{ρ_o} of g. Then x_o is a point of attraction of the algorithms (3.3) and (4.1) with additional normalization on the level ρ_o.

We did not need any approximation property as is the case for most multi-grid methods. If such a property is satisfied, however, this leads to an acceleration as we shall show. Let $H_o = H^{(\ell)}(x_o)$ as in (1.8). Using the definition of H_o we may rewrite $\psi'(x_o)P_{x_o}$ as

$$P_{x_o}[L_\omega^\nu x_o\lambda'(x_o)^T/\lambda_o + L_\omega^\nu(\tilde{H}_o-H_o)(-F_o+\lambda_oB+Bx_o\lambda'(x_o)^T)+(E_n-L_\omega^\nu)B^{-1}F_o/\lambda_o]P_{x_o}$$

from which, comparing with (5.3), (5.4), we conclude that the spectral radius becomes smaller when $\tilde{H}_o$ is an approximation to H_o. Although the algorithm will in general only be linearly convergent in contrary to (1.8) we expect it to have a small convergence factor if the algorithm uses points $x^{(o)}$, $x_o^{(1)}$ for which $\|H_o-\tilde{H}_o\|$ is small.

6. Numerical Results and Conclusions

The methods of the preceding sections will now be utilized to solve finite-difference discretizations of problem (o.1) - (o.3). We assume that Ω is the unit square and that the usual five-point difference approximation is chosen on square meshes with $h^{(i)} = 2^{-i-2}$, $i = o,1,\ldots,3$, yielding matrices $B^{(i)}$. On grid $G^{(i)}$ we use the discretization

$$B^{(i)}x^{(i)} = \mu b^{(i)}, \ \mu = \lambda^{-1} , \tag{6.1}$$

where the j-th component of the vector $b^{(i)}$ is given by $F(x_j^{(i)})$. We shall compare the performance of the algorithms (1.8), (3.3) and (4.1). For a comparison of (1.8) and (2.4) with other algorithms we again refer to [13].

At first we consider following the path for the case $\varepsilon = o$. The solution curve has a single point and we follow the curve taking ρ as parameter. The ρ-sequence 2,4,...,2o was used on the coarsest grid $G^{(o)}$ for algorithm (4.1). For the first ρ-value $e^{(o)} = (1,\dots,1)^T/\|(1,\dots,1)^T\|_b$ was taken as starting vector while for the following values the last solution after renormalization was used. In the first column of Table 1 are given the ρ-values obtained on the finest grid $G^{(3)}$ divided by 8 and in the second column the corresponding μ-values. In the next two columns the numbers k of multi-grid cycles for algorithm (4.1) are listed and the approximate computing time in seconds. For the sequence of ρ-values obtained on $G^{(2)}$ the two-grid method (3.3) was used on grids $G^{(2)}$, $G^{(3)}$ and the next two columns contain the number of two-grid cycles and the computing time. Finally, the numbers in the last columns are the number of steps for algorithm (1.8) on $G^{(3)}$ and the corresponding time. The choice of starting vectors for (3.3) and (1.8) was the same as that for (4.1).

$\rho^{(3)}/8$	$\mu^{(3)}$	multi-grid		two-grid		gen. inv. iter.	
		k	time	k	time	k	time
1.978	2.319	5	19	3	44	3	132
3.941	4.o45	5	19	3	36	3	86
5.93o	5.294	5	19	4	45	2	51
7.985	6.142	5	19	4	43	2	5o
1o.15	6.637	6	22	4	44	2	52
12.49	6.8o6	6	22	4	41	2	5o
15.15	6.658	7	25	4	41	2	49
18.13	6.178	8	31	5	46	2	48
21.79	5.345	9	33	5	48	2	51
26.3o	4.178	9	34	5	48	2	53

Table 1 Comparison of multi-grid methods and generalized inverse iteration. $\varepsilon = o$, $h^{(3)} = 1/32$.

The termination criterion was $|\mu_{k+1} - \mu_k| < 1o^{-5}$ and the number of smoothing steps was $\nu^{(1)} = 3$, $\nu^{(2)} = 4$ and $\nu^{(3)} = 6$ with $\omega = 1$. The relatively high number of smoothing steps was caused by an optimization with respect to computing time. In the two-grid method the iterative solution of the linear system on $G^{(i-1)}$ took more time than a smoothing step on $G^{(i)}$ while in the multi-grid method the correction from $G^{(o)}$ was not so good leading also to an increase of $\nu^{(3)}$. For the solution of the linear systems the conjugate gradient algorithm (without preconditioning) proposed in [1] was chosen. An alternative would have been the subroutine SYMMLQ (cf. [15]). It should be noted that the structure of the matrices does not increase the order of magnitude of operations needed by these algorithms compared with a solution of a linear system with matrix B.

The interpolation operator with weights

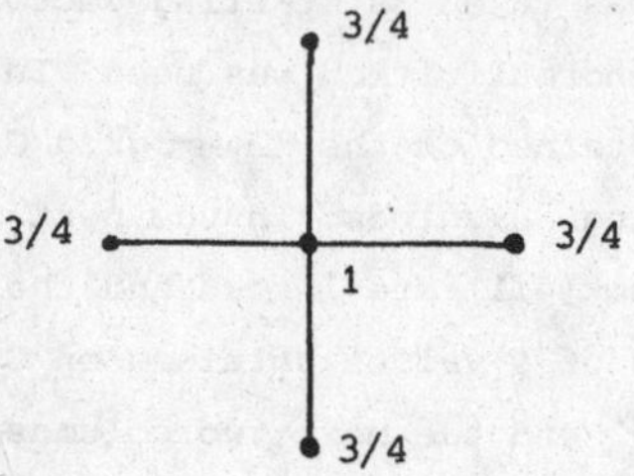

was used for points on $G^{(i+1)}$ but not on $G^{(i)}$ and the restriction operator was taken as its adjoint. The computations were performed by A. Kosubek in single-precision ALGOL-W on the IBM 37o-158 at the computing centre of the University of Dortmund.

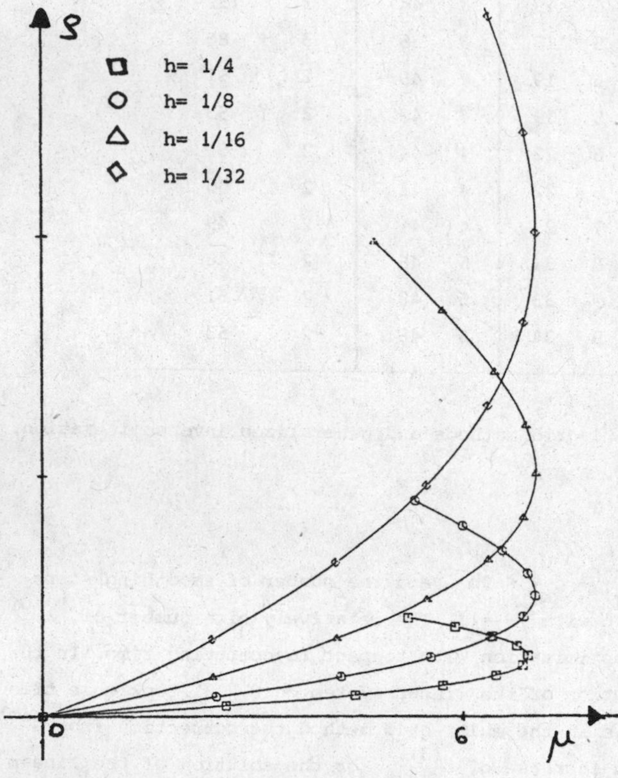

Figure 1

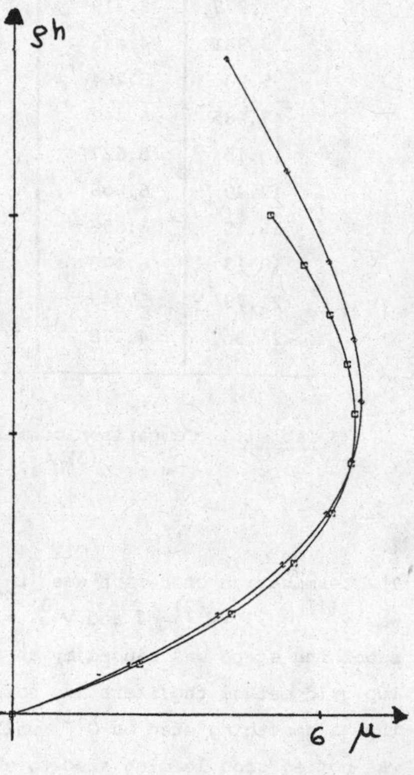

Figure 2

The solution curves in the ρ-μ diagram are plotted in Figure 1 where the computed points are marked. In order to check the correspondence between the points on the curves those for h = 1/4 and h = 1/32 were plotted in Figure 2 after multiplying the ρ-values by a multiple of h to make the norm $\|\cdot\|_B$ independent of h.

The computing times show that algorithm (4.1) is faster than (3.3) which seems to be only slightly faster than (1.8). This impression, however, is deceptive. If the ρ-steps are taken larger then (1.8) needs more steps or may even not converge at all without damping, while the given times remain more or less the same for the multi-grid methods which get their starting values from the coarser grids. In order to illustrate the behaviour of the latter methods we have added Table 2.

Having found points in the neighbourhood of the turning point for $G^{(o)}$ algorithms (3.3) and (4.1) were used to compute the corresponding points for $G^{(3)}$. Algorithm (1.8) diverged for the first ρ-value when started with $x_1^{(3)} = e^{(3)}$.

$\rho^{(3)}/8$	$\mu^{(3)}$	multi-grid		two-grid		gen. inv. iter.
		k	time	k	time	
11.27	6.758	3	13	4	58	no conv.
11.86	6.792	4	15	4	38	
12.47	6.8o5	4	16	4	38	
13.o9	6.799	4	16	4	38	
13.72	6.773	5	2o	4	39	

Table 2 Computation of a few points close to the turning point. $\varepsilon = o$, $h^{(3)} = 1/32$.

The results show that with the proposed algorithms and for the example chosen it is possible, as suggested by Brandt (cf. [2]), to find a certain point with coarse-grid computations only and to switch then to the multi-grid algorithm to obtain the corresponding solution on finer grids.

But the results of Table 1 and Figure 1 and 2 seem to indicate that an increasing distance between the solution curves above the turning point and a decreasing correspondence between the computed points causes the growing amount of work needed by the multi-grid algorithms especially for (4.1). This is confirmed by the corresponding results for problem (o.1) - (o.3) with $\varepsilon = .2$ given in Tables 3,4 and Figures 3,4.

$\rho^{(3)}/8$	$\mu^{(3)}$	multi-grid		two-grid		gen. inv. iter.	
		k	time	k	time	k	time
4.9o1	4.82o	5	19	3	43	3	127
29.73	9.o95	4	17	4	49	3	117
54.o1	8.319	2	12	3	39	2	47
77.25	7.72o	3	14	3	41	2	45
99.85	7.377	2	11	3	4o	2	48
122.o	7.195	2	11	3	38	2	42
143.8	7.113	2	14	3	38	2	4o
165.4	7.o79	3	14	3	4o	2	46
186.8	7.126	3	14	3	41	2	45
2o8.1	7.187	3	14	3	44	2	46
229.3	7.272	3	14	3	38	2	45
25o.4	7.374	3	14	3	42	2	48
271.4	7.49o	3	14	3	41	2	44

Table 3 Comparison of multi-grid methods and generalized inverse iteration. $\varepsilon = .2$, $h^{(3)} = 1/32$.

Here the curve has two turning points and in Table 4 the neighbourhood of the second turning point is approached directly. Only the algorithm (4.1) converged with a crude initial guess. The sequence of ρ-values on $G^{(o)}$ was 5, 3o, 55,...,28o.

$\rho^{(3)}/8$	$\mu^{(3)}$	multi-grid		two-grid	gen. inv. iter.
		k	time		
149.9	7.1o3	3	14	no conv.	no conv.
155.9	7.o97	3	14		
161.9	7.o96	3	14		
168.o	7.o98	3	14		
174.o	7.1o4	3	14		

Table 4 Computation of a few points close to the second turning point. $\varepsilon = .2$, $h^{(3)} = 1/32$.

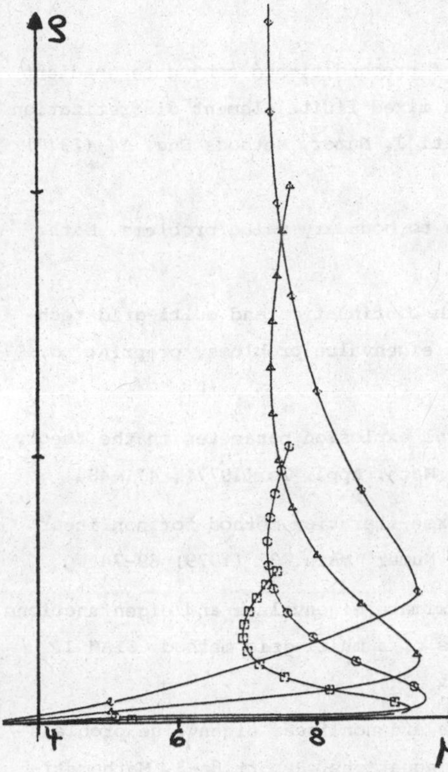

Figure 3

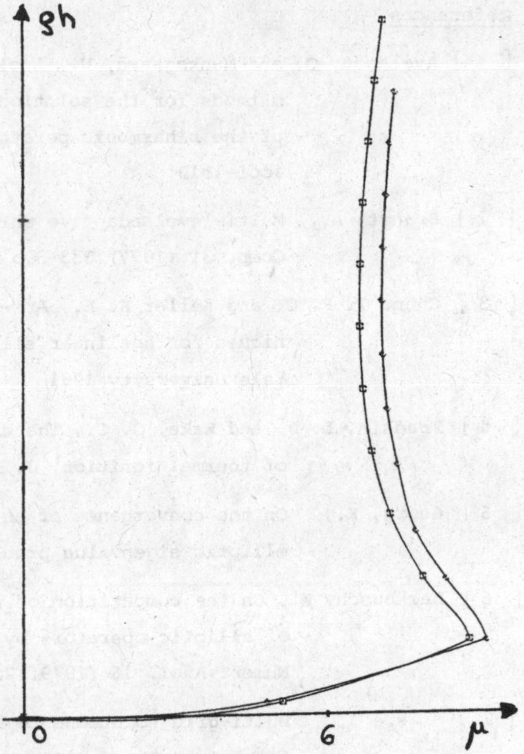

Figure 4

The generalized inverse iteration (1.9) and in the form (2.4) for variational inequalities has proved in [13] to be an efficient algorithm for the bifurcation problems under consideration. For a typical problem the multi-grid methods (3.3) and especially (4.1) are not only more efficient but they also allow to find a certain point on the solution curve for the finest grid without having to follow the path up to that point with fine-grid computations. Although the above multi-grid algorithms are rather simple versions problem (0.1)-(0.3) with $\varepsilon=.2$ is solved by (4.1) quite satisfactorily. In the case $\varepsilon=0$ already a suitable multiplication of the correction in Step 4 of (4.1), i. e. under- or overinterpolation (see, for example, [3]), gave a better correspondence and a higher efficiency. Several ways to accelerate (4.1) have not been tested here as well as multi-grid methods which use smoothing operations on all intermediate grids.

Recently several workers have considered the application of multi-grid methods to turning-point problems, see, for example, contributions in this volume. The discussion of the relative merits of these approaches has to be postponed since the full content of these papers is not known to the author. The above method definitely has advantages, for example, over the regularization by fixing a value at a certain point especially in the neighbourhood of bifurcation points.

References

[1] Axelsson, O. and Munksgaard, N., A class of preconditioned conjugate gradient methods for the solution of a mixed finite element discretization of the biharmonic perator. Int. J. Numer. Methods Eng. 14 (1979) 1oo1-1o19

[2] Brandt, A., Multi-level adaptive solution to boundary value problems. Math. Comp. 31 (1977) 333-39o

[3] Chan, T. F. C. and Keller H. B., Arc-length continuation and multi-grid techniques for nonlinear elliptic eigenvalue problems, preprint no.197 Yale University 1981

[4] Fradkin, L. J. and Wake, G. C., The critical explosion parameter in the theory of thermal ignition. J. Inst. Math. Appl. 2o (1977), 471-484

[5] Georg, K., On the convergence of an inverse iteration method for nonlinear elliptic eigenvalue problems. Numer. Math. 32 (1979) 69-74

[6] Hackbusch, W., On the computation of approximate eigenvalues and eigenfunctions of elliptic operators by means of a multi-grid method. SIAM J. Numer. Anal. 16 (1979) 2o1-215

[7] -.- Multi-grid solutions to linear and nonlinear eigenvalue problems for integral and differential equations. Report 8o-3, Mathematisches Institut, Universität zu Köln 198o

[8] Hackbusch, W. and Mittelmann, H. D., On multi-grid methods for variational inequalities, preprint Nr. 57 (Angewandte Mathematik), Universität Dortmund 1981

[9] Kosubek, A., Numerische Lösung indefiniter Gleichungssysteme mit Mehrgittermethoden. Diplomarbeit, Universität Dortmund 1981

[1o] McCormick, S. F., A mesh refinement method for Ax=λBx. Math. Comp. 36 (1981) 485-498

[11] Miersemann, E., Verzweigungsprobleme für Variationsungleichungen. Math. Nachr. 65 (1975) 187-2o9

[12] Mittelmann, H. D., Bifurcation problems for discrete variational inequalities, to appear in Math. Meth. Appl. Sci.

[13] -.- An efficient algorithm for bifurcation problems of variational inequalities, manuscript NA-81-14, Computer Science Dept., Stanford University 1981

[14] Mittelmann, H. D. and Weber,H., A bibliography on numerical methods for bifurcation problems. preprint Nr. 56 (Angewandte Mathematik), Universität Dortmund

[15] Paige, C. C. and Saunders, M. A., Solution of sparse indefinite systems of linear equations. SIAM J. Numer. Anal. 12 (1975) 617-629

[16] Paumier, J.-C., Une méthode numérique pour le calcul des points de retournement. Application à un problème aux limites non-linéare. I., Numer. Math. 37 (1981), 433-444, II., Numer. Math. 37 (1981) 445-452

[17] Peters, G. and Wilkinson, J. H., Inverse iteration, ill-conditioned equations and Newton's method. SIAM Review 21 (1979) 339-36o

[18] Rippl, H., Konjugierte-Gradienten-Verfahren zur Lösung indefiniter symmetrischer linearer Gleichungssysteme. Diplomarbeit, Universität Dortmund 1981

[19] Varga, R., Matrix iterative analysis. Prentice Hall, Englewood Cliffs, New Jersey 1962

[2o] Wilkinson, J. H., The algebraic eigenvalue problem. Clarendon Press, Oxford 1965

USE OF THE MULTIGRID METHOD FOR LAPLACIAN PROBLEMS IN THREE DIMENSIONS

Zenon P. Nowak
Warsaw Technical University
00-665 Warszawa, Nowowiejska 24

1. Introduction

The Dirichlet and the Neumann boundary value problems for irregular domains in three dimensions are usually solved by integral equation methods. The integral formulation is without a competitor in fluid mechanics, where it serves as a basis for all currently used methods for calculating the three dimensional flows around complicated bodies (see, e.g., Hess [11]). The success of these methods, at least in the case of the exterior Neumann problem, is due to the fact that the boundary condition is easily expressed and the condition at the infinity is inherent in the integral formulation.

For practical calculations the boundary surface is usually approximated by a system of plane polygonal elements ("panels"). The density of the single or double-layer potential is assumed to be constant on each panel. The values of these constants are determined from the requirement that the potentials satisfy the boundary condition at certain preassigned points of the boundary surface. The resulting system of linear algebraic equations represents a discrete counterpart of the Fredholm integral equation problem.

In the case of the exterior Neumann problem the integrals over panels of simple shape may be calculated analytically (Hess and Smith [10]). However, the exact formulas are time consuming to compute. Therefore, the integrands are usually approximated by a few terms of a power series expansion or set equal to constants. In the last case, i.e., when the midpoint rule is used for all panels, we obtain the method of point singularities: simple poles or dipoles distributed on the surface.

The nonsparse systems of algebraic equations, arising from such discretizations, are usually solved by iteration (e.g., Brakhage [2], Atkinson [1], Hemker and Schippers [9]). An efficient iterative technique, called the multigrid (MG) method has recently evolved from the works of Brandt [3], Hackbusch [6], Nicolaides [12], Wesseling [15] and others (for an extensive bibliography see [7]).

The method was primarily devised for solving sparse systems, arising from the discretization of partial differential equations. The application of the MG method to nonsparse systems was first considered by Hackbusch [8], who gave proofs of convergence of MG iterates for very general Fredholm equation problems. The results of Hackbusch were further developed in the work of Hemker and Schippers [9]. The MG schemes of papers [8, 9] require $O(m^2)$ arithmetic operations, where m denotes the number of equations to be solved. This represents a marked improvement over earlier algorithms of Brakhage [2] and Atkinson [1], which require $O(m^3)$ and, respectively, $O(m^2 \log m)$ operations to obtain sufficiently converged solution.

In the present paper we are concerned with the solution of the exterior Neumann and the interior Dirichlet problems, with boundary condition defined on a single closed Lyapounov's surface. At the cost of certain complications in the proofs, present results could be extended to the case of several disjoint surfaces. For three dimensions the Lyapounov conditions can not be avoided, at least in the present state of the potential theory (Riesz and Sz.-Nagy [13], p. 223).

In Sect. 4 of this paper the Fredholm equation counterparts of the boundary value problems are discretized by using a certain variant of the method of panels, which seems to be convenient in practice. The boundary surface is approximated by a polyhedron, consisting of the triangular panels satisfying certain regularity assumptions. Using a general convergence result (Lemma 1) we prove (see Theorem 1) that the discrete solutions of Fredholm equation problems converge to the exact solutions at the rate $O(m^{-b/2})$, where m is the number of panels and b is a fixed positive number less than the smoothness exponent of the surface (denoted by a in (1), see Sect. 2). It is also proved that the single or double-layer potentials, arising from the discrete solutions, tend uniformly together with their derivatives of arbitrarily high order to the exact potentials and their corresponding derivatives on closed subdomains of the region. The same results are also obtained for the method of point singularities, derived from the "exact" method of panels by midpoint integrations. It is evident that each method of intermediate accuracy has the same properties.

In Sect. 5 we discuss the convergence of the simplest one-stage (two-level) variant of the MG method, when applied to the discrete systems of Sect. 4. From a general convergence result (Lemma 2), related to the results of Hackbusch [8], we infer (see Theorem 2) that the second iterate of the one-stage MG method differs from the converged solution by $O(m^{-2b/3})$, which is less than the error $O(m^{-b/2})$ of the discretization method. This result is valid for certain simple and convenient choices of the prolongation and restriction operators, and when the number of equations at the lower level is: $\hat{m} = m^{2/3}$. The MG calculation takes about $14m^2/3$ multiplications and $17\,m^2/3$ additions. Since the MG process must be preceded by an introductory stage costing $O(m^2)$ operations, where O is rather "big" , it seems that little benefit would be derived by introducing further levels in the MG iteration step.

During the final phase of preparation of the present paper it came to the author's knowledge that the convergence of the MG process for the exterior Neumann problem in three dimensions has been recently analyzed by Wolff [16] (loc.cit.[14]).

2. Boundary value problems and their integral equation counterparts.

Let S be a closed smooth surface. We shall always assume that S satisfies Lyapounov's conditions:

(L_1) The angle $\alpha(\bar{n}_p, \bar{n}_q)$ between the exterior normals at points $p, q \in S$ satisfies the inequality

(1) $\alpha(\bar{n}_p, \bar{n}_q) \leq M\cdot(r(p,q))^a, \quad 0 < a \leq 1,$

where M, a are constants and $r(p,q)$ denotes the distance between p and q;

(L_2) There exists a constant number $c > 0$, such that every line parallel to the normal $\bar{n}_p$, $p \in S$, cuts not more than once that part of S which lies inside the sphere with radius c and center p .

The open regions interior and exterior to S will be denoted by R_i and R_e, respectively.

We consider the following interior Dirichlet and exterior Neumann problems :

(D_i) Find the function $U(p)$, $p \in R_i$, satisfying the Laplace equation $\Delta U = 0$ in R_i, and such that

(2) $U(p) = f(p)$ for $p \in S$,

where $f(p)$ is given on S ;

(N_e) Find the function $U(p)$, $p \in R_e$, satisfying $\Delta U = 0$ in R_e and such that

(3) $\partial U(p)/\partial n_p = -f(p)$ for $p \in S$, and

$U(p) \longrightarrow 0$ when p tends to the infinity.

Let us denote

$$(4)\qquad K(p,q)=\frac{1}{2\pi}\,\frac{\cos(\alpha(\bar n_q,\ \bar r_{qp}))}{r(p,q)^2}\ ,\quad q\in S\ ,\ p\neq q\ ,$$

where $\alpha(\bar n_q, \bar r_{qp})$ is the angle between the exterior normal $\bar n_q$ at q and the vector $\bar r_{qp}$ pointing from q to p.

It can be shown (see, e.g., Günter [5]) that if f is continuous on S then the solutions of problems D_i and N_e can be expressed as

$$(5)\qquad U(p)=\begin{cases} -\iint\limits_S K(p,q)\ u(q)\ dS_q\ , & p\in R_i,\ \text{for problem } D_i,\\[2mm] \dfrac{1}{2\pi}\iint\limits_S \dfrac{u(q)}{r(p,q)}\,dS_q\ , & p\in R_e,\ \text{for problem } N_e,\end{cases}$$

where u satisfies the Fredholm integral equations of the second kind

$$(6)\qquad \begin{aligned} u(p) &= \iint\limits_S K(p,q)\ \ u(q)\ dS_q + f(p)\ ,\quad p\in S,\quad \text{for problem } D_i,\ \text{and}\\ u(p) &= \iint\limits_S K(q,p)\ \ u(q)\ dS_q + f(p)\ ,\quad p\in S,\quad \text{for problem } N_e. \end{aligned}$$

We shall introduce the integral operator L, transforming bounded functions $u(p)$, defined on S, into other such functions Lu, defined at $p\in S$ by

$$(7)\qquad Lu(p)=\begin{cases}\iint\limits_S K(p,\ q)\ u(q)\ dS_q & \text{for problem } D_i,\\ \iint\limits_S K(q,\ p)\ u(q)\ dS_q & \text{for problem } N_e\ .\end{cases}$$

Now the integral equations (6) are led to the common form

$$(8)\qquad u = Lu + f$$

3. Auxiliaries

3.1. Properties of Lyapounov's surface. From condition (L_2) it follows that the part of Lyapounov's surface, lying inside the sphere with center $q\in S$ and radius c (briefly, inside $s(q;c)$) can be represented in the form

$$(9)\qquad n = F(s,t)\ ,$$

where n,s,t are the coordinates of the Cartesian system with origin at q and the n-axis pointing in the direction of the normal vector $\bar n_q$ (see Fig. 1).

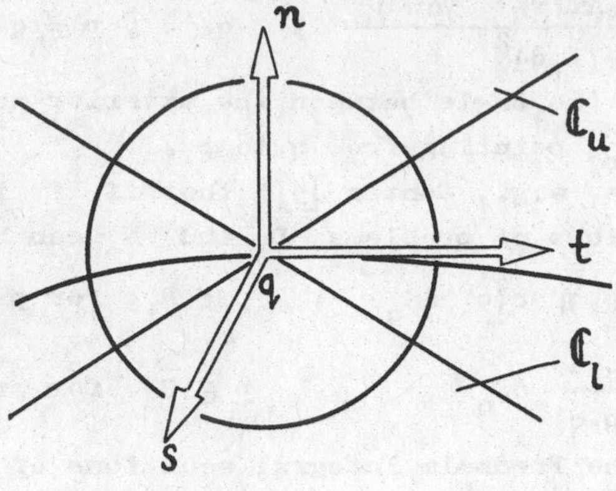

Fig. 1

If the point $(s,t,F(s,t))$ lies inside Lyapounov's sphere $s(q;c)$ then

(10) $$|F(s,t)| \leqq \tilde{M}(s^2 + t^2)^{(1+a)/2} \quad \text{for a constant } \tilde{M},$$

and, consequently, $|F(s,t)| \leqq \tilde{M}\, c^a\, (s^2 + t^2)^{1/2}$. This means that the surface S lies between the two cones $\mathbb{C}_l$ and $\mathbb{C}_u$, defined by $n = \pm \tilde{M}_1(s^2 + t^2)^{1/2}$, where $\tilde{M}_1 = \tilde{M}c^a$. Each line cutting $\mathbb{C}_l$ and $\mathbb{C}_u$ inside Lyapounov's sphere must also cut S (in fact, the continuous function $n - F(s,t)$ is nonpositive at the intercept on $\mathbb{C}_l$ and nonnegative at the intercept on $\mathbb{C}_u$, and must therefore vanish at a certain point of that line).

3.2. <u>o,O and O_+ symbols.</u> We agree that all $o(m)$, $o_1(m),\dots,o_4(m)$, etc., tend to zero when $m \to \infty$, and that $|O(m)| \leqq C\cdot m$, $|O_+(m)| \geqq C\cdot m$, where C is a positive constant, which may depend on other constants or functions characterizing the problem or the method of its solution (e.g., M, a in (1), f in (2,3), $\hat{M}_1$, $\hat{M}_2$ in Sect. 4.2, etc.). The o's may also depend on these constants or functions.

3.3. <u>Classes C_o, C_b, C_{om}, C_{bm}.</u> Let C_o denote the class of piecewise continuous bounded functions $g(p)$, $p \in S$, with the norm

$$\|g\|_{C_o} = \sup_{p\in S} |g(p)|.$$

A function $g(p)$, $p\in S$, will be said to belong to the class C_b for some $0 < b \leqq 1$ if

$$\| g \|_{C_b} = \max_{q \in S} |g(q)| + \sup_{p,q \in S} \left(|g(p) - g(q)| / (r(p,q))^b \right) < \infty .$$

Let C_{om} and C_{bm} be the classes of discrete functions $g_m(k)$, defined for $k = 1,\dots,m$, and endowed with the analogous norms

$$\|g_m\|_{C_{om}} = \max(|g_m(k)| : k = 1,\dots,m) ,$$

$$\|g_m\|_{C_{bm}} = \|g_m\|_{C_{om}} +$$

(11) $$\max(| g_m(i) - g_m(j)| / (r(p_{im}, p_{jm}))^b : i,j = 1,\dots,m; i \neq j),$$

where $p_{im} \in S$ for $i = 1,\dots,m$, and $p_{im} \neq p_{jm}$ if $i \neq j$.

4. Discrete counterparts of the problem $u = Lu + f$.

4.1. Convergence of the discrete solutions.

Lemma 1. For each $m \geqq m_o$ let R_m be a linear operator acting from C_o into C_{om}, and let P_m be a linear operator acting from C_{om} into C_o. Suppose that

(12) $$M_1 \| g_m \|_{C_{om}} \leqq \|P_m g_m\|_{C_o} \leqq M_2 \|g_m\|_{C_{om}} \quad \text{for all } g_m \in C_{om} ,$$

(13) $$\|R_m\|_{C_o \to C_{om}} \leqq M_3 ,$$

(14) $$\|P_m R_m - I\|_{C_b \to C_o} = o_1(m) ,$$

where M_1, M_2, M_3, b are positive constants and I denotes the identity operator on C_o. Suppose also that a linear operator L, acting from C_o into C_b, satisfies the conditions

(15) $$\|L\|_{C_o \to C_b} = M_4 < \infty ,$$

(16) $$\|(I - L)^{-1}\|_{C_o \to C_o} = M_5 < \infty .$$

Assume further that a linear operator l_m acting from C_{om} into C_{om} satisfies consistency condition

(17) $$\| 1_m - R_m L P_m \|_{C_{om} \to C_{om}} = o_2(m).$$

Then for sufficiently large m the stability condition follows:

(18) $$\| (i_m - 1_m)^{-1} \|_{C_{om} \to C_{om}} \leqq M_6 < \infty ,$$

where i_m denotes the identity operator on C_{om}. Moreover, let u be the solution of the problem $u = Lu + f$ for some $f \in C_b$, and let

(19) $$\| f_m - R_m f \|_{C_{om}} = o_3(m)$$

for certain $f_m \in C_{om}$. Then the solutions u_m of the problem $u_m = 1_m u_m + f_m$ satisfy the convergence condition

(20) $$\| P_m u_m - u \|_{C_o} \leqq M' o_1(m) + M'' o_2(m) + M''' o_3(m) \equiv o_4(m) ,$$

where M', M'', M''' are constants.

<u>Proof</u>. Let us choose any $g_m \in C_{om}$. Identity

$$P_m (i_m - 1_m) = (I - L) P_m + (I - P_m R_m) LP_m + P_m (R_m LP_m - 1_m),$$

and condition (12) give

$$M_2 \| (i_m - 1_m) g_m \|_{C_{om}} \geqq \| (I-L) P_m g_m \|_{C_o} - \| (I - P_m R_m) L P_m g_m \|_{C_o}$$
$$- \| P_m (R_m LP_m - 1_m) g_m \|_{C_o}$$

By (12, 16) we have

(21) $$\| (I - L) P_m g_m \|_{C_o} \geqq \frac{1}{M_5} \| P_m g_m \|_{C_o} \geqq \frac{M_1}{M_5} \| g_m \|_{C_{om}} .$$

Further, (14, 15, 12) give

$$\| (I - P_m R_m) LP_m g_m \|_{C_o} \leqq$$

(22) $$\leqq \| I - P_m R_m \|_{C_b \to C_o} \| L \|_{C_o \to C_b} \| P_m \|_{C_{om} \to C_o} \| g_m \|_{C_{om}}$$
$$\leqq o_1(m) M_4 M_2 \| g_m \|_{C_{om}} , \text{ and}$$

$$\|P_m (R_m L P_m - 1_m) g_m\|_{C_o} \leqq \|P_m\|_{C_{om} \to C_o} \|R_m L P_m - 1_m\|_{C_{om} \to C_{om}} \|g_m\|_{C_{om}}$$

$$\leqq M_2 \cdot o_2(m) \|g_m\|_{C_{om}} .$$

These estimates imply

$$(23) \quad \|(i_m - 1_m) g_m\|_{C_{om}} \geqq \left(\frac{M_1}{M_2 M_5} - o'(m)\right) \|g_m\|_{C_{om}} .$$

For sufficiently large m we have $M_1/(M_2 M_5) - o'(m) \geqq M_7 > 0$. Then $\|(i_m - 1_m) g_m\|_{C_{om}} = 0$ implies $\|g_m\|_{C_{om}} = 0$. Consequently, the operator $(i_m - 1_m)^{-1}$ exists and the stability (18) follows with $M_6 = 1/M_7$. To prove (20) we first note that

$$(24) \quad P_m u_m - u = P_m(u_m - R_m u) + (P_m R_m - I) u ,$$

and that

$$u_m - R_m u = (i_m - 1_m)^{-1}((1_m - R_m L P_m) R_m + R_m L(P_m R_m - I)) u$$

$$+ (i_m - 1_m)^{-1} (f_m - R_m f) .$$

If $f \in C_b$ then by (15, 16) also $u \in C_b$. Further,

$$\|L\|_{C_o \to C_o} \leqq M_4 \quad \text{and} \quad \|R_m\|_{C_b \to C_{om}} \leqq M_3 .$$

Consequently,

$$(25) \quad \|u_m - R_m u\|_{C_{om}} \leqq M_6(o_2(m) \cdot M_3 + M_3 \cdot M_4 \cdot o_1(m)) \|u\|_{C_b}$$

$$+ M_6 \cdot o_3(m) \equiv o''(m) .$$

Now, (24, 25, 12, 14) give (20) with $o_4(m) = M_2 o''(m) + o_1(m) \|u\|_{C_b}$. Proof is complete.

4.2. <u>Examples</u>: the methods of panels and point singularities.

For each integer $m \geqq m_o$ the surface S will be approximated by a polyhedron $\hat{S}_m$, consisting of the plane triangular elements $\hat{S}_{1m}, \ldots, \hat{S}_{mm}$.

Let d_{im} be the size of $\hat{S}_{im}$, that is: $d_{im} = \max(r(p,q) : p, q \in \hat{S}_{im})$, and let r_{im} be the radius of the circle inscribed into $\hat{S}_{im}$. Let $d_m = \max(d_{im} : i = 1, \ldots, m)$.

We shall always assume that

(i) $d_m = o(m)$

(ii) $d_{im}/d_m \geqq \hat{M}_1 > 0$ for $m \geqq m_o$; $i=1,\dots,m$;

(iii) $r_{im}/d_{im} \geqq \hat{M}_2 > 0$ for $m \geqq m_o$; $i=1,\dots,m$;

(iv) Vertices of all the triangles $\hat{S}_{im}$, $i=1,\dots,m$, lie on S;

(v) If $\hat{S}_{im} \cap \hat{S}_{jm}$ (where $i,j=1,\dots,m$; $i \neq j$) is not empty then it is either a vertex or the entire side of $\hat{S}_{im}$ and $\hat{S}_{jm}$; each side of any triangle is shared with only one other triangle;

(vi) The normal to $\hat{S}_{im}$, $i=1,\dots,m$, at a point $p \in \hat{S}_{im}$ can not cut any $\hat{S}_{jm}$, $j \neq i$, inside the sphere $s(p;d_m)$.

Let us introduce the operator $\hat{P}_m$, transforming the functions $g_m \in C_{om}$ into functions $\hat{P}_m g_m$ defined on $\hat{S}_m$ by the formula

(26) $$\hat{P}_m g_m(q) = g_m(i) \text{ when } q \in \hat{S}_{im},\ i=1,\dots,m .$$

(ambiguities occurring at boundaries of $\hat{S}_{im}$ may be resolved in an arbitrary manner).

Let $\hat{n}_q$ be a vector, normal to $\hat{S}_m$ at $q \in \hat{S}_m$, and pointing into R_e. In analogy to (4) and (7) we define

(27) $$\hat{K}_m(p,q) = \frac{1}{2\pi} \frac{\cos(\alpha(\hat{n}_q, \bar{r}_{qp}))}{r(p,q)^2}, \quad q \in \hat{S}_m,\ p \neq q ,$$

and introduce the operator l_m, transforming the functions $g_m \in C_{om}$ into functions $l_m g_m \in C_{om}$ defined by

(28) $$l_m g_m(i) = \begin{cases} \iint_{\hat{S}_m} \hat{K}_m(\hat{p}_{im},q)\, \hat{P}_m g_m(q)\, d\hat{S}_m(q) & \text{for problem } D_i , \\ \iint_{\hat{S}_m} \hat{K}_m(q,\hat{p}_{im})\, \hat{P}_m g_m(q)\, d\hat{S}_m(q) & \text{for problem } N_e , \end{cases}$$

where $i=1,\dots,m$; $\hat{p}_{im}$ is the center of the circle inscribed into $\hat{S}_{im}$. By (26) we have

(29) $$l_m g_m(i) = \sum_{j=1}^{m} A_{ij} g_m(j), \text{ where}$$

(30) $$A_{ij} = \begin{cases} \iint_{\hat{S}_{jm}} \hat{K}_m(\hat{p}_{im},q)\, d\hat{S}_m(q) & \text{for problem } D_i , \\ \iint_{\hat{S}_{jm}} \hat{K}_m(q, \hat{p}_{im})\, d\hat{S}_m(q) & \text{for problem } N_e . \end{cases}$$

Let q_{im}, $m \geq m_o$, $i=1,\dots,m$, be the points of S, chosen so that

$r(\hat{p}_{im}, q_{im}) = O(d_m)$. Let f be the function in (2) or (3) and let f_m be defined by

(31) $\quad f_m(i) = f(q_{im}), \quad m \geq m_o; \; i=1,\dots,m$.

In the method of panels the problem (8) is discretized by introducing the system of linear algebraic equations $u_m = l_m u_m + f_m$, with l_m and f_m just defined. In analogy to (5), the potential $U(p)$ can be approximated by

$$(32) \quad U_m(p) = \begin{cases} -\iint_{\hat{S}_m} \hat{K}_m(p,q)\, \hat{P}_m u_m(q)\, d\hat{S}_m(q), & p \in R_i, \text{ for problem } D_i, \\ \frac{1}{2\pi} \iint_{\hat{S}_m} \frac{1}{r(p,q)}\, \hat{P}_m u_m(q)\, d\hat{S}_m(q), & p \in R_e, \text{ for problem } N_e. \end{cases}$$

In the method of point singularities all the coefficients A_{ij} are calculated from the midpoint formula

$$(33) \quad A_{ij} = \begin{cases} \hat{K}_m(\hat{p}_{im}, \hat{p}_{jm})\, |\hat{S}_{jm}| & \text{for problem } D_i, \\ \hat{K}_m(\hat{p}_{jm}, \hat{p}_{im})\, |\hat{S}_{jm}| & \text{for problem } N_e, \end{cases}$$

where $|\hat{S}_{jm}|$ denotes the area of $\hat{S}_{jm}$. As previously, l_m will be defined by (29) and f_m by (31). The midpoint rule, when applied to (32), gives the following approximation to the potential:

$$(34) \quad U_m(p) = \begin{cases} -\sum_{j=1}^{m} \hat{K}_m(p, \hat{p}_{jm})\, u_m(j)\, |\hat{S}_{jm}|, & p \in R_i, \text{ for problem } D_i, \\ \frac{1}{2\pi} \sum_{j=1}^{m} \frac{1}{r(p, \hat{p}_{jm})}\, u_m(j)\, |\hat{S}_{jm}|, & p \in R_e, \text{ for problem } N_e. \end{cases}$$

Using Lemma 1 we shall now prove the following

<u>Theorem 1</u>. Suppose that S is a Lyapounov surface and that the polyhedra $\hat{S}_m$, $m=m_o, m_o+1, \dots$, satisfy conditions (i)-(vi). Suppose also that $f \in C_a$, where a is such as in (1). Then for sufficiently large m the systems $u_m = l_m u_m + f_m$ of the method of panels (l_m as in (29,30), f_m as in (31)) possess the solutions u_m. The values $u_m(i)$, $i=1,\dots,m$, when distributed on S by using an operator P_m, give functions $P_m u_m(p)$, $p \in S$, such that

$$(35) \quad \| P_m u_m - u \|_{C_o} = O(m^{-b/2}),$$

where u is the solution of (8) and b is a fixed number: $0 < b < a$. Further, the functions (32) and their derivatives of arbitrarily high order tend uniformly on closed subregions of R_i (or, resp., R_e) to the solution $U(p)$ of problem D_i (resp., N_e) and to its corresponding derivatives. The analogous results are valid for the method of point singularities (l_m as in (29,33), f_m as in (31), U_m as in (34)).

<u>Proof</u> will be split into parts A, B and C. In part A we define a transformation T_m of $\hat{S}_m$ onto S. In part B this transformation serves to define certain operators P_m, R_m and to verify the assumptions of Lemma 1. In part C we prove the convergence of U_m and of its derivatives.

(A) If m is sufficiently large then, as a consequence of (i), each $\hat{S}_{im}$, $i=1,..,m$, lies inside Lyapounov's sphere $s(q;c)$ with center at a vertex q of $\hat{S}_{im}$. By (iii) each angle of $\hat{S}_{im}$ is larger than $2\hat{M}_2 > 0$. Using assumption (iv) and estimate (10) we can show by a simple calculation that the angle $\alpha(\hat{n}_q, \bar{n}_q)$ between the normal to $\hat{S}_{im}$ and the normal to S at a vertex $q \in \hat{S}_{im}$ is less than $\tilde{M}_2\, d^a_{im} \leqq \tilde{M}_2\, d^a_m$, where $\tilde{M}_2$ depends only on $\tilde{M}$, a and $\hat{M}_2$ in (10) and (iii), respectively (briefly, $\alpha(\hat{n}_q, \bar{n}_p) = O(d^a_m)$). As a consequence, if $\hat{S}_{im}$ and $\hat{S}_{jm}$, $j \neq i$, have a common side, then the angle between the normals to $\hat{S}_{im}$ and $\hat{S}_{jm}$ is $O(d^a_m)$. We thus see that the angle formed by $\hat{S}_{im}$ and $\hat{S}_{jm}$ is either $O(d^a_m)$ or $\pi - O(d^a_m)$. The first possibility is excluded by (vi), which would then be violated at points of $\hat{S}_{im}$, lying close to the common side of $\hat{S}_{im}$ and $\hat{S}_{jm}$. Hence, the angle between $\hat{S}_{im}$ and the plane bisecting the corner formed by $\hat{S}_{im}$ and $\hat{S}_{jm}$ is $\frac{\pi}{2} - O(d^a_m)$.

Let Π_{im} be the pyramid, or the parallelepiped, formed by such bisecting planes passing through the sides of S_{im} (see Fig.2), and let v_{im} be the vertex of Π_{im} (if Π_{im} is a parallelepiped then v_{im} lies at the infinity).

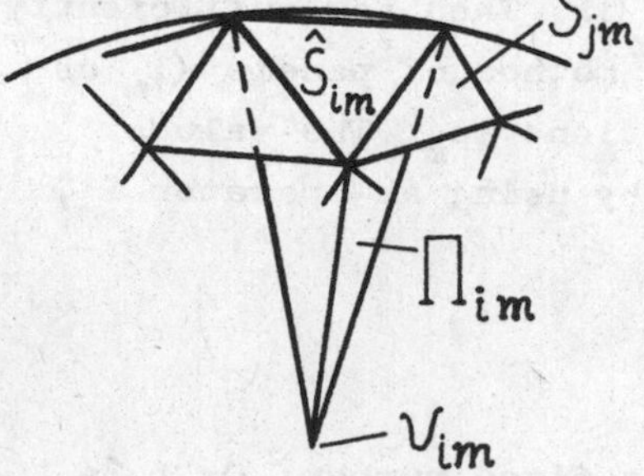

Fig.2

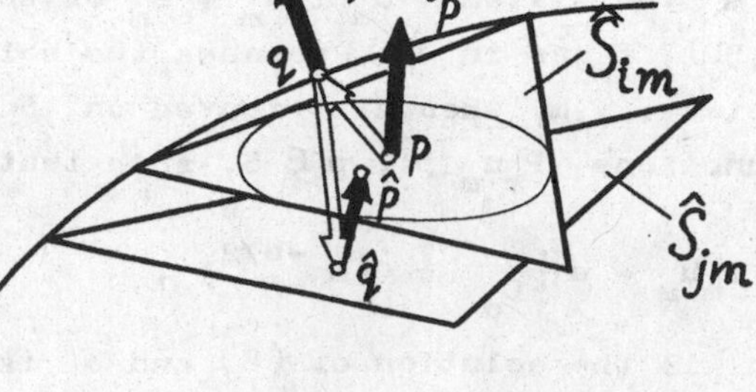

Fig.3

Suppose that the three angles between $\hat{S}_{im}$ and the bisecting planes are equal. In such case the height of Π_{im} is $r(v_{im},\hat{p}_{im})$, which by our estimate of these angles is larger than $r_{im}/O(d_m^a)$. In other situations the height is further increased. Hence, in view of (iii) and (ii) the distance between v_{im} and any point $p \in \hat{S}_{im}$ is always larger than $\tilde{M}_3 d_m^{1-a}$, where $\tilde{M}_3 \equiv \tilde{M}_3(\tilde{M},a,\hat{M}_1,\hat{M}_2)>0$ (briefly, $r(p, v_{im}) = O_+(d_m^{1-a})$).

Let q be a vertex of $\hat{S}_{im}$ and let $\mathbb{C}_l(q)$ and $\mathbb{C}_u(q)$ be the cones, defined as in Sect. 3 (Fig.1). The angle between $\bar{n}_q$ and any vector $\bar{r}(v_{im},p)$, $p\in\hat{S}_{im}$, is $O(d_m^a)$ and, consequently, the size of that part of Π_{im} which lies outside both cones is $O(d_m)$. Since the distance between v_{im} and $\hat{S}_{im}$ is $O_+(d_m^{1-a})$ and $\hat{S}_{im}$ lies outside both cones (in fact, the angle between $\hat{S}_{im}$ and $\bar{n}_q$ is $\frac{\pi}{2} - O(d_m^a)$), we see that v_{im} lies inside one of these cones. Hence, each axis pointing from v_{im} to $p\in\hat{S}_{im}$ cuts the surface of both cones inside the sphere $s(q,O(d_m))$. As we have seen in Sect.3, it must therefore cut S in at least one point $T_{im}(p)\in s(q,O(d_m))\subset s(q;c)$. It can not cut S twice inside $s(q;c)$ when m is so large that the angle $O(d_m^a)$ between $\bar{n}_q$ and $\bar{r}(v_{im},p)$ is less than a certain number $\tilde{M}_4(M,a)>0$ (Günter [5], p.3).

We have thus defined a transformation $T_{im}(p)$, acting from $\hat{S}_{im}$ into S. Let S_{im} be the image of $\hat{S}_{im}$ under T_{im}. If $p,q\in\hat{S}_{im}$ and $p\neq q$ then $T_{im}(p)\neq T_{im}(q)$. Otherwise, $v_{im} = T_{im}(p) = T_{im}(q)\in S$ which, as we have seen, is not possible. Hence, there exists the inverse transformation T_{im}^{-1} of S_{im} onto $\hat{S}_{im}$.

Since the angle between the adjacent $\hat{S}_{im}$ and $\hat{S}_{jm}$ is $\pi - O(d_m^a)$, it follows from (v) that each Π_{im} is surrounded by three adjacent Π_{jm}. Consequently, each S_{im} is surrounded by three adjacent S_{jm}. Hence, $S = \bigcup_{i=1}^{m} S_{im}$, since otherwise no S_{im}, adjacent to the boundary of $\bigcup_{i=1}^{m} S_{im}$, would have a neighbour on the opposite side of that boundary.

Now, let us suppose that certain S_{im} and S_{jm} $(j\neq i)$ overlap, i.e., that the set $\Sigma_{im} \equiv S_{im}\cap\bigcup_{j\neq i} S_{jm}$ has an interior point. Then $\Sigma_{im} \equiv S_{im}$, since otherwise certain parts of the boundary of Σ_{im} would lie inside S_{im}, and no S_{jm} $(j\neq i)$, adjacent to these parts, would have a neighbour on the opposite side of that boundary. Let $p=\hat{p}_{im}$ be the center of the circle inscribed into $\hat{S}_{im}$ (see Fig.3), and let $q= T_{im}(p)$. Since $\Sigma_{im} \equiv S_{im}$, the point p belongs to a certain S_{jm}, $j\neq i$. Let $\hat{q}=T_{jm}^{-1}(q)$. We shall see that, contrary to (vi), the normal $\hat{n}_{\hat{q}}$ at $\hat{q}\in\hat{S}_{jm}$ cuts $\hat{S}_{im}$ at a point $\hat{p}\in s(\hat{q}, O(d_m^{1+a}))$. In fact,

$r(p,q) = O(d_m^{1+a})$ and, similarly, $r(q,\hat{q}) = O(d_m^{1+a})$, Hence $r(p,\hat{q}) = O(d_m^{1+a})$. Further, $\alpha(\hat{n}_p,\hat{n}_{\hat{q}}) \leqq \alpha(\hat{n}_p,\bar{n}_q) + \alpha(\bar{n}_q,\hat{n}_{\hat{q}}) = O(d_m^a)$. Therefore,

$$r(\hat{q},\hat{p}) \leqq \frac{r(p,\hat{q})}{\cos(\alpha(\hat{n}_p,\hat{n}_{\hat{q}}))} = O(d_m^{1+a}) \text{ , i.e., } \hat{p} \in s(\hat{q}, O(d_m^{1+a})) \text{ .}$$

Now, $r(p,\hat{p}) \leqq r(p,\hat{q}) + r(\hat{q},\hat{p}) = O(d_m^{1+a})$. From (ii) and (iii) it follows that $r_{im} = O_+(d_m)$. Hence, $r(p,\hat{p}) < r_{im}$, and $\hat{p} \in \hat{S}_{im}$. This proves that no S_{im}, S_{jm}, $i \neq j$, may overlap.

Let us introduce the function T_m defined on $\hat{S}_m$ by

$$T_m(q) = T_{im}(q) \quad \text{if} \quad q \in \hat{S}_{im}, \ i=1,\ldots,m \text{ .}$$

From the properties of T_{im}, $i=1,\ldots,m$, it follows that T_m is a one-to-one transformation of $\hat{S}_m$ onto S.

Let $d\hat{S}_{im}$ be the surface element of $\hat{S}_{im}$ at p and let dS_{im} be its image at $q = T_{im}(p)$ (see Fig. 4).

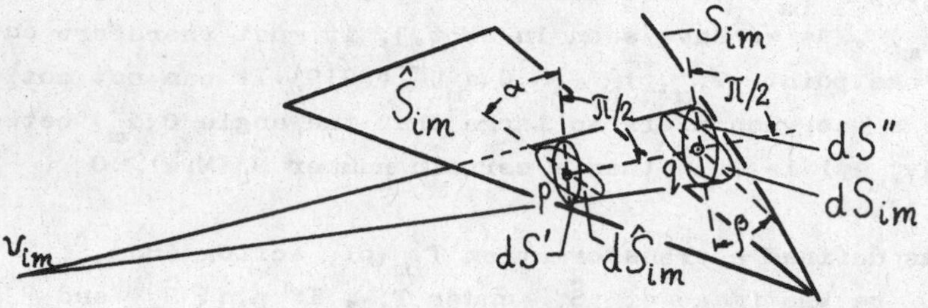

Fig. 4

Using the estimates $r(v_{im},p) = O_+(d_m^{1-a})$, $r(p,q) = O(d_m^{1+a})$, $\alpha = O(d_m^a)$, $\beta = O(d_m^a)$ we find

$$J_{im}(q) = \frac{d\hat{S}_{im}}{dS_{im}} = \frac{d\hat{S}_{im}}{dS'} \frac{dS'}{dS''} \frac{dS''}{dS_{im}} =$$

$$(36) \qquad \frac{\cos\beta}{\cos\alpha}\left(\frac{r(v_{im},p)}{r(v_{im},p) + r(p,q)}\right)^2 = 1 + O(d_m^{2a}) \text{ .}$$

Let us denote

$$J_m(q) = J_{im}(q) \quad \text{for} \quad q \in S_{im}, \ i=1,\ldots,m \text{ .}$$

The area of $\hat{S}_m$ is

$$(37) \quad |\hat{S}_m| = \iint_S J_m(q)\, dS_q = (1+O(d_m^{2a}))\,|S|$$

The assumptions (ii) and (iii) imply: $|\hat{S}_m| \geqq m\,\pi\,\hat{M}_1^2\,\hat{M}_2^2\,d_m^2$, whence by (37) it follows that $d_m = O(m^{-1/2})$.

(B) For all $g_m \in C_{om}$ we put

(38) $$P_m g_m(p) = g_m(i) \text{ if } p \in S_{im},\ i=1,\dots,m .$$

For $g \in C_o$ we put

(39) $$R_m g(i) = g(p_{im}),\ i=1,\dots,m ,$$

where $p_{im} = T_m(\hat{p}_{im})$.

It is obvious that such P_m, R_m satisfy (12, 13) with $M_1=M_2=M_3=1$. Further, for any $g \in C_b$ and $p \in S_{im}$ we have

$$|P_m R_m g(p) - g(p)| = |g(p_{im}) - g(p)| \leq \|g\|_{C_b} (r(p_{im},p))^b$$
$$= \|g\|_{C_b} O(d_m^b),$$

where the last relation follows from the estimate

$$r(p_{im},p) \leq r(\hat{p}_{im},T_m^{-1}(p)) + r(\hat{p}_{im},p_{im}) + r(p, T_m^{-1}(p)) \leq$$
$$\leq d_m + O(d_m^{1+a}) .$$

Hence, (14) holds for $o_1(m) = O(d_m^b) = O(m^{-b/2})$. If $f \in C_a$ then (19) follows similarly with

$$o_3(m) \leq \|f\|_{C_a} \cdot (r(q_{im},p_{im}))^a = O(d_m^a) = O(m^{-a/2}) .$$

If S is Lyapounov's surface with the smoothness exponent a, then the operator L, defined by (7), satisfies (15) and (16) for any $0 < b < a$ (Günter [5], pp. 47-53, 212 for problem D_i and pp. 58-62, 168 for problem N_e). Further, if $f \in C_a$ then there exists the solution $u \in C_b$ of the problem (8) (Günter [5], p. 212 for problem D_i and p. 170 for problem N_e). Now, (35) will follow from (20) if we verify that (17) holds for $o_2(m) = O(m^{-b/2})$.

Let $g_m \in C_{om}$. If l_m is defined as in (29,30) then

(40) $$\delta_m(i) \equiv l_m g_m(i) - R_m L P_m g_m(i) = \iint_{\hat{S}_m} \hat{K}_m(\hat{p}_{im},q)\ \hat{P}_m g_m(q)\, d\hat{S}_m(q)$$
$$- \iint_S K(p_{im},q)\ P_m g_m(q)\, dS_q .$$

Using definitions (26,27,38) and integrating by substitution we find

$$\delta_m(i) = \iint_S (J_m(q)\ E_{im}(q) + (J_m(q) - 1) K(p_{im},q))\, P_m g_m(q)\, dS_q ,$$

where

(41) $$E_{im}(q) = \hat{K}_m(\hat{p}_{im}, T_m^{-1}(q)) - K(p_{im},q) .$$

Noting that $\hat{K}_m(\hat{p}_{im}, T_m^{-1}(q)) = 0$ if $q \in S_{im}$, we obtain

$$\delta_m(i) = \sum_{j=1;\,j\neq i}^{m} g_m(j) \iint_{S_{jm}} J_m(q)\, E_{im}(q)\, dS_q$$

$$(42) \qquad + \sum_{j=1;\,j\neq i}^{m} g_m(j) \iint_{S_{jm}} (J_m(q)-1)\, K(p_{im},q)\, dS_q$$

$$- \; g_m(i) \iint_{S_{im}} K(p_{im},q)\, dS_q = t_1+t_2-t_3, \text{ say.}$$

In the method of point singularities the same formulas apply if we put

$$(43) \qquad E_{im}(q) = \hat{K}_m(\hat{p}_{im},\hat{p}_{jm}) - K(p_{im},q) \quad \text{for} \quad q \in S_{jm}\,.$$

For problem N_e we only reverse the arguments of $\hat{K}_m$ and K (e.g., $\hat{K}_m(\hat{p}_{im},\hat{p}_{jm})$ becomes $\hat{K}_m(\hat{p}_{jm},\hat{p}_{im})$).

In order to estimate $|E_{im}(q)|$ at $q \in S_{jm}$, $j\neq i$, we shall consider a general situation, represented in Fig.5, where $\hat{p} \in \hat{S}_{im}$, $\hat{q} \in \hat{S}_{jm}$, $p \in S_{im}$, $q \in S_{jm}$ (the triangles $\hat{S}_{im}$ and $\hat{S}_{jm}$ may not touch).

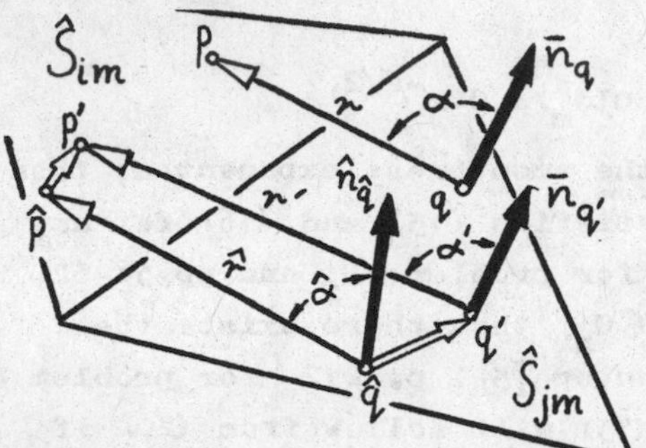

Fig.5

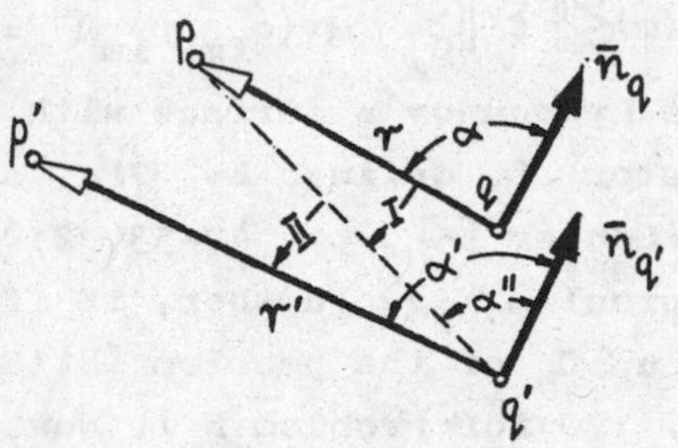

Fig.6

Let us suppose that

$$(44) \qquad r(p,q) = O_+(d_m) \text{ and } r(\hat{p},\hat{q}) = O_+(d_m).$$

Using the notation explained in Fig.5 we can write

$$2\pi(\hat{K}_m(\hat{p},\hat{q}) - K(p,q)) = \cos\hat{\alpha}/\hat{r}^2 - \cos\alpha/r^2 =$$
$$= \cos\alpha(r-\hat{r})(r+\hat{r})/(\hat{r}^2 r^2) + (\cos\hat{\alpha} - \cos\alpha)/\hat{r}^2 \equiv \tau_1 + \tau_2, \text{ say.}$$

Since $|\hat{r}-r| = O(d_m)$, we see that

$$r/\hat{r} = (\hat{r}+O(d_m))/\hat{r} = 1 + O(d_m)/O_+(d_m) = O(1), \text{ and}$$
$$(r+\hat{r})/\hat{r} = 1 + r/\hat{r} = O(1).$$

For Lyapounov's surface we have: $\cos\alpha = O(r^{a})$ and, consequently,

$$\tau_1 = O(d_m)\, r^{a-3} = O(d_m^a)\, r^{-2} .$$

Let $p' = T_m(\hat{p})$ and $q' = T_m(\hat{q})$. Passing from $\hat{r}$ to r' (see Fig.5) we displaced the ends of the segment $\hat{r}$ by $O(d_m^{1+a})$. Since $\hat{r} = O_+(d_m)$, the angle between r and $\hat{r}$ is $O(d_m^a)$. We have also seen that $\alpha(\bar{n}_{q'}, \hat{n}_{\hat{q}}) = O(d_m^a)$. Hence, $|\hat{\alpha} - \alpha| = O(d_m^a)$. Now let us note that the pair $(\bar{n}_{q'}, r')$ can be obtained from $(\bar{n}_q, r)$ in the two stages I and II, represented in Fig. 6. For Lyapounov's surface it can be easily shown that

$$|\cos\alpha - \cos\alpha''| = O(d_m^a), \quad |\cos\alpha'' - \cos\alpha'| = O(d_m^a) ,$$

which gives $|\cos\alpha - \cos\alpha'| = O(d_m^a)$. Since also

$$|\cos\hat{\alpha} - \cos\alpha'| = O(d_m^a), \text{ we obtain: } |\cos\hat{\alpha} - \cos\alpha| = O(d_m^a).$$

Consequently, $\tau_2 = O(d_m^a)\, r^{-2}$. The estimates of τ_1 and τ_2 imply

$$\hat{K}_m(\hat{p},\hat{q}) - K(p,q) = O(d_m^a)\,(r(p,q))^{-2} . \tag{45}$$

Now, if $q \in S_{jm}$, $j \neq i$, then the arguments of $\hat{K}_m$ and K in(41) or (43), when identified with the corresponding arguments $\hat{p},\hat{q},p,q$ in(45) satisfy the restrictions (44). Reversing the arguments does not alter this situation. Hence, for both problems D_i and N_e, and for both methods we have

$$|E_{im}(q)| = O(d_m^a)\,(r(p_{im},q))^{-2} . \tag{46}$$

Since $J_m(q) = O(1)$ and the set $\bigcup_{j\neq i} S_{jm}$ lies outside a sphere $s(p_{im};\, O(d_m))$, we can estimate the first term on the right hand side of (42) as

$$|t_1| = O(d_m^a)\cdot\|g_m\|_{C_{om}} \iint_{S-s(p_{im};O(d_m))} (r(p_{im},q))^{-2}\, dS_q =$$

$$= O(d_m^a)\cdot O(|\log d_m|)\|g_m\|_{C_{om}} = O(d_m^b)\|g_m\|_{C_{om}}$$

for fixed $0 < b < a$.

Using (36) and the estimates: $K(p_{im},q) = K(q,p_{im}) = O((r(p_{im},q))^{a-2})$ we obtain

$$|t_2| = O(d_m^{2a})\,\|g_m\|_{C_{om}} \iint_{S-s(p_{im};O(d_m))} (r(p_{im},q))^{a-2}\, dS_q = O(d_m^{2a})\|g_m\|_{C_{om}}$$

Since S_{im} belongs to a sphere $s(p_{im};O(d_m))$ we find

$$|t_3| = \|g_m\|_{C_{om}} \cdot \iint\limits_{S\cap s(p_{im};O(d_m))} O(r(p_{im}, q))^{a-2})\, dS_q = O(d_m^a)\,\|g_m\|_{C_{om}} .$$

Hence, for both problems and both methods we have

(47) $|\delta_m(i)| = |l_m g_m(i) - R_m LP_m g_m(i)| = O(d_m^b)\|g_m\|_{C_{om}}$, $i = 1,\dots,m$,

which is the same as (17) for $o_2(m) = O(d_m^b) = O(m^{-b/2})$. Hence, (35) is proved for $o_4(m) = O(m^{-b/2})$.

(C) For both problems D_i and N_e let Q denote the appropriate operator on the right hand sides of (5), so that (5) reads: $U=Q(u)$. For all $v \in C_o$ we have (Günter [5], pp. 22,40):

(48) $|Q(v)(p)| \leqq M' \|v\|_{C_o}$ if $p \in R_i$ or, resp., $p \in R_e$,

where M' is a constant. Hence, (20) implies

(49) $|Q(P_m u_m)(p) - U(p)| \leqq M' o_4(m)$ for $p \in R_i$ (resp., $p \in R_e$),

where, as we have seen, $o_4(m) = O(d_m^b)$ for both problems and both methods under consideration.

Let us introduce the set $B(\varepsilon) \equiv \{p: r(p,q) \geqq \varepsilon$ for all $q \in S\}$. The uniform convergence $U_m \to U$ on closed subregions will follow if we prove that for some $o(m)$ the estimate holds

(50) $|U_m(p) - U(p)| = O(d_m^b)$ when $p \in R_i \cap B(o(m))$ (resp., $R_e \cap B(o(m))$).

By (49) it is enough to prove that

(51) $|U_m(p) - Q(P_m u_m)(p)| = O(d_m^b)$ when $p \in R_i \cap B(o(m))$ (resp., $R_e \cap B(o(m))$).

Using the definitions of Q and U_m and integrating by substitution we find

(52) $$U_m(p) - Q(P_m u_m)(p) = Q((J_m - 1)\cdot P_m u_m)(p) + \iint\limits_S F_m(p,q)\, J_m(q)\, P_m u_m(q)\, dS_q \equiv \vartheta_1 + \vartheta_2 \text{, say,}$$

where ϑ_1 is the appropriate potential with density $(J_m(q) - 1)\cdot P_m u_m(q)$, $q \in S$, and the function $F_m(p,q)$ in ϑ_2 is :

(53) $F_m(p,q) = K(p,q) - \hat{K}_m(p, T_m^{-1}(q))$, if U_m is defined by (32; D_i)

(54) $F_m(p,q) = K(p,q) - \hat{K}_m(p, \hat{p}_{jm})$ for $q \in S_{jm}$, $j=1,\dots,m$, in case (34; D_i)

(55) $F_m(p,q) = \frac{1}{2\pi}\left(1/r(p, T_m^{-1}(q)) - 1/r(p,q)\right)$ in case (32; N_e)

(56) $F_m(p,q) = \frac{1}{2\pi}\,(1/r(p,\hat{p}_{jm}) - 1/r(p,q))$ for $q \in S_{jm}, j=1,\dots,m$, in case $(34; N_e)$

By (36,48, 20) the term ϑ_1 can be estimated as

(57) $|\vartheta_1| \leqq M'(\|u\|_{C_0} + o_4(m))\, O(d_m^{2a}) = O(d_m^{2a})$.

For ϑ_2 we have

$$|\vartheta_2| \leqq (1+O(d_m^{2a}))(\|u\|_{C_0} + o_4(m)) \iint_S |F_m(p,q)|\, dS_q$$

(58) $$\leqq M'' \iint_S |F_m(p,q)|\, dS_q \, .$$

Using symbols explained in Fig.7, we can put (53) into the form

$$F_m(p,q) = \frac{1}{2\pi}\,(\cos\alpha/r^2 - \cos\hat{\alpha}/\hat{r}^2) = \frac{1}{2\pi}((\cos\alpha - \cos\hat{\alpha})/r^2 + \cos\hat{\alpha}\cdot(\hat{r}-r)(\hat{r}+r)/(r^2\,\hat{r}^2)).$$

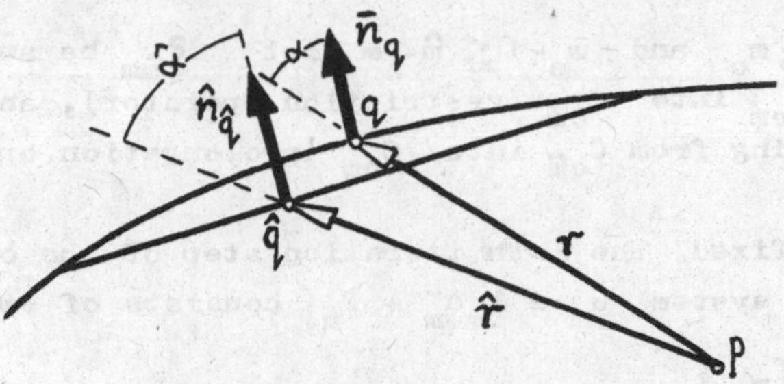

Fig.7

Let $p \in R_i \cap B(d_m)$, i.e., $r \equiv r(p,q) \geqq d_m$ for all $q \in S$. The estimates $|\alpha - \hat{\alpha}| = O(d_m^a)$, $|r-\hat{r}| = O(d_m^{1+a})$ give

$$|F_m(p,q)| = O(d_m^a)/(r(p,q))^2 + O(d_m^{1+a})/(r(p,q))^3$$

and, consequently,

(59) $\iint_S |F_m(p,q)|\, dS_q = O(d_m^a\,|\log d_m|) + O(d_m^a) = O(d_m^b)$.

Now, (59,58,57,52) imply (51) and thus also (50) for $o(m) = d_m$.

In case (54) the point $\hat{q}$ in Fig.7 should be replaced by $\hat{p}_{jm}$. Then $|r-\hat{r}| = O(d_m)$, and (59) follows for all $p \in R_i \cap B(d_m^{1-b})$. Hence, (50) remains valid if we choose $o(m) = d_m^{1-b}$.

In the cases (55,56) the analogous calculations give (50) for $o(m) = d_m$.

Now, let G be a closed region contained in R_i (resp., R_e), and let $\varepsilon = \min(\ r(p,q) : p \in G, q \in S\ )$. For sufficiently large $m \geq m_\varepsilon$ we have $o(m) < \varepsilon/2$ and, consequently, the sphere with center at $p \in G$ and radius $\varepsilon/2$ is contained in $R_i \cap B(o(m))$ (resp., $R_e \cap B(o(m))$), where (50) holds. Since $U_m(p) - U(p)$ are harmonic functions, we obtain (Courant and Hilbert [4], p. 274)

$$|\partial U_m(p)/\partial x - \partial U(p)\ /\partial x| \leqq O(d_m^b)/\varepsilon\ , \quad p \in G,\ m > m_\varepsilon\ ,$$

where x is a space coordinate. This proves the last assertion of the theorem for the sequence $\partial U_m/\partial x$. The case of higher-order derivatives is similar. Proof is complete.

5. One-stage MG method for $u_m = l_m u_m + f_m$.

5.1. General presentation. For reader's convenience we present a short description of the one-stage MG iteration of the second kind (Hackbusch[8]).

For each integer $m \geq m_o$ and $m_o - 1 \leq \hat{m} < m$ let $\rho_{\hat{m}m}$ be an operator acting from C_{om} into $C_{o\hat{m}}$ (restriction operator), and let $\pi_{m\hat{m}}$ be an operator acting from $C_{o\hat{m}}$ into C_{om} (prolongation operator).

Let m and $\hat{m}$ be fixed. The i-th iteration step of the one stage MG method for the system $u_m = l_m u_m + f_m$ consists of smoothing

(60) $\quad \tilde{u}_m = l_m u_m^{(i-1)} + f_m\ ,$

followed by correction

(61) $\quad u_m^{(i)} = \tilde{u}_m - \pi_{m\hat{m}}\, c_{\hat{m}}\ ;$

$c_{\hat{m}} \in C_{o\hat{m}}$ is obtained by solving the auxiliary system

(62) $\quad c_{\hat{m}} = \hat{l}_{\hat{m}}\, c_{\hat{m}} + g_{\hat{m}}\ ,$

where

(63) $\quad g_{\hat{m}} = \rho_{\hat{m}m}\, l_m\, (u_m^{(i-1)} - \tilde{u}_m)\ ,$

and $\hat{l}_{\hat{m}}$ is a linear operator acting from $C_{o\hat{m}}$ into $C_{o\hat{m}}$, such that $(i_{\hat{m}} - \hat{l}_{\hat{m}})^{-1}$ exists (otherwise, the correction would be unfeasible).

The iteration process may be started with $u_m^{(o)} = 0 \in C_{om}$.

5.2. Convergence of the MG iterates.

Lemma 2. Assume that for certain $m \geqq m_o$ and $m_o - 1 \leq \hat{m}(m) < m$ the operators $\rho_{\hat{m}m}$, $\pi_{m\hat{m}}$, $\hat{l}_{\hat{m}}$ and l_m satisfy the conditions

(64) $\quad \rho_{\hat{m}m}\, \pi_{m\hat{m}} = i_{\hat{m}}$

(65) $\bar{M}_1 \| g_{\hat{m}} \|_{C_{o\hat{m}}} \leqq \| \Pi_{m\hat{m}} g_{\hat{m}} \|_{C_{om}} \leqq \bar{M}_2 \| g_{\hat{m}} \|_{C_{o\hat{m}}}$ for all $g_{\hat{m}} \in C_{o\hat{m}}$,

(66) $\| \rho_{\hat{m}m} \|_{C_{om} \to C_{o\hat{m}}} \leqq \bar{M}_3$,

(67) $\| i_m - \Pi_{m\hat{m}} \rho_{\hat{m}m} \|_{C_{bm} \to C_{om}} = o(m)$,

(68) $\hat{l}_{\hat{m}} = \rho_{\hat{m}m} l_m \Pi_{m\hat{m}}$,

(69) $\| l_m \|_{C_{om} \to C_{bm}} \leqq \bar{M}_4$,

(70) $\| (i_m - l_m)^{-1} \|_{C_{om} \to C_{om}} \leqq \bar{M}_5 < \infty$

where $\bar{M}_1, .., \bar{M}_5$ are constants. Then for sufficiently large m the operators $(i_{\hat{m}} - \hat{l}_{\hat{m}})^{-1}$ exist. If $u_m^{(i)}$ is the result of the i-th iteration step, consisting of (60) and (61), and if $u_m = l_m u_m + f_m$ then

(71) $\| u_m^{(i)} - u_m \|_{C_{om}} \leqq (\hat{o}(m))^i \| u_m^{(o)} - u_m \|_{C_{om}}$, $i=1,2,...$,

where $\hat{o}(m) = \bar{M}_6 \cdot o(m)$ for a certain constant $\bar{M}_6$.

<u>Proof</u>. By (68) we have

$$\Pi_{m\hat{m}} (i_{\hat{m}} - \hat{l}_{\hat{m}}) = (i_m - l_m) \Pi_{m\hat{m}} + (i_m - \Pi_{m\hat{m}} \rho_{\hat{m}m}) l_m \Pi_{m\hat{m}} .$$

Proceeding as in (21, 22) and using (70,65,67,69) we obtain the following counterpart of (23):

$$\| (i_{\hat{m}} - \hat{l}_{\hat{m}}) g_{\hat{m}} \|_{C_{o\hat{m}}} \geqq (\bar{M}_7 + \tilde{o}(\hat{m})) \| g_{\hat{m}} \|_{C_{o\hat{m}}} \geqq \bar{M}_8 \| g_{\hat{m}} \|_{C_{o\hat{m}}} \quad \text{for all } g_{\hat{m}} \in C_{o\hat{m}}$$

Hence, $(i_{\hat{m}} - \hat{l}_{\hat{m}})^{-1}$ exists and $\| (i_{\hat{m}} - \hat{l}_{\hat{m}})^{-1} \|_{C_{o\hat{m}} \to C_{o\hat{m}}} \leqq 1/\bar{M}_8$.

Now, using (60,61,64,68) we obtain

$$u_m^{(i)} - u_m = Q_m (u_m^{(i-1)} - u_m), \quad \text{where}$$

$$Q_m = (i_m + \Pi_{m\hat{m}} (i_{\hat{m}} - \hat{l}_{\hat{m}})^{-1} \rho_{\hat{m}m} l_m)(i_m - \Pi_{m\hat{m}} \rho_{\hat{m}m}) l_m .$$

From (65,66,69,67) it follows that

$$\| Q_m \|_{C_{om} \to C_{om}} \leqq (1 + \bar{M}_2 \bar{M}_3 \bar{M}_4 / \bar{M}_8) \, o(m) \, \bar{M}_4 \equiv \bar{M}_6 \cdot o(m) \equiv \hat{o}(m).$$

This implies (71). Proof is complete.

5.3. <u>The use in the calculation of the potential</u> by the methods of Sect. 4.2.

For convenience we shall assume that m is such that $k=m^{1/3}$ is an integer number (e.g., m=1000, 1331, 1728,...) We choose $\hat{m}(m)=m^{2/3}$. We assume also that the triangles $\hat{S}_{im}$, $i=1,\dots,m$, are ordered in such a manner that all the groups $(\hat{S}_{jk+1,m},\dots,\hat{S}_{(j+1)k,m})$, $j=0,1,\dots,\hat{m}-1$, satisfy the condition

(72) $r(\hat{p}_{im}, \hat{p}_{lm}) = O(d_m \cdot k^{1/2})$ for all $jk+1 \leqq i,l \leqq (j+1)k$.

This condition is satisfied if the members of each group are sufficiently tightly packed.

A simple choice of the restriction operator is

(73) $\rho_{\hat{m}m} g_m(j) = \frac{1}{k} \sum_{i=1}^{k} g_m((j-1)k+i)$, $j=1,\dots,\hat{m}$; $g_m \in C_{om}$

The prolongation operator will be defined by

(74) $\pi_{m\hat{m}} g_{\hat{m}}(i) = g_{\hat{m}}\left(\left[\frac{i+k-1}{k}\right]\right)$, $i=1,\dots,m$; $g_{\hat{m}} \in C_{o\hat{m}}$,

where $[\cdot]$ denotes the integer part of a real number.

We shall now prove the following

<u>Theorem</u> 2. Suppose that

(a) The polyhedra $\hat{S}_m$, where $m=k_o^3, (k_o+1)^3, \dots$, satisfy conditions (i) - (vi);

(b) f belongs to C_a on Lyapounov's surface S;

(c) The operator l_m is defined by (29, 30) or (29,33) and the function f_m is given by (31);

(d) $\hat{m}(m)=m^{2/3}$; the triangles $\hat{S}_{im}$, $i=1,\dots,m$, are ordered in such a manner that (72) holds, the operators $\rho_{\hat{m}m}$, $\pi_{m\hat{m}}$ are defined by (73,74); $\hat{l}_{\hat{m}} = \rho_{\hat{m}m} l_m \pi_{m\hat{m}}$.

Then

(75) $\|u_m^{(2)} - u_m\|_{C_{om}} = O(m^{-2b/3})$, and

(76) $\|P_m u_m^{(2)} - u\|_{C_o} = O(m^{-b/2})$,

where $u_m^{(2)}$ is the second iterate of the one-stage MG method, $u_m = l_m u_m + f_m$, u is the solution of (8), P_m is defined by (38), and b is a fixed number: $0<b<a$. The two iterations cost $14\, m^2/3 + m^{4/3} + O(m)$ multiplications and $17\, m^2/3 + m^{5/3} + O(m)$ additions.

<u>Proof</u>. If $\mathcal{S}_{\hat{m}m}$ and $\pi_{m\hat{m}}$ are defined by (73,74) then (64) is obvious, (65,66) are satisfied for $\bar{M}_1=\bar{M}_2=\bar{M}_3=1$ and (67) holds with $o(m) = O(d_m^b \; k^{b/2}) = O(m^{-b/3})$, as it follows easily from (72) if the points p_{im} in the definition (11) are: $p_{im} = T_m(\hat{p}_{im})$, $i=1,..,m$. Since the stability (70) follows from Lemma 1, it remains to prove (69).

We have seen in the proof of Theorem 1 that all the considered operators l_m satisfy (17) if P_m and R_m are defined by (38,39). Using (12,13,15) we obtain $\|l_m\|_{C_{om}\to C_{om}} < \infty$. In view of definition (11) it now remains to prove that

$$(77) \quad |l_m g_m(i) - l_m g_m(j)| \leq \bar{M}_9 \, \|g_m\|_{C_{om}} \cdot (r(p_{im},p_{jm}))^b$$

$$\text{for } i,j=1,\ldots,m; i\neq j; g_m \in C_{om}.$$

By (40,47) we have

$$(78) \quad \begin{aligned} |l_m g_m(i) - l_m g_m(j)| &\leq |\delta_m(i) - \delta_m(j)| + |R_m LP_m g_m(i) - R_m LP_m g_m(j)| \leq \\ &\leq O(d_m^b) \, \|g_m\|_{C_{om}} + \|R_m LP_m g_m\|_{C_{bm}} (r(p_{im},p_{jm}))^b. \end{aligned}$$

It is obvious that $\|R_m\|_{C_b\to C_{bm}} = 1$. Hence, (12,15) imply

$$(79) \quad \|R_m LP_m g_m\|_{C_{bm}} \leq \|R_m\|_{C_b\to C_{bm}} \|L\|_{C_o\to C_b} \|P_m g_m\|_{C_o} \leq M_4 \, M_2 \, \|g_m\|_{C_{om}}.$$

If $i\neq j$ then $O(d_m^b) = O(r(p_{im},p_{jm})^b)$. Hence, (78,79) imply (77). We thus see that (71) holds for $\hat{o}(m) = O(m^{-b/3})$. If $i=2$ then (71) gives (75), which in view of (12,35) implies

$$\|P_m u_m^{(2)} - u\|_{C_o} \leq \|P_m u_m^{(2)} - P_m u_m\|_{C_o} + \|P_m u_m - u\|_{C_o} \leq$$

$$\leq M_2 \cdot O(m^{-2b/3}) + O(m^{-b/2}) = O(m^{-b/2}).$$

Direct solution of (62) requires about $\hat{m}^3/3 = m^2/3$ multiplications and the same number of additions; (60) and (63) require about $2\,m^2$ multiplications and $2\,m^2$ additions. Hence, the two iterations cost about $14\,m^2/3$ multiplications and additions. The preliminary calculations of the coefficient matrix of $\hat{l}_{\hat{m}}$ costs about $m(m + m^{2/3})$ additions and $m^{4/3}$ multiplications. Proof is complete.

We see that only two steps of the one-stage MG method are sufficient to get the iteration error (75) which is less than the final error (76) (the estimate (76) is exact in the sense that, in the general case, the exponent -b/2 can not be replaced by -c/2, where $c \geqq a$).

Finally, let us note that Theorem 2 remains valid if the triangles $\hat{S}_{im}$, i=1,...,m , are divided into unequal groups, satisfying condition (72) (the operators $\varrho_{\hat{m}m}$ and $\pi_{\hat{m}m}$ must then be defined by an obvious extension of (73,74)).

The results of calculations based on the methods of this paper will be presented elsewhere.

REFERENCES

[1] Atkinson, K.E. : Iterative variants of the Nyström method for the numerical solution of integral equations, Numer. Math. 22, 17-31 (1973).

[2] Brakhage, H. : Über die numerische Behandlung von Integralgleichungen nach der Quadraturformelmethode, Numer. Math. 2, 183-196 (1960).

[3] Brandt, A. : Multi-level adaptive solutions to boundary value problems, Math. Comp. 31, 333-390 (1977).

[4] Courant, R. and Hilbert, D. : Methods of mathematical physics, vol. II, New York-London-Sydney (1966).

[5] Günter, N.M. : Die Potentialtheorie und ihre Anwendung auf Grundaufgaben der mathematischen Physic , Leipzig (1957).

[6] Hackbusch, W. : On the multi-grid method applied to difference equations, Computing 20, 291-306 (1978).

[7] Hackbusch, W. : On the convergence of multi-grid iterations, Beiträge Numer. Math. 9, 213-239 (1981).

[8] Hackbusch, W.: Die schnelle Auflösung der Fredholmschen Integralgleichung zweiter Art, Beiträge Numer. Math. 9, 47-62 (1981).

[9] Hemker, P.W. and Schippers, H. : Multiple grid methods for the solution of Fredholm integral equations of the second kind, Math. Comp. 36, 215-232 (1981).

[10] Hess, J.L. and Smith, A.M.O. : Calculation of potential flow about arbitrary bodies, Progress in Aeronautical Sciences 8, 1-138 , New York (1966).

[11] Hess, J.L. : Review of integral equation techniques for solving potential flow problems with emphasis on the surface-source method, Comp. Meth. in Appl. Mech. and Eng. 5, 145-196 (1975).

[12] Nicolaides, R.A.: On multigrid convergence in the indefinite case, Math. Comp. 32, 1082-1086 (1978).

[13] Riesz, F. and Sz.-Nagy, B. : Leçons d'analyse fonctionelle, Budapest (1972).

[14] Schippers, H. : Application of multigrid methods for integral equations to two problems from fluid mechanics, MC Report, Math. Centrum, Amsterdam (1981).

[15] Wesseling, P. : A convergence proof for a multiple grid method, Report NA-21, Dept. of Math., Univ. of Technology, Delft (1978).

[16] Wolff, H. : Multigrid method for the calculation of potential flow around 3-D bodies, MC Report, Math. Centrum, Amsterdam (1981).

APPLICATIONS OF MULTI-GRID METHODS FOR TRANSONIC FLOW CALCULATIONS

Wolfgang Schmidt
Dornier GmbH
D-7990 Friedrichshafen

Antony Jameson
Princeton University
Princeton, NY 08544

Abstract

Multiple grid methods are discussed on the basis of the Poisson equation. In the second part, routine-type application of a multi-grid solver for the full potential equation in combination with a boundary layer integral method is demonstrated. Finally, the importance of multi-grid techniques for systems of partial differential equations (Euler, Navier Stokes) is discussed and some first results are presented for a scheme presently under consideration.

Introduction

Very fast and accurate methods for transonic flow computations are needed by design engineers to improve present day and to develop future transport as well as military aircraft. Such methods permit detailed transonic flow studies within a short time at low cost. Multiple grid methods have been demonstrated to be very powerful for non-elliptic two-dimensional, transonic flow problems by South and Brandt[1] as well as Jameson[2]. Since practical flow analysis implies contour-conformal grid systems, which in general will be non-orthogonal and stretched, fast flow solvers must be insensitive against mesh spacing.

In order to understand the basic properties and the efficiency of multi-grid methods, a study of such methods has been performed on the basis of the Poisson equation. These numerical experiments indicated the combination of multi-grid plus an ADI-scheme to be the most robust procedure. Therefore the combination of such a solver for the full potential equation as proposed by Jameson[2] is presented in the second part of the present paper. Comparison with measurements is done after inclusion of viscous effects by means of integral boundary layer methods.

While such potential flow solutions prove to be useful for flows with moderate shock strength, the isentropic and irrotational flow assumption does lead to significant errors in the upper transonic speed range. Additional deficiencies occur in transonic lifting flow as pointed out in Ref. 3 due to the Kutta condition for potential flow theory. These problems can be overcome by solving either the time dependent exact inviscid equations (Euler equations) or the time dependent viscid equations (Navier Stokes equations). However, techniques being developed for the full potential equation (quasi-linear second order equation) cannot be simply applied for the Euler or Navier Stokes equations, since they represent systems of first order partial differential equations of hyperbolic nature in time. Since standard solution methods for these equations require very large computer time, efficient multi-grid solvers are highly desirable for such equations. In the last part of the present paper two different approaches for multi-grid schemes for the Euler equations are discussed. For the one similar for the Ni-scheme[4] first results are presented.

Multi-Grid Study for the Poisson Equation

The model case which will be considered is a test function

$$f = c_1 (x+y)^{d_1} \sin(a_1 x) \sin(b_1 y) + c_2 (x+y)^{d_2} \sin(a_2 x) \sin(b_2 y) \tag{1}$$

which should be equal to the solution g of the two-dimensional Poission equation

$$Lg = W \tag{2}$$

where the operator L on the solution is the central second order difference approximation in each point (i,j)

$$(Lg)_{i,j} = g_{i+1,j} - 2g_{i,j} + g_{i-1,j} + (\Delta x/\Delta y)^2 (g_{i,j+1} - 2g_{i,j} + g_{i,j-1}) \tag{3}$$

and the right hand side is the same approximation of the test function f

$$W_{i,j} = f_{i+1,j} - 2f_{i,j} + f_{i-1,j} + (\Delta x/\Delta y)^2 (f_{i,j+1} - 2f_{i,j} + f_{i,j-1}) \qquad (4)$$

such that in the converged state

$$g = f$$

The multi-grid equation is formulated as follows. Let

$$L_h g - W = 0 \qquad (5)$$

be the equation with the grid width h. Then L_h approximates the linear differential operator L on the grid with a spacing proportional to the parameter h. Let U be the present estimate of g, and let V be the required correction to U such that U+V satisfies (5). Then the basis of the multi-grid method is to replace (5) and determine V by

$$L_{2h} V + I_{2h}^{h} (L_h U - W) = 0 \qquad (6)$$

where L_{2h} is the same approximation to L on a grid in which the spacing has been doubled, and I_{2h}^{h} is an operator which transfers to each grid point of the coarse grid the residual $L_h U-W$ of the coincident point of the fine mesh. After the solution of Eq. (6) the approximation on the fine grid is updated by interpolating the correction calculated on the coarse grid to the fine grid, so that U is replaced by

$$U^{new} = U + I_{h}^{2h} V \qquad (7)$$

where I_{h}^{2h} is an interpolation operator.

Equation (6) can in turn be solved by introducing an approximation on a yet coarser grid, so that a multiple sequence of grids may be used, leading to a rapid solution procedure for two reasons. First, the number of operations required for a relaxation sweep on one of the coarse grids is much smaller than the number required on the fine grid. Second, the rate of convergence is faster on a coarse grid, reflecting the fact that corrections can be propagated from one end of the grid to the other in a smaller number of steps.

To extend this idea to nonlinear equations and to avoid a perturbation form of the equation, Eq. (6) is reorganised by adding and subtracting the current solution U to give

$$L_{2h}(U+V) - L_{2h}U + I_{2h}^{h}(L_h U-W) = 0$$

or $$L_{2h}\overline{U} - \overline{W} = 0 \tag{8}$$

where $\overline{U}$ is the improved estimate of the solution to be determined on the coarse grid, and $\overline{W}$ is an appropriately modified right-hand side

$$\overline{W} = L_{2h}U - I_{2h}^{h}(L_h U-W) \tag{9}$$

The updating formula (7) now becomes

$$U^{new} = U + I_h^{2h}(\overline{U}-U) \tag{10}$$

As updating operator I_h^{2h} fourth order interpolation formulas are used (third order at boundaries).

The success of the multiple grid method generally depends on the use of a relaxation algorithm which rapidly reduces the high frequency components of error on any given grid. To analyse the influence of this smoothing algorithm, three different relaxation schemes were tested:

- horizontal line relaxation (SLOR)
- alternating direction scheme AF1[5]

  $$(\alpha - \delta_x^2)\left(\frac{\alpha}{r} - \delta_y^2\right)\delta g = \omega\frac{\alpha}{r}Lg \tag{11}$$

  where

  $r = (\Delta x/\Delta y)^2$, α is a parameter to be chosen, ω is an over-relaxation factor, and the residual Lg is calculated using the result of the previous iteration

- alternating direction scheme AF2[6]

  $$(\alpha r\delta_y^- - \delta_x^2)(\alpha - \delta_y^+)\delta g = \omega\alpha\, Lg \tag{12}$$

  using one-sided difference operators δ_y^- and δ_y^+.

On an orthogonal mesh with equidistant spacing in Δx and Δy the computational results indicate that horizontal line relaxation and AF2 work well in the multi-grid mode when

$$\Delta x < \Delta y \quad (128 \times 32 \text{ grid})$$

but not so well when

$$\Delta x > \Delta y \quad (32 \times 128 \text{ grid})$$

On an equally spaced grid ($\Delta x = \Delta y$) all three schemes worked, but AF1 was best. The average rate of reduction of residual per work unit (measured as fine grid iteration) that could be achieved with AF1 in multi-grid mode was

- 0.230 on a 128 x 128 grid
- 0.259 on a 64 x 64 grid
- 0.272 on a 32 x 32 grid

It is interesting that the observed convergence rate gets faster as the grid gets finer. The result on the 128x128 grid represents a reduction from an initial average residual of $0.108 \cdot 10^{-1}$ to a final average residual of $0.576 \cdot 10^{-13}$ in eleven multi-grid cycles (17.66 work units).

The average error $|f-g|$ was reduced from $0.573 \cdot 10^{-1}$ to $0.554 \cdot 10^{-13}$ for a mean rate of reduction of 0.162 per work unit. For comparison the rate of convergence with line relaxation on a 64x64 grid (using 1 grid) was 0.982 for the residual and 0.993 for the error. In multi-grid mode line relaxation gave rates of 0.576 for the residual and 0.511 for the error.

An extensive study was made for the rate of convergence of the AF1 scheme in multigrid mode with different choices of the parameters α and ω. Fig. 1 shows curves of the rate of convergence for the 128x 128 grid plotted against ω for different values of α. Optimal convergence rates can be obtained for certain combinations of α, ω only, however these combinations do depend on grid spacing and the optimum is very narrow. In production-type codes automatic adjustment for optimal α-ω combination would be highly desirable, however the logics for that are not understood yet.

MAD-Full Potential Solver

In the case of transonic flow we have to allow for a change from elliptic to hyperbolic type as the flow becomes locally supersonic. In the model problem

$$a\,\Phi_{xx} + b\,\Phi_{yy} = 0 \qquad (13)$$

this corresponds to one of the coefficients, say a, to become negative. The classical alternating direction scheme AF1

$$(\alpha - A\delta_x^2)\,(\alpha - B\delta_y^2)\,\delta\Phi = \omega\alpha L\Phi \qquad (14)$$

then has the disadvantage[7] that if one regards the iterations to represent time steps Δt in an artificial time direction t, it simulates the time dependent equation

$$\alpha\,\Delta t\,\Phi_z = \alpha\,\Phi_{xx} + b\,\Phi_{yy} \qquad (15)$$

When $\alpha < 0$ and Cauchy data is given at $x = 0$ - corresponding to supersonic inflow - this leads to an ill posed problem which admits oscillatory solutions which are undamped in time and grow in the x-direction.

Therefore Jameson[2] proposed the following generalised alternating direction scheme. Let the scalar parameter α in Eq. (14) be replaced by a difference operator

$$S \equiv \alpha_o + \alpha_1\,\delta_x^- + \alpha_2\,\delta_y^- \qquad (16)$$

where δ_x^- and δ_y^- denote one sided difference operators in the x and y directions. This yields the scheme

$$(S - A\delta_x^2)\,(S - B\delta_y^2)\,\delta\Phi = \omega SL\Phi \qquad (17)$$

in which the residual $L\Phi$ is differenced by the operator S. The corresponding time dependent equation is now a hyperbolic one of the form

$$\beta_o\,\Phi_t + \beta_1\,\Phi_{xt} + \beta_2\,\Phi_{y}t = a\,\Phi_{xx} + b\,\Phi_{yy} \qquad (18)$$

where the coefficients β_o, β_1, β_2 depend on the paramters α_o, α_1, α_2.

Two remarks should be made to this scheme:

- There are additional error terms because the operator S does not commute with $a\delta_x^2$, $b\delta_y^2$, and the order of the factors may matter.

- Applied on a single grid it is necessary to use a sequence of the parameters α_o, α_1, α_2 to reduce all frequency bands of the error.

From the different possible strategies a very simple one has been chosen. Each cycle begins on the fine grid. The alternating direction iteration is performed once on each grid until the coarsest grid is reached. Then it is performed once on each grid going back up to the second finest grid, and the cycle terminates with the interpolation of the correction from the second finest to the fine grid. For viscous flow computations, this result is being used for a boundary layer computation, the displacement thickness of which is entering the next multi-grid cycle to provide finally a converged viscid solution. The additional effort for the viscid analysis is less than the time of one MAD-cycle per cycle, without increasing the total amount of MAD-cycles needed for convergence.

Some typical results are presented here. Further results are given in Ref. 8. All the examples were calculated on a circular domain generated by conformal mapping of the airfoil to a unit circle, with 160x32 in the Θ and r direction on the fine grid. Five grids were used in the multi-grid scheme, giving a coarse grid of 10x2 cells; Fig. 2 shows a typical result for the RAE 2822 airfoil including viscous effects. The result was obtained after 12 MAD-cycles reducing the average residual from $0.1194 \cdot 10^{-3}$ to $0.7355 \cdot 10^{-5}$. Total CPU-time on an IBM 3031 computer for this case is 120 sec.

Fig. 3 shows similar results for the DO-A1 airfoil (CAST 7). Again there is very good agreement with the experimental data. Finally, Fig. 4 portrays inviscid results for the cylinder flow at M = 0.50 after 29 cycles and a reduction in average residual from $0.7876 \cdot 10^{-2}$ to $0.71 \cdot 10^{-6}$.

Multiple Grid Schemes for Systems of First Order Partial Differential Equations

The example which will be considered is two-dimensional unsteady inviscid flow which is described by the unsteady two-dimensional Euler equations

$$\frac{\partial\vec{U}}{\partial t} + \frac{\partial\vec{F}}{\partial x} + \frac{\partial\vec{G}}{\partial y} = 0 \tag{19}$$

where
$$\vec{U} = \begin{Bmatrix} \rho \\ \rho u \\ \rho v \\ \rho E \end{Bmatrix}, \quad \vec{F} = \begin{Bmatrix} \rho u \\ \rho u^2+p \\ \rho uv \\ \rho uH \end{Bmatrix}, \quad \vec{G} = \begin{Bmatrix} \rho v \\ \rho vu \\ \rho v^2+p \\ \rho vH \end{Bmatrix} \tag{20}$$

and p, ρ, u, v, E and H denote the static pressure, density, Cartesian velocity components, total energy and total enthalpy. For a perfect gas is

$$E = \frac{p}{(\gamma-1)\rho} + \frac{1}{2}\,(u^2+v^2)\ , \quad H = E + {}^{p}/_{\rho} \tag{21}$$

If only homoenergetic steady flow ($H = H_o$ = const) is of interest, the transient phase does not have to be time-accurate, thus allowing for accelerating techniques. Two possible ones are described in Ref. 9, the local time stepping and a forcing term proportional to $H-H_o$. Both acceleration techniques basically permit the additional use of multiple grid techniques.

All three schemes presently under multi-grid evaluation are based on the integral conservation formulation of Eq. (19)

$$\iint \frac{\partial}{\partial t}\,\vec{U}\,dvol + \oint \bar{\bar{H}}\,\hat{n}\,ds = 0 \tag{22}$$

describing the change of the flow vector $\vec{U}$ in time in one volume to be equal to fluxes of the flow in mass, momentum, energy through the surfaces of this volume.

The three schemes differ in the approximation of the flow in each cell

- central point Runge Kutta Stepping (Ref. 9)
  where the flow is discretised by values in the volume center

- nodal point Runge Kutta Stepping
  where the flow is discretised by values in the volume corners
- one step distribution formula scheme (Ref.4)

The nodal point and the distribution formula schemes are basically better suited to multi-grid techniques since after mesh halving points will remain mesh points.

Best results in multi-grid so far have been obtained using the one step distribution formula scheme. Therefore this procedure is described briefly.

The basic idea again is to use the coarser grids to propagate the fine grid connections properly and rapidly throughout the field, thus improving convergence rate to steady state while maintaining low truncation errors by using the fine grid discretisation.

The changes ΔU_{2h} in the coarse grid, obtained by removing every other line from the fine grid, are determined by

$$\Delta U_{2h} = I_{2h}^{h} \delta U_{h} \tag{23}$$

Where I_{2h}^{h} is an operator which transfers to each control volume of the coarse grid the correction δU_h of the fine grid using specific distribution formulas.

After computing the corrections δU_{2h} on all coarse grid points, the flow properties at the finest grid are updated by

$$U^{new} = U + I_{h}^{2h} \delta U_{2h} \tag{24}$$

where I_h^{2h} is a linear interpolation operator which interpolates the coarse grid corrections to give the corrections at each fine grid point of the finest mesh.

The scheme is quite sensitive to the choice of distribution formulas and the boundary condition treatment. Also, compared with the Runge-Kutta stepping schemes in Ref. 9, a fairly large dissipation has to be incorporated to make multi-grid converging efficiently. Fig. 5 portrays the converged pressure distribution C_p for the NACA 0012 airfoil at $M_\infty = 0{,}80$ and $\alpha = 0^o$. The added convergence history with an average reduction rate per cycle for 500 cycles of 0.9686 is quite good. Fig. 6 shows similar results for the cylinder at $M_\infty = 0.35$. Here, the reduction in $\partial\rho/\partial t$ is equal to 0.9785. However, both results show effects of numerical viscosity. The cylinder-case gives no fully symmetric solution, the NACA airfoil case shows errors in the shock region compared with fully converged solutions of the very accurate Runge Kutta stepping scheme in Ref. 9.

Conclusions

Engineering application of computational methods requires fast and reliable numerical schemes. Multi-grid techniques are very well suited to meet those requirements, and have proved to be very useful for problems governed by quasi-linear equations. However, more has to be done to develop fast and accurate schemes for systems of first order partial differential equations since these equations are the most relevant ones for computational fluid mechanics.

References

1 South, J.C., and Brandt, A.: Application of a multi-level grid method to transonic flow calculations. Transonic Flow Problems in Turbomachinery, edited by T.C. Adamson and M.F. Platzer, Hemisphere, Washington, 1977

2 Jameson, A.: Acceleration of transonic potential flow calculations on arbitrary meshes by the multiple grid method. AIAA-Paper 79-1458, AIAA 4th CFD Conference, Williamsburg, July 1979

3 Schmidt, W., Jameson, A., Whitfield, D.: Finite volume solution for the Euler equations for transonic flow over airfoils and wings including viscous effects. AIAA-Paper 81-1265, AIAA 14th FPD Conference, Palo Alto, June 1981

4 Ni, R.H.: A multiple grid scheme for solving the Euler equations. 5th AIAA CFD Conference, AIAA-Paper 81-1025, June 1981, Palo Alto

5 Jameson, A.: An alternating direction method for the solution of the transonic small disturbance equation. New York Univ., ERDA Report COO-3077-96, 1975.

6 Ballhaus, W.F., Jameson, A., Albert, J.: Implicit approximate factorization schemes for the efficient solution of steady transonic flow problems. 3rd AIAA CFD Conference, Albuquerque, June 1977.

7 Jameson, A.: Iterative solution of transonic flows over airfoils and wings, including flows at Mach 1. Comm. Pure Appl. Math., Vol. 27, 1974, pp. 283-309.

8 Longo, J., Schmidt, W., Jameson, A.: Viscous transonic airfoil flow simulation. DGLR Symposium "Strömung mit Ablösung", Stuttgart, 1981.

9 Jameson, A., Schmidt, W., Turkel, E.: Numerical Solutions of the Euler Equations by Finite Volume Methods Using Runge-Kutta Time-Stepping Schemes. AIAA-Paper 81-1259, AIAA 14th FPD Conference, Palo Alto, June 1981.

FIG. 1: INFLUENCE OF PARAMETERS α, ω ON CONVERGENCE

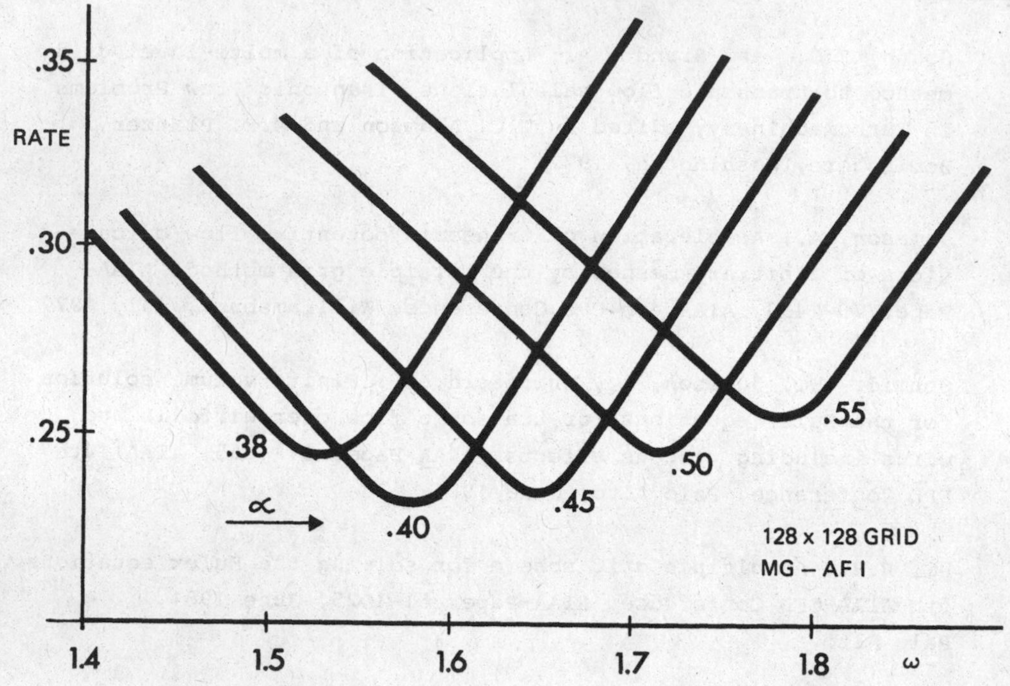

FIG. 2: RAE AIRFOIL 2822

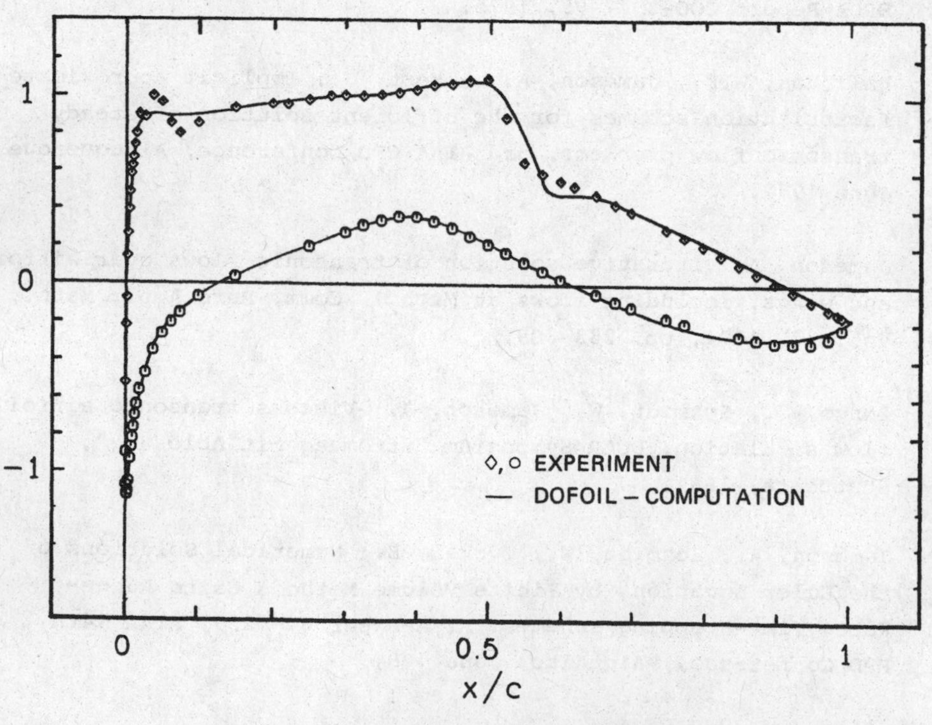

FIG. 3: DO-A1 (CAST 7) AIRFOIL

●○ EXPERIMENT CL = .110 CD = .0102 CM =-.118
M∞ = .760 ± .001 α = −2.0 $R_e = 6 \cdot 10^6$ (Ref. AGARD AR-138)
+x DOFOIL COMPUTATION

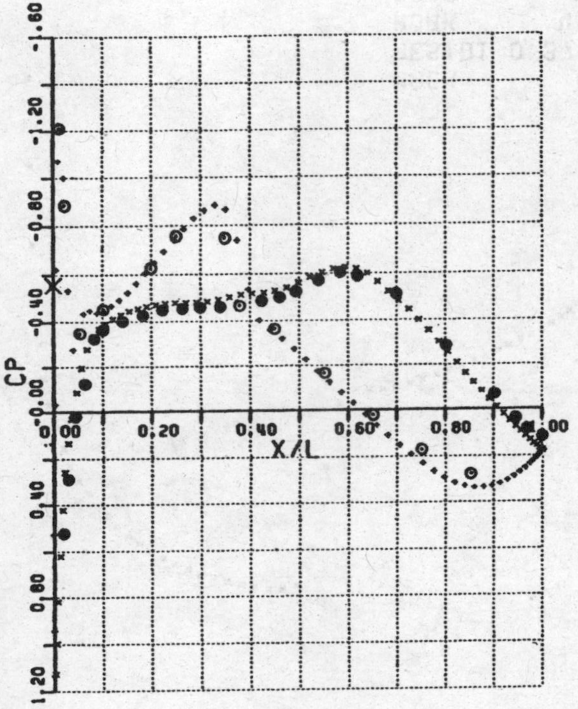

CL= 0.134053 CD= 0.010449 CM= -0.123160

FIG. 4: FULL POTENTIAL MAD – SOLUTION

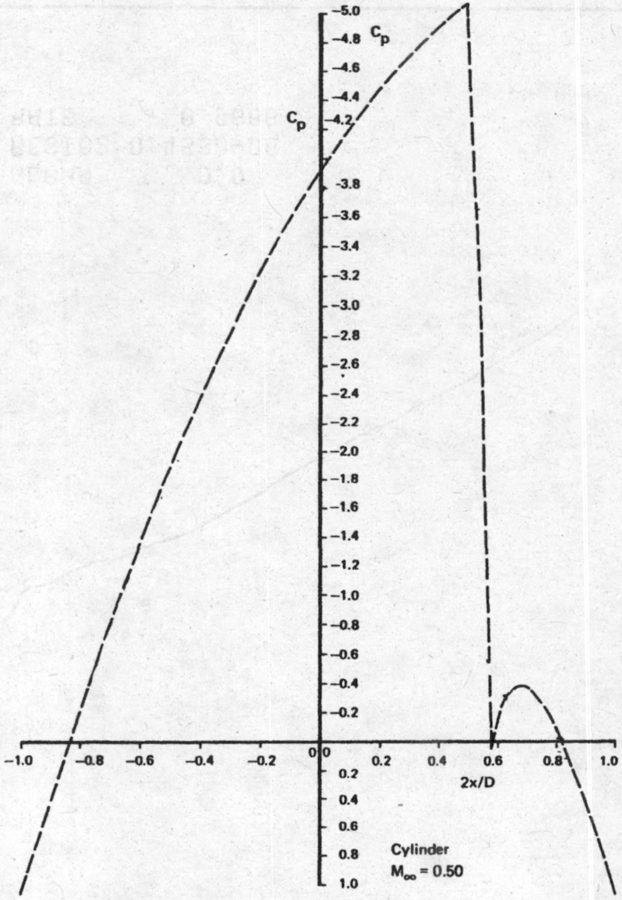

FIG. 5: NACA 0012

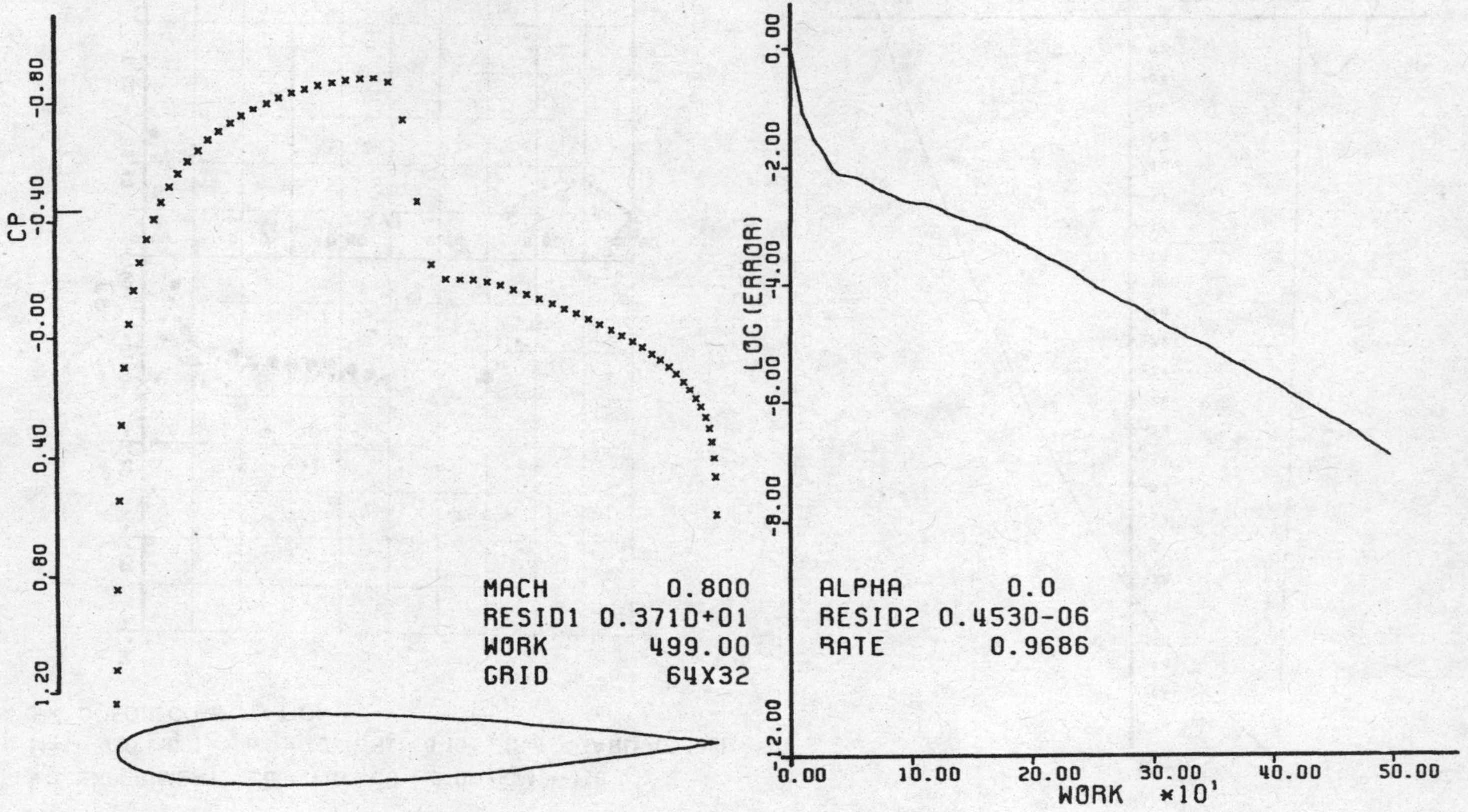

FIG. 6: CIRCLE

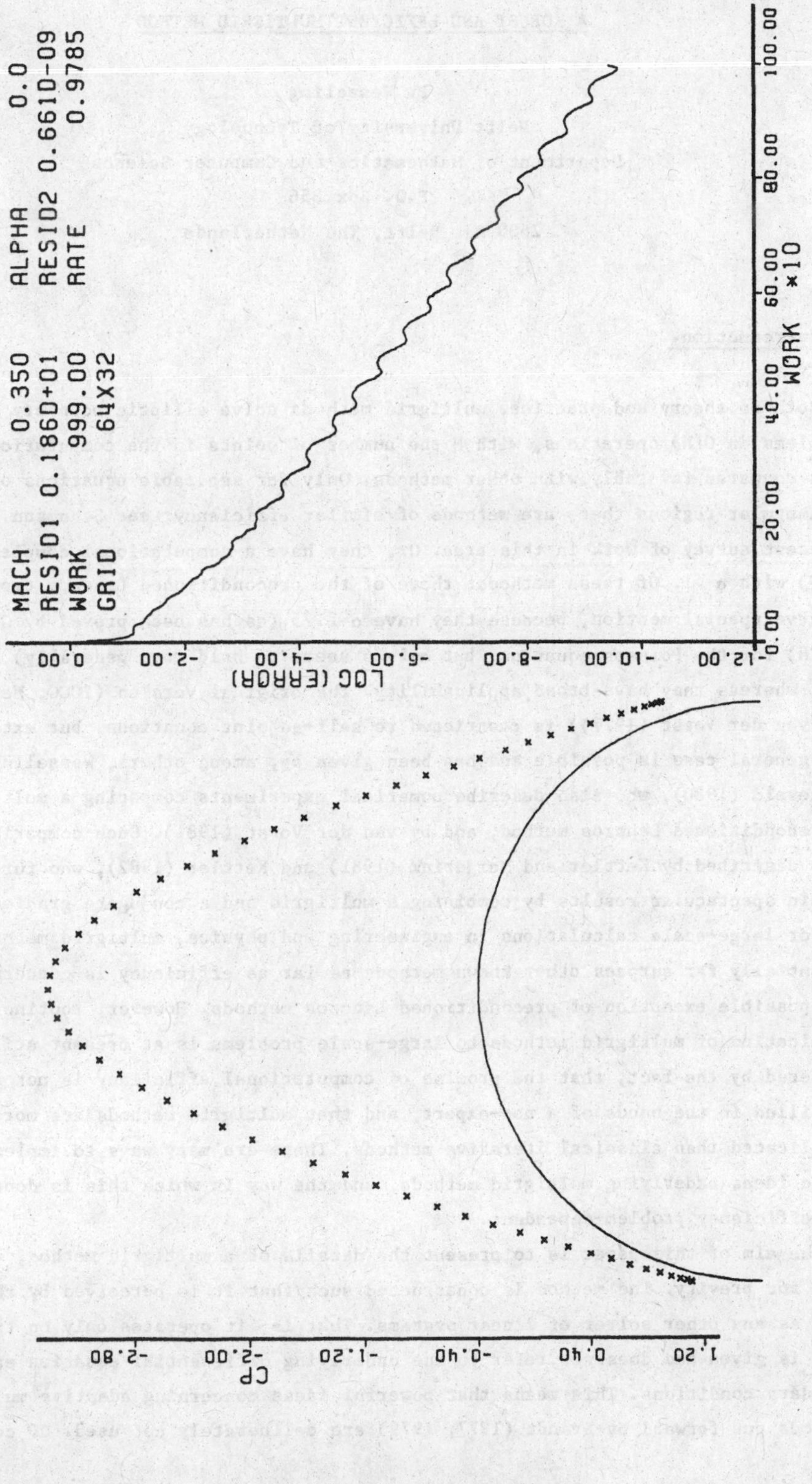
MACH 0.350
RESID1 0.186D+01
WORK 999.00
GRID 64X32
ALPHA 0.0
RESID2 0.661D-09
RATE 0.9785
LOG(ERROR)
0.00
-2.00
-4.00
-6.00
-8.00
-10.00
-12.00
0.00
20.00
40.00
60.00
80.00
100.00
WORK ×10
CP
-2.80
-2.00
-1.20
-0.40
0.40
1.20

A ROBUST AND EFFICIENT MULTIGRID METHOD

P. Wesseling
Delft University of Technology
Department of Mathematics and Computer Science
P.O. Box 356
2600 AJ Delft, The Netherlands

1. Introduction.

Both in theory and practice, multigrid methods solve elliptic boundary value problems in O(N) operations, with N the number of points in the computational grid. This compares favorably with other methods. Only for separable equations on rectangular regions there are methods of similar efficiency; see Schumann (1980) for a recent survey of work in this area. Or, they have a computational complexity of $O(N^\alpha)$ with $\alpha > 1$. Of these methods, those of the preconditioned Lanczos type deserve special mention, because they have $\alpha=1.25$ (as has been proved by Gustafsson (1978) for the Poisson equation, but $\alpha=1.25$ seems to hold more generally), which is low, whereas they have broad applicability. The original version (ICCG, Meijerink and van der Vorst (1977)) is restricted to self-adjoint equations, but extension to the general case is possible and has been given by, among others, Wesseling and Sonneveld (1980), who also describe numerical experiments comparing a multigrid and a preconditioned Lanczos method, and by van der Vorst (1981). Such comparisons are also described by Kettler and Meijerink (1981) and Kettler (1982), who furthermore obtain spectacular results by combining a multigrid and a conjugate gradient method.

For large-scale calculations in engineering and physics, multigrid methods potentially far surpass other known methods as far as efficiency is concerned, with the possible exception of preconditioned Lanczos methods. However, routine application of multigrid methods to large-scale problems is at present still hampered by the fact, that the promise of computational efficiency is not always fulfilled in the hands of a non-expert, and that multigrid methods are more complicated than classical iterative methods. There are many ways to implement the basic ideas underlying multigrid methods, and the way in which this is done may make the efficiency problem-dependent.

The aim of this paper is to present the details of a multigrid method, called MGD1 for brevity. The method is constructed such that it is perceived by the user just as any other solver of linear systems. That is, it operates only on the matrix that is given and does not refer to the underlying differential equation and boundary conditions. This means that powerful ideas concerning adaptive multigrid methods put forward by Brandt (1977, 1979) are deliberately not used. Of course, the basic multigrid methodology is employed, as typified in various ways by the work of Fedorenko (1962), Bakhvalov (1966), Astrachancev (1971), Brandt (1973), Frederickson (1975), Wachspress (1975), Hackbusch (1978a), Wesseling (1977),and become widely known and appreciated by Brandt (1977). It depends on the user (is he familiar with multigrid methods or not) and on the problem, whether an adaptive or a non-adaptive approach, such as presented here, is to be preferred.

The method MGD1 is fast for a large class of elliptic boundary value problems, as will be made plausible and demonstrated experimentally. Operation counts and rates of convergence are given, and a comparison of computational efficiency with other methods is made.

The main characteristics of the method are the use of incomplete LU-decomposition for smoothing, 7-point prolongation and restriction operators, coarse-grid Galerkin approximation, and the use of a fixed multigrid strategy, that will be called the sawtooth cycle. It does not need to be adapted to the problem, and is expected to be useful to the non-specialist. A FORTRAN program will be described in a forthcoming report.

2. A multigrid method.

A multigrid method will be presented for the solution of an elliptic boundary value problem on a rectangle, discretized by finite differences. A computational grid Ω^ℓ and a corresponding set of grid-functions U^ℓ are defined as follows:

$$\Omega^\ell \equiv \{(x_0,y_0) \cup (x_i,y_j),\ x_i=x_0+ih_x,\ y_j=y_0+jh_y,\ i,j=1(1)2^\ell\},$$
$$U^\ell \equiv \{u^\ell:\Omega^\ell \to \mathbb{R}\}. \tag{2.1}$$

The formulation (2.1) allows elimination of physical boundaries where Dirichlet boundary conditions are given. For example, if all boundary conditions are of Dirichlet type and the region is the unit square one may choose: $h_x=h_y=h=(2^\ell+2)^{-1}$, $x_0=y_0=h$. It is not really necessary to have the number of x- and y-gridlines equal. By coordinate stretching one can generate a non-equidistant mesh in the physical plane. The linear algebraic system generated by the difference scheme is denoted by:

$$A^\ell u^\ell = f^\ell. \tag{2.2}$$

The multigrid method makes use of a hierarchy of computational grids Ω^k and corresponding sets of grid-functions U^k, $k=\ell-1(-1)1$, defined by (2.1) with ℓ replaced by k. On the coarser grids (i.e. grids with larger step-size, hence smaller k) equation (2.2) is approximated by:

$$A^k u^k = f^k,\quad k = \ell-1(-1)1. \tag{2.3}$$

Furthermore, let there be given a restriction operator r^k and a prolongation operator p^k:

$$r^k : U^k \to U^{k-1} \quad , \quad p^k : U^{k-1} \to U^k. \tag{2.4}$$

A^k, f^k, r^k, p^k will be specified later.

For the so-called smoothing process use is made of incomplete LU-(ILU-) decomposition. Temporarily suppressing the superscript k, we assume that we have a lower and an upper triangular matrix L and U respectively, such that

$$LU = A + C. \tag{2.5}$$

L, U and C will be specified later. Consider the following iterative process for solving (2.2) or (2.3):

$$x := x + (LU)^{-1}\,(f-Ax)\ ,$$

or

$$x := (LU)^{-1}\,(f+Cx). \tag{2.6}$$

The multigrid method presented in the following quasi-Algol program can be regarded as a method to accelerate the iterative process (2.6):

multigrid method MGD1:

<u>begin</u> $f^{\ell-1} := r^\ell C^\ell(u^\ell - \bar{u}^\ell)$;

<u>for</u> k := ℓ-1(-1)2 <u>do</u> $f^{k-1} := r^k f^k$;

$u^1 := (L^1U^1)^{-1} f^1$;

<u>for</u> k := 2(1)ℓ-1 <u>do</u> $u^k := (L^kU^k)^{-1}(C^kp^ku^{k-1} + f^k)$;

$\bar{u}^\ell := u^\ell + p^\ell u^{\ell-1}$;

$u^\ell := (L^\ell U^\ell)^{-1}(C^\ell\bar{u}^\ell + f^\ell)$

<u>end</u> of one iteration with MGD1;

Using (2.5), one may verify that $C^\ell(u^\ell - \bar{u}^\ell) = f^\ell - A^\ell u^\ell$ is the residue associated with the current iterand u^ℓ. As we shall see, C^ℓ has only two non-zero diagonals, hence $C^\ell(u^\ell - \bar{u}^\ell)$ is a fast way to compute the residue $f^\ell - A^\ell u^\ell$. When starting, $\bar{u}^\ell$ and u^ℓ are not available. A cheap way to get started is to use the initial estimate u^ℓ=0, and replace $C^\ell(u^\ell - \bar{u}^\ell)$ by $f^\ell - A^\ell u^\ell = f^\ell$. In this case the computation effectively starts on the coarsest grid, which is advantageous, because no effort is wasted in correcting a perhaps unfortunate first guess. If one has a good initial estimate available, one may generate u^ℓ and $\bar{u}^\ell$ as follows:

$\bar{u}^\ell$:= initial estimate;

$u^\ell := (L^\ell U^\ell)^{-1}\ (C^\ell\bar{u}^\ell + f^\ell)$;

In this case the computation starts with a smoothing step on the finest grid, which costs little more than a straightforward residue calculation.

The method is not recursive, and is easily implemented in FORTRAN. Each coarse grid is visited only once, and one smoothing step is performed after each coarse grid correction. This multigrid strategy may be depicted graphically as follows (for 4 grids):

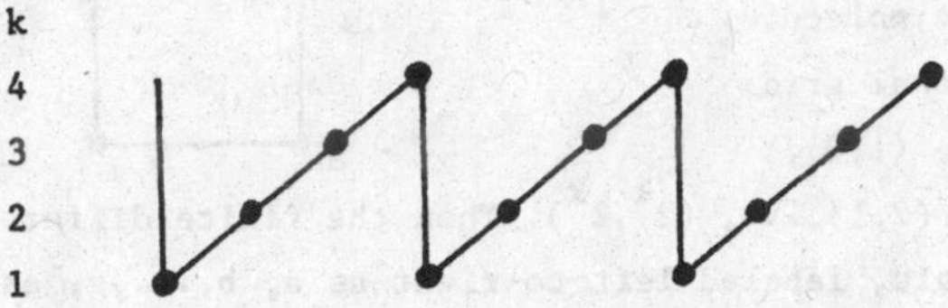

Each dot represents a smoothing operation. This diagram suggests "sawtooth cycle" as an appropriate name for this strategy. This is probably the simplest multigrid strategy that one can think of; cf. Brandt (1977), Hackbusch (1981), Hemker (1981) for a description of many possible multigrid strategies.

A measure of the computational cost of one execution of multigrid method MGD1 may be obtained by counting the arithmetic operations that are visible in the mathematical formulae. The following table gives the operation count per grid-point of Ω^k for various parts of the algorithm.

C^k	r^k	$(L^kU^k)^{-1}$	p^k
3	2	13	1.5

The results for $(L^kU^k)^{-1}$, r^k and p^k follow from subsequent sections. With $n_k=(2^k+1)^2$ the number of grid-points of Ω^k, we obtain the following total count:

$$6n_\ell + 2\sum_{k=2}^{\ell-1} n_k + 13n_1 + 18.5\sum_{k=2}^{\ell-1} n_k + 2.5n_\ell + 17n_\ell \cong 30.4^\ell ,$$

or about 30 operations per grid-point of Ω^ℓ.

The decisions made in the design of MGD1 are based on comparative experiments described by Wesseling (1980) and Mol (1981).

3. Incomplete LU-decomposition.

ILU- decomposition was used by Meijerink and van der Vorst (1977) as an effective preconditioning for conjugate gradient methods, and was introduced by Wesseling and Sonneveld (1980) as a smoothing process for multigrid methods. For experiments with and analysis of various ILU smoothing processes, see Wesseling and Sonneveld (1980), Hemker (1980a), Mol (1981), Kettler and Meijerink (1981), and the extensive treatment by Kettler (1982).

The general second order elliptic differential operator can be approximated by

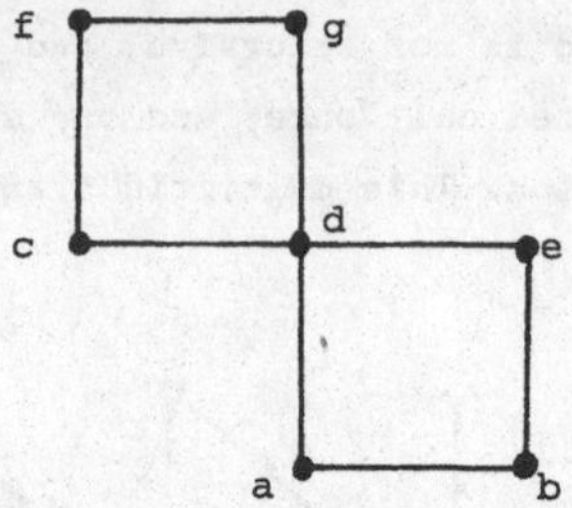

central or one-sided differences using the 7-point finite difference molecule depicted here. If no mixed derivative is present atoms b and f are superfluous, and we have the familiar 5-point molecule. Let the points of the computational grid Ω^k be ordered as follows: (0,0), (1,0), (2,0),..., (2,0), (0,1), (1,1), (2,1),..., $(2^k,2^k)$. Then the finite difference matrix A^k has 7 non-zero diagonals, labeled left-to-right as a, b,...,g, each of which corresponds with the atom with the same label. By non-zero we mean: possibly non-zero.

The ILU-decomposition to be employed here can be described as follows. In the same locations as a, b, c, L is prescribed to have non-zero diagonals α, β, γ respectively. The main diagonal of L is specified to be unity. At locations d, e, f, g, U has non-zero diagonals δ, ε, ζ, η, respectively. The rest of L and U is zero. L and U can be conveniently computed by Crout-like formulae, as follows, on an m*n grid. Subscript k is the row-number.

$$\alpha_k = a_k/\delta_{k-m}, \qquad \beta_k = (b_k-\alpha_k\varepsilon_{k-m})/\delta_{k-m+1},$$
$$\gamma_k = (c_k-\alpha_k\zeta_{k-m})/\delta_{k-1}, \qquad \delta_k = d_k-\gamma_k\varepsilon_{k-1}-\beta_k\zeta_{k-m+1}-\alpha_k\eta_{k-m}, \tag{3.1}$$
$$\varepsilon_k = e_k-\beta_k\eta_{k-m+1}, \qquad \zeta_k = f_k-\gamma_k\eta_{k-1}, \quad \eta_k = g_k.$$

Quantities that are not defined are to be replaced by 0.
The error matrix C = LU-A has only two non-zero diagonals ψ and λ, located next to and inside b and f, respectively, and given by:

$$\psi_k = \beta_k\varepsilon_{k-m+1}, \quad \lambda_k = \gamma_k\zeta_{k-1}. \tag{3.2}$$

The solution of LU = q is obtained by back-substitution:

$$u_i := q_i - \gamma_i u_{i-1} - \beta_i u_{i-m+1} - \alpha_i u_{i-m}, \quad i=1(1)mn;$$
$$u_i := (u_i - \varepsilon_i u_{i+1} - \zeta_i u_{i+m-1} - \eta_i u_{i+m})/\delta_i, \quad i=mn(-1)1. \tag{3.3}$$

It is easily verified, that the construction of L and U, the construction of C, and the solution of LUu=q takes 17,2 and 13 arithmetic operations per grid-point, respectively.
The computation of Cu+f takes 4 operations, because C has only 2 non-zero diagonals, hence a complete smoothing step $u:=(LU)^{-1}(Cu+f)$ takes 17 operations. L and U can be stored in the space for A, because A is not needed. If A has only 5 non-zero diagonals, this requires an extra storage of 2 reals per grid point. Storage of C also requires 2 reals per grid-point, but C can also be computed instead of being stored, in which case the cost of a smoothing step increases from 17 to 19. The

memory requirement for the quantities defined on the coarse grid is about (1/4+1/16+..)=1/3 of the memory requirement on the finest grid.

We will not dwell upon the existence and stability of the ILU-decomposition just described. Meijerink and van der Vorst (1977) have shown existence if A is a symmetric M-matrix, i.e. $a_{ij}=a_{ji}$, $a_{ij}<0$ for i=j and $A^{-1}>0$. They also note that one has existence under much more general circumstances. We have found experimentally, that the ILU-decomposition exists and provides an efficient smoothing process, if in the non-self-adjoint case a sufficient amount of artificial viscosity is introduced on the finest grid. This is also necessary for all other smoothing processes that we know of.

In order to make Fourier methods applicable, in smoothing (and two-level) analysis it must be assumed that the values of the elements of L and U do not vary along diagonals; if (3.1) is used this is usually not the case near boundaries, and in certain strongly anisotropic diffusion problems the influence of the boundaries on the ILU-decomposition extends inwards over many meshes. In such cases smoothing and two-level analysis are not realistic for the present method MGD1.

4. Prolongation and restriction.

Let the value of the grid-function u^k in the point $(s.2^{-k},t.2^{-k})$ be denoted by u^k_{st}. Prolongation and restriction are defined by:

$$(p^k u^{k-1})_{2s,2t} = u^{k-1}_{st}, \qquad (p^k u^{k-1})_{2s+1,2t} = \frac{1}{2}(u^{k-1}_{st} + u^{k-1}_{s+1,t}),$$
$$(p^k u^{k-1})_{2s,2t+1} = \frac{1}{2}(u^{k-1}_{st} + u^{k-1}_{s,t+1}), \qquad (p^k u^{k-1})_{2s+1,2t+1} = \frac{1}{2}(u^{k-1}_{st} + u^{k-1}_{s+1,t+1}). \tag{4.1}$$

$$(r^k u^k)_{st} = u^k_{2s,2t} + \frac{1}{2}(u^k_{2s+1,2t} + u^k_{2s,2t+1}$$
$$+ u^k_{2s-1,2t} + u^k_{2s,2t-1} + u^k_{2s+1,2t-1} + u^k_{2s-1,2t+1}). \tag{4.2}$$

The following diagrams may clarify the structure of p^k and r^k.

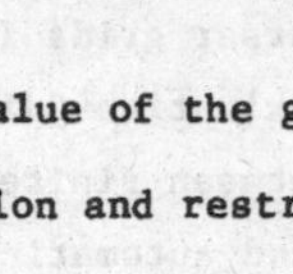
p^k

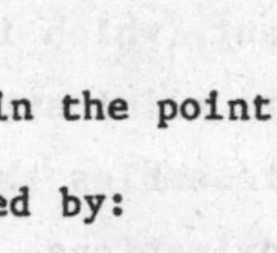
r^k

This p^k provides linear interpolation, while having a sparser matrix representation than all other linear interpolation operators, and r^k is its adjoint, in the sense that

$$(p^k u^{k-1}, v^k)_k = (u^{k-1}, r^k v^k)_{k-1}, \ \forall\, v^k \in U^k, \tag{4.3}$$

with $(u^k, v^k)_k \equiv \sum_{\Omega^k} u^k_{ij} v^k_{ij}$. Because r^k is a weighted average of 7 points, we call this 7-point prolongation and restriction.

5. Galerkin coarse grid approximation.

The coarse grid operators A^k are defined as follows:

$$A^{k-1} = r^k A^k p^k , \quad k = \ell(-1)2. \tag{5.1}$$

We call this a Galerkin approximation because (5.1) implies:

$$(A^k p^k u^{k-1}, p^k v^{k-1})_k = (A^{k-1} u^{k-1}, v^{k-1}) , \quad \forall v^{k-1} \in U^{k-1} , \tag{5.2}$$

if (4.3) holds; hence, we have a case of projection in a lower-dimensional subspace.

A more obvious way to generate A^{k-1} is to use a finite difference method. Under (5.1), if r^k and p^k are as in section 4, standard central equidistant finite difference approximations of the operators $\partial^2/\partial x^2$, $\partial^2/\partial y^2$, $\partial^2/\partial x \partial y$ using the 7-point difference molecule of section 3 are invariant, hence Galerkin and finite difference approximation are identical. When lower derivatives occur, or the coefficients are variable, or the mesh non-equidistant (which is the case on coarser grids if on the finest grid Dirichlet boundaries are eliminated), the two approximations differ. Their mutual relationship closely resembles the relationship between finite difference and finite element approximations. The Galerkin method automatically generates accurate approximations and takes care of special circumstances, such as changing mesh-size or varying coefficients, but the numerical analyst can always achieve the same accuracy with an ably designed finite difference approximation. The transformation of upwind differences by (5.1) is interesting; the difference molecules on the finest and five coarser grids are given below:

$$\begin{array}{ccc} 0 & 0 & \\ -1 & 1 & 0 \\ & 0 & 0 \end{array} \qquad \begin{array}{ccc} -1 & 1 & \\ -5 & 4 & 1 \\ & -1 & 1 \end{array} (*2^{-2}) \qquad \begin{array}{ccc} -5 & 5 & \\ -15 & 8 & 7 \\ & -5 & 5 \end{array} (*2^{-3}) \qquad \begin{array}{ccc} -21 & 21 & \\ -51 & 16 & 35 \\ & -21 & 21 \end{array} (*2^{-4})$$

$$\begin{array}{ccc} -85 & 85 & \\ -187 & 32 & 155 \\ & -85 & 85 \end{array} (*2^{-5}) \qquad \begin{array}{ccc} -341 & 341 & \\ -715 & 64 & 651 \\ & -341 & 341 \end{array} (*2^{-6})$$

For the derivation of these molecules, the formulae given by Mol (1981) have been used. Apparently, upwind differencing is gradually replaced by central differencing, plus a higher order truncation error containing a mixed third derivative. Diagonal dominance is lost, but in practice we have never encountered numerical "wiggles" or instability of the ILU-decomposition; note that as the grid gets coarser, the ILU-decomposition becomes more exact. Later, succesfull experiments with the upwind-discretized convection-diffusion equation will be reported.

One way of programming (5.1) is as follows. It is based on a datastructure used earlier by Frederickson (1975). In this section, Greek subscripts are 2-tuples identifying points of the computational grid, e.g. $\alpha=(\alpha_1,\alpha_2)$ indicates the grid-point with indices (α_1,α_2) (cf. the ordering introduced in section 3). To the atoms of the difference molecule 2-tuples are assigned according to the accompanying diagram; these are also identified by Greek subscripts.

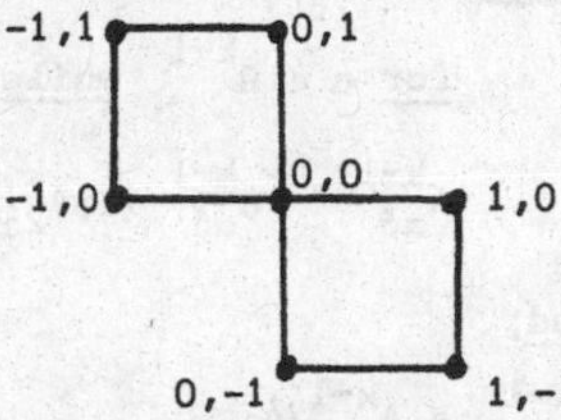

By $A^k_{\alpha\beta}$ we denote the element of the matrix A^k in row number $1+\alpha_1+m\alpha_2$ and column $1+\alpha_1+m\alpha_2+\beta_1+m\beta_2$, with m the number of grid-points of Ω^k in the x-direction. For example, $\beta=(1,-1)$ corresponds with the b-diagonal of section 3. If α is outside Ω^k or β is outside the molecule then $A^k_{\alpha\beta}$ is defined to be zero.

With these conventions, matrix-vector multiplication can be formulated as follows:

$$(A^k u^k)_\alpha = \sum_\beta A^k_{\alpha\beta}\, u^k_{\alpha+\beta}, \tag{5.3}$$

with range $\beta = Z \times Z$. Restriction and prolongation can be represented as follows:

$$(r^k u^k)_\alpha = \frac{1}{2} \sum_\beta \mu_\beta u^k_{2\alpha+\beta}\,, \tag{5.4}$$

$$(p^k u^k)_\alpha = \frac{1}{2} \sum_\beta \mu_{\alpha-2\beta} u^{k-1}_\beta\,, \tag{5.5}$$

with the weight factors μ_β defined by (cf. (4.1) and (4.2)): if β is inside the molecule, then $\mu_\beta=1$, except $\mu_{0,0}=2$; outside the molecule, $\mu_\beta=0$. Eq. (5.5) is not a convenient way to compute $p^k u^k$, but it can be used to derive a useful formula for $r^k A^k p^k$. From (5.3)-(5.5) it follows that for any u^{k-1}:

$$(r^k A^k p^k u^{k-1})_\alpha = \frac{1}{2} \sum_\beta \mu_\beta \sum_\gamma A^k_{2\alpha+\beta,\gamma} \frac{1}{2} \sum_\delta \mu_{2\alpha+\beta+\gamma-2\delta}\, u^{k-1}_\delta \tag{5.6}$$

Since the range of β, γ and δ may be taken to be all of $Z \times Z$, change of variables is easy. Let $\delta'=\alpha+\delta$, $\gamma'=\beta+\gamma-2\delta'$, then (5.6) takes on a form from which we may conclude (omitting primes):

$$(r^k A^k p^k)_{\alpha\delta} = \frac{1}{4} \sum_{\beta,\gamma} \mu_\beta \mu_\gamma A^k_{2\alpha+\beta,2\delta+\gamma-\beta} \tag{5.7}$$

It is found that if A^k has a general 7 (or fewer)-point structure, then A^{k-1} has a 7-point structure. An important point is that $2\delta+\gamma-\beta$ is only 95(61) times inside the molecule of A^k if A^k is a 7-(5-)point molecule, for all 7^3=343 possible combinations of β,γ and δ. This is exploited by putting the α-loop inside the δ-, β- and γ-loops, and arrange the computation as follows:

A^{k-1} := 0;
for $\delta \in$ molecule **do**
for $\beta \in$ molecule **do**
for $\gamma \in$ molecule **while** $2\delta+\gamma-\beta \in$ molecule **do**
begin $\mu = \mu_\beta \mu_\gamma$;
  for $\alpha \in \Omega^{k-1}$ **while** $2\alpha+\beta \in \Omega^k$ **do**
  $A^{k-1}_{\alpha\delta} := A^{k-1}_{\alpha\delta} + \mu A^k_{2\alpha+\beta,2\delta+\gamma-\beta}$
end;
$A^{k-1} := A^{k-1}/4$;

The cost of the inner loop is 2 operations, hence the total cost is 2*95(2*61) for a 7-(5-) point operator A^k. The division by 16 adds one operation, so that our final conclusion is, that the construction of A^{k-1} takes 191 (123) operations per grid-point of Ω^{k-1} for a 7-(5-) point operator A^k. The total work for the construction of A^k, k=ℓ-1(-1)1 in operations per grid-point of Ω^ℓ is about 191/3≈64(7-point A^ℓ), or 123/4+191/12≈47(5-point A^ℓ). This has to be done only once, before the multigrid iterations start. It depends on the complexity of A^ℓ whether computation of A^k, k < ℓ is cheaper with the finite difference method or with the above Galerkin method. But more important is the fact, that the Galerkin method enables us to work only with the given matrix A^ℓ and not to refer to the underlying problem (equation and boundary conditions), and that always good coarse grid approximations are obtained automatically.

6. Numerical experiments

Standardized test-problems are useful for demonstrating and comparing the applicability and performance of multigrid methods. If the coefficients are constant, smoothing analysis (cf. Brandt (1977)) can help to understand and construct efficient smoothing processes. Two-level analysis (cf. Brandt and Dinar (1979), Foerster et al. (1981), Ries et al. (1981)) realistically predicts the rate of convergence of certain (not all, for example not for the sawtooth cycle described previously) multigrid methods; for a certain method applied to the Poisson equation Braess (1981) has given a rigorous prediction. Such analyses use Fourier methods; see Hemker (1980b) for an introduction to the Fourier analysis of multigrid methods. Where Fourier analysis is not applicable, one can, given sufficient computer time, compute the spectral norm and spectral radius (i.e. the asymptotic rate of convergence) numerically, or just observe the rate of convergence and use heuristic arguments in order to verify the soundness and efficiency of a multigrid method. In practice, quite often the observed rate of converge of a good multigrid method is considerably better than the asymptotic rate, which is not reached because already after a few iterations discretization or even machine accuracy is obtained.

The test problems should be standardized, so that results reported by different authors can be easily compared and reproduced. The fact that constant coeffient test problems have a very special type of spectrum (cf. Curtis (1981)) makes them somewhat exceptional; therefore test problems with variable coefficients should also be included. Unless one wishes to design a method especially for a specific problem the special properties of a given test problem, such as for example a coefficient being constant, should not be exploited, and not be taken into account in operation counts.

We have looked for suitable test problems that have already been treated by other authors, and will report results for the following problems:

(i) $\phi_{xx} + \phi_{yy} = 4$, (6.1)

(ii) $\phi_{xx} + 0.01\,\phi_{yy} = 2.02$, (6.2)

(iii) $0.01\,\phi_{xx} + \phi_{yy} = 2.02$, (6.3)

(iv) $\phi_{xx} + 1.7\,\phi_{xy} + \phi_{yy} = 4$, (6.4)

(v) $u\phi_x + v\phi_y = 0.001(\phi_{xx}+\phi_{yy}) - 1$, (6.5)

$(u,v) = (1,0),\ (0,1),\ (1,1),\ (1,-1),$
(a) (b) (c) (d)

(vi) $(a\phi_x)_x + (a\phi_y)_y = 0, \quad a = |\sin kx \sin ky|$. (6.6)

The region is $\Omega = (0,1) \times (0,1)$. The computational grid is equidistant. For problems (i) - (iv) the boundary condition is $\phi|_{\partial\Omega} = x^2+y^2$, exact solution: $\phi = x^2+y^2$. For (v) and (vi), $\phi|_{\partial\Omega} = 0$, exact solution not known for (v), zero for (vi). Lack of time prevented us from including cases with Neumann boundary conditions, but this should certainly be included, because the rate of convergence, of some multigrid methods (but not MGD1, except for (iii)) is affected by the type of boundary condition. Standard 5-point central differencing is used on the finest grid except for (v), where upwind differencing according to Il'in (1969) is employed. There is no particular reason for using Il'in discretization in the present context; any other form of upwind discretization would lead to roughly the same results (this is an example where Galerkin coarse grid discretization is cheaper than finite differences, because of the cost of the Il'in coefficients). The initial guess of the solution is $\phi \equiv 0$, except for (vi), where the initial ϕ is uniformly randomly distributed. The boundary conditions are not eliminated from the equations.

Besides being important for many applications in its own right (but fast solvers of Fourier analysis/cyclic reduction type (see e.g. Schumann (1980)) exist already for some time, and are competitive with multigrid methods) problem (i) typifies self-adjoint elliptic equations with smoothly varying coefficients. Problems (ii) and (iii) represent anisotropic diffusion problems or problems with strong coordinate stretching in one direction. Although mathematically almost identical, the rate of convergence of some methods (including MGD1) may differ significantly between (ii) and (iii). In (iv) we have a mixed derivative, which does not occur often in mathematical physics, but which we include for completeness. Non-orthogonal coordinate transformations give rise to mixed derivatives. The convection-diffusion problem (v) represents a singular perturbation problem of a type that is ubiquitous in fluid dynamics. Finally, (vi) represents problems with slowly or rapidly but continuously varying coefficients.

Because some or even all of the specific difficulties represented by these test problems may occur simultaneously in a given application, a method should perform well for more than one problem. The method MGD1 described here performs efficiently for all six test problems, although for (iii) it is somewhat less efficient than for the other five. For fluid mechanical applications (i) and (v) should be mastered, for reservoir engineering the method should be able to handle all problems, and in addition the more difficult test problems of Stone and Kershaw (see Kettler and Meijerink (1981), cf. Kettler (1982)), in which the coefficients are discontinuous, and the region non-rectangular. The inclusion of a Navier-Stokes test problem seems desirable; see Wesseling and Sonneveld (1980) for some results.

In the following table we list publications that give results for the test-problems above. In some cases, the right-hand-side and boundary conditions may be different.

Test-problem	(i)	(ii)	(iii)	(iv)	(v)	(vi)
Brandt (1977)	*					
Hackbusch (1978b)	*,Δ	*		*		
Nicolaides (1979)	*,Δ					*
W,S,M	*	*	*	*	*	*
F,R,B,S	*,Δ	*	*		∇	

Table 6.1 Publications on test-problems (i) - (vi)

Publications that consider related methods are grouped together. W,S,M stands for Wesseling and Sonneveld (1980), Wesseling (1980), Mol (1981); F,R,B,S for Foerster et al. (1981), Ries et al. (1981), Börgers (1981), Stüben et al. (1982). The bulk of the results for MGD1 quoted here are taken from W,S,M. In these publications small inconsequential differences occur due to the fact, that M eliminates the boundary conditions, whereas W does not. Where the symbols vary along a row in table 6.1, different algorithms were used for different test-problems, otherwise the same algorithm was used. Where no entry occurs, this means that we have no results at our disposal; the method may or may not be applicable. Some publications treat other test-problems, not discussed here, as well.

The following table gives results for method MGD1.

	(i)	(ii)	(iii)	(iv)
M,ℓ	8,6	10,6	4,4	7,6
ρ,t	0.033, 20	0.15, 36	0.0016, 11	0.025, 19
	(va)	(vb)	(vc)	(vd)
M,ℓ	3,4	2,4	1,4	4,4
ρ,t	0.0030, 12	$7*10^{-5}$, 7	$3*10^{-9}$, 4	0.040, 21

Table 6.2 Method MGD1 applied to test-problems (i) - (v).

In table 6.2, M is the number of iterations that were carried out; ℓ determines the number of grid-points $(2^{\ell}+1)*(2^{\ell}+1)$ of the finest grid; ρ is the average reduction factor, defined by: ρ^M=quotient of Euclidean norms of residues Ax-f before and after M iterations; $t=-30/{}^{10}\log\rho$ is the number of operations per grid-point of the finest grid for 0.1 reduction of the residual. For (v) the boundary conditions were eliminated; this is of no consequence.

Of course, ρ depends on M and on the initial guess, hence, ρ is afflicted with a certain arbitrariness, which the spectral norm and radius lack. For (i) it has been determined numerically, that the spectral radius ρ_∞=0.090 (t=29).

Clearly, for (ii) the results are worse than for the other cases. For this case the smoothing factor of ILU goes to 1 as ε→0, and with Neumann boundary conditions along x=0, x=1 MGD1 does not work (in that case the differential problem (6.2) is badly posed). Kettler (1982) gives and ILU-decomposition which does not suffer from this defect. With Dirichlet boundary conditions, however, MGD1 seems to be dependable for (ii), as is suggested by the following results of more extensive experiments with (ii); ρ_∞ is the spectral radius.

ℓ\ε	0.5	10^{-1}	10^{-2}	10^{-4}
3	.038	.11	.042	.001
4	.091	.22	.19	.003
5	.10	.26	.41	.017
6	.10	.27	.55	.068

Table 6.3 Estimated ρ_∞ for $\phi_{xx}+\varepsilon\phi_{yy} = 2+2\varepsilon$, $\phi\big|_{\partial\Omega} = x2+y2$.

Comparison with the work of Hackbusch (1978) is relatively straightforward, because his method is similar to MGD1, the main differences being the use of two checkerboard (CH) or zebra-Gauss-Seidel relaxations (Z) for smoothing, and the use of 9-point prolongation and restriction. The main difference in computational work will be due to the difference in smoothing strategy. Assuming variable coefficients, two applications of CH take 18 operations per grid-point. Solution of a tri-diagonal system takes 8 operations, or 5 if the LU-decompositions are stored at a cost of 2 reals per grid-point. Including the cost of the right-hand-sides we arrive at a cost of two applications of Z of 24, or 18 with extra storage, except for test-problem (iv), where the cost is 32 or 26. CH is not applicable to (ii), (iii), (iv). Recalling that the cost of ILU is 17 and the total cost of MGD1 is 30, we estimate the total cost of the methods used by Hackbusch to be 32 with CH, 42 with Z or 32 with extra storage, and 56 or 46 for test-problem (iv). From Hackbusch (1978b) we then deduce the following table.

	(i), CH	(i), Z	(ii), Z	(iv), Z
M,ℓ	6,6	8,6	8,6	8,6
ρ,t	.048,24	.038,30(23)	.063,35(27)	.199,60(46)

Table 6.4 Computational cost of methods of Hackbusch (1978b).
CH: checkerboard-Gauss-Seidel; Z: zebra-Gauss-Seidel.
Between brackets: with extra storage.

Z is not applicable to (iii), because the lines are chosen in the x-direction. Also including y-lines would perhaps change the efficiency for (i) and (iv) little, but almost double the cost for (ii) and (iii). Because MGD1 and the methods of Hackbusch have much in common this is mainly a practical comparison of Gauss-Seidel and ILU smoothing methods.

The results of Brandt (1977) will not be discussed, because this early work has been extended by F,R,B,S, who give results of two-level analysis for a variety of combinations of restrictions, prolongations and smoothing processes. In the so-called W-cycle (with double the two-level cost) two-level analysis usually gives a good estimate for ρ_∞. For (i) the best method (using CH smoothing and total-reduction concepts) in Ries et al. (1981) results in: ρ_∞=0.074, cost=23.5, t=21. In the cost estimate the special values of the Poisson coefficients are exploited. The best method in Stüben et al. (1982) (using alternating direction Z smoothing) has for (i), (ii) and (iii):

ρ_∞=0.023, cost=51.6, t=31.5, (i),
ρ_∞=0.119, cost=51.6, t=55.8, (ii) and (iii),

with the cost estimate valid for variable coefficients. For (v) a detailed analysis is included by Börgers (1981); for the best method (using CH smoothing) for (v) with $\varepsilon=10^{-5}$ on a 65×65 grid the following results emerge:

$\rho=0.41$, cost=28, t=72, (va),

$\rho=0.37$, cost=28, t=65, (vc),

with ρ the average reduction factor over the last 14 of 20 iterations. When comparing with table 6.2 one has to keep in mind, that table 6.2 gives only the initial rate of convergence, observed during the first few iterations. For (va,b,c) this is irrelevant, because here MGD1 is almost exact. For comparison we give the following numerical estimate for (vd) with $\varepsilon=10^{-5}$ on a 65×65 grid: $\rho_\infty=0.29$, t=56. This concludes our comparison with F,R,S,B.

Nicolaides (1979) reports experiments with two multigrid-finite-element methods. One of these, using "linear elements", results for (i) in the same system of equations on the finest grid that we are considering here. Prolongation, restriction and coarse grid approximation are much the same as in MGD1, but smoothing consists of a few Gauss-Seidel relaxations before and after coarse grid correction. CPU-time measurements are given in units of a Gauss-Seidel relaxation, for which we take a cost of 9 (assuming variable coefficients). On a 64*64 grid t=35.1 (cf. table 6.2) is reported, with a smooth initial guess, for (i) with right-hand-side zero. For (vi) Nicolaides (1979) reports experiments with bilinear elements only corresponding to a 9-point discretization. Here the cost of one relaxation is 17 (for variable coefficients). The following results are obtained.

k	2	4	8	16	32
N	76	82	99	107	104
MGD1	25	25	25	26	26

<u>Table 6.5</u> Values of t for test-problem (vi)

Three multigrid iterations were carried out on a 65×65 grid; k is the coefficient in (6.6), t as in table 6.2, N stands for Nicolaides. The initial guess is uniformly randomly distributed. Because in this case there are singificant differences between coarse-grid Galerkin and difference approximation, we have also tried MGD1 with finite differences; this is denoted as MGD1*.
On a 33×33 grid we have estimated the spectral radius ρ_∞, and obtain the following result.

k	8	16	32
MGD1	.31	.18	.13
MGD1*	.30	.30	>1

<u>Table 6.4</u> Values of ρ_∞ for test-problem (iv).

As is to be expected, with coarse-grid finite difference approximation the method deteriorates as the rate of variation of the coefficients in the differential equation increases. This can probably be remedied by taking a suitable average of the coefficients when constructing coarse-grid finite difference approximations. The Galerkin approximation used in MGD1 does this automatically.

7. Final remarks.

A multigrid method (MGD1) has been presented, that is perceived by the user as any other linear systems solver, and requires no insight in the properties of multigrid methods. The user has to specify the matrix and the right-hand-side only. This is made possible by using Galerkin coarse-grid approximations and a fixed multigrid strategy, the so-called sawtooth cycle. The matrix should represent a 5- or a 7-point discretization of an elliptic equation on a rectangle in the usual way.

In cases with rapidly varying coefficients, the Galerkin method has the additional advantage of providing better coarse-grid approximations than straightforward finite differences.

MGD1 works well for a large variety of problems, including non-self-adjoint singularly perturbed equations, equations with a mixed derivative, and strongly anisotropic diffusion problems, except for a certain combination of anisotropic direction and Neumann boundary conditions. This restriction is removed, and the method is generalized to more general domains and discontinuous coefficients by Kettler and Meijerink (1981), cf. Kettler (1982).

Comparison with available results for other methods shows, that the robustness of MGD1 and the abandoning of any form of adaptivity is not paid for in terms of efficiency. In fact, at the moment MGD1 seems generally to be somewhat faster than other, more specialized methods, except when these exploit special values of the coefficients.

In the near future even more efficient methods may evolve, because it seems unlikely that the full potential of the various methods is already fully exhausted. The combination of conjugate gradient and multigrid methods (Kettler and Meijerink (1981)) is very promising, cf. Kettler (1982). The structure of multigrid methods is very rich. Various elements of the method can be chosen in many ways; smoothing process, prolongation, restriction, coarse-grid approximation, multigrid cycle. One can speak of a multigrid "philosophy", or perhaps better of a multigrid methodology. No single "best" or "standard" method is likely to emerge, just as in other areas of numerical mathematics.

References

G.P. Astrachancev, An iterative method of solving elliptic net problems. USSR Comp. Math. Math. Phys. 11, 2, 171-182, 1971.

N.S. Bakhvalov, On the convergence of a relaxation method with natural constraints on the elliptic operator, USSR Comp. Math. Math. Phys. 6, no.5, 101-135, 1966.

C. Börgers, Private communication: Mehrgitterverfahren für eine Mehrstellendiskretisierung der Poisson-Gleichung und für eine zweidimensionale singular gestörte Aufgabe. Diplomarbeit (prof. Trottenberg), 1981.

D. Braess, The contraction number of a multigrid method for solving the Poisson equation. Num. Math. 37, 387-404, 1981.

A. Brandt, Multi-level adaptive technique (MLAT) for fast numerical solution to boundary value problems. In: Proc. Third Int. Conf. Num. Meth. Fluid Dyn., Paris, 1972, Lect. Notes in Phys. 18, 82-89, Springer-Verlag, Berlin etc. 1973.

A. Brandt, Multi-level adaptive solutions to boundary-value problems, Math. Comp. 31 333-390, 1977.

A. Brandt, Multi-level adaptive solutions to singular perturbation problems. In: Numerical analysis of singular perturbation problems (P.W. Hemker, J.J. Miller, eds.), 53-142, Academic Press, New York, 1979.

A. Brandt and N. Dinar: Multi-grid solution to elliptic flow problems. In: Numerical methods for partial differential equations (S.V. Parter, ed.), 53-149. New York etc., Academic Press, 1979.

A.R. Curtis, On a property of some test equations for finite difference or finite element methods. IMA J. Num. Anal. 1, 369-375, 1981.

R.P. Fedorenko, A relaxation method for solving elliptic difference equations, USSR Comp. Math. Math. Phys. 1, 1092-1096, 1962.

H. Foerster, K. Stüben, U. Trottenberg, Non-standard multigrid techniques using checkered relaxation and intermediate grids. In: Elliptic problem solvers (M. Schulz, ed.), 285-300, Academic Press, New York etc., 1981.

P.O. Frederickson, Fast approximate inversion of large sparse linear systems, Mathematics report 7-75, Lakehead University, Thunder Bay, Canada, 1975.

I. Gustafsson, A class of first order factorization methods. BIT 18, 142-156, 1978.

W. Hackbusch, A fast iterative method for solving Poisson's equation in a general region. In: Numerical treatment of differential equations, Oberwolfach 1976 (R. Bulirsch , R.D. Grigorieff, J. Schröder, eds.). Lecture Notes in Math. 631, Springer-Verlag, Berlin etc., 1978a.

W. Hackbusch, On the multigrid method applied to difference equations, Computing 20 pp. 291-306, 1978b.

W. Hackbusch, On the convergence of multi-grid iterations. Beiträge zur Numerischen Mathematik 9, 213-239, 1981.

P.W. Hemker, The incomplete LU-decomposition as a relaxation method in multi-grid algorithms. In: Boundary and interior layers - computational and asymptotic methods, Proceedings, Dublin 1980, (J.J.H. Miller, ed.), Boole-Press, Dublin 1980a.

P.W. Hemker, Fourier analysis of gridfunctions, prolongations and restrictions. Mathematical Centre, 413 Kruislaan, Amsterdam, Report NW 93/80, 1980b.

P.W Hemker, Introduction to multigrid methods, Nieuw Archief v. Wiskunde 29, 71-101, 1981.

A.M. Il'in, Differencing scheme for a differential equation with a small parameter affecting the highest derivative. Math. Notes Acad. Sc. USSR 6, 596-602, 1969.

R. Kettler, in the present volume (1982).

R. Kettler and J.A. Meijerink, A multigrid method and a combined multigrid-conjugate gradient method for elliptic problems with strongly discontinuous coefficients in general domains. KSEPL, Volmerlaan 6, Rijswijk, The Netherlands, Publication 604, 1981.

J.A. Meijerink and H.A. van der Vorst, An iterative solution method for linear systems of which the coefficient matrix is a symmetric M-matrix, Math. Comp. 31 pp. 148-162, 1977.

W.J.A. Mol, On the choice of suitable operators and parameters in multigrid methods, Report NW 107/81, Mathematical Centre, Amsterdam, 1981.

R.A. Nicolaides, On some theoretical and practical aspects of multigrid methods. Math. Comp. 33, pp. 933-952, 1979.

M. Ries, U. Trottenberg, G. Winter, A note on MGR methods. Universität Bonn, preprint no. 461, 1981.

U. Schumann, Fast elliptic solvers and their application in fluid dynamics. In: W. Kollmann (ed.), Computational fluid dynamics, Hemisphere, Washington etc., 1980.

K. Stüben, C.A. Thole, U. Trottenberg, private communication, 1982.

H.A. van der Vorst, Iterative solution methods for certain sparse linear systems with a non-symmetric matrix arising from PDE-problems. J. Comp. Phys. 44, 1-20, 1981.

E.L. Wachspress, A rational finite element basis, chapter 10. Academic Press, New York, 1975.

P. Wesseling, Numerical solution of the stationary Navier-Stokes equations by means of a multiple grid method and Newton iteration. Report NA-18, Delft University of Technology, 1977.

P. Wesseling, Theoretical and practical aspects of a multigrid method, Report NA-37, Delft University of Technology, 1980.

P. Wesseling and P. Sonneveld, Numerical experiments with a multiple grid and a preconditioned Lanczos type method. In: Approximation methods for Navier-Stokes problems, Proceedings, Paderborn 1979 (R. Rautmann, ed.), Lecture Notes in Math. 771, 543-562, Springer-Verlag, Berlin etc. 1980.

MULTIGRID BIBLIOGRAPHY

Kurt Brand
GMD - IMA, Postfach 1240
D - 5205 St. Augustin 1
F. R. Germany

This bibliography is under permanent compilation at the Institut für Mathematik of the Gesellschaft für Mathematik und Datenverarbeitung (GMD - IMA), St. Augustin / F. R. Germany. A first version of the bibliography is due to Heinz Reutersberg. The intention is to collect all papers dealing with multigrid methods. In doing so, there are some insufficiencies, more or less inevitable:

What is an adequate definition of the multigrid principle? Error smoothing processes and the transfer between a sequence of grids should be involved. Which related approaches should be taken into account? We decided not to include, e.g., papers only dealing with reduction type or defect correction methods. For some multigrid - related papers see the supplement.

We try to include new papers as fast as possible. Therefore we have some preprints and technical reports (sometimes preliminary ones). This can result in bibliographical problems when the paper appears later on (and maybe slightly changed or even under a different title) in some journal.

Of course, this bibliography is by no means complete or free from errors. For further updating, readers are kindly requested to send any corrections or additions to the author.

Acknowledgments: I would like to thank Reinhild Schwarz and Horst Schwichtenberg for a lot of technical work in compiling and automatically processing the entries of this bibliography. I am also indebted to Hartmut Foerster for fruitful discussions concerning form and contents and to Ulrich Trottenberg for creating a relaxed (and sometimes rollicking) working atmosphere.

[1] Agarwal, R.K.:
Unigrid and multigrid algorithms for the solution of coupled, partial-differential equations using fourth-order-accurate compact differencing. Report, Symposium on Numerical Boundary Condition Procedures and Multigrid Methods. NASA-Ames Research Center, Moffett Field, CA, 1981.

[2] Alcouffe, R.E.; Brandt, A.; Dendy, J.E. (Jr.); Painter, J.W.:
The multi-grid methods for the diffusion equation with strongly discontinuous coefficients. SIAM J. Sci. Stat. Comput., 2, pp. 430-454, 1981.

[3] Allgower, E.L.; Böhmer, K.; McCormick, S.F.:
Discrete correction methods for operator equations. Numerical Solution of Nonlinear Equations. Proceedings, Bremen 1980 (E.L. Allgower, K. Glashoff, H.O. Peitgen, eds.). Lecture Notes in Mathematics, 878, pp. 30-97. Springer-Verlag, Berlin, 1981.

[4] Arlinger, B.:
Multigrid technique applied to lifting transonic flow using full potential equation. Report L-0-1 B439, Saab-Scania, 1978.

Askar, A.:
cf. [74].

[5] Asselt, E.J. van:
Application of the Osher-Engquist difference scheme and the full multi-grid method to a two dimensional nonlinear elliptic model equation. Preprint NW 103/81, Dept. of Numerical Mathematics, Mathematical Centre, Amsterdam, 1981.

[6] Asselt, E.J. van:
The multi grid method and artificial viscosity. Multigrid Methods. Proceedings of the Conference Held at Köln-Porz, November 23-27, 1981 (W. Hackbusch, U. Trottenberg, eds.). Lecture Notes in Mathematics. Springer-Verlag, Berlin, 1982.

[7] Astrakhantsev, G.P.:
An iterative method of solving elliptic net problems. U.S.S.R. Computational Math. and Math. Phys., 11 no. 2, pp. 171-182, 1971.

[8] Astrakhantsev, G.P.; Rukhovets, L.A.:
A relaxation method in a sequence of grids for elliptic equations with natural boundary condition. Z. Vycisl. Mat. i Mat. Fiz., 21 no. 4, pp. 926-944, 1981.

[9] Auzinger, W.; Stetter, H.J.:
Defect correction and multigrid iterations. Multigrid Methods. Proceedings of the Conference Held at Köln-Porz, November 23-27, 1981 (W. Hackbusch, U. Trottenberg, eds.). Lecture Notes in Mathematics. Springer-Verlag, Berlin, 1982.

[10] Axelsson, O.:
On multigrid methods of the two-level type. Multigrid Methods. Proceedings of the Conference Held at Köln-Porz, November 23-27, 1981 (W. Hackbusch, U. Trottenberg, eds.). Lecture Notes in Mathematics. Springer-Verlag, Berlin, 1982.

[11] Axelsson, O.; Gustafsson, I.:
Preconditioning and two-level multigrid methods of arbitrary degree of approximation. Report 8120, Mathematisch Instituut, Katholieke Universiteit, Nijmegen, 1981.

[12] Bakhvalov, N.S.:
On the convergence of a relaxation method with natural constraints on the elliptic operator. U.S.S.R. Computational Math. and Math. Phys., 6 no. 5, pp. 101-135, 1966.

[13] Bank, R.E.:
A comparison of two multi-level iterative methods for nonsymmetric and indefinite elliptic finite element equations. SIAM J. Numer. Anal., 18, pp. 724-743, 1981.

[14] Bank, R.E.:
A multi-level iterative method for nonlinear elliptic equations. Elliptic Problem Solvers (M.H. Schultz, ed.), pp. 1-16. Academic Press, New York, NY, 1981.

[15] Bank, R.E.:
Analysis of a multi-level inverse iteration procedure for eigenvalue problems. Research Report, Comp. Sci. Dept., Yale University, New Haven, CT, 1981.

[16] Bank, R.E.; Dupont, T.F.:
Analysis of a two-level scheme for solving finite element equations. Report CNA-159, Center for Numerical Analysis, University of Texas at Austin, 1980.

[17] Bank, R.E.; Dupont, T.F.:
An optimal order process for solving finite element equations. Math. Comp., 36, pp.35-51, 1981.

[18] Bank, R.E.; Rose, D.J.:
Analysis of a multi-level iterative method for nonlinear finite element equations. Technical Report, Dept. of Mathematics, Yale University, New Haven, CT, 1981.

[19] Bank, R.E.; Sherman, A.H.:
A comparison of smoothing iterations for multi-level methods. Advances in Computer Methods for Partial Differential Equations III (R. Vichnevetsky, R.S. Stepleman, eds.), pp. 143-147. IMACS, New York, NY, 1979.

[20] Bank, R.E.; Sherman, A.H.:
Algorithmic aspects of the multi-level solution of finite element equations. Sparse Matrix Proceedings 1978 (I.S. Duff, G.W. Stewart, eds.), pp. 62-89. SIAM, Philadelphia, 1979.

[21] Bank, R.E.; Sherman, A.H.:
PLTMG user's guide - July 1979 version. Report CNA-152, Center for Numerical Analysis, University of Texas at Austin, 1979.

[22] Bank, R.E.; Sherman, A.H.:
The use of adaptive grid refinement for badly behaved elliptic partial differential equations. Math. Comput. Simulation XXII, pp. 18-24, 1980.

[23] Bank, R.E.; Sherman, A.H.:
An adaptive multi-level method for elliptic boundary value problems. Computing, 26, pp. 91-105, 1981.

[24] Becker, K.:
Mehrgitterverfahren zur Lösung der Helmholtz-Gleichung im Rechteck mit Neumannschen Randbedingungen. Diplomarbeit, Institut für Angewandte Mathematik, Universität Bonn, 1981.

Bercovier, M.:
cf. [195].

Böhmer, K.:
cf. [3].

[25] Börgers, C.:
Mehrgitterverfahren für eine Mehrstellendiskretisierung der Poissongleichung und für eine zweidimensionale singulär gestörte Aufgabe. Diplomarbeit, Institut für Angewandte Mathematik, Universität Bonn, 1981.

[26] Boerstoel, J.W.:
A multigrid algorithm for steady transonic potential flows around aerofoils using Newton iteration. Multigrid Methods. NASA Conference Publication 2202, Ames Research Center, pp. 151-172, Moffett Field, CA, 1981.

[27] Boerstoel, J.W.:
A fast-solver algorithm for steady transonic potential-flow computations with Newton iteration and multigrid relaxation. Proceedings of the Fourth GAMM-Conference on Numerical Methods in Fluid Mechanics (H. Viviand, ed.), pp. 21-41. Vieweg, Braunschweig, 1982.
[28] Braess, D.:
The contraction number of a multigrid method for solving the Poisson equation. Numer. Math., 37, pp. 387-404, 1981.
[29] Braess, D.:
The convergence rate of a multigrid method with Gauss-Seidel relaxation for the Poisson equation. Multigrid Methods. Proceedings of the Conference Held at Köln-Porz, November 23-27, 1981 (W. Hackbusch, U. Trottenberg, eds.). Lecture Notes in Mathematics. Springer-Verlag, Berlin, 1982.
[30] Braess, D.:
The convergence rate of a multigrid method with Gauss-Seidel relaxation for the Poisson equation (revised). Preprint, Institut für Angewandte Mathematik, Ruhr-Universität Bochum, 1982.
Branca, H.W.:
cf. [152].
[31] Brand, K., McCormick, S.F. (eds.):
Multigrid Newsletter (starting Jan., 1981). IMA, Gesellschaft für Mathematik und Datenverarbeitung, St. Augustin; Dept. of Mathematics, Colorado State University, Fort Collins, CO, 1981.
Brandt, A.:
cf. [2], [72], [190].
[32] Brandt, A.:
Multi-level adaptive technique (MLAT) for fast numerical solution to boundary value problems. Proceedings Third International Conference on Numerical Methods in Fluid Mechanics, Paris 1972 (H. Cabannes, R. Teman, eds.). Lecture Notes in Physics, 18, pp. 82-89. Springer-Verlag, Berlin, 1973.
[33] Brandt, A.:
Multi-level adaptive techniques (MLAT). I. The multi-grid method. Research Report RC 6026, IBM T.J. Watson Research Center, Yorktown Heights, NY, 1976.
[34] Brandt, A.:
Multi-level adaptive solutions to boundary-value problems. Math. Comp., 31, pp. 333-390, 1977.
[35] Brandt, A.:
Multi-level adaptive techniques (MLAT) for partial differential equations: ideas and software. Mathematical Software III. (J.R. Rice, ed.), pp. 277-318. Academic Press, New York, NY, 1977.
[36] Brandt, A.:
Multi-level adaptive finite-element methods. I. Variational problems. Special Topics of Applied Mathematics (J. Frehse, D. Pallaschke, U. Trottenberg, eds.), pp. 91-128. North-Holland Publishing Company, Amsterdam, 1979.
[37] Brandt, A.:
Multi-level adaptive techniques (MLAT) for singular-perturbation problems. Numerical Analysis of Singular Perturbation Problems (P.W. Hemker; J.J.H. Miller, eds.), pp. 53-142. Academic Press, London, 1979.
[38] Brandt, A.:
Multi-level adaptive computations in fluid dynamics. AIAA J., 18, pp. 1165-1172, 1980.

[39] Brandt, A.:
Numerical stability and fast solutions to boundary value problems. Boundary and Interior Layers - Computational and Asymptotic Methods (J.J.H. Miller, ed.), pp. 29-49. Boole Press, Dublin, 1980.

[40] Brandt, A.:
Stages in developing multigrid solutions. Numerical Methods for Engineering I (E. Absi, R. Glowinski, P. Lascaux, H. Veysseyre, eds.), pp. 23-45. Dunod, Paris, 1980.

[41] Brandt, A.:
Multigrid solutions to steady-state compressible Navier-Stokes equations. I. Preprint no. 492, Sonderforschungsbereich 72, Universität Bonn, 1981.

[42] Brandt, A.:
Multigrid solvers for non-elliptic and singular-perturbation steady-state problems. Research Report, Dept. of Applied Mathematics, Weizmann Institute of Science, Rehovot, 1981.

[43] Brandt, A.:
Multigrid solvers on parallel computers. Elliptic Problem Solvers (M.H. Schultz, ed.), pp. 39-84. Academic Press, New York, NY, 1981.

[44] Brandt, A.:
Guide to multigrid development. Multigrid Methods. Proceedings of the Conference Held at Köln-Porz, November 23-27, 1981 (W. Hackbusch, U. Trottenberg, eds.). Lecture Notes in Mathematics. Springer-Verlag, Berlin, 1982.

[45] Brandt, A.; Cryer, C.W.:
Multigrid algorithms for the solution of linear complementarity problems arising from free boundary problems. MRC Report no. 2131, Mathematics Research Center, University of Wisconsin, Madison, WI, 1980.

[46] Brandt, A.; Dinar, N.:
Multi-grid solutions to elliptic flow problems. Numerical Methods for Partial Differential Equations (S.V. Parter, ed.), pp 53-147. Academic Press, New York, NY, 1979.

[47] Brandt, A.; Ophir, D.:
Language for processes of numerical solutions to differential equations. Second annual report. U.S. Army Contract DAJA37-79-C-0504, 1981.

[48] Brandt, A.; Ta'asan, S.:
Multi-grid methods for highly oscillatory problems. Research Report, Dept. of Applied Mathematics, Weizmann Institute of Science, Rehovot, 1981.

[49] Brandt, A.; Dendy, J.E. (Jr.); Ruppel, H.:
The multigrid method for semi-implicit hydrodynamics codes. J. Comput. Phys., 34, pp. 348-370, 1980.

[50] Brandt, A.; McCormick, S.F.; Ruge, J.:
Multigrid methods for differential eigenproblems. Report, Dept. of Mathematics, Colorado State University, Ft. Collins, CO, 1981.

[51] Brown, J.J.:
A multigrid mesh-embedding technique for three-dimensional transonic potential flow analysis. Multigrid Methods. NASA Conference Publication 2202, Ames Research Center, pp. 131-150, Moffett Field, CA, 1981.

[52] Camarero, R.; Younis, M.:
Efficient generation of body-fitted coordinates for cascades using multigrid. AIAA J., 18 no. 5, pp. 487-488, 1980.

Caughey, D.A.:
cf. [188].

[53] Chan, T.F.C.; Keller, H.B.:
Arc-length continuation and multi-grid techniques for nonlinear elliptic eigenvalue problems. Technical Report no.197, Computer Science Dept., Yale University, New Haven, CT, 1981.

Clemm, D.S.:
cf. [76].

Cryer, C.W.:
cf. [45].

[54] Deconinck, H.; Hirsch, C.:
A multigrid method for the transonic full potential equation discretized with finite elements on an arbitrary body fitted mesh. Multigrid Methods. NASA Conference Publication 2202, Ames Research Center, pp. 61-82, Moffett Field, CA, 1981.

[55] Deconinck, H.; Hirsch, C.:
A multitgrid finite element method for the transonic potential equation. Multigrid Methods. Proceedings of the Conference Held at Köln-Porz, November 23-27, 1981 (W. Hackbusch, U. Trottenberg, eds.). Lecture Notes in Mathematics. Springer-Verlag, Berlin, 1982.

Dendy, J.E. (Jr.):
cf. [2], [49].

[56] Dendy, J.E. (Jr.):
Black box multigrid. Multigrid Methods. NASA Conference Publication 2202, Ames Research Center, pp. 249-274, Moffett Field, CA, 1981.

[57] Dendy, J.E. (Jr.); Hyman, J.M.:
Multi-grid and ICCG for problems with interfaces. Elliptic Problem Solvers (M.H. Schultz, ed.), pp. 247-253. Academic Press, New York, NY, 1981.

Dinar, N.:
cf. [46].

[58] Dinar, N.:
Fast methods for the numerical solution of boundary-value problems. PH.D. Thesis, Dept. of Applied Mathematics, Weizmann Institute of Science, Rehovot, 1978.

[59] Donovang, M.:
Defektkorrekturen nach Stetter und Pereyra und MG-Extrapolation nach Brandt: Beziehungen und Anwendung auf elliptische Randwertaufgaben. Diplomarbeit, Institut für Angewandte Mathematik, Universität Bonn, 1981.

Dupont, T.F.:
cf. [16], [17].

[60] Fedorenko, R.P.:
A relaxation method for solving elliptic difference equations. U.S.S.R. Computational Math. and Math. Phys., 1 no. 5, pp. 1092-1096, 1962.

[61] Fedorenko, R.P.:
The speed of convergence of an iterative process. U.S.S.R. Computational Math. and Math. Phys., 4 no. 3, pp. 227-235, 1964.

[62] Fix, G.J.; Gunzburger, M.D.:
On numerical methods for acoustic problems. Comput. Math. Appl., 6, pp. 265-278, 1980.

[63] Foerster, H.; Witsch, K.:
On efficient multigrid software for elliptic problems on rectangular domains. Math. Comput. Simulation XXIII, pp. 293-298, 1981.

[64] Foerster, H.; Witsch, K.:
Multigrid software for the solution of elliptic problems on rectangular domains: MG00 (Release 1). Multigrid Methods. Proceedings of the Conference Held at Köln-Porz, November 23-27, 1981 (W. Hackbusch, U. Trottenberg, eds.). Lecture Notes in Mathematics. Springer-Verlag, Berlin, 1982.

[65] Foerster, H.; Stüben, K.; Trottenberg, U.:
Non standard multigrid techniques using checkered relaxation and intermediate grids. Elliptic Problem Solvers (M.H. Schultz, ed.), pp. 285-300. Academic Press, New York, NY, 1981.

[66] Forester, C.K.:
Advantages of multi-grid methods for certifying the accuracy of PDE modeling. Multigrid Methods. NASA Conference Publication 2202, Ames Research Center, pp. 23-45, Moffett Field, CA, 1981.

Fray, J.M.J.:
cf. [170].

[67] Fuchs, L.:
A Newton-multi-grid method for the solution of nonlinear partial differential equations. Boundary and Interior Layers - Computational and Asymptotic Methods (J.J.H. Miller, ed.). Boole Press, Dublin, 1980.

[68] Fuchs, L.:
Multi-grid solution of the Navier-Stokes equations on non-uniform grids. Multigrid Methods. NASA Conference Publication 2202, Ames Research Center, pp. 84-101, Moffett Field, CA, 1981.

[69] Fuchs, L.:
Transonic flow computation by a multi-grid method. Numerical Methods for the Computation of Inviscid Transonic Flows with Shock Waves (A. Rizzi, H. Viviand, eds.), pp. 58-65. Vieweg, Braunschweig, 1981.

[70] Gary, J.:
The multigrid iteration applied to the collocation method. SIAM J. Numer. Anal., 18, pp. 211-224, 1981.

[71] Gary, J.:
On higher order multigrid methods with application to a geothermal reservoir model. Internat. J. Numer. Meth. in Fluids, to appear.

[72] Gaur, S.P.; Brandt, A.:
Numerical solution of semiconductor transport equations in two dimensions by multi-grid method. Advances in Computer Methods for Partial Differential Equations II (R. Vichnevetsky, ed.), pp. 327-329. IMACS (AICA), New Brunswick, NJ, 1977.

Ghia, K.N.:
cf. [73].

[73] Ghia, U.; Ghia, K.N.; Shin, C.T.:
Solution of incompressible Navier-Stokes equations by coupled strongly-implicit multi-grid method. Report, Symposium on Numerical Boundary Condition Procedures and Multigrid Methods. NASA-Ames Research Center, Moffett Field, CA, 1981.

[74] Grinstein, F.F.; Rabitz, H.; Askar, A.:
The multigrid method for accelerated solution of the discretized Schrödinger equation. Report, Dept. of Mathematics, Princeton University, 1982.

[75] Grosch, C.E.:
Performance analysis of Poisson solvers on array computers. Super-Computers: 2, pp. 147-181. Infotech International, Maidenhead, 1979.

[76] Guderley, K.G.; Clemm, D.S.:
Eigenvalue and near eigenvalue problems solved by Brandt's multigrid method. Flight Dyn. Report AFFDL-TR-79-3147, Wright-Patterson Air-Force Base, 1979.

Gunzburger, M.D.:
cf. [62].

Gustafsson, I.:
cf. [11].

[77] Gustavson, F.G.:
Implementation of the multi-grid method for solving partial differential equations. Research Report RA 82, IBM T.J. Watson Research Center, pp. 51-57, Yorktown Heights, NY, 1976.

[78] Hackbusch, W.:
Ein iteratives Verfahren zur schnellen Auflösung elliptischer Randwertprobleme. Report 76-12, Institut für Angewandte Mathematik, Universität Köln, 1976.

[79] Hackbusch, W.:
A fast numerical method for elliptic boundary value problems with variable coefficents. Proceedings of the Second GAMM-Conference on Numerical Methods in Fluid Mechanics (E.H. Hirschel, W. Geller, eds.), DFVLR, pp. 50-57. DFVLR, Köln, 1977.

[80] Hackbusch, W.:
A multi-grid method applied to a boundary problem with variable coefficients in a rectangle. Report 77-17, Institut für Angewandte Mathematik, Universität Köln, 1977.

[81] Hackbusch, W.:
On the convergence of a multi-grid iteration applied to finite element equations. Report 77-8, Institut für Angewandte Mathematik, Universität Köln, 1977.

[82] Hackbusch, W.:
A fast iterative method for solving Helmholtz's equation in a general region. Fast Elliptic Solvers (U. Schumann, ed.), pp. 112-124. Advance Publications, London, 1978.

[83] Hackbusch, W.:
A fast iterative method for solving Poisson's equation in a general region. Numerical Treatment of Partial Differential Equations (R. Bulirsch, R.D. Grigorieff, J. Schröder, eds.). Proceedings of a Conference Held at Oberwolfach, July 4-10, 1976. Lecture Notes in Mathematics, 631, pp. 51-62. Springer-Verlag, Berlin, 1978.

[84] Hackbusch, W.:
On the multigrid method applied to difference equations. Computing, 20, pp. 291-306, 1978.

[85] Hackbusch, W.:
On the computation of approximate eigenvalues and eigenfunctions of elliptic operators by means of a multi-grid method. SIAM J. Numer. Anal., 16, pp. 201-215, 1979.

[86] Hackbusch, W.:
On the fast solution of parabolic boundary control problems. SIAM J. Control Optim., 17, pp. 231-244, 1979.

[87] Hackbusch, W.:
On the fast solutions of nonlinear elliptic equations. Numer. Math., 32, pp. 83-95, 1979.

[88] Hackbusch, W.:
A note on the penalty correction method. Report 80-6, Institut für Angewandte Mathematik, Universität Köln, 1980.

[89] Hackbusch, W.:
Analysis and multigrid solutions of mixed finite element and mixed difference equations. Preprint, Institut für Angewandte Mathematik, Ruhr-Universität Bochum, 1980.

[90] Hackbusch, W.:
Convergence of multi-grid iterations applied to difference equations. Math. Comp., 34, pp. 425-440, 1980.

[91] Hackbusch, W.:
Multi-grid solutions to linear and nonlinear eigenvalue problems for integral and differential equations. Report 80-3, Institut für Angewandte Mathematik, Universität Köln, 1980.

[92] Hackbusch, W.:
Numerical solution of nonlinear equations by the multigrid iteration of the second kind. Numerical Methods for Nonlinear Problems. Proceedings of the International Conference Held at the University College Swansea (C. Taylor, ed.), pp. 1041-1050. Pineridge Press, 1980.

[93] Hackbusch, W.:
On the fast solving of elliptic control problems. J. Optim. Theory Appl., 31, pp. 565-581, 1980.

[94] Hackbusch, W.:
Regularity of difference schemes. II. Regularity estimates for linear and nonlinear problems. Report 80-13, Institut für Angewandte Mathematik, Universität Köln, 1980.

[95] Hackbusch, W.:
Survey of convergence proofs for multigrid iterations. Special Topics of Applied Mathematics (J. Frehse, D. Pallaschke, U. Trottenberg, eds.), pp. 151-164. North-Holland Publishing Company, Amsterdam, 1980.

[96] Hackbusch, W.:
The fast numerical solution of very large elliptic difference schemes. J. Inst. Math. Appl., 26, pp. 119-132, 1980.

[97] Hackbusch, W.:
Bemerkungen zur iterierten Defektkorrektur und zu ihrer Kombination mit Mehrgitterverfahren. Rev. Roumaine Math. Pures Appl., 26, pp. 1319-1329, 1981.

[98] Hackbusch, W.:
Die schnelle Auflösung der Fredholmschen Integralgleichung zweiter Art. Beiträge Numer. Math., 9, pp. 47-62, 1981.

[99] Hackbusch, W.:
Error analysis of the nonlinear multigrid method of the second kind. Apl. Mat., 26, pp. 18-29, 1981.

[100] Hackbusch, W.:
Numerical solution of linear and nonlinear parabolic control problems. Optimization and Optimal Control (A. Auslender, W. Oettli, J. Stoer, eds.). Lecture Notes in Control and Information Sciences, 30., pp. 179-185. Springer-Verlag, Berlin, 1981.

[101] Hackbusch, W.:
On the convergence of multi-grid iterations. Beiträge Numer. Math., 9, pp. 213-239, 1981.

[102] Hackbusch, W.:
On the regularity of difference schemes. Ark. Mat., 19, pp. 71-95, 1981.

[103] Hackbusch, W.:
Optimal Hp,p/2 error estimates for a parabolic Galerkin method.** SIAM J. Numer. Anal., 18, pp. 681-692, 1981.

[104] Hackbusch, W.:
The fast numerical solution of time periodic parabolic problems. SIAM J. Sci. Stat. Comput., 2, pp. 198-206, 1981.

[105] Hackbusch, W.:
Multi-grid convergence theory. Multigrid Methods. Proceedings of the Conference Held at Köln-Porz, November 23-27, 1981 (W. Hackbusch, U. Trottenberg, eds.). Lecture Notes in Mathematics. Springer-Verlag, Berlin, 1982.

[106] Hackbusch, W.:
Multi-grid solution of continuation problems. Iterative Solution of Nonlinear Systems (R. Ansorge, T. Meis, W. Törnig, eds.). Lecture Notes in Mathematics. Springer-Verlag, Berlin, 1982.

[107] Hackbusch, W.:
On multi-grid iterations with defect correction. Multigrid Methods. Proceedings of the Conference Held at Köln-Porz, November 23-27, 1981 (W. Hackbusch, U. Trottenberg, eds.). Lecture Notes in Mathematics. Springer-Verlag, Berlin, 1982.

[108] Hackbusch, W.:
Introduction to multi-grid methods for the numerical solution of boundary value problems. Computational Methods for Turbulent, Transonic and Viscous Flows (J.A. Essers, ed.). Hemisphere, to appear.

[109] Hackbusch, W.; Hofmann, G.:
Results of the eigenvalue problem for the plate equation. Z. Angew. Math. Phys., 31, pp. 730-739, 1980.

[110] Hackbusch, W.; Mittelmann, H.D.:
On multi-grid methods for variational inequalities. Report no. 57, Abt. Mathematik, Universität Dortmund, 1981.

[111] Hemker, P.W.:
Fourier analysis of gridfunctions, prolongations and restrictions. Preprint NW 93/80, Dept. of Numerical Mathematics, Mathematical Centre, Amsterdam, 1980.

[112] Hemker, P.W.:
Multi-grid bibliography. Colloqium Numerical Integration of Partial Differential Equations (J. Verwer, ed.), Dept. of Numerical Mathematics, Mathematical Centre, Amsterdam, 1980.

[113] Hemker, P.W.:
On the structure of an adaptive multi-level algorithm. BIT, 20, pp. 289-301, 1980.

[114] Hemker, P.W.:
The incomplete LU-decomposition as a relaxation method in multigrid algorithms. Boundary and Interior Layers - Computational and Asymptotic Methods (J.J.H. Miller, ed.), pp. 306-311. Boole Press, Dublin, 1980.

[115] Hemker, P.W.:
A note on defect correction processes with an approximate inverse of deficient rank. Preprint NN 23/81, Dept. of Numerical Mathematics, Mathematical Centre, Amsterdam, 1981.

[116] Hemker, P.W.:
Algol 68 Fourier analysis program. Programlisting, Dept. of Numerical Mathematics, Mathematical Centre, Amsterdam, 1981.

[117] Hemker, P.W.:
Algol 68 multigrid library. Programlisting, Dept. of Numerical Mathematics, Mathematical Centre, Amsterdam, 1981.

[118] Hemker, P.W.:
Introduction to multigrid methods. Nieuw Archief voor Wiskunde (3), 29, pp. 71-101, 1981.

[119] Hemker, P.W.:
Mixed defect correction iteration for the accurate solution of the convection diffusion equation. Multigrid Methods. Proceedings of the Conference Held at Köln-Porz, November 23-27, 1981 (W. Hackbusch, U. Trottenberg, eds.). Lecture Notes in Mathematics. Springer-Verlag, Berlin, 1982.

[120] Hemker, P.W.; Schippers, H.:
Multiple grid methods for the solution of Fredholm integral equations of the second kind. Math. Comp., 36, pp. 215-232, 1981.

Hirsch, C.:
cf. [54], [55].

Hofmann, G.:
cf. [109].

[121] Holland, W.; McCormick, S.F.; Ruge, J.:
Unigrid methods for boundary value problems with nonrectangular domains. Multigrid Methods. NASA Conference Publication 2202, Ames Research Center, pp. 235-247, Moffett Field, CA, 1981.

[122] Houwen, P.J. van der; Sommeijer, B.P.:
Analysis of Richardson iteration in multigrid methods for nonlinear parabolic differential equations. Report NW 105/81, Dept. of Numerical Mathematics, Mathematical Centre, Amsterdam, 1981.

[123] Houwen, P.J. van der; Vries, H.B. de:
Preconditioning and coarse grid corrections in the solution of the initial value problem for nonlinear partial differential equations. Preprint NW 95/80, Dept. of Numerical Mathematics, Mathematical Centre, Amsterdam, 1980.

Hussaini, M.Y.:
cf. [136], [206].

Huynh, Q.:
cf. [136].

Hyman, J.M.:
cf. [57].

Jameson, A.:
cf. [186].

[124] Jameson, A.:
Acceleration of transonic potential flow calculations on arbitrary meshes by the multiple grid method. Paper AIAA-79-1458, AIAA Fourth Computational Fluid Dynamics Conference, New York, NY, 1979.

[125] Jespersen, D.C.:
Multilevel techniques for nonelliptic problems. Multigrid Methods. NASA Conference Publication 2202, Ames Research Center, pp. 1-22, Moffett Field, CA, 1981.

[126] Jespersen, D.C.:
Multigrid methods for partial differential equations. Studies in Numerical Analysis (G. Golub, ed.), to appear.

[127] Johnson, G.M.:
Multiple-grid acceleration of Lax-Wendroff algorithms. Report, Lewis Research Center, Cleveland, OH, 1981.

Keller, H.B.:
cf. [53].

[128] Kettler, R.:
Analysis and comparison of relaxation schemes in robust multigrid and preconditioned conjugate gradient methods. Multigrid Methods. Proceedings of the Conference Held at Köln-Porz, November 23-27, 1981 (W. Hackbusch, U. Trottenberg, eds.). Lecture Notes in Mathematics. Springer-Verlag, Berlin, 1982.

[129] Kettler, R.; Meijerink, J.A.:
A multigrid method and a combined multigrid-conjugate gradient method for elliptic problems with strongly discontinous coefficients in general domains. Shell Publication 604, KSEPL, Rijswijk, 1981.

[130] Kneile, K.:
Accelerated convergence of structured banded systems using constrained corrections. Multigrid Methods. NASA Conference Publication 2202, Ames Research Center, pp. 285-303, Moffett Field, CA, 1981.

[131] Kroll, N.:
Direkte Anwendungen von Mehrgittertechniken auf parabolische Anfangsrandwertaufgaben. Diplomarbeit, Institut für Angewandte Mathematik, Universität Bonn, 1981.

[132] Lee, H.N.; Meyers, R.E.:
On time dependent multi-grid numerical techniques. Comput. Math. Appl., 6, pp. 61-65, 1980.

Lehmann, H.:
cf. [153].

[133] Linden, J.:
Mehrgitterverfahren für die Poisson-Gleichung in Kreis und Ringgebiet unter Verwendung lokaler Koordinaten. Diplomarbeit, Institut für Angewandte Mathematik, Universität Bonn, 1981.

[134] Linden, J.; Trottenberg, U.; Witsch, K.:
Multigrid computation of the pressure of an incompressible fluid in a rotating spherical gap. Proceedings of the Fourth GAMM-Conference on Numerical Methods in Fluid Mechanics (H. Viviand, ed.), pp. 183-193. Vieweg, Braunschweig, 1982.

[135] Lomax, H.; Pulliam, J., Jespersen, D.C.:
Eigensystem analysis techniques for finite-difference equations. I. Multigrid techniques. Paper AIAA-81-1027, AIAA Fifth Computational Fluid Dynamics Conference, pp. 55-80, New York, NY, 1981.

[136] Lustman, L.R.; Huynh, Q.; Hussaini, M.Y.:
Multigrid method with weighted mean scheme. Multigrid Methods. NASA Conference Publication 2202, Ames Research Center, pp. 47-59, Moffett Field, CA, 1981.

[137] Maitre, J.F.; Musy, F.:
The contraction number of a class of two-level methods; an exact evaluation for some finite element subspaces and model problems. Multigrid Methods. Proceedings of the Conference Held at Köln-Porz, November 23-27, 1981 (W. Hackbusch, U. Trottenberg, eds.). Lecture Notes in Mathematics. Springer-Verlag, Berlin, 1982.

[138] Mansfield, L.:
On the solution of nonlinear finite element systems. SIAM J. Numer. Anal., 17, pp. 752-765, 1980.

[139] Mansfield, L.:
On the multi-grid solution of finite element equations with isoparametric elements. Numer. Math., 37, pp. 423-432, 1981.

[140] McCarthy, D.R.; Reyner, T.A.:
A multigrid code for the three-dimensional transonic potential flow about axisymmetric inlets at angle of attack. Paper AIAA-80-1365, AIAA Thirteenth Fluid and Plasma Dynamics Conference, New York, NY, 1980.

McCormick, S.F.:
cf. [3], [50], [121].

[141] McCormick, S.F.:
An algebraic interpretation of multigrid methods. Technical Report, Lawrence Livermore Laboratory, Livermore, CA, 1979.

[142] McCormick, S.F.:
Multigrid methods: An alternate viewpoint. Report UCID-18487, Lawrence Livermore Laboratory, Livermore, CA, 1979.

[143] McCormick, S.F.:
Mesh refinement methods for integral equations. Numerical Treatment of Integral Equations (J. Albrecht, L. Collatz, eds.), pp. 183-190. Birkhäuser Verlag, Basel, 1980.

[144] McCormick, S.F.:
A mesh refinement method for A*x=Lambda*B*x. Math. Comp., 36, pp. 485-498, 1981.

[145] McCormick, S.F.:
Multigrid short course notes. Report, Lockheed Palo Alto Research Center, 1981.

[146] McCormick, S.F.:
Numerical software for fixed point microprocessor applications and for fast implementation of multigrid techniques. ARO Report 81-3, Proceedings of the Army Numerical Analysis and Computers Conference, 1981.

[147] McCormick, S.F.:
Multigrid methods for variation problems: The V-cycle. Research Report, Dept. of Mathematics, Colorado State University, Ft. Collins, CO, 1982.

[148] McCormick, S.F.; Rodrigue, G.H.:
Multigrid methods for multiprocessor computers. Technical Report, Lawrence Livermore Laboratory, Livermore, CA, 1979.

[149] McCormick, S.F.; Ruge, J.:
Multigrid methods for variational problems. Preprint, Dept. of Mathematics, Colorado State University, Ft. Collins, CO, 1981.

[150] McCormick, S.F.; Ruge, J.:
Unigrid for multigrid simulation. Internal Report, Dept. of Mathematics, Colorado State University, Ft. Collins, CO, 1981.

[151] McCormick, S.F.; Thomas, J.W.:
Multigrid methods applied to water wave problems. Internal Report, Dept. of Mathematics, Colorado State University, Ft. Collins, CO, 1981.

Meijerink, J.A.:
cf. [129].

[152] Meis, T.; Branca, H.W.:
Schnelle Lösung von Randwertaufgaben. Z. Angew. Math. Mech., 62, 1982.

[153] Meis, T.; Lehmann, H.; Michael, H.:
Application of the multigrid method to a nonlinear indefinite problem. Multigrid Methods. Proceedings of the Conference Held at Köln-Porz, November 23-27, 1981 (W. Hackbusch, U. Trottenberg, eds.). Lecture Notes in Mathematics. Springer-Verlag, Berlin, 1982.

[154] Merriam, M.:
Formal analysis of multigrid techniques applied to Poisson's equation in three dimensions. AIAA-81-1028, Proceedings AIAA Fifth Computational Fluid Dynamics Conference, New York, NY, 1981.

Meyers, R.E.:
cf. [132].

Michael, H.:
cf. [153].

Mittelmann, H.D.:
cf. [110].

[155] Mittelmann, H.D.:
A fast solver for nonlinear eigenvalue problems. Iterative Solution of Nonlinear Systems (R. Ansorge, T. Meis, W. Törnig, eds.). Lecture Notes in Mathematics. Springer-Verlag, Berlin, 1982.

[156] Mittelmann, H.D.:
Multi-grid methods for simple bifurcation problems. Multigrid Methods. Proceedings of the Conference Held at Köln-Porz, November 23-27, 1981 (W. Hackbusch, U. Trottenberg, eds.). Lecture Notes in Mathematics. Springer-Verlag, Berlin, 1982.

[157] Mol, W.J.A.:
A multigrid method applied to some simple problems. Memorandum no. 287, Dept. of Numerical Mathematics, Twente University of Technology, Twente, 1979.

[158] Mol, W.J.A.:
Numerical solution of the Navier-Stokes equations by means of multigrid method and Newton-iteration. Preprint NW 92/80, Dept. of Numerical Mathematics, Mathematical Centre, Amsterdam, 1980.

[159] Mol, W.J.A.:
Computation of flows around a Karman-Trefftz profile. Preprint NW 114/81, Dept. of Numerical Mathematics, Mathematical Centre, Amsterdam, 1981.

[160] Mol, W.J.A.:
On the choice of suitable operators and parameters in multigrid methods. Report NW 107/81, Dept. of Numerical Mathematics, Mathematical Centre, Amsterdam, 1981.

Musy, F.:
cf. [137].

[161] NI, R.H.:
A multiple grid scheme for solving Euler equations. Paper AIAA-81-1025, AIAA Fifth Computational Fluid Dynamics Conference, New York, NY, 1981.

[162] Nicolaides, R.A.:
On multiple grid and related techniques for solving discrete elliptic systems. J. Comput. Phys., 19, pp. 418-431, 1975.

[163] Nicolaides, R.A.:
On the l2 convergence of an algorithm for solving finite element equation.** Math. Comp., 31, pp. 892-906, 1977.

[164] Nicolaides, R.A.:
On multi-grid convergence in the indefinite case. Math. Comp., 32, pp. 1082-1086, 1978.

[165] Nicolaides, R.A.:
On the observed rate of convergence of an iterative method applied to a model elliptic difference equation. Math. Comp., 32, pp. 127-133, 1978.

[166] Nicolaides, R.A.:
On finite element multigrid algorithms and their use. The Mathematics of Finite Elements and Applications III, MAFELAP 1978 (J.R. Whiteman, ed.), pp. 459-466. Academic Press, London, 1979.

[167] Nicolaides, R.A.:
On some theoretical and practical aspects of multigrid methods. Math. Comp., 33, pp. 933-952, 1979.

[168] Nowak, Z.P.:
Use of the multigrid method lor Laplacian problem in three dimensions. Multigrid Methods. Proceedings of the Conference Held at Köln-Porz, November 23-27, 1981 (W. Hackbusch, U. Trottenberg, eds.). Lecture Notes in Mathematics. Springer-Verlag, Berlin, 1982.

Ophir, D.:
cf. [47].

[169] Ophir, D.:
Language for processes of numerical solutions to differential equations. PH.D. Thesis, Dept. of Applied Mathematics, Weizmann Institute of Science, Rehovot, 1978.

[170] Oskam, B.; Fray, J.M.J.:
General relaxation schemes in multigrid algorithms for higher order singularity methods. Multigrid Methods. NASA Conference Publication 2202, Ames Research Center, pp. 217-234, Moffett Field, CA, 1981.

Painter, J.W.:
cf. [2].

[171] Painter, J.W.; Symm, G.T.:
Multigrid experience with the neutron diffusion equation. Report LA-UR 79-1634, Los Alamos Scientific Laboratory, Los Alamos, NM, 1979.

[172] Poling, T.C.:
Numerical experiments with multi-grid methods. M.A. Thesis, Dept. of Numerical Mathematics, College of William and Mary, Williamsburg, VA, 1978.

Pulliam, J., Jespersen, D.C.:
cf. [135].

Rabitz, H.:
cf. [74].

Reyner, T.A.:
cf. [140].

[173] Ries, M.:
Lösung elliptischer Randwertaufgaben mit approximativen und iterativen Reduktionsverfahren. Dissertation, Institut für Angewandte Mathematik, Universität Bonn, 1981.

[174] Ries, M.; Trottenberg, U.:
MGR - ein blitzschneller elliptischer Löser. Preprint no. 277, Sonderforschungsbereich 72, Universität Bonn, 1979.

[175] Ries, M.; Trottenberg, U.; Winter, G.:
A note on MGR methods. Linear Algebra Appl., Bonn, to appear 1982.

Rodrigue, G.H.:
cf. [148].

Rose, D.J.:
cf. [18].

[176] Rubin, S.G.:
Incompressible Navier-Stokes and parabolized Navier-Stokes procedures and computational techniques. Computational Fluid Dynamics. Lecture Series 1982-04, von Karman Institute for Fluid Dynamics, Rhode-Saint-Genese, 1982.

Ruge, J.:
cf. [50], [121], [149], [150].

[177] Ruge, J.W.:
Multigrid methods for differential eigenvalue and variational problems and multigrid simulation. PH.D. Thesis, Dept. of Mathematics, Colorado State University, Ft. Collins, CO, 1981.

Rukhovets, L.A.:
cf. [8].

Ruppel, H.:
cf. [49].

[178] Schaffer, S.:
High order multi-grid methods to solve the Poisson equation. Multigrid Methods. NASA Conference Publication 2202, Ames Research Center, pp. 275-284, Moffett Field, CA, 1981.

[179] Schaffer, S.:
High-order multigrid methods. PH.D. Thesis, Dept. of Mathematics, Colorado State University, Ft. Collins, CO, 1982.

Schippers, H.:
cf. [120].

[180] Schippers, H.:
Multigrid techniques for the solution of Fredholm integral equations of the second kind. MCS 41, Colloqium on the Numerical Treatment of Integral Equations. (H.J.J. Te Riele, ed.), Dept. of Numerical Mathematics, Mathematical Centre, pp. 29-49, Amsterdam, 1979.

[181] Schippers, H.:
Multiple grid methods for oscillating disc flow. Boundary and Interior Layers - Computational and Asymptotic Methods (J.J.H. Miller, ed.), Dublin, 1980.

[182] Schippers, H.:
The automatic solution of Fredholm integral equations of the second kind. Report NW 99/80, Dept. of Numerical Mathematics, Mathematical Centre, Amsterdam, 1980.

[183] Schippers, H.:
Application of multigrid methods for integral equations to two problems from fluid dynamics. Multigrid Methods. NASA Conference Publication 2202, Ames Research Center, pp. 193-216, Moffett Field, CA, 1981.

[184] Schippers, H.:
Multiple grid methods for equations of the second kind with applications in fluid mechanics. Proefschrift, Dept. of Numerical Mathematics, Mathematical Centre, Amsterdam, 1982.

[185] Schippers, H.:
On the regularity of the principal value of the double layer potential. J. Engrg. Math., 1982.

[186] Schmidt, W.; Jameson, A.:
Applications of multi-grid methods for transonic flow calculations. Multigrid Methods. Proceedings of the Conference Held at Köln-Porz, November 23-27, 1981 (W. Hackbusch, U. Trottenberg, eds.). Lecture Notes in Mathematics. Springer-Verlag, Berlin, 1982.

Sherman, A.H.:
cf. [19], [20], [21], [22], [23].

[187] Shiftan, Y.:
Multi-grid methods for solving elliptic difference equations. M.SC. Thesis, Dept. of Applied Mathematics, Weizmann Institute of Science, Rehovot, 1972.

Shin, C.T.:
cf. [73].

[188] Shmilovich, A.; Caughey, D.A.:
Application of the multi-grid method to calculations of transonic potential flow about wing-fuselage combinations. Multigrid Methods. NASA Conference Publication 2202, Ames Research Center, pp. 101-130, Moffett Field, CA, 1981.

[189] Solchenbach, K.; Stüben, K.; Trottenberg, U.; Witsch, K.:
Efficient solution of a nonlinear heat conduction problem by use of fast reduction and multigrid methods. Preprint no. 421, Sonderforschungsbereich 72, Universität Bonn, 1980.

Sommeijer, B.P.:
cf. [122].

Sonneveld, P.:
cf. [203].

[190] South, J.C. (Jr.); Brandt, A.:
Application of a multi-level grid method to transonic flow calculations. Transonic Flow Problems in Turbomachinery (T.C. Adamson, M.F. Platzer, eds.). Hemisphere, Washington, DC, 1977.

Stetter, H.J.:
cf. [9].

[191] Strakhovskaya, L.G.:
An iterative method for evaluating the first eigenvalue of an elliptic operator. U.S.S.R. Computational Math. and Math. Phys., 17 no. 3, pp. 88-101, 1977.
Stüben, K.:
cf. [65], [189].
[192] Stüben, K.:
MG01: A multi-grid program to solve Delta U - c(x,y)U = f(x,y) (on Omega), U=g(x,y) (on dOmega), on nonrectangular bounded domains Omega. IMA-Report no. 82.02.02, Gesellschaft für Mathematik und Datenverarbeitung, St. Augustin, 1982.
[193] Stüben, K.; Trottenberg, U.:
Multigrid methods: Fundamental algorithms, model problem analysis, and applications. Multigrid Methods. Proceedings of the Conference Held at Köln-Porz, November 23-27, 1981 (W. Hackbusch, U. Trottenberg, eds.). Lecture Notes in Mathematics. Springer-Verlag, Berlin, 1982.
[194] Stüben, K.; Trottenberg, U.:
On the construction of fast solvers for elliptic equations. Computational Fluid Dynamics. Lecture Series 1982-04, von Karman Institute for Fluid Dynamics, Rhode-Saint-Genese, 1982.
Symm, G.T.:
cf. [171].
Ta'asan, S.:
cf. [48].
[195] Tal-Nir, A.; Bercovier, M.:
Implementation of the multigrid method "MLAT" in the finite element method. Report no. 54, Institut National de Recherche en Informatique et en Automatique (INRIA), Le Chesnay, 1981.
Thomas, J.W.:
cf. [151].
Trottenberg, U.:
cf. [65], [134], [174], [175], [189], [193], [194].
[196] Trottenberg, U.:
Schnelle Lösung partieller Differentialgleichungen - Idee und Bedeutung des Mehrgitterprinzips. Jahresbericht 1980/81, Gesellschaft für Mathematik und Datenverarbeitung, pp. 85-95, Bonn, 1981.
[197] Verfürth, R.:
The contraction number of a multigrid method with mesh ratio 2 for solving Poisson's equation. Report, Institut für Angewandte Mathematik, Ruhr-Universität Bochum, 1982.
Vries, H.B. de:
cf. [123].
[198] Wesseling, P.:
Numerical solution of stationary Navier-Stokes equations by means of a multiple grid method and Newton iteration. Report NA-18, Dept. of Mathematics, Delft University of Technology, Delft, 1977.
[199] Wesseling, P.:
A convergence proof for a multiple grid method. Report NA-21, Dept. of Mathematics, Delft University of Technology, Delft, 1978.
[200] Wesseling, P.:
The rate of convergence of a multiple grid method. Numerical Analysis. Proceedings, Dundee 1979 (G.A. Watson, ed.). Lecture Notes in Mathematics, 773, pp. 164-184. Springer-Verlag, Berlin, 1980.

[201] Wesseling, P.:
Theoretical and practical aspects of a multigrid method. Report NA-37, Dept. of Mathematics, Delft University of Technology, Delft, 1980.

[202] Wesseling, P.:
A robust and efficient multigrid method. Multigrid Methods. Proceedings of the Conference Held at Köln-Porz, November 23-27, 1981 (W. Hackbusch, U. Trottenberg, eds.). Lecture Notes in Mathematics. Springer-Verlag, Berlin, 1982.

[203] Wesseling, P.; Sonneveld, P.:
Numerical experiments with a multiple grid and a preconditioned Lanczos method. Approximation Methods for Navier-Stokes Problems. Proceedings, Paderborn 1979 (R. Rautmann, ed.). Lecture Notes in Mathematics, 771, pp. 543-562. Springer-Verlag, Berlin, 1980.

Winter, G.:
cf. [175].

Witsch, K.:
cf. [63], [64], [134], [189].

[204] Wolff, H.:
Multi-grid techniek voor het oplossen van fredholm-integraalvergelijkingen van de tweede soort. Report NN 19/79, Dept. of Numerical Mathematics, Mathematical Centre, Amsterdam, 1979.

[205] Wolff, H.:
Multiple grid methods for the calculation of potential flow around 3-d bodies. Preprint NW 119/82, Dept. of Numerical Mathematics, Mathematical Centre, Amsterdam, 1982.

Wong, Y.S.:
cf. [206].

Younis, M.:
cf. [52].

[206] Zang, T.A.; Wong, Y.S.; Hussaini, M.Y.:
Spectral multi-grid methods for elliptic equations. Multigrid Methods. NASA Conference Publication 2202, Ames Research Center, pp. 173-191, Moffett Field, CA, 1981.

Supplement.

This supplement to the multigrid bibliography provides a (very incomplete) collection of some papers dealing with multigrid forerunners and related subjects, e.g., hierarchical relaxation, coarse grid acceleration, nested iterations and reduction type methods.

[1] Astrakhantsev, G.P.:
The iterative improvement of eigenvalues. U.S.S.R. Computational Math. and Math. Phys., 16 no. 1, pp. 123-132, 1976.

[2] Atkinson, K.:
Iterative variants of the Nyström method for the numerical solution of integral equations. Numer. Math., 22, pp. 17-31, 1973.

Aziz, K.:
cf. [18].

Beelen, T.:
cf. [12].

[3] Brakhage, H.:
Über die numerische Behandlung von Integralgleichungen nach der Quadraturformelmethode. Numer. Math., 2, pp. 183-196, 1960.

Dahlquist, G.:
cf. [9].

Förster, H.:
cf. [4], [5].

[4] Foerster, H.; Förster, H.; Trottenberg, U.:
Modulare Programme zur schnellen Lösung elliptischer Randwertaufgaben mit Reduktionsverfahren. TR2D01, TR2D02: Programme zur Lösung der Helmholtz-Gleichung mit Dirichletschen Randbedingungen im Rechteck. Preprint no. 216, Sonderforschungsbereich 72, Universität Bonn, 1978.

[5] Foerster, H.; Förster, H.; Trottenberg, U.:
Modulare Programme zur schnellen Lösung elliptischer Randwertaufgaben mit Reduktionsverfahren. Algorithmische Details der Programme TR2D01, TR2D02. Preprint no. 420, Sonderforschungsbereich 72, Universität Bonn, 1980.

[6] Frederickson, P.O.:
Fast approximate inversion of large sparse linear systems. Mathematics Report no. 7-75, Dept. of Mathematical Sciences, Lakehead University, Ontario, 1975.

Hemker, P.W.:
cf. [12].

[7] Hyman, J.M.:
Mesh refinement and local inversion of elliptic partial differential equations. J. Comput. Phys., 23, pp. 124-134, 1977.

[8] Kronsjö, L.:
A note on the "nested iterations" method. BIT, 15, pp. 107-110, 1975.

[9] Kronsjö, L.; Dahlquist, G.:
On the design of nested iterations for elliptic difference equations. BIT, 11, pp. 63-71, 1972.

[10] McCormick, S.F.:
A revised mesh refinement strategy for Newton's method applied to nonlinear two-point boundary value problems. Numerical Treatment of Differential Equations in Applications. Proceedings, Oberwolfach (R. Ansorge, W. Törnig, eds.). Lecture Notes in Mathematics, 679, pp. 15-23. Springer-Verlag, Berlin, 1978.

[11] Miranker, W.L.:
Hierarchical relaxation. Computing, 23, pp. 267-285, 1977.

[12] Polak, S.J.; Wachters, A.; Beelen, T.; Hemker, P.W.:
A mesh-parameter-continuation method. Elliptic Problem Solvers (M.H. Schultz, ed.), pp. 383-390. Academic Press, New York, NY, 1981.

Reutersberg, H.:
cf. [16].

[13] Reutersberg, H.:
Reduktionsverfahren zur Lösung diskreter elliptischer Randwertaufgaben im R3.** Z. Angew. Math. Phys., 60, pp. 313-314, 1980.

[14] Reutersberg, H.:
Reduktionsverfahren zur Lösung elliptischer Differenzengleichungen im R3.** Bericht Nr. 121 der Gesellschaft für Mathematik und Datenverarbeitung. R. Oldenbourg, München, 1980.

[15] Schröder, J.; Trottenberg, U.:
Reduktionsverfahren für Differenzengleichungen bei Randwertaufgaben I. Numer. Math., 22, pp. 37-68, 1973.

[16] Schröder, J.; Trottenberg, U.; Reutersberg, H.:
Reduktionsverfahren für Differenzengleichungen bei Randwertaufgaben II. Numer. Math., 26, pp. 429-459, 1976.

[17] Schröder, J.; Trottenberg, U.; Witsch, K.:
On fast Poisson solvers and applications. Numerical Treatment of Partial Differential Equations (R. Bulirsch, R.D. Grigorieff, J. Schröder, eds.). Proceedings of a Conference Held at Oberwolfach, July 4-10, 1976. Lecture Notes in Mathematics, 631, pp. 153-187. Springer-Verlag, Berlin, 1978.

[18] Settari, A.; Aziz, K.:
A generalization of the additive correction methods for the iterative solution of matrix equations. SIAM J. Numer. Anal., 10, pp. 506-521, 1973.

[19] Southwell, R.V.:
Relaxation Methods in Theoretical Physics. Clarendon Press, Oxford, 1946.

[20] Stiefel, E.:
Über einige Methoden der Relaxationsrechnung. Z. Angew. Math. Phys., 3, pp. 1-33, 1952.

Trottenberg, U.:
cf. [4], [5], [15], [16], [17].

[21] Trottenberg, U.:
Reduction methods for solving discrete elliptic boundary value problems - an approach in matrix terminology. Fast Elliptic Solvers (U. Schumann, ed.). Advance Publications, London, 1977.

[22] Trottenberg, U.; Witsch, K.:
Zur Kondition diskreter elliptischer Randwertaufgaben. GMD-Studien no. 60, Gesellschaft für Mathematik und Datenverarbeitung, St. Augustin, 1981.

[23] Vallee Poussin, F. de la:
An accelerated relaxation algorithm for iterative solution of elliptic equations. SIAM J. Numer. Anal., 5, pp. 340-351, 1968.

Wachters, A.:
cf. [12].

Witsch, K.:
cf. [17], [22].

PARTICIPANTS

Abou El-Seoud, M. Samir	TH Darmstadt, West Germany
Asselt, Evert Jan van	Mathematisch Centrum, Amsterdam, Netherlands
Auzinger, Winfried	TU Wien, Austria
Axelsson, Owe	University of Nijmegen, Netherlands
Becker, Klaus	Universität Bonn, West Germany
Berger, Peter	Universität Stuttgart, West Germany
Bock, Hans Georg	Universität Bonn, West Germany
Böhmer, Klaus	Universität Marburg, West Germany
Börgers, Christoph	GMD-IMA, St. Augustin, West Germany
Börsch-Supan, Wolfgang	Joh. Gutenberg-Universität, Mainz, West Germany
Boerstoel, Jan W.	National Aerospace Lab., Amsterdam, Netherlands
Braess, Dietrich	Ruhr-Universität Bochum, West Germany
Brakhagen, Franz	GMD-IMA, St. Augustin, West Germany
Brand, Kurt	GMD-IMA, St. Augustin, West Germany
Brandt, Achi	Weizmann Institute of Science, Rehovot, Israel
Bredif, Marc	O.N.E.R.A., Chatillon, France
Bunse, Wolfgang	Universität Bielefeld, West Germany
Caillat, Michel	CNAM, Paris, France
Carlo, Antonio Di	University of Rome, Italy
Cube, Markus von	GMD-IMA, St. Augustin, West Germany
Dahmen, Wolfgang	Universität Bonn, West Germany
Dauphin, Yves	Université libre de Bruxelles, Belgium
Deconinck, Herman	Vrije Universiteit Brussel, Belgium
Deuflhard, Peter	Universität Heidelberg, West Germany
Dick, Erik	State University at Gent, Belgium
Donovang, Margarete	GMD-IMA, St. Augustin, West Germany
Duff, Iain, S.	AERE Harwell, Oxfordshire, Great Britain
Engels, Hermann	RWTH Aachen, West Germany
Esser, Rüdiger	Universität Köln, West Germany
Favini, Bernado	University of Rome, Italy
Finger, Karl-Heinz	Interatom, Berg.-Gladbach, West Germany
Foerster, Hartmut	GMD-IMA, St. Augustin, West Germany
Fromm, Jens	DFVLR, Köln, West Germany
Gentzsch, Wolfgang	DFVLR, Göttingen, West Germany
Gietl, Horst	Leibniz RZ, München, West Germany
Gipser, Michael	TH Darmstadt, West Germany
Gleue, Jörg	Hahn-Meitner-Institut, Berlin, West Germany
Gorenflo, Rudolf	FU Berlin, West Germany
Grigorieff, Rolf Dieter	TU Berlin, West Germany
Hackbusch, Wolfgang	Ruhr-Universität Bochum, West Germany
Häggblad, Bo	ASEA, Västeras, Sweden
Hänel, Dieter	RWTH Aachen, West Germany
Hanke, Heinz-Günter	Ruhr-Universität Bochum, West Germany
Hedstrom, Gerald	Lawrence Livermore Nat. Lab., Livermore, CA, U.S.A.
Hemker, Pieter W.	Mathematisch Centrum, Amsterdam, Netherlands
Henke, Horst	RWTH Aachen, West Germany
Hoffmann, Jobst	RWTH Aachen, West Germany
Hofmann, Götz	Ruhr-Universität Bochum, West Germany
Honig, G.	Universität Mainz, West Germany
Hoppe, Stephan	Universität Essen, West Germany
Jain, Romesh Kumar	DFVLR, Braunschweig, West Germany
Jarausch, Helmut	RWTH Aachen, West Germany
Joly, Pascal	Université P. et M. Curie, Paris, France
Kaspar, Bernhard	TH Darmstadt, West Germany
Kettler, Rob	Delft University of Technology, Netherlands
Krämer-Eis, Peter	Universität Bonn, West Germany
Kroll, Norbert	DFVLR, Braunschweig, West Germany
Kuban, Angelika	FU Berlin, West Germany

Kutsche, Ralf-Detlef	FU Berlin, West Germany
Lehmann, Helge	Universität Köln, West Germany
Linden, Johannes	Universität Bonn, West Germany
Lorentz, Rudolf	GMD-IMA, St. Augustin, West Germany
Mackens, Wolfgang	RWTH Aachen, West Germany
Maitre, Jean Francois	Ecole Centrale de Lyon, France
Meinguet, Jean	Universitê catholique de Louvain, Belgium
Meis, Theodor	Universität Köln, West Germany
Merschen, Toni	RWTH Aachen, West Germany
Michael, Harald	Universität Köln, West Germany
Mittelmann, Hans D.	Universität Dortmund, West Germany
Mülthei, Heinrich	Universität Mainz, West Germany
Munz, Claus-Dieter	Universität Kalrsruhe, West Germany
Musy, Francois	Ecole centrale de Lyon, France
Nahrgang, Christoph	Universität Bielefeld, West Germany
Nowak, Zenon	ITLi MS Polit. Warszawa, Poland
Oertel, Klaus-Dieter	GMD-IMA, St. Augustin, West Germany
Ophir, Dan	GMD-IMA, St. Augustin, West Germany
Pallaske, Ulrich	Bayer AG, Leverkusen, West Germany
Phillips, Timothy N.	Oxford Univ. Comp. Lab., Great Britain
Projahn, Ulrich	TH Darmstadt, West Germany
Qun, Lin	Academia Sinica, Beijing, R.P. China
Raemaekers, Hubert	Mathematisch Centrum, Amsterdam, Netherlands
Reutersberg, Heinz	GMD-IMA, St. Augustin, West Germany
Rieger, Herbert	TH Darmstadt, West Germany
Ries, Manfred	Universität Trier, West Germany
Schloeder, Johannes	Universität Bonn, West Germany
Schmidt, Geert H.	Shell Research, Rijswijk, Netherlands
Schmidt, Wolfgang	Dornier, Friedrichshafen, West Germany
Schneider, Hans-Jürgen	Solingen, West Germany
Schrauf, Geza	Universität Bonn, West Germany
Schreiber-Martens, Alwine	Krupp Forschungsinstitut, Essen,West Germany
Schröder, Johann	Universität Köln, West Germany
Schulte, Dieter	Krupp Forschungsinstitut, Essen, West Germany
Schwarz, Reinhild	GMD-IMA, St. Augustin, West Germany
Schwarz, Roland	Universität Köln, West Germany
Schwichtenberg, Horst	GMD-IMA, St. Augustin, West Germany
Solchenbach, Karl	GMD-IMA, St. Augustin, West Germany
Steffen, Bernhard	KFA-ZAM, Jülich, West Germany
Stehle, Burghart	KfK-INR, Karlsruhe, West Germany
Stephany, Peter	GMD-IMA, St. Augustin, West Germany
Stetter, Hans-J.	TU Wien, Austria
Strack, Klaus-Günther	RWTH Aachen, West Germany
Strasdas, Wolfgang	Universität Essen, West Germany
Stüben, Klaus	GMD-IMA, St. Augustin, West Germany
Tatone, Amabile	Universita de l'Aquila, Italy
Thiebes, Joe	Universität Bonn, West Germany
Thole, Clemens-August	GMD-IMA, St. Augustin, West Germany
Trottenberg, Ulrich	Universität Bonn, West Germany
Unger, Heinz	Universität Bonn, West Germany
Verfürth, Rüdiger	Ruhr-Universität Bochum, West Germany
Weber, Anton	DFVLR, Köln, West Germany
Wesseling, Pieter	TH Delft, Netherlands
Widlund, Olof	Courant Institute, New York, U.S.A.
Winter, Gerd	Universität Bonn, West Germany
Witsch, Kristian	Universität Düsseldorf, West Germany
Wittman, Helga	DFVLR, Köln, West Germany
Wohlrab, Ottwin	Universität Bonn, West Germany
Yu-Fa, Guo	GMD-IMA, St. Augustin, West Germany
Zimare, Hartmut	Fa. Kahle, Düsseldorf, West Germany

Lecture Notes in Mathematics

continuation on page 655

Vol. 817: L. Gerritzen, M. van der Put, Schottky Groups and Mumford Curves. VIII, 317 pages. 1980.

Vol. 818: S. Montgomery, Fixed Rings of Finite Automorphism Groups of Associative Rings. VII, 126 pages. 1980.

Vol. 819: Global Theory of Dynamical Systems. Proceedings, 1979. Edited by Z. Nitecki and C. Robinson. IX, 499 pages. 1980.

Vol. 820: W. Abikoff, The Real Analytic Theory of Teichmüller Space. VII, 144 pages. 1980.

Vol. 821: Statistique non Paramétrique Asymptotique. Proceedings, 1979. Edited by J.-P. Raoult. VII, 175 pages. 1980.

Vol. 822: Séminaire Pierre Lelong–Henri Skoda, (Analyse) Années 1978/79. Proceedings. Edited by P. Lelong et H. Skoda. VIII, 356 pages, 1980.

Vol. 823: J. Král, Integral Operators in Potential Theory. III, 171 pages. 1980.

Vol. 824: D. Frank Hsu, Cyclic Neofields and Combinatorial Designs. VI, 230 pages. 1980.

Vol. 825: Ring Theory, Antwerp 1980. Proceedings. Edited by F. van Oystaeyen. VII, 209 pages. 1980.

Vol. 826: Ph. G. Ciarlet et P. Rabier, Les Equations de von Kármán. VI, 181 pages. 1980.

Vol. 827: Ordinary and Partial Differential Equations. Proceedings, 1978. Edited by W. N. Everitt. XVI, 271 pages. 1980.

Vol. 828: Probability Theory on Vector Spaces II. Proceedings, 1979. Edited by A. Weron. XIII, 324 pages. 1980.

Vol. 829: Combinatorial Mathematics VII. Proceedings, 1979. Edited by R. W. Robinson et al.. X, 256 pages. 1980.

Vol. 830: J. A. Green, Polynomial Representations of GL_n. VI, 118 pages. 1980.

Vol. 831: Representation Theory I. Proceedings, 1979. Edited by V. Dlab and P. Gabriel. XIV, 373 pages. 1980.

Vol. 832: Representation Theory II. Proceedings, 1979. Edited by V. Dlab and P. Gabriel. XIV, 673 pages. 1980.

Vol. 833: Th. Jeulin, Semi-Martingales et Grossissement d'une Filtration. IX, 142 Seiten. 1980.

Vol. 834: Model Theory of Algebra and Arithmetic. Proceedings, 1979. Edited by L. Pacholski, J. Wierzejewski, and A. J. Wilkie. VI, 410 pages. 1980.

Vol. 835: H. Zieschang, E. Vogt and H.-D. Coldewey, Surfaces and Planar Discontinuous Groups. X, 334 pages. 1980.

Vol. 836: Differential Geometrical Methods in Mathematical Physics. Proceedings, 1979. Edited by P. L. García, A. Pérez-Rendón, and J. M. Souriau. XII, 538 pages. 1980.

Vol. 837: J. Meixner, F. W. Schäfke and G. Wolf, Mathieu Functions and Spheroidal Functions and their Mathematical Foundations Further Studies. VII, 126 pages. 1980.

Vol. 838: Global Differential Geometry and Global Analysis. Proceedings 1979. Edited by D. Ferus et al. XI, 299 pages. 1981.

Vol. 839: Cabal Seminar 77 – 79. Proceedings. Edited by A. S. Kechris, D. A. Martin and Y. N. Moschovakis. V, 274 pages. 1981.

Vol. 840: D. Henry, Geometric Theory of Semilinear Parabolic Equations. IV, 348 pages. 1981.

Vol. 841: A. Haraux, Nonlinear Evolution Equations- Global Behaviour of Solutions. XII, 313 pages. 1981.

Vol. 842: Séminaire Bourbaki vol. 1979/80. Exposés 543–560. IV, 317 pages. 1981.

Vol. 843: Functional Analysis, Holomorphy, and Approximation Theory. Proceedings. Edited by S. Machado. VI, 636 pages. 1981.

Vol. 844: Groupe de Brauer. Proceedings. Edited by M. Kervaire and M. Ojanguren. VII, 274 pages. 1981.

Vol. 845: A. Tannenbaum, Invariance and System Theory: Algebraic and Geometric Aspects. X, 161 pages. 1981.

Vol. 846: Ordinary and Partial Differential Equations, Proceedings. Edited by W. N. Everitt and B. D. Sleeman. XIV, 384 pages. 1981.

Vol. 847: U. Koschorke, Vector Fields and Other Vector Bundle Morphisms – A Singularity Approach. IV, 304 pages. 1981.

Vol. 848: Algebra, Carbondale 1980. Proceedings. Ed. by R. K. Amayo. VI, 298 pages. 1981.

Vol. 849: P. Major, Multiple Wiener-Itô Integrals. VII, 127 pages. 1981.

Vol. 850: Séminaire de Probabilités XV. 1979/80. Avec table générale des exposés de 1966/67 à 1978/79. Edited by J. Azéma and M. Yor. IV, 704 pages. 1981.

Vol. 851: Stochastic Integrals. Proceedings, 1980. Edited by D. Williams. IX, 540 pages. 1981.

Vol. 852: L. Schwartz, Geometry and Probability in Banach Spaces. X, 101 pages. 1981.

Vol. 853: N. Boboc, G. Bucur, A. Cornea, Order and Convexity in Potential Theory: H-Cones. IV, 286 pages. 1981.

Vol. 854: Algebraic K-Theory. Evanston 1980. Proceedings. Edited by E. M. Friedlander and M. R. Stein. V, 517 pages. 1981.

Vol. 855: Semigroups. Proceedings 1978. Edited by H. Jürgensen, M. Petrich and H. J. Weinert. V, 221 pages. 1981.

Vol. 856: R. Lascar, Propagation des Singularités des Solutions d'Equations Pseudo-Différentielles à Caractéristiques de Multiplicités Variables. VIII, 237 pages. 1981.

Vol. 857: M. Miyanishi. Non-complete Algebraic Surfaces. XVIII, 244 pages. 1981.

Vol. 858: E. A. Coddington, H. S. V. de Snoo: Regular Boundary Value Problems Associated with Pairs of Ordinary Differential Expressions. V, 225 pages. 1981.

Vol. 859: Logic Year 1979–80. Proceedings. Edited by M. Lerman, J. Schmerl and R. Soare. VIII, 326 pages. 1981.

Vol. 860: Probability in Banach Spaces III. Proceedings, 1980. Edited by A. Beck. VI, 329 pages. 1981.

Vol. 861: Analytical Methods in Probability Theory. Proceedings 1980. Edited by D. Dugué, E. Lukacs, V. K. Rohatgi. X, 183 pages. 1981.

Vol. 862: Algebraic Geometry. Proceedings 1980. Edited by A. Libgober and P. Wagreich. V, 281 pages. 1981.

Vol. 863: Processus Aléatoires à Deux Indices. Proceedings, 1980. Edited by H. Korezlioglu, G. Mazziotto and J. Szpirglas. V, 274 pages. 1981.

Vol. 864: Complex Analysis and Spectral Theory. Proceedings, 1979/80. Edited by V. P. Havin and N. K. Nikol'skii, VI, 480 pages. 1981.

Vol. 865: R. W. Bruggeman, Fourier Coefficients of Automorphic Forms. III, 201 pages. 1981.

Vol. 866: J.-M. Bismut, Mécanique Aléatoire. XVI, 563 pages. 1981.

Vol. 867: Séminaire d'Algèbre Paul Dubreil et Marie-Paule Malliavin. Proceedings, 1980. Edited by M.-P. Malliavin. V, 476 pages. 1981.

Vol. 868: Surfaces Algébriques. Proceedings 1976–78. Edited by J. Giraud, L. Illusie et M. Raynaud. V, 314 pages. 1981.

Vol. 869: A. V. Zelevinsky, Representations of Finite Classical Groups. IV, 184 pages. 1981.

Vol. 870: Shape Theory and Geometric Topology. Proceedings, 1981. Edited by S. Mardešić and J. Segal. V, 265 pages. 1981.

Vol. 871: Continuous Lattices. Proceedings, 1979. Edited by B. Banaschewski and R.-E. Hoffmann. X, 413 pages. 1981.

Vol. 872: Set Theory and Model Theory. Proceedings, 1979. Edited by R. B. Jensen and A. Prestel. V, 174 pages. 1981.

Vol. 927: Y. Z. Flicker, The Trace Formula and Base Change for GL (3). XII, 204 pages. 1982.

Vol. 928: Probability Measures on Groups. Proceedings 1981. Edited by H. Heyer. X, 477 pages. 1982.

Vol. 929: Ecole d'Eté de Probabilités de Saint-Flour X - 1980. Proceedings, 1980. Edited by P.L. Hennequin. X, 313 pages. 1982.

Vol. 930: P. Berthelot, L. Breen, et W. Messing, Théorie de Dieudonné Cristalline II. XI, 261 pages. 1982.

Vol. 931: D.M. Arnold, Finite Rank Torsion Free Abelian Groups and Rings. VII, 191 pages. 1982.

Vol. 932: Analytic Theory of Continued Fractions. Proceedings, 1981. Edited by W.B. Jones, W.J. Thron, and H. Waadeland. VI, 240 pages. 1982.

Vol. 933: Lie Algebras and Related Topics. Proceedings, 1981. Edited by D. Winter. VI, 236 pages. 1982.

Vol. 934: M. Sakai, Quadrature Domains. IV, 133 pages. 1982.

Vol. 935: R. Sot, Simple Morphisms in Algebraic Geometry. IV, 146 pages. 1982.

Vol. 936: S.M. Khaleelulla, Counterexamples in Topological Vector Spaces. XXI, 179 pages. 1982.

Vol. 937: E. Combet, Intégrales Exponentielles. VIII, 114 pages. 1982.

Vol. 938: Number Theory. Proceedings, 1981. Edited by K. Alladi. IX, 177 pages. 1982.

Vol. 939: Martingale Theory in Harmonic Analysis and Banach Spaces. Proceedings, 1981. Edited by J.-A. Chao and W.A. Woyczyński. VIII, 225 pages. 1982.

Vol. 940: S. Shelah, Proper Forcing. XXIX, 496 pages. 1982.

Vol. 941: A. Legrand, Homotopie des Espaces de Sections. VII, 132 pages. 1982.

Vol. 942: Theory and Applications of Singular Perturbations. Proceedings, 1981. Edited by W. Eckhaus and E.M. de Jager. V, 363 pages. 1982.

Vol. 943: V. Ancona, G. Tomassini, Modifications Analytiques. IV, 120 pages. 1982.

Vol. 944: Representations of Algebras. Workshop Proceedings, 1980. Edited by M. Auslander and E. Lluis. V, 258 pages. 1982.

Vol. 945: Measure Theory. Oberwolfach 1981, Proceedings. Edited by D. Kölzow and D. Maharam-Stone. XV, 431 pages. 1982.

Vol. 946: N. Spaltenstein, Classes Unipotentes et Sous-groupes de Borel. IX, 259 pages. 1982.

Vol. 947: Algebraic Threefolds. Proceedings, 1981. Edited by A. Conte. VII, 315 pages. 1982.

Vol. 948: Functional Analysis. Proceedings, 1981. Edited by D. Butković, H. Kraljević, and S. Kurepa. X, 239 pages. 1982.

Vol. 949: Harmonic Maps. Proceedings, 1980. Edited by R.J. Knill, M. Kalka and H.C.J. Sealey. V, 158 pages. 1982.

Vol. 950: Complex Analysis. Proceedings, 1980. Edited by J. Eells. IV, 428 pages. 1982.

Vol. 951: Advances in Non-Commutative Ring Theory. Proceedings, 1981. Edited by P.J. Fleury. V, 142 pages. 1982.

Vol. 952: Combinatorial Mathematics IX. Proceedings, 1981. Edited by E. Billington, S. Oates-Williams, and A.P. Street. XI, 443 pages. 1982.

Vol. 953: Iterative Solution of Nonlinear Systems of Equations. Proceedings, 1982. Edited by R. Ansorge, Th. Meis, and W. Törnig. VII, 202 pages. 1982.

Vol. 954: S.G. Pandit, S.G. Deo, Differential Systems Involving Impulses. VII, 102 pages. 1982.

Vol. 955: G. Gierz, Bundles of Topological Vector Spaces and Their Duality. IV, 296 pages. 1982.

Vol. 956: Group Actions and Vector Fields. Proceedings, 1981. Edited by J.B. Carrell. V, 144 pages. 1982.

Vol. 957: Differential Equations. Proceedings, 1981. Edited by D.G. de Figueiredo. VIII, 301 pages. 1982.

Vol. 958: F.R. Beyl, J. Tappe, Group Extensions, Representations, and the Schur Multiplicator. IV, 278 pages. 1982.

Vol. 959: Géometrie Algébrique Réelle et Formes Quadratiques. Proceedings, 1981. Edité par J.-L. Colliot-Thélène, M. Coste, L. Mahé, et M.-F. Roy. X, 458 pages. 1982.

Vol. 960: Multigrid Methods. Proceedings, 1981. Edited by W. Hackbusch and U. Trottenberg. VII, 652 pages. 1982.